GUIDED WAVE PHOTONICS

PHOTONICS

Fundamentals
and Applications
with MATLAB®

Optics and Photonics

Series Editor

Le Nguyen Binh

Monash University, Clayton, Victoria, Australia

GUIDED WAVE PHOTONICS

PHOTONICS

Fundamentals
and Applications
with MATLAB®

Le Nguyen Binh

CRC Press
Taylor & Francis Group
Boca Raton   London   New York

CRC Press is an imprint of the
Taylor & Francis Group, an **informa** business

CRC Press
Taylor & Francis Group
6000 Broken Sound Parkway NW, Suite 300
Boca Raton, FL 33487-2742

© 2012 by Taylor & Francis Group, LLC
CRC Press is an imprint of Taylor & Francis Group, an Informa business

First issued in paperback 2019

No claim to original U.S. Government works

ISBN 13: 978-0-367-45223-0 (pbk)
ISBN 13: 978-1-4398-2855-7 (hbk)

Library of Congress Cataloging-in-Publication Data

Binh, Le Nguyen.
 Guided wave photonics : fundamentals and applications with MATLAB / Le Nguyen Binh.
  p. cm.
 "A CRC title."
 Includes bibliographical references and index.
 ISBN 978-1-4398-2855-7 (hardcover : alk. paper)
 1. Optical wave guides--Computer simulation. 2. Photonics--Mathematics. 3. MATLAB. I. Title.

TK8305.B46 2012
621.36--dc23   2011031719

Visit the Taylor & Francis Web site at
http://www.taylorandfrancis.com

and the CRC Press Web site at
http://www.crcpress.com

To the memory of my father

To my mother

To Phuong and Lam

Contents

Preface

This book presents the theory and simulation of optical waveguides and wave propagations in a guided environment, the guided wave photonics. It consists of a unified treatment of three distinct but related topics of formulation of wave equations in the transverse plane and time-dependent propagation directions, the coupling of the guided wave systems, and nonlinear behavior of the guided waves in such waveguiding systems.

A departure from the convention followed by most books in the field is the omission of introductory chapters on the fundamentals of the theory of guided waves. Rather, the book as a whole is aimed at readers with a background on the propagation of electromagnetic waves, offered in a semester course covering essential matters on the guiding of lightwaves in waveguide structures. However, essential background materials are given in the appendices. These guided wave devices play the most important roles in the engineering of optical communications transmission at ultra-high speed.

Therefore, guiding of lightwaves in planar and three-dimensional structures is described in Chapters 2 and 3, while Chapters 4 and 5 describe the guiding of lightwaves in circular optical waveguides, the single-mode optical fibers. Chapters 6 and 7 describe the coupling phenomena via the scalar and full coupled-mode theories. Chapter 8 gives a brief treatment of nonlinear optical waveguides and Chapter 10 a treatment of optical fibers and their applications in performing a number of photonic manipulating functions such as phase conjugation and time demultiplexing in optical transmission systems. Nonlinear effects in such guided wave devices are also treated, wherever appropriate in sections of the chapters.

The motivation and materials for this book have been provided mainly by research conducted in the University of Western Australia, Monash University, Siemens Central Research Laboratories at Otto Hahn Ring, Munich, Germany and Nortel Networks Advanced Technology Research Centers at Harlow, England with which the author has been associated as a technical and academic member.

Many other people contributed significantly to this book. Research students at Monash University contributed over the years; in particular, my research scholars and doctoral graduates Dr. Su-Vun Chung, Dr. Shu Zheng, Dr. Wenn Jing Lai, Dr. X. Wang and Dr. Nguyen Duc Nhan. Undergraduate students attending the courses in advanced photonics and optical fiber communications and electromagnetic wave propagations of the Department of Electrical and Computer Systems Engineering of Monash University, Faculty of Engineering of the Christian Albrechts University of Kiel, Germany and Network Technology Research Center of Nanyang Technological University of Singapore, have questioned and challenged several aspects of wave guiding in optical waveguides, I thus thank them for their exchanges of ideas and curiosity.

I wish to thank Ashley Gasque of CRC Press for her encouragement and assistance in the formulation of this book. I extend special thanks to the wife Nguyen Thi Phuong and my son Le Nguyen Lam for putting up with the intrusion on family time as a result of all the days and nights spent on the chapters of this book. Last and not least, I thank my mother Nguyen Thi Huong and late father for giving their children the best education philosophy over many years and their teaching of "learning for life."

Le Nguyen Binh
Glen Iris, Australia

MATLAB® and Simulink® are trademarks of the Math Works, Inc. and are used with permission. The Mathworks does not warrant the accuracy of the text or exercises in this book. This book's use or discussion of MATLAB® and Simulink® software or related products does not constitute endorsement or sponsorship by the Math Works of a particular pedagogical approach or particular use of the MATLAB® and Simulink® software. For product information, please contact:

The MathWorks, Inc.
3 Apple Hill Drive
Natick, MA 10760-2098 USA
Tel: 508-647-7000
Fax: 508-647-7001
E-mail: info@mathworks.com
Web: www.mathworks.com

Author

Le Nguyen Binh received a B.E. (Hons) and PhD in electronic engineering and integrated photonics in 1975 and 1980 respectively, both from the University of Western Australia, Nedlands, Western Australia. In 1980 he joined the Department of Electrical Engineering of Monash University after a three-year period with CSIRO Australia as a research scientist.

He was appointed as reader of Monash University in 1995. He has worked for the Department of Optical Communications of Siemens AG Central Research Laboratories in Munich and the Advanced Technology Centre of Nortel Networks in Harlow, U.K. He was a visiting professor of the Faculty of Engineering of Christian Albrechts University of Kiel, Germany.

Dr. Binh has published more than 300 papers in leading journals, refereed conferences, and has written two books in the fields of photonic signal processing and digital optical communications for CRC Press. His current research interests are in advanced modulation formats for long-haul optical transmission, electronic equalization techniques for optical transmission systems, ultra-short pulse lasers and photonic signal processing, optical transmission systems and network engineering.

He is currently the technical director of the European Research laboratories of Hua Wei Technologies GmbH in Munich, Germany. His research interests are now focused on 100G and beyond optical transmission and integrated optical technology and digital signal processing for extremely high transmission bit rate.

List of Abbreviations and Notations

A	Area, constant
A_{eff}	Effective area—of fiber evaluated as the area of the guided mode ($= \pi r_0^2$)—see also mode spot size
a	Fiber core radius—see also mode spot size and mode field diameter
a_k	Integer "1" or "0"
AM	Amplitude Modulation
ASE	Amplified Spontaneous Emission
ASK	Amplitude Shift Keying
α_L	Attenuation (linear scale)
α_{dB}	Attenuation factor in dB
B_e	3 dB Bandwidth—electrical
B_o	3 dB Bandwidth—optical
B	Normalized propagation constant—see also β
b_0, b_1	Energy of a "0" or "1" transmitted and received at the front end of a photo detector
β	Propagation constant of fiber (see also effective index of guided mode)
β_1	First order differentiation of the propagation with respect to the angular frequency—propagation delay
β_2	Second order differentiation of the propagation constant—group velocity dispersion (GVD)
β_3	Third order differentiation of the propagation constant—related to dispersion slope.
BDPSK	Binary Differential Phase Shift Keying
C	Constant or capacitance
c	Velocity of light in vacuum—note: use exact value of c for calculation of wavelength grid for DWDM; $c = 2.998e^8$ m/s
CD	Chromatic Dispersion
CMT	Couple Mode Theory
CPFSK	Continuous Phase Frequency Shift keying
CPM	Continuous Phase Modulation
CS-RZ	Carrier Suppressed Return-to-Zero format
D	Dispersion factor (usual unit = ps/nm/km)
D_w	Waveguide dispersion factor (see also material dispersion factor M)
D_T	Total Dispersion factor—chromatic dispersion
d	Fiber diameter ($= 2a$)
DBM	Duo-Binary Modulation
DCF	Dispersion Compensating Fiber
DCM	Dispersion Compensating Module
DD	Direct Detection

Demux	Demultiplexer
δ	Delta factor—equivalent to the Q-factor optical carrier frequency (center)
Δ	Relative refractive index difference
DFB	Distributed Feedback (Laser)
DI	Delay Interferometer
DPSK	Differential Phase Shift Keying
DQPSK	Differential Quadrature Phase Shift Keying
DRA	Distributed Raman Amplifier
DSF	Dispersion Shifted Fiber
DSP	Digital Signal Processing (processor)
DuoB	Duo-Binary
EDFA	Erbium-Doped Fiber Amplifier (Er:doped fiber amplifier)
EO	Electro-Optic
EOM	Electro-Optic Modulators
ESI	Equivalent Step Index
η	Quantum efficiency
$\mathcal{F}$	Fourier transform
f	Frequency
f_o or ν	Optical frequency
FDTWEA	Finite Difference Traveling Wave Electrodes Analysis
FWM	Four-Wave Mixing
GVD	Group Velocity Dispersion
H	Plank's constant ($h = 6.624e^{-34}$ J-s).
I, i	Current large signal or bias current and small signal
IF	Intermediate Frequency
IM	Intensity Modulation/Modulator
IM/DD	Intensity Modulation/Direct Detection
IO	Integrated optics
I-Q	In-phase and Quadrature
ITU	International Telecommunications Union
k_B	Boltzmann's constant ($1.38e^{-23}$ J/°K)
L	Length of optical fiber
L_{eff}	Effective length, length along which the linear dispersion or nonlinear effects are effective
L_D	Dispersion length
LO	Local Oscillator
MADPSK	Multi-level (M-ary) Amplitude-Differential Phase Shift Keying
MI	Modulation Instability
MMF	Multi-Mode Optical Fibers
MOF	Microstructure Optical Fibers
MSK	Minimum Shift Keying
Mux	Multiplexers or Multiplexing
MZDI	Mach-Zehnder Delay Interferometer

MZI	Mach-Zehnder Interferometer
MZIM	Mach-Zehnder Interferometer Modulator or Mach Zehnder Interferometric Intensity Modulator
NA	Numerical Aperture
NL	Nonlinear
n_{eff}	Effective refractive index of guided mode ($= \beta/\kappa$)
NLPN	Nonlinear Phase Noise
NLSE	Non-Linear Schroedinger Equation
NRZ	Non-Return-to-Zero pulse shaping
NZDSF	Non-Zero Dispersion Shifted Fiber (ITU-655)
OA	Optical Amplifier (*see also* EDFA and ROA for lumped and distributed amplification)
O-DPSK	Offset Differential Phase Shift Keying
OFDM	Orthogonal Frequency Division Multiplexing
OOK	On-Off Keying or Amplitude Shift keying (ASK)
OPLL	Optical Phase Locked Loop
OSNR	Optical Signal-to-noise Ratio
P_0	Optical power output
$p_s(t)$	Signal power
Pdf or pdf	Probability density function
PLL	Phase Locked Loop
PM	Phase Modulator
PMD	Polarization Mode Dispersion in unit of $ps/\sqrt{km}$
PMF	Polarization Maintaining Fiber
Q	Electronic charge; q = 1.6e^{-16} C
QAM	Quadrature Amplitude Modulation
R	Responsivity
r_0	Mode spot size—e^{-1} position from the mode center under Gaussian mode profile, *see also* effective area of single mode fiber
ROA	Raman Optical Amplifier
RZ	Return-to-Zero pulse shaping
RZ33	RZ pulse of width of 33% of bit period format
RZ50	RZ pulse of width of 50% of bit period format
RZ67	RZ pulse of width of 67% of bit period format (normally CS-RZ)
SDH	Synchronous Digital Hierarchy
SI	Spectral noise density of a bias current I
SMF	Single-Mode Fiber
SPM	Self Phase Modulation—nonlinear effects
SR	Spectral noise density of the thermal noise of a resistor R
SSMF	Standard Single-Mode Fiber (ITU-652)
Star-QAM	Star Quadrature Amplitude Modulation, signal constellation distributed on circle like a star
T_b	Bit period
T_f	Fall time

T_r	Rise time
T	Absolute temperature (°K)
TOD	Third Order Dispersion
V	Normalized frequency parameter
V_e	Equivalent step index V-parameter
ν	Frequency of lightwaves
VSTF	Volterra Series Transfer Function
XPM	Cross Phase Modulation
Z_T	Trans-impedance of an electronic amplifier

1

Introduction

Current advances in ultra-high speed optical communications would not be possible if lightwaves could not be guided over long loss and low dispersive fiber over a very long distance. Originally proposed as a dielectric waveguide by Kao and Hockham [1] in an article published in the *Proceedings of the IEE* in 1966, this has been developed extensively in circular waveguide, the optical fibers and then in planar waveguide structure, integrated optical circuits. Both these technological developments have played crucial roles in the practical implementation of lasers, single optical fibers, and optical modulators. This chapter thus gives an overview of optical wave guiding in dielectric media.

An optical guided wave structure performs two principal functions. Firstly, it must satisfy the bounded conditions such that the wave profile in the transverse plane must be oscillation. Secondly, the wave traveling speed along the propagation direction motion of such guided structures must be optimized according to the purposes of the applications, for example, single mode and low dispersion fibers, high dispersion for compensation fibers, highly nonlinear and low dispersive rib waveguides, etc. Therefore, in the chapters following this introductory chapter, we describe the important parameters and design of guided wave photonic structures to reach these objectives.

1.1 Historical Overview of Integrated Optics and Photonics

The term "integrated optics" was first coined by Miller in 1969 [2], as an analogy of lightwave to the electronic integrated circuits. Indeed the present term used in industry is planar lightwave circuits (PLC). Since that proposed term there has been tremendous progress. Various integrated optical devices have been researched, developed and deployed in practical optical transmission systems and networks. Extensive surveys have been given over the years [3,4] and even defined and described on the Internet [5,6]. Both linear and nonlinear integrated optics [7] have been exploited.

Four decades ago, G.R. Moore made a prediction, which is now the famous Moore's law. That means that the number of integrated transistors on a chip would be doubled every 18 months. This is still true now and it is expected to be valid for another decade at least. Moore's law is applied not only to the growth of the semiconductor industry but also to the growth of data traffic [8]. This is mainly driven by the emergence of high bandwidth consuming applications and services such as Internet Protocol Television (IPTV), file sharing, and high definition television (HDTV). Besides that, the expansion of the major telecommunications infrastructure in the Far East (China, India) and in developing countries also brings about huge data traffic.

Rapid deployment of fiber to the x (FTTx) further pushes the demand for network capacity for delivering the data traffic. Recently, the Federal Government of Australia initiated a project to design, build and operate the active infrastructure of Australia's Next

Generation National Broadband Network [9]. Similarly in South East Asia, Singapore has also introduced its National Broadband Networks [10–12] for the twenty-first century. Both national-scale projects aim at providing broadband access links, which can scale up to 100 Mb/s, to 50% of Australian and Singapore residents by 2012 and then 1 Gb/s soon after that period. This is considered as the digital economy of the twenty-first century. The delivery of information at this rate is unheard of in the history of human communications. It would not have been possible if optical fibers, especially single mode fibers, were not invented and exploited over the last three decades. For this invention, Dr. Charles Kao was awarded the Nobel Prize for Physics in 2009 for his proposed dielectric waveguides, as was G. Hockham in 1966. Since then huge investments were made by Corning, Schott Glass, Sumitomo and several glass-based global corporations to the fabrication and manufacturing of guided circular wave structures in order to dominate the new and emerging technology over the last quarter of the twentieth century and thence this century.

Even at this Gb/s to the home the consumption of the capacity has still not been fully used, the available bandwidth possibly supplied by a single mode fiber of about 25 THz would be sufficient for the near future expansion of global optical networks. The principal point is how to deliver effectively and economically both in the core systems/networks and the last kilometer distribution networks. In addition, societies have reached the age of creativity and the trend of usage of video transmission in global communities has risen tremendously indicating that the bit rate or capacity of the backbone networks must be increased significantly to respond to these demands. The video of the future will include super HDTV, high definition holographic television and image transmission whose compressed bit rate is in the order of gigabits per second. The increase in the capacity–distance product in one single fiber over the years is shown in Figure 1.1 indicating a decade multiplication factor every four years.

The development of optical communications after the invention of optical fiber has fulfilled demand in the past and even much longer into the future of human communications. But in order to keep up with the exponential growth of the data traffic in the near and medium range future, more and more hardware components and transmission technologies have to be developed. Wavelength-division multiplexing (WDM), the technology of combining a number of wavelengths into the same fiber, successfully increases the fiber capacity by 32 times, or even by hundreds of times if its advanced version, dense wavelength-division multiplexing (DWDM), is employed. However, the bottleneck is at the routers and switches where hundreds of channels must be demultiplexed for O/E (optical-to-electrical conversion), routing, E/O (electrical-to-optical conversion) and then multiplexing [13,14]. The whole network speed can further be significantly increased if the optical signal can be routed/switched directly in the optical domain without the need of O/E. All-optical switching, signal processing and optical time division multiplexing (OTDM) are promising solutions [15–21] and they can be performed in a nonlinear optical waveguide. This is described in the last chapter of this book. The speed of OTDM channels has been demonstrated to terabit per second range [22].

As one can observe about the bit rate per channel in 1974, it was not higher than 1 Mb/s transmission over 1 km of optical fibers which were of multimode type. At that stage the modulation of the laser is directly manipulating the driving current injected to the laser cavity. Because the length-bandwidth of the multimode fiber is small (less than 1.0 GHz/km) then direct modulation is sufficient. As expected the dispersion of optical fibers can be reduced when only one optical mode is guided in a circular fiber, direct modulation of the laser cavity injection current would not be sufficiently high. External modulation

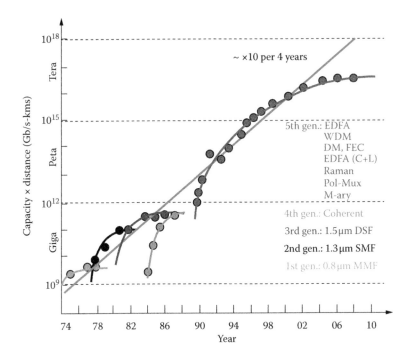

FIGURE 1.1
Capacity-distance product in Gb/s-kms achieved over the last 35 years in one single mode fiber.

of the lightwaves generated from a narrowband laser could push the operating bandwidth higher than 10 Gb/s which is well above the maximum bit rate of 8 Gb/s for the case of direct modulation.

The external modulation is possible if lightwaves are guided in planar or three dimensional waveguide structures such as $LiNbO_3$ or polymeric rib-structure. Thus, intensive research works over the last two decades on multi-GHz bandwidth integrated optical modulators. It is noted that the optical pulses must be very fast. That means that the low frequency roll-off must be in the range of less than 100 MHz and high frequency must be in the range of multi-GHz range. The total insertion loss of these integrated optical modulators (IOM-typically about 6 dB) limit the operating of optical transmission till the invention of optical amplifiers, the Er-doped fiber amplifiers (EDFA) that leads to an explosion of research and development of IOMs, especially $Ti:LiNbO_3$ structures [23].

In OTDM systems, data is encoded using ultra-short optical pulses occupying N time slots in the OTDM time frame. Each channel is assigned a time slot and its data can be accessed with the aid of an optical clock pulse train corresponding to that time slot. Hence, generation of ultra-short optical pulses with multiple gigabits repetition rate is critical for ultra-high bit rate optical communications, particularly for the next generation of terabits per second, optical fiber systems. The modulation of lightwaves at this ultra bit rate is critical for the realization of such optical sequences. Thus, optical modulators realized in integrated optical structures are critical for ultra-broadband optical communication systems and networks. Furthermore, the field of optical packet switching has gained recognition in recent years and requires ultra-short and high peak power pulse generators to provide all-optical switching [24–30]. Hence the generation of an ultra-short pulse sequence is important and mode-locked fibers incorporating wideband phase or amplitude modulators are needed as described by Binh and Ngo [31].

The ultra-fast operations of optical fiber communications systems would not be possible if optical waveguides had not been extensively developed so that the lightwaves could be modulated, transmitted and received and as routing and switching.

1.2 Why Analysis of Optical Guided Wave Devices?

Over the past few decades or so, there has been a rapid growth of research, development and applications of lightwave technology, photonic telecommunication and high-speed all-optical integrated systems. This has led to the need to further study and develop certain optical guided-wave couplers based on single mode optical fibers, optical thin film, slab and channel waveguides. Whilst single mode fiber and its directional coupler remain the most important components in the field of light transmission and routing, other optical guided-wave couplers (especially those consisting of composite or hybrid waveguides) are playing a new and increasingly important role in opening up new areas of applications, especially all-optical integrated devices.

Dielectric optical waveguides with widths or diameters from micrometers to millimeters have also attracted applications in many fields, such as optical communication, optical sensing and optical power delivery systems [32]. In this book, based on exact solutions of Maxwell's equations and numerical calculations, the basic guiding properties of lightwaves in guided wave structures are described.

In the process of material selection, system design and device fabrication, there are always several optical and structural parameters available for selection and possible manipulation, the variation of which can lead to slight variation or even a drastic change of behavior of the guide-wave coupler. It is therefore crucial in certain cases and very helpful generally, to have a simple, practical and accurate theoretical analysis of any optical guided-wave coupler systems. This will reveal their operational characteristics, such as power-coupling properties, and help to choose the appropriate materials and their processing techniques before manufacture, and allow the design parameters and fabrication procedures of the devices to be optimized.

A number of optical material systems can be used for creating optical waveguides for passive or active photonic devices. Silica and doped silica have been demonstrated to be the best material combination for the core and cladding regions of circular dielectric waveguides or single mode optical fibers in modern optical transmission systems and networks. Silica on silicon combined materials has also been employed as planar optical waveguiding media for passive photonic devices in order to match the guided modes of the circular and buried rib planar waveguides.

Anisotropic crystals can also be selected for fabrication of active integrated photonic devices such as electro-optical and acousto-optic modulators. The most common optical modulators employed extensively for optical transmission systems are lithium niobate of x-, y- or z-cut orientation which is then doped with titanium using diffusion or ion exchange techniques.

Compound semiconductor materials are also used for modulators such as GaAs, InGaAsP, etc. The electro-absorption property is the principal effect employed for such modulators.

One can also choose fused silica (SiO$_2$) and single crystal silicon (Si) as typical dielectric materials for waveguiding under the following considerations: (i) fused silica and single crystal silicon are among the most important photonic and opto-electronic materials within the visible and near-infrared ranges; (ii) both silica and silicon waveguides or wires with submicrometer- or nanometer-diameters have recently been successfully fabricated; and (iii) their optical and physical properties are well known, and they have typical values of moderate and high refractive indices (about 1.45 for silica and 3.5 for silicon) [33]. Photonic device applications using these photonic materials for realizing photonic devices can benefit from minimizing the width of the waveguides. However the fabrication of low-loss optical waveguides with sub-wavelength diameters remains very challenging because of high precision requirement. Recently, several types of dielectric submicrometer- and nanometer-diameter wires of optical qualities can be achieved. These wires whose diameters are less than a micrometer, can be tens to thousands times thinner than the commonly used micrometer-diameter waveguides. They can be used as air-clad wire-waveguides with sub-wavelength-diameter cores, and building blocks in the future micro- and nano-photonic devices [34].

1.3 Principal Objectives

The principal objectives of the book thus are: (i) to describe the fundamental principles of the guiding of lightwaves in planar and circular dielectric waveguide structures; (ii) to present theoretical and numerical techniques for the design and implementation of optical waveguides and systems of optical waveguides so as to form a network of optical guided wave components; (iii) to describe the coupling phenomena of lightwaves from one waveguide to the others, thence the coupled mode theory in scalar and vectorial approaches; (iv) to present the nonlinear effects in guided wave devices and associated phenomena so that their effects in the transmission of optical signals through optical fibers can be evaluated and methods to overcome these unwanted impairments, can be developed; (v) to illustrate the design of planar rib nonlinear optical waveguides to generate optical phase conjugated signals so as to completely compensate the distorted signals; and finally (v) to describe the generation of optical amplification through parametric conversion effects and its uses in the demultiplexing of ultra-high speed OTDM signals in the optical domain.

At the beginning of this book, a fundamental presentation of the guiding of lightwaves in optical waveguide structures is given. Thence the design of a number of waveguide structures is treated with effective index method and finite difference methods. The coupling of lightwaves from one waveguide to the others is then treated with the coupled mode theory under scalar approximation and vectorial approaches. Certain discrepancies are found among a few different formulations, especially when the wave guidance and coupling are not so weak or the polarization effects are of concern. The situation demanded further investigations on the application of coupled mode theory (CMT), especially when the coupled guided-modes (two or more) are non-degenerate (i.e., non-identical). A possible unified view on CMT is also expected. A number of novel devices are described utilizing asymmetric waveguide and coupler structures, such as a composite fiber-slab system, are also treated. In particular, the following observations were made and proved to be the basis for this treatment: (i) a general, coupled-mode and compound-mode analysis was lacking for composite (hybrid), optical guided-wave couplers consisting of a single mode fiber and an asymmetric slab waveguide (linear or nonlinear, with or without gratings); (ii) further analytical solutions are still desired in terms of the power-coupling properties, even for the well-known coupled-mode equations (usually solved numerically by computer) for two-mode coupler systems consisting of a dual-mode waveguide or two single mode waveguides; (iii) a comparative study of the major coupled-mode formulations was also desired when applied to the composite, asymmetric guided-wave coupler systems.

This was necessary to assess the simplicity, applicability and accuracy of each of the major formulations. Thus objectives of this treatment of the lightwave coupling systems are given by: (a) the formulation of scalar CMT with respect to a novel asymmetric, composite fiber-slab guided-wave couplers, as well as the conventional two-mode couplers; (b) to achieve analytical solutions in exact or closed forms wherever possible, including all the new coupling coefficients, power conservation laws, and power coupling or redistribution features of the above coupler systems; and (c) to develop simple, robust and effective algorithms and computer programs to numerically simulate or solve other cases where analytical solutions are not possible or are difficult to obtain; (d) to reveal the main features of the power coupling and redistribution, as well as possible effects of the variation of a particular design parameter, such as the structural or optical material constants; and finally (e) the format of our improved and generalized formulations (e.g., the coupled-mode equations and the analytical solutions) is made as close as possible to that of previous work, allowing comparison of special cases. The structural and optical constants, coupling coefficients and other constants are replaced by more general expressions.

For completeness, circular optical waveguides now commonly known as optical fibers are described. Only *single mode* optical fibers are presented as these fibers are considered to be the most important fibers for modern optical communications systems and networks due to its low loss and low dispersion properties and are currently deployed throughout the global information core networks. Geometrical and profile structures of these fibers are described coupled with the conditions for single mode guiding. The properties of the fiber attenuation and dispersion of information signals carrying through the fibers are given in detail. It is noted here that the mode size of the single modes guided through the fibers follows a close profile of that of a Gaussian profile, thus the solution of the guide mode by solving the eigenvalue equation obtained from the boundary conditions of the wave equation is no longer extremely important but by substituting the guided profile into the wave equation gives us the conditions for which the propagation constants of the mode can be found.

Nonlinear optical waveguides are also considered and a numerical technique, the finite element method, is given so as to illustrate the behavior of the guided mode in such structures. The nonlinearity of optical waveguides are also treated with the four wave mixing effects and applications of such results in the generation of phase conjugation, self phase modulation and triple correlation so that bispectrum can be generated for the design of ultra-sensitive optical receivers.

1.4 Chapters Overview

In Chapter 2, an introduction to planar optical waveguides is described, that is the lightwaves are confined in the vertical direction and no restriction in the lateral direction or the waveguide is structured infinitely long in this dimension. This direction is then restricted to form a three dimensional optical waveguide and is described in Chapter 3. Both analytical and numerical techniques are given for the solution of the wave equations to find the eigenvalue solutions of the propagation constant which determines the propagation velocity of the guided mode. It is noted that the wave equation can normally be solved in the transverse plane to obtain the guided conditions and thus the dimension of the waveguides and the index profile of different regions. The propagation constants are also found so that one can determine the propagation velocity of the guided mode along the propagation direction. The propagation constant must be maximized so that the mode would propagate fastest.

Chapters 4 and 5 are dedicated to the treatment of optical waveguides having circular sectional area, the optical fibers. Only single mode optical fibers are described. Even more specifically, the refractive index difference between the core and cladding regions is very small. Thus the weak guiding conditions are applied, that is the mode is gently guided in the core region and only about 70% of the mode power is distributed in the core and thus quite a substantial amount of mode power is propagating in the cladding region. Although analytical solutions of the eigenvalue equations are given, it is not necessary to solve the wave equations because the mode profile can be measured and a Gaussian profile is usually obtained. Thus if one knows the profile of the guided mode this can be substituted into the wave equation and then optimization can be performed to obtain the propagation constant of the guided mode. Because there is only one mode this is quite simple. This is essential so as to explain to practice engineers in the field of optical communication engineering. The impacts of the mode size on attenuation and dispersion, thence pulse broadening, are described. Some modern fibers such as microstructure optical fibers are also analyzed to illustrate the guiding and anti-guiding phenomena in such fibers.

Chapters 6 and 7 describe in detail the couple mode theory in scalar and vectorial modes for analyses of guided wave coupling systems. These two chapters present theoretical and analytical treatment of such coupling systems. A series of advanced analysis and modeling of both the novel, asymmetric and composite fiber-slab guided-wave coupler systems, as well as the conventional two-mode coupler system is presented. This is described with increasing complexity in terms of analytical formulations and numerical calculations. Chapter 6 presents the simplified (also called first-order or conventional) scalar coupled-mode formulations, analysis and modeling whilst Chapter 7 provides the full coupled mode theory counterparts. This is to specifically address the effect of the field-overlap (butt coupling) typical for asymmetric and non-degenerate guided-mode coupler systems which is neglected in the simplified analysis. Chapter 7 also describes the necessary vector corrections required to take account of problems such as polarization effects (e.g., coupling between the transverse magnetic [TM] modes). An example of the coupling in defect dual core microstructure optical fibers is also given to demonstrate the effectiveness of the couple mode theory.

Chapter 8 then gives some typical structures of nonlinear optical waveguides in which some regions, either the core or cladding regions, are formed with materials whose nonlinear Kerr coefficient is quite high so that switching of the guided modes, eigensolutions, happen. A distinction of this treatment is that a conjugate eigen-pair of solutions can be obtained leading to the observation of bistability phenomena in such nonlinear guided wave structures.

Chapter 9 gives an introduction to optical modulators fabricated using integrated guided wave structures. The principles of modulation of the phase and intensity of the guided wave are simplified by the presentation of the phasor diagram. Its corresponding dynamics can be related to the design of the waveguides. These modulators are important for optical communications systems operating in ultra-high speed regions.

Chapter 10 then illustrates important applications of nonlinearity in guided wave devices to achieve phase conjugation, the parametric amplification and four wave mixing processes to form the optical triple correlation for the design of ultra-sensitive optical receivers. Raman amplification and scattering in microstructure optical fibers are also studied and the nonlinear property is given.

In brief, we must note that there are several types of guided wave devices which can be included in the chapters of this book. However we elected to present the fundamentals of guiding lightwaves and coupling of lightwaves from one path to the other rather than go deeply into operations of optical devices formed on an integrated optical platform. We also illustrate only optical modulators and some nonlinear applications to demonstrate the effectiveness of guided wave devices in modern optical communication systems and networks.

References

1. C.K. Kao and G.A. Hockham, "Dielectric-fibre surface waveguides for optical frequencies," *Optoelectronics, IEE Proceedings J.*, 133(3), 191–198, 1966.
2. S.E. Miller, "Integrated optics: An introduction," *Bell Sys. Tech. J.*, 48, 2059–2069, 1969.
3. C.R. Doerr and K. Okamoto, "Advances in silica planar lightwave circuits," *IEEE J. Lightwave Tech.*, 24(12), 4763–4770, 2006.
4. P.S. Chung, "Waveguide modes, coupling techniques, fabrication and losses in optical integrated optics," *J. Elect. Electron. Australia*, 5, 201–214, 1985.
5. http://electron9.phys.utk.edu/optics421/modules/m10/integrated_optics.htm. Access date: June 2010.
6. R.G. Hunsperger, "Integrated Optics Theory and Technology," originally published in the series: *Advanced Texts in Physics*, 6th Ed., 2009, XXVIII, Springer, Berlin.
7. L.N. Binh and S.V. Chung, "Nonlinear interactions in thin film structures," Proceedings of the 8th Australian Workshop on Optical Communications, Adelaide 1983, Session VII.
8. R.E. Wagner, J.R. Igel, R. Whitman, M.D. Vaughn, A.B. Ruffin, and S. Bickharn, "Fiber-based broadband-access deployment in the United States," *J. Lightwave Technol.*, 24, 4526–4540, 2006.
9. Department of Broadband, Communications and Digital Economy, Government of Australia http://www.dbcde.gov.au/communications/national_broad band_network. Access date: June 2009.
10. H.Y.Khoong, "NextGenerationNationalBroadbandNetworkforSingapore(NextGenNBN)," http://www.ida.gov.sg/doc/News%20and%20Events/News_and_Events_Level2/20080407164702/OpCoRFP7Apr08.pdf. Access date: May 2008.
11. Singapore IDA, "IDA To Pre-Qualify Interested Parties To Be Operating Company For Singapore's Next Generation National Broadband Network," 2008, http://www.ida.gov.sg/News%20and%20Events/20080303140126.aspx. Access date: May 2008.
12. H. Kogelnik, "Perspectives on optical communications," The Optical Fiber Communication Conference & Exposition and the National Fiber Optic Engineers Conference (OFC/NFOEC), 2008. OFC 2008, San Diego, CA, 2008.
13. A. Bogoni, L. Poti, P. Ghelfi, M. Scaffardi, C. Porzi, F. Ponzini, G. Meloni, G. Berrettini, A. Malacarne, and G. Prati, "OTDM-based optical communications networks at 160 Gbit/s and beyond," *Optical Fiber Technol.*, 13, 1–12, 2007.
14. K. Vlachos, N. Pleros, C. Bintjas, G. Theophilopoulos, and H. Avramopoulos, "Ultrafast time-domain technology and its application in all-optical signal processing," *J. Lightwave Technol.*, 21, 1857–1868, 2003.
15. M.M. Mosso, W.R. Ruziscka, F.S. da Silva, and C.F.C. da Silva, "OTDM quasi-all-optical demultiplexing techniques comparative analysis," *Proc. Int. Microw. Optoelect. Conf.*, Vols 1 and 2, 692–697, 1997.

16. A. Bogoni, L. Poti, R. Proietti, G. Meloni, E. Ponzini, and P. Ghelfi, "Regenerative and reconfigurable all-optical logic gates for ultra-fast applications," *Electron. Lett*, 41, 435–436, 2005.

17. H.J.S. Dorren, A.K. Mishra, Z.G. Li, H.K. Ju, H. de Waardt, G.D. Khoe, T. Simoyama, H. Ishikawa, H. Kawashima, and T. Hasama, "All-optical logic based on ultrafast gain and index dynamics in a semiconductor optical amplifier," *IEEE J. Sel. Topics Quant. Elect.*, 10, 1079–1092, 2004.

18. M. Scaffardi, N. Andriolli, G. Meloni, G. Berrettini, F. Fresi, P. Castoldi, L. Poti, and A. Bogoni, "Photonic combinatorial network for contention management in 160 Gb/s-interconnection networks based on all-optical 2×2 switching elements," *IEEE J. Sel. Topics Quant. Elect.*, 13, 1531–1539, 2007.

19. A. Bogoni, P. Ghelfi, M. Scaffardi, C. Porzi, F. Ponzini, and L. Poti, "Demonstration of feasibility of a complete 160 Gbit/s OTDM system including all-optical 3R," *Optics Communications*, 260, 136–139, 2006.

20. T. Houbavlis, K.E. Zoiros, M. Kalyvas, G. Theophilopoulos, C. Bintjas, K. Yiannopoulos, N. Pleros et al., "All-optical signal processing and applications within the ESPRIT project DO_ ALL," *IEEE J. Lightwave Technol.*, 23, 781–801, 2005.

21. H. Kogelnik, "Perspectives on optical communications," The Optical Fiber Communication Conference & Exposition and the National Fiber Optic Engineers Conference (OFC/NFOEC), 2008. OFC 2008, San Diego, CA, 2008.

22. T.D. Vo, H. Hu, M. Galili, E. Palushani, J. Xu, L.K. Oxenløwe, S.J. Madden et al., "Photonic chip based transmitter optimization and error-free receiver demultiplexing of 1.28 Tbaud Data," Proc. Int. Workshop on Nonlinear Syst. Adv. Sig. Proc., HCM City, Vietnam, Sept. 2010, Univ Press, Univ. of Science, HCM City.

23. L.N. Binh, "Lithium niobate optical modulators: Devices and applications," *J. Crystal Growth*, 288, 180–187, 2006.

24. J. Herrera, O. Raz, E. Tangdiongga, Y. Liu, H.C.H. Mulvad, F. Ramos, J. Marti et al., "160-Gb/s all-optical packet switching over a 110-km field installed optical fiber link," *IEEE J. Lightwave Technol.*, 26, 176–182, 2008.

25. V. Eramo, M. Listanti, and A. Germoni, "Cost evaluation of optical packet switches equipped with limited-range and full-range converters for contention resolution," *IEEE J. Lightwave Technol.*, 26, 390–407, 2008.

26. S.N. Fu, P. Shum, N.Q. Ngo, C.Q. Wu, Y.J. Li, and C.C. Chan, "An enhanced SOA-based double-loop optical buffer for storage of variable-length packet," *IEEE J. Lightwave Technol.*, 26, 425–431, 2008.

27. D. Klonidis, C.T. Politi, R. Nejabati, M.J. O'Mahony, and D. Simeonidou, "OPSnet: design and demonstration of an asynchronous high-speed optical packet switch," *IEEE J. Lightwave Technol.*, 23, 2914–2925, 2005.

28. P. Zhou and O. Yang, "How practical is optical packet switching in core networks?" Global Telecommunications Conference, 2003. Proc. GLOBECOM '03. IEEE Global Telecomm. Conf., 2003. GLOBECOM '03. IEEE Vol. 5, 2003.

29. M.J. O'Mahony, D. Simeonidou, D.K. Hunter, and A. Tzanakaki, "The application of optical packet switching in future communication networks," *IEEE Comms Magazine*, 39, 128–135, 2001.

30. T.S. El-Bawab and J.-D. Shin, "Optical packet switching in core networks: Between vision and reality," *IEEE Communications Magazine*, 40, 60–65, 2002.

31. L.N. Binh and Q.N. Ngo, "*Ultra-Short Pulse Fiber Lasers*," CRC Press, Taylor & Francis Group, Boca Raton, FL, 2010.

32. C. Manolatou, S.G. Johnson, S. Fan, P.R. Villeneuve, H.A. Haus, and J.D. Joannopoulos, "High-density integrated optics," *IEEE J. Lightwave Tech.*, 17, 1682–1692, 1999.

33. K.K. Lee, D.R. Lim, H.C. Luan, A. Agarwal, J. Foresi, and L.C. Kimerling, "Effect of size and roughness on light transmission in a Si/SiO_2 waveguide: experiments and model," *Appl. Phys. Lett.* 77, 1617–1619, 2000. Erratum: *Appl. Phys. Lett.*, 77, 2258, 2000.

34. L. Tong, J. Lou, and E. Mazur, "Single-mode guiding properties of sub-wavelength-diameter silica and silicon wire waveguides," *Opt. Exp.*, 12(6), 1025–1035, 2004.

2

Single-Mode Planar Optical Waveguides

2.1 Introduction

The term "integrated optics" was first coined by Miller in 1969 [1], as an analogy of lightwave to the electronic integrated circuits. Indeed the present term used in industry is planar lightwave circuits (PLC). Since that proposed term there has been tremendous progress. Various integrated optical devices have been researched, developed and deployed in practical optical transmission systems and networks. Extensive surveys have been given over the years [2,3] and even defined and described on the Internet [4,5]. Both linear and nonlinear integrated optics [6] have been exploited.

Recent experimental demonstrations have pushed the information transmission bit rate per channel to 100 Gb/s [7–9] with the multiplexing of several wavelength channels reaching to tens of terabits/s [10–12]. At such speed we need to have the following optical functionality: optical modulation, switching, optical pre-processing using nonlinearity of integrated optical devices and the compensation of dispersion of single mode optical fibers using fiber or integrated optic components and equalization of the losses of fibers by optical amplification. Prior to the availability of Er-doped fiber amplifiers (EDFA), attempts to increase the span distance between repeaters of amplitude modulated single fiber transmission systems by employing coherent techniques [13,14] in which a narrow line width and high power is used to mix with the received signals to improve the receiver sensitivity were made. This line width requirement on the local oscillator laser limits the deployment of such coherent transmission in real practice. Only recently, coherent optical communications have attracted much interest once again as a possible technique to further increase the transmission spans [15,16]. Naturally the modulation formats, such as amplitude shift keying, phase shift keying and frequency shift keying, are employed to reduce the effective signal band width in order to minimize the dispersion effects in single mode optical fibers. The detection of ultra-high speed optical signals can thus be in the direct or coherent detection. This attracts the employment of integrated optical devices in the optical transmitters, receivers and online components.

The basic component of these integrated photonic devices is the optical waveguide formed by a thin film or diffused waveguiding layer structure on some substrate. From the mode propagation point of view, design optimization requires accurate estimation of the propagation constant thence dispersion characteristics, mode size and group velocities, depending on the types of applications. These requirements led us to develop simple, accurate and efficient methods of analysis or single mode waveguides. This is the motivation of the first chapter of the book.

Most optical waveguides with a graded index profile, especially lithium niobate types, are fabricated by a diffusion process that is commonly formed by the diffusion of impurities into various substrate ferroelectric materials such as lithium niobate and/ or lithium tantalate. On the other hand the proton exchange process can also be used so that the Li ions can be exchanged by the hydrogen ion. In both processes, diffusion and ion exchange, a crystal stress is established and a change of the refractive index is created, thus graded index profile distribution. Complementary error function is usually used to represent the distribution of the impurity from the surface of the substrate into the depth of the substrate [17] and the time of diffusion of the metallic impurities. A Gaussian profile is expected when the diffusion time is sufficiently long [18] which is used to fit the experimental values of Se into CdS crystal. In addition various profiles can be used to form optical waveguides using molecular beam epitaxy (MBE), metallic organic chemical vapor deposition (MOCVD) techniques as in the waveguiding structures for laser diodes of separate confinement of the heterojunction [19].

Exact analytical solutions are available for the step [20], exponential [21], hyperbolic secant [22], clad-linear [23], clad-parabolic [24] and Fermi [25] profiles. In general, approximate analytical or exact numerical methods are required to analyze general classes of profiles. For some practical profiles, universal charts describing the mechanism of waveguiding have already been presented by several authors. These have been obtained by variational analysis [26], Wentzel–Kramers–Brillouin (WKB) [27], and multilayer staircase [28]. However these curves are only accurate for multimode waveguides, with the exception of the last two references. Single mode planar optical waveguides are very important for integrated optic circuits for applications in advanced ultra-high speed optical communications, see for example, the pioneering works by Korotky et al. [29]. However the methods of analysis are confined predominantly to those originally used in multimode waveguides. In the single mode regime, the variational and WKB methods are expected to perform poorly. In the former, the solutions are strongly affected by the choice of trial fields. In the later, more accurate prediction of the phase changes at the turning point are required. In the Runge–Kutta outward integration method, instability is caused by the solution and error increasing at large "x" [17]. This problem is resolved by approximating the fields at sufficient depth in the waveguide by an evanescent field [30]. However this requires the knowledge of the location of solution matching.

Thus this chapter recognizes that any general technique of analysis must be numerical in nature due to the more stringent accuracy requirements for single mode waveguides. In the case that the single mode waveguide is used as a nonlinear interaction medium then the phase matching is very important and thus an accurate estimation of the dispersion curves plays a very important part in the conversion efficiency [6,31].

For diffused waveguides or graded index profile distribution of the refractive index from the surface of the waveguide to the deepest position, two widely used methods for waveguide modal analysis are given. The variational method with a simple Hermite-Gaussian field was first introduced by Korotky [29] to calculate the mode spot size in a diffused channel waveguide. In Section 2.4 we show that the method to estimate the modal characteristics of all diffused waveguide profiles is inaccurate and computationally intensive for the calculation of the dispersion characteristics, but is a very close estimation of the mode spot size.

This chapter is organized as follows: Section 2.2 describes the formation of planar optical waveguides with a derivation of the wave equation for the guiding of either transverse electric (TE) or transverse magnetic (TM) mode or both. Single mode planar optical waveguides are the main focus and thus all the guiding conditions are established along this line of theoretical treatment. Section 2.3 outlines an approximate analytical method for the solutions of asymmetric index profile waveguides including the variational and WKB techniques [32,33]. Optical waveguides with a symmetric planar index profile are then treated in Section 2.4. The equations required for the derivation of the wave equation are given in Appendix A and an exact analysis of an asymmetric waveguide structure are given in Appendix B.

2.2 Formation of Planar Single-Mode Waveguide Problems

A planar dielectric waveguide with the geometry shown in Figure 2.1 can support modes with two polarizations. These are the TE and TM field guided modes. In practice either polarized mode can be excited. Provided certain boundary conditions are met, these modes are bounded and propagated along the z-axis, each with a unique effective phase velocity. If we allow for the uniformity of the refractive index and geometrical dimension in the propagation direction, the phase velocity is only a function of the transverse index profile. In this section we consider the simplest configuration with index variation only in the −x direction. Note that the notation of the coordinate system follows the right hand rule (RHS).

2.2.1 Transverse Electric/Transverse Magnetic Wave Equation

The wave equation for the guided modes stems from the Maxwell equations are given in Appendix A together with the associate constituent relations. We consider the steady state solutions in the dielectric medium free from any sources and losses. By omitting the common factor $e^{j(\omega t - \beta z)}$ from the equations we can write, for a medium characterized with the refractive index $n(x)$, the well-known wave equation:

$$\frac{d^2 E_y}{dx^2} + \left[k_0^2 n^2(x) - \beta^2\right] E_y = 0 \tag{2.1}$$

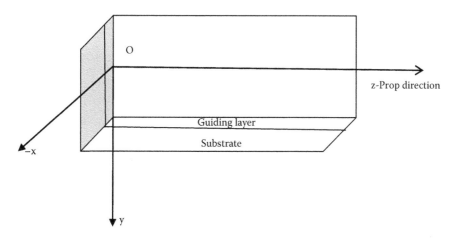

FIGURE 2.1
Schematic structure of a planar dielectric waveguide.

and

$$\frac{d^2 H_y}{dx^2} + \left[k_0^2 n^2(x) - \beta^2 \right] H_y = \frac{1}{n^2(x)} \frac{d^2 n^2(x)}{dx^2} \frac{dH_y}{dx}$$

where k_0 = wave number in free space; β = propagation constant of the wave along the z-axis; $E_y(x)$ = TE field; H_y = TM field.

2.2.1.1 Continuity Requirements and Boundary Conditions

In practical waveguides, it is common to expect that at least one region of dielectric continuity is encountered by the optical fields. At the location of the discontinuity, the wave equations are not valid. However the identity of the modes is preserved by matching the fields and their derivatives on either side of the dielectric discontinuity. For the TE modes these boundary conditions impose the continuity E_y; dE_y/dy across the interface. For the TM modes we require the continuity H_y; $(1/n^2(x))$ (dH_y/dy). In addition the bound modes satisfy the conditions that E_y, H_y vanishes at $x = \infty$. Together they give rise to the eigenvalue equation from which the propagation constant can be calculated.

Note that the principal object of the eigenvalue equation is to estimate the maximum value of the propagation constant so that the dependence of this propagation parameter on the optical frequency/wavelength is minimum, so that there is minimum dispersion of the waves at different wavelengths that is normally found for the other spectral components of a modulated lightwave channel in optical communication systems. A maximum value of the propagation constant along the z direction means that the direction of the wave vector is close to the propagation axis.

2.2.1.2 Index Profile Construction

For the purpose of computation and analysis it is customary to write the refractive index profile in the general form as

$$n^2(x) = \begin{cases} n_s^2 \left[1 + 2\Delta S\left(\dfrac{x}{d} \right) \right] & x \geq 0 \\[2mm] n_c^2 & x < 0 \end{cases}$$

where (2.2)

$$\Delta = \frac{n_0^2 - n_s^2}{2 n_s^2}$$

Δ is the profile height and n_c, n_0 and n_s are the refractive indices of the cover, the guiding layer and the substrate, respectively. $S(x)$ is the profile shape function and d is the diffusion depth of a graded index distribution. Figure 2.2 shows a typical representation of the graded index profile. It turns out that further normalization of the shape profile can be represented as

$$S\left(X = \frac{x}{d} \right) = \frac{n^2(x) - n_s^2}{n_0^2 - n_s^2}$$

(2.3)

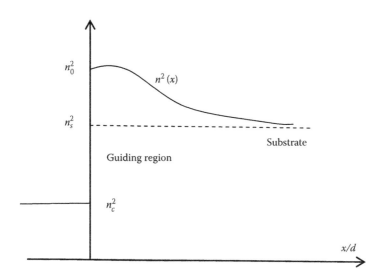

FIGURE 2.2
Square of the refractive index distribution profile for an asymmetrical waveguide.

This definition of the profile shape is unaffected by the symmetry of the waveguide. In a symmetric waveguide structure with an axis of symmetry at $x = 0$ these equations are equally valid.

2.2.1.3 Normalization and Simplification

The presence of the non-zero term in the RHS of Equation 2.1 complicates the analysis. It is identically zero for a step index profile. The exact solutions of this equation are available for the exponential, hyperbolic secant and an inverted-x profile [34]. However for smooth profiles normally encountered in practice, several authors [35] found by perturbation analysis that the RHS of the equation can be neglected. However the fundamental mode of an infinite parabolic profile faces a 44% error in the group velocity of the profile shape of the guided mode.

Following Kolgenik and Ramaswamy [36] we can introduce the normalized parameter for the waveguide as follows

$$V = \frac{dk_0}{n_0^2 - n_s^2}$$

$$A = \frac{n_s^2 - n_c^2}{n_0^2 - n_s^2};$$

and

$$b = \frac{n_e^2 - n_s^2}{n_0^2 - n_s^2} \tag{2.4}$$

where A is defined as the asymmetry factor, b is the normalized propagation constant, $n_e = \beta/k_0$ is the effective refractive index of the guided mode along the propagation axis, and V is the normalized frequency.

For a guided mode it requires

$$n_s < n_e < n_0 \tag{2.5}$$

thus the normalized propagation constant must satisfy $B < 1$. The real advantages of normalization comes from the analysis and design optimization point of view. Substituting the normalized parameters into the wave equation, Equation 2.1, we obtain:

$$\frac{d^2\varphi}{dx^2} + V^2[S(X) - b]\varphi = 0$$

where

$$\varphi(X) \equiv E_y; H_y \tag{2.6}$$

for the TE and TM modes, respectively. The propagation for the TM modes are accurately represented by those of the TE modes, except at cutoff or for waveguide at large symmetry [31]. However if extreme accuracy or mode splitting is required then Equation 2.6 can be modified to the changes involving only a slight modification of the boundary condition.

2.2.1.4 Modal Parameters of Planar Optical Waveguides

The solution of the wave equation, Equation 2.6, together with the boundary conditions, enables the determination of various optical parameters. The following are the most commonly used for the design of single optical guided-wave devices:

2.2.1.4.1 Mode Size

The mode size Γ_a for an asymmetrical field is defined as the full-width half-maximum (FWHM) power intensity. For a full description of the field the peak position of the intensity I_p and the field asymmetrical factor Γ_1/Γ_2, defined with respect to I_p are required as shown in Figure 2.3. The knowledge of the mode size is critical to match that of the single mode optical fiber for inline integration with fiber transmission systems. These parameters are defined in terms of the optical power of the field due to practical reasons because the mode size is normally monitored using a charge-coupled device (CCD) camera through which the intensity of the mode field is converted into the charge current and displayed or digitized for data processing.

2.2.1.4.2 Propagation Constant and Effective Refractive Index

The variation of the normalized propagation constant b as a function of the normalized frequency parameter V is normally required for the design and characterization of optical waveguide. For example, this relationship specifies the diffusion depth required once the mode index at a specific operating wavelength is given. The purpose of this section is to present the relation of the modal field to the propagation constant.

If we integrate Equation 2.6 with respect to X then we have

$$\int_{-\infty}^{+\infty} \left\{ \frac{d^2\varphi}{dx^2} + V^2[S(X) - b]\varphi \right\} dX = 0$$

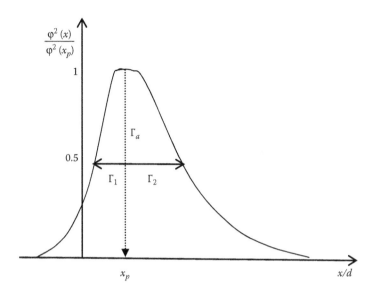

FIGURE 2.3
The minimum set of parameters required for characterization of fundamental mode field in an asymmetrical planar waveguide.

or

$$\int_{-\infty}^{+\infty} \varphi'' \, dX + V^2 \int_{-\infty}^{+\infty} \left\{ [S(X) - b] \varphi(X) \right\} dX \qquad (2.7)$$

where the dashes denote the derivative with respect to X. If we further impose the condition that $\varphi(-\infty) = \varphi(+\infty) = 0$, the integral in the left becomes zero and thus we obtain

$$b = \frac{\displaystyle\int_{-\infty}^{+\infty} S(X)\varphi(X)\, dX}{\displaystyle\int_{-\infty}^{+\infty} \varphi(X)\, dX} \qquad (2.8)$$

This equation indicates that the dispersion characteristics for an arbitrary index profile must be a smooth curve. The rule of refractive index used by Tien [37] to explain a host of new wave phenomena in integrated optical waveguides is that light tends to propagate in the region where the refractive index is largest. In the context of planar optical waveguides with an arbitrary index profile, this rule suggests that for a given profile at a given frequency, the mode field adjusts itself so that maximum value of b is achieved. This corresponds to the minimum phase velocity allowed for the mode. This rule is indeed a direct statement of Fermat's law in ray optics and is a special case of a generalized rule in quantum mechanics formulated in the form of the well-known Feynman's path integral of which Maxwell's equations are also satisfied [38].

2.2.1.4.3 Waveguide Dispersion and Spot Size
A second and potentially useful relation between b and the modal field can be established from the stationary expression for b. This can be obtained from Equation 2.6 after

multiplication by φ and taking the integration with respect to X from negative to positive infinitive. After integrating the results by parts and imposing the boundary conditions $\varphi(-\infty)\varphi'(-\infty) = \varphi(+\infty)\varphi'(\infty) = 0$ we obtain the well-known stationary relation

$$bV^2 = \frac{V^2 \int_{-\infty}^{+\infty} S(X)\varphi^2(X)\,dX - \int_{-\infty}^{+\infty} \varphi^2(X)\,dX}{\int_{-\infty}^{+\infty} \varphi^2(X)\,dX} \tag{2.9}$$

This is the basic equation for the variational analysis. It possesses the unique property that for any trial fields which satisfy the boundary conditions above, the quotient remains stationary provided the mismatch between the trial and actual fields is small [39].

Thus we can write

$$\frac{d(bV^2)}{dV^2} = \frac{\int_{-\infty}^{+\infty} S(X)\varphi^2(X)\,dX}{\int_{-\infty}^{+\infty} \varphi^2(X)\,dX} \tag{2.10}$$

i.e.,

$$\frac{1}{2}\left(bV + V\frac{d(bV)}{dV}\right) = V\frac{\int_{-\infty}^{+\infty} S(X)\varphi^2(X)\,dX}{\int_{-\infty}^{+\infty} \varphi^2(X)\,dX} \tag{2.11}$$

Taking the derivative a second time and using Equations 2.9 and 2.11 we obtain

$$\frac{1}{2}\left(V\frac{d^2(bV)}{dV^2} + 2\frac{d(bV)}{dV}\right) = V\frac{d}{dV}\left(b + \frac{2}{V^2 W_m}\right) + \frac{1}{2}\left(b + \frac{d(bV)}{dV}\right) \tag{2.12}$$

where we define a new spot size parameter as

$$W_m = \frac{2\int_{-\infty}^{+\infty} \varphi^2(X)\,dX}{\int_{-\infty}^{+\infty} [\varphi'(X)]^2\,dX} \tag{2.13}$$

Further algebraic manipulation leads to the simple relationship

$$V\frac{d^2(bV)}{dV^2} = 4\frac{d}{dV}\left(\frac{2}{VW_m^2}\right) \tag{2.14}$$

This relation is analogous to the relation between Petermann's spot size and the waveguide dispersion in single mode optical fibers [40]. The preceding analysis was first performed by Sansonetti [41] which inspired Petermann to define a new spot size in the characterization of single mode optical fibers from spot size measurement.

The relation by Equation 2.14 can be found for optical waveguide with a profile follow-ing a Hermite-Gaussian variational field of

$$\varphi(X) = \begin{cases} A_0 \alpha_0^{1/2} e^{-\alpha_0 X^2/2} ; X \geq 0 \\ 0; X < 0 \end{cases} \tag{2.15}$$

$$W_m^2 = \frac{4}{\alpha_0^2} = \Gamma_a^2 \tag{2.16}$$

where α_0 is the variational spot size parameter and A_0 is a constant. The RHS of Equation 2.16 corresponds to Γ_a^2 as defined by Korotky [29] which has been defined above. Although α_0 is an approximate mode spot size, several experiments [29] show excellent agreement between the theoretical and experimental values W_m and Γ_a.

2.3 Approximate Analytical Methods of Solution

Despite the availability of direct numerical integration methods for the analysis of optical waveguides, approximate analytical solutions are still being used, improved and sought after. We have to strike the balance between accuracy and simplification. Three well-known methods of analysis are described in this section. They are valid for single mode planar optical waveguides.

The variational method [29] is applicable only to the fundamental mode of the asym-metrical waveguide due to the form of the trial field. The equivalent profile method is valid only for symmetrical waveguides because it requires the field to be monotonously decreasing. The WKB method can be used in both cases. We thus group the methodologi-cal approaches into symmetry and asymmetry. In Section 2.3.1 the analytical formulae for a number of widely used profiles are obtained. We explore the improvements to the WKB method and limitations. In Section 2.3.2 we compare the accuracy of the equivalent profile-moment methods using a step and a cosh reference profile [42,43]. The WKB method may not work at all for the analysis of the single mode optical waveguides.

2.3.1 Asymmetrical Waveguides

2.3.1.1 Variational Techniques

The variational method is based on the substitution of a TE_0 mode lookalike trial field into the stationary expression of the normalized propagation constant b given in Equation 2.8. The shape of the field is then adjusted to maximize b for all values of V (see Equation 2.9). The mathematical procedure is given in Snyder and Love, 1983 [24].

Following the field profile defined by Korotky et al. [29] and Riviere et al. [28], a trial solu-tion can be proposed which closely fits the form of the TE_0 field

$$\varphi(X) = \begin{cases} \sqrt{\alpha_0} e^{-\frac{\alpha_0 X^2}{2}} X \geq 0 \\ 0 ..... X < 0 \end{cases} \tag{2.17}$$

Note that one drawback of the form of this field is that it vanishes at $X = 0$. For single mode optical waveguides, this condition is very nearly only for guides with very large asymmetry. However only a single parameter needs to be optimized, thus the optimization scheme is simple. In the following it is shown that there exist closed form formulae for several profiles.

2.3.1.1.1 Eigenvalue Equation

If we substitute Equation 2.17 and the derivative of this trial field into Equation 2.9 a simpler expression is obtained, after some tedious algebra, as

$$b = I_1 - \frac{3\alpha_0}{2V^2} \tag{2.18}$$

where

$$I_1 = 4\alpha_0 \left(\frac{\alpha_0}{\pi}\right)^{1/2} \int_0^\infty S(X)X^2 e^{-\alpha_0 X^2} dX$$

is the only profile dependent expression. The correct value of α_0 is obtained by noting that b must be stationary with respect to α_0, thus

$$\frac{db}{d\alpha_0} = 0 = \frac{dI_1}{d\alpha_0} - \frac{3}{2V^2} \tag{2.19}$$

Substituting this α_0 into Equation 2.18 we obtain the eigenvalues given in Table 2.1.

2.3.1.1.2 Fundamental Mode Cutoff Frequency

The lowest order mode in an asymmetric optical waveguide has a non-zero cutoff frequency for $A \neq 0$. Thus we can set $b = 0$ and $V = V_c$ in Equations 2.18 and 2.19 to obtain the desired cutoff frequencies. Since V_c appears in both equations, one can solve simultaneously for α_0 initially before substituting back to obtain the cutoff value for the V-parameter, V_c. It happens that the cutoff V_c for a Gaussian profile is given as $V_c = 1.9741$. Table 2.2 tabulates the analytical expressions for the cutoff frequency of the exponential and complementary error function profiles.

TABLE 2.1

Optimum Value of α_0 for Selected Asymmetrical Clad-diffused Waveguide

Profile	$S(X)$	α_0
Gaussian	e^{-X^2}	$(\alpha_0 + 1)\left[\dfrac{(\alpha_0 + 1)}{\alpha_0}\right]^{1/2} - V = 0$
Exponential	e^{-X}	$4V^2\sqrt{\alpha_0} - 3\pi(\alpha_0 + 1)^2 = 0$
Complementary error function erfc	$\mathrm{erfc}(X)$	$2(4\alpha_0 + 1)\sqrt{\dfrac{\alpha_0}{\pi}} - (1 + 6\alpha_0)\left[1 - \mathrm{erfc}\dfrac{1}{2}\sqrt{\alpha_0}\right]e^{\frac{1}{4\alpha_0}} - \dfrac{12\alpha_0^3}{V^2}$

TABLE 2.2

Optimum Value of α_0 and Fundamental Cutoff Normalized Frequency
Parameters of exp and erfc Profiles

Profile	Parameter	Equation
Exponential	α_0	$\dfrac{\left(2\alpha_0^2+5\alpha_0-2\right)}{16\alpha_0\left(2\alpha_0+1\right)}\left[1-\text{erfc}\dfrac{1}{2\sqrt{\alpha_0}}\right]+e^{\frac{1}{4\alpha_0}}+1=0$
	V_c	$\left(\dfrac{3\alpha_0}{2}+\dfrac{1}{\sqrt{\alpha_0}}\right)\left[\left(1+\dfrac{1}{2\alpha_0}\right)e^{\frac{1}{4\alpha_0}}\left(1-\text{erfc}\dfrac{1}{2\sqrt{\alpha_0}}\right)\right]=V_c^2$
Complementary error function erfc	α_0	$\dfrac{\left(\alpha_0+1\right)}{\sqrt{\alpha_0}}\tan^{-1}\sqrt{\alpha_0}-\dfrac{\alpha_0}{\left(\alpha_0+1\right)}-1=0$
	V_c	$V_c^2=\dfrac{3\pi\left(\alpha_0+1\right)^2}{4\sqrt{\alpha_0}}$

The computation of the propagation constant and the mode cutoff frequency for TE_0 mode that requires numerical integration would be tedious. Fortunately the commonly encountered graded profile waveguide shown in Table 2.1 only involves root-search for the estimation of α_0. There is no existing method to estimate the probable range of α_0 as a function of V. Thus the consuming process in the computation of the parameter V is the correct estimation of the interval for root search algorithm. Nevertheless one would be interested in the instigation of the accuracy of the results over a selected range of V. The estimated values of the propagation constant b are calculated and tabulated for a number of profile distributions with a set of specific parameters $n_c = 1.0$, $n_s = 2.177$ and $\Delta = 0.043$. This is a typical profile structure for air cover diffused waveguide profile in $LiNbO_3$ or $LiTaO_3$ substrate. The trial field distribution is Hermite-Gaussian. The corresponding cutoff frequencies at the cutoff limit are given for TE_0 in Table 2.4.

The values of the propagation constants are as expected. The accuracy of the variational field fit to the actual field improves with increasing frequency. At large V, the field in the cover decreases rapidly. Similarly, the evanescent field in the substrate follows a similar trend. Thus the mode field is confined within the guiding region and its shape is accurately modeled by a Hermite-Gaussian function. This behavior was first observed experimentally by Kiel and Auracher [44]. This observation lead to the motivation of Korotky et al.'s [29] pioneering work in the use of this simple trial field. An earlier method by Taylor requires up to 21 terms in the variational field involving parabolic cylinder functions [26]. A simple relationship between α_0 and the mode spot size Γ_a can be obtained as [28]:

$$\Gamma_a = \frac{1.555}{\sqrt{\alpha_0}} \tag{2.20}$$

Thus the mode size Γ_a can be obtained directly without using numerical computing. However Korotky et al. found good agreement between experimental and theoretical results [29], our analytical results, given here in Table 2.3, show that there are substantial discrepancies in the propagation constant for single mode optical waveguides. This means

TABLE 2.3

b-V Data for Selected Profiles Calculated with a Hermite-Gaussian Trial Field (Variational Method; Exact = Analytical Expression)

V	Exponential		Gaussian		Complementary Error Function	
	b (Var.)	*b* (Exact)	*b* (Var.)	*b* (Exact)	*b* (Var.)	*b* (Exact)
2	0.066	0.105	—	—	—	—
3	0.193	0.299	0.216	0.275	0.015	0.068
4	0.289	0.321	0.370	0.413	0.121	0.169
5	0.362	0.390	0.476	0.510	0.213	0.255
10	0.560	0.578	0.719	0.732	0.477	0.497
100	—	0.897	0.970	0.971	0.883	0.885

that the mode spot size Γ_a may not be so dependent on the frequencies. Experimentally speaking the variation of the wavelength of the guiding waveguide and detection is due the sensitivity of the spot size image monitoring device. This must be taken into account for the measurements of the FWHM of the image. To investigate this possibility, Γ_a is plotted versus the normalized frequency V parameter shown in Figure 2.4. This step is also taken to examine the behavior of α_0 near the cutoff frequency of the fundamental mode. The difficulty in the calculation of the cutoff frequencies given in Table 2.4 can be observed. This is due to the volatility of the confinement of the mode near cutoff. This is a well known phenomenon in optical fibers [45]. Figure 2.4 shows the variation of Γ_a with respect to V over a range of frequencies including at $V = V_c$ for the profiles of exponential, Gaussian and complementary error shapes. Two sets of data and curves are given so as to notice the method of using the root-search algorithm to compute the optimum spot size parameters α_0. More than one root can exist in the search interval. The smooth set of curves given in Figure 2.3 is obtained by choosing only the negative going cross-over of the curves given in Figure 2.4. The kinks observed in Figure 2.5 are obtained when choosing the smaller and incorrect root. The propagation constant computed from this false zero is much smaller and can be negative. Thus it is preferred to operate the waveguide far from the cutoff region so that the mode spot size is not strongly dependent on the V-parameter. This scenario is important for the case when planar optical waveguides are used as an optical amplifier [46], e.g., Er:LiNbO$_3$ waveguide, the wavelength of the pump beam is far from the operating wavelength region and may be close to the cutoff. This must be taken into account. If not then the fluctuation of the mode spot size would alter the amplification gain of the amplifier (Figure 2.6). In practice, the refractive index profile could never be modeled by any form of analytical function and an equivalent profile may

TABLE 2.4

Cutoff Frequencies of the TE$_0$ Mode

Profile	Exponential		Gaussian		Complementary Error Function	
	V_c (Var.)	V_c (Exact)	V_c (Var.)	V_c (Exact)	V_c (Var.)	V_c (Exact)
	1.563	1.087	1.974	1.433	2.839	2.085
α_0 at V_c	—	0.143	—	0.500	—	0.697

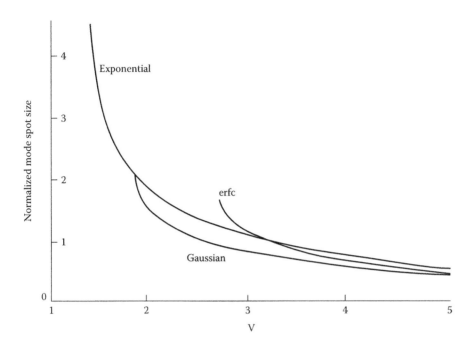

FIGURE 2.4
Mode spot size calculated using Hermite-Gaussian trial field for different profiles.

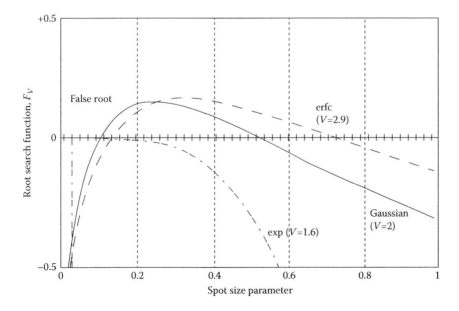

FIGURE 2.5
Multiple roots of the root search function of the variational method.

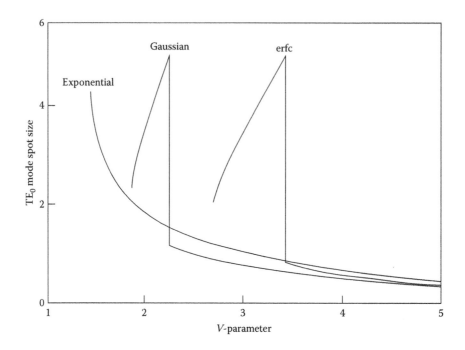

FIGURE 2.6
Mode spot size of TE_0 mode as a function of V for profiles of exponential, Gaussian and complementary error function estimated using variational method.

be used. The correct spot size behavior computed numerically using the numerical algorithm given here follows a similar trend as shown in Figure 2.7. The variational spot size is superimposed on these curves for comparison, the agreement is remarkable. The wavy curves in Figure 2.8 are caused by numerical noises. The tolerance on each plot is >1%. Such accuracy is achieved due to the definition described here for the spot size that does

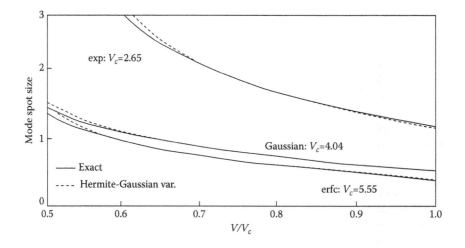

FIGURE 2.7
Spectral variation of mode spot size: accuracy of Hermite-Gaussian trial field fitting for single mode diffused clad profiles. Single mode diffused-clad profiles $n_c = 1.0$; $n_s = 2.177$; $\Delta = 0.043$ ($A = 20$).

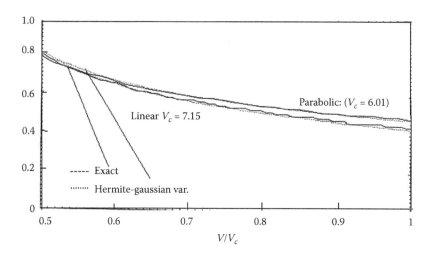

FIGURE 2.8
Accuracy of Hermite-Gaussian trial field fitting for single mode clad power law profiles. Single mode clad-power-law profiles with $n_c = 1.0$; $n_s = 2.177$; $\Delta = 0.043$ ($A = 20$).

not take into account the tails of the field. This is where serious agreement between the exact and the Hermite-Gaussian function occurs. This explains the discrepancies in the normalized propagation constant b as estimated by this method.

2.3.1.2 Wentzel–Kramers–Brilluoin Method

The WKB method was first developed by Jeffery [47] and applied to the calculation of energy eigenvalues in quantum mechanics by Wentzel [48], Kramers [49] and Brilluoin [50]. Due to the similarities between problems involving the quantum mechanical potential well and the refractive index profiles of optical waveguides [37] the method can be easily adapted for use in guided wave optics. Marcuse first used the method for proving the eigenvalue equation of asymmetrical graded index optical waveguides [23]. A simplified derivation by Hocker and Burns [51] based on ray optics confirmed Marcuse's results. This is due to the equivalence of the WKB and ray optics formalism [52].

The central problem of the techniques lie with the connection of oscillatory and evanescent fields at the turning point where the original WKB solutions are singular. Langer solved the problem by approximating the actual fields there by Airy functions [53]. This is equivalent to replacing the actual profile locally by a linear segment. Its slope and position are implicitly related to the propagation constant in the eigenvalue equation. We have examined the turning point phenomena in detail (see Appendix C). We thus can state that the WKB method is not limited by the inaccurate phase prediction at the turning point. A more serious limitation is caused by the neglect of the cladding. Coupling effects between the turning point and cladding have been studied in detail by Arnold [54]. He found that the cladding effects can be isolated and built into the eigenvalue equation. However the corrections involved a complicated nest of Airy functions and the analytic simplicity of the method is lost.

We took a simpler and more practical approach to account for cladding effects and studied the behavior of the WKB errors and found that it is minute for asymmetrical waveguides provided a simple correction is added. More discussions on the improvement of the method presented will be given in appropriate sections.

2.3.1.2.1 Derivation of the Wentzel–Kramers–Brilluoin Eigenvalue Equation

In an asymmetrical planar optical waveguide, the solution of the WKB eigenvalue equation can be obtained by matching the field and its derivative at the dielectric interfaces. For the WKB method this is complicated by the fact that the solutions must be matched correctly at the turning point. The Jeffery's solution can be referred to as the 0th order WKB method and Langer's method with turning point correction as the 1st order WKB method (see Figure 2.9). Following Gordon's [55] and Marcuse's [23] articles, one can write for a graded index asymmetrical waveguide as

$$\varphi(X) = \begin{cases} a_0 e^{V\sqrt{A+bX}} \,; X \leq 0 \\ p^{-1/2}(X)\cos\left[\phi(X) - \pi/4\right]; 0 \leq X < X_t^- \\ \left(\dfrac{2\pi\phi}{3p}\right)^{-1/2}\left[J_{1/3}(\phi) + J_{-1/3}(\phi)\right]; X = X_t^- \\ \left(\dfrac{2\pi\phi}{3p}\right)^{+1/2}\left[I_{1/3}(\phi) + I_{-1/3}(\phi)\right]; X = X_t^+ \\ \left(\dfrac{p(X)}{4}\right)^{-1/2} e^{-\phi(X)} \,; X_t^+ \leq X < \infty \end{cases} \tag{2.21}$$

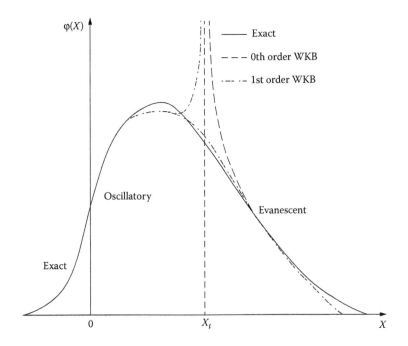

FIGURE 2.9

An illustration of regions of validity of the WKB solutions. The turning point is given by $S(X_t) = B$. The WKB eigenvalue equation is obtained by ensuring that the WKB solutions in the guide are matched to the exact field in the cover (superstrate).

where

$$\phi(X) = V \int_X^{X_t} \left| \sqrt{S(X) - b} \right| dX$$

and

$$p(X) = \left[\frac{S(X) - b}{1 - b} \right]^{1/2};$$

a_0 is a constant; I and J are the Bessel's functions representation of the Airy solutions at the turning point, X_t.

The turning point is defined such that

$$S(X_t) = b \tag{2.22}$$

A general proof is given in Appendix C which shows that at a turning point, the approximation of the exact field by Airy function is extremely good if

$$S'(X_t) \approx 0 \text{ and } S''(X_t) \approx 0 \tag{2.23}$$

Furthermore, the oscillatory solution for $0 \le X \le X_t^-$ and the evanescent field for $X > X_t^+$ are just asymptotic expansions of Bessel's solutions for $\phi (X) \gg 1$; i.e., $V \gg 1 >$ these are just the 0th order WKB solutions (see Appendix C). They have to be used in these forms with the correct phase arguments to ensure uniformity of the WKB solutions in both the guide and the substrate are already correctly matched.

The eigenvalue equation follows by ensuring the smooth matching of the WKB solution and the exact evanescent field at $X = 0$. The continuity of $\varphi(0)$ and $\varphi'(0)$ gives, after some algebraic manipulations

$$V \int_0^{X_t} \left| \sqrt{S(X) - b} \right| dX = \left(m + \frac{1}{4} \right) \pi + \tan^{-1} \left(\frac{\sqrt{A + b}}{1 - b} + \delta \right) \tag{2.24}$$

where

$$\delta = \frac{S'(0)}{4V\sqrt{1 - b}}.$$

If setting $d = 0$ then the WKB eigenvalue equation becomes

$$2V \int_0^{X_t} \left| \sqrt{S(X) - b} \right| dX - \frac{\pi}{2} - 2\tan^{-1} \left(\frac{\sqrt{A + b}}{1 - b} + \delta \right) = 2m\pi \tag{2.25}$$

as obtained by Marcuse [23]. For practical multimode waveguides, Hocker and Burns [51] claimed that

$$\tan^{-1} \left(\frac{\sqrt{A + b}}{1 - b} \right) \approx \pi / 2$$

since $A \gg 1$. Thus leading to the following relationship

$$V \int_0^{X_t} \left| \sqrt{S(X) - b} \right| dX = \left(m + \frac{3}{4} \right) \pi \qquad (2.26)$$

The third term of the RHS of Equation 2.25 exists due to the phase shift undergone by the modal field by the discontinuity at $X = 0$. Indeed Equation 2.25 is just the mathematical statement of the similar phase resonance condition of ray optics [56] which states that the phase accumulated along the ray path over one period including reflection as it traverses the guide from $X = 0$ to $X = X_t$ must be a multiple of 2π, if constructive interference is to occur. Constructive interference is essential for maintaining a stable modal pattern. In fact the phase changes at turning point and the dielectric discontinuity are just

$$-\frac{\pi}{2} \text{ and } -2\tan^{-1}\left(\frac{\sqrt{A+b}}{1-b} \right).$$

Figure 2.10 illustrates the ray path of the process. To solve b for a given V, the turning point is tuned until the phase resonance condition is met. This result has been derived without the use of WKB formalism by Hocker and Burns [51].

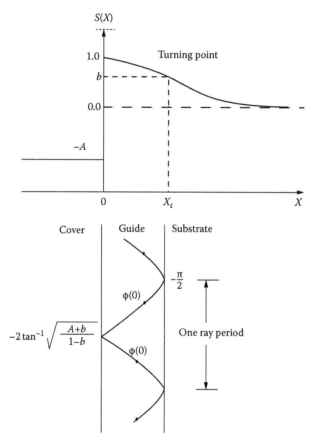

FIGURE 2.10
Ray optic derivation of the WKB eigenvalue equation.

2.3.1.2.2 Limitation of the Wentzel–Kramers–Brilluoin Method

Three sources of errors are inherent in the eigenvalue equation, Equation 2.25. It can be easily shown that Equation 2.24 can reduce to the eigenvalue equation for the TE modes of a step index profile waveguide providing the phase changes can be obtained correctly. For a graded index profile this may not be the case, as described in the next section. Thus the phase accumulated in the guided region as predicted by the WKB is only an approximation. This is due to the representation of the field by an equivalent cosine-like field.

Is the change at a dielectric interface dependent on the slope of the refractive index profile in the second medium at the interface? It is believed that in general $\delta \neq 0$ judging from the exact analysis of the linear-clad profile. This factor was omitted from the results of Marcuse [23]. Furthermore the phase change at the turning point is estimated from Equation 2.23 without resorting to the correctness of the evanescent field representation of the actual field beyond $X = X_t$. The question is whether one can lump together all sources of errors in the phase into a single error parameter. Thus it is possible to propose, in general

$$V \int_0^{X_t} \sqrt{S(X) - b} \cdot dX = (m + \gamma)\pi \tag{2.27}$$

where γ is the total accumulated phase change at the turning point, γ is normally equal to 3/4.

2.3.1.2.3 Profiles with Analytical Wentzel–Kramers–Brilluoin Solutions

The WKB integral given in Equation 2.26 is integrable for the step, clad-linear, clad-parabolic, exponential, and cosh graded-index profiles. With the help of the integration formulae given in [58] the results for the normalized propagation constant b and mode cutoff frequencies are presented in Table 2.5. For other profiles, the integral has to

TABLE 2.5

Equation for Calculating b and V_c via the WKB Method

Profile	b	V_c
Clad-linear	$1 - \dfrac{A_0}{V^{2/3}}\qquad A_0 = \left[\dfrac{3\pi}{2}(m+\gamma)\right]^{2/3}$	$\dfrac{3\pi}{2}(m+\gamma)$
Clad-parabolic	$1 - \dfrac{4(m+\gamma)}{V}$	$4(m+\gamma)$
Exponential	$b : \sqrt{1-b} - \sqrt{b}\, \tan^{-1}\left(\sqrt{\dfrac{1-b}{b}}\right) = \dfrac{\pi(m+\gamma)}{2V}$	$\dfrac{\pi}{2}(m+\gamma)$
cosh-2	$\left[1 - \dfrac{2}{V}(m+\gamma)\right]^2$	$2(m+\gamma)$
Step	$1 - \left[\dfrac{2(m+\gamma)}{V}\right]^2$	$(m+\gamma)\pi$
Gaussian	–	$(m+\gamma)\pi$

be integrated numerically. Although with modern computing facilities with ultra-high speed processors, analytical solutions would give us some insight and understanding of the behavior of the wave solution. If numerical integration is conducted then for each trial value of the normalized propagation constant b, the turning point changes, especially when V becomes very small, b approaches zero and the turning point value becomes very large. Thus the WKB were not popular before but are now with modern and ultra-high speed computing systems.

2.3.1.2.4 Ordinary Wentzel–Kramers–Brilluoin Results

We are faced with two forms of the WKB eigenvalue equation in Equations 2.24 and 2.27. This section studies the performance of this equation over a wide variety of profiles for representative values of V. The effect of the asymmetry factor on the dispersion was identified by Ramaswamy and Lagu [17] in which they found that for $A > 10$, the error is negligible. However their conclusion is only valid for multimode waveguides. In modern optical waveguides for advanced optical communications systems, single mode optical waveguides are mainly the guided wave media for applications. The results obtained for single mode optical waveguides prove otherwise when $A = 20$ is employed with the waveguide parameters $n_s = 2.177$ (lithium niobate as substrate) cover layer $n_c = 1.0$ (air) and $\Delta = 0.043$. Equation 2.26 reaches its asymptotic value when $A \rightarrow \infty$ the exact numerical results have been obtained with an integration step of 0.01. The profile truncation point is set at $X = 10$ for diffused profiles and for $X = 1$ for the clad power law profiles. The values of b and V for different profiles are calculated and tabulated in Tables 2.6 and 2.7. The improvement of the values of the normalized propagation constant and the V-parameter can be observed and is self-explanatory. The value of V is set at a region closed to that of the cutoff of the guided mode, TE_0.

2.3.1.2.5 Enhanced Wentzel–Kramers–Brilluoin Method

There are serious drawbacks to both the variational and WKB methods in the computing of the dispersion characteristics of the diffused clad single mode planar waveguides. The cosine-exponential trial field can substantially improve the accuracy of the variational analysis [58]. However it requires optimization.

For good field distribution and immunity to the bending of the waveguide, it is anticipated that the waveguide is operating in the region close to the cutoff of the TE_1 mode. Figures 2.11 and 2.12 show the range of applicability of each method for diffused clad as

TABLE 2.6

b-V for Clad-linear Profile and Clad-parabolic Profile and Exponential Profile

V	Clad-linear Profile			Clad-parabolic Profile			Exponential Profile		
	b (WKB)		b (Exact)	b (WKB)		b (Exact)	b (WKB)		b (Exact)
	Enhanced	Ordinary		Enhanced	Ordinary		Enhanced	Ordinary	
2	–	–	–	–	–	–	0.1086	0.0831	0.1050
3	–	–	0.0335	0.981	0.000	0.1577	0.23331	0.2054	0.2292
4	0.1333	0.0792	0.1479	0.3081	0.2500	0.3262	0.3249	0.2992	0.3212
5	0.2498	0.2045	0.2500	0.4417	0.4000	0.4475	0.3939	0.3705	0.3903
10	0.5218	0.5001	0.5182	0.7149	0.7000	0.7153	0.5809	0.5658	0.5781
100	0.8945	0.8923	0.8939	0.9705	0.9700	–	0.8974	0.8954	0.8968

TABLE 2.7

b-V for Gaussian Profile, erfc Profile and cosh-2 Profile

	Gaussian Profile			erfc Profile			Cosh-2 Profile		
	b (WKB)		b (Exact)	b (WKB)		b (Exact)	b (WKB)		b (Exact)
V	Enhanced	Ordinary		Enhanced	Ordinary		Enhanced	Ordinary	
2	0.0452	0.0104	0.817	–	–	–	0.1001	0.0625	0.1231
3	0.2538	0.2071	0.2750	0.0575	0.0281	0.0677	0.2908	0.2500	0.3074
4	0.4008	0.3630	0.4133	0.1651	0.1293	0.1695	0.4244	0.3906	0.4357
5	0.3939	0.3705	0.3903	0.5013	0.4712	0.5095	0.2539	0.2198	0.2552
10	0.5809	0.5658	0.5781	0.7301	0.7173	0.7323	0.4991	0.4776	0.4971
100	0.9706	0.9702	0.9707	0.8959	0.8835	0.8852	0.9707	0.9702	0.9708

well as clad power-law profiles. Except for the Gaussian profile, all other profiles show that the enhanced WKB method is sufficiently accurate if the operating point lies in the range $0.75 < V/V_c < 1.0$. The discrepancies in the Gaussian profile are caused by the steepness of the profile. Errors caused by the presence of the uniform substrate index should taper off in the stated range of validity. The variational method with a Hermite-Gaussian field cannot be used for the calculation of b for any of these profiles due to the poor overlapping between the trial field and the actual field. The waveguide designs should use an appropriate method for a particular application.

2.3.2 Symmetrical Waveguides

In this section we deal mainly with symmetrical optical waveguides and dwell mainly on the equivalent moment methods described above. The WKB method is also treated briefly. It involves only a slight modification of the previous equations and the entries given in Table 2.8. As presented the WKB method performs poorly in these kind of waveguides.

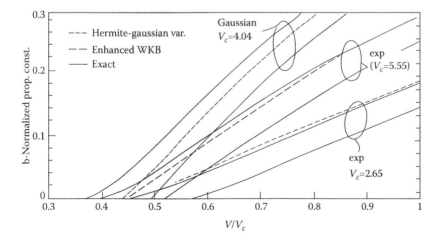

FIGURE 2.11

Dispersion characteristics: Comparison of methods of analysis of single mode clad-power-law with $n_s = 2.17$; $\Delta = 0.043$ of index profile of Gaussian and exponential shape.

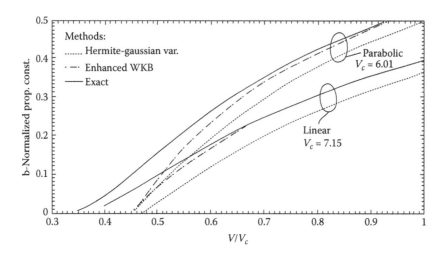

FIGURE 2.12
Dispersion characteristics: Comparison of methods of analysis of single mode clad-power-law with $n_s = 2.17$; $\Delta = 0.043$ of index profile of parabolic and linear shape.

On the other hand the moment method (based on cosh-2 profile) is accurate in the range of frequencies for single mode operation.

2.3.2.1 Wentzel–Kramers–Brilluoin Eigenvalue Equation

For a symmetrical optical waveguide Equation 2.27 becomes

$$2V \int_0^{X_t} \sqrt{S(X) - b} \, dX = (m + \gamma_s)$$

(2.28)

$$\gamma_s = \frac{1}{2} \quad \text{diffused waveguides}$$

TABLE 2.8

WKB Calculated Cutoff Frequencies Two Lowest Order Modes TE_0 and TE_1

Profile	Vc_1, Vc_2 (WKB) Enhanced	Vc_1, Vc_2 (WKB) Ordinary	Vc_1, Vc_2 (Exact)
Clad-linear	3.2	3.53	2.46
	7.92	8.24	7.15
Clad-parabolic	2.72	3.0	1.96
	6.72	7.0	6.01
Exponential	1.07	1.18	1.09
	2.64	2.75	2.65
Gaussian	2.14	2.56	1.43
	5.28	5.96	4.04
erfc	2.32	2.56	2.09
	5.72	5.96	5.55
Cosh-2	1.36	1.50	1.24
	3.36	3.50	3.32

If the turning point coincides with a dielectric discontinuity, the correct phase shift formula is to be used. For buried modes the complicated expression is given in Appendix B. This section is limited to the profiles following diffused or clad-power shape. The factor of 2 in Equation 2.28 is accounted for the WKB-defined effective guide width now extended from $-X_t$ to X_t, the turning points on both sides of the guiding region. Thus the formulae in Table 2.5 for the estimation of b and V can be translated to symmetrical optical waveguides by transforming $V \to 2V$ and $\gamma \to \gamma_s$.

2.3.2.2 Two-Parameter Profile-Moment Method

The profile-moment method is related to the variational formalism of optical waveguide problems [59]. The trial field is derived from that of a reference profile where an exact analytical expression is available. It is known that the field distribution of the fundamental mode follows a bell-shape like trend. Thus by adjusting the V-parameter, a close match to the modal field can be obtained. This condition can be satisfied by monotonously varying the variational parameters.

2.3.2.2.1 Theoretical Basis

The starting point is that for two symmetrical waveguides having the same substrate index, normalized mode propagation constants are related by [60]

$$\beta^2 - \beta_r^2 = \frac{k_0^2 \int_{-\infty}^{\infty} \left[n^2(x) - n_r^2(x) \right] \varphi(x) \varphi_r(x) \, dx}{\int_{-\infty}^{\infty} \varphi(x) \varphi_r(x) \, dx} \tag{2.29}$$

where the subscript r indicates the quantities belong to the reference waveguide. Then the condition for the two waveguides to be equivalent is $\beta = \beta_r$, thus we have

$$\int_{-\infty}^{\infty} \left[n^2(x) - n_r^2(x) \right] \varphi(x) \varphi_r(x) \, dx = 0 \tag{2.30}$$

One can express the product of the field of this equation as a series as

$$\varphi(x) \varphi_r(x) = \sum_{l=0}^{\infty} c_l(k_0) x^{2l} \tag{2.31}$$

where c_l are the frequency-dependent coefficients of the series. Thus Equation 2.30 becomes

$$\sum_{l=0}^{\infty} c_l(k_0) \int_{0}^{\infty} \left[n^2(x) - n_r^2(x) \right] x^{2l} \, dx = 0 \tag{2.32}$$

Since we impose the condition that these waveguides have the same substrate index we can write Equation 2.32 in terms of the profile shape function $S(X)$ leading to

$$\sum_{l=0}^{\infty} c_l(k_0) \left[N_{2l} - N_{2lr} \right] = 0 \tag{2.33}$$

in which N_{2l} can be identified by

$$N_{2l} = 2\left[n_0^2 - n_s^2\right]d^{2(l+1)}\Omega_{2l} = 0 \tag{2.34}$$

where Ω_{2l} is defined as

$$\Omega_{2l} = \int_0^\infty S(X)X^{2l}\,dX = 0 \tag{2.35}$$

For profiles which are nearly identical, one can assume that for $\beta = \beta_r$ over the range of k_0 for which the fields are slowly varying, it is sufficient to retain only two terms in the series. Thus we have

$$N_0 = N_{0r}$$
$$\tag{2.36}$$
$$N_2 = N_{2r}$$

Expanding these two terms we have

$$\left[n_0^2 - n_s^2\right]d\Omega_0 = \left[n_{0r}^2 - n_s^2\right]d_r\,\Omega_{0r}$$
$$\tag{2.37}$$
$$\left[n_0^2 - n_s^2\right]d^3\Omega_2 = \left[n_{0r}^2 - n_s^2\right]d_r\,\Omega_{2r}$$

which can also be expressed in terms of the normalized frequency V-parameter as

$$\frac{V}{V_r} = \left\{\frac{\Omega_0\Omega_2}{\Omega_{0r}\Omega_{2r}}\right\}^{1/4} \tag{2.38}$$

2.3.2.2.2 Estimation of Normalized Propagation Constant

The normalized propagation constant b can be expressed in terms of V and β as

$$b(V) = \left(\frac{d}{V}\right)^2 \frac{n_e^2 - n_s^2}{k_0^2} \tag{2.39}$$

where $n_e = \beta/k_0$ is the effective refractive index of the guided mode or the refractive index of the guided medium as seen by the mode along the z direction. Since $n_e = n_{er}$ thence

$$\frac{b(V)}{b(V_r)} = \frac{\Omega_0}{\Omega_{0r}}\left\{\frac{\Omega_0\Omega_2}{\Omega_{0r}\Omega_{2r}}\right\}^{1/2} \tag{2.40}$$

This equation states that the propagation $b(V)$ of an arbitrary waveguide can be derived from that of a reference waveguide provided that the profile moments and the dispersion relation for $b_r(V_r)$ are known. Table 2.9 lists the three lowest moments of profiles having analytical forms of their shape functions. The profiles listed in this table having step, clad-linear, exponential and cosh have exact analytical solutions for their propagation constant. Thus any of these profiles can be employed as a reference profile.

We select two profiles, the step and cosh profiles for two case studies as follows:

TABLE 2.9

Profile Moments of Selected Profile Shape

Profile	Ω_0	Ω_2	Ω_4	Ω_4/Ω_2	SDF [see Equation 2.51]
Step	1	0.333	0.2	0.6	1.0
Clad-linear	0.5	0.0833	0.0033	0.40	0.67
Clad-parabolic	0.667	0.133	0.571	0.43	0.72
Exponential	1	2	24	12	10.08
Gaussian	0.866	0.443	2.659	6	5.04
erfc(x)	0.564	0.188	0.226	1.2	1.01
Cosh-2(x)	1	0.693	0.823	1.19	1.00

2.3.2.2.2.1 Step Reference Profile [20]

$$V_r\sqrt{1-b_r} = m\frac{\pi}{2} - \tan^{-1}\left(\sqrt{\frac{b_r}{1-b_r}}\right) \qquad m = 0,1,2... \tag{2.41}$$

where

$$V_r = \left[3\Omega_0\Omega_2\right]^{1/4} V$$

$$b = \left[\frac{\Omega_0^3}{3\Omega_2}\right]^{1/2} b_r \tag{2.42}$$

2.3.2.2.2.2 Cosh Reference Profile [20]

$$b_r = \left\{\left(1+\frac{1}{4V_r^2}\right)^{1/2} - \frac{1}{V_r}\left(m+\frac{1}{2}\right)\right\}^2 \qquad m = 0,1,2,... \tag{2.43}$$

where

$$V_r = \left[\frac{12\Omega_0\Omega_2}{\pi^2}\right]^{1/4} V$$

$$b = \frac{\pi}{2}\left[\frac{\Omega_0^3}{3\Omega_2}\right]^{1/2} b_r \tag{2.44}$$

These equations are required for calculation of the dispersion relation characteristics.

2.3.2.2.3 Estimation of TE₁ Mode Cutoff

The next higher order mode is TE$_1$. The cutoff frequency of this mode is the upper limit of the single mode operation. We can write the product of the guided waves of the reference waveguide and the one to be analyzed as

$$\varphi(x)\varphi_r(x) = \sum_{l=0}^{\infty} c_l(k_0)x^{2l+2} \tag{2.45}$$

and similarly

$$N_2 = N_{2r}$$
$$N_4 = N_{4r} \tag{2.46}$$

Thus the relationship between the profile moments and the refractive index can be obtained as

$$\left[n_0^2 - n_s^2 \right] d^3 \Omega_2 = \left[n_{0r}^2 - n_s^2 \right] d_r^{\,3} \Omega_2$$
$$\left[n_0^2 - n_s^2 \right] d^5 \Omega_4 = \left[n_{0r}^2 - n_s^2 \right] d_r^{\,5} \Omega_4 \tag{2.47}$$

The estimation of non-profile moment terms using the definition of V for each guide gives the desired mode-cutoff relation after setting $b = 0$.

2.3.2.2.3.1 Step Reference Profile [20]

$$V_c = \frac{\pi}{2} \left(\frac{5\Omega_4}{27\Omega_2} \right) \tag{2.48}$$

2.3.2.2.3.2 Cosh Reference Profile

$$V_c = \sqrt{2} \left(\frac{5\pi^2 \Omega_4}{252\Omega_2^3} \right)^{1/4} \tag{2.49}$$

where the cutoffs for the reference profiles have been derived from Equations 2.41 and 2.43.

The propagation constant and the cutoff frequency V_c are tabulated in Tables 2.10 and 2.11 for two typical profiles, the step and clad-linear types, and Tables 2.12 through 2.15 are for clad-parabolic, exponential, Gaussian and erfc(x) profiles, respectively. Note that the moments of the complementary error function are not listed in Table 2.9 as there are no close form solutions.

2.3.2.2.4 Choice of Methods

There are some interesting insights as observed from the tables.

In the clad-power law profiles, the moment-equivalent step index profile (moment-ESI) method consistently gives better results for both the propagation constant and the cutoff frequencies of TE_0 mode.

On the other hand, diffused waveguides characterized by non-decreasing higher order moments of Table 2.9 are more accurately modeled by the cosh profile.

At low frequencies both approaches are asymptotically exact.

TABLE 2.10

b-V Data for Step Profile

| V | b (Moment) | | b (WKB) | b (Exact) |
	Step Reference	Cosh Reference		
0.5	b = b (exact)	0.192	<0	0.189
1.0	For all V	0.481	0.383	0.454
1.5		0.697	0.726	0.628
2.0		0.848	0.846	0.725
3.0		>1	0.931	0.849
4.0		>1	0.961	0.902

TABLE 2.11

b-V Data for Clad-Linear Profile

| V | b (Moment) | | b (WKB) | b (Exact) |
	Step Reference	Cosh Reference		
0.5	0.0560	0.0563	<0	0.0561
1.0	0.173	0.177	<0	0.174
1.5	0.286	0.300	0.149	0.290
2.0	0.375	0.404	0.297	0.384
3.0	0.491	0.558	0.464	0.515
4.0	0.558	0.660	0.557	0.579

TABLE 2.12

b-V Data for Clad-parabolic Profile

| V | b (Moment) | | b (WKB) | b (Exact) |
	Step Reference	cosh Reference		
0.5	0.0951	0.0959	<0	0.0952
1.0	0.270	0.280	<0	0.272
1.5	0.419	0.448	0.333	0.423
2.0	0.525	0.580	0.500	0.535
3.0	0.653	0.762	0.667	0.673
4.0	0.721	0.877	0.750	0.751

TABLE 2.13

b-V Data for Exponential Profile

| V | b (Moment) | | b (WKB) | b (Exact) |
	Step Reference	Cosh Reference		
0.5	0.142	0.148	0.0205	0.152
1.0	0.263	0.294	0.205	0.317
1.5	0.320	0.387	0.337	0.424
2.0	0.350	0.431	0.426	0.498
3.0	0.377	0.491	0.539	0.593
4.0	0.389	0.525	0.609	0.653

TABLE 2.14

b-V Data for Gaussian Profile

| V | *b* (Moment) | | *b* (WKB) | *b* (Exact) |
	Step Reference	Cosh Reference		
0.5	0.146	0.148	<0	0.147
1.0	0.342	0.363	0.207	0.354
1.5	0.466	0.520	0.421	0.498
2.0	0.541	0.628	0.549	0.594
3.0	0.620	0.763	0.688	0.709
4.0	0.657	0.842	0.762	0.774

The eigenvalue equation, Equation 2.41, can be reduced to

$$b(V \to 0) = \left[\left(1 + \frac{1}{4V^2} \right)^{1/2} - \frac{1}{2V} \right]^2 \tag{2.50}$$

which is just the eigenvalue equation for the cosh profile. This comparison is valid only when the profiles have equal volume Ω_0.

For large value V their dispersion curve splits and higher order moment scaling factors in Equations 2.41 and 2.43 have to be used. However due to different properties of the higher order moments of the step and cosh profiles, neither one can be used to predict each other's dispersion characteristics accurately.

The WKB method gives consistently better results at large frequencies. To give an idea of the asymptotic range of applicability of the WKB and moment methods, the dispersion curves for the diffused, as well as the clap-power law profile, are plotted in Figures 2.13 through 2.16.

2.3.2.2.5 A New Method for Profile Classification

Table 2.15 and Figure 2.16 indicate that the dispersion characteristic of the complementary error function profile is well above the expected accuracy of the moment method as calculated from a cosh reference profile. Even at $V = 4.0$ a near perfect agreement is obtained. To account for such observations a shape derivation factor (SDF) can be proposed as

TABLE 2.15

b-V Data for erfc(*x*) Profile

| V | *b* (Moment) | | *b* (WKB) | *b* (Exact) |
	Step Reference	Cosh Reference		
0.5	0.0672	0.0679	<0	0.0678
1.0	0.187	0.194	0.0281	0.193
1.5	0.286	0.306	0.177	0.304
2.0	0.355	0.394	0.293	0.391
3.0	0.436	0.512	0.443	0.509
4.0	0.479	0.586	0.534	0.584

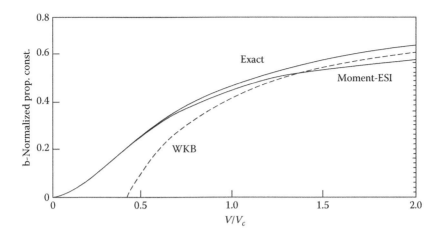

FIGURE 2.13
Dispersion characteristics: range of applicability of WKB method and moment method. Symmetric clad-linear profile with $A = 20$; V_c (TE$_1$ mode) = 2.7995.

$$SDF = \frac{\Omega_4 / \Omega_2}{\Omega_{4r} / \Omega_{2r}} \qquad (2.51)$$

where the subscript r is referred to the reference profile. For clad profiles one can chose the step profile as a reference whereas for diffused waveguides the cosh profile offers a much better fit. This SDF parameter is thus entered in Table 2.10. We could see the benefit of this factor for erfc(x) profile which has an SDF factor of 1.01 as compared to 10 for an exponential profile. Thus the former method offers better accuracy.

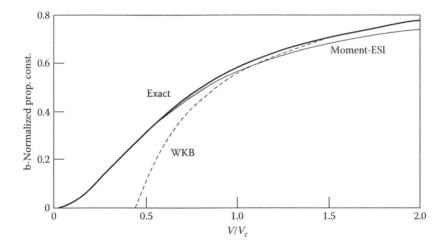

FIGURE 2.14
Dispersion characteristics: range of applicability of WKB method and moment method. Symmetric clad-parabolic profile with $A = 20$; V_c (TE$_1$ mode) = 2.330; $n_c = 2.177$; $\Delta = 0.043$.

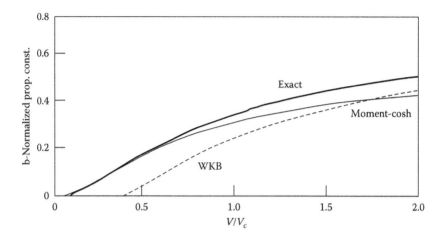

FIGURE 2.15
Dispersion characteristics: range of applicability of WKB method and moment—cosh method. Symmetric exponential profile with $A = 20$; V_c (TE$_1$ mode) = 1.2024; $n_c = 2.177$; $\Delta = 0.043$.

2.3.2.3 New Equivalence Relation for Planar Optical Waveguides

By sketching the spatial distribution of the modal field of the TE$_1$ mode in a symmetrical waveguide and the TE$_0$ field in an asymmetrical waveguide, we find out why the profile-moment method could not work in both cases. However there is some surprise when $X \geq 0$ the distribution is similar if $A \to \infty$ as shown in Figure 2.17. For the same profile, if the field distributions are identical then there exists a relation between the dispersion curves of both structures. Figure 2.18 illustrates the correspondence between the modes of both waveguide structures. One can postulate that

$$V_{ca1} = V_{cs2}$$

$$V_{ca2} = V_{cs4}$$

$$(2.52)$$

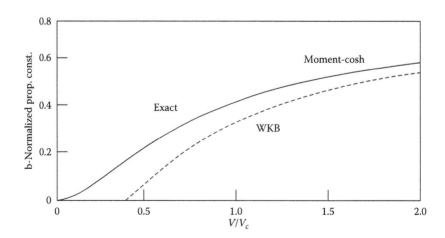

FIGURE 2.16
Dispersion characteristics: Range of applicability of WKB method and moment-cosh method. Symmetric erfc(x) profile with $A = 20$; V_c (TE$_1$ mode) = 2.3187; $n_c = 2.177$; $\Delta = 0.043$.

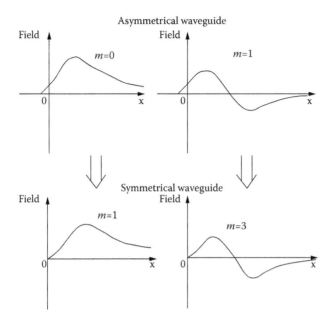

FIGURE 2.17
Correspondence between the modal fields of asymmetrical and symmetrical waveguides with the field distribution of odd modes.

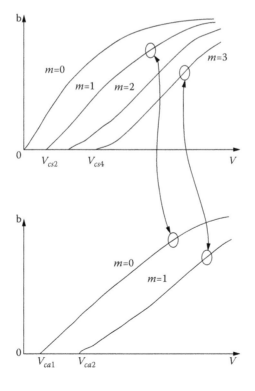

FIGURE 2.18
The m-mode dispersion curves of asymmetrical and symmetrical waveguides from the $(2m + 1)$th mode dispersion curve of the corresponding symmetrical waveguide.

TABLE 2.16

Ratio of TE_0 and TE_1 Mode Cutoff
V-parameter in Asymmetrical Waveguides
(Obtained for Profiles Without Analytical
Solutions by Forward Recurrence Algorithm
With 0.01 Step Size)

Profile	$\dfrac{V_{ca2}}{V_{ca1}}$	
	$A \sim 20$	$A \to 20$
Step	3.33	3.0
Clad-linear	2.91	2.67
Clad-parabolic	3.07	2.78
Exponential	2.43	2.30
Gaussian	2.83	2.57
erfc(x)	2.65	2.47
Cosh-2	2.68	2.45

where the left hand side (LHS) denotes the cutoffs of the TE_0 and TE_1 modes in the asymmetrical waveguide. V_{cs2}; V_{cs4} are the cutoffs of the TE_1 and TE_3 modes of the corresponding symmetrical waveguide.

Furthermore, it is noted that the separation of the b-V characteristic curves at cutoff of the symmetrical waveguide is nearly uniform. For the step profile, this separation equals $\pi/2$ whereas for graded profiles, this is only approximately true. Thus we can write

$$V_{cs4} \simeq 3V_{cs2} \tag{2.53}$$

Combining Equations 2.52 and 2.53 leads to

$$\left.\frac{V_{ca2}}{V_{ca1}}\right|_{A\to\infty} \simeq 3 \tag{2.54}$$

This equation allows us to conduct preliminary tests on the postulation above. Table 2.16 tabulates the ration of the V-parameters at cutoff of the modes TE_0 and TE_1 for $A \sim 20$ and $A \to \infty$ for the profiles listed. It shows that Equation 2.54 does not satisfy for the step profile. This may be contributed to by the error in the assumptions about the separation of the b-V curves at cutoff in Equation 2.53. To see if Equation 2.52 can be satisfied one can compare the TE_0 mode cutoffs of the symmetrical waveguide as $A \to \infty$. The cutoffs are so close in the last two columns of the Table 2.17 as to be considered exactly equal. Thus this is the new corresponding relationship between the m-modes of an asymmetrical waveguide and the odd $(2m + 1)$th modes of the corresponding symmetrical waveguide.

2.3.2.3.1 Solution of a Simple Symmetric Waveguide Solution

2.3.2.3.1.1 Structure   A symmetric slab or planar optical waveguide consists of a slab (or core) of dielectric "transparent" material of refractive index n_1, embedded between two layers of materials of index n_2 acting as the substrate and superstrate layers as shown in Figure 2.1. The refractive index of the core is higher than those of the substrate and superstrate (cladding) in order for lightwaves to be guided (see Figure 2.19).

TABLE 2.17

Prediction of Odd Mode Cutoffs in Symmetrical Waveguides from the
Mode Cutoffs of the Corresponding Waveguides

Profile	V_{ca1} (Asymmetrical Waveguide TE_0)		V_{ca4} (Symmetrical TE_1)
	$A \sim 20$	$A \to 20$	
Step	1.35	1.57	1.57
Clad-linear	2.46	2.80	2.80
Clad-parabolic	1.96	2.26	2.25
Exponential	1.09	1.20	1.20
Gaussian	1.43	1.64	1.64
erfc(x)	2.09	2.35	2.37
Cosh-2	1.24	1.41	1.41

Assuming that the structure is extended to infinite in y and z directions, and a guiding thickness of 2a and that the materials are isotropic and lossless (i.e., permittivities are real and scalar) and nonmagnetic.

2.3.2.3.1.2 Numerical Aperture If we assume at the moment that total internal reflections at the boundaries are required for guiding, what is the acceptance angle such that light-waves can be launched? The ray path entering the optical fiber core (in the axial plane) or a slab waveguide for total internal reflection is shown in Figure 2.20.

Applying Snell's law at the air-core and core-cladding boundaries of the dielectric waveguide, the total internal reflection can take place only if:

$$n_0 \sin \theta_0 \leq n_1 \cos \theta_c \qquad (2.55)$$

where θ_c is the critical angle such that

$$n_1 \sin \theta_c = n_2 \sin 90 = n_2 \qquad (2.56)$$

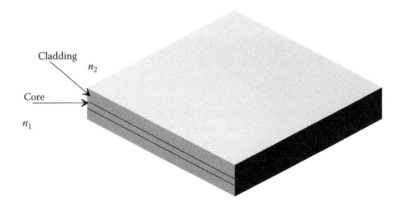

FIGURE 2.19
Cross section of a slab optical waveguide. The optical waveguide is assumed to be confined in the vertical direction x and extended infinitely in the lateral direction y. Lightwaves are guided and propagating along the z direction.

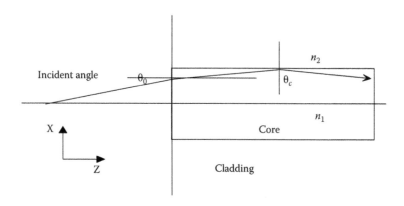

FIGURE 2.20
Numerical aperture of a dielectric waveguide. Lightwaves are approximated as light rays. This is true for the case when several lightwaves propagate in the wave guide. Light rays entering the waveguide interface are refracted and then totally reflected at the core-cladding boundaries. The numerical aperture can be determined by calculating the maximum angle of the incident ray at the entry face of the fiber and air.

Thus the numerical aperture (NA) which is defined as the maximum value of $\sin \theta_0$ is given by Equation 2.55.

$$NA = (\sin \theta_0)_{max} = n_1 \cos \theta_c$$

thus we have

$$NA = (n_1^2 - n_2^2)^{1/2} \tag{2.57}$$

2.3.2.3.1.3 Modes of the Symmetric Dielectric Slab Waveguides   Consider a monochromatic (i.e., single ω or λ) wave propagating in the z direction with its electric field

$$E(x, y, z) = E(x)e^{j(\omega t - \beta z)} \tag{2.58}$$

i.e., field dependent on x, and uniform along the y direction, β is the propagation constant along the z direction, then in the absence of charges and currents. Maxwell's equations representing the fields of the optical magnetic waves of the optical guided modes following a wave equation are given as

$$\nabla^2 E + \frac{1}{c^2}\frac{\partial^2 E}{\partial t^2} = 0 \tag{2.59a}$$

with the time dependent of the electric field as in Equation 2.58 we have $d^2/dt^2 = -\omega^2$ and $d^2/dz^2 = \beta^2$ and substituting into Equation 2.59a and using $\omega^2/c^2 = n^2(\omega)k_0^2$ we have the wave equation:

$$\nabla_t^2 E + (\beta^2 - n^2(\omega)k_0^2)E = 0 \tag{2.59b}$$

where

$$V_t^2 = \frac{\partial^2}{\partial x^2} + \frac{\partial^2}{\partial y^2} \tag{2.59c}$$

Therefore for a planar optical waveguide with an infinite extension in the y direction we have $d/dy = 0$ and for TE modes only E_y is significant, the wave equation becomes:

$$\frac{d^2 E_y}{dx^2} + (\beta^2 - \omega^2 \mu \varepsilon) E_y = 0 \tag{2.60a}$$

where μ and ε are the permeability and permittivity of medium n_1 or n_2. [$\mu = \mu_0$ and $\varepsilon = \varepsilon_r \varepsilon_0$], nonmagnetic and (glass) dielectric material. Similarly a wave equation involved H_y is given by

$$\frac{d^2 H_y}{dx^2} + (\beta^2 - \omega^2 \mu \varepsilon) H_y = 0 \tag{2.60b}$$

Equation 2.60a and b can be rewritten using $k = \omega/c$ and $c = (\mu_0 \varepsilon_0)^{-1/2}$ is the light velocity in vacuum and k is the wave number in vacuum, as

$$\frac{d^2 E_y}{dx^2} + (\beta^2 - k^2 n_j^2) E_y = 0 \tag{2.60c}$$

$$\frac{d^2 H_y}{dx^2} + (\beta^2 - k^2 n_j^2) H_y = 0 \tag{2.60d}$$

where $n_j = n_1$ or n_2 depending on whether the equations are applied in the core or cladding regions; $n_1 = (\varepsilon_{r1})^{1/2}$, $n_2 = (\varepsilon_{r2})^{1/2}$ with ε_{r1} and ε_{r2} are the relative permittivities of the core and cladding regions, respectively. From Equation 2.60c and d we observe that the variation of the fields along the axis Ox as

- *Sinusoidal* behavior when $k^2 n_j^2 > \beta^2$ or guided waves inside the core
- *Exponential* (decay) behavior when $k^2 n_j^2 < \beta^2$, i.e., no radiation in cladding region

In other words, for a properly designed optical waveguide the optical field is oscillating in regions where the longitudinal propagation constant is smaller than the plane-wave propagation constant and "evanescent" with exponential-like behavior elsewhere. Thus the lightwaves are "trapped" in the core region and guided through the waveguide length. In the next section we analyze the wave equation so that conditions for guiding lightwaves are established.

2.3.2.3.1.4 Guided-Modes
(a) General Solutions and the Eigenvalue Equation   Optical waves are guided along the waveguide when their electromagnetic (EM) fields are oscillatory in the slab waveguide region and exponentially decay in the cladding region, that is

$$kn_2 \le \beta \le kn_1 \tag{2.61}$$

We now define a transverse propagation constant **u/a** and transverse decay constant **v/a** as

$$\frac{u^2}{a^2} = k^2 n_1^2 - \beta^2 \tag{2.62a}$$

$$\frac{v^2}{a^2} = -k^2 n_2^2 + \beta^2 \tag{2.62b}$$

Adding Equation 2.62a and b gives:

$$\frac{u^2}{a^2} + \frac{v^2}{a^2} = k^2(n_1^2 - n_2^2) \tag{2.63}$$

or alternatively

$$V^2 = u^2 + v^2 = k^2 a^2 (n_1^2 - n_2^2) \tag{2.64}$$

We observe that the Equation 2.62a and b represent the propagation constant in the transverse direction as illustrated in Figure 2.21.

In order for the lightwaves to be guided or effectively oscillating in the transverse direction we can see that the transverse propagation constant u/a must be positive in the core region and negative in the cladding region.

The parameter V is defined as the normalized frequency (V) which is dependent only on the guide thickness and light frequency (i.e., wavelength) and the refractive index difference between the core and cladding regions (the slab and superstrate or substrate regions for planar optical waveguides).

The field E_y for TE modes and H_y for TM modes are a linear combination of cos(ux/a) and sin(ux/a) inside the core layer (i.e., when $|x| \leq a$) and exponentially decay in the cladding region (that is when $|x| > a$) with exp($-vx/a$) and exp($+vx/a$) in the superstrate or substrate. We therefore have a continuum of optical guided modes depending on whether the solution function follows a symmetrical or anti-symmetric pattern (e.g., cosine or sine functions).

Mathematically the general solution of the wave equations given in Equation 2.60c and d would be a combination of the sine and cosine or even and odd functions, respectively. In the following sections we split the solution into two parts, the even and odd modes corresponding with the even and odd functions. We can write a combination of these solutions as we have usually seen done in mathematics of differential equations.

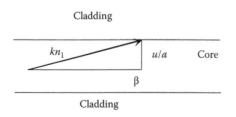

FIGURE 2.21
Representation of the propagation constant along the "ray" direction, the propagation direction z and the transverse direction.

Even TE Modes (for Modes with Solution Function Cosine)   For $|x| \leq a$, that is inside the core region the only significant component for TE mode is E_y and H_x

$$E_y(x) = A \cos \frac{ux}{a} \tag{2.65a}$$

$$H_y = 0 \quad \text{and} \quad H_z = A \sin \frac{ux}{a} \tag{2.65b}$$

For $|x| > a$, that is the field portion of the lightwaves are in the cladding region

$$E_y = Ce^{-\frac{v}{a}(x-a)} \tag{2.66}$$

The arbitrary constants A and C can be found by applying the boundary conditions as follows:

We have the value of E_y at $x = a^+$ and $x = a^-$ must be equal; using Equation 2.65a and b:

$$A \cos u = Ce^{-\frac{v}{a}(x-a)} \tag{2.67}$$

evaluated at $x = a$ this equality becomes

$$C = A \cos u \tag{2.68}$$

The coefficient A (thus C) can then be found by using H_z and one of Maxwell's equation as H_z (at core $x = a^+$) = H_z (at cladding $x = a^-$) at $x = a^+$ in core we have

$$H_z = \frac{1}{j\omega\mu_0} \frac{dE_y}{dz} = \frac{1}{j\omega\mu_0} \left(-\frac{u}{a} A \sin \frac{u}{a} x \right) \tag{2.69}$$

and in cladding at $x = a^-$

$$H_z = \frac{1}{j\omega\mu_0} \left(-C \frac{v}{a} e^{-\frac{v}{a}(x-a)} \right) \tag{2.70}$$

Therefore equating these boundary conditions (Equations 2.69 and 2.70) we obtain

$$C = A \frac{u}{v} \sin u \tag{2.71}$$

The eigenvalue equation for the even modes can be achieved by equating Equations 2.71 and 2.68:

$$v = u \tan u \tag{2.72}$$

This equation is called the eigenvalue equation which can be solved to find the propagation constant β along the z direction. The number of guided modes that can be supported by the slab optical waveguide can then be easily determined. The number of possible values of β gives the number of guided even TE modes, thus whether the waveguide is a single mode or multimode waveguide pending on the number of odd modes possibly supported by the waveguide. This is investigated in the next part.

Odd TE Modes   Similarly for odd TE modes, the solution function follows a sine function. Writing the solution for E_y in the core region and the evanescent field in the cladding regions then applying the boundary conditions at the core-cladding interface we obtain the eigenvalue equation for odd modes as:

$$v = -\frac{u}{\tan u} \tag{2.73}$$

Graphical Solutions   Combining Equations 2.72 and 2.73 we observe that the waveguides can support only discrete modes and the propagation constant β related to u and v parameters can be found by solving graphically the intersection between circles of V and curves representing Equations 2.64, 2.72 or 2.73. These solutions are illustrated in Figure 2.4.

(b) Cutoff Properties   From Figure 2.22 we observe that

- $V = 0$, i.e., zero optical frequency or λ is zero, that is there exists no lightwaves. Thus we observe that we always have at least one guided (even) mode, TE_0
- $V < \pi/2$, there exists only one guided mode TE_0 (fundamental even mode)
- $V > \pi/2$, odd TE_0 mode appears (second mode—fundamental odd mode)
- $V = \pi$, third mode (TE_1, first order even mode)

That is, each time V reaches a multiple integer of $\pi/2$, a new TE mode reaches its cutoff. The fundamental TE even mode is cut off only when $V = 0$, that is when there is no waveguide.

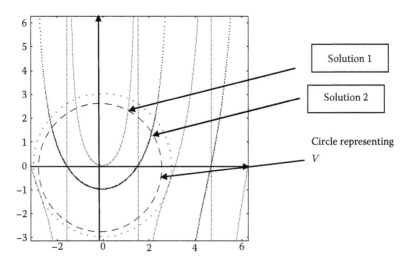

FIGURE 2.22
(See Color Insert) Graphical solution of Equations 2.64, 2.72, or 2.73: _____ $v = u \tan u$, __ __ __ $v = -u/\tan u$, and ------- $V^2 = u^2 + v^2$.

2.3.3 Concluding Remarks

The variational method incorporating a Hermite-Gaussian trial field is inaccurate for calculating the *b-V* dispersion characteristics of single mode optical waveguides. However it is an accurate and convenient tool for mode spot size calculations provided that the mode is well confined in the single-spectral range.

The WKB approach is numerically straightforward without any divergence. The enhanced WKB formulae with the correct phase connection at the dielectric interfaces yield very accurate results even for single mode optical waveguides. This is in contrast with a number of published works.

The profile-moment method is applicable only to symmetrical profiles. An SDF is defined to allow a decision on the choice of reference profile to obtain optimum performance of this method. For diffused profiles, the hyperbolic cosh reference profile is required to yield accurate results. For clad profiles, the step reference profile offers accurate results and is significantly better when SDF → 1.

However the profile-moment method does not cover the entire single mode region for all profiles. At larger *V* its accuracy deteriorates significantly. Neither does the WKB formulae cover the single mode region, a hybrid profile-moment-WKB method has to be considered. A simple profile characterization factor is presented to assess the applicability of the profile-moment method.

From the analysis of the mode field distribution and numerical computation, the *m*-modes of an asymmetrical waveguide and the $(2m + 1)$ odd modes of its corresponding asymmetrical waveguide are directly related. Computations of their mode cutoffs allow us to establish the exact correspondence for an asymmetrical waveguide with infinite asymmetrical factor.

2.4 Appendix A: Maxwell Equations in Dielectric Media

2.4.1 Maxwell Equations

The general Maxwell equations can be written as

$$\nabla \times E = -j\omega\mu H \tag{2.74}$$

$$\nabla \times H = J + j\omega\varepsilon E \tag{2.75}$$

$$\nabla \cdot D = \sigma/\varepsilon_0 \tag{2.76}$$

$$\nabla \cdot B = 0 \tag{2.77}$$

Note that in the above equations: $j = \sqrt{-1}$; $\mu \cong \mu_0$ is the magnetic permeability for non-magnetic materials which normally constitute an optical waveguide; $\varepsilon = \varepsilon_0 n^2$ is the dielectric constant of the material where ε_0 is the dielectric constant of free space and n is the refractive index of the materials; J is the current density and σ is the surface charge density, which are possible sources. The displacement vector D is related to the electric field via $D = \varepsilon E$ and the magnetic field induction B is related to the magnetic field via $B = \mu H$.

In practice, problems of optical waveguide and couplers are often analyzed in the regions that are free of the above sources, i.e., $J = 0$, and $\sigma = 0$. In these cases, we have

$$\nabla \times E = -j\omega\mu H \tag{2.78}$$

$$\nabla \times H = j\omega\varepsilon E \tag{2.79}$$

$$\nabla \cdot (\varepsilon E) = 0 \tag{2.80}$$

$$\nabla \cdot H = 0 \tag{2.81}$$

2.4.2 Wave Equation

With the usual expression of the time dependent lightwave carrier modulated signals $e^{j\omega t}$, the wave equation can thus be obtained as:

$$\nabla \times \nabla \times \vec{E} - \varepsilon(\omega)\frac{\omega^2}{c^2}\vec{E} = 0 \tag{2.82}$$

The refractive index of the medium can be related to the permittivity including the nonlinear third order effects. Using the relation:

$$\nabla \times \nabla \times \vec{E} = \nabla(\nabla \cdot \vec{E}) - \nabla^2\vec{E} = -\nabla^2\vec{E}$$
$$\because \nabla \cdot \vec{D} = 0 \tag{2.83}$$

then the wave equation

$$\nabla^2\vec{E} - n^2(\omega)\frac{\omega^2}{c^2}\vec{E} = 0$$

2.4.3 Boundary Conditions

In the regions free of the sources, we have the following boundary conditions

- Continuity of the magnetic field and the component of the electric field tangential to the interface, i.e.,

$$H^{(1)} = H^{(2)}, E^{(1)} /\!/ = E^{(2)} /\!/ ; \tag{2.84}$$

- Continuity of the normal component of the displacement vector, i.e.,

$$D_\perp^{(1)} = D_\perp^{(2)} \quad \text{or} \quad n_1{}^2 E_\perp^{(1)} = n_2{}^2 E_\perp^{(2)} \tag{2.85}$$

2.4.4 Reciprocity Theorems

2.4.4.1 General Reciprocity Theorem

From the above source-free Maxwell equations, we have for two optical media of dielectric constants ε_1 and ε_2:

$$\nabla \times (\nabla \times E) = \omega^2 \varepsilon \mu E \tag{2.86}$$

$$\nabla \times (\nabla \times H) = \omega^2 \varepsilon \mu H \tag{2.87}$$

Using the above equations and the identity $\nabla \cdot (A \times B) = B \cdot (\nabla \times A) - A \cdot (\nabla \times B)$, we have

$$\nabla \cdot (E_1 \times H_2 - E_2 \times H_1) = j\omega(\varepsilon_2 - \varepsilon_1)E_1 \cdot E_2 \tag{2.88}$$

and its integral equivalence

$$\frac{\partial}{\partial z} \iint_{A^\infty} (E_1 \times H_2 - E_2 \times H_1) \cdot \hat{z}\, dA = j\omega \iint_{A_\infty} [\varepsilon_2(x,y) - \varepsilon_1(x,y)]E_1 \cdot E_2\, dA \tag{2.89}$$

2.4.4.2 Conjugate Reciprocity Theorem

The conjugate reciprocity theorem can be obtained in a similar way as above, except for using the conjugate form of filed expressions. This is particularly convenient in constructing the formulation for the lossless waveguides or couplers, in particular the expression of the power conservation. Following some algebra, we have

$$\nabla \cdot \left(E_\mu \times H_\nu^* + E_\nu^* \times H_\mu\right) = -j\omega\left(\varepsilon_\mu - \varepsilon_\nu\right)E_\mu \cdot E_\nu^* \tag{2.90}$$

$$\frac{\partial}{\partial z} \iint_{A_\infty} (E_\mu \times H_\nu^* + E_\nu^* \times H_\mu) \cdot \hat{z}\, dA = -j\omega \iint_{A_\infty} (\varepsilon_\mu - \varepsilon_\nu)E_\mu \cdot E_\nu^*\, dA \tag{2.91}$$

2.5 Appendix B: Exact Analysis of Clad-Linear Optical Waveguides

The exact analysis of TE modes guided in an optical waveguide whose refractive index profile follows a clad-linear shape. The profile was first analyzed exactly by Marcuse [23], then treated in full by Adams [61] and applied to the study of low threshold current laser diode by Walpole et al. [62]. The results presented here are different from these published formulae as they are expressed in terms of Bessel functions of real positive order. Starting with the eigenvalue equation we derive the propagation constant and the cutoffs of the waveguide. The treatments of symmetrical and asymmetrical profiles are given separately.

2.5.1 Asymmetrical Clad-Linear Profile

2.5.1.1 Eigenvalue Equation

The eigenvalue equation is given by [61]

$$\frac{Ai'(\alpha_0) - V^{1/3}\sqrt{A+b}\,Ai'(-\alpha_0)}{Bi'(\alpha_0) - V^{1/3}\sqrt{A+b}\,Bi'(-\alpha_0)} = \frac{Ai'(\alpha_1) - V^{1/3}\sqrt{b}\,Ai(\alpha_1)}{Bi'(\alpha_1) - V^{1/3}\sqrt{A+b}\,Bi(-\alpha_1)} \tag{2.92}$$

where $\alpha_0 = (1-b)V^{2/3}$; $\alpha_1 = bV^{2/3}$; Ai and Bi are Airy functions and the dash denotes the derivatives with respect to the argument.

Using the relations between the Airy and Bessel functions of Ref. [63], we can convert Equation 2.92 into an immediate form

$$\frac{J_{-2/3}(\gamma_0) - J_{2/3}(\gamma_0) + Q[J_{-1/3}(\gamma_0) + J_{1/3}(\gamma_0)]}{J_{-2/3}(\gamma_0) - J_{2/3}(\gamma_0) + Q[J_{-1/3}(\gamma_0) - J_{1/3}(\gamma_0)]}$$

$$= \frac{-I_{-2/3}(\gamma_1) + I_{2/3}(\gamma_1) + I_{-1/3}(\gamma_1) - I_{1/3}(\gamma_1)}{I_{-2/3}(\gamma_1) + I_{2/3}(\gamma_1) + I_{-1/3}(\gamma_1) + I_{1/3}(\gamma_1)} \tag{2.93}$$

where the arguments γ_0; γ_1 of the Bessel functions and Q are given as

$$\gamma_0 = \frac{2}{3}V(1-b)^{3/2}$$

$$\gamma_1 = \frac{2}{3}V(b)^{3/2} \tag{2.94}$$

$$Q = \left[\frac{A+b}{1-b}\right]^{3/2}$$

In arriving at Equation 2.93 we have also used[*]

$$Ai'(\alpha_1) = -\frac{\alpha_1}{3}(I_{-2/3} - I_{2/3}) \tag{2.95}$$

Finally using the recurrence relations for the J and I functions results in the required version of the eigenvalue equation

$$\frac{(2+3\gamma_0 Q)J_{1/3}(\gamma_0) - 3\gamma_0 J_{4/3}(\gamma_0)}{(4Q+3\gamma_0)J_{2/3}(\gamma_0) - 3\gamma_0 J_{5/3}(\gamma_0)} = \frac{(2+3\gamma_1 Q)I_{1/3}(\gamma_0) - 3\gamma_1 I_{4/3}(\gamma_1)}{(4+3\gamma_1)I_{2/3}(\gamma_1) - 3\gamma_1 I_{5/3}(\gamma_1)} \tag{2.96}$$

---

[*] *Note*: The Equation 2.95 given in Abramowitz and Stegun [63] does not have the negative sign—a vital error.

2.5.1.2 Mode Cutoff

At cutoff we have $b = 0$ and $Q = [A]^{1/2}$; $\gamma_0 = 2/3V$; $\gamma_1 = 0$. The RHS of Equation 2.96 $\to \infty$ and a substitution of the asymptotic formula for the Bessel functions gives the mode cutoff for modes with large V_C

$$\frac{2}{3}V_C \sim \left(m + \frac{1}{12}\right)\pi - \tan^{-1}\left(\sqrt{A}\right) \quad m = 0, 1, 2... \tag{2.97}$$

2.5.2 Symmetrical Waveguide

2.5.2.1 Eigenvalue Equation

The eigenvalue equations for the odd and even modes are given in Ref. [61] and can be derived from Equation 2.92. Only the LHS is affected and the eigenvalue equations for these modes are given as

$$\frac{Ai'(-\alpha_0)}{Bi'(-\alpha_0)} = \frac{Ai'(\alpha_1) - V^{1/3}\sqrt{b}\,Ai\,(\alpha_1)}{Bi'(\alpha_1) - V^{1/3}\sqrt{A+b}\,Bi\,(-\alpha_1)} \quad \text{even mode}$$

$$\frac{Ai\,(-\alpha_0)}{Bi(-\alpha_0)} = \frac{Ai'(\alpha_1) - V^{1/3}\sqrt{b}\,Ai\,(\alpha_1)}{Bi'(\alpha_1) - V^{1/3}\sqrt{A+b}\,Bi\,(-\alpha_1)} \quad \text{odd mode} \tag{2.98}$$

Similarly using the relations between the Airy and Bessel functions of Ref. [63], we can convert Equation 2.98 into an immediate form

$$\frac{-2J_{-2/3}(\gamma_0) + 3\gamma_0 J_{4/3}(\gamma_0)}{3\gamma_0 J_{2/3}(\gamma_0)} = \frac{-I_{-2/3}(\gamma_1) + I_{2/3}(\gamma_1) + I_{-1/3}(\gamma_1) - I_{1/3}(\gamma_1)}{I_{-2/3}(\gamma_1) + I_{2/3}(\gamma_1) + I_{-1/3}(\gamma_1) + I_{1/3}(\gamma_1)}$$

$$\frac{-3\gamma_0 J_{1/3}(\gamma_0)}{3\gamma_0 J_{5/3}(\gamma_0) - 4J_{2/3}(\gamma_0)} = \frac{-I_{-2/3}(\gamma_1) + I_{2/3}(\gamma_1) + I_{-1/3}(\gamma_1) - I_{1/3}(\gamma_1)}{I_{-2/3}(\gamma_1) + I_{2/3}(\gamma_1) + I_{-1/3}(\gamma_1) + I_{1/3}(\gamma_1)} \tag{2.99}$$

2.5.2.2 Mode Cutoff

The RHS of Equation 2.96 goes to infinite at cutoff so we have

$$J_{2/3}\left(\frac{2V_C}{3}\right) = 0 \quad m = 0, 2... \text{even modes}$$

$$J_{2/3}\left(\frac{2V_C}{3}\right) - V_C J_{5/3}\left(\frac{2V_C}{3}\right) = 0 \quad m = 1, 3... \text{odd modes} \tag{2.100}$$

Alternatively these equations can be obtained from Equation 2.97.

2.6 Appendix C: Wentzel–Kramers–Brilluoin Method, Turning Points and Connection Formulae

2.6.1 Introduction

Consider the scalar wave equation

$$\frac{d^2\varphi(x)}{dx^2} + K^2\varphi(x) = 0 \tag{2.101}$$

where $K^2 = k_0^2 n^2(x) - \beta^2$ is the transverse propagation constant and $\varphi(x)$ is the modal field in the transverse plane. The characteristic mode factor $e^{j(\omega t - \beta z)}$ is omitted. This version of the wave equation is selected to present the turning points in the subsequent analysis. The turning point is defined by $x_t \rightarrow K(x_t) = 0$.

When the refractive index function $n(x)$ has a certain simple form, Equation 2.101 can be solved explicitly for $\varphi(x)$, the good behavior of this function restricting the axial propagation constant β to discrete values. These are the characteristics of the bound modes. However in most practical optical waveguides, explicit solutions of the fields are not available and approximation methods of solutions must be developed.

The WKB method is based on an asymptotic expansion in k_0^{-1}, the first term of which leads to geometrical optic results, or the 0th order WKB solutions, and higher order terms lead to exact modal solutions. The principal concern of this method lies in the transitional region which connects the oscillatory fields and its evanescent neighbors. These are the turning points of the problem where the semi-classical approximation breaks down. The way in which the WKB solution is valid on either side of the turning point connects, remains as the central problem of the method.

Before proceeding to derive the WKB solutions we assign the following symbols: $\phi(x)$; $\varphi(x)$; $\Phi(x)$ = exact modal field, WKB solution and approximate modal field valid at the turning point.

2.6.2 Derivation of the Wentzel–Kramers–Brilluoin Approximate Solutions

Following established tradition, we postulate a solution of Equation 2.101 in the form of

$$\phi(x) = Ae^{jk_0 S(x)}$$

$$j = \sqrt{-1}; A = \text{constant} \tag{2.102}$$

Thus Equation 2.101 can be transformed into the Riccati equation

$$j\frac{1}{k_0}\frac{d^2 S(x)}{dx^2} - \left(\frac{dS(x)}{dx}\right)^2 + \left(n^2(x) - n_e^2\right) = 0 \tag{2.103}$$

$$n_e \equiv \text{effective index of waveguide mode}$$

Now let $y = S'$ and assume that y admits of a formal series expansion of the form:

$$y = \sum_{n=0}^{\infty} k_0^{-n} y_n \qquad (2.104)$$

Thence

$$y' = S'' = \sum_{n=0}^{\infty} k_0^{-n} y_n' \qquad (2.105)$$

and

$$y^2 = S'^2 = \left\{ \sum_{n=0}^{\infty} k_0^{-n} y_n' \right\}^2 = y_0^2 \left\{ 1 + k_0^{-1} \frac{2y_1}{y_0} + k_0^{-2} \left(\frac{2y_1}{y_0} + \frac{y_1^2}{y_0^2} \right) + \cdots \right\} \qquad (2.106)$$

Therefore substituting Equations 2.105 and 2.106 into Equation 2.103 and equating the coefficients of the like-power of k_0 leads to the following recurrence relations:

$$y_0^2 = n^2 - n_e^2 = \frac{K^2}{k_0^2}$$

$$jy_0^1 = 2y_1 y_0 \qquad (2.107)$$

$$jy_1' = 2y_2 y_0 + y_1^2$$

Thus we can obtain

$$y_0' = \pm \frac{K}{k_0}$$

$$\pm j \frac{K'}{k_0} = \pm \frac{2K}{k_0} y_1 \qquad (2.108)$$

which can be integrated into the form

$$\frac{j}{2k_0} \ln K + C = \frac{1}{k_0} \int_{x_0}^{x} y_1 \, dx \qquad (2.109)$$

where C is the integration constant.

Hence

$$\phi(x) = Ae^{jk_0 S(x)} = Ae^{jk_0 \int_{x_0}^{x} \sum_{n=0}^{\infty} k_0^{-n} 0 y_n \, dx} = Ae^{jk_0 \left\{ \int_{x_0}^{x} y_0 \, dx + k_0^{-1} \int_{x_0}^{x} y_1 \, dx + k_0^{-2} \int_{x_0}^{x} y_2 \, dx + \cdots \right\}}$$

$$= A_{\pm} e^{\pm j \int_{x_0}^{x} K \, dx + jk_0 C}$$

$$\rightarrow \varphi(x) = A_{\pm} K^{-1/2} e^{\pm j \int_{x_0}^{x} K \, dx} \tag{2.110}$$

where $A_{\pm}$ denotes the constants corresponding to the $\pm$ solutions, respectively. For $K^2 > 0$ we have

$$\varphi_1(x) = DK^{-1/2} \cos\left(\pm j \int_{x_0}^{x} K \, dx + \delta \right) \tag{2.111}$$

$$D, \delta \equiv \text{arb. constants}$$

For $K^2 < 0$ we have

$$\varphi_2(x) = B_{\pm} \left| K \right|^{-1/2} e^{\left(\pm \int_{x_0}^{x} |K| \, dx \right)} \tag{2.112}$$

Clearly at the turning point defined by $K^2(x) = 0$ both the oscillatory and evanescent fields diverge. Hence neither form can be retained during the transition from one interval to the other in which K^2 changes sign.

Furthermore a back substitution of the WKB solutions, for example φ into the original Equation 2.101 produces an inhomogeneous equation

$$\frac{d^2\varphi(x)}{dx^2} + K^2\varphi(x) = W(x)$$

$$W(x) = \frac{3}{4}\left(\frac{K'}{K}\right)^2 - \frac{1}{2}\frac{K''}{K} \tag{2.113}$$

For this equation, any point where $K^2(x)$ vanishes is a singular point. However far from the singularity, a higher order WKB solution can be obtained with Equation 2.113 as the starting point and incorporating $XW(x)$ into K^2. This method contrasts with the straightforward idea using high-order recurrence relation to obtain higher order terms [64].

2.6.3 Turning Point Corrections

2.6.3.1 *Langer's Approximate Solution Valid at Turning Point*

For a refractive index profile which is unbounded we can write, for an nth order zero at $x_0 = 0$ as

$$K^2(x) = (x - x_n)^n f(x)$$ (2.114)

where

$$f(x) = \sum_{i=0}^{\infty} C_i (x - x_i)^i$$

is a non-vanishing polynomial of x at $x_0 = 0$. For simplicity and without loss of generality, let $x_0 = 0$. Thus for values of x close to zero we can write

$$K^2(x) = C_0 (x)^n$$ (2.115)

Therefore in the vicinity of the turning point, we can represent the wave equation by an approximate differential equation

$$\frac{d^2\phi(x)}{dx^2} + C_0 x^n \phi(x) = 0$$ (2.116)

where $\phi \equiv \varphi$ at $x = 0$.

The solutions to Equation 2.116 are Bessel functions. Thus it would be necessary to transform the wave equation before setting the condition as shown in Equation 2.115. We can now introduce the Liouville transform

$$\xi = \int_0^x K(x) dx \quad \text{and} \quad \phi = K^{1/2}(x)v$$ (2.117)

After some algebra the equation becomes

$$\frac{d^2 v}{d\xi^2} + \left\{ \frac{3}{4} K^{-4} \left(\frac{d^2 K}{dx^2} \right)^2 - \frac{1}{2} K^{-3} \frac{d^2 K}{dx^2} + 1 \right\} v = 0$$ (2.118)

Now using the form of the approximate transverse propagation constant in Equation 2.115 and evaluating the terms in the bracket reducing Equation 2.118 to the required intermediate form

$$\frac{d^2v}{d\xi^2} + \left\{1 + C_0^{-1}x^{-(n+2)}\left(\frac{n^2+4n}{16}\right)\right\}v = 0 \tag{2.119}$$

Thence using 2.117 we obtain

$$\frac{1}{\xi^2} = C_0^{-1}x^{-(n+2)}\left(\frac{n^2+2}{2}\right)^2 \tag{2.120}$$

Then substituting into Equation 2.119 we arrive at

$$\frac{d^2v}{d\xi^2} + \left\{1 + \frac{1}{\xi^2}\cdot\left(\frac{n(n+4)}{4(n+4)^2}\right)\right\}v = 0 \tag{2.121}$$

Now changing the variable $v \to \xi^{1/2}W$ leads to

$$\frac{d^2v}{d\xi^2} = \xi^{1/2}\frac{d^2W}{d\xi^2} + \xi^{-1/2}\frac{dW}{d\xi} - \frac{1}{4}\xi^{-3/2}W \tag{2.122}$$

Finally substituting Equation 2.122 back into Equation 2.121 and multiplying by $\xi^{3/2}$ throughout, the resultant equation is transformed to the desired canonical form:

$$\frac{d^2W}{d\xi^2} + \xi\frac{dW}{d\xi} + \left\{\xi^2 - \frac{1}{(n+2)^2}\right\}W = 0 \tag{2.123}$$

This is the Bessel equation of order $1/(n+2)$ with independent solutions denoted by plus and minus signs

$$W = \left\{\frac{\pi^2}{4}A_\pm\right\}J_{\pm m}(\xi) \tag{2.124}$$

with

$$m = \frac{1}{n+2}$$

Thus from $v \to \xi^{1/2}W$ and Equation 2.117 we can recover the solutions of the original form Equation 2.116 at the turning point as

$$\phi = K^{-1/2}\xi^{1/2}W(\xi) = \left\{\frac{\pi^{1/2}}{2^{1/2}}A_\pm\right\}\left(\frac{\xi}{K}\right)^{1/2}J_{\pm m}(\xi) \tag{2.125}$$

where $A_\pm$ is an arbitrary constant and ξ is related to $K(x)$ via the Liouville transformation. The relationship between Equations 2.123 and 2.116 can be found in the standard textbook on Bessel's function [57]. We include here for interested readers:

To study the behavior of the approximate solutions at the turning point, we obtain the DE satisfied by ϕ. To do this we represent Equation 2.125 into the form

$$\phi = G(x)\xi^m J_{\pm m}(\xi)$$

with

$$G(x) = \left\{ \frac{\pi^{1/2}}{2^{1/2}} A_\pm \right\} K^{-1/2}(x)\xi^{1/2-m} \tag{2.126}$$

Differentiating Equation 2.126 with respect to x we obtain:

$$\frac{d\phi(x)}{dx} = \frac{\left\{ \frac{\pi^{1/2}}{2^{1/2}} A_\pm \right\}\xi^{1/2-m}}{G(x)} \left\{ \xi^m J'_{\pm m} + m\xi^m J'_{\pm m} \right\} + G'(x)\xi^m J_{\pm m} \tag{2.127}$$

and

$$\frac{d^2\phi(x)}{dx^2} = \xi^m J_{\pm m} G''(x) - \frac{\pi^2}{4} A_\pm^4 G^{-3}(x)\xi^{2-3m} J_{\pm m}$$

Using Equation 2.126 we can write in more compact form

$$\frac{d^2\phi(x)}{dx^2} = \frac{G''(x)}{G(x)}\phi(x) - K^2\phi(x) \tag{2.128}$$

Thus we have the relation

$$K(x) = \frac{d\xi(x)}{dx} = \frac{\pi}{2} A_\pm^2 \xi^{1-2m} G^{-2}(x) \tag{2.129}$$

and the Bessel identity

$$\xi^2 J'' + \xi J' = \left(m^2 - \xi^2 \right) J(\xi) \tag{2.130}$$

2.6.3.2 Behavior of Turning Point

Equation 2.128 can be recast into

$$\frac{d^2\phi(x)}{dx^2} + \left\{ K^2(x) - \theta(x) \right\}\phi(x) = 0 \tag{2.131}$$

where $(x) = \dfrac{G''(x)}{G(x)}$.

This is the differential equation that satisfies the approximate Langer's solution ϕ at the turning point. To prove the validity of ϕ, we only have to show that $G(x)$ is bounded at $x = 0$, so that the coefficient of ϕ does not possess singularity. To do this we write

$$\frac{G(x)}{\left\{\dfrac{\pi^{1/2}}{2^{1/2}} A_{\pm}\right\}} = K^{-1/2}(x)\xi^{1/2-m} = x^{-n/4} f^{-1/2}(x)\left[I_1(x)\right]^{1/2-m} \tag{2.132}$$

Using the expression for K in Equation 2.115 then

$$I_1(x) = \int_0^x x^{n/2} f^{1/2}(x)\, dx \tag{2.133}$$

Thence integration by parts gives the expanded form

$$I_1(x) = \frac{x^{n/2+1} f^{1/2}}{n/2+1} - \left\{ 1 - \frac{\displaystyle\int_0^x x^{n/2+1} f'(x) f^{-1/2}(x)\, dx}{2x^{n/2+1} f^{1/2}(x)} \right\} \tag{2.134}$$

Substituting Equation 2.134 into 2.132 we obtain

$$\frac{G(x)}{\left\{\dfrac{\pi^{1/2}}{2^{1/2}} A_{\pm}\right\}} = \frac{f^{-m/2}}{(n/2+1)^{1/2-m}}\left[1 - I_2(x)\right]^{1/2-m} \tag{2.135}$$

where the intermediate integral is given by

$$I_2(x) = \frac{\displaystyle\int_0^x x^{n/2+1} f'(x) f^{-1/2}(x)\, dx}{x^{n/2+1} f^{1/2}(x)} \tag{2.136}$$

Now applying l'Hospital's rule shows that

$$\lim_{x\to 0} I_2(x) = 0 \tag{2.137}$$

whilst we observe that $f(0) \neq 0$ by definition. Hence $G(x) \neq 0$ at the turning point, the proposition has been proved.

2.6.3.3 Error Bound for ϕ Turning Point

To investigate the accuracy with which the differential equation (DE) for error bound for ϕ represents the exact wave equation at the turning point, one only has to compute the value of $\theta(x) = 0$ at $x = 0$. Thus we find

$$\theta(x) = \frac{G''(x)}{G(x)} = \frac{3}{4}\left(\frac{K'}{K}\right)^2 - \frac{1}{2}\left(\frac{K''}{K}\right) + (m-1/4)\frac{K^2}{\xi^{1/2}} \tag{2.138}$$

Note that the first two terms of this equation is the function $W(x)$ defined earlier. Thus for $m = 1/2$ (i.e., $n = 0$) we have the turning point of order zero and $\phi \to \varphi$. Therefore it can be concluded that in a region removed from any turning point the WKB solutions are of the same form as Φ and vice versa. This observation is substantially proven by investigations of the asymptotic behavior of ϕ away from the turning point. The proof of this result paves the way for the important connection formulae which is the eventual purpose of this exercise.

Now, using the form of $K^2(x)$ given in Equation 2.114, we can put $\theta(x)$ in term of $f(x)$ as

$$\frac{3}{4}\left(\frac{K'}{K}\right)^2 = \frac{3}{16}\left\{\left(\frac{f'}{f}\right)^2 + \frac{2n}{x}\left(\frac{f'}{f}\right) + \frac{n^2}{x^2}\right\} - \frac{1}{2}\left(\frac{K''}{K}\right)$$

$$= \frac{1}{8}\left(\frac{f'}{f}\right)^2 - \frac{n}{4}\left(\frac{n}{2}-1\right)x^{-2} - \frac{n}{4}\left(\frac{f'}{f}\right)x^{-1} - \frac{1}{4}\left(\frac{f''}{f}\right) \tag{2.139}$$

and

$$\left(m^2 - \frac{1}{4}\right)K^2\xi^{-2} = \left(m^2 - \frac{1}{4}\right)x^n f(x)\frac{1}{I_1^2(x)} \tag{2.140}$$

Thus we can rewrite Equation 2.134 in a more suitable form

$$I_1(x) = \left\{x^{n/2+1}f^{1/2} - \frac{1}{2}\int_0^x x^{n/2+1}f'(x)f^{-1/2}(x)\,dx\right\}(n/2+1)^{-1} \tag{2.141}$$

Since the integral $\to 0$ in the limit of $x \to 0$ we can expand I_1^{-2} in a binomial series as

$$I_1^{-2} = \left(\frac{n}{2}+1\right)^2\left(x^{\frac{n}{2}+1}f^{1/2}\right)^{-2}\left\{1 + \frac{I_3}{x^{\frac{n}{2}+1}f^{-2}} + \frac{3}{4}\frac{I_3^2}{\left(x^{\frac{n}{2}+1}f^{1/2}\right)^2} + \cdots\right\} \tag{2.142}$$

with $I_3 = \int_0^x f^{1/2}f'x^{\frac{n}{2}+1}\,dx$.

Integrating by parts again we can expand I_3 into the form

$$I_3 = \frac{1}{\left(\frac{n}{2}+2\right)}\left\{x^{\frac{n}{2}+2}f^{-1/2}f' - \int_0^x x^{\frac{n}{2}+2}f^{-1}\left(f^{1/2}f'' - \frac{1}{2}(f')^2 f^{-1/2}\right)dx\right\}$$

$$\tag{2.143}$$

$$\therefore I_3^2 = \left(\frac{n}{2}+2\right)^{-2}\left\{x^{\frac{n}{2}+4}f^{-1}(f')^2 - 2x^{\frac{n}{2}+2}f^{-1/2}f'I_4 + (I_4)^2 + \cdots\right\}$$

where I_4 is the integral expression given in Equation 2.142. Finally substituting Equations 2.143 and 2.142 into 2.139 we obtain

$$\left(m^2 - \frac{1}{4}\right)K^2\xi^{-2} = -\frac{n}{16x^2}\left(\frac{n+4}{2}\right) - \frac{2}{8}\frac{1}{x}\left(\frac{f}{f'}\right) - \frac{3}{16}\frac{n}{n+4}\left(\frac{f'}{f}\right)^2$$

$$+ \left[\frac{n}{8}\frac{1}{x^2}\left(\frac{f}{f'}\right) + \frac{3}{8}\frac{n}{n+4}\frac{1}{x}\left(\frac{f'}{f}\right)^2\right]\frac{I_4(x)}{x^{n/2+1}f^{1/2}(x)} + \cdots \qquad (2.144)$$

Therefore substituting Equations 2.144 and 2.139 into 2.138 we arrive at

$$\theta(x) = \frac{3}{2}\frac{1}{n+6}\left(\frac{n+5}{n+4}\right)\left(\frac{f'}{f}\right)^2 - \left(\frac{f''}{f}\right) + \text{terms in } x \text{ and higher} \qquad (2.145)$$

Equation 2.145 shows that if $f(x)$ is a relatively slowly varying function at $x = 0$ then $\theta(0) \to 0$ and the Langer's approximate solution Φ is a good approximation to the actual mode field at the turning point. This is a general result of which the expression obtained by Marcuse [23] is a special case.

2.6.4 Correction Formulae

It is shown above that the DE satisfied by Langer's approximate solution Φ is non-singular. Therefore unlike the WKB solution Φ is single valued. It is not restricted to yield representations of the solution of the wave equation, Equation 2.101, in the intervals on one to the other side of the turning point. Therefore in problems involving a single turning point no question of connection formulae arises in association with it. However in actual waveguides, the simplest scalar wave equation contains two turning points at the minimum. Thus a single function such as Langer's approximate solution, Φ, valid at a turning point cannot possibly describe the modal fields since essentially there are two regions of evanescent field behavior connected by a region where the field is oscillatory.

Thus as a first step to obtain a connection formulae, we introduce a new variable

$$t = -\int_0^x |K|(x)\,dx = j\xi \qquad (2.146)$$

$$j = \sqrt{-1}$$

so that we can express Φ as ϕ_1 or ϕ_2 to denote the solutions in regions in which $K^2 > 0$ and $K^2 < 0$, respectively. We now have

$$\phi_1 = \left\{\left(\frac{\pi}{2}\right)^{1/2} A_\pm \left[\xi/K\right]^{1/2}\right\} J_{\pm m}(\xi); x \geq 0$$

$$\phi_2 = \left\{\left(\frac{\pi}{2}\right)^{1/2} B_\pm\right\}\left[t/K\right]^{1/2} I_{\pm m}(\xi); x < 0 \qquad (2.147)$$

The asymptotic solution of the Bessel function of large argument are well known and we can write [63]

$$\phi_1^{as} = \phi_1\big|_{\xi \to \infty} = A_\pm K^{-1/2} \cos\left(\xi \mp \frac{m\pi}{2} - \frac{\pi}{4}\right)$$

$$\phi_2^{as} = \phi_2\big|_{t \to \infty} = B_\pm |K|^{-1/2} e^{-t + j2\pi\left(\mp m - \frac{1}{4}\right)}$$

(2.148)

These relations confirm our earlier hypothesis that ϕ_1^{as}, ϕ_2^{as} are just linear combinations of the WKB solutions $\varphi_1 = A_\pm K^{-1/2} e^{\pm j\xi}$ and $\varphi_2 = B_\pm |K|^{-1/2} e^{-t}$.

We consider the case in which the two turning points are sufficiently far apart so that we can use a turning point of order $n = 1$. Now that the solutions for the linear turning point are just the Airy functions or Bessel functions of order 1/3. These are

$$\Phi_1 = \left(\frac{\pi}{2}\right)^{1/2} A_\pm \left(\frac{\xi}{K}\right)^{1/2} J_{\pm 1/3}(\xi); x \geq 0$$

$$\Phi_2 = \left(\frac{\pi}{2}\right)^{1/2} B_\pm \left(\frac{t}{|K|}\right)^{1/2} I_{\pm 1/3}(t); x < 0$$

(2.149)

These expressions are just alternative ways of writing the same solutions. They must be identical. Continuity requirements at $x = 0$ gives

$$B_+ = -A_-$$

$$B_- = A_-$$

(2.150)

Thus the asymptotic forms can be written down as

$$\phi_1^{as} = A_\pm K^{-1/2} \cos\left(\xi \mp \frac{\pi}{6} - \frac{\pi}{4}\right)$$

$$\phi_2^{as} = \mp \frac{A_\pm |K|^{-1/2}}{2} e^{-t - j\pi\left(\frac{1}{2} \mp \frac{2}{3}\right)}$$

(2.151)

To derive the first connection formula we follow the procedure established by Langer [53] and write

$$\phi_{2+}^{as} + \phi_{2-}^{as} + \phi_{1+}^{as} + \phi_{1-}^{as}$$

$$A_+ = A_-$$

(2.152)

thus $\to |K|^{-1/2} e^{-t} \to 2|K|^{-1/2} \cos\left(\xi - \frac{\pi}{4}\right)$.

The arrow indicates that the asymptotic solution on the left goes into the expression on the right as one crosses the turning point. These arrows are irreversible or a small error in the phase of the cosine would be magnified by the positive exponential on the LHS of Equation 2.152.

Similarly we can write down another set of connection formula as

$$|K|^{-1/2} e^{-t} \leftarrow 2K^{1/2} \cos\left(\xi + \frac{\pi}{4}\right) \tag{2.153}$$

These formulae suffice for applications involving two turning points.

2.6.5 Application of Correction Formulae

There are three distinct categories of problems encountered in wave propagation in slab dielectric waveguides where the connection formulae obtained above can be applied to yield eigenvalue equations for bound modes. In the WKB context, a bound mode corresponds to a solution which is oscillatory between two turning points beyond which the solution is evanescent. These are treated separately.

2.6.5.1 Ordinary Turning Point Problem

The refractive index profile is illustrated in Figure 2.23a where the turning points are at x and x_2. Thus for $x_1 < x \leq x_2$, $K^2 > 0$ and the field is oscillatory. Elsewhere, the field is

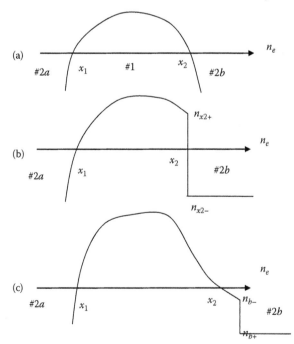

FIGURE 2.23

(a) Ordinary turning point with caustics at $x = x_1$ and x_2 where $n_2(x) = ne_2$. (b) Step discontinuity at $x = x_2$. (c) Buried modes with mode index n_e close to step discontinuity at $x = b$.

evanescent. The connection formulae connect solutions on both sides of a turning point and these must be applied at both x_1 and x_2. Correct phase matching of the oscillatory fields yields the WKB eigenvalue equation. This equation contains the implicit prescription of the required propagation constant.

Thus in region #2a and #2b we have, respectively,

$$\varphi_{2a} = A_a |K|^{-1/2} e^{-t_1}$$

$$\varphi_{2b} = A_b |K|^{-1/2} e^{-t_2} \tag{2.154}$$

with $t_1 = \int_{x_1}^{x} |K| \, dx; .t_2 = \int_{x_2}^{x} |K| \, dx.$

By virtue of the first connection formulae Equation 2.152, the solution in the region #1 which connects with region #2a is

$$\varphi_{1a} = 2A_a K^{-1/2} \cos\left(\xi_1 - \frac{\pi}{4}\right) \tag{2.155}$$

with $\xi_1 = \int_{x_1}^{x} K \, dx.$

And that which connects the solution in region #2b is

$$\varphi_{1b} = 2A_b K^{-1/2} \cos\left(\xi_2 - \frac{\pi}{4}\right) \tag{2.156}$$

with $\xi_2 = \int_{x}^{x_2} K \, dx.$

Now with φ_{1a} and φ_{1b} represent one and only one solution in #1. Thus this consistency condition implies that

$$A_a = A_b$$

$$\xi_1 - \frac{\pi}{4} = \pm\left(\xi_2 - \frac{\pi}{4}\right) \tag{2.157}$$

The RHS of Equation 2.157 can be written as, taking the minus sign

$$-\left(\xi_2 - \frac{\pi}{4}\right) = \xi_1 - \frac{\pi}{4} - \eta$$

$$\eta = \int_{x_1}^{x_2} K \, dx - \frac{\pi}{2} \tag{2.158}$$

which must be a multiple of π to ensure the correct phase matching of the WKB oscillatory solution. Equation 2.158 is expanded to give the familiar form of the WKB eigenvalue equation

$$\int_{x_1}^{x_2} \left[k_0^2 n^2(x) - \beta^2 \right]^{1/2} = \left(m + \frac{1}{2} \right) \pi; \quad m = 0, 1, 2, \dots \tag{2.159}$$

This result fails when $x_1 = x_2$ or when there is an index discontinuity in the vicinity of the turning point.

2.6.5.2 Effect of an Index Discontinuity at a Turning Point

Figure 2.23b illustrates a typical example of a thin film waveguide deposited on a substrate. The turning point at $x = x_2$ coincides with the film/air interface. At $x = x_2$ it is erroneous to apply the connection formulae, since they are derived on the assumption of a linear variation of $K^2(x_2)$. We resort, instead, to the boundary conditions imposed on the fields at $x = x_2$. If we denote $n(x_{2-})$ and $n(x_{2+})$ as the values of the refractive index just before and after the step then the standard phase shift at the discontinuity [61] is proportional to δ_{x_2} given by

$$\delta_{x_2} = \tan^{-1} \left(\frac{\beta^2 - k_0^2 n^2(x_{2+})}{-\beta^2 + k_0^2 n^2(x_{2-})} \right)^{1/2} \tag{2.160}$$

and the WKB eigenvalue equation transforms to

$$\int_{x_1}^{x_{2-}} \left[k_0^2 n^2(x) - \beta^2 \right]^{1/2} = \left(m + \frac{1}{4} \right) \pi + \delta_{x_2}; \quad m = 0, 1, 2, \dots \tag{2.161}$$

Note that in the large step $\delta_{x_2} \to \pi/2$ as in the case of a strongly asymmetrical waveguide.

2.6.5.3 Buried Modes near an Index Discontinuity at a Turning Point

The analysis in the previous sub-section treats only the turning point at $x = x_2$ and fails directly on top of the index discontinuity. Thus strictly speaking Equation 2.161 is only accurate for certain order modes (value of m) which satisfy this condition.

We denote the refractive index just before and after the step at $x = b$ by $n(b^-)$ and $n(b^+)$, respectively. Then in the region $x_2 < x < b$ the step at $x = b$ causes significant reflection of energy. Thus it is no longer accurate to represent the fields in the region by a single decay exponential. Therefore we include both decaying and growing exponentials and write

$$\varphi_1 = AK^{-1/2} \cos \left(\xi_1 - \frac{\pi}{4} \right); x_1 < x < x_2$$

$$\varphi_{2a} = A|K|^{-1/2} \cos \left(\int_{x_1}^{x_2} K dx \right) e^{t_2} + \frac{1}{2} c \sin \left(\int_{x_1}^{x_2} K dx \right) e^{-t_2}; x_2 < x < b \tag{2.162}$$

$$\varphi_{2b} = A|K|^{-1/2} e^{\left(\int_{xb}^{x} |K| dx \right)}; x < b$$

where A is a constant. The coefficients of the growing and decaying fields in Equation 2.162 have been chosen to satisfy the connection formulae at $x = x_2$. Thus the equation of φ_1

in Equation 2.162 is redundant as far as the eigenvalue equation is concerned. Continuity requirements on φ_{2a} and φ_{2b} at $x = b$ gives the required eigenvalue equation

$$\int_{x1}^{x2-} \left[k_0^2 n^2(x) - \beta^2 \right]^{1/2} = \left(m + \frac{1}{4} \right) \pi + \delta_b ; m = 0, 1, 2.... \quad (2.163)$$

where

$$\delta_b = \tan^{-1} \left\{ \frac{\left| |K(b^-)| + |K(b^+)| \right|}{\left| |K(b^-)| - |K(b^+)| \right|} \right\} e^{2\int_{x2}^{b-} |K| dx}$$

is the corrected overall phase shift of the buried modes [65].

2.7 Appendix D: Design and Simulation of Planar Optical Waveguides

2.7.1 Introduction

As we have outlined in the series of lectures on optical waveguides in which we treat a slab optical waveguide as the extension of an optical fiber where the structure is restricted to one-dimension and the other dimension of its cross section has been extended to infinity. The optical waveguides considered in this experiment are not a step index slab type as considered in the theoretical section above but rather they are graded-indexed, i.e., the refractive index is gradually decreased from the core to the cladding region.

This introductory experiment in optical waveguides aims to familiarize potential optical communications engineers with the structures and behavior of the optical field distribution in a number of guided optical wave structures such as straight, bend and Y-junction, etc. The computer experiment is written in such a way that you can read and perform the preliminary work and experiment in stages.

The objectives of this section are

- To design parameters of slab optical waveguides so that they can support a certain number of guided modes in the single or multimode regions
- To use the fundamental mode of the optical waveguide for observation and measurement of optical fields of several waveguide composite structures
- To propagate the fundamental optical field through a number of optical waveguide structures such as straight, bend and Y-junction optical guided waves structures.
- Programs for simulation of the propagation of guided lightwaves are listed in Appendix 9.

2.7.2 Theoretical Background

2.7.2.1 Structures and Index Profiles

Optical waveguides are the fundamental element in modern optical communications and photonic signal processing systems. Optical fibers are the guiding medium for optical signal transmission and are formed by a circular core inside a circular cladding region. The mathematics required to represent the electric and magnetic field components of the guided waves in optical fibers would involve Bessel's functions. These waveguides would be treated in detail in the fourth year of a course of optical communications engineering. A simplified version of optical fibers is the slab optical waveguide whose structure is shown in Figure 2.1. The cladding regions are the superstrate and substrate and would have an identical constant refractive index. The guiding region is a slab or thin film layer sandwiched between the cladding regions. The refractive index of the slab region must be higher than that of the cladding and its thickness must be sufficiently thick to support confined (bound) optical guided modes.

The refractive index profile of the step-slab region can be uniform, i.e., constant throughout, **or graded where n(x) decreases gradually** from the center of the slab to the cladding. For the sake of simplicity to obtain an analytical solution of the wave equation representing the guided field, the index profile of our slab structures would have a "cosh^{-2}" distribution (graded index profile) given by:

$$n^2(x) = n_s^2 + \frac{2n_s \Delta n}{\cosh^2 \dfrac{2x}{h}} \tag{2.164}$$

or approximately

$$n(x) = n_s + \frac{\Delta n}{\cosh^2 \dfrac{2x}{h}} \tag{2.165}$$

where n_s = cladding refractive index (for both the superstrate and substrate), Δn = refractive index difference between the cladding and slab regions and h is the total thickness of the guiding layer (slab thickness). It is convenient to define a normalized parameter V as

$$V = kh(2n_s \Delta n)^{1/2} \tag{2.166}$$

where $k = 2\pi/\gamma$ and γ is the operating wavelength of the optical waves in vacuum.

Note: the expression of the parameter V is identical with that of a circular optical fiber. However in this experiment we are dealing with "planar" optical waveguide structures.

2.7.2.2 Optical Fields of the Guided Transverse Electronic Modes

Normally the optical fields in a slab waveguide would consist of two quasi-polarizations, TE and TM where the non-zero electromagnetic field components are (E_y, H_x, H_z) and (E_x, E_z, H_y) for TE and TM modes, respectively. In this experiment we consider only the behavior of TE modes in slab optical waveguides.

The wave equation for TE modes can be derived from Maxwell's equations; in the case when the refractive index difference is small the wave equation can be approximated to have a scalar form as

$$\frac{d^2 E_y}{dx^2} - (\beta^2 - n^2 k^2)E_y = 0 \tag{2.167}$$

when the refractive index distribution of the waveguide structure $n(x)$ has a cosh^{-2} profile in the slab region and constant in the cladding regions, the field solution of Equation 2.167 would have an analytical form of

$$E_y(x) = \frac{u_v(2x/h)}{\cosh^2(2x/h)} \tag{2.168}$$

where $v = 0,1,2,3.....$ are subject to the boundary conditions that the field must vanish at a distance very far from the slab-cladding interface. The function $u_v(2x/h)$ would take the following forms:

2.7.2.2.1 For Even Transverse Electronic Modes, $v = 0,2,4,...$

$$u_v(2x/h) = 1 - \frac{1}{2}v(2s-v)\frac{\sinh^2(2x/h)}{1.1!}$$

$$+ \frac{1}{4}v(v-2)(2s-v)(2s-v-2)\frac{\sinh^4(2x/h)}{(1.3.2!)} + \cdots \tag{2.169}$$

with $s = 0.5\{(1 + V^2)^{1/2} - 1\}$ the **total number of guided even modes** that this kind of optical waveguide can support.

2.7.2.2.2 For Odd Transverse Electronic Modes, $v = 1,3,5,...$

$$u_v(2x/h) = \sinh(2x/h)$$

$$\cdot \left[\begin{array}{l} 1 - \frac{1}{2}(v-1)(2s-v-1)\frac{\sinh^2(2x/h)}{3.1!} \\[2mm] + \frac{1}{4}(v-1)(v-3)(2s-v-3)\frac{\sinh^4(2x/h)}{(3.5..2!)} + \cdots \end{array} \right] \tag{2.170}$$

with $s = 0.5[(1 + V^2)^{1/2} - 1]$ **is** the **maximum number of guided odd mode**s.
  For lower order modes, we have

$$u_0 = 1 \tag{2.171a}$$

$$u_1 = \sinh\frac{2x}{h} \tag{2.171b}$$

$$u_2 = 1 - 2(s-1)\ \sinh^2\frac{2x}{h} \tag{2.171c}$$

$$u_3 = \sinh\frac{2x}{h}\left[1 - \frac{2}{3}(s-2)\sinh^2\frac{2x}{h}\right] \tag{2.171d}$$

The propagation constant β_v and the effective indices ($n_{eff} = \beta_v/k$) of the vth order modes are given by

$$\beta_v^2 = n_s^2 k^2 + 4(s-v)^2/h^2 \tag{2.172}$$

$$n_{\text{eff}}^2 = n_s^2 + (s-v)^2(\lambda/\beta_v)^2 \tag{2.173a}$$

and the normalized propagation constant b is defined as

$$b = \frac{n_{\text{eff}}^2 - n_s^2}{n^2 - n_s^2} \tag{2.173b}$$

with n as the refractive index at the center of the guide or approximately

$$b = \frac{n_{\text{eff}}^2 - n_s^2}{2n_s \Delta n} \tag{2.173c}$$

thus the optical field of a slab optical waveguide can be found if we can specify the following parameters: the slab thickness h; the cladding refractive index; the refractive index difference Δn; and the operating wavelength.

2.7.2.3 Design of Optical Waveguide Parameters: Preliminary Work

Choose the parameters n_s, Δn and h of your waveguide. Some typical refractive indices of certain transparent materials for superstrate and substrate of optical waveguides are

- $n_s = 1.447$ for silica glass at the operating wavelength of 1300 nm. This is also the base material for modern telecommunication optical fibers.
- $n_s = 3.6$ for GaAs semiconductor waveguide at 1300 nm. This is the base material for optical waveguides formed in the resonant cavity and waveguide of semiconductor lasers for optical fiber communications.
- $n_s = 2.2$ for lithium niobate crystal at certain crystal axis as seen by the TE waves. This material is also the base material for optical modulators for optical communications.

Make sure that the chosen parameters would form an optical waveguide which would support no more than four guided modes at the operating wavelength of 1300 nm. Can you design an optical waveguide such that it supports only one guided mode which is the only fundamental mode of the optical waveguide?

In fact we can plot the b-V curve for $v = 0, 1, 2, ...$ and from this diagram we can design a mono-mode, 2-mode ... optical waveguides. Notice that only TE modes are considered here.

2.7.3 Simulation of Optical Fields and Propagation in Slab Optical Waveguide Structures

A computer simulation program has been written to study the evolution of optical fields in slab optical waveguides which form the basic component for several optical wave guiding devices.

To study numerically the behavior of optical waves, particularly the fundamental mode field, in these structures, the whole waveguide region including the slab and cladding regions, i.e., W and L, are sliced into several intervals along the propagation direction z as well as in the vertical direction for numerical calculations.

The field in the first plane, i.e., at $z = 0$, can be found by Equation 2.170. This field would then be propagated to the next plane through a discredited equation by applying the finite difference method to the wave equation with the z-dependence in Maxwell's equation. This para-axial wave propagation equation is given by:

$$2jkn_s \frac{\partial E_y}{\partial z} = \frac{\partial^2 E_y}{dx^2} + k^2 \left[n^2(x,y) - n_s^2 \right] E_y \tag{2.174}$$

which can be written using the center-finite-difference technique as:

$$jkn_s \frac{E_{i,k+1} - E_{i,j}}{\Delta z} = \frac{E_{i-1} - 2E_i + E_{i+1}}{\Delta x^2} + k^2 \left[n_i(x,y) - n_s^2 \right] E_i \tag{2.175}$$

where $j = (-1)^{1/2}$, the subscripts i and j denote the variation of E with respect to the x and z directions, respectively. That is the cross section plane is partitioned into several layers with order i and the propagation steps along the z direction are assigned with order j. The obtained results of the field at location j would then be used as the field initial distribution for propagating through the structure to obtain the optical field of the next plane $j + 1$ and so on. Thus we can employ an analytical method to obtain the field solution for the optical field in the transverse plane. A numerical method (the finite difference) is used to study the evolution of the optical field propagating along the optical waveguide structure.

In this section we are not going to study the finite difference method but the evolution of the optical field in optical waveguides. In the following parts we would examine the optical field behavior in the structures illustrated in Figure 2.24.

An additional dimension, the z axis, is now added to the optical waveguide devices. These structures are shown in Figure 2.25 through 2.27.

The full list of all programs in FORTRAN language for simulations of a number of waveguide structures such straight, bend, taper, Y-junction and inteferometric Mach-Zehnder type is given in Appendix 9.

2.7.3.1 Lightwaves Propagation in Guided Straight Structures

Please do not hesitate to seek assistance from the demonstrators for procedures in running the simulation package. Typical steps for simulation are: run MATLAB; go to directory

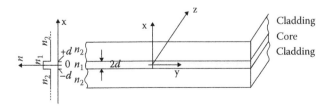

FIGURE 2.24
Schematic structure of the slab optical waveguide: The guiding layer is sandwiched between the superstrate (upper cladding region) and substrate (lower cladding region) of identical refractive indices.

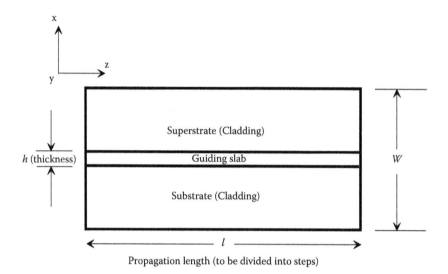

FIGURE 2.25
Side view of a slab optical waveguide in a straight optical device structure. h is the waveguide thickness, W is the total width of the structure in the transverse plane to be specified for numerical simulation, L is the total length of the device. W is to be divided into several equi-spacing layers for numerical simulation. The length L along the propagation direction is also spitted into several steps for propagation from one plane to the other and so on.

STRAIGHT; run FD1, the program for the beam propagation method; choose parameters as prompted by the program such as: (i) waveguide region to be analyzed; (ii) operating wavelength; (iii) slab thickness; (iv) the number of x intervals for optical field and number of propagation steps in the z direction; (v) the refractive indices of slab and cladding, i.e., cladding index and the index difference; and (vi) propagation distance.

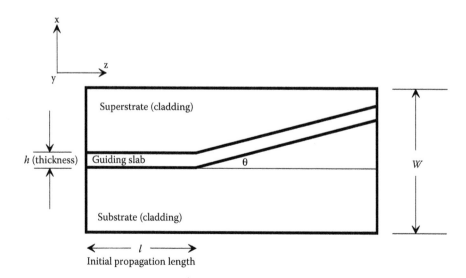

FIGURE 2.26
Side view of the Y-structure using slab optical waveguide. θ is the half Y-junction angle, l is the straight initial section before splitting.

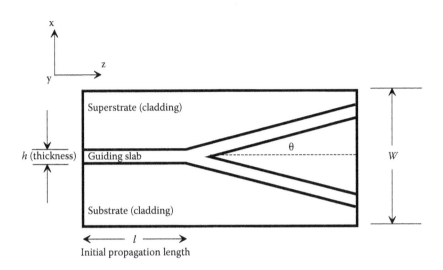

FIGURE 2.27
Side view of the interferometric structure using slab optical waveguide. θ is the half Y-splitting angle, l is the straight initial section before splitting.

a. After successfully obtaining the guiding of optical waves, keep one or two parameter constants (such as the waveguide thickness), vary stepwise other parameters, e.g., index difference, wavelength, etc. Observe the evolution of the optical field and plot the 3D guided wave field profile and the field contour. Note: when specifying the number of planes to be plotted by MATLAB the product of the number of intervals in x and z directions must not exceed the MATLAB limit which is 8188 depending on the available computer memory and MATLAB version.

b. Observe the field evolution with respect to the change in refractive index difference, waveguide thickness etc.

An example of the wave guided and propagation in a straight waveguide is shown in Figure 2.28.

Note: The graphical facilities are dependent on the computer networking at the time of the experiment.

2.7.3.2 Lightwaves Propagation in Guided Bent Structures

Similar to the steps as above: (i) choose the most suitable optical waveguide structures of the STRAIGHT structure to enter into the bend structure parameters; (ii) the BEND directory is to be evoked; (iii) additional parameters for this structure are the bend angle θ and the length l of the straight section. Start with a bend angle of about 0.5 degree of arc; (iv) now vary the bend angle in step of 0.5 or 1 degree of arc to about 10 degree of arc; (v) observe the radiation of the guided field at the bend section and REPORT the guided and radiated optical fields; (vi) vary the refractive index difference and run the program for the bend angle of about 2–4 degrees and observe the confinement and radiation of the optical field at straight and bend sections.

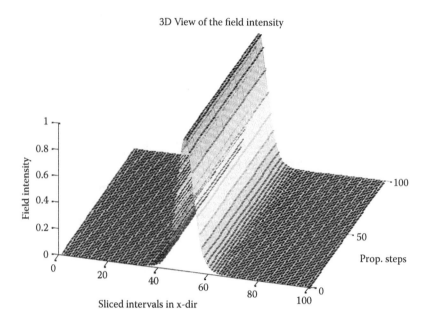

FIGURE 2.28
Guided mode propagation in a straight slab optical waveguide. Waveguide parameters: cladding refractive index = 2.2, refractive index change = 0.02, propagation length = 100 micron, step size = 0.5 micron. Step size in x-direction = 0.02 micron.

2.7.3.3 *Lightwaves Propagation in Y-Junction (Splitter) and Interferometric Structures*

As illustrated in Figure 2.2c the Y-junction or optical splitter is considered as a combination of two identical bend sections. Note that the section right at the Y-junction has a width that is wider than that of the straight section. Thus the number of guided modes would be higher for this very short section: (i) choose the half angle θ of the Y-junction from 0.5 to 5 degrees of arc, run the appropriate program see Appendix 11, to observe the field evolution and measure the field strength, i.e., "intensity" or optical power, distribution across the whole device; (ii) you can vary the refractive index difference for a small half angle at the Y-junction to observe the splitting effect. Report the field behavior at these Y-junctions; (iii) use the programs provided for simulation of the propagation of lightwaves through an interferometric optical waveguide structure the "Mach-Zehnder" in which the input lightguide is spitted into two paths and the combined into one output port.

2.7 Problems

1. a. A slab optical waveguide has a symmetrical step refractive index profile based on pure silica as the cladding material with a refractive index of 1.480 at 1550 nm free space wavelength.

    i.   The scalar wave equation for the z-propagating transverse electric monochromatic lightwaves is given by:

$$\frac{d^2 E_y}{dx^2} + (k^2 n_j^2 - \beta^2) E_y = 0,$$

where n_j ($j = 1$ or 2) represents the refractive index in either the core or the cladding regions, $k = 2\pi/\lambda$ is the free space wave number, λ is the free space wavelength, β is the propagation constant of the lightwaves along the z-direction.

Write down the wave solutions for both the even and odd TE-guided modes in the core and cladding regions. The two corresponding eigenvalue equations of these guided modes are given by

$$v = u \tan u \text{ for even TE-guided modes}$$

$$v = -\frac{u}{\tan u} \text{ for odd TE-guided modes}$$

where u and v are defined by:

$$u^2 = a^2(k^2 n_1^2 - \beta^2)$$

$$v^2 = a^2(-k^2 n_2^2 + \beta^2).$$

You are required to prove just one of the above eigenvalue equations (either one of your choice).

ii. Obtain an expression for the normalized frequency V-parameter, and thence design a slab optical waveguide so that it can guide two TE optical-guided modes at an operating free space wavelength of 1550 nm. It is recommended that the following parameters of the optical waveguide should be specified: the slab core thickness, the cladding thickness, the relative refractive index difference between the core and cladding regions, and the cutoff wavelength of TE modes of order higher then the two TE modes.

iii. Sketch the electric field and intensity distribution of the fundamental guided TE mode in the transverse plane of the designed waveguide in Part (ii).

b. Describe briefly the attenuation or loss of silica planar waveguide layer and the dispersion curve of such planar waveguide whose refractive index difference is smaller than 1%.

2. A slab optical waveguide has a symmetrical refractive index based on pure silica as the substrate material. The core refractive index is 1.50 and its thickness is 4.00 μm. The cladding thickness is 20 μm. The refractive index difference is 0.09. The operating wavelength is 1300.0 nm.

a. Is the operating wavelength in the UV, visible, near infrared, infrared, or far infrared region?

b. Is the silica material less lossy at 1300 nm than at 1550 nm? Give reasons.

c. Find the normalized frequency parameter V for the planar optical waveguide. Thence find the number of odd and even guided modes. If possible, write a procedure in MATLAB to calculate the propagation constants of these guided modes.

d. If the cladding refractive index is 1.515, would the optical waveguide support any guided mode at 1300 nm wavelength? If it does, find the number of guided modes for this structure. Sketch the field and intensity distribution of these modes across the waveguide cross section.

e. Using the refractive index profile as in Part (d), design the geometrical structure of the waveguide so that it can support only one TE-even mode.

3. Assuming that the refractive indices at 1550 nm wavelength are the same, repeat Problem 2 with an operating optical wavelength of 1550 nm.

4. The structure of an optical planar waveguide is a splitting junction Y. Sketch this structure with a Y-angle of 2 degrees of arc. The requirement is that the output optical fields at the output ports of the Y-junction must be that of a single TE-even mode. Using silica as the substrate material of a refractive index of 1.500 at 1530 nm wavelength in vacuum, design the planar optical waveguide sections of the Y-junction. Designers should assume that the splitting tilted junction area would support the same guided modes as that of the output straight branches.

References

1. S.E. Miller, "Integrated optics: An introduction," *Bell Sys. Tech. J.*, 48, 2059–2069, 1969.
2. C.R. Doerr and K. Okamoto, "Advances in silica planar lightwave circuits," *IEEE J. Lightwave Tech.*, 24(12), 4763–4770, December 2006.
3. P.S. Chung, "Waveguide modes, coupling techniques, fabrication and losses in optical integrated optics," *J. Elkect. Electron. Australia*, 5, 201–214, 1985.
4. http://electron9.phys.utk.edu/optics421/modules/m10/integrated_optics.htm. Access date: July 2010.
5. R.G. Hunsperger, "Integrated optics theory and technology," Originally published in the series: *Advanced Texts in Physics*, 6th ed., 2009, XXVIII, Springer, Berlin.
6. L.N. Binh and S.V. Chung, "Nonlinear interactions in thin film structures," Proceedings of the 8th Australian Workshop on Optical Communications, Adelaide 1983, Session VII.
7. K. Uchiyama and T. Morioka, "All-optical time-division demultiplexing experiment with simultaneous output of all constituent channels from 100Gbit/s OTDM signal," *Electron. Lett.* 37(10), 642–643, May 2001.
8. M. Nakazawa, T. Yamamoto, and K.R. Tamura, "1.28Tbit/s–70km OTDM transmission using third- and fourth-order simultaneous dispersion compensation with a phase modulator," *Electron. Lett.* 36(24), 2027–2029, November 2000.
9. C. Schubert, R.H. Derksen, M. Möller, R. Ludwig, C.-J. Weiske, J. Lutz, S. Ferber, A. Kirstädter, G. Lehmann, and C. Schmidt-Langhorst, "Integrated 100-Gb/s ETDM receiver," *IEEE J. Lightw. Tech.*, 25(1), 122–129, January 2007.
10. H. Suzuki, M. Fujiwara, and K. Iwatsuki, "Application of super-DWDM technologies to terrestrial terabit transmission systems," *IEEE J. Lightw. Tech.*, 24(5), 1998–2005, May 2006.
11. J.P. Turkiewicz, E. Tangdiongga, G. Lehmann, H. Rohde, W. Schairer, Y.R. Zhou, E. S. R. Sikora et al., "160 Gb/s OTDM networking using deployed fiber," *IEEE J. Lightw. Tech.*, 23(1), 225–234, January 2005.
12. R. Nagarajan, M. Kato, V.G. Dominic, C.H. Joyner, R.P. Schneider, Jr., A.G. Dentai, T. Desikan et al., "400 Gbit/s (10 channel x40 Gbit/s) DWDM photonic integrated circuits," *Electron. Lett.* 41(6), March 2005.

13. T. Okoshi, "Recent advances in coherent optical fiber communications," *IEEE J. Lightw. Tech.*, LT-5, 44–52, 1987.
14. L.G. Karzovky, and O.K. Tonguz, "ASK and FSK coherent lightwave systems: A simplified approximate analysis," *IEEE J. Lightw. Tech.*, 8(3), 338–351, March 1990.
15. S. Tsukamoto, D.-S. Ly-Gagnon, K. Katoh, and K. Kikuchi, "Coherent demodulation of 40-Gbit/s polarization-multiplexed QPSK signals with 16-GHz spacing after 200-km transmission", paper: PDP29, Proc. CLEOS 2005, San Jose, 2005.
16. D.-S. Ly-Gagnon, S. Tsukamoto, K. Katoh, and K. Kikuchi, "Coherent detection of optical quadrature phase-shift keying signals with carrier phase estimation," *IEEE J. Lightw. Tech.*, 24(1), 12–21, January 2006.
17. V. Ramaswamy and R.K. Lagu, "Numerical field solution for an arbitrary asymmetrical gradient index planar waveguide," *IEEE J. Lightw. Tech.*, LT-1, 408–417, 1983.
18. R.V. Schmidt and I.P. Kaminow, "Metal-diffused optical waveguides in LiNbO$_3$," *Appl. Phys. Letters.*, 25, 458–460, 1974.
19. W. Streifer, R.D. Burnham, and D.R. Scifres, "Modal analysis of separate-confinement heterojunction lasers with inhomegeneous cladding layers," *Opt. Lett.*, 8, 283–285, 1981.
20. D. Marcuse, *"Light Transmission Optics,"* Chapter 8, Van Nostrand, New York, 1972.
21. E.M. Conwell, "Modes in optical waveguides formed by diffusion," *Appl. Phys. Lett.*, 26, 328–329, 1973.
22. H. Kirchoff, "The solution of Maxwell equations for inhomogeneous dielectric slabs," *A.E.U.*, 26, 537–541, 1972.
23. D. Marcuse, "TE modes of graded index – slab waveguides," *IEEE J. Quant. Elect.*, QE-9, 1000–1006, 1973.
24. A.W. Snyder and J.D. Love, *"Optical Waveguide Theory,"* Chapter 12, Chapman and Hall, London, 1983.
25. T.R. Chen and Z.L. Yang, "Modes of a planar waveguide with Fermi index profile," *Appl. Optics*, 24, 2809–2812, 1985.
26. H.F. Taylor, "Dispersion characteristics of diffused channel waveguides," *IEEE J. Quant Electronics*, QE-12, 748–752, 1976.
27. G.B. Hocker and W.K. Burns, "Mode dispersion in diffused channel waveguides by the effective index method," *Appl. Optics*, 16, 113–118, 1975.
28. L. Riviere, A. Yi-Yan, and H. Carru, "Properties of single mode planar with Gaussian index profile," *IEEE J. Lightw. Tech.*, LT-3, 368–377, 1986.
29. S. Korotky, W. Minford, L. Buhl, M. Divino, and R. Alferness, "Mode size and method for estimating the propagation constant of single mode Ti:LiNbÒ strip waveguides," *IEEE Trans. Microwave Th. and Tech.*, MTT-30, 1784–1789, 1982.
30. A.N. Kaul, S.L. Hossain, and K. Thyagarajan, "A simple numerical method for studying the propagation characteristics of single-mode graded-index planar optical waveguides," *IEEE Trans. Microw. Th. and Tech.*, MTT_34, 288–292, 1986.
31. L.N. Binh and S.V. Chung, "Design considerations for second harmonic generation in thin film grown by ion beam assisted deposition method," Proc. 3rd Laser Conf., Melbourne, Session 10A, 1983.
32. J. Killingbeck, "A pocket calculator determination of energy eigenvalues," *J. Phys A.*, 10, L09–L103, 1977.
33. R.A. Sammut and C. Pask, "Simplified numerical analysis of optical fibers and planar waveguides," *Elect. Lett.*, 17, 105–106, 1981.
34. J.D. Love and A.K. Ghatak, "Exact solutions for TM modes in graded index slab waveguides," *IEEE J. Quant Elect.*, QE 15, 14–16, 1979.
35. D. Marcuse, "The effects of the $\nabla^2 n$ term on the modes of an optical square-law medium," *IEEE J. Quant. Elect.*, QE-9, 958–960, 1973.
36. H. Kolgenik and V. Ramaswamy, "Scaling rule for thin film optical waveguides," *Appl. Opt.*, 13, 1857–1862, 1974.

37. P.K. Tien, "Integrated optics and new wave phenomena," *Rev. Mod. Physics*, 49, 361–420, 1977.
38. A. Watson, "Physics—Where the action is," *New Scientist*, 42–44, 30th January, 1986.
39. L. Mammel and L.G. Cohen, "Numerical prediction of fiber transmission characteristics from arbitrary refractive index profiles," *Appl. Opt.*, 21, 699–703, 1982.
40. K. Petermann, "Constraints for fundamental mode-size for broadband dispersion-compensated single mode fibers," *Elect. Lett.*, 19, 712–714, 1983.
41. P. Sansonetti, "Modal dispersion in single mode-fibers: simple approximation issued from mode spot size spectral behavior," *Elect. Lett.*, 18, 647–648, 1982.
42. R.J. Black and C. Pask, "Slab waveguides characteristics by moments of refractive index profiles," *IEEE J. Quant. Elect.*, QE-20, 996–999, 1984.
43. S. Ruschin, "Approximate formula for the propagation constant of the basic mode in slab waveguides of arbitrary index profiles," *Appl. Opt.*, 24, 4189–4191, 1985.
44. R. Kiel and F. Auracher, "Coupling of single-mode Ti:diffused LiNbO$_3$ waveguides to single mode fibers," *Opt. Comm.*, 30, 23–28, 1979.
45. W.A. Gambling and H. Matsumara, "Propagation in radially inhomogeneous single-mode fiber," *Opt. Quantum Electronics*, 10, 31–40, 1978.
46. E. Desurvire, "*Erbium-Doped Fiber Amplifiers, Principles and Applications,*" Wiley, New York, 1994.
47. J. Jeffery, "On certain approximate solutions of linear differential equations of second order," *Proc. London Math. Soc.*, 23, 428–436, 1923.
48. G. Wentzel, "Eine Verrallgemeirneung der Quanttenbedingungen fur die Zwecke der Wellemechanik," *Z. Phys.*, 38, 518–529, 1926.
49. H.A. Kramers, "Wellenmechanik und halbzahlig Quantisierung," *Z. Phys.*, 39, 828–840, 1926.
50. L. Brillouin, "Remarque's sur la mechanique ondulatoire," *J. de Phys.*, 7, 353–368, 1926.
51. G.B. Hocker and W.K. Burns, "Mode dispersion in diffused channel waveguides of arbitrary index profiles," *IEEE J. Quant Elect.*, QE-11, 270–276, 1975.
52. A. Ankiewicz, "Comparison of wave and ray techniques for solution of graded index optical waveguide problems," *Optica Acta*, 25, 361–373, 1978.
53. R.E. Langer, "On the connection formulae and the solutions of the wave equation", *Phys Rev.*, 51, 669–676, 1937.
54. J.M. Arnold, "Asymptotic analysis of planar and cylindrical inhomogeneous waveguides," *Radio Sciences*, 16, 511–518, 1981.
55. J.P. Gordon, "Optics of general guiding media," *Bell Syst. Tech. J.*, 45, 321–332, 1966.
56. P.K. Tien and R. Ulrich, "Theory of prism-coupler and thin film lightguides," *J. Opt. Soc. Am.*, 60, 1325–1337, 1966.
57. I.S. Gradshteyn and I.M. Ryzhik, "*Table of Integrals, Series and Products,*" Academic Press, Boston, USA, 1965.
58. P.K. Mitra and A. Sharma, "Analysis of single mode inhomogeneous planar waveguides," *IEEE J. Lightwave Tech.*, LT-4, 204–212, 1986.
59. R.J. Black and C. Pask, "Developments in the theory of equivalent-step-index fibers," *J. Opt. Soc. Am.*, A1, 1129–1131, 1984.
60. A.W. Snyder and R.A. Sammut, "Fundamental (HE11) modes of graded optical fibers," *J. Opt. Soc. Am.*, 69, 1663–1671, 1979.
61. M.J. Adams, "*Introduction to Optical Waveguides,*" Wiley, New York, 1981.
62. J.N. Walpole, J.P. Donnelly, P.J. Taylor, L.J. Missaggia, C.T. Harris, R.J. Bailey, A. Napoleone, S.H. Groves, S.R. Chinn, R. Huang, and J. Plant, "Slab-coupled 1.3-mm semiconductor laser with single spatial large diameter mode," *IEEE Photonics Technol. Lett.*, 14(6), 756–758, June 2002.
63. M. Abramowitz and I.A. Stegun, "*Handbook of Mathematical Functions,*" Dover Publications, New York, 1972.
64. J.L. Dunham, "The energy levels of a rotating vibrator," *Phys. Rev.*, 41, 721–731,1932.
65. J. Plesingr and J. Cryroky, "Computer simulation of lightwave propagation in photonic waveguide structures," *Radioengineering*, 2(1), 1978, April 1993.

3

3D Integrated Optical Waveguides

Following the fundamentals of planar optical waveguide, this chapter describes the three dimensional (3D) optical waveguides in which the waveguide is restricted in both transverse directions. A simplified analysis of these waveguides, the effective index method (EIM) and numerical techniques, the finite difference method (FDM), are described and examples are given. In this chapter, we analyze the modes which are guided by 3D waveguides with rectangular geometries using mainly Marcatili's method and the EIM as analytical techniques.

Thence on the numerical method we select the simple FDM as the principal technique because it is simple and gives accurate results for optical waveguides operating in the linear region. We thus chose the FDM to study the quasi-transverse electronic (TE) and quasi-transverse magnetic (TM) polarized waveguide modes due to its simplicity and plausible accuracy. We have employed the semivectorial analysis which automatically takes full account of the discontinuities in the normal electric field components across any arbitrary distribution of internal dielectric interfaces. The eigenmodes of the Helmholtz equation is solved by the application of the shifted inverse power iteration method. This method warrants both the mode size and its relevant propagation constant, which are both important parameters to the design of optical waveguide. The grid size is non-uniform to maximize the accuracy of the optical guided modes and their propagation constants. Diffused waveguides and rib waveguides are designed with different parameters to demonstrate the effectiveness of the method and lead to an optimum design of waveguides of optical modulation and micro-ring resonators.

3.1 Introduction

To achieve an efficient design of high speed modulators and switches, especially micro-ring resonators, the fabrication of rib waveguides and the Ti:NbO$_3$ waveguide with suitable mode size is essential to minimize waveguide insertion loss and also to maximize the overlap integral between the guided optical field and the applied modulating field. Furthermore, the bending or radius of curvature is so important for the ring resonator to keep the ring size as small as possible. Extensive studies have been devoted in recent decades to fabricating Ti:diffused LiNbO$_3$ waveguides which couple efficiently to single-mode fibers [1–5]. A major milestone was achieved when a total fiber-waveguide-fiber insertion loss of 1 dB was achieved for z-cut LiNbO$_3$ at 1.3 μm [4]. Such low loss was achieved by choosing fabrication parameters to yield a relatively deep, clean diffusion, which simultaneously minimized the fiber waveguide mode mismatch loss and the propagation loss. Suchoski and Ramaswamy [6] have reported on the optimization of fabrication parameters to obtain Ti:LiNbO$_3$ single mode waveguides which exhibit both minimum mode size and low propagation loss at 1.3 μm.

All these design requirements have led to the significance of the analysis of polarized modes in channel waveguides.

In general, the optical mode of the waveguide is acquired by solving the Helmholtz equation. However, only a few simple waveguide structures can be solved analytically. Therefore, extensive attempts have been made to obtain numerical solutions for a two-dimensional (2D) cross-section of optical waveguides [7–23]. One method is the approximate modeling of 2D slab waveguide solution successively in both directions, following either the method of Marcatili [5] or the EIM [24]. However, these methods are not applicable to arbitrarily shaped optical waveguides, neither do they handle waveguide mode near the cutoff region efficiently. A significant number of numerical methods have been proposed to obtain rigorous solutions to the wave equation with pertinent boundary conditions. The popular techniques are the FDM [10], the finite element method (FEM) [22] or the beam propagating method (BPM) [14]. The application of different techniques based on the above methods such as semivectorial *E*-field FDM [12], semivectorial *H*-field FDM [25], and Rayleigh quotient solution [26], have been studied and reported. These methods are applicable to arbitrarily shaped optical waveguides. In FEM and FDM, partial differential equations are discretized and then transformed into matrix equations. The calculations of mode indices and optical field distributions are then equivalent to obtaining eigenvalues and eigenfuctions of the coefficient matrices.

In this chapter, we first treat the 3D optical waveguide from an analytical point of view with the representation of a two-dimensional distribution of the refractive index profile by two effective planar profiles. The propagation and mode guiding conditions obtained for these planar waveguides are thus combined for the 3D waveguide.

Sections 3.2 and 3.3 describe the analytical estimation of guided mode using the Marcatili method and the EIM. Section 3.4 then outlines the numerical formulation of the non-uniform finite difference (FD) scheme. Both quasi-TE and quasi-TM polarized modes are addressed. We also assess the accuracy of the numerical result of this scheme by computing the effective refractive index of rib and slab dielectric waveguides. The effect of grid spacing is also investigated. The effectiveness of the variable grid spacing in dealing with waveguide mode near the cutoff region is also given. Sections 3.4 and 3.5 then give the treatment of the 3D optical waveguides by the FDM for uniform index regions, the rib waveguide and diffused index profiles, the diffused optical channel waveguides. Section 3.4 describes the modeling of the 3D optical waveguide with a graded index profile such as the Ti:LiNbO$_3$ channel waveguides. The effects of various waveguide fabrication parameters such as the diffusion time, diffusion temperature, thickness and width of the titanium strips are studied. The accuracy of the numerical model is assessed by comparing our simulations with experimental and simulation results that are reported in several literatures. Section 3.5 describes the modeling of rib optical waveguides using the same FDM.

3.2 Marcatili's Method

The cross-section of typical 3D waveguides are shown in Figure 3.1 including a raised strip or channel waveguide, strip-loaded, rib or ridge and embedded structures with a substrate and an overlay region.

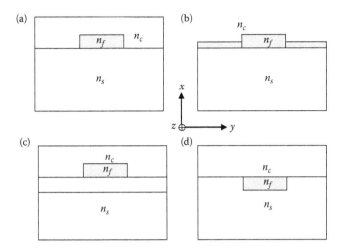

FIGURE 3.1
Channel 3D optical waveguides: (a) raised strip or channel; (b) ridge or rib; (c) strip-loaded; (d) embedded channel.

Usually the raised channel waveguide is formed by depositing a thin film layer, e.g., by the molecular chemical vapor deposition (MOCVD) or by sputtering, then if we remove the film material in the outer regions by some means, such as dry reactive etching, while keeping the film layer in the central portion intact, we have the *raised stripe* or *channel waveguides*. The *ridge* or *rib waveguides* are similar to the raised strip waveguides except that the film layer on the two sides is partially removed as shown in Figure 3.1b. If we place a dielectric strip on the top of the film layer, as shown schematically in Figure 3.1c, we have the *strip-loaded waveguides*. By embedding a high-index bar in the substrate region, we have the *buried* or *embedded strip waveguides* (Figure 3.1d). Channel, ridge, strip-loaded, and buried strip waveguides are 3D waveguides with rectangular boundaries. Circular and elliptical fibers, discussed in Chapters 4 and 5, are 3D waveguides with curved boundaries. The refractive index of the 3D waveguide can vary with respect to the distance of depth. In this case we have a graded index channel waveguide such as diffused channel optical waveguides formed by diffusion of impurity into $LiNbO_3$ substrate at a temperature around $1000°C$.

In this section, we analyze the modes which are guided by 3D waveguides with rectangular geometries using mainly the Marcatili method, and the EIM. The chapter consists of five sections. Since fields of 3D waveguides are complicated and difficult to analyze, we begin with a qualitative description.

3.2.1 Field and Modes Guided in Rectangular Optical Waveguides

3.2.1.1 Mode Fields of H_x Modes

In 2D waveguides, one of the dimensions transverse to the direction of propagation is very large in comparison to the operating wavelength. This is the y direction in Figure 3.1. The waveguide width in this direction is treated as infinitely large. As a result, fields guided by 2D dielectric waveguides can be classified as TE or TM modes as discussed in Chapter 2. For TE modes, the longitudinal electric field component, E_z is zero, and all other field components can be expressed in terms of H_z. For TM modes, H_z vanishes and all other field components can be expressed in terms of E_z. In 3D optical waveguides, the waveguide width and height are comparable to the operating wavelength. Neither the width nor height can be treated as infinitely large. Thus neither E_z nor H_z vanish, except in some special cases. As a result, modes guided by 3D optical waveguides are neither TE nor TM modes except for the special cases. In general, they are *hybrid modes*. A complicated scheme is needed to designate the hybrid modes. Since all field components are present, the analysis for hybrid modes is very complicated. Intensive numerical computations are often required [27]. The description of the EIM given here briefly follows the explanations given in Chen, 2007 [28]

In many dielectric waveguide structures, the index difference is small. As a result, one of the transverse electric field components is much stronger than the other transverse electric field component. Goell has suggested a physically intuitive scheme to describe hybrid modes [29]. In Goell's scheme, a hybrid mode is labeled by the direction and distribution of the strong transverse electric field component. If the dominant electric field component is in the x (or y) direction and if the electric field distribution has $p - 1$ nulls in the x direction and $q - 1$ nulls in the y direction, then the hybrid mode is identified as $E_{x,pq}$ (or $E_{y,pq}$) modes. The superscript denotes the direction of the *dominant transverse electric field component.*

Now considering a weakly guiding rectangular optical waveguide with a core of index n_1 and surrounded with lower indices n_j with $j = 2, 3, 4$ and 5. The waveguide cross-section is shown in Figure 3.2.

The rectangular waveguide can be considered to be equivalent to two slab waveguides, one extended in the x direction, termed as the *H*-waveguide and one in the y direction termed as the *V*-waveguide. That means that the field is confined as a mode in the y direction and the other in the x direction. This is normally called the hybrid mode. Thus we can write the field component H_x in the five regions as portioned in Figure 3.2 as follows:

$$H_{x1} = C_1 \cos\left(\kappa_{x1}x + \phi_{x1}\right)\cos\left(\kappa_{y1}y + \phi_{y1}\right)e^{-j\beta z}; \text{region 1}$$

$$H_{x2} = C_2 \cos\left(\kappa_{x2}x + \phi_{x2}\right)e^{-j\kappa_{y2}y}e^{-j\beta z}; \text{region 2}$$

$$H_{x3} = C_3 e^{-j\kappa_{x3}x} \cos\left(\kappa_{y3}y + \phi_{y3}\right)e^{-j\beta z}; \text{region 3} \tag{3.1}$$

$$H_{x4} = C_4 e^{-j\kappa_{y4}y} \cos\left(\kappa_{x4}x + \phi_{x4}\right)e^{-j\beta z}; \text{region 4}$$

$$H_{x5} = C_5 e^{-j\kappa_{x5}x} \cos\left(\kappa_{y5}y + \phi_{y5}\right)e^{-j\beta z}; \text{region 5}$$

where C_j, ϕ_{xj}, ϕ_{yj} are the constants to be determined using the boundary conditions, κ_{xj}, κ_{yj} are the propagation constants effective in the x and y transverse directions, respectively. For each region the propagation constants in the x, y and z directions. κ_{xj}, κ_{yj}, β must satisfy

$$\kappa_{xj}^2 + \kappa_{yj}^2 + \beta^2 = \kappa^2 n_j^2; j = 1, 2, 3, 4, 5 \tag{3.2}$$

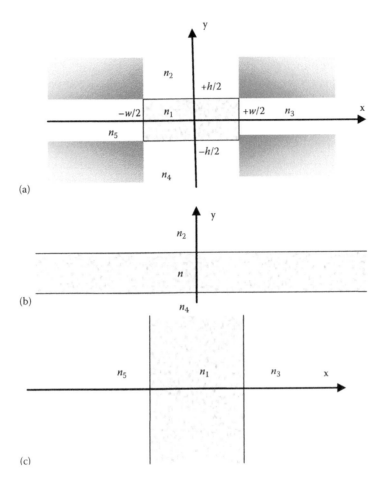

FIGURE 3.2
Model used to analyze E_y modes of: (a) rectangular waveguide, (b) H-waveguide, and (c) V-waveguide.

There are no additional constraints on the transverse propagation constant. In fact the transverse propagation constant in regions 2–5 are imaginary, that is the fields must decay to zero in these regions except in the rectangular section.

When expressed in terms of κ_{xj}, κ_{yj}, ϕ_x, ϕ_y, Equation 3.1 can be simplified to

$$H_{x1} = C_1 \cos\left(\kappa_{x1}x + \phi_x\right)\cos\left(\kappa_{y1}y + \phi_y\right)e^{-j\beta z} \text{ ; region 1}$$

$$H_{x2} = C_2 \cos\left(\kappa_{x2}x + \phi_x\right)e^{-j\kappa_{y2}y}e^{-j\beta z} \text{ ; region 2}$$

$$H_{x3} = C_3 e^{-j\kappa_x x}\cos\left(\kappa_y y + \phi_y\right)e^{-j\beta z} \text{ ; region 3} \qquad (3.3)$$

$$H_{x4} = C_4 e^{-j\kappa_y y}\cos\left(\kappa_x x + \phi_x\right)e^{-j\beta z} \text{ ; region 4}$$

$$H_{x5} = C_5 e^{-j\kappa_x x}\cos\left(\kappa_y y + \phi_y\right)e^{-j\beta z} \text{ ; region 5}$$

3.2.1.2 Boundary Conditions at the Interfaces

3.2.1.2.1 Horizontal Boundary $y = \pm h/2$; $|x| < w/2$

Along the horizontal boundaries, the tangential components are E_x, E_z, H_x, H_z, the x-components are ignored as their amplitudes are extremely small compared with the other components. Using Maxwell's equation we can observe that

- E_z is continuous at the boundary and tangential implying that $(1/n_j^2)\,(\partial H_x/\partial y)$.
- Being tangential to the horizontal lines, H_x must be continuous everywhere along the horizontal lines. Thence the tangential derivative $\partial H_x/\partial x$ and therefore H_z must also be continuous on the horizontal lines. In other words, if H_x is continuous at the horizontal lines, so is H_z.

Thus all the boundary conditions are met if we have the continuity of the term $(1/n_j^2)\,(\partial H_x/\partial y)$.

3.2.1.2.2 Vertical Boundary $x = \pm w/2$; $|y| < h/2$

Along this boundary the tangential components are in the y and z direction and the normal direction is x. Only the components E_y, H_x are significant and E_y is continuous if H_x is continuous. Applying these conditions for the field components at $x = \pm\, w/2$ we obtain:

$$E_{z1} - E_{z3} = \frac{j\eta_0}{k}\left(\frac{1}{n_1^2}\frac{\partial H_{x1}}{\partial y} - \frac{1}{n_3^2}\frac{\partial H_{x3}}{\partial y}\right) + O\left(\partial^2\right)$$

$$= \frac{j\eta_0}{k}\frac{1}{n_1^2}\left(\frac{\partial\left(H_{x1} - H_{x3}\right)}{\partial y}\right) - \frac{j\eta_0}{n_3}\frac{n_1^2 - n_3^2}{n_1^2}\frac{1}{kn_3}\frac{\partial\left(H_{x3}\right)}{\partial y} \tag{3.4}$$

The second term of Equation 3.4 can be ignored due to the very small difference in the refractive index terms. Thus it can be written as:

$$E_{z1} - E_{z3} = \frac{j\eta_0}{k}\frac{1}{n_1^2}\left(\frac{\partial\left(H_{x1} - H_{x3}\right)}{\partial y}\right) + O\left(\partial^2\right) \tag{3.5}$$

In other words the component E_z is continuous if H_x is continuous there.

3.2.1.2.3 Transverse Vector κ_x, κ_y

The transverse momentum vector κ_x can now be determined from the boundary conditions discussed above. One would seek an oscillating behavior of the waves in the waveguide region and exponentially decay to zero in the cladding regions. At $y = \pm h/2$ the continuity of H_x and $(1/n_j^2)(\partial H_x/\partial y)$ leads to

$$C_1\cos\left(\frac{1}{2}\kappa_y h + \phi_y\right) = C_2 e^{-j\kappa_{y2}h/2}$$

$$-\frac{\kappa_y}{n_1^2}C_1\sin\left(\frac{1}{2}\kappa_y h + \phi_y\right) = -\frac{j\kappa_{y2}}{n_2^2}C_2 e^{-j\kappa_{y2}h/2} \tag{3.6}$$

Combining these equations we obtain the relation

$$\tan\left(\frac{1}{2}\kappa_y h + \phi_y\right) = -\frac{j\kappa_{y2}n_1^2}{\kappa_y n_2^2} \tag{3.7}$$

From Equation 3.2 we can deduce that

$$j\kappa_{y2} = \sqrt{k^2(n_1^2 - n_2^2) - \kappa_y^2} \tag{3.8}$$

Thus Equation 3.7 becomes

$$\tan\left(\frac{1}{2}\kappa_y h + \phi_y\right) = \frac{\sqrt{k(n_1^2 - n_2^2) - \kappa_y^2}}{\kappa_y n_2^2} \tag{3.9}$$

Or alternatively we have

$$\frac{1}{2}\kappa_y h + \phi_y = m\pi + \tan^{-1}\left(\frac{\sqrt{k^2(n_1^2 - n_2^2) - \kappa_y^2}}{\kappa_y n_2^2}\right)$$

$$\frac{1}{2}\kappa_y h + \phi_y = n\pi + \tan^{-1}\left(\frac{\sqrt{k^2(n_1^2 - n_4^2) - \kappa_y^2}}{\kappa_y n_4^2}\right); \quad \text{at} \quad y = -h/2 \tag{3.10}$$

with q, m, n as integers. Then eliminating ϕ_y we can rewrite as

$$\kappa_y h_y = q\pi + \tan^{-1}\left(\frac{\sqrt{k^2(n_1^2 - n_2^2) - \kappa_y^2}}{\kappa_y n_2^2}\right) + \tan^{-1}\left(\frac{\sqrt{k^2(n_1^2 - n_4^2) - \kappa_y^2}}{\kappa_y n_4^2}\right) \tag{3.11}$$

This is the dispersion relation for the TM modes guided in the channel waveguide and is also similar to that for a planar waveguide. The last two terms on the right-hand side (RHS) of Equation 3.11 represent the phase shift, normally called the Goos-Hanchen shift for the "rays" penetrating into the cladding of the guided fields. Thus, similar to this boundary condition and the dispersion relationship, the dispersion characteristics for the transverse vector κ_y can be written as

$$\kappa_x w = p\pi + \tan^{-1}\left(\frac{\sqrt{k^2(n_1^2 - n_3^2) - \kappa_x^2}}{\kappa_x}\right) + \tan^{-1}\left(\frac{\sqrt{k^2(n_1^2 - n_5^2) - \kappa_x^2}}{\kappa_x}\right) \tag{3.12}$$

with p as an integer.

3.2.2 Mode Fields of E_y Modes

Similar to the analysis given for the H_x modes, the E_x modes can be found with the dispersion relation by using the continuity properties of the field components H_y; $\partial H_y/\partial y$. We then obtain

$$\kappa_y h = q\pi + \tan^{-1}\left(\frac{\sqrt{k^2(n_1^2 - n_2^2) - \kappa_y^2}}{\kappa_y}\right) + \tan^{-1}\left(\frac{\sqrt{k^2(n_1^2 - n_4^2) - \kappa_y^2}}{\kappa_y}\right) \tag{3.13}$$

$$\kappa_x w = p\pi + \tan^{-1}\left(\frac{n_1^2\sqrt{k^2(n_1^2-n_3^2)-\kappa_x^2}}{\kappa_x n_3^2}\right) + \tan^{-1}\left(\frac{n_1^2\sqrt{k^2(n_1^2-n_5^2)-\kappa_x^2}}{\kappa_x n_5^2}\right) \quad (3.14)$$

Equations 3.13 and 3.14 specify the dispersion relationship for the TM modes with a planar waveguide thickness of W. The terms involving the arctan are the Goos-Hanchen phase shifts due to the reflection and penetration of the lightwave fronts into some distance at the interface between the two different dielectric interfaces [7].

Thus Marcatili's method is modeled for two equivalent planar waveguides in the horizontal and vertical directions. It corresponds to the dispersion relation equations, Equation 3.11 and 3.12, for TM modes guided by planar *waveguide of thickness* W. The dominant electric field of E_x modes is in parallel with the horizontal boundaries. Thus, we use the dispersion equation of TE modes guided by waveguide H to determine κ_y. The dominant electric field component of E_x modes is perpendicular to the vertical boundaries of waveguide W. Therefore, we use the dispersion for TM modes guided by the 2D waveguide to evaluate κ_x. With κ_x, κ_y known, the propagation constant can be determined from Equation 3.2.

3.2.3 Dispersion Characteristics

As an example, we consider a dielectric bar of index n_1 immersed in a medium with index n_2 as shown in Figure 3.3 with uniform refractive indices in the regions surrounding the channel waveguiding region. To facilitate comparison, we define the normalized frequency parameter V and the normalized guide index b, or normalized propagation constant in terms of n_1, n_2, h

$$V = kh\sqrt{n_1^2-n_2^2} \simeq \frac{2\pi}{\lambda}hn\sqrt{2\Delta}$$

$$b = \frac{\beta^2 - k^2 n_2^2}{k^2\left(n_1^2 - n_2^2\right)} \quad (3.15)$$

Thus the normalized effective refractive index can be evaluated as a function of the normalized-frequency parameter V to give the dispersion curves as shown in Figure 3.4 in which the curves obtained from FEM and the Marcatili methods are also contrasted with agreement.

A numerical evaluation is given here for silica doped with a GeO$_2$ waveguide and a cladding region of pure silica. The relative refractive index of the core and the pure silica cladding is 0.3% or 0.5%, then using the single mode operation given in Figure 3.4, we can select $V = 1$ and using Equation 3.15, then the cross-section of the rectangular waveguide is 3×3 μm² for 0.5% relative refractive index and for 0.3% of the dimension is 6×6 μm², the refractive index of pure silica is 1.448 for an operating wavelength of 1550 nm.

3.3 Effective Index Method

3.3.1 General Considerations

Similar to the Marcatili method discussed in the last section, the *EIM* is also an approximate method for analyzing rectangular waveguides. In the Marcatili method, a 3D waveguide

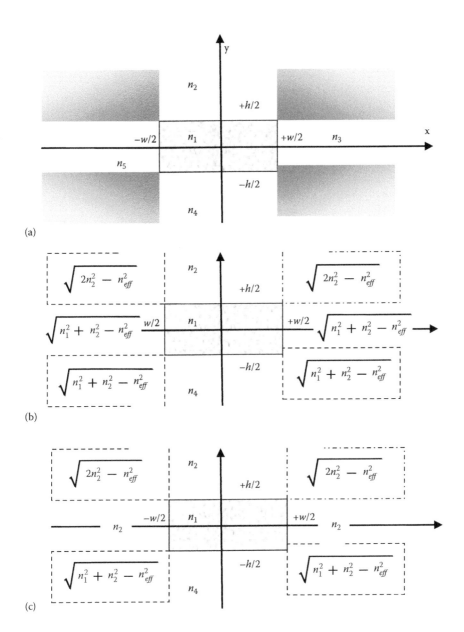

FIGURE 3.3
An embedded channel optical waveguide: (a) waveguide structure, and its representation using (b) EIM, and (c) model of the waveguide using the Marcatili method.

(see Figure 3.1) is replaced by two 2D waveguides: waveguides H and W depicted in Figure 3.3. The two 2D waveguides are *mutually independent* in that the waveguide parameters of the two 2D waveguides come directly from the original 3D waveguide.

To provide a theoretical basis for the EIM, in lieu of the original 3D waveguide, considering a *pseudo-waveguide* that can be considered to be a superposition of two equivalent waveguides H and V or H' and V'. The pseudo-waveguide is chosen such that waveguides H and V, or H' and V', can be easily identified and analyzed. The dispersion of waveguide

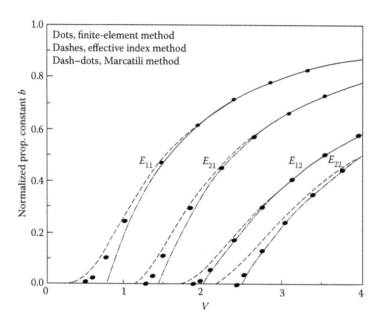

FIGURE 3.4

Dispersion characteristics, dependence of the normalized propagation constant of the guided modes as a function of the parameter V, the normalized frequency, comparison of three numerical, analytical methods for rectangular optical waveguides consisting of uniform core and cladding.

V, or V', is used as an approximation for β of the original 3D waveguide. The structures of these waveguides are shown in Figure 3.5.

Considering the E_y modes guided by a 3D waveguide shown in Figure 3.5a, all field components of E_y modes can be expressed in terms of H_x which can be written as $h_x(x, y)$ $e^{-j\beta z}$, with $h_x(x, y)$ is the field distribution in the transverse plane and $n(x, y) = n_j; j = 1 - 5$. The wave equation in the transverse plane can be obtained as

$$\left[\frac{\partial^2}{\partial x^2} + \frac{\partial^2}{\partial y^2} + k^2 n^2(x, y) - \beta^2 \right] h_x(x, y) = 0 \tag{3.16}$$

Instead of considering the 3D waveguide problem we can now modify the refractive index distribution so that a pseudo-waveguide structure can be obtained as a planar 2D planar structure in the x and then the y direction, as shown in Figure 3.5. Then a superposition of the guiding condition can be achieved. The refractive index of the pseudo-waveguide can be written as

$$n_{ps}^2 = n_x^2(x) + n_y^2(y) \tag{3.17}$$

Then one can determine $n(x, y)$ by the common method of separation of variables, that is the distribution $h_x(x, y)$ can be represented as the product of two field distribution functions $E_y(x)$ and $E_x(y)$, the wave equation in the transverse plane can thus be written as:

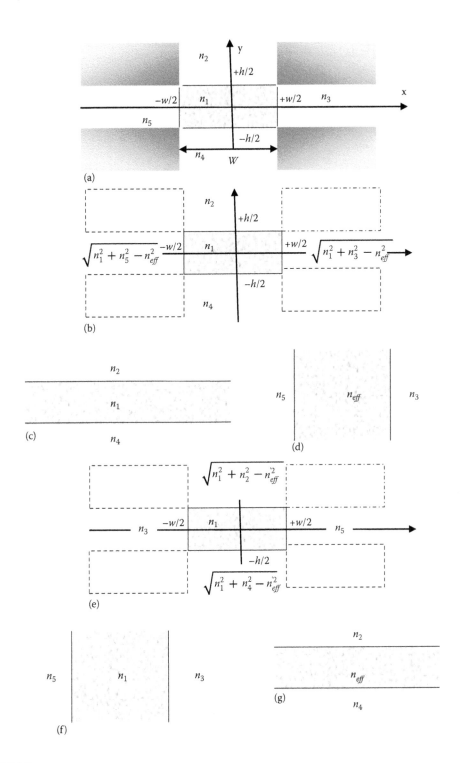

FIGURE 3.5
Model of pseudo-waveguides used in the EIM. (a) Rectangular channel waveguide, (b) pseudo-waveguide, (c) planar waveguide I, (d) planar waveguide II, (e) alternate pseudo-waveguide, (f) planar waveguide V', and (g) planar waveguide H'.

$$\frac{1}{E_y(x)}\frac{\partial^2 E_y(x)}{\partial x^2} + \frac{1}{E_x(y)}\frac{\partial^2 E_x(y)}{\partial y^2} + \left[k^2 n_x^2(x) + k^2 n_y^2(y) - \beta^2\right] h_x(x,y) = 0$$

or   (3.18)

$$\frac{1}{E_x(y)}\frac{\partial^2 E_x(y)}{\partial y^2} + k^2 n_y^2(y) = -\frac{1}{E_y(x)}\frac{\partial^2 E_y(x)}{\partial x^2} - k^2 n_x^2(x) - \beta^2$$

The solutions of this equation can only be physically possible when the two fields approach zero at a distance very far from the center. That means that the boundary is equivalent to a perfect metallic wall. So we have

$$\frac{1}{E_x(y)}\frac{\partial^2 E_x(y)}{\partial y^2} + k^2\left[n_y^2(y) - n_{eff}^2\right] = 0$$

                                                                                                            (3.19)

$$-\frac{1}{E_y(x)}\frac{\partial^2 E_y(x)}{\partial x^2} - k^2\left[n_x^2(x) + n_{eff}^2\right] - \beta^2 = 0$$

These two equations can be solved subject to the boundary conditions to arrive with the propagation constant along the z direction of the waveguide. Thus the complete solution is the product of the two field functions $E_y(x)$ and $E_x(y)$. The phase term represents the propagation of the field along the z direction.

3.3.2 A Pseudo-Waveguide

Consider the waveguide structure shown in Figure 3.5. The refractive index distribution of the channel waveguide core and cladding is shown and given as:

$$n_{ps}^2(x,y) = \begin{cases} n_1^2; & \text{region 1} \\ n_2^2; & \text{region 2} \\ n_1^2 + n_3^2 - n_{eff}^2; & \text{region 3} \\ n_4^2; & \text{region 4} \\ n_1^2 + n_5^2 - n_{eff}^2; & \text{region 5} \end{cases}$$ (3.20)

This distribution can be considered as the superposition of two distributed functions:

$$n_y^2(y) = \begin{cases} n_2^2; & \text{region } y > h/2 \\ n_1^2; & \text{region } -h/2 \le y \le h/2 \\ n_4^2; & \text{region } y < -h/2 \end{cases}$$

                                                                                                            (3.21)

$$n_x^2(x) = \begin{cases} n_3^2 - n_{eff}^2; & \text{region } x > w/2 \\ 0; & \text{region } -w/2 \le x \le w/2 \\ n_5^2 - n_{eff}^2; & \text{region } x < -w/2 \end{cases}$$

Thence similar to the method obtained in the Marcatili method, the dispersion relation can be obtained as:

$$kh\sqrt{n_1^2 - n_{\text{eff}}^2} = q\pi + \tan^{-1}\left(\frac{n_1^2}{n_2^2} \frac{\sqrt{n_{\text{eff}}^2 - n_2^2}}{\sqrt{-n_{\text{eff}}^2 + n_1^2}}\right) + \tan^{-1}\left(\frac{n_1^2}{n_4^2} \frac{\sqrt{n_{\text{eff}}^2 - n_4^2}}{\sqrt{-n_{\text{eff}}^2 + n_1^2}}\right) \qquad (3.22)$$

$$kw\sqrt{n_{\text{eff}}^2 - N} = p\pi + \tan^{-1}\left(\frac{\sqrt{N^2 - n_3^2}}{\sqrt{-n_{\text{eff}}^2 + N^2}}\right) + \tan^{-1}\left(\frac{\sqrt{N^2 - n_5^2}}{\sqrt{n_{\text{eff}}^2 - N^2}}\right) \qquad (3.23)$$

Using these dispersion relations the dispersion characteristics of an embedded channel waveguide with cladding, as shown in Figure 3.3, can be obtained very close to that given in Figure 3.4 as the dashed curves.

3.3.3 Finite Difference Numerical Techniques for 3D Waveguides

The main purpose of selecting the FDM to study the quasi-TE and quasi-TM polarized waveguide modes is due to its simplicity and plausible accuracy. We have employed the semivectorial analysis [12,23,25] which automatically takes full account of the discontinuities in the normal electric field components across any arbitrary distribution of internal dielectric interfaces. The semivectorial FDM, despite its simplicity and being free from troublesome spurious solutions, has two major disadvantages of being computationally intensive and requiring large amounts of memory. Hence, it is necessary to introduce the discretization scheme on the non-uniform mesh, in which mesh intervals can be changed arbitrarily depending on waveguide structures. For this reason, we have modeled the waveguide mode with FDM which employs a non-uniform discretization scheme [9,23]. Such a discretization scheme enables us to increase the size of the problem space so that the field component at the boundary can be assumed to have vanished. The grid spacing increases monotonically with increasing distance from the guiding region. The grid lines can also be aligned with the boundaries of the step index changes in conventional structures such as rib, ridge and strip-loaded waveguides as well as quantum well structures. Furthermore, by judiciously placing the grid lines and corresponding cell structure efficiently, we can reduce the required matrix size and hence redundant computer calculations, while preserving the accuracy of the calculations. The non-uniform discretization scheme also enables us to handle waveguide mode near the cutoff region with a relative simple boundary condition. The eigenmodes of the Helmholtz equation is solved by the application of the shifted inverse power iteration method. This method warrants both the mode size and its relevant propagation constant, which are both important parameters to the design of optical waveguide.

Apart from being able to access the accuracy of the final product of our work, which is the SVMM (Semivectorial Mode Modeling) computer program, we also present an overview of its application in modeling the Ti:LiNbO$_3$ channel waveguide for optical devices such as modulators and switches.

3.4 Non-Uniform Grid Semivectorial Polarized Finite Difference Method for Optical Waveguides with Arbitrary Index Profile

3.4.1 Propagation Equation

For harmonic wave propagation in the z direction along a rib or channel waveguide, we consider the following fields

$$E(x, y, z) = (E_x, E_y, E_z) \exp j(\omega t - \beta z) \tag{3.24}$$

$$H(x, y, z) = (H_x, H_y, H_z) \exp j(\omega t - \beta z) \tag{3.25}$$

$$D = \varepsilon(x, y)E, \quad B = \mu H \tag{3.26}$$

where the dielectric permittivity $\varepsilon(x, y)$ is piecewise constant and the magnetic permeability μ is completely constant throughout the solution domain. The components of the electric and magnetic fields in Equation 3.1 are functions of x and y only. Then, applying the Maxwell equations in the magnetic and charge-free media and appropriate algebra we obtain the wave equation

$$\nabla \times (\nabla \times E) = \nabla(\nabla \cdot E) - \nabla^2 E = \omega^2 \varepsilon \mu E = k^2 n^2 E \tag{3.27}$$

in which $k = \omega(\varepsilon_0 \mu_0)^{1/2} = 2\pi/\lambda$ and $\varepsilon = \varepsilon_0 n^2(x, y)$ with λ being the free space wavelength. With the divergence of $\nabla \cdot D = 0$ and $\nabla \log_e \varepsilon = \nabla \varepsilon / \varepsilon$, we get

$$\nabla \cdot E = -E \cdot \nabla \log_e \varepsilon = -E \cdot \nabla n^2/n \tag{3.28}$$

This may be substituted into Equation 3.4 to yield the wave equation

$$\nabla^2 E + k^2 E + \nabla(E \cdot \nabla n^2/n) = 0 \tag{3.29}$$

As $n(x, y)$ is piecewise constant, $\nabla n^2/n = 0$ and it should be noted that $\nabla n^2/n$ is undefined at internal dielectric interfaces where $n(x, y)$ is discontinuous. With the assumption that the fields are polarized either perpendicular (quasi-TM) to or parallel (quasi-TE) to the crystal surface and that the major field components of the modes are perpendicular to the direction of the propagation, Equation 3.7 can be reduced to

$$\left(\nabla_t^2 + k^2 n^2\right) E = \beta^2 E \tag{3.30}$$

in which

$$\nabla_T^2 = \frac{\partial^2}{\partial x^2} + \frac{\partial^2}{\partial y^2},$$

the transverse Laplacian and β is the propagation constant. This is essentially the Helmholtz wave equation.

3.4.2 Formulation of Non-Uniform Grid Difference Equation

Figure 3.6a shows the grid lines used in the FDM formulation. The grid lines are chosen in such a way that denser grids are allocated around the guiding region while coarser grids are assigned to regions further away from the waveguide. Boundaries of abrupt index changes are straddled by the grid lines wherever necessary. Figure 3.6b shows the magnified view of a portion of the grid for a more detailed illustration. Each cell point is located in the center of each rectangular cell. h_i and h_j are the horizontal and vertical grid sizes. The refractive index within each cell is assumed to be uniform. $n_{i,j}$ and

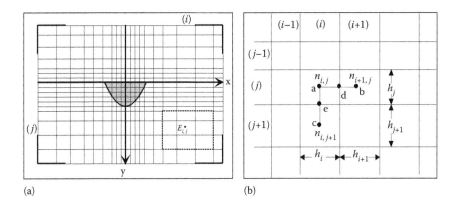

FIGURE 3.6
Non-uniform discretized grid for FDM scheme. (a) and (b) A magnified portion of grid lattice and cell structure of point i, j.

$n_{i+1, j}$ represent the values of the refractive index of each small cell as an approximation, which are taken from the continuous refractive index profile $n(x, y)$. Non-uniform spacing of the grid lines provides some flexibility in setting up the non-uniform grid FDM. The non-uniform discretization with increasing spacing away from the guiding region permits sufficient extension of the boundary. This enables us to assume a Dirichlet boundary condition (metal box) where all fields have vanished.

3.4.2.1 Quasi-Transverse Electronic Mode

For quasi-TE polarized mode, E_y is assumed to be zero. E_x is continuous across the horizontal interfaces but discontinuous across vertical interfaces. Therefore, the quasi-TE modes are the eigensolutions of the equation

$$\nabla_t^2 E_x + k^2 n^2 E_x = \beta^2 E_x \tag{3.31}$$

The discontinuity across the vertical interface will need to be taken into account when formulating the difference equation.

Figure 3.7 illustrates the quasi-TE field discontinuity at the boundary between cells (i, j) and $(i + 1, j)$. Consider the points a, d and b, with d being at the boundary of the dielectric interface. The horizontal axis is the x-axis while the vertical axis is the electric field amplitude of the respective position of the cell. Assume that the x-axis is pointing towards the east. So, E_E and E_W are the field amplitudes just to the east and the west of the boundary between the cells (i, j) and cell $(i + 1, j)$. $E_{i,j}^v$ is the virtual field in cell (i, j) which is the extension of the actual field $E_{i+1,j}$. In other words, $E_{i,j}^v$ is the field seen by the cell $(i + 1, j)$. Similarly, $E_{i+1,j}^v$ is the extension of $E_{i,j}$. n_E and n_W are the refractive indices just to the east and the west of the boundary. Since we consider a slowly varying index distribution, we assume that n_E and n_W are approximately equal to $n_{i,j}$ and $n_{i+1,j}$, respectively. The boundary conditions between the cells (i, j) and $(i + 1, j)$ are given as follows:

$$n_E^2 E_E = n_W^2 E_W \Rightarrow n_{i,j}^2 E_E = n_{i+1,j}^2 E_W \tag{3.32}$$

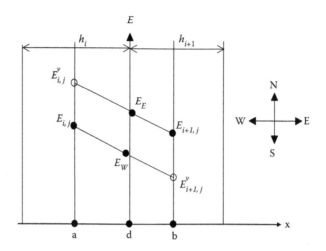

FIGURE 3.7

Quasi-TE electric field discontinuity at the boundary between cells (i, j) and cell $(i+1, j)$. Solid lines are the actual field profiles along the x-axis while $E_{i,j}^v$ and $E_{i+1,j}^v$ are virtual fields.

$$\frac{\partial E_E}{\partial x} = \frac{\partial E_W}{\partial x} = p^+ \tag{3.33}$$

where p^+ represents the field gradient at the boundary between the cells. We can then use the approximate relationship between $E_{i, j}$, $E_{i+1, j}$, $E_{i,j}^v$, $E_{i+1,j}^v$, and obtain the following equations for E_E and E_W:

$$E_{i+1,j} \approx E_E + (h_{i+1}/2) \cdot p^+$$

$$E_{i,j}^v \approx E_E - (h_i/2) \cdot p^+$$

$$E_{i+1,j}^v \approx E_W + (h_{i+1}/2) \cdot p^{+'} \tag{3.34}$$

$$E_{i,j}^v \approx E_W - (h_i/2) \cdot p^+$$

where h_i and h_{i+1} are the horizontal lengths of the cells (i, j) and $(i+1, j)$. The four equations above are in fact redundant. Therefore we need only to consider either $E_{i,j}^v$ or $E_{i+1,j}^v$, which we choose $E_{i+1,j}^v$ in our case. The following shows the algebraic manipulation (Equation 3.11):

$$p^+ = 2(E_{i+1,j}^v - E_{i,j})/(h_i + h_{i+1})$$

$$E_{i+1,j}^v = E_{i+1,j} + (E_W - E_E) \tag{3.35}$$

$$h_{i+1}(E_W - E_{i,j}) = h_i(E_{i+1,j} - E_E)$$

and then

$$E_{i+1,j}^v = \frac{n_{i+1,j}^2(h_i + h_{i+1})E_{i+1} + h_{i+1}(n_{i+1,j}^2 - n_{i,j}^2)E_{i,j}}{(n_{i,j}^2 h_i + n_{i+1,j}^2 h_{i+1})} \tag{3.36}$$

With similar procedure between cells (i, j) and $(i+1, j)$, we can obtain

$$p^- = 2\frac{E_{i,j} - E_{i-1,j}^v}{h_i + h_{i-1}} \tag{3.37}$$

$$E_{i-1,j}^v = \frac{n_{i-1,j}^2(h_i + h_{i-1})E_{i-1,j} + h_{i-1}(n_{i-1,j}^2 - n_{i,j}^2)E_{i,j}}{(n_{i,j}^2 h_i + n_{i-1,j}^2 h_{i-1})}$$

where p^- is now the field gradient at the boundary between the cells $(i-1, j)$ and (i, j). Note that the quasi-TE electric field is continuous in terms of y direction even if there are discontinuities in the refractive index. Therefore $E_{i,j+1}^v = E_{i,j+1}$, $E_{i,j-1}^v = E_{i,j-1}$.

The second derivative can be derived as

$$\frac{\partial^2 E_{i,j}}{\partial x^2} = \frac{1}{h_i}[p^+ - p^-] = \frac{1}{h_i}\left[\frac{2(E_{i+1,j}^v - E_{i,j})}{h_{i+1} + h_i} - \frac{2(E_{i,j} - E_{i-1,j}^v)}{h_i + h_{i-1}}\right] \tag{3.38a}$$

$$\frac{\partial^2 E_{i,j}}{\partial y^2} = \frac{1}{h_j}[p^+ - p^-] = \frac{1}{h_j}\left[\frac{2(E_{i,j+1} - E_{i,j})}{h_{j+1} + h_j} - \frac{2(E_{i,j} - E_{i,j-1})}{h_j + h_{j-1}}\right] \tag{3.38b}$$

Thence we get the discrete wave equation as

$$\frac{\partial^2 E_{i,j}}{\partial x^2} = \frac{2n_{i-1,j}^2}{h_i(n_{i,j}^2 h_i + n_{i-1,j}^2 h_{i-1})}E_{i-1,j} + \frac{2n_{i+1,j}^2}{h_i(n_{i,j}^2 h_i + n_{i+1,j}^2 h_{i+1,j})}E_{i+1,j}$$

$$-\left[\frac{2n_{i,j}^2}{h_i(n_{i,j}^2 h_i + n_{i-1,j}^2 h_{i-1})} + \frac{2n_{i,j}^2}{h_i(n_{i,j}^2 h_i + n_{i+1,j}^2 h_{i+1})}\right]E_{i,j} \tag{3.39a}$$

$$\frac{\partial^2 E_{i,j}}{\partial y^2} = \frac{1}{h_j}\left[\frac{2(E_{i,j+1} - E_{i,j})}{h_{j+1} + h_j} - \frac{2(E_{i,j} - E_{i,j-1})}{h_j + h_{j-1}}\right] \tag{3.39b}$$

Substituting these equations into the Helmholtz equation,

$$C_{i-1,j}E_{i-1,j} + C_{i+1,j}E_{i+1,j} - C_{i,j}E_{i,j} + C_{i,j-1}E_{i,j-1} + C_{i,j+1}E_{i,j+1} = \beta^2 E_{i,j} \tag{3.40}$$

where

$$C_{i-1,j} = \frac{2n_{i-1}^2}{h_i(n_{i,j}^2 h_i + n_{i-1,j}^2 h_{i-1})}$$

$$C_{i+1,j} = \frac{2n_{i+1}^2}{h_i(n_{i,j}^2 h_i + n_{i+1,j}^2 h_{i+1})}$$

$$C_{i,j-1} = \frac{2}{h_j(h_j + h_{j-1})} \tag{3.41}$$

$$C_{i,j+1} = \frac{2}{h_j(h_j + h_{j+1})}$$

$$C_{i,j} = C_{i-1,j} + C_{i+1,j} + C_{i,j-1} + C_{i,j+1} - k^2 n_{i,j}^2$$

The above equation is essentially an eigenvalue equation of

$$C_{TE}E_{TE} = \beta_{TE}^2 E_{TE} \tag{3.42}$$

in which C_{TE} is a non-symmetric band matrix which contains the coefficient of the above equations, β_{TE}^2 is the TE propagation eigenvalue, and E_{TE} is the corresponding normalized eigenvector representing the field profile $E_x(x, y)$.

3.4.2.1.1 *Quasi-transverse Magnetic Mode*

The quasi-TM mode can be formulated in a similar fashion. The only difference is that for the quasi-TM polarized mode, E_x is assumed to be zero and E_y is continuous across the vertical interfaces but discontinuous across horizontal interfaces. Essentially, the quasi-TM modes are the eigensolutions of the equation

$$\nabla_t^2 E_y + k^2 n^2 E_y = \beta^2 E_y \tag{3.43}$$

The detailed derivation of the equation can be found in Ref. [23]. The following are the derivatives and its relevant difference equations:

$$\frac{\partial^2 E_{i,j}}{\partial y^2} = \frac{2n_{i,j-1}^2}{h_i(n_{i,j}^2 h_i + n_{i,j-1}^2 h_{j-1})} E_{i,j-1} + \frac{2n_{i,j+1}^2}{h_i(n_{i,j}^2 h_i + n_{i,j+1}^2 h_{j+1})} E_{i,j+1}$$

$$- \left[\frac{2n_{i,j}^2}{h_i(n_{i,j}^2 h_i + n_{i,j-1}^2 h_{j-1})} + \frac{2n_{i,j}^2}{h_i(n_{i,j}^2 h_i + n_{i,j+1}^2 h_{j+1})} \right] E_{i,j} \tag{3.44a}$$

$$\frac{\partial^2 E_{i,j}}{\partial x^2} = \frac{1}{h_i} \left[\frac{2(E_{i+1,j} - E_{i,j})}{h_{i+1} + h_i} - \frac{2(E_{i,j} - E_{i-1,j})}{h_i + h_{i-1}} \right] \tag{3.44b}$$

Substituting these into the Helmholtz equation we get

$$C_{i-1,j}E_{i-1,j} + C_{i+1,j}E_{i+1,j} - C_{i,j}E_{i,j} + C_{i,j-1}E_{i,j-1} + C_{i,j+1}E_{i,j+1} = \beta^2 E_{i,j} \tag{3.45}$$

with

$$C_{i-1,j} = \frac{2}{h_i(h_i + h_{i-1})}$$

$$C_{i+1,j} = \frac{2}{h_i(h_i + h_{i+1})}$$

$$C_{i,j-1} = \frac{2n_{j-1}^2}{h_j(n_{i,j}^2 h_j + n_{i,j-1}^2 h_{j-1})} \tag{3.46}$$

$$C_{i,j+1} = \frac{2n_{j+1}^2}{h_j(n_{i,j}^2 h_j + n_{i,j+1}^2 h_{j+1})}$$

$$C_{i,j} = C_{i-1,j} + C_{i+1,j} + C_{i,j-1} + C_{i,j+1} - k^2 n_{i,j}^2$$

3.4.2.1.2 Eigenvalue Matrix

To solve the difference equation, we need first to discretize the problem space. We assume that the space is sliced into NX pieces along the x direction and NY pieces along the y direction. This will give us a total of N $(=NX \times NY)$ grid points. The refractive index of each cell is then allocated according to the relevant index distribution.

When the FD wave equation is evaluated at a grid point, say $E_{i,j}$, it will yield a five-point linear equation in terms of the E field of the immediate neighbors, namely $E_{i-1,j}$, $E_{i+1,j}$, $E_{i,j-1}$, $E_{i,j+1}$, each with its relevant coefficient as shown in Equations 3.11 and 3.12. For a cross-sectional area of a waveguide with N such grid points, we would end up with N linearly dependant algebraic equations.

We will now scan through the grid points row after row, at the same time re-labeling the subscripts of E from 1 to N. Consider the original grid point (i, j). Assuming that the new sequence number is k, then Equation 3.45 can be rewritten as

$$p_k E_k + l_k E_{k-1} + r_k E_{k+1} + t_k E_{k-Nx} + b_k E_{k+Nx} = \beta^2 E_k \tag{3.47}$$

where p_k, l_k, r_k, t_k, b_k, are the coefficients $C_{i,j}$, $C_{i-1,j}$, $C_{i+1,j}$, $C_{i,j-1}$, $C_{i,j+1}$, respectively. We can then collect terms and write the equations in a matrix form.

For a 3×3 grid of the refractive index profile, we can write the matrix equations as the eigenvalue equation of the form $[C] \cdot [E] = \beta^2[E]$ in which $[C]$ is a non-symmetric band matrix which contains the coefficient of the above equations, β^2 is the propagation eigenvalue, and $[E]$ is the corresponding normalized eigenvector representing the field profile $E(i, j)$. In the next section we will discuss the approach that we adopt in solving the eigenvalue problem given as

$$
\begin{bmatrix}
p_1 & r_1 & 0 & b_1 & 0 & 0 & 0 & 0 & 0 \\
l_2 & p_2 & r_2 & 0 & b_2 & 0 & 0 & 0 & 0 \\
0 & l_3 & p_3 & r_3 & 0 & b_3 & 0 & 0 & 0 \\
t_4 & 0 & l_4 & p_4 & r_4 & 0 & b_4 & 0 & 0 \\
0 & t_5 & 0 & l_5 & p_5 & r_5 & 0 & b_5 & 0 \\
0 & 0 & t_6 & 0 & l_6 & p_6 & r_6 & 0 & b_6 \\
0 & 0 & 0 & t_7 & 0 & l_7 & p_7 & r_7 & 0 \\
0 & 0 & 0 & 0 & t_8 & 0 & l_8 & p_8 & r_8 \\
0 & 0 & 0 & 0 & 0 & t_9 & 0 & l_9 & p_9
\end{bmatrix}
\begin{bmatrix}
E_1 \\ E_2 \\ E_3 \\ E_4 \\ E_5 \\ E_6 \\ E_7 \\ E_8 \\ E_9
\end{bmatrix}
= \beta^2
\begin{bmatrix}
E_1 \\ E_2 \\ E_3 \\ E_4 \\ E_5 \\ E_6 \\ E_7 \\ E_8 \\ E_9
\end{bmatrix}
\tag{3.48}
$$

There are a few major features of the matrix equation above: (i) this type of matrix is often referred to as a tridiagonal matrix with fringes. The order of the matrix is $N \times N$, the square of the total number of grid points. Most of terms in the matrix are zeros. (ii) The matrix is non-symmetrical relative to the diagonal term. (iii) The central three diagonal terms always exist and are always non-zero. (iv) The coefficients p, l, r, t, b make up the five bands of the matrix, with p being the main diagonal, l and r being the subdiagonal while t and b are the superdiagonal. (v) The subdiagonal diagonal terms are just one term away from the main diagonal while the superdiagonal terms are NX terms away from the main diagonal. The distance between the main diagonal and the last non-zero superdiagonal band is commonly referred to as the half bandwidth of a band matrix. (vi) Terms such as l_1,

r_N, $t_1 - t_{NX}$, $b_{N-Nx} - b_N$ are missing. This is so since the evaluations of these terms require the E values outside the boundary area, and these values have been assumed zero. Therefore they need not be represented.

3.4.2.2 Inverse Power Method

The properties and characteristics of the eigenvalue problem are well known and have been addressed rather extensively in many text books [30,31]. This section would only provide a brief overview to highlight the more specific points related to our particular approach.

An $N \times N$ matrix A is said to have an *eigenvector* x and a corresponding *eigenvalue* λ if the following condition is satisfied:

$$A \cdot x = \lambda x \tag{3.49}$$

There can be more than one distinct eigenvalue and eigenvector corresponding to a given matrix. The zero vector is not considered to be an eigenvector at all. The above equation holds only if

$$\det | A - \lambda I | = 0 \tag{3.50}$$

which is known as the characteristic equation of the matrix. If this is expanded, it becomes an Nth degree polynomial in λ whose roots are the eigenvalues. This is an indication that there are always N, though not necessarily distinct, eigenvalues. Equal eigenvalues coming from multiple roots are called degenerate. Root-searching in the characteristic equation however, is usually a very poor computational method for finding eigenvalues. There are many more efficient algorithms available in locating the eigenvalues and their corresponding vectors.

Unfortunately there is no universal method for solving all matrix types. For certain problems, either the eigenvalues or eigenvectors are needed, while others require both. Furthermore, some problems may only need a small number of solutions out of the total N solutions available, while others need all. To complicate the matter even further, the eigensolutions could be complex, and some matrices can be so ill-behaved that round-off errors in computing can lead to a non-convergence of the solution. Therefore it is of vital importance to be able to choose the right approach in solving an eigenproblem. Choosing an algorithm often involves the classification of matrixes into types like symmetry, non-symmetry, tridiagonal, banded, positive definite, definite, Heisenberg, sparse, random, etc. The matrix in our problem is a non-symmetric banded matrix with bandwidth equal to twice the number of columns in the grid profile. It has great sparsity for most of the elements are zero. Also, we need only a few eigenvalues that correspond to the guided modes of the waveguide. In other words, there are only a limited number of guided modes, hence the number of eigenvalue λ. The number of eigensolutions required is small compared with the size of the matrix (often in the order of tens of thousands). All these different factors have led to the choice of the approach called the inverse iteration method [30,31].

The basic idea behind the inverse iteration method is quite simple. Let y be the solution of the linear system

$$(A - \tau I) \cdot y = b \tag{3.51}$$

where b is a random vector and τ is close to some eigenvalue λ of A. Then the solution y will be close to the eigenvector corresponding to λ. The procedure can be iterated: replace b by y and solve for a new y, which will be even closer to the true eigenvector. We can see why this works by expanding both y and b as linear combinations of the eigenvectors x_j of A:

$$y = \sum_j \alpha_j x_j \quad \text{and} \quad b = \sum_j \beta_j x_j \tag{3.52}$$

Then we have

$$\sum_j \alpha_j (\lambda_j - \tau) x_j = \sum_j \beta_j x_j \tag{3.53}$$

so that

$$\alpha_j = \frac{\beta_j}{\lambda_j - \tau} \quad \text{and} \quad y = \sum_j \frac{\beta_j x_j}{\lambda_j - \tau} \tag{3.54}$$

If τ is close to λ_n, say, then provided β_n is not accidentally too small, y will be approximately x_n, up to a normalization. Moreover, the iteration of this procedure gives another power of $\lambda_j - \tau$ in the denominator of Equation 3.43. Thus the convergence is rapid for well-separated eigenvalues.

Suppose at the ith stage of iteration we are solving the equation

$$(A - \lambda_i I) \cdot y = x_i \tag{3.55}$$

where x_i and λ_i are our current guesses for some eigenvector and eigenvalue of interest (we shall see below how to update λ_i). The exact eigenvector and eigenvalue satisfy

$$A \cdot x = \lambda x \rightarrow (A - \lambda_i I) \cdot x = (\lambda - \lambda_i) x \tag{3.56}$$

Since y of Equation 3.31 is an improved approximation to x, we normalize it and set

$$x_{i+1} = \frac{y}{|y|} \tag{3.57}$$

We get an improved estimate of the eigenvalue by substituting our improved guess y in Equation 3.56. By Equation 3.34, the left-hand side is x_i, so calling λ our new value λ_{i+1}, we find

$$\lambda_{i+1} = \lambda_i + \frac{|x|^2}{|x_i \cdot y|}.$$

Although the formulae of the inverse iteration method seem to be rather straightforward, the actual implementation can be quite tricky. Most of the computational load occurs in solving the linear system of equations. It would be advantageous if we could solve Equation 3.32 quickly. It is to be remembered that the size of the matrix in our case is dependent upon the total grid size of the problem space. For a typical grid size of 100×100 for example, the coefficient matrix would be $10,000 \times 10,000$. The core memory required in a digital computer to store the entire matrix would be phenomenal. Linear system solvers such as the routines that are available in LINPACK employs a common LU (L = Lower triangular matrix; U = Upper triangular matrix) factorization (Gaussian elimination) plus a backward substitution combination algorithm, much like the manual way of solving linear equations. There is extensive coverage on this topic in most numerical text books [31]. We will therefore not discuss it further except to mention that the LU factorization needs only to be done before the first iteration. When the iteration starts, we already have the steps involved in elimination stored away in an array and only backward substitution is necessary. This approach, even with a storage optimized mode in the LINPACK routine still has a storage requirement of about $3 \times$ (bandwidth of matrix × matrix size). Even though this would mean a considerable reduction in memory storage, it still amounts to a rather substantial memory size.

Also, the pre-conditioner that employs the incomplete Cholesky conjugate gradient method [32] and the Orthomin [32] accelerator have been found to be the most stable and converge most quickly for our matrix. On average, the combination of the pre-conditioner and accelerator enable us to complete a simulation of a typical waveguide in 3 to 5 min on a Pentium 4 PC. The same simulation that incorporates the LINPACK LU decomposition routine would take 25 min on the same computer with a substantially greater amount of memory. Since the zero elements are no longer involved in the calculations, it is understandable that the LINPACK iterative method will perform more efficiently.

By incorporating the LINPACK numerical solver and the inverse iterative method, we have successfully implemented a mode modeling program, SVMM, capable of modeling the channel waveguide of an arbitrary index profile. The inverse iterative method also enables us to model the higher order modes that are supported by the waveguide structure.

3.4.3 Ti:LiNbO₃ Diffused Channel Waveguide

The modeling of the Ti:LiNbO₃ channel waveguide, a graded index waveguide, plays a significant role in the design of the optical modulators and switches. Efficient design of such optical devices requires good knowledge of the modal characteristics of the relevant channel waveguide. In Ref. [7], we have outlined the general overview of the waveguide fabrication process. In this section, we will attempt to employ our SVMM program to simulate the waveguide mode of the Ti:LiNbO₃ waveguide and compare the results with that of published experimental results. Our objectives, apart from assessing the usefulness of SVMM, are also to understand the key features in the fabrication of the Ti:LiNbO₃ waveguide for Mach-Zehnder optical modulator.

To achieve our purpose, a good knowledge of the refractive index profile of the diffused waveguide is required. Over the past decades much work has been done in fabricating low loss, minimum mode size Ti diffused channel waveguide [6,33]. From these references, we can gather our knowledge of the diffusion process involved in the fabrication of the LiNbO₃ waveguide and its relevant diffusion profile. Based on this knowledge, we can then profess to model the modal characteristics of the waveguide by SVMM. The following section shows how SVMM can be used for the design of practical waveguides.

3.4.3.1 Refractive Index Profile of the Ti:LiNbO₃ Waveguide

When Ti metal is diffused, the Ti-ion distribution spreads more widely than the initial strip width. The profiles can be described by the sum of an error function, while the Ti-ion distributions perpendicular to the substrate surface can be approximated by a Gaussian function [1,6,34]. This of course is true only if the diffusion time is long enough to diffuse all the Ti metal into the substrate. We consider this case as having the finite diffusant source. However, if the total diffusion time is shorter than needed to exhaust the Ti source, the lateral diffusion profile would take up the sum of the complementary error function while the depth index profile is given by the complementary function [35]. This case is considered to have had an infinite diffusant source [23]. In our study, we would assume that there is sufficient time for the source to be fully diffused because in most practical waveguides, it is undesirable to have Ti residue deposited on the surface of the waveguide because this will increase the propagation loss [36]. This increase in propagation loss is a result of stronger interaction with the LiNbO$_3$ surface (and thus an increased scattering loss) as the modes become more weakly guided.

In general, the refractive index distribution of a weakly guiding channel waveguide is

$$n(x, y) = n_b + \Delta n(x, y) = n_b + \Delta n_0 \cdot f(x) \cdot g(y) \tag{3.58}$$

where n_b is the refractive index of the bulk (substrate) and $\Delta n(x, y)$ is the variation of the refractive index in the guiding region. $\Delta n(x, y)$ in our diffusion model is essentially a separable function where $f(x)$ and $g(y)$ are the functions that describe the lateral and perpendicular diffusion profile while Δn_0 is known as the surface index change after diffusion. The surface index change is defined as the change of the refractive index on the substrate just below the center of the Ti strip. In other words, it is the refractive index when both $f(x)$ and $g(y)$ assume the value of unity.

The variation of the refractive index can be modeled as below [23]

$$\Delta n(x, y) = \frac{dn}{dc} \tau \int_{-w/2}^{w/2} \frac{2}{d_y \sqrt{\pi}} \exp\left[-\left(\frac{y}{d_y}\right)^2\right] \frac{1}{d_x \sqrt{\pi}} \cdot \exp\left[-\left(\frac{x-u}{d_x}\right)^2\right] du$$

$$= \Delta n_0 \cdot f(x) \cdot g(y) \tag{3.59}$$

where

$$f(x) = \frac{1}{2}\left[\text{erf}\left(\frac{x+\frac{w}{2}}{d_x}\right) - \text{erf}\left(\frac{x-\frac{w}{2}}{d_x}\right)\right] \Bigg/ \text{erf}\left(\frac{w}{2d_x}\right) \tag{3.60}$$

$$g(x) = \exp\left[-\left(\frac{y}{d_y}\right)^2\right] \tag{3.61}$$

and

$$\Delta n_0 = \frac{dn}{dc} \frac{2}{\sqrt{\pi}} \frac{\tau}{d_y} \text{erf}\left(\frac{w}{2d_x}\right) \tag{3.62}$$

with

$$d_x = 2\sqrt{D_x t}, \quad d_y = 2\sqrt{D_y t} \tag{3.63}$$

In the above expressions, t is the total diffusion time, c is the Ti concentration, d_x and d_y are the diffusion lengths, and D_x and D_y are the diffusion constants in each direction. τ and w are the initial Ti strip thickness. dn/dc is the change of index per unit change in Ti metal concentration. The change of surface index would approach the value, where

$$\Delta n_0 = \frac{dn}{dc} \frac{2}{\sqrt{\pi}} \frac{\tau}{d_y}.$$

Any increase in the surface index will have to come from a thicker Ti strip, or a decrease in diffusion depth, d_y, which involves an increase or decrease in diffusion temperature. According to the work of Fukuma et al. [1], the diffusion length is very close to one another in both lateral and depth direction (isotropic diffusion) at 1025°C for z-cut crystal. An increase in temperature greater than that would result in a higher diffusion constant in the depth direction and a lower value for lateral diffusion and vice versa for a diffusion temperature lower than 1025°C. The diffusion length can also be changed by monitoring the diffusion time. Essentially, longer diffusion time would mean a lower surface index change as most of the Ti source would be diffused deeper into the substrate. Again, model 2 depicts a higher change of surface index since not all the Ti metal is exhausted. The graphs shown in Figures 3.8 through 3.10 show the variation of the diffusion profile as we vary both the initial titanium width and the diffusion time. The fabrication condition and parameters are assumed to have $T = 1025°C$, $\tau = 1100Å$, $dn/dc = 0.625$, and $dx = dy = 2\ \mu m$.

We can see in the graphs shown in Figures 3.8 through 3.10 that by controlling the width of the initial Ti strip width we can vary the change of the refractive index and the relative size of the channel waveguide, thus enabling us to control the number of mode that can be supported by the waveguide.

In general, a narrow initial Ti-film strip width would give a near cutoff mode due the very small change of the refractive index change. The optical mode would be weakly

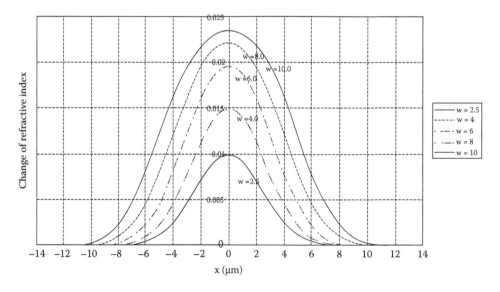

FIGURE 3.8
Lateral diffusion variation with increasing Ti strip width.

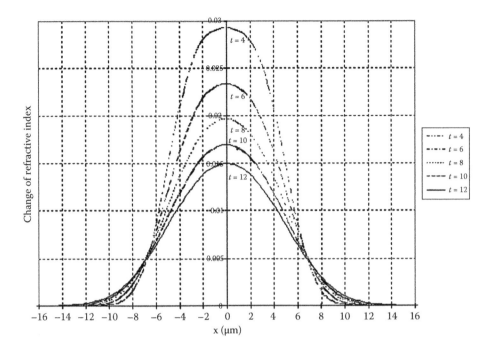

FIGURE 3.9
Lateral diffusion variation with increasing diffusion time.

confined, thus giving a larger mode size. As we increase the Ti width, the refractive index change would be higher and the waveguide mode would be better confined and have a smaller mode size. However, the mode size would increase with a further increase in Ti width due to a larger physical size of the waveguide. The change of surface index can also be controlled by varying the thickness of the Ti strip. As Equation 3.53 implied,

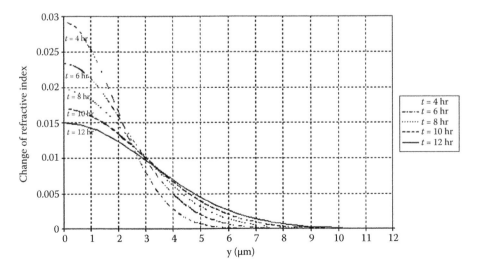

FIGURE 3.10
Depth index variation with increasing diffusion time.

the surface index change is proportional to the strip thickness, τ. The Ti thick film can be diffused at 1000–1050°C for 6 h and would be around 500–800 Å [6]. If the Ti strip is too thin, the refractive index change approaches cutoff conditions. All these characteristics will be illustrated by the next section when we model the waveguide mode with SVMM.

3.4.3.2 Numerical Simulation and Discussion

With the above knowledge of the diffusion profile, we are now in a good position to feed these models into the SVMM program to investigate the modal characteristics of the Ti:LiNbO₃ waveguide. In this section, we attempt to simulate the experimental work reported in Suchoski and Ramaswamy, 1987 [6] in fabricating minimum mode size low loss Ti:LiNbO₃ channel waveguide. We will restrict our analysis to the z-cut *y* propagating material since this would be the substrate cut for the optical modulator. For this particular substrate cut, the relevant optical field would be TM polarized, which corresponds to the polarization along the extraordinary index axis of the crystal. Hence, the change of refractive index concern would be the extraordinary index, n_e.

In Suchoski's work [6], the TM polarized mode width and depth, which is defined as 1/*e* intensity full width and full depth, are measured for the Ti:LiNbO₃ waveguides fabricated under the condition where T = 1025°C for 6 h. The sample waveguides have Ti thickness ranging from 500 to 1100 Å, and Ti strip widths ranging from 2.5 to 10 μm.

The laser source wavelength is assumed to be at 1.3 μm. The graphs shown in Figures 3.11 through 3.15 are extracted from their work. In view of these experimental results, we can see that the mode size increases as the Ti strip width is decreased from 4 to 2.5 μm. This increase is more pronounced, especially with the thinner Ti films, because the waveguides become closer to cutoff, as thinner Ti film results in a lower value of Δ*n*. The TM mode depth and width decrease as the Ti thickness is increased from 500 to 800 Å. However, for 4 μm strip widths, the mode size does not decrease further for Ti films thicker than 800 Å. This is an indication that it is not possible to diffuse any more Ti into the substrate for Ti

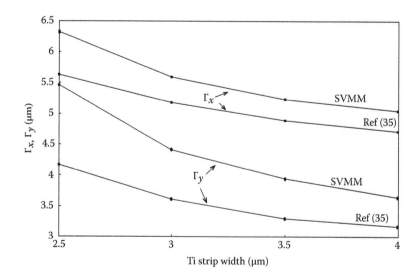

FIGURE 3.11
Simulation of mode sizes with nominal diffusion parameters.

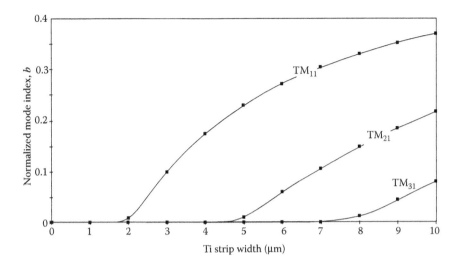

FIGURE 3.12
Typical modal field distribution of a diffused channel waveguide, μm.

thickness of more than 800 Å for 6 h diffusion time. We now proceed to simulate the above experiment with our program. We will focus on Ti thickness that ranges between 700 to 800 Å because it is the thickness that gives minimum mode sizes, which is ideal for the design of the optical modulator for maximizing the overlap integral between guided optical modes and the applied modulating field. To achieve that, we must first work out the suitable diffusion parameter to be used in our program. Various values of *dn/dc* have been reported [1]. Measurements reported by Minakata et al. [36] shows the change of extraordinary index n_e per Ti concentration as:

$$\frac{dn_e}{dc} = 0.625$$

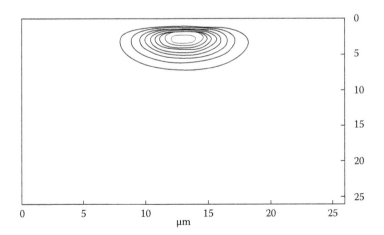

FIGURE 3.13
Contour plot of the modal field of a diffused channel waveguide—vertical and horizontal dimensions in micrometers.

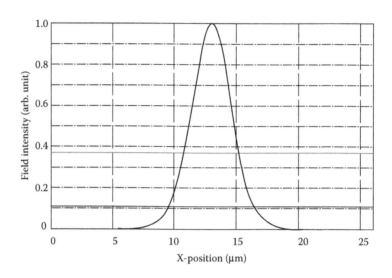

FIGURE 3.14
Horizontal mode profile of a diffused channel waveguide.

The nominal values for diffusion constant, D_x and D_y from the work of Fukuma and Noda [1] which were both measured to be 1.2×10^{-4} $\mu m^2/s$ at the nominated temperature which is 1025°C. This makes both diffusion length of d_x and d_y the value of 2 μm.

With these nominal parameters, we simulate the waveguide with a Ti thickness, τ of 700 Å. Figures 3.11 through 3.15 are the results of our simulation compared to the experimental one and some illustrations of the TM mode profile.

As it turns out, the simulated results appear to have overestimated both Γ_x and Γ_y. Such discrepancy is anticipated as fabrication of the diffused waveguide is subjected to many changes. Various reports [1,6,33,35–38,40–55] have shown that even though the nominal diffusion condition can be very much the same, the measured diffusion parameter can

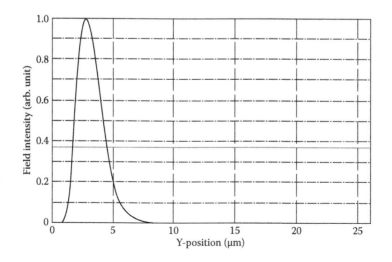

FIGURE 3.15
Vertical mode profile of a diffused channel waveguide.

still differ greatly from one another due to possible differences in stoichiometry between different crystals and measurements techniques. Therefore, there would certainly be some uncertainties that lie in fabrication parameters and also the application of the refractive index model described in Equation 3.49. Such uncertainty can be compensated by adjusting the value of dn/dc and also D_x and D_y. We find that by adjusting the following diffusion parameters where $dn/dc = 0.8$; $D_x = 1.4 \times 10^{-4} \, \mu m/s^2$; and $D_y = 1.1 \times 10^{-4} \, \mu m/s^2$; our simulation results correspond well within the design limit with the experimental work done by Suchoski [6] for the case where the waveguide is well guided. The result is shown in Figure 3.16.

Having found the suitable diffusion parameter, the simulation of another experimental result from Ref. [6] is conducted for $\tau = 750$ Å and Ti strip width, w ranges from 2.5 to 10 μm. Figure 3.17 depicts the comparison of both simulated and experiment results, the mode size variation with respect to the width of the Ti strip.

The results in Figure 3.17 show that the mode width, Γ_x corresponds well to the experimental result with differences of less than 3%. The mode depth, Γ_y, however matches only to within 8%. Despite the slight discrepancy, the SVMM's result still shows the qualitative characteristic of the diffused waveguide. We can also observe that the modal width, Γ_x starts at a large value and then decreases with wider Ti strip width. Effectively, the larger initial mode size is due to the lower refractive index change resulting from a much narrower Ti width, thus causing the optical mode to be less confined. As the Ti strip becomes wider, it gives a higher change of refractive index, hence a better confined optical mode. The mode width however, would increase further as we increase the Ti width simply because of the increase in the physical width of the waveguide. At the same time, the larger physical width would enable the waveguide to support higher order mode.

Figure 3.18 depicts the variation of the normalized mode index b defined as [39] $b = (n_{eff}^2 - n_s^2)/(2\Delta n \cdot n_s)$ with respect to the variation of the width of the Ti strip. The waveguide becomes more strongly guided, i.e., higher effective index, as the Ti width is increased. At the same time, higher order modes begin to appear as the strip width gets

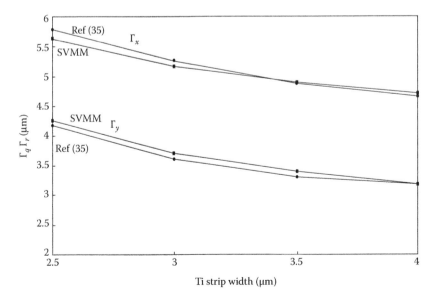

FIGURE 3.16
Comparison of simulated and experimental mode sizes for $\tau = 700$ Å.

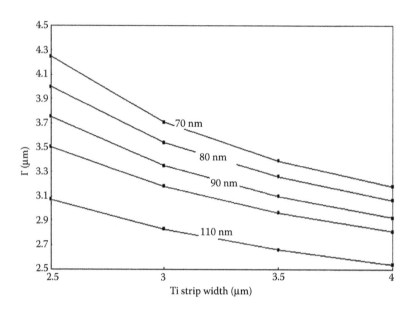

FIGURE 3.17
Experimental mode sizes for $\tau = 700$ Å with the Ti film thickness as a parameter.

significantly larger than 6 μm. Figure 3.19 shows the distribution of higher order modes of a waveguide diffused with a 10 μm Ti strip width.

The modal depth, however, decreases with wider Ti strip width because any wider Ti strip width does not affect the diffusion depth, but lateral mode distribution would support higher order modes. The surface index would increase with thicker Ti film, thus leading to smaller modal depth. The surface index however, only reaches a maximum value as we increase w. Therefore, by increasing the Ti strip width to a certain point, the modal

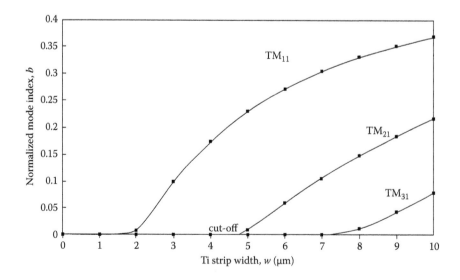

FIGURE 3.18
Normalized mode index, b as a function of Ti strip width, w.

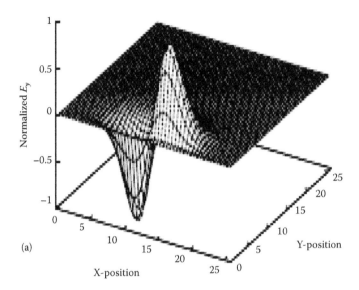

(a)

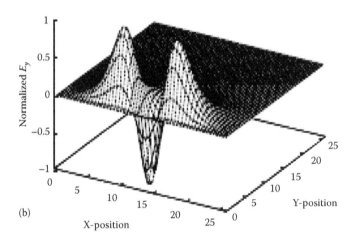

(b)

FIGURE 3.19

(a) TM_{21} mode: $\tau = 750$ Å, $w = 10$ μm; (b) TM_{31} mode: $\tau = 750$ Å, $w = 10$ μm.

depth would cease to decrease further, as observed by both the experimental and simulated results. At this point, lateral diffusion would dominate. It is worth remembering that the limiting case of increasing width in Ti width is a planar waveguide.

Figure 3.20 shows simulation results with waveguides of the same diffusion parameter but with the Ti thickness for the diffusion of 800 Å. It shows that SVMM overestimates the mode size of the diffused waveguide. As a matter of fact, the thickness of 800 Å, as mentioned before, corresponds to the case where the Ti thickness has just depleted. In other words, the change of surface index is at its highest point for that diffusion time. The mode size would therefore appear to be much smaller compared to those waveguides of which the Ti has been diffused sufficiently longer than the time needed to just deplete all the Ti. This explains why the simulated modal width and depth are larger than the practical one. Figures 3.21 and 3.22 summarize the simulated results for waveguide for a range of Ti thickness.

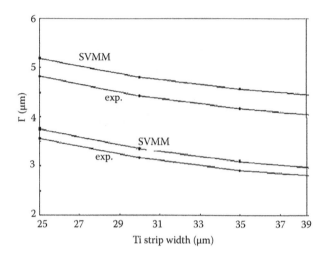

FIGURE 3.20
Comparison of simulated and experimental mode sizes for $\tau = 800$ Å.

From the curves of Figures 3.21 and 3.22, we can see that the modal width and depth increase monotonically with Ti thickness. It is not difficult to see from the experimental results that the diffusion model and diffusion parameters are no longer valid when the Ti film exceeds the thickness which is fully diffused at around 800 Å. For any thickness beyond that, we will need to resort to another diffusion model. In our case this isn't necessary because having Ti film thicker than the thickness of diffusible quantity would lead to scattering loss, thus increasing the total insertion loss of the device.

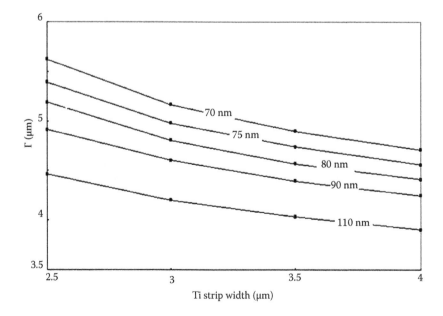

FIGURE 3.21
Simulated modal width Γ_x (*in horizontal direction*) for a range of Ti thickness, τ.

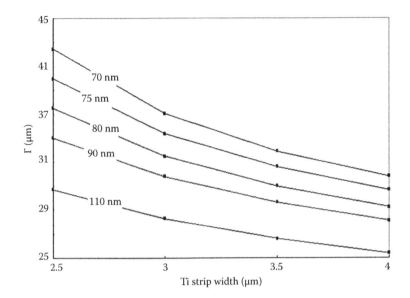

FIGURE 3.22
Simulated modal width Γ_y (*in vertical direction*) for a range of Ti thickness, τ.

In this section thus far, we have demonstrated how SVMM can be used apart from simulating rib waveguide, to simulate diffused channel waveguide. As a matter of fact, the FDM can be employed to obtain reasonable accuracy of the mode index and its distribution as well as the evolution with different diffusion parameters for optical waveguides having an arbitrary index profile. Simulation of the Ti:LiNbO$_3$ waveguide however, is not a straightforward matter because fabrication of such a waveguide is subjected to many changes such as differences in crystal quality, diffusion process, density variations of the deposited titanium films and also differences in measurement techniques. As a result of that, we can see inconsistencies in published literature. Fouchet et al. [35] had shown the relation between refractive index change $\Delta n_{e,0}(Z)$ and Ti concentration $C(Z)$ in the mathematical form of

$$\Delta n_{e,0}(Z) = A_{e,0}(C_0, \lambda) \cdot (C(Z))^{\alpha_{e,0}} \qquad (3.64)$$

The expression shows that the proportionality coefficient $A_{e,0}$ depends not only on the wavelength λ, but also on the diffusion parameters which is characterized by C_0, the Ti surface concentration.

In other words, the diffusion model that we used in our simulation is only a crude representation of the diffused waveguide. To enhance the accuracy of the simulation, we will need to provide a more accurate diffusion model which takes into account the dispersion relationship of the change in the refractive index profile in Ti:LiNbO$_3$, despite being a crude representation of the diffusion process, is still sufficient to demonstrate the credibility of SVMM in modeling diffused waveguide. The simulations that we did in this particular section have not only shown the usefulness of SVMM, but have also provided a qualitative overview of the design of the Ti:LiNbO$_3$ waveguide. At this point, this program can surely be calibrated against diffusion data that are measured in house and be used as a tool in the design of the Ti:LiNbO$_3$ waveguide for optical modulators.

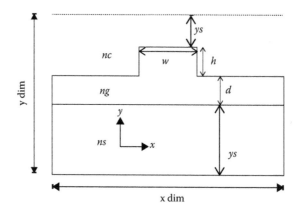

FIGURE 3.23
Typical structure of rib waveguide.

3.5 Mode Modeling of Rib Waveguides

In every FD approach, a few approximations are made and will therefore introduce some error into the final result. The following are a few approximations that are likely to introduce some error in our calculation: (i) the approximation of the full vectorial wave equation by the semivectorial one; (ii) the replacement of the differential equation with the difference equations; (iii) discretization error; (iv) round-off error; and (v) the errors that are introduced by the LINPACK numerical solver itself.

To assess the accuracy, capability and limitation of our program, we have calculated fundamental mode indices of three well-known rib waveguides that are often used as the waveguide modeling benchmark. Results of polarized modes have been published [10–22]. The geometry of the rib waveguide is shown in Figure 3.23. Parameters include width of the rib w, height of the rib h, thickness of the guiding layer underneath the rib d, index of the substrate n_s, and index of the guiding layer n_g and are listed in Table 3.1. The refractive index of the air cladding region, n_c is unity.

The three waveguides each have a different characteristic. Structure 1 has relatively large vertical refractive index steps ($\Delta n = 2.44$ and 0.1) which could, for example, correspond to a GaAs guiding layer bound by air and a $Ga_{0.75}Al_{0.25}As$ confining layer. In the lateral direction, the rib height is large and the width narrow. This structure, with strong light confinement in both lateral and vertical direction, is useful for curved guides as radiation loss is minimized. This structure does not allow the application of EIM because the slab outside the rib is cutoff.

Structure 2 shows a weakly guiding feature. In this case the rib height is much less, allowing the mode to extend laterally. This is particularly useful for directional coupler structures, as strong coupling between adjacent guides will result in short coupling lengths. The guiding

TABLE 3.1

Parameters of Rib Waveguide for Calculation Benchmark

Guide	n_g	n_s	d (μm)	h (μm)	w (μm)
1	44	34	0.2	1.1	2
2	44	36	0.9	0.1	3
3	44	435	5	2.5	4

layer thickness is made small to give a thin mode shape in the vertical direction, and thus low voltage operation. Essentially, this structure is tightly confined vertically and weakly confined horizontally. Such features enable the application of EIM [1,16,24] because the small etch step and large width to height ratio are the conditions of validity of this approximate method.

Structure 3 gives a good coupling to an optical fiber. Insertion loss is a crucial parameter for most waveguide devices, and is determined by propagation loss and losses due to mode mismatch. Fresnel reflection loss is also important, but can be reduced to insignificant levels by using $\lambda/4$ anti-reflection coatings. Mode profiles of a circularly symmetric optical fiber and a waveguide will, in general, be different, due to the differing refractive indices of the semiconductor and the fiber, and also the differing shapes of the modes. The effects of both these factors may be alleviated by the use of appropriate waveguide designs. In structure 3 the guiding layer is relatively thick, and the stripe width and height are adjusted to give a more symmetric mode shape. In this structure the slab mode is near cutoff. Again, because the rib height is nearly twice the slab thickness and the rib width is less than the rib height, the accuracy of the EIM is expected to be poor. Figures 3.24 through 3.27 show the contour

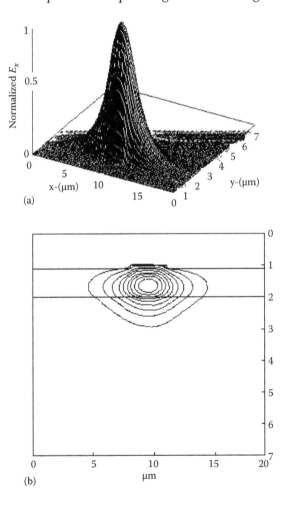

FIGURE 3.24
(a) 3D plot of TE polarized mode profile for waveguide structure with low rib. (b) Contour plot of TE polarized mode profile.

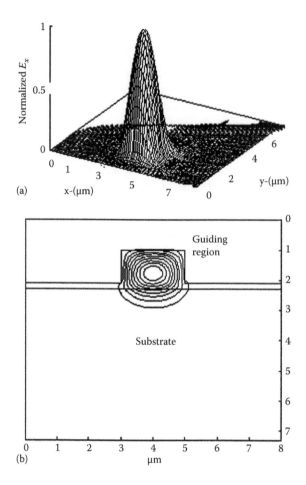

FIGURE 3.25
(a) 3D plot of TE polarized mode profile for waveguide structure 1. (b) Contour plot of TE polarized mode profile for waveguide structure 1.

plot and 3D plot of the TE polarized mode of the three waveguide structure calculated by the SVMM program.

The grid size h_x and h_y are 0.1. Since we assume that the field value around the computational boundary is zero, it would mean that we require a much larger computational window for both structure 2 and structure 3 so that the assumption would be valid. This, however, would mean that we can either use a coarser grid leading to a reduction in computing accuracy, or maintain the grid size but end up with a huge eigenmatrix to solve. For that reason, the variable grid size comes in handy. We can avoid a severe storage penalty by judiciously placing the denser mesh around the area the higher field values are assumed and the coarser mesh at a region of a much lower field value. This would thus allow us to extend the boundary of the computation without incurring a severe storage problem while preserving the accuracy of the computation. The choice of grid size and its influence on the accuracy of the final results shall be illustrated in the next section.

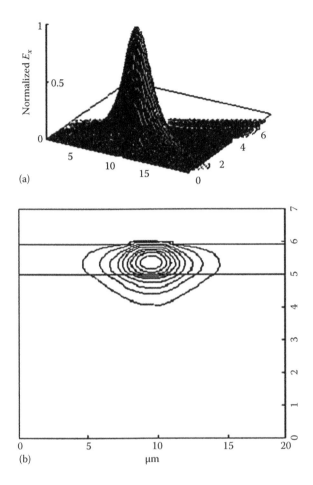

FIGURE 3.26
(a) 3D plot of TE polarized mode profile for waveguide structure 2. (b) Contour plot of TE polarized mode profile for waveguide structure 2.

3.5.1 Choice of Grid Size

A variable grid size, e.g., dense in the region close to the center of the guided mode, would give more accurate results. To assess the effect of grid size on the accuracy of our simulation program, we compute the effective index for the TE polarized mode of structure 1 by varying the grid size in both x and y directions, namely h_x and h_y. We can compare these simulated results with those reported by Lusse et al. [11] who used a dense mesh of 508×394 mesh points with their full vectorial FDM.

In Table 3.2 the simulation results of the rib waveguide structures using the SVMM program described above are shown. In simulation 1–6, the value of $h_y = 0.1$ is kept constant while reducing h_x from 0.5 down to 0.025. As we can see, as h_x reaches 0.025, we can no longer get a significant improvement on the accuracy. A further reduction of grid size down to 0.01 would be highly impractical because we would end up with 800 grid points along the x direction, thus paying a high penalty in terms of computer memory. In simulation 7–9, we keep h_x at 0.025 while reducing h_y from 0.1 down to 0.025, another significant

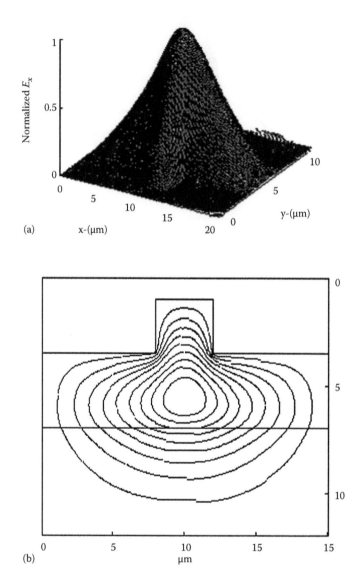

(a)

(b)

FIGURE 3.27
(a) 3D plot of TE polarized mode profile for waveguide structure 3. (b) Contour plot of TE polarized mode profile for waveguide structure 3.

improvement in accuracy is shown and the results get very close to the one simulated by Lusse et al. [11] with both h_x and h_y equal to 0.025, a grid size of 320 × 292, the difference of our calculated effective index with that of Lusse et al. 1994 [11] is 2.78 × 10^{-5}. Simulations 10 and 11 show how the non-uniform scheme could economize storage usage while preserving the desired accuracy. By placing denser grid mesh around the region where higher field values, the center of the guided mode, and coarser mesh for region further away from the expected center, the mesh size can be reduced from 320 × 292 down to 240 × 226 (a total reduction of 39200 points) without significant loss in the result accuracy as observable in Figure 3.28. The non-uniform grid allocation scheme has in this particular case shown its usefulness. (It is to be remembered that each reduction of grid size needs to be multiplied

TABLE 3.2

Calculation of Effective Index with Different Choice of Grid Size

Simulation No.	h_x (μm)	h_y (μm)	x dim (μm)	y dim (μm)	Total Grid	Effective Index
1	0.5	0.1	8.0	7.3	16 × 73	3913474
2	0.25	0.1	8.0	7.3	32 × 73	3899896
3	0.125	0.1	8.0	7.3	64 × 73	3895512
4	0.1	0.1	8.0	7.3	80 × 73	3894906
5	0.05	0.1	8.0	7.3	160 × 73	3894048
6	0.025	0.1	8.0	7.3	320 × 73	3893836
7	0.025	0.05	8.0	7.3	320 × 146	3888583
8	0.025	0.025	8.0	7.3	320 × 292	3887148
9	0.0–2.0:0.1 2.0–2.5:0.05 2.5–0:0.025 0–4.0:0.05 4.0–5.5:0.025 5.5–6.0:0.05 6.0–8.0:0.1	0.0025	8.0	7.3	240 × 292	3887162
10	0.0–2.0:0.1 2.0–2.5:0.05 2.5–0:0.025 0–4.0:0.05 4.0–5.5:0.025 5.5–6.0:0.05 6.0–8.0:0.1	0.0–4.0:0.025 4.0–7.3:0.05	8.0	7.3	240 × 226	3887165
P. Lusse	–	–	–	–	508 × 394	88687

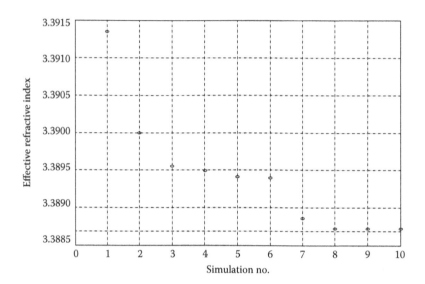

FIGURE 3.28
Refractive index variation with simulation number to obtain converged solution.

by 26 for that is the amount of workspace required by the coefficient matrix, eigenvector and the LINPACK numerical solver).

3.5.2 Numerical Results

Tables 3.3a and b show the values of the propagation constants of both TE and TM polarized modes for all three waveguides. The results are compared with several published results. The bolded entries of the tables are results of our work. We can see from Table 3.3 that our results compare favorably with all the other published results.

The numerical results presented so far have indicated the order of accuracy of the SVMM program. We can observe that the exemplar results are compared well with other published results.

3.5.3 Higher Order Modes

In our earlier discussion, we indicated that the inverse power method can be used to work out the other eigenmodes of the waveguide. To illustrate that, we simulate the waveguide mode of the waveguide structure published by Rahman and Davies [22]. Table 3.4 outlines the parameters of the waveguide structure. Figures 3.29 and 3.30 show the fundamental mode and the leading asymmetric mode of the TE polarized field.

The leading asymmetric mode of Figure 3.30 can be obtained with an initial eigenvalue that is close to the eigenvalue of the leading asymmetric mode. One way to acquire a good initial guess for an independent eigenvalue is by perturbing the last few significant digits of the last calculated eigenvalue. In our case, the eigenvalue of the fundamental mode (see Figure 3.29) was calculated to be 347.78889. We then proceeded to the calculation of the asymmetric mode with an initial guess of 346. Other eigenmodes can also be worked out in a similar fashion. However, we need to remember that there are only a limited number of eigenmodes that are supported by certain waveguide structures. A good indication that the particular eigenmode is physically not feasible is an effective index which is lower than

TABLE 3.3

Comparisons of Effective Indices and Normalized Indices at $\lambda = 1.55$

a) TE Polarized Mode

Methods	Guide 1		Guide 2		Guide 3	
	n_{eff}	b	n_{eff}	b	n_{eff}	B
SVMM	3887148	0.4835	3953612	0.4391	4368918	0.3782
Sv-BPM [14]	388711	0.4834	395471	0.4405	436805	0.3608
Helmholtz [15]	388764	0.4839	395560	0.4416	436808	0.3614
SI [19]	38874	0.4837	39506	0.4354	43688	0.3759
SV [25]	3869266	0.4656	3954	0.4401	4368112	0.3621
FD [26]	3882623	0.4789	3952147	0.4373	436804	0.3611

b) TM Polarized Mode

Methods	Guide 1		Guide 2		Guide 3	
	n_{eff}	b	n_{eff}	b	n_{eff}	B
SVMM	3879173	0.4755	390647	0.3803	4368434	0.3685
Sv-BPM [14]	387924	0.4756	390693	0.3809	436772	0.3543
Helmholtz [15]	387990	0.4762	390712	0.3811	346772	0.3543
SI [19]	38788	0.4752	39032	0.3763	43684	0.3669
SV [25]	3867447	0.4638	3905927	0.3796	4367719	0.3542
FD [26]	3875430	0.4718	3905701	0.3794	4367751	0.3549

TABLE 3.4

Parameters of Rib Waveguide of Ref. [16] ($\lambda = 1.15$ μm)

Guide	n_g	n_s	d (μm)	h (μm)	w (μm)
Ref. [22]	44	40	0.5	0.5	3

that of the refractive index of the substrate, thus giving a negative value of the normalized index. This is illustrated in Figure 3.31.

The third order mode distribution depicted in Figure 3.31 is acquired by further reducing the initial guess of the eigenvalues from 346 to 345. As a result of that, we get an effective index of 398 which is lower that the refractive index of the substrate which is 40 in this case. This results in a normalized index b of −0.047. As shown in the contour plot, most of the field is radiated into the substrate of the waveguide.

This feature of SVMM that enables us to work out the higher order modes is extremely important to find out if the designed waveguide can support multimode operation. We will see in the next chapter how such a feature can be exploited in the design of a single mode waveguide.

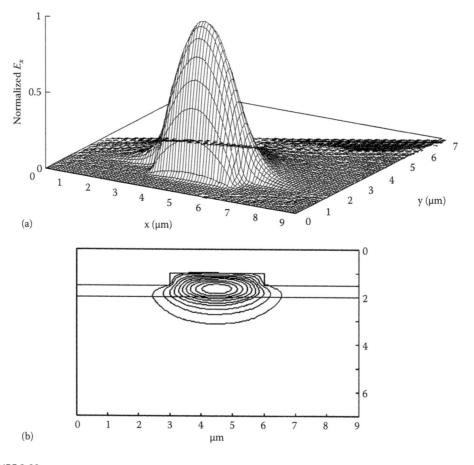

FIGURE 3.29

(a) 3D plot of fundamental mode of waveguide from Rahman and Davies, 1985 [22]. (b) Contour plot of the fundamental mode (TE-polarized) of the waveguide from Rahman and Davies, 1985 [22].

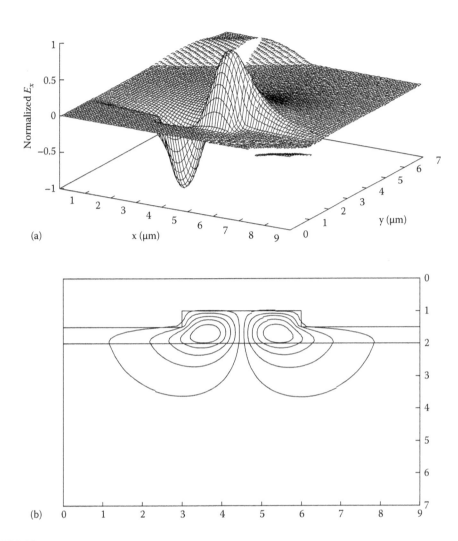

FIGURE 3.30
(a) 3D plot for the leading asymmetric mode of waveguide in Rahman and Davies, 1985 [22]. Calculated effective index = 4025302. (b) Contour plot of leading asymmetric mode (TE polarized) of waveguide in Rahman and Davies, 1985 [22].

3.6 Conclusions

In this chapter, the simplified approach for the analytical study of channel waveguides, the 3D version, is described using the Marcatili method and effective index techniques. Simplified analytical dispersion relations have been obtained for these 3D waveguides. An example of the design of a GeO_2 doped core rectangular channel waveguide is given.

Further, we have successfully developed numerical techniques based on a semivectorial FD analysis to solve the Helmholtz equation. The numerical model that we have formulated can accurately and effectively model the guided modes in optical waveguides of arbitrary index profile distribution. A non-uniform mesh allocation scheme is employed in the formulation of the difference equations to free more computer

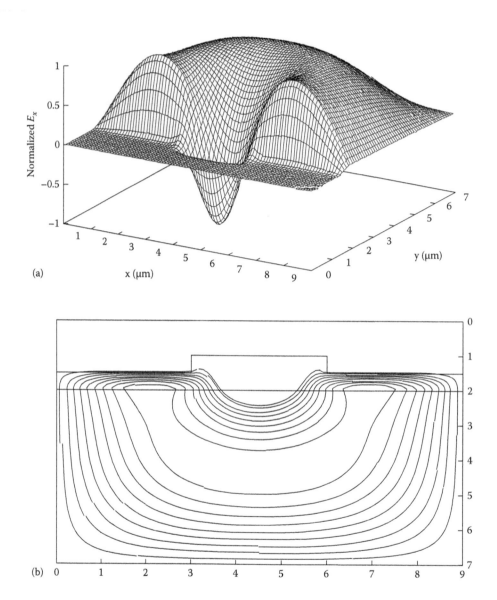

(a)   x (µm)

(b)

FIGURE 3.31
(a) 3D plot of the third order mode which is not supported by the waveguide structure. Calculated effective index = 3980958, normalized index = −0.047314. (b) Contour plot of radiated mode.

memory for the computation of waveguide regions that bear greater significance. The accuracy of our computer program, SVMM, is assessed by computing the propagation constants and the effective indices of several rib waveguides that have been known to be excellent benchmark waveguide structures. The results presented here in this chapter as a test of the formulation given are compared favorably with other published results [14,15,19,25,26]. We then continue to simulate the optical guided modes of diffused optical waveguides in LiNbO₃. The computed mode sizes are consistent with published experimental values. The numerical simulations however, have shown the inadequacies of the adopted diffusion model for its inability to model the diffused waveguide in a more robust sense. It is suggested that highly refined and robust representation of the refractive index profile of the Ti:LiNbO$_3$ diffused waveguide should be employed. Despite the shortcoming of the diffusion model, SVMM can be used as an analytical and design tool for integrated optical waveguides.

3.7 Problems

1. Effective refractive index method

   A number of channel 3D optical waveguides are to be formed on a glass substrate including: (a) raised strip or channel, (b) ridge or rib, (c) strip-loaded, and (d) embedded channel. The refractive index of the glass substrate is silica on silicon with a refractive index of 1.488 at 1550 nm wavelength.

   a. Refer to Figure 3.1 of Chapter 3 and sketch the waveguide structures.

   b. Propose the materials for the waveguide structure deposited on top of the substrate. Determine the refractive index of the waveguiding region and the width and thickness of the channel waveguide so that the waveguide can support two modes.

   c. Repeat (2) so that the waveguide can support only one polarized mode. Determine whether the mode is TE or TM polarized.

2. Refer to Figure 3.4 on the dispersion relation between the normalized propagation constant b and the V-parameter for 2D optical waveguide.

   a. Determine the value of V so that only one mode is guided. What mode is this?

   b. Determine the cutoff wavelength of the waveguide quantitatively.

   c. Similar to (b), design an optical waveguide so that it can support three modes.

3. Give a brief description of the finite difference method or computing the parameters of optical waveguides.

References

1. M. Fukuma and J. Noda, "Optical properties of titanium-diffused LiNbO$_3$ strip waveguides and their coupling to a fiber characteristics," *Appl. Optics*, 20(4), 591–597, 15 February 1980.
2. C.H. Bulmer, S.K. Sheem, R.P. Moeller, and W.K. Burns, "High efficiency flip-chip coupling between single mode fibers and LiNbO$_3$ channel waveguides," *Appl. Phys. Lett.*, 37, 351–355, 1981.
3. V. Ramaswamy et al., "High efficiency single mode fiber to Ti:LiNbO$_3$ waveguide coupling," *Elect. Lett.*, 10, 30–31, 1982.
4. R.C. Alferness et al., "Efficient single mode fiber to titanium diffused lithium niobate waveguide coupling for l=1.32μm," *IEEE J. Quant. Elect.*, QE-18, 1807–1811, 1982.
5. E.A.J. Marcatili, "Dielectric rectangular waveguide and directional couplers for integrated optics," *Bell Syst. Tech. J.*, 48, 2071–2102, 1969.
6. P.G. Suchoski and R.V. Ramaswamy, "Minimum mode size low loss Ti:LiNbO$_3$ channel waveguides for efficient modulator operation at 1.3 μm," *IEEE J. Quant. Elect.*, QE-23(10), 1673–1679, October 1987.

7. K.S. Chiang, "Review of numerical and approximate methods for the modal analysis of general optical dielectric waveguides," *Opt. Quant. Elect.*, 26, S113–S134, 1994.
8. M. Saad, "Review of numerical methods for the analysis of arbitrary shaped microwave and optical dielectric waveguides," *IEEE Trans. Microwave Theory and Tech.*, MTT-33(10), 894–899, October 1985.
9. S. Seki et al., "Two dimensional analysis of optical waveguides with a non-uniform finite difference method," *IEE Proceedings Part J Optoelect.*, 138(2), 123–127, April 1991.
10. M.J. Robertson et al., "Semiconductor waveguides: Analysis of optical propagation in single rib structures and directional couplers," *IEE Proc., Part J Optoelect.*, 132(6), 336–342, December 1985.
11. P. Lusse et al., "Analysis of vectorial mode fields in optical waveguides by a new finite difference method," *IEEE J. Lightwave Tech.*, 12(11), 487–493, March 1994.
12. M.S. Stern, "Semivectorial polarized **H** field solutions for dielectric waveguides with arbitrary index profiles," *IEE Proceedings Part J Optoelect.*, 135(5), 333–338, October 1988.
13. W. Huang and H.A. Haus, "A simple variational approach to optical rib waveguides," *IEEE J. Lightwave Tech.*, 9(1), 56–61, January 1991.
14. Pao-Lo Liu and Bing-Jin Li, "Semivectorial beam propagation method for analysing polarized modes of rib waveguides," *IEEE J. Quant. Elect.*, 28(4), 778–782, April 1992.
15. Pao-Lo Liu and Bing-Jin Li, "Semivectorial Helmholtz beam propagation by Lanczos reduction," *IEEE J. Quant. Elect.*, 29(8), 2385–2389, August 1993.
16. T.M. Benson et al., "Rigorous effective index method for semiconductor rib waveguides," *IEE Proceedings Part J Optoelect.*, 139(1), 67–70, February 1992.
17. P.C. Kendall et al., "Advances in rib waveguide analysis using weighted index method or the method of moments," *IEE Proceedings Part J Optoelect.*, 137(1), 27–29, February 1990.
18. S.V. Burke, "Spectral index method app. lied to rib and strip-loaded directional couplers," *IEE Proceedings Part J Optoelect.*, 137(1), 7–10, February 1990.
19. M.S. Stern, "Analysis of the spectral index method for vector modes of rib waveguides," *IEE Proceedings Part J Optoelect.*, 137(1), 21–26, February 1990.
20. G. Ronald Hadley and R.E. Smith, "Full vector waveguide modeling using an iterative finite difference method with transparent boundary conditions," *IEEE J. Lightwave Tech.*, 13(3), 465–469, March 1995.
21. T.M. Benson et al., "Polarisation correction app. lied to scalar analysis of semiconductor rib waveguides," *IEE Proceedings Part J Optoelect.*, 139(1) 39–41, February 1992.
22. B.M.A. Rahman and J.B. Davies, "Vectorial-H finite element solution of GaAs/GaAlAs rib waveguides," *IEE Proceedings Part J Optoelect.*, 132, 349–353, 1985.
23. Chang Min Kim and R.V. Ramaswamy, "Modeling of graded index channel waveguides using non-uniform finite difference method," *IEEE J. Lightwave Tech.*, 7(10), 1581–1589, October 1989.
24. G.B. Hocker and W.K. Burns, "Mode dispersion in diffused channel waveguides by the effective index method," *Appl. Optics.*, 16(1), 113–118.
25. M.S. Stern, "Semivectorial polarized finite difference method for optical waveguides with arbitrary index profiles," *IEE Proceedings Part J Optoelect.*, 135(1), 56–63, February 1988.
26. M.S. Stern, "Rayleigh quotient solution of semivectorial field problems for optical waveguides with arbitrary index profiles," *IEE Proceedings Part J Optoelect.*, 138, 185–190, 1990.
27. K. Ogusu, "Numerical analysis of the rectangular dielectric waveguide and its modifications," *IEEE Trans. Microwave Theory Technol.*, MTT-25(11), 874–885, 1977.
28. C-L. Chen, "Foundations for Guided wave optics," J. Wiley, New Jersey, 2007.
29. J.E. Goell, "A circular-harmonic computer analysis of rectangular dielectric waveguide," *Bell. Systems. Tech. J.*, 48, 2133–2160, 1969.
30. R.L. Burden and J.D. Faires, *"Numerical Analysis,"* 4th edition, PWS-Kent Pub., 492–505.
31. W.H. Press et al., *"Numerical Recipes-the Art of Scientific Computing,"* Cambridge Univ. Press, 377–379.
32. T.C. Opp et al., "LINPACK user's guide version 1.0-a package for solving large sparse linear systems by various iterative methods," Center for Numerical Analysis, The University of Texas, Austin.

33. S.K. Korotky et al., "Mode size and method for estimating the propagation constant of single mode Ti:LiNbO₃ strip waveguides," *IEEE J. Quant. Elect.*, QE-18(10), 1796–1801, October 1982.
34. M. Minakata, S. Shaito, and M. Shibata, "Two dimensional distribution of refractive index changes in Ti diffused LiNbO₃ waveguides," *J. Appl. Phys.*, 50(5), 3063–3067, May 1979.
35. S. Fouchet et al., "Wavelength dispersion of Ti induced refractive index change in LiNbO₃ as a function of diffusion parameters," *IEEE J. Lightwave Tech.*, LT-5(5), 700–708, May 1987.
36. M. Minikata et al., "Precise determination of refractive index changes in Ti-diffused LiNbO₃ optical waveguides," *J. Appl. Phys*, 49(9), 4677–4682, September 1978.
37. W.K. Burns et al., "Ti diffusion in Ti:LiNbO₃ planar and channel optical waveguides," *J. Appl. Phys.*, 50(10), 6175–6182, October 1979.
38. M.D. Feit et al., "Comparison of calculated and measured performance of diffused channel-waveguide couplers," *J. Opt. Soc. Am.*, 73(10), 1296–1304, October 1983.
39. Anurag Sharma and Pushpa Bindal, "Analysis of diffused planar and channel waveguides," *IEEE J. Quant. Elect.*, 29(1), 150–157, January1993.
40. R.K. Lagu and R.V. Ramaswamy, "A variational finite-difference method for analysing channel waveguides with arbitrary index profiles," *IEEE J. Quant.Elect.*, QE-22(6), 968–976, June 1986.
41. E. Strake et al., "Guided Modes of Ti:LiNbO₃ channel waveguides: a novel quasi analytical technique in comparison with the scalar finite-element method," *IEEE J. Lightwave Tech.*, 6(6), 1126–1135, June 1988.
42. N. Schulz et al., "Finite difference method without spurious solutions for the hybrid-mode analysis of diffused channel waveguides," *IEEE Trans. Microwave Theory Tech.*, 38(6), 722–729, June 1990.
43. K.T. Koai and Pao-Lo Liu, "Modeling of Ti:LiNbO₃ waveguide devices: Part I-Directional Couplers," *IEEE J.Lightwave Tech.*, 7(3), 533–539, March 1989.
44. K.T. Koai and Pao-Lo Liu, "Modeling of Ti:LiNbO₃ waveguide devices: Part II-S-shaped channel waveguide bends," *IEEE J. Lightwave Tech.*, 7(7), 1016–1022, July 1989.
45. M. Valli and A. Fioretti, "Fabrication of good quality Ti:LiNbO₃ planar waveguides by diffusion in dry and wet O2 atmospheres," *J. Modern Optics*, 35(6), 885–890, 1988.
46. D.S. Smith et al., "Refractive Indices of Lithium Niobate," *Optics Comm.*, 17(3), 332–335, June 1976.
47. D.F. Nelson and R.M. Mikulyak, "Refractive indices of congruently melting lithium niobate," *J. Appl. Physics*, 45(8), 3688–3689, August 1974.
48. R.C. Alferness et al., "Characteristics of Ti diffused lithium niobate optical directional couplers," *Appl. Optics*, 18(23), 4012–4016, 1 December 1979.
49. G.B. Hocker and W.K. Burns, "Modes in diffused optical waveguides of arbitrary index profile," *IEEE J.Quant. Elect.*, QE-11(6), 270–1975, June 1975.
50. Murray R. Spiegel, "Mathematical Handbook of Formulas and Tables," Shaum's Outline Series, McGraw Hill, 1990.
51. Emis Datareviews, "Properties of Lithium Niobate," Series no. 5, INSPEC publication, 131–146, 1991.
52. Anurag Sharma and Pushpa Bindal, "An accurate variational analysis of single mode diffused channel waveguides," *Opt. Quant. Elect.*, 24, 1359–1371, 1992.
53. Anurag Sharma and Pushpa Bindal, "Variational analysis of diffused planar and channel waveguides and directional couplers," *J. Opt. Soc. Am. A.*, 11(8), 2244–2248, August 1994.
54. J. Ctyroky et al., "3-D analysis of LiNbO₃: Ti channel waveguides and directional couplers," *IEEE J. Quant. Elect.*, QE. 20(4), 400–409, April 1984.
55. S.V. Chung, R.A. Lee, and L.N. Binh, "An exact ray representation of bound modes in a step-profile planar slab waveguide," *J. Modern optics*, 32(7), 779– 791, 1987.

4

Single-Mode Optical Fibers: Structures and Transmission Properties

4.1 Optical Fibers

4.1.1 Brief History

We have, as described in the previous chapters, analyzed the planar optical waveguides in which the guiding region is a slab imbedded between a substrate and a superstrate having an identical refractive index. The guiding of lightwaves in an optical fiber is similar to that of the planar waveguide except the lightwaves are guiding through a circular guiding structure.

Within the context of this book on guided wave photonics, fiber optics would be most relevant as circular optical waveguides, we should point out the following development in optical fiber communications systems so that we will be able to focus on modern optical systems engineers based on the fundamental understanding of electromagnetic fields theory

- The step index and graded index multimode optical fibers find very limited applications in systems and networks for long-haul applications.
- The single mode optical fibers (SMF) have been achieved for very a small difference in the refractive indices between the core and cladding regions. Thus the guiding in modern optical fiber for telecommunications is called "weakling" guiding. This development was intensively debated and agreed by the optical fiber communications technology community during the late 1970s.
- The invention of optical amplification in rare-earth doped SMF in the late 1980s has transformed the design and deployment of optical fiber communications systems and networks in the last decade and the coming decades of the twenty-first century. The optical loss of the fiber and the optical components in the optical networks can be compensated for by using these fiber in-line optical amplifiers.
- Therefore the pulse broadening of optical signals during transmission and distribution in the networks become much more important for system design engineers.

Due to the above development we shall focus the theoretical approach on the understanding of optical fibers towards the practical aspects of the design of optical fibers with minimum dispersion or for a specified dispersion factor. This can be carried out, from practical measurements, so that the optical field distribution would follow a Gaussian distribution. Having known the field distribution, one would be able to obtain the propagation constant of the single guided mode and hence the spot size of this mode, thus the energy concentration inside the core of the optical fiber. From the basic concept of optical dispersion by using the definition of group velocity and group delay, we would be able to derive the chromatic dispersion in SMFs. After arming ourselves with the basic equations for dispersion we would be able to embark on the design of optical fibers with a specified dispersion factor.

4.1.2 Optical Fiber: General Properties

4.1.2.1 Geometrical Structures and Index Profile

An optical fiber consists of two concentric dielectric cylinders. The inner cylinder, or core, has a refractive index of $n(r)$ and radius a. The outer cylinder, or cladding, has index n_2 with $n(r) > n_2$ and a larger outer radius. A core of about 4–9 μm and a cladding diameter of 125 μm are the typical values for silica-based SMF. A schematic diagram of an optical fiber is shown in Figure 4.1.

The refractive index n of an optical waveguide is usually changed with radius r from the fiber axis ($r = 0$) and is expressed by

$$n^2(r) = n_2^2 + \mathrm{NA}^2 s\left(\frac{r}{a}\right) \tag{4.1}$$

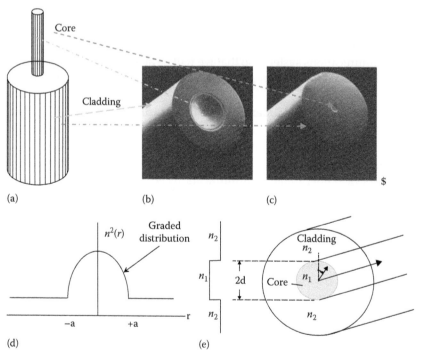

(a)　　　　(b)　　　　(c)

(d)　　　　(e)

FIGURE 4.1

(a) Schematic diagram of the step index fiber: coordinate system, structure. The refractive index of the core is uniform and slightly larger than that of the cladding. For silica glass the refractive index of the core is about 1.478 and that of the cladding about 1.47 at 1550 nm wavelength region. (b) Cross section of an etched fiber–multimode type 50 nm diameter. (c) SMF etched cross section. (d) Graded index profile. (e) Fiber cross section and step index profile $d = a$ = radius of fiber.

where NA is the numerical aperture evaluated at the core axis, while $s(r/a)$ is the profile function which characterizes any profile shape ($s = 1$ at maximum) with a scaling parameter (usually the core radius).

4.1.2.1.1 Step Index Profile

In a step index profile the refractive index remains constant in the core region, thus

$$s\left(\frac{r}{a}\right) = 1 \text{ for } r \leq a \tag{4.2}$$

$$s\left(\frac{r}{a}\right) = 0 \text{ for } r > a \tag{4.3}$$

So we have for a step index profile

$$n^2\left(r\right) = n_1^2 \text{ for } r < a \tag{4.4}$$

and

$$n^2\left(r\right) = n_2^2 \text{ for } r > a \tag{4.5}$$

Exercise 1

Refer to the technical specification of the SMF Corning SMF-28. State whether the index profile of the fiber is a perfect step index profile or graded index profile? Is it true that the profile is a perfect step index distribution? If not then what is the real manufactured profile?

4.1.2.1.2 Graded Index Profile

We consider hereunder the two most common types of graded index profiles: power-law index and the Gaussian profile.

4.1.2.1.2.1 Power-law Index Profile The core refractive index of optical fiber usually follows a graded profile. In this case the refractive index rises gradually from the value n_2 of the cladding glass to the value n_1 at the fiber axis. Therefore $s(r/a)$ can be expressed as

$$s\left(\frac{r}{a}\right) = \left\{1 - \left(\frac{r}{a}\right)^\alpha \text{ for } r \leq a \text{ and } = 0 \text{ for } r < a \right. \tag{4.6}$$

with α = power exponent. Thus the index profile distribution $n(r)$ can be expressed in the usual way as (by using Equations 4.6 and 4.2, by substituting $NA^2 = n_1^2 - n_2^2$.

$$n^2\left(r\right) = \begin{cases} n_1^2\left[1 - 2\Delta\left(\frac{r}{a}\right)^\alpha\right] & \text{for } r \leq a \\ n_2^2 \text{ for } r > a \end{cases} \tag{4.7}$$

with $\Delta = NA^2/n_1^2$ is the elative refractive difference. Observing the equation describing the profile shape in Equation 4.7, there are three special cases

- $\alpha = 1$: the profile function $s(r/a)$ is linear and the profile is called a triangular profile
- $\alpha = 2$: the profile is a quadratic function with respect to the radial distance and the profile is called the parabolic profile
- $\alpha = \infty$: then the profile is a step type

4.1.2.1.2.2 Gaussian Index Profile In the Gaussian index profile the refractive index changes gradually from the core center to a distance very far away from it and $s(r)$ can be expressed as

$$s\left(\frac{r}{a}\right) = e^{-(r/a)^2} \tag{4.8}$$

4.1.3 Fundamental Mode of Weakly Guiding Fibers

The electric and magnetic fields $E(r, \phi, z)$ and $H(r, \phi, z)$ of the optical fibers in cylindrical coordinates can be found by solving Maxwell's equations. However, only the lower order modes of ideal step index fibers are important for modern optical fiber communications systems. The fact is that $\Delta < 1\%$, thus optical waves are weakly guided and E and H are then approximate solutions of the scalar wave equation in a cylindrical coordinate system (x, θ, ϕ)

$$\left[\frac{\delta^2}{\delta r^2} + \frac{1}{r}\frac{\delta}{\delta r} + k^2 n_j^2\right]\varphi(r) = \beta^2\varphi(r) \tag{4.9}$$

where $n_j = n_1, n_2$, and $\varphi(r)$ is the spatial field distribution of the nearly transverse electromagnetic (EM) waves

$$E_x = \psi(r)e^{-i\beta z}$$

$$H_y = \left(\frac{\varepsilon}{\mu}\right)^{1/2} E_x = \frac{n_2}{Z_0} E_x \tag{4.10}$$

with E_y, E_z, H_x, H_z negligible, $\varepsilon = n_2^2\varepsilon_o$ and $Z_0 = (\varepsilon\mu)^{1/2}$ is the vacuum impedance. That is the waves can be seen as a plane wave traveling down along the fiber tube. These plane waves are reflected between the dielectric interfaces, in other words, it is trapped and guided in and along the core of the optical fiber.

4.1.3.1 Solutions of the Wave Equation for Step Index Fiber

The field spatial function $\varphi(r)$ would have the form of Bessel functions (from Equation 4.9) as

$$\varphi(r) = A\frac{J_0(ur/a)}{J_0(u)} \quad \text{for} \quad 0 < r < a \tag{4.11}$$

$$\varphi(r) = A\frac{K_0(vr/a)}{K_0(v)} \quad \text{for } r > a \tag{4.12}$$

where J_0 and K_0 are Bessel functions of the first kind and modified of the second kind, respectively, and u and v are defined as

$$\frac{u^2}{a^2} = k^2 n_1^2 - \beta^2 \tag{4.13a}$$

$$\frac{v^2}{a^2} = -k^2 n_2^2 + \beta^2 \tag{4.13b}$$

Thus following Maxwell's equations relation, we can find that E_z can take two possible solutions which are orthogonal as

$$E_z = -\frac{A}{kan_2}\begin{pmatrix} \sin\phi \\ \cos\phi \end{pmatrix} \left| \begin{array}{ll} \dfrac{uJ_1\left(u\dfrac{r}{a}\right)}{J_0(u)} & \text{for } 0 \le r < a \\[4mm] \dfrac{vK_1\left(\dfrac{vr}{a}\right)}{K_0(v)} & \text{for } r > a \end{array} \right. \tag{4.14}$$

The terms u and v must satisfy simultaneously two equations

$$u^2 + v^2 = V^2 = ka\left(n_1^2 - n_2^2\right)^{1/2} = kan_2\left(2\Delta\right)^{1/2} \tag{4.15}$$

$$u\frac{J_1(u)}{J_0(u)} = v\frac{K_1(v)}{K_0(v)} \tag{4.16}$$

where Equation 4.16 is obtained by applying the boundary conditions at the interface $r = a$ (E_z is the tangential component and must be continuous at this dielectric interface). Equation 4.16 is usually called the eigenvalue equation. The solution of this equation would give the values of β, which would take discrete values and are the propagation constants of the guided lightwaves.

Equation 4.15 shows that the longitudinal field is in the order of u/kan_2 with respect to the transverse component. In practice $\Delta \ll 1$, by using Equation 4.15, we observe that this longitudinal component is negligible compared with the transverse component. We thus consider the mode as *transversely polarized*. The fundamental mode is then usually denominated as LP_{01} mode (LP = linearly polarized) for which the field distribution is shown in Figure 4.2 and the graphical representation of the eigenvalue equation (Equation 4.16) calculated as the variation of $\beta/k = b$ as the normalized propagation constant and the V-parameter is shown in Figure 4.3c.

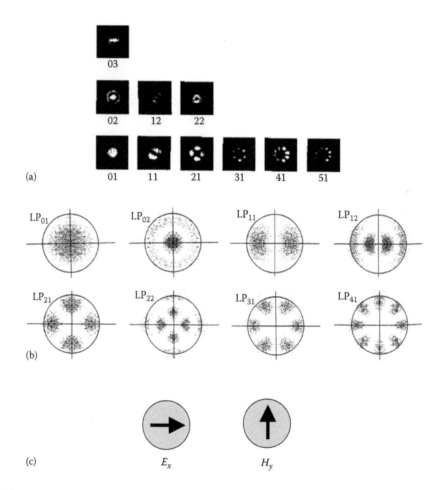

FIGURE 4.2
(a) Spectrum of guided modes in a multimode fiber, the numbers under the image or intensity distribution is the mode order of the guided waves. (b) Calculated intensity distribution of LP guided modes in a step index optical fibers with $V = 7$. (c) Electric and magnetic field distribution of an LP_{01} mode polarized along O of the fundamental mode of an SM fiber.

4.1.3.2 Gaussian Approximation

4.1.3.2.1 Fundamental Mode Revisited

We note again that E and H are approximate solutions of the scalar wave equation and the main properties of the fundamental mode of weakly guiding fibers that can be observed as follows

- The propagation constant β (in z direction) of the fundamental mode must lie between the core and cladding wave numbers. This means the effective refractive index of the guided mode lies with the range of the cladding and core refractive indices.

- Accordingly, the fundamental mode must be nearly a transverse EM wave as described by Equation 4.10.

$$\frac{2\pi n_2}{\lambda} < \beta < \frac{2\pi n_1}{\lambda} \tag{4.17}$$

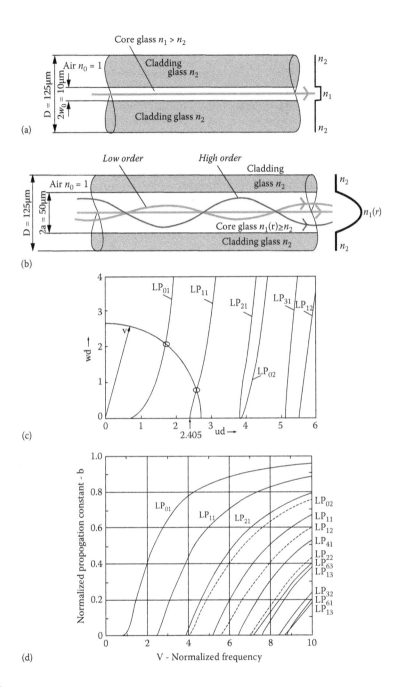

FIGURE 4.3
(a) Guided modes as seen in the transverse plane of a circular optical fiber. (b) "Ray" model of lightwave propagating in SM fiber. (c) Ray model for multimode graded index fiber. (d) Graphical illustration of solution for eigenvalues (propagation constant–wave number of optical fibers). (e) *b-V* characteristics of guided fibers.

- The spatial dependence $\psi(r)$ is a solution of the scalar wave equation, Equation 4.9.

4.1.3.2.1.1 Gaussian Approximation The main objectives are to find a good approximation for the field $\psi(r)$ and the propagation constant β. These can be found through the eigenvalue equation and Bessel's solutions as shown in the previous section. It is desirable if we can approximate the field to a good accuracy to obtain simple expressions to have a clearer understanding of light transmission on SMF, without going through graphical or numerical methods. Furthermore, experimental measurements and numerical solutions for step and power-law profiles show that $\psi(r)$ is approximately Gaussian in appearance. We thus approximate the field of the fundamental mode as

$$\varphi(r) \cong Ae^{-1/2(r/r_0)^2} \tag{4.18}$$

where r_0 is defined as the spot size, i.e., at which the intensity equals to e^{-1} of the maximum. Thus if the wave equation, Equation 4.9, is multiplied by $r\psi(r)$ and using the identity

$$r\varphi\frac{\delta^2\varphi}{\delta r^2} + \varphi\frac{\delta\varphi}{\delta r} = \frac{\delta}{\delta r}\left(r\varphi\frac{\delta\varphi}{\delta r}\right) - r\left(\frac{\delta\varphi}{\delta r}\right)^2 \tag{4.19}$$

then, by integrating from 0 to infinitive and using $\left[r\Psi\dfrac{d\Psi}{dr}\right]_0^\infty = 0$ we have

$$\beta^2 = \frac{\displaystyle\int_0^\infty\left[-\left(\frac{\delta\varphi}{\delta r}\right)^2 + k^2n^2\left(r\right)\varphi^2\right]r\delta r}{\displaystyle\int_0^\infty r\varphi^2\delta r} \tag{4.20}$$

The procedure to find the spot size is then followed by substituting $\psi(r)$ (Gaussian) in Equation 4.18 into Equation 4.20 then differentiating and setting $\delta^2\beta/\delta r$ evaluated at r_0 to zero, that is the propagation constant β of the fundamental mode *must* give the largest value of r_0.

Knowing r_0 and β the fields E_x and H_y (Equation 4.10) are fully specified.

Case 1: Step Index Fiber
Substituting the step index profile given by Equation 4.10 and $\psi(r)$ into Equation 4.18 and then Equation 4.20 leads to an expression for β in terms of r_0 given by

$$V = k\cdot a\mathrm{NA} \tag{4.21}$$

The spot size is thus evaluated by setting

$$\frac{\delta^2\beta}{\delta r_0} = 0 \tag{4.22}$$

and r_0 is then given by

$$r_0^2 = \frac{a^2}{\ln V^2} \tag{4.23}$$

Substituting Equation 4.23 into Equation 4.21 we have

$$(a\beta)^2 = (akn_1)^2 - \ln V^2 - 1 \tag{4.24}$$

This expression is physically meaningful only when $V > 1$ (r_0 is positive)

Case 2: Gaussian Index Profile Fiber
Similarly for the case of a Gaussian index profile, by following the procedures for step index profile fiber we can obtain

$$(a\beta)^2 = (an_1 k)^2 - \left(\frac{a}{r_0}\right)^2 + \frac{V^2}{\left(\frac{a}{r_0} + 1\right)} \tag{4.25}$$

and

$$r_0^2 = \frac{a^2}{V - 1} \text{ by using } \frac{\delta^2 \beta}{\delta r_0} = 0 \tag{4.26}$$

That is maximizing the propagation constant of the guided waves. The propagation constant is at maximum when the "light ray" is very close to the horizontal direction. Substituting Equation 2.26 into Equation 2.25 we have

$$(a\beta)^2 = (akn_1)^2 - 2V + 1 \tag{4.27}$$

thus Equations 2.26 and 2.27 are physically meaningful only when $V > 1$ ($r_0 > 0$).

It is obvious from Equation 2.28 that the spot size of the optical fiber with a V-parameter of 1 is extremely large. This is very important that the V-parameter of a single mode optical fiber must not be close to unity. In practice we observe that the spot size is large but finite (observable). In fact if V is smaller than 1.5 the spot size becomes large. This will be investigated in detail in the next chapter.

4.1.3.3 Cutoff Properties

Similarly to the case of planar dielectric waveguides, from Figure 4.3 we observe that when we have $V < 2.405$, only the fundamental LP_{01} exists. Thus we have Figure 4.4.

Note: For SM operation the V-parameter must be less than or equal to 2.405, however in practice $V < 3$ is acceptable.

In fact the value 2.405 is the first zero of the Bessel function $J_0(u)$. In practice one cannot really distinguish between the V value between 2.3 and 3.0. Experimentally observation

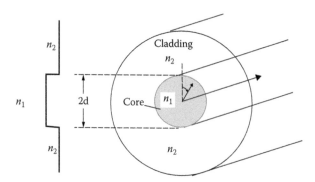

FIGURE 4.4

(a) Intensity distribution of the LP_{01} mode. (b) Variation of the spot size—field distribution with radial distance r with V as a parameter.

shows that optical fiber can still support only one mode. Thus designers do usually take the value of V as 3.0 or less to design an SMF.

The V-parameter is inversely proportional with respect to the optical wavelength. Thus if an optical fiber is launched with lightwaves whose optical wavelength is smaller than the operating wavelength at which the optical fiber is SM, then the optical fiber is supporting more than one mode. The optical fiber is said to be operating in a multimode region.

Thus one can define the cutoff wavelength for optical fibers as follows: the wavelength (λ_c) *above which* only the fundamental mode is guided in the fiber, is called the *cutoff wavelength* λ_c. This cutoff wavelength can be found by using the V-parameter as $V_c = V$ (at cutoff) = 2.405, thus

$$\lambda_c = \frac{2\pi a NA}{V_c} \tag{4.29}$$

Exercise 2

An optical fiber has the following parameters: a core refractive index of 1.46, a relative refractive index difference of 0.3% and a cladding diameter of 125 μm and a core diameter of 8.0 μm. (a) Find the fiber NA and hence the fiber acceptance angle. (b) What is the cutoff wavelength of this fiber? (c) What is the number of optical guided modes which can be supported if the optical fiber is excited with lightwaves of a wavelength of 810 nm? (d) If the cladding diameter is reduced to 50 and 20 μm, comment on the field distribution of the guided SM.

In practice the fibers tend to be effectively SM for larger values of V, say $V < 3$ for the step profile, because the higher order modes suffer radiation losses due to fiber imperfections. Thus if $V = 3$, from Equation 4.15 we have $a < 3\lambda/2NA$, in this case that $\lambda = 1$ μm and the numerical aperture NA must be very small ($<<1$) for radius a to have some reasonable dimension. Usually Δ is about 1% or less for SSMFs (standard single mode optical fiber) in telecommunications systems.

4.1.3.4 Power Distribution

The axial power density or intensity profile $S(r)$, the z-component of Poynting's vector is given by

$$S(r) = \frac{1}{2} E_x H_y^*$$ (4.30)

Substituting Equation 4.10 into Equation 4.30, we have

$$S(r) = \frac{1}{2} \left(\frac{\varepsilon}{\mu} \right)^{1/2} e^{-(r/r_0)^2}$$ (4.31)

The total power is then given by

$$P = 2\pi \int_0^\infty r S(r) dr = \frac{1}{2} \left(\frac{\varepsilon}{\mu} \right)^{1/2} r_0^2$$ (4.32)

and hence the fraction of power $\eta(r)$ within 0 to r across the fiber cross section is given by

$$\eta(r) = \frac{\displaystyle\int_0^r r S(r) dr}{\displaystyle\int_0^\infty r S(r) dr} = 1 - e^{-\left(r^2/r_0^2 \right)}$$ (4.33)

Table 4.1 gives the expressions for P and $\eta(r)$ of the step index and Gaussian profile fibers (by substituting the appropriate values of r_0 into Equations 3.32 and 3.33).

As a rule of thumb and experimental confirmation that an optical fiber is best for guided mode is that the optical power contained in the core is about 70%–80% of the total power.

Exercise 3

Using Gaussian approximation for the intensity distribution of the fundamental mode of the SMF with $V = 2$, find the fraction of power in the core region with $a = 4$ μm.

TABLE 4.1

Analytical expressions for total optical guided power and its fractional power confined inside the core region for step index and Gaussian index profiles

	Step index	Gaussian
$S(r/a)$ for $V > 1$	$\frac{1}{2} \left(\frac{\varepsilon}{\mu} \right)^{1/2} e^{-(r/a)^2 \ln V^2}$	$\frac{1}{2} \left(\frac{\varepsilon}{\mu} \right)^{1/2} e^{-(r/a)(V-1)}$
Power P for $V > 1$	$\frac{1}{2} \left(\frac{\varepsilon}{\mu} \right)^{1/2} \frac{a^2}{\ln V^2}$	$\frac{1}{2} \left(\frac{\varepsilon}{\mu} \right)^{1/2} \frac{a^2}{V-1}$
	$\frac{1}{2} \left(\frac{\varepsilon}{\mu} \right)^{1/2} \frac{a^2}{\ln V^2}$	$\frac{1}{2} \left(\frac{\varepsilon}{\mu} \right)^{1/2} \frac{a^2}{V-1}$
$\eta(r)$ for $V > 1$ $\eta(r)$ for $V > 1$	$1 - e^{-(r/a)^2 \ln V^2} = 1 - \frac{1}{V^2}$ for $r = a$	$1 - e^{-(r/a)^2(V-1)}$

Exercise 4

Find the radius a for maximum confinement of light power, i.e., maximum r_0, for step index and parabolic profile optical fibers.

4.1.3.5 Approximation of Spot Size r_0 of a Step Index Fiber

As stated above, spot size r_0 would play a major role in determining the performance of an SM fiber. It is useful if we can approximate the spot size as long as the fiber is operating over a certain wavelength. When an SM fiber is operating above its cutoff wavelength, a good approximation (greater than 96% accuracy) for r_0 is given by

$$\frac{r_0}{a} = 0.65 + 1.619V^{-3/2} + 2.879V^{-6} = 0.65 + 0.434\left(\frac{\lambda}{\lambda_c}\right)^{+3/2} + 0.0419\left(\frac{\lambda}{\lambda_c}\right)^{+6} \qquad (4.34)$$

for $0.8 \leq \dfrac{\lambda}{\lambda_c} \leq 2.0$ single mode.

Exercise 5

What is the equivalent range for the V-parameter of Equation 2.34? Inspect the b-V and $V^2(d^2(Vb)/dV^2)$ versus V and b, if possible do a curve fitting, to obtain the approximate relationship for r/a and V (MATLAB procedure is recommended).

Exercise 6

Refer to the technical specification of Corning SMF-28 and LEAF

a. State the core diameter of the fibers, the spot size or mode field diameters (MFDs) of the fibers.
b. Thence estimate the effective areas of these fibers.
c. What is the ration of the effective area and the physical area of the cores of the fibers?

4.1.4 Equivalent Step Index (ESI) Description

As we can observe there are two possible orthogonally polarized modes (E_x, H_y) and (E_y, H_x) which can be propagating *simultaneously*. These modes are usually approximated by a single LP mode. These modes' properties are well-known and well understood for step index optical fibers and analytical solutions are also readily available.

Unfortunately, practical SMFs never have a perfect step index profile due to the variation of the dopant diffusion and polarization. These non-step index fibers can be approximated, under some special conditions, by an ESI profile technique.

A number of index profiles of modern SM fibers, e.g., non-zero dispersion shifted fibers (NZDSF) are shown in Figure 4.5. The ESI profile is determined by approximating the fundamental mode electric field spatial distribution $\psi(r)$ by a Gaussian function as described in Section 2.2.2.2(b). The electric field can thus be totally specified by the e^{-1} width of this function or *mode spot size* (r_0). Alternatively the term MFD is also used and is equivalent to twice the size of the mode spot size r_0.

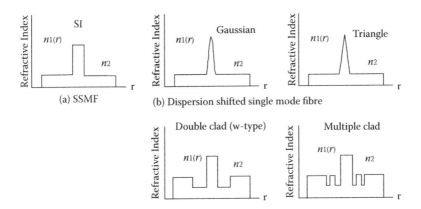

FIGURE 4.5
Index profiles of a number of modern fibers, e.g., dispersion shifted SM fibers.

4.1.4.1 Definitions of Equivalent Step Index Parameters

The ESI description can be used to design single mode (SM) fiber with graded index, W⁻ or segmented core profiles (under some limitations). These non-step index profiles can be described by ESI parameters denoted as followed:

V_e = effective or equivalent V-parameter
a_e = ESI core radius
λ_{ec} = ESI cutoff wavelength
Δ_e = equivalent relative index difference

These parameters are related to two moments M_0 and M_1 defined as

$$M_n = \int_0^\infty \left[n^2(r) - n^2(a)\right] r^n dr \tag{4.35}$$

For $n = 1, 2$. The effective V_e parameter and effective core radius r_e are given by

$$V_e^2 = 2k^2 \int_0^\infty \left[n^2(r) - n^2(a)\right] r dr \tag{4.36}$$

$$V_e^2 = 2k^2 M_1 \quad \text{and} \quad a_e = 2\frac{M_1}{M_0} \tag{4.37}$$

It follows from Equation 4.36 and 4.37, the parameters λ_{ec} and Δ_e by setting

$$V_e^2 = 2k^2 a_e^2 n_1^2 \Delta_e \tag{4.38}$$

and $V_e = 2.405$ (cutoff condition for step index). Therefore the cutoff wavelength for an ESI profile fiber is:

$$\lambda_{ec} = \frac{2\pi\sqrt{2M_1}}{2.405} \tag{4.39}$$

It is noteworthy that V_e as given in Equation 2.36 is equivalent to the mode *volume*. Physically the significance of V_e can be compared to the *average density* of a disk with a local density equal to $[n^2(r) - n^2(a)]$.

4.1.4.2 Accuracy and Limits

The ESI approximation is generally accurate to within 2% at least over the wavelength range $0.8 < \lambda/\lambda_c < 1.5$. For most practical purposes this range is the operating wavelength to minimize the dispersion property of SMFs.

4.1.4.3 Examples on Equivalent Step Index Techniques

4.1.4.3.1 Graded Index Fibers

These index profiles of graded fibers are given by Equation 4.7. We thus have

$$n^2(r) - n^2(a) = s(r/a) = 1 - \left(\frac{r}{a}\right) \tag{4.40}$$

Substituting Equation 4.40 into Equation 4.36 gives

$$\frac{V_e}{V} = \left(\frac{\alpha}{\alpha+2}\right)^{1/2} \tag{4.41}$$

where $V^2 = k^2 a^2$ NA is the V-parameter of a step index fiber with the core index at the fiber axis of n_1. Hence we have

$$\lambda_{ec} = \frac{V}{2.405}\left(\frac{\alpha}{\alpha+2}\right)^{1/2} \tag{4.42}$$

Exercise 7

For an SMF with a triangular profile index distribution whose equivalent V-parameter is equal to 2 at 1550 nm wavelength, what is the V-parameter value at the center of the core of the fiber? If the diameter of the core of the two fibers are kept identical then what is the ration of the refractive indices at the core center of the fibers? Repeat for a parabolic profile.

4.1.4.3.2 Graded Index Fiber with a Central Dip

The fiber index profile with a central dip and gradually graded increase to the outer cladding is shown in Figure 4.6.

Similarly to Equation 4.6 for a graded index fiber with maximum index at the core axis we have

$$S(r/a) = 1 - \gamma(1-x)^\alpha \text{ for } 0 < r < a \tag{4.43}$$

where γ is the depth and $0 < \gamma < 1$. When $r = 0$ we have a step index profile and when $r = 1$ we have the central axis refractive index equal to the cladding index.

Using Equations 4.35 and 4.36, V_e can be easily found and given by:

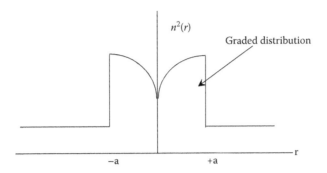

FIGURE 4.6
Refractive index profile of a graded index fiber with a central dip. This is a typical profile of manufactured fiber if a good collapsing of the fiber perform is not achieved.

$$\frac{V_e^2}{V^2} = 1 - \frac{2\gamma}{(\alpha+1)(\alpha=2)} \tag{4.44}$$

4.1.4.4 General Method

The general technique to find the ESI parameters for optical fibers can be started by rising the stationary expression in Equation 4.20 for expressing β of the actual fiber as compared to *its* equivalent propagation constant β_e as:

$$\beta^2 = \beta_e^2 + k^2 \frac{\int_0^\infty \left[n^2(r) - n_e^2(r)\right] r\psi^2(r)dr}{\int_0^\infty r\psi^2(r)dr} \tag{4.45}$$

where $n_e^2(r)$ is the equivalent counterpart of $n(r)$ when the fiber is expressed in its equivalent step form. The field expression $\psi(r)$ is assumed to be similar for the actual fiber and its step equivalence. Once the field $\psi(r)$ can be replaced by the approximate exact field shape, we can find V_e and a_e that minimize $\beta^2 - \beta_e^2$ in Equation 4.45. Generally these parameters are functions of both V and a, thus it is impossible to get one ESI technique applicable to a wide range of wavelengths and complicated numerical calculations are required.

4.2 Nonlinear Optical Effects

In this section the nonlinear effects are described. These effects play important roles in the transmission of optical pulses along SMFs. The nonlinear effects can be classified into three types: the effects that change the refractive index of the guided medium due to the intensity of the pulse, the self phase modulation (SPM); the scattering of the lightwave to other frequency-shifted optical waves when the intensity reaches over a certain threshold, the Brillouin and Raman scattering phenomena; and the mixing of optical waves to generate a fourth wave, the degenerate four wave mixing (FWM). Besides these nonlinear effects there is also photorefractive effect which is due to the change of the refractive index of silica due to the intensity of ultraviolet (UV) optical waves. This phenomenon is used to fabricate grating whose spacing between dark and bight regions satisfies Bragg's diffraction condition. These are fiber Bragg gratings and would be used as optical filters and dispersion compensator when the spacing varies or is chirped.

4.2.1 Nonlinear Self Phase Modulation Effects

All optical transparent materials are subject to the change of the refractive index with the intensity of the optical waves, the optical Kerr effect. This physical phenomenon is originated from the harmonic responses of electrons of optical fields leading to the change of the material susceptibility. The modified refractive index $n_{1,2}^K$ of the core and cladding regions of the silica-based material can be written as:

$$n_{1,2}^K = n_{1,2} + \bar{n}_2 \frac{P}{A_{\mathit{eff}}} \tag{4.46}$$

where n_2 is the nonlinear index coefficient of the guided medium, the average typical value of n_2 is about 2.6×10^{-20} m²/W. P is the average optical power of the pulse and A_{eff} is the effective area of the guided mode. The nonlinear index changes with the doping materials in the core. Although the nonlinear index coefficient is very small, the effective area is also very small, about 50–70 µm², and the length of the fiber under the propagation of optical signals is very long and the accumulated phase change is quite substantial. This leads to the SPM and cross phase modulation (XPM) effects in the optical channels.

4.2.2 Self Phase Modulation

Under a linear approximation we can write the modified propagation constant of the guided LP mode in an SMF as

$$\beta^K = \beta + k_0 \bar{n}_2 \frac{P}{A_{\mathit{eff}}} = \beta + \gamma P \tag{4.47}$$

where

$$\gamma = \frac{2\pi \bar{n}_2}{\lambda A_{\mathit{eff}}}$$

is an important nonlinear parameter of the guided medium taking an effective value from 1–5 kmW⁻¹ depending on the effective area of the guided mode and operating wavelength. Thus the smaller the mode spot size or MFD the larger the nonlinear SPM effect. For dispersion compensating fiber the effective area is about 15 µm² while for SSMF and NZDSF the effective area ranges from 50–80 µm². Thus the nonlinear threshold power of dispersion compensating fiber (DCF) is much lower than that of SSMF and NZDSF. We will see later that the maximum launched power into DCF is limited at about 0 dBm or 1 mW in order to avoid a nonlinear distortion effect, while limited at about 5 dBm for SSMF.

The accumulated nonlinear phase changes due to the nonlinear Kerr effect over the propagation length L is given by:

$$\phi_{NL} = \int_0^L \left(\beta^K - \beta\right) dz = \int_0^L \gamma P(z) dz = \gamma P_{in} L_{eff}$$

(4.48)

with $P(z) = P_{in} e^{-\alpha z}$

defined as the representation of the attenuation of the optical signals along the propagation direction z. In order to consider the nonlinear SPM effect as small compared with the linear chromatic dispersion effect, one can set $\phi_{NL} \ll 1$ or $\phi_{NL} = 0.1$ rad. and the effective length of the propagating fiber is set at $L_{eff} = 1/\alpha$ with optical losses equalized by cascaded optical amplification sub-systems. Then the maximum input power to be launched into the fiber can be set at

$$P_{in} < \frac{0.1\alpha}{\gamma N_A}$$

(4.49)

for $\gamma = 2$ (W · km)$^{-1}$ and $N_A = 10$, $\alpha = 0.2$ dB/km (or 0.0434×0.2 km^{-1}) then $P_{in} < 2.2$ mW or about 3 dBm and accordingly 1 mW for DCF. In practice due the randomness of the arrival of "1" and "0", this nonlinear threshold input power can be set at about 10 dBm as the total average power of all wavelength multiplexed optical channels launched into the propagation fiber.

4.2.3 Cross Phase Modulation

The change of the refractive index of the guided medium as a function of the intensity of the optical signals can also lead to the phase of optical channels in a different spectral region close to that of the original channel. This is XPM effects. This is critical in wavelength division multiplexed (WDM) channels, and even more critical in dense WDM when the frequency spacing between channels is 50 GHz or even narrower. In such systems the nonlinear phase shift of a particular channel depends not only on its power but also on that of other channels. The phase shift of the ith channel can be written as [1]:

$$\phi_{NL}^i = \gamma L_{eff} \left(P_{in}^i + 2 \sum_{j \neq i}^{M} P_j \right)$$

(4.50)

with M = number of multiplexed channels. The factor 2 in Equation 4.50 is due to the bipolar effects of the susceptibility of silica materials. The XPM thus depends on the bit pattern and the randomness of the synchronous arrival of the "1". It is hard to estimate so the numerical simulation would normally be employed to obtain the XPM distortion effects by numerical simulation using the wave propagation of the signal envelop via the nonlinear Schrodinger equation (NLSE). The evolution of slow varying complex envelopes A(z, t) of optical pulses along an SMF is governed by the NLSE [3]:

$$\frac{\partial A(z,t)}{\partial z} + \frac{\alpha}{2} A(z,t) + \beta_1 \frac{\partial A(z,t)}{\partial t} + \frac{j}{2} \beta_2 \frac{\partial^2 A(z,t)}{\partial t^2} - \frac{1}{6} \beta_3 \frac{\partial^3 A(z,t)}{\partial t^3}$$

$$= -j\gamma \left| A(z,t) \right|^2 A(z,t)$$

(4.51)

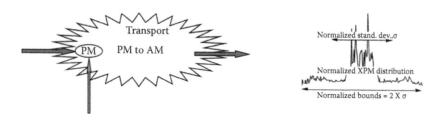

FIGURE 4.7
Illustration of XPM effects—phase modulation conversion to amplitude modulation and hence interference between adjacent channels.

where z is the spatial longitudinal coordinate, α accounts for fiber attenuation, β_1 indicates differential group delay (DGD), β_2 and β_3 represent second and third order factors of fiber chromatic dispersion (CD), and γ is the nonlinear coefficient. This equation is described in detail in Chapter 3. The phase modulation due to nonlinear phase effects is then converted to amplitude modulation and thence the cross talk to other adjacent channels. This is shown in Figure 4.7.

4.2.4 Stimulated Scattering Effects

Scattering of lightwave by impurities can happen due to the absorption and vibration of the electrons and dislocation of molecules in silica-based materials. The back scattering and absorption is commonly known as Rayleigh scattering losses in fiber propagation in whose phenomena the frequency of the optical carrier does not change. While other scattering processes in which the frequency of the lightwave carrier is shifted to another frequency region are commonly known as inelastic scattering, as Raman scattering and Brillouin scattering. In both cases the scattering of photons to a lower energy level photon with energy difference between these levels decreases with the energy of phonons. Optical phonons result from the electronic vibration for Raman scattering while acoustic phonons or mechanical vibration of the linkage between molecules lead to Brillouin scattering. At high power, when the intensity reaches over a certain threshold then the number of scattered photons is exponentially grown, then the phenomena is a simulated process. Thus the phenomena can be called as stimulated Brillouin scattering (SBS) and a stimulated Raman scattering (SRS). SRS and SBS were first observed in the 1970s [2–4].

4.2.4.1 Stimulated Brillouin Scattering

Brillouin scattering comes from the compression of the silica materials in the presence of an electric field, the electrostriction effect. Under the pumping of an oscillating electric field of frequency f_p, an acoustic wave of frequency F_a is generated. Spontaneous scattering is an energy transfer from the pump wave to the acoustic wave and then a phase matching to transfer a frequency shifted optical wave of frequency as a sum of the optical signal waves and the acoustic wave. This acoustic wave frequency shift is around 11 GHz with a bandwidth of around 50–100 MHz (due to the gain coefficient of the SBS) and a beating envelope would be modulating the optical signals. Thus jittering of the received signals at the receiver would be formed, hence the closure of the eye diagram in the time domain.

Once the acoustic wave is generated it beats with the signal waves to generate the side band components. This beating beam acts as a source and further transfers the signal beam energy into the acoustic wave energy and further amplifies this wave to generate further jittering effects. The Brillouin scattering process can be expressed by the following coupled equations [5].

$$\frac{dI_p}{dz} = -g_B I_p I_s - \alpha_p I_p$$

$$-\frac{dI_s}{dz} = +g_B I_p I_s - \alpha_s I_s$$

(4.52)

The SBS gain g_B is frequency dependent with a gain bandwidth of around 50–100 MHz for pump wavelength at around 1550 nm. For silica fiber g_B is about $5e^{-11}$ mW^{-1}. The threshold power for the generation of SBS can be estimated (using Equation 4.52) as

$$g_B P_{th\_SBS} \frac{L_{eff}}{A_{eff}} \approx 21$$

(4.53)

with the effective length

$$L_{eff} = \frac{1 - e^{-\alpha L}}{\alpha}$$

where I_p = intensity of pump beam, I_p = intensity of signal beam, g_B = Brillouin scattering gain coefficient, and α_s, α_p = losses of signal and pump waves.

For the standard SSMF, this SBS power threshold is about 1.0 mW. Once the launched power exceeds this power threshold level the beam energy is reflected back. Thus the average launched power is usually limited to a few dBm due to this low threshold power level.

4.2.4.2 Stimulated Raman Scattering

SRS occurs in silica-based fiber when a pump laser source is launched into the guided medium, the scattering light from the molecules and dopants in the core region would be shifted to a higher energy level and then jump down to a lower energy level, hence amplification of photons in this level. Thus a transfer of energy from a different frequency and energy level photons occurs. The stimulated emission happens when the pump energy level reaches above the threshold level. The pump intensity and signal beam intensity are coupled via the coupled equations:

$$\frac{dI_p}{dz} = -g_R I_p I_s - \alpha_p I_p$$

$$-\frac{dI_s}{dz} = +g_R I_p I_s - \alpha_s I_s$$

(4.54)

where I_p = intensity of pump beam, I_p = intensity of signal beam, g_R = Raman scattering gain coefficient, and α_s, α_p = losses of signal and pump waves.

The spectrum of the Raman gain depends on the decay lifetime of the excited electronic vibration state. The decay time is in the range of 1 ns and Raman-gain-bandwidth is about 1 GHz. In SMFs the bandwidth of the Raman gain is about 10 THz. The pump beam wavelength is usually about 100 nm below the amplification wavelength region. Thus in order to extend the gain spectra a number of pump sources of different wavelengths are used. Polarization multiplexing of these beams is also used to reduce the effective power launched in the fiber so as to avoid the damage of the fiber. The threshold for stimulated Raman gain is given by

$$g_R P_{th\_SRS} \frac{L_{eff}}{A_{eff}} \approx 16$$

with the effective length

$$L_{eff} = \frac{1 - e^{-\alpha L}}{\alpha} \quad \text{or} \approx 1/\alpha \quad \text{for long length} \tag{4.55}$$

For SSMF with an effective area of 50 μm², $g_R \sim 1e^{-13}$ mW then the threshold power is about 570 mW near the C-band spectral region. This would require at least two pump laser sources which should be polarization multiplexed into one combined pump beam. The SRS is used frequently in modern optical communications systems, especially when no undersea optical amplification is required, the distributed amplification of SRS offers significant advantages as compared with lumped amplifiers, such as Erbium-doped Fiber Amplifier (EDFA). The broadband gain and low gain ripple of SRS is also another advantage for dense wavelength division multiplexing (DWDM) transmission.

4.2.4.3 Four-Wave Mixing

FWM is considered as a scattering process in which three photons are mixed to generate the fourth wave. This happens when the momentum of the four waves satisfy a phase matching condition. That is the condition of maximum power transfer. Figure 4.8 illustrates the mixing of different wavelength channels to generate inter-channel cross talk.

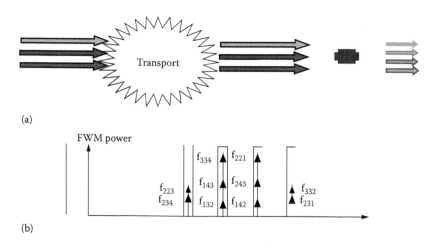

(a)

(b)

FIGURE 4.8
Illustration of FWM of optical channels. (a) Momentum vectors of channels. (b) Frequencies resulting from mixing of different channels.

The phase matching can be represented by a relationship between the propagation constant along the z-direction in an SMF as:

$$\beta(\omega_1) + \beta(\omega_2) - \beta(\omega_3) - \beta(\omega_4) = \Delta(\omega) \tag{4.56}$$

with $\omega_1, \omega_2, \omega_3, \omega_4$ the frequencies of the first to fourth waves and Δ is the phase mismatching parameter. In the case that the channels are equally spaced with a frequency spacing of Ω, as in DWDM optical transmission, then we have $\omega_1 = \omega_2$; $\omega_3 = \omega_1 + \Omega$; $\omega_4 = \omega_1 - \Omega$. One can use the Taylor's series expansion around the propagation constant at the center frequency of the guide carrier β_0 then we can obtain [6]:

$$\Delta(\omega) = \beta_2 \Omega^2 \tag{4.57}$$

The phase matching is thus optimized when β_2 is zero that means that in the region where there is no dispersion, thus the FWM effect is largest due to perfect phase matching and thus the inter-channel crosstalk is biggest. This is the reason why dispersion shifted fiber is not commonly used when the zero dispersion wavelength is fallen in the spectral region of operation of channel. In modern transmission fiber the zero dispersion wavelength is shifted to outside the C-band, say 1510 nm, so that there is a small dispersion factor at 1550 nm and the C-band ranging from 2–6 ps/nm · km, for example, Corning LEAF or NZDSF. This small amount of dispersion is sufficient to avoid the FWM with a channel spacing of 100 GHz or 50 GHz.

The XPM signal is proportional to instantaneous signal power. Its distribution is bounded <5 channels and otherwise effectively unbounded. Thus the Link budgets include XPM evaluated at maximum outer bounds.

4.3 Optical Fiber Manufacturing and Cabling

This section is devoted to a brief description of the manufacturing of optical fibers and the cabling of several fibers for optical communications systems. The manufacturing techniques and cabling process affect the transmission and physical properties of the fibers. We focus on these aspects for a general understanding of optical transmission systems.

As we have described in previous sections, the SSMF structure is a cylindrical core with a refractive index slightly higher than that of the cladding region. For optical communications operating in the 1300 nm and 1700 nm wavelength regions the silica material is the base material. A "pure" silica tube is the starting structure and a combination of silica, germanium oxide GeO_2 and P_2O_5 are then deposited inside the tube. Other dopants such as B_2O_3 and fluoride can also be used to reduce the refractive index of some small regions of the core. These are the segmented core and W-type fibers which will be described in the next chapter.

Once the deposition of the impurities are done (see Figure 4.9) the tube is collapsed to produce a fiber perform as shown in Figure 4.10. Also shown in this figure is a schematic of the fiber drawing machine and the fiber drawing tower. The refractive index of the fiber perform and its details are shown in Figure 4.11.

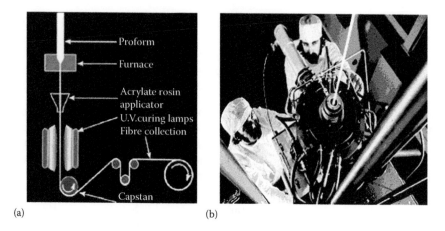

(a)   (b)

FIGURE 4.9
(See Color Insert) (a) Schematic of fiber drawing machine. (b) Picture of fiber microwave furnace and diameter monitoring and feedback control.

4.4 Concluding Remarks

This chapter has introduced the fundamental concepts of optical slab waveguide and then the circular optical waveguide or optical fibers. The basic properties of the fiber structures, its profile, the spot size, the cutoff wavelength, and the Gaussian approximation are described. The Gaussian approximation makes the understanding of the optical guided mode simple. It also allows us to obtain directly the optical mode distribution and thus several other approximations required to obtain the simplest form of the important parameters of SMFs.

Once the basic properties of an SMF are found, they form the basic set of parameters so that optical fibers whose effective index profiles are non-step can be found, based on the ESI technique that converts the parameters to an equivalent step-like profile and hence other optical properties.

Only structural and wave properties of lightwave signals traveling in optical fibers are presented here. As optical communications systems engineers, we have to understand and develop techniques for analyzing and identifying the transmission of digital and analogue signals through optical fibers, the attenuation and broadening of optical signals after transmission through the transmission medium, namely attenuation and broadening via dispersion of lightwave pulses. These topics will be treated in the next section.

Furthermore an appendix (Appendix 4.1 of Section 4.11) listed at the end of this chapter, gives detailed examples of how to design fibers for dispersion flattening and compensating, especially when they are used for Raman optical amplification medium in distributed or lumped configurations.

4.5 Signal Attenuation and Dispersion

This section describes the mechanism and properties of lightwave modulated signals when propagating in optical fibers, SMF only, in particular SSMF and NZDSF and dispersion

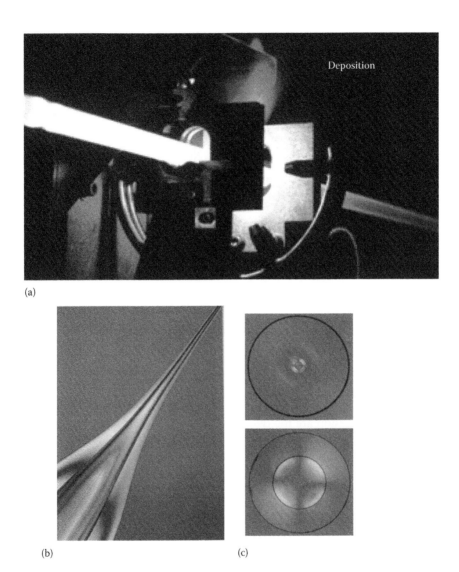

FIGURE 4.10
(See Color Insert) (a) Schematic of a fiber deposition and fabrication of a fiber perform, deposition of core material and collapsing. (b) Fiber perform, before drawing into fiber stands. (c) Cross section of a fiber perform with refractive index profile exactly the same as the fiber index profile of SM (upper) and multimode (lower) types.

compensating fibers. Attenuation and dispersion effects, the two principal phenomena in the design of transmission systems, are described in detail.

4.5.1 Introductory Remarks

In Chapter 2 the basic structure and fundamental aspects of lightwaves propagating in planar optical waveguides were treated. The SMFs are the basic structure for standard communication transmission systems and are introduced in Section 4.2. This section deals with the transmission of optical signals over optical fibers, mainly the loss and spreading of optical signals transmitted through optical fibers namely the attenuation and dispersion effects.

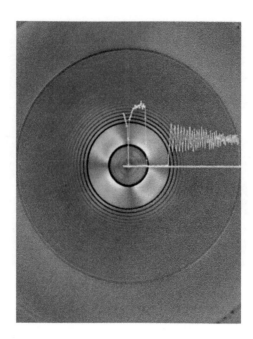

FIGURE 4.11
(See Color Insert) Real index profile across an SM fiber perform. Note: non-step like profile—so why modeled as step index structure?

Attenuation and dispersion are the two most important effects that play major parts in optical fiber transmission systems. The attenuation of optical signal would limit the availability of optical power along the transmission path, and for very low attenuation, dispersion limit the repeater spacing below what would be possible from the attenuation factor.

The fiber loss has been reduced from 100 dB/km (i.e., transmission possible over only a few meters) at 1300 nm in 1970 to about 0.25 and 0.15 dB/km which is very close to the theoretical possible transparent limit and transmission over several hundred kilometers of fibers, for 1300 and 1550 nm wavelength region, respectively, in 1980.

The dispersion and pulse broadening of optical fibers has also been reduced due to "smart design" of optical fiber structures. In early 1970 we saw a remarkable development of theories for understanding lightwaves guiding in optical fibers of multimode types. The breakthrough in the reduction of loss in optical fibers and the ability of manufacturing of optical fibers with a very small core diameter lead to the design of single optical fibers. The remarkable theoretical development of optical waveguiding in "weakly guiding" (i.e., a very small difference between the core and cladding regions) fiber structure leads to a plane wave-like transmission of lightwaves. Furthermore, the availability of narrow linewidth lasers allows systems engineers to design and implement several high speed long distance fiber optic communications systems.

The attenuation that arises from intrinsic material properties and from waveguide properties is described and a general attenuation coefficient is derived. The chromatic dispersion for SM fiber in linear limit, that means that we assume that the optical power launched into the fiber to be less than the threshold for nonlinear effects, is then treated. The effects of optical waveguide parameters on the dispersion factors are analyzed. The balance of the opposite signed dispersion factors between material and waveguides factors can be obtained so that a minimum dispersion factor can be designed for optical fibers with a dispersion compensated or shifted characteristics can be achieved.

In later chapters of this book we will treat optical amplifiers and techniques to compensate for the broadening of optical pulses so that an ultra-long ultra-high speed optical fiber communication system can be designed and implemented.

4.5.2 Signal Attenuation in Optical Fibers

Optical loss in optical fibers is one of the two main fundamental limiting factors as it reduces the average optical power reaching the receiver. The optical loss is the sum of three major components: intrinsic loss, microbending loss and splicing loss.

4.5.2.1 Intrinsic or Material Attenuation

Intrinsic loss consists mainly of absorption loss due to OH impurities and Rayleigh scattering loss. The intrinsic loss is a function of λ^{-6}. Thus the longer the operating wavelength, the lower the loss. However it also depends on the transparency of the optical materials that are used to form the optical fibers. For silica fiber the optical material loss is low over the wavelength range 0.8 μm 1.8 μm. Over this wavelength range there are three optical windows in which optical communication are utilized. The first window over the central wavelength 810 nm is about 20.0 nm bandwidth over the central wavelength. The second and third windows most commonly used in present optical communications are over 1300 nm and 1550 nm with a range of 80 nm and 40 nm, respectively. The intrinsic losses are about 0.3 and 0.15 dB/km at 1550 nm and 1300 nm regions respectively.

This is a few hundred thousand times improvement over the original transmission of signal over 5.0 m with a loss of about 60 dB/km. Most communication fibers systems are operating at 1300 nm due to the minimum dispersion at this range. "Power hungry" systems or extra-long systems should operate at 1550 nm.

4.5.2.2 Absorption

The absorption loss in silica glass is composed mainly of UV and infra-red (IR) absorption tales of pure silica. The IR absorption tale of *pure silica* has been shown due to the *vibration of the basic* tetrahedron and thus *strong resonances* occurs around 8–13 μm with a loss of about 10^{-10} dB/km. This loss is shown in Curve IR of Figure 4.1. Overtones and combinations of these vibrations lead to various absorption peaks in the low wavelength range as shown by Curve UV.

Various impurities that also lead to spurious absorption effects in the wavelength range of interest (1.2–1.6 μm) are transition metal ions and water in the form of OH ions. These sources of absorptions have been practically reduced in recent years.

4.5.2.3 Rayleigh Scattering

The Rayleigh scattering loss, L_R, which is due to microscopic non-homogeneities of the material, shows a λ^{-4} dependence and is given by

$$L_R = (0.75 + 4.5\Delta)\lambda^{-4}\ \mathrm{dB/km} \tag{4.58}$$

where Δ is the relative index difference as defined above and λ is the wavelength in μm. Thus to minimize the loss Δ should be made as low as possible.

4.5.2.4 Waveguide Loss

The losses due to waveguide structure arise from power leakage, bending, microbending of the fiber axis and defects and joints between fibers. The power leakage is significant only for depressed cladding fibers.

4.5.2.5 Bending Loss

When a fiber is bent the plane wave fronts associated with the guided mode are pivoted at the center of curvature and their longitudinal velocity along the fiber axis increases with the distance from the center of curvature. As the fiber is bent further over a critical curve, the phase velocity would exceed that of plane wave in the cladding and radiation occurs.

The bend loss L_B for a radius R (radius of curvature) is given by

$$L_B = -10Log_{10}\left(1 - 890\frac{r_0^6}{\lambda^4 R^2}\right) \text{for silica}$$

(4.59)

4.5.2.6 Microbending Loss

Microbending loss results from power coupling from the guided fundamental mode of the fiber to radiation modes. This coupling takes place when the fiber axis is bent *randomly* in a *high* spatial frequency. Such bending can occur during packing of the fiber and during the cabling process as shown in Figure 4.12.

The microbending loss of an SM fiber is a function of the fundamental mode spot size r_0. Fibers with large spot size are extremely sensitive to microbending. It is therefore desirable to design the fiber to have as small a spot size as possible to minimize bending loss. The microbending loss can be expressed by the relation

(a)   (b)   (c)   (d)

FIGURE 4.12
(See Color Insert) Installation of fiber cables. (a) Installation of fiber cable by hanging. (b) Installation of fiber cable by ploughing. (c) Installation of undersea fiber cable. (d) Splicing two optical fibers.

$$L_m = 2.15 \times 10^{-4} r_0^6 \lambda^{-4} L_{mm} \text{ dB/km} \tag{4.60}$$

where L_{mm} is the microbending loss of a 50 μm core multimode fiber having an NA of 0.2.

4.5.2.7 Joint or Splice Loss

Ultimately the fibers will have to be spliced together to form the final transmission link. With fiber cable that averages 0.4–0.6 dB/km (Figure 4.13), splice loss in excess of 0.2 dB/splice drastically reduces the non-repeatered distance that can be achieved. It is therefore extremely important that the fiber be designed such that splicing loss be minimized.

Splice loss is mainly due to axial misalignment of the fiber core as shown in Figure 4.14.

Splicing techniques, which rely on aligning the outside surface of the fibers, require extremely tight tolerances on core to outside surface concentricity. Offsets of the order of 1 μm can produce significant splice loss. This loss is given by

$$L_s = \frac{10}{\ln 10} \left(\frac{d}{r_0}\right)^2 \text{dB} \tag{4.61}$$

where d is the *axial* misalignment of the fiber cores. It is obvious that minimizing optical loss involves making trade-offs between the different sources of loss. It is advantageous to have a large spot size to minimize both Rayleigh and splicing losses whereas minimizing bending and microbending losses requires a small spot size. In addition, as will be described in the next section, the spot size plays a significant role in the chromatic dispersion properties of SM fibers.

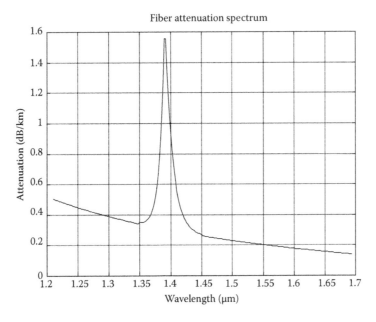

FIGURE 4.13
Attenuation of optical signals as a function of wavelength. The minimum loss at wavelength: at $\lambda = 1.3$ μm about 0.3 dB/km and at $\lambda = 1.5$ μm loss of about 0.13 dB/km. For cabled fibers the attenuation factor at 1550 nm is 0.25 dB/km.

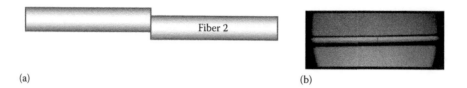

(a)   (b)

FIGURE 4.14
(a) Misalignment in splicing two optical fibers generating losses. (b) Aligned spliced fibers.

4.5.2.8 Attenuation Coefficient

Under general conditions of power attenuation inside an optical fiber the attenuation coefficient of the optical power P can be expressed as

$$\frac{dP}{dz} = -\alpha P \tag{4.62}$$

where α is the attenuation coefficient. This attenuation coefficient can include all effects of power loss when signals are transmitted though the optical fibers.

Considering optical signals with an average optical power entering at the input of the fiber length L is P_{in} and P_{out} is the output optical power, then we have P_{in} and P_{out} related to the attenuation coefficient α as

$$P_{out} = P_{in}\, e^{(-L)} \tag{4.63}$$

It is customary to express α in dB/km by using the relation

$$\alpha(dB/km) = -\frac{10}{L}\log_{10}\left(\frac{P_{out}}{P_{in}}\right) = 4.343\alpha \tag{4.64}$$

Standard optical fibers with a small Δ (the relative refractive index difference between the core and cladding region) would exhibit a loss of about 0.2 dB/km, i.e., that the purity of the silica is very high. Such purity of a bar of silica would allow us to see though a 1 km glass bar the person standing at the other end without distortion! The attenuation curve for silica glass is shown in Figure 4.1.

4.6 Signal Distortion in Optical Fibers

4.6.1 Basics on Group Velocity

Consider a *monochromatic* field given by

$$E_x = A\cos(\omega t - \beta z) \tag{4.65}$$

where A is the wave amplitude, ω is the radial frequency and β is the propagation constant along the z direction. If setting ($\omega t - \beta z$) constant then the wave phase velocity is given by

$$v_p = \frac{dz}{dt} = \frac{\omega}{\beta} \tag{4.66}$$

Now we consider the propagating wave consists of *two* monochromatic fields of frequencies $\omega + \delta\omega$ and $\omega - \delta\omega$ of

$$E_{x1} = A \cos\left[\left((\omega + \delta\omega)t - (\beta + \delta\beta)z\right)\right] \tag{4.67}$$

$$E_{x2} = A \cos\left[\left((\omega - \delta\omega)t - (\beta - \delta\beta)z\right)\right] \tag{4.68}$$

The total field is then given by:

$$E_x = E_{x1} + E_{x2} = 2A \cos\left(\omega t - \beta z\right) \cos\left(\delta\omega t - \delta\beta z\right) \tag{4.69}$$

If $\omega \gg \delta\omega$ then $\cos(\omega t - \beta z)$ varies much faster than $\cos(\delta\omega t - \delta\beta z)$. Now by setting ($d\omega t - d\beta z$) constant and define the *group velocity* as

$$v_g = \frac{d\omega}{d\beta} \rightarrow v_g^{-1} = \frac{d\beta}{d\omega} \tag{4.70}$$

The group delay t_g per unit length (setting L at 1.0 km) is thus given to be

$$t_g = \frac{L\left(\text{of } 1\,\text{Km}\right)}{v_g} = \frac{d\beta}{d\omega} \tag{4.71}$$

The pulse spread $\Delta\tau$ per unit length due to group delay of light sources of spectral width σ_λ (i.e., the full-width half-mark [FWHM] of the optical spectrum of the light source) is

$$\Delta\tau = \frac{dt_g}{d\lambda}\sigma_\lambda \tag{4.72}$$

i.e., the spread of the group delay due to the spread of source wavelength in ps/km. Thus the linewidth of the light source makes a great difference in the distortion of the optical signal transmitted through the optical fiber. The narrower the source linewidth the less dispersed the optical pulses are. Typical linewidth of Fabry-Perot semiconductor lasers is about 1–2.0 nm while the DFB (distributed feedback) laser would exhibit a linewidth of 100 MHz (how many nm is this 100 MHz optical frequency equivalent to?)

Optical signal traveling along a fiber becomes increasingly distorted. This distortion is a consequence of *intermodal* delay effects and *intramodal* dispersion. Intermodal delay effects are significant in multimode optical fibers due to each mode having a different value of group velocity at a specific frequency, while intermodal dispersion is pulse spreading that occurs within an SM. It is the result of the group velocity being a function of the wavelength λ and is therefore referred to as chromatic dispersion.

Two main causes of intermodal dispersion are

- Material dispersion which arises from the variation of the refractive index $n(\lambda)$ as a function of wavelengths. This causes a wavelength dependence of the group velocity of any given mode.
- Waveguide dispersion, which occurs because the mode propagation constant $\beta(\lambda)$ is a function of wavelength λ and core radius a and the refractive index difference.

The group velocity associated with the fundamental mode is frequency dependent because of chromatic dispersion. As a result different spectral components of the light pulse travel at different group velocities, a phenomenon referred to as *the group-velocity-dispersion* (GVD), intramodal dispersion or as material dispersion and waveguide dispersion.

4.6.2 Group Velocity Dispersion

4.6.2.1 Material Dispersion

The refractive index of silica as a function of wavelength is shown in Figure 4.15. The refractive index is plotted over the wavelength region of 1.0–2.0 µm which is the most important range for silica based optical communications systems as the loss is lowest at 1300 nm and 1550 nm windows (Figure 4.13).

The propagation constant β of the fundamental mode guided in the optical fiber can be written as

$$\beta(\lambda) = \frac{2\pi n(\lambda)}{\lambda} \tag{4.73}$$

The group delay t_{gm} per unit length of Equation 3.9 can be obtained

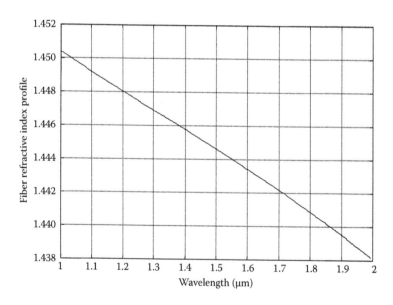

FIGURE 4.15
Variation in the refractive index as a function of optical wavelength of silica.

$$t_{gm} = \frac{d\beta}{d\omega} \tag{4.74}$$

where we can use

$$d\omega = d\left(\frac{2\pi c}{\lambda}\right) = -\frac{2\pi c}{\lambda^2} d\lambda \tag{4.75}$$

Thus

$$t_{gm} = -\frac{\lambda^2}{2\pi c} \frac{d\beta}{d\lambda} \tag{4.76}$$

Substituting Equation 4.73 to 4.76 we have

$$t_{gm} = \frac{1}{c}\left[n(\lambda) - \frac{\lambda dn(\lambda)}{d\lambda} \right] \tag{4.77}$$

Figures 4.16 through 4.18 show the time signal and its envelope and equivalent spectrum as well as the evolution of the carrier phasor leading to the dispersion of the signals propagating through the guided medium.

Thus the pulse dispersion per unit length $\Delta\tau_m/\Lambda\lambda$ due to material (using Equation 4.77) for a source having root mean square (RMS) spectral width σ_λ of

$$\Delta\tau_m = -\frac{\lambda}{c} \frac{d^2 n}{d\lambda^2} \sigma_\lambda \tag{4.78}$$

if setting $\Delta\tau_m = M(\lambda)\sigma_\lambda$, thence

$$M(\lambda) = -\frac{\lambda}{c} \frac{d^2 n}{d\lambda^2} \tag{4.79}$$

$M(\lambda)$ is assigned as *the material dispersion factor or "material dispersion parameter"*, its unit is commonly expressed in ps/(nm · km).

Thus if the refractive index can be expressed as a function of the optical wavelength then the material dispersion can be calculated. In fact in practice optical material engineers have to characterize all optical properties of new materials. The refractive index $n(\lambda)$ can usually be expressed in Sellmeier's dispersion formula as

$$n^2(\lambda) = 1 + \sum_k \frac{G_k \lambda^2}{\left(\lambda^2 - \lambda_k^2\right)} \tag{4.80}$$

where G_k are Sellmeier's constants and k is an integer and normally taken as a range of $k = 1 - 3$. In the late 1970s several silica based glass materials have been manufactured and their properties are measured. The refractive indices are usually expressed using Sellmeier's coefficients. These coefficients for several optical fiber materials are given in Table 4.1.

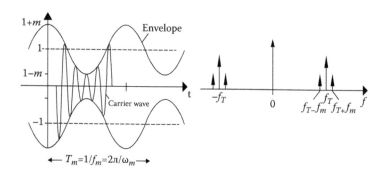

FIGURE 4.16
Time signal and spectrum.

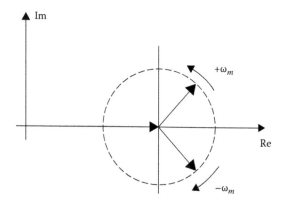

FIGURE 4.17
Vector phasor diagram of the complex envelope.

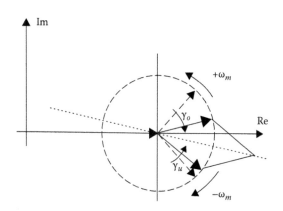

FIGURE 4.18
Magnitude of the complex envelope when not sinusoidal, the envelope subject to nonlinear distortions.

TABLE 4.2

Sellmeier's coefficients for several optical fiber silica based materials with germanium doped in the core region

Sellmeier's constants	Germanium concentration, C (mole %)			
	0 (pure silica)	3.1	5.8	7.9
G_1	0.6961663	0.7028554	0.7088876	0.7136824
G_2	0.4079426	0.4146307	0.4206803	0.4254807
G_3	0.8974794	0.8974540	0.8956551	0.8964226
λ_1	0.0684043	0.0727723	0.0609053	0.0617167
λ_2	0.1162414	0.1143085	0.1254514	0.1270814
λ_3	9.896161	9.896161	9.896162	9.896161

By using curve fitting, the refractive index of pure silica $n(\lambda)$ can be expressed as:

$$n(\lambda) = c_1 + c_2\lambda^2 + c_3\lambda^{-2} \tag{4.81}$$

where $c_1 = 1.45084$, $c_2 = -0.00343$ μm^{-2} and $c_3 = 0.00292$ μm². Thus from Table 4.2 and Equation 4.81, we can use Equation 4.79 to determine the material dispersion factor for certain wavelength ranges.

For the doped core of the optical fiber the Sellmeier's expression (Equation 4.80) can be approximated by using a curve fitting technique to approximate it to the form in Equation 4.81. The material dispersion factor $M(\lambda)$ becomes zero at wavelengths around 1350 nm and about −10 ps/(nm · km) at 1550 nm (Figure 4.19). However the attenuation at 1350 nm is about 0.4 dB/km compared with 0.2 dB/km at 1550 nm as shown in Table 4.2.

4.6.2.2 Waveguide Dispersion

The effect of waveguide dispersion can be approximated by assuming that the refractive index of the material is independent of wavelength. Let us now consider the group delay, i.e., the time required for a mode to travel along a fiber of length L. This kind of dispersion depends strongly on Δ and *V-parameters*. To make the results of fiber parameters, we define a *normalized propagation constant b* as

$$b = \frac{\frac{\beta^2}{k^2} - n_2^2}{n_1^2 - n_2^2} \tag{4.82}$$

for small Δ. We note that the β/k is in fact the *"effective"* refractive index of the guided optical mode propagating along the optical fiber, that is the guided waves traveling the axial direction of the fiber "sees" it as a medium with a refractive index of an equivalent "effective" index.

In the case that the fiber is a weakly guided waveguide with the effective refractive index takes a value significantly closer to that of the core or cladding index, Equation 4.82 can then be approximated by

$$b \cong \frac{\frac{\beta}{k} - n_2}{n_1 - n_2} \tag{4.83}$$

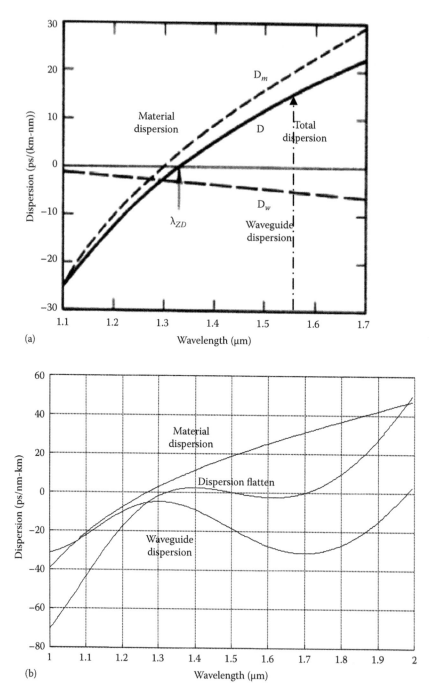

FIGURE 4.19

Chromatic dispersion factor of: (a) SSMF, (b) dispersion flatten fiber: plotted curves representing the material dispersion factor as a function of optical wavelength for silica base optical fiber (yellow curve) with a zero dispersion wavelength at 1290 nm. This curve is generated as an example. For standard SMFs which are currently installed throughout the world the total dispersion is around +17 ps/(nm · km) at 1550 nm and almost zero at 1310 nm. Can we estimate the waveguide dispersion curve for the standard SM optical fiber at around 1300 nm and 1550 nm windows?

solving Equation 4.83 for β, we have

$$\beta = n_2 k (b\Delta + 1) \tag{4.84}$$

the group delay for waveguide dispersion is then given by (per unit length)

$$t_{wg} = \frac{d\beta}{d\omega} = \frac{1}{c} \frac{d\beta}{dk} \tag{4.85}$$

$$t_{wg} = \frac{1}{c}\left[n_1 + n_2\Delta \frac{d(bk)}{dk} \right] = \frac{1}{c}\left[n_1 + n_2\Delta \frac{d(bk)}{dk} \right] = \frac{1}{c}\left[n_1 + n_2\Delta \frac{d(bV)}{dV} \right] \tag{4.86}$$

Equation 4.86 can be obtained from Equation 4.85 by using the expression of *V*. Thus the pulse spreading $\Delta\tau_\omega$ due to the waveguide dispersion per unit length by a source having an optical bandwidth (or linewidth σ_λ) is given by

$$\Delta\tau_\omega = \frac{dt_{gw}}{d\lambda}\sigma_\lambda = -\frac{n_2\Delta}{c\lambda} V \frac{d^2(Vb)}{dV^2}\sigma_\lambda \tag{4.87}$$

and the *waveguide dispersion factor or "waveguide dispersion parameter"* (similar to the material dispersion factor) is then defined as:

$$D(\lambda) = -\frac{n_2(\lambda)\Delta}{c\lambda} V \frac{d^2(Vb)}{dV^2} \tag{4.88}$$

in unit of ps/(nm · km). In the range of $0.9 < \lambda/\lambda c < 2.6$, the factor $V(d^2(Vb)/dV^2)$ can be approximated (to $<5\%$ error) by

$$V \frac{d^2(Vb)}{dV^2} \cong 0.080 + 0.549(2.834 - V)^2 \tag{4.89}$$

or alternatively using the definition of cutoff wavelength and the expression of the *V*-parameter we obtain

$$V \frac{d^2(Vb)}{dV^2} \cong 0.080 + 3.175\left(1.178 - \frac{\lambda_c}{\lambda} \right) \tag{4.90}$$

Note: Readers/students to prove the equivalent equation of Equations 4.85 through 4.90. Furthermore, the sign assignment of the material and the waveguide dispersion factors must be the same. Otherwise a negative and positive of these dispersion factors would create confusion. Can you explain what would happen to the pulse if it is transmitted through an optical fiber having a total negative dispersion factor?

Thus from Equations 4.90 and 4.89 we can calculate the waveguide dispersion factor and hence the pulse dispersion factor for a particular source spectral width σ_λ. It is noted that the dispersion considered in this chapter is for step index fiber only. For grade index fiber, ESI parameters must be found and the chromatic dispersion can then be calculated.

4.6.2.3 Alternative Expression for Waveguide Dispersion Parameter

Alternatively the waveguide dispersion parameter can be expressed as a function of the propagation constant β by using $\omega = 2\pi c/\lambda$ and Equation 4.90, then the waveguide dispersion factor can be written as:

$$D(\lambda) = -\frac{2\pi c}{\lambda^2}\beta_2 = -\frac{2\pi c}{\lambda^2}\frac{d\beta^2}{d\omega^2} \tag{4.91}$$

Thus the waveguide dispersion factor is directly related to the second order derivative of the propagation constant with respect to the optical radial frequency (Figure 4.20).

An example of a design of an optical fiber operating in the SM region is given in Figure 4.19. The cladding material is pure silica. Shown in this figure are the curves of the material dispersion (yellow), waveguide dispersion (red) and total dispersion for an SMF with non-uniform refractive index profile in the core.

4.6.2.4 Higher Order Dispersion

We observe also from Figure 4.6 that the bandwidth–length product of the optical fiber can be extended to infinitive if the system is operating at the wavelength such that the total dispersion is zero. However the dispersive effects do not disappear completely at this zero-dispersion wavelength. Optical pulses still experience broadening because of higher order dispersion effects. It is easily imagined that the total dispersion factor cannot be made zero "flatten" over the optical spectrum. This is higher order dispersion which is governed by the slope of the total dispersion curve, called the dispersion slope $S = d(D + M)/d\lambda$; $S(\lambda)$ can thus be expressed as:

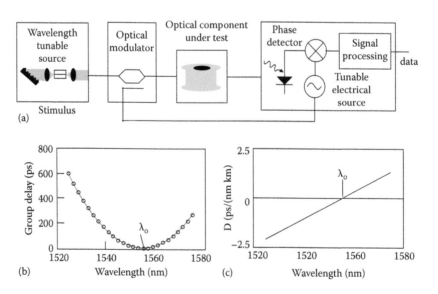

FIGURE 4.20
(a) Chromatic dispersion measurement of two-port optical device. (b) Relative group delay versus wavelength. (c) Dispersion parameter versus wavelength.

$$S(\lambda) = \left(\frac{2\pi c}{\lambda^2}\right)^2 \frac{d^3\beta}{d\lambda^3} + \left(\frac{4\pi c}{\lambda^3}\right)\frac{d^2\beta}{d\lambda^2} \tag{4.92}$$

S is also known as the differential-dispersion parameter.

4.6.2.5 Polarization Mode Dispersion

The delay between two principal state of polarization (PSP) is normally negligibly small at 10 Gb/s. However, at high bit rate and in ultra-long-haul transmission, PMD severely degrades the system performance [7–10]. The instantaneous value of DGD ($\Delta\tau$) varies along the fiber and follows a Maxwellian distribution [11–13] (see Figures 4.21 through 4.23).

The Maxwellian distribution is governed by the following expression:

$$f(\Delta\tau) = \frac{32(\Delta\tau)^2}{\pi^2(\Delta\tau)^3} \exp\left\{-\frac{4(\Delta\tau)^2}{\pi(\Delta\tau)^2}\right\} \quad \Delta\tau \geq 0 \tag{4.93}$$

The mean DGD value $\langle\Delta\tau\rangle$ is commonly termed "fiber PMD" and provided in the fiber specifications. The following expression gives an estimate of the maximum transmission limit L_{max} due to the PMD effect [30]:

$$L_{max} = \frac{0.02}{(\Delta\tau)^2 \cdot R^2} \tag{4.94}$$

where R is the bit rate. Based on Equation 4.94, L_{max} for both old fiber vintage and contemporary fibers are obtained as follows

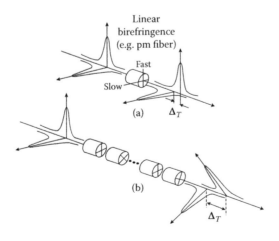

FIGURE 4.21
Conceptual model of PMD: (a) simple birefringence device; (b) randomly concatenated birefringence.

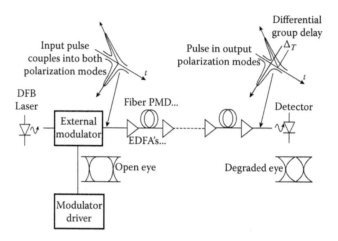

FIGURE 4.22
Effect of PMD in a digital optical communication system, degradation of the received eye diagram.

- $\langle \Delta\tau \rangle = 1$ ps/km (old fiber vintages): for bit rate of $R = 40$ Gb/s then the maximum distance $L_{max} = 12.5$ km; for $R = 10$ Gb/s then $L_{max} = 200$ km.

- $\langle \Delta\tau \rangle = 0.1$ ps/km (contemporary fiber for modern optical systems): then if the bit rate $R = 40$ Gb/s then the maximum transmission distance is $L_{max} = 1{,}250$ km; for $R = 10$ Gb/s; $L_{max} = 20{,}000$ km.

Question: Inspect the technical specifications of Corning fibers given in the appendix of Chapter 2, extract the values of the PMD. Explain the difference between the values of the fibers. What is the standard value allowable for PMD in modern fibers?

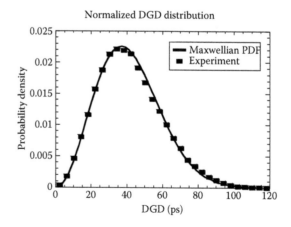

FIGURE 4.23
Maxwellian distribution of PMD random process.

4.7 Transfer Function of Single Mode Fibers

4.7.1 Linear Transfer Function

The treatment of the propagation of modulated lightwaves through SM fiber in the linear and nonlinear regimes has been well documented [14–19]. For completeness of the transfer function of SMFs, in this section we restrict our study to the frequency transfer function and impulse responses of the fiber to the linear region of the media. Furthermore, the delay term in the NLSE can be ignored, as it has no bearing on the size and shape of the pulses. From NLSE we can thus model the fiber simply as a quadratic phase function. This is derived from the fact that the nonlinear term of NLSE can be removed and Taylor's series approximation around the operating frequency (central wavelength) can be obtained and the frequency and impulse responses of the SM fiber can be obtained [17–19]. The input-output relationship of the pulse can therefore be depicted. Equation 4.1 expresses the time-domain impulse response $h(t)$ and the frequency domain transfer function $H(\omega)$ as a Fourier transform pair:

$$h(t) = \sqrt{\frac{1}{j4\pi\beta_2}} \exp\left(\frac{jt^2}{4\beta_2}\right) \leftrightarrow H(\omega) = \exp\left(-j\beta_2\omega^2\right) \tag{4.95}$$

where β_2 is well known as the GVD parameter. The input function $f(t)$ is typically a rectangular pulse sequence and β_2 is proportional to the length of the fiber. The output function $g(t)$ is the dispersed waveform of the pulse sequence. The propagation transfer function in Equation 4.95 is an exact analogy of diffraction in optical systems (see item 1, Table 2.1, p.14 [20]). Thus, the quadratic phase function also describes the diffraction mechanism in one-dimensional optical systems, where distance x is analogous to time t. The establishment of this analogy allows us to borrow many of the imageries and analytical results that have been developed in the diffraction theory. Thus, we may express the step response $s(t)$ of the system $H(\omega)$ in terms of the Fresnel cosine and sine integrals as follows

$$s(t) = \int_0^t \sqrt{\frac{1}{j4\pi\beta_2}} \exp\left(\frac{jt^2}{4\beta_2}\right) dt = \sqrt{\frac{1}{j4\pi\beta_2}} \left[C\left(\sqrt{1/4\beta_2}\, t\right) + jS\left(\sqrt{1/4\beta_2}\, t\right) \right] \tag{4.96}$$

with

$$C(t) = \int_0^t \cos\left(\frac{\pi}{2}\tau^2\right) d\tau$$

$$S(t) = \int_0^t \sin\left(\frac{\pi}{2}\tau^2\right) d\tau \tag{4.97}$$

where $C(t)$ and $S(t)$ are the Fresnel cosine and sine integrals.

Using this analogy, one may argue that it is always possible to restore the original pattern $f(x)$ by refocusing the blurry image $g(x)$ (e.g., image formation, item 5, Table 2.1 [20]). In the electrical analogy, it implies that it is possible to compensate the quadratic phase media perfectly. This is not surprising. The quadratic phase function $H(\omega)$ in Equation 4.95 is an all-pass transfer function, thus it is always possible to find an inverse function to recover $f(t)$. One can express this differently in information theory terminology, the quadratic phase channel has a theoretical bandwidth of infinity; hence its information capacity is infinite. Shannon's channel capacity theorem states that there is no limit on the reliable rate of transmission through the quadratic phase channel. Figure 4.24 shows the pulse and impulse responses of the fiber. It is noted that only the envelope of the pulse

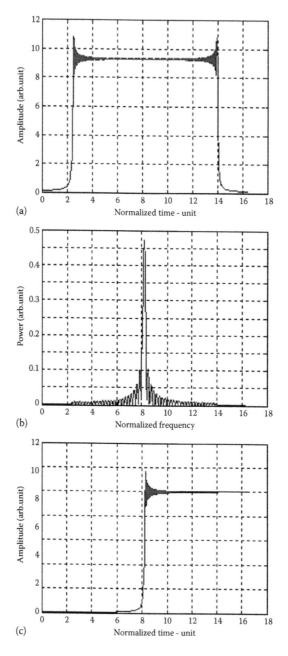

FIGURE 4.24

Rectangular pulse transmission of through an SMF: (a) pulse response, (b) frequency spectrum, (c) step response of the quadratic-phase transmittance function. Note: horizontal scale in a normalized unit of time.

is shown and the phase of the lightwave carrier is included as the complex values of the amplitudes. As observed the chirp of the carrier is significant at the edges of the pulse. At the center of the pulse, the chirp is almost negligible at some limited fiber length, thus the frequency of the carrier remains nearly the same as at its original starting value. One could obtain the impulse response quite easily but in this work we believe that the pulse response is much more relevant in the investigation of the uncertainty in the pulse sequence detection. Rather the impulse response is much more important in the process of equalization.

The uncertainty of the detection depends on the modulation formats and detection process. The modulation can be implemented by manipulation of the amplitude, the phase or the frequency of the carrier or both amplitude and phase or multi-sub-carriers such as the orthogonal frequency division multiplexing (OFDM) [21]. The amplitude detection would be mostly affected by the ripples of the amplitudes of the edges of the pulse. The phase of the carrier is mostly affected near the edge due to the chirp effects. However if differential phase detection is used then the phase change at the transition instant is the most important and the opening of the detected eye diagram. For frequency modulation the uncertainty in the detection is not very critical provided that the chirping does not enter into the region of the neighborhood of the center of the pulse in which the frequency of the carrier remains almost constant.

The picture changes completely if the detector/decoder is allowed only a finite time window to decode each symbol. In the convolution coding scheme for example, it is the decoder's constraint length that manifests due to the finite time window. In the adaptive equalization scheme, it is the number of equalizer coefficients that determines the decoder window length. Since the transmitted symbols have already been broadened by the quadratic phase channel, if they are next gated by a finite time window, the information received could be severely reduced. The longer the fiber, the more the broadening of the pulses is widened, hence the more uncertain it becomes in the decoding. It is the interaction of the pulse broadening on one hand, and the restrictive detection time window on the other, that gives rise to the finite channel capacity.

It is observed that the chirp occurs mainly near the edge of the pulses when it is in the near field region, about a few kilometers for standard SM fibers. In this near field distance the accumulation of nonlinear effects is still very weak and thus this chirp effects dominate the behavior of the SM fiber. The nonlinear Volterra transfer function presented in the next section would thus have minimum influence. This point is important for understanding the behavior of lightwaves circulating in shot length fiber devices in which both the linear and nonlinear effects are to be balanced, such as active mode locked soliton and multi-bound soliton lasers [21,22]. In the far field the output of the fiber is Gaussian like for the square pulse launched at the input. In this region the nonlinear effects would dominate over the linear dispersion effect as they have been accumulated over a long distance.

The linear time variant system such as the SM fiber would have a transfer function of

$$H(f) = |H(f)| e^{-j\alpha(f)} \tag{4.98}$$

where

$$\alpha = \pi^2 \beta_2 L = \frac{\left(-\pi D L \lambda^2\right)}{(2c)}$$

is proportional to the length L and the dispersion factor D (ps/nm/km). The phase of the frequency transfer response is a quadratic function of the frequency thus the group delay would follow a linear relationship with respect to the frequency as observed in Figure 4.25. The frequency response in amplitude terms is infinite and is a constant while the phase response is a quadratic function with respect to the frequency of the base band signals. The carrier is chirped accordingly as observed in Figures 4.26 and 4.27. The chirping effect is very significant near the edge of the rectangular pulse and almost nil at the center of the pulse, in the near field region of less than 1 km of standard SM fiber. In the far field region the pulse becomes Gaussian like. Thus the response of the fiber in the linear region can be seen as shown in Figure 4.28 for a Gaussian pulse input to the fiber. The output pulse is also Gaussian by taking the Fourier transform of the input pulse and multiplying by the fiber transfer function. Thence an inverse Fourier would indicate the output pulse shape follows a Gaussian profile.

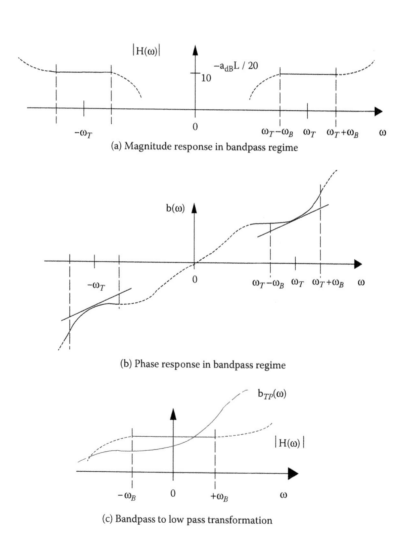

(a) Magnitude response in bandpass regime

(b) Phase response in bandpass regime

(c) Bandpass to low pass transformation

FIGURE 4.25
Frequency response of an SMF: (a) magnitude; (b) phase response in bandpass regime; (c) baseband equivalence.

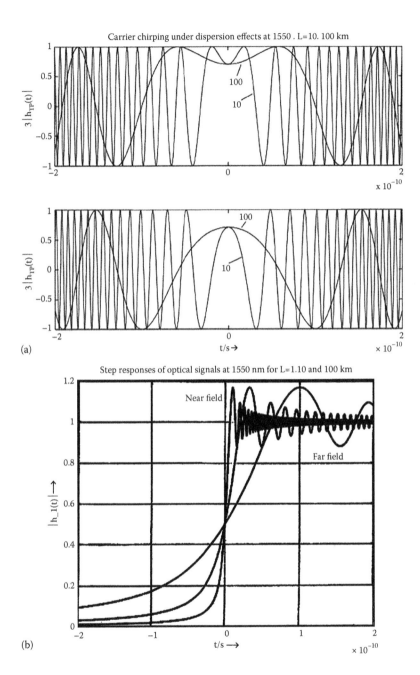

FIGURE 4.26
(a) Carrier chirping effects and (b) step response of an SMF of L = 1, 10 and 100 km.

This leads to a rule of thumb for consideration of the scaling of the bit rate and transmission distance as: *"Given that a modulated lightwave of a bit rate B can be transmitted over a maximum distance L of SMF with a bit error rate (BER) of error-free level, than if the bit rate is halved then the transmission distance can be increase by four times."* For example for 10 Gb/s amplitude shift keying modulation format signals can be transmitted over 80 km of standard SMF then at 40 Gb/s only 5 km can be transmitted for a BER of 10^{-9}.

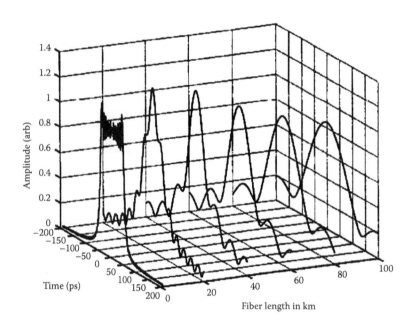

FIGURE 4.27
Pulse response from near field (~ < 2 km) to far field (> 80 km).

4.7.2 Nonlinear Fiber Transfer Function

The weakness of most of the recursive methods in solving the NLSE is that they do not provide much useful information to help the characterization of nonlinear effects. The Volterra series model provides an elegant way of describing a system's nonlinearities, and enables the designers to see clearly where and how the nonlinearity affects the system performance. Although Refs. [2–4] have given an outline of the kernels of the transfer function using the Volterra series, it is necessary for clarity and physical representation of these functions to give brief derivations here on the nonlinear transfer functions of an optical fiber operating under nonlinear conditions.

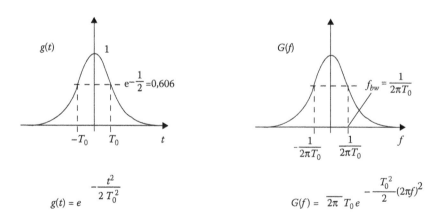

FIGURE 4.28
Fiber response to Gaussian pulse. Gaussian→ Gaussian!

The Volterra series transfer function (VSTF) of a particular optical channel can be obtained in the frequency-domain as a relationship between the input spectrum $X(\omega)$ and the output spectrum $Y(\omega)$, as

$$Y(\omega) = \sum_{n=1}^{\infty} \int_{-\infty}^{\infty} \cdots \int_{-\infty}^{\infty} H_n(\omega_1, \cdots, \omega_{n-1}, \omega - \omega_1 - \cdots - \omega_{n-1})$$

$$\times X(\omega_1) \cdots X(\omega_{n-1}) X(\omega - \omega_1 - \cdots - \omega_{n-1}) d\omega_1 \cdots d\omega_{n-1}$$

(4.99)

where $H_n(\omega_1, \ldots, \omega_n)$ is the nth-order frequency domain Volterra kernel including all signal frequencies of orders 1 to n. The wave propagation inside a single-mode fiber can be governed by a simplified version of the NLS wave equation [1] with only the SPM effect is included as

$$\frac{\partial A}{\partial z} = -\frac{\alpha_0}{2} A - \beta_1 \frac{\partial A}{\partial t} - j\frac{\beta_2}{2}\frac{\partial^2 A}{\partial t^2} - \frac{\beta_3}{6}\frac{\partial^3 A}{\partial t^3} + j\gamma |A|^2 A$$

(4.100)

where $A = A(t, z)$. The proposed solution of the NLSE can be written with respect to the VSTF model of up to the fifth order as

$$A(\omega, z) = H_1(\omega, z) A(\omega) + \int_{-\infty}^{\infty} \int_{-\infty}^{\infty} H_3(\omega_1, \omega_2, \omega - \omega_1 + \omega_2, z)$$

$$\times A(\omega_1) A^*(\omega_2) A(\omega - \omega_1 + \omega_2) d\omega_1 d\omega_2$$

$$+ \int_{-\infty}^{\infty}\int_{-\infty}^{\infty}\int_{-\infty}^{\infty}\int_{-\infty}^{\infty} H_5(\omega_1, \omega_2, \omega_3, \omega_4, \omega - \omega_1 + \omega_2 - \omega_3 + \omega_4, z)$$

(4.101)

$$\times A(\omega_1) A^*(\omega_2) A(\omega_3) A^*(\omega_4)$$

$$\times A(\omega - \omega_1 + \omega_2 - \omega_3 + \omega_4) d\omega_1 d\omega_2 d\omega_3 d\omega_4$$

where $A(\omega) = A(\omega, 0)$, that is the amplitude envelop of the optical pulses at the input of the fiber. Taking the Fourier transform of Equation 4.3 and assuming $A(t, z)$ is of sinusoidal form we have

$$\frac{\partial A(\omega, z)}{\partial z} = G_1(\omega) A(\omega, z) \int_{-\infty}^{\infty}\int_{-\infty}^{\infty} G_3(\omega_1, \omega_2, \omega - \omega_1 + \omega_2) A(\omega_1, z) A^*(\omega_2, z)$$

$$\times A(\omega - \omega_1 + \omega_2, z) d\omega_1 d\omega_2$$

(4.102)

where

$$G_1(\omega) = -\frac{\alpha_0}{2} + j\beta_1\omega + j\frac{\beta_2}{2}\omega^2 - j\frac{\beta_3}{6}\omega_3$$

and $G_3(\omega_1, \omega_2, \omega_3) = j\gamma$. ω is taking the values over the signal bandwidth and beyond in overlapping the signal spectrum of other optically modulated carriers while $\omega_1..\omega_3$ are all also taking values over similar range as that of ω. For general expression the limit of integration is indicted over the entire range to infinitive.

Substituting Equation 4.101 into 4.102 and equating both sides, the kernels can be obtained after some algebraic manipulations

$$\frac{\partial}{\partial z}\Bigg[H_1(\omega,z)A(\omega)+\int_{-\infty}^{\infty}\int_{-\infty}^{\infty}H_3(\omega_1,\omega_2,\omega-\omega_1+\omega_2,z)A(\omega_1)A^*(\omega_2)A(\omega-\omega_1+\omega_2)d\omega_1d\omega_2$$

$$+\int_{-\infty}^{\infty}\int_{-\infty}^{\infty}\int_{-\infty}^{\infty}\int_{-\infty}^{\infty}H_5(\omega_1,\omega_2,\omega_3,\omega_4,\omega-\omega_1+\omega_2-\omega_3+\omega_4,z)$$

$$\times A(\omega_1)A^*(\omega_2)A(\omega_3)A^*(\omega_4)A(\omega-\omega_1+\omega_2-\omega_3+\omega_4)d\omega_1d\omega_2d\omega_3d\omega_4$$

$$=G_1(\omega)\Bigg[H_1(\omega,z)A(\omega)+\int_{-\infty}^{\infty}\int_{-\infty}^{\infty}H_3(\omega_1,\omega_2,\omega-\omega_1+\omega_2,z)$$

$$\times A(\omega_1)A^*(\omega_2)A(\omega-\omega_1+\omega_2)d\omega_1d\omega_2$$

$$+\int_{-\infty}^{\infty}\int_{-\infty}^{\infty}\int_{-\infty}^{\infty}\int_{-\infty}^{\infty}H_5(\omega_1,\omega_2,\omega_3,\omega_4,\omega-\omega_1+\omega_2-\omega_3+\omega_4,z)A(\omega_1)A^*(\omega_2)A(\omega_3)A^*(\omega_4)$$

$$\times A(\omega-\omega_1+\omega_2-\omega_3+\omega_4)d\omega_1d\omega_2d\omega_3d\omega_4\Bigg]+\int_{-\infty}^{\infty}\int_{-\infty}^{\infty}G_3(\omega_1,\omega_2,\omega-\omega_1+\omega_2)$$

$$\times\Bigg[H_1(\omega_1,z)A(\omega_1)+\int_{-\infty}^{\infty}\int_{-\infty}^{\infty}H_3(\omega_{11},\omega_{12},\omega_1-\omega_{11}+\omega_{12},z)$$

$$\times A(\omega_{11})A^*(\omega_{12})A(\omega_1-\omega_{11}+\omega_{12})d\omega_{11}d\omega_{12}$$

$$+\int_{-\infty}^{\infty}\int_{-\infty}^{\infty}\int_{-\infty}^{\infty}\int_{-\infty}^{\infty}H_5(\omega_{11},\omega_{12},\omega_{13},\omega_{14},\omega_1-\omega_{11}+\omega_{12}-\omega_{13}+\omega_{14},z)$$

$$\times A(\omega_{11})A^*(\omega_{12})A(\omega_{13})A^*(\omega_{14})\times A(\omega_1-\omega_{11}+\omega_{12}-\omega_{13}+\omega_{14})d\omega_{11}d\omega_{12}d\omega_{13}d\omega_{14}\Bigg]$$

$$\times\Bigg[H_1(\omega_1,z)A(\omega_1)+\int_{-\infty}^{\infty}\int_{-\infty}^{\infty}H_3(\omega_{11},\omega_{12},\omega_1-\omega_{11}+\omega_{12},z)$$

$$\times A(\omega_{11})A^*(\omega_{12})A(\omega_1-\omega_{11}+\omega_{12})d\omega_{11}d\omega_{12}$$

$$+\int_{-\infty}^{\infty}\int_{-\infty}^{\infty}\int_{-\infty}^{\infty}\int_{-\infty}^{\infty}H_5(\omega_{21},\omega_{22},\omega_{23},\omega_{24},\omega_2-\omega_{21}+\omega_{22}-\omega_{23}+\omega_{24},z)$$

$$\times A(\omega_{21})A^*(\omega_{22})A(\omega_{23})A^*(\omega_{24})$$

$$\times A(\omega_2-\omega_{21}+\omega_{22}-\omega_{23}+\omega_{24})d\omega_{21}d\omega_{22}d\omega_{23}d\omega_{24}\Bigg]^*$$

$$\times\Big[H_1(\omega-\omega_1+\omega_2,z)A(\omega-\omega_1+\omega_2)\Big]$$

$$+\int_{-\infty}^{\infty}\int_{-\infty}^{\infty}H_3(\omega_{31},\omega_{32},\omega-\omega_1+\omega_2-\omega_{31}+\omega_{32},z)$$

$$\times A(\omega_{31})A^*(\omega_{32})A(\omega-\omega_1+\omega_2-\omega_{31}+\omega_{32})d\omega_{31}d\omega_{32}$$

$$+\int_{-\infty}^{\infty}\int_{-\infty}^{\infty}\int_{-\infty}^{\infty}\int_{-\infty}^{\infty}H_5(\omega_{31},\omega_{32},\omega_{33},\omega_{34},\omega-\omega_1+\omega_2-\omega_{31}+\omega_{32}-\omega_{33}+\omega_{34},z)$$

$$\times A(\omega_{31})A^*(\omega_{32})A(\omega_{33})A^*(\omega_{34})$$

$$\times A(\omega-\omega_1+\omega_2-\omega_{31}+\omega_{32}-\omega_{33}+\omega_{34})\times d\omega_{31}d\omega_{32}d\omega_{33}d\omega_{34}$$

$$(4.103)$$

Equating the first order terms on both sides we obtain

$$\frac{\partial}{\partial z} H_1(\omega, z) = G_1(\omega) H_1(\omega, z) \tag{4.104}$$

Thus the solution for the first order transfer function, Equation 4.104, is then given by

$$H_1(\omega, z) = e^{G_1(\omega)z} = e^{\left(-\frac{\alpha_0}{2} + j\beta_1\omega + j\frac{\beta_2}{2}\omega^2 - j\frac{\beta_3}{6}\omega^3\right)z} \tag{4.105}$$

This is, in fact, the linear transfer function of an SMF with the dispersion factors β_2 and β_3 as already shown in the previous section.
Similarly for the third order terms we have

$$\frac{\partial}{\partial z} \int_{-\infty}^{\infty} \int_{-\infty}^{\infty} H_3(\omega_1, \omega_2, \omega - \omega_1 + \omega_2, z)$$

$$\times A(\omega_1) A^*(\omega_2) A(\omega - \omega_1 + \omega_2) d\omega_1 d\omega_2$$

$$= \int_{-\infty}^{\infty} \int_{-\infty}^{\infty} G_3(\omega_1, \omega_2, \omega - \omega_1 + \omega_2) H_1(\omega_1, z) A(\omega_1) H_2^*(\omega_2, z) \tag{4.106}$$

$$\times A(\omega_2) H_1(\omega - \omega_1 + \omega_2) A(\omega - \omega_1 + \omega_2) d\omega_1 d\omega_2$$

Now letting $\omega_3 = \omega - \omega_1 + \omega_2$ then it follows

$$\frac{\partial H_3(\omega_1, \omega_2, \omega_3, z)}{\partial z} = G_1(\omega_1 - \omega_2 + \omega_3) H_3(\omega_1, \omega_2, \omega_3, z)$$

$$+ G_3(\omega_1, \omega_2, \omega_3) H_1(\omega_1, z) H_1^*(\omega_2, z) H_1(\omega_3, z) \tag{4.107}$$

The third kernel transfer function can be obtained as

$$H_3(\omega_1, \omega_2, \omega_3, z) = G_3(\omega_1, \omega_2, \omega_3)$$

$$\times e^{\frac{\left(G_1(\omega_1) + G_1^*(\omega_2) + G_1(\omega_3)\right)z - e\, G_1(\omega_1 - \omega_2 + \omega_3)z}{G_1(\omega_1) + G_1^*(\omega_2) + G_1(\omega_3) - G_1(\omega_1 - \omega_2 + \omega_3)}} \tag{4.108}$$

The fifth order kernel can similarly be obtained as

$$H_5(\omega_1, \omega_2, \omega_3, \omega_4, \omega_5, z)$$

$$= \frac{H_1(\omega_1, z)H_1^*(\omega_2, z)H_1(\omega_3, z)H_1^*(\omega_4, z)H_1(\omega_5, z) - H_1(\omega_1 - \omega_2 + \omega_3 - \omega_4 + \omega_5, z)}{G_1(\omega_1) + G_1^*(\omega_2) + G_1(\omega_3) + G_1^*(\omega_4) + G_1(\omega_5) - G_1(\omega_1 - \omega_2 + \omega_3 - \omega_4 + \omega_5)}$$

$$\times \left[\frac{G_3(\omega_1, \omega_2, \omega_3 - \omega_4 + \omega_5)G_3(\omega_3, \omega_4, \omega_5)}{G_1(\omega_3) + G_1^*(\omega_4) + G_1(\omega_5) - G_1(\omega_3 - \omega_4 + \omega_5)} \right.$$

$$+ \frac{G_3(\omega_1, \omega_2 - \omega_3 + \omega_4, \omega_5)G_3^*(\omega_2, \omega_3, \omega_4)}{G_1^*(\omega_2) + G_1(\omega_3) + G_1^*(\omega_4) - G_1^*(\omega_2 - \omega_3 + \omega_4)}$$

$$\left. + \frac{G_3(\omega_1 - \omega_2 + \omega_3, \omega_4, \omega_5)G_3(\omega_1, \omega_2, \omega_3)}{G_1(\omega_1) + G_1^*(\omega_2) + G_1(\omega_3) - G_1(\omega_1 - \omega_2 + \omega_3)} \right]$$

$$- \frac{G_3(\omega_1, \omega_2, \omega_3 - \omega_4 + \omega_5)G_3(\omega_3, \omega_4, \omega_5)}{G_1(\omega_3) + G_1^*(\omega_4) + G_1(\omega_5) - G_1(\omega_3 - \omega_4 + \omega_5)}$$

$$\times \frac{H_1(\omega_1, z)H_1^*(\omega_2, z)H_1(\omega_1 - \omega_2 + \omega_3, z) - H_1(\omega_1 - \omega_2 + \omega_3 - \omega_4 + \omega_5, z)}{G_1(\omega_1) + G_1^*(\omega_2) + G_1(\omega_3 - \omega_4 + \omega_5) - G_1(\omega_1 - \omega_2 + \omega_3 - \omega_4 + \omega_5)} \quad (4.109)$$

$$- \frac{G_3(\omega_1, \omega_2 - \omega_3 + \omega_4, \omega_5)G_3^*(\omega_2, \omega_3, \omega_4)}{G_1^*(\omega_2) + G_1(\omega_3) + G_1^*(\omega_4) - G_1^*(\omega_2 - \omega_3 + \omega_4)}$$

$$\times \frac{H_1(\omega_1, z)H_1^*(\omega_2 - \omega_3 + \omega_4, z)H_1(\omega_5, z) - H_1(\omega_1 - \omega_2 + \omega_3 - \omega_4 + \omega_5, z)}{G_1(\omega_1) + G_1^*(\omega_2 - \omega_3 + \omega_4) + G_1(\omega_5) - G_1(\omega_1 - \omega_2 + \omega_3 - \omega_4 + \omega_5)}$$

$$- \frac{G_3(\omega_1 - \omega_2 + \omega_3, \omega_4, \omega_5)G_3(\omega_1, \omega_2, \omega_3)}{G_1(\omega_1) + G_1^*(\omega_2) + G_1(\omega_3) - G_1(\omega_1 - \omega_2 + \omega_3)}$$

$$\times \frac{H_1(\omega_1 - \omega_2 + \omega_3, z)H_1^*(\omega_4, z)H_1(\omega_5, z) - H_1(\omega_1 - \omega_2 + \omega_3 - \omega_4 + \omega_5, z)}{G_1(\omega_1 - \omega_2 + \omega_3)G_1(\omega_4) + G_1(\omega_5) - G_1(\omega_1 - \omega_2 + \omega_3 - \omega_4 + \omega_5)}$$

Higher order terms can be derived with ease if higher accuracy is required. However in practice such higher order would not exceed the fifth rank. We can understand that for a length of a uniform optical fiber the first to nth order frequency spectrum transfer can be evaluated indicating the linear to nonlinear effects of the optical signals transmitting through it. Indeed the third and fifth order kernel transfer functions based on the Volterra series indicate the optical filed amplitude of the frequency components which contribute to the distortion of the propagated pulses. An inverse of these higher order functions would give the signal distortion in the time domain. Thus the VSTFs allow us to conduct distortion analysis of optical pulses and hence an evaluation of the bit-error-rate of optical fiber communications systems.

The superiority of such Volterra transfer function expressions allow us to evaluate each effect individually, especially the nonlinear effects so that we can design and manage the optical communications systems under linear or nonlinear operations. Currently this linear-nonlinear boundary of operations is critical for system implementation, especially for optical systems operating at 40 Gb/s where linear operation and carrier suppressed return-to-zero format is employed. As a norm in series expansion the series needs to be converged to a final solution. It is this convergence that would allow us to evaluate the limit of nonlinearity in a system.

4.7.3 Transmission Bit Rate and the Dispersion Factor

The effect of dispersion on the system bit rate B_r is obvious and can be estimated by using the criterion:

$$B_r \cdot \Delta t < 1 \tag{4.110}$$

where $\Delta\tau$ is the total pulse broadening. When the fiber length, the total dispersion $D_T(= M(\lambda) + D(\lambda))$ and a source linewidth σ_λ the criterion becomes:

$$B_r \cdot L \cdot |D_T| \sigma_\lambda \le 1 \tag{4.111}$$

For a total dispersion factor of 1 ps/(nm · km) and a semiconductor laser of linewidth of 2–4 nm, the bit rate-length product cannot exceed 100 Gb/s-km. That is if a 100 km transmission distance is used then the bit rate cannot be higher than 1.0 Gb/s (Figures 4.29 and 4.30).

4.8 Fiber Nonlinearity

The nonlinear effects in optical fibers were described in Chapter 2. This section revisits these effects and their influence on the propagation of optical signals over long length of fibers. The nonlinearity and linear effects in optical fibers can be classified as shown in Figure 4.31.

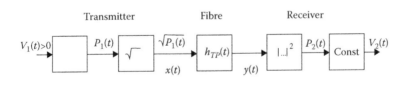

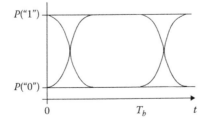

FIGURE 4.29
Schematic of an optical transmission system and its equivalent transfer functions.

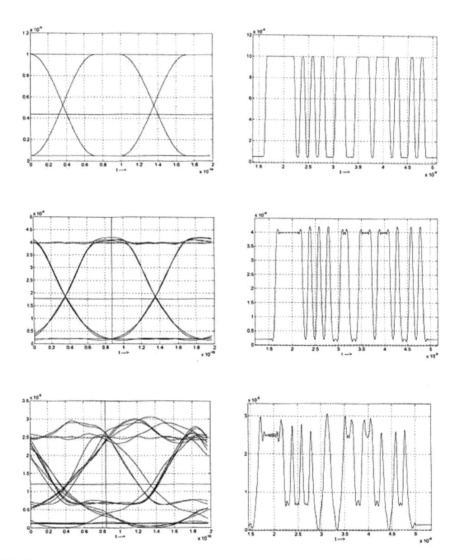

FIGURE 4.30
Eye diagram of time signals at 10 Gb/s transmission over SSMF after 0, 20, 80 km length.

Fiber RI is dependent on both operating wavelengths and lightwave intensity. This intensity-dependent phenomenon is known as the Kerr effect and is the cause of fiber nonlinear effects.

4.8.1 SPM, XPM Effects

The power dependence of refractive index (RI) is expressed as [3]:

$$n' = n + \bar{n}_2 \left(P / A_{eff} \right)$$

(4.112)

where P is the average optical power of the guided mode, $\bar{n}_2$ is the fiber nonlinear coefficient and A_{eff} is the effective area of the fiber.

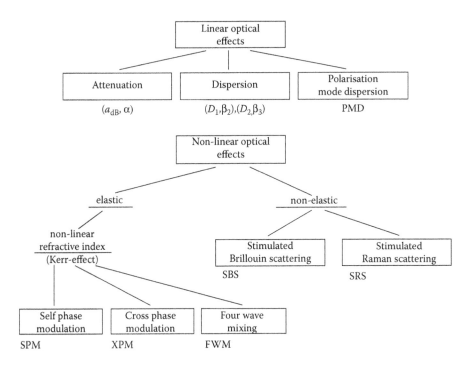

FIGURE 4.31
Linear and nonlinear fiber properties in SMFs.

Fiber nonlinear effects include intra-channel SPM, inter-channel XPM, FWM, SRS and SBS. SRS and SBS are not the main degrading factors as their effects are only getting noticeably large with very high optical power. On the other hand, FWM severely degrades the performance of an optical system with the generation of ghost pulses only if the phases of optical signals are matched with each other. However, with high local dispersions such as in SSMF, effects of FWM become negligible [3,31]. In terms of XPM, its effects can be considered to be negligible in a DWDM system in the following scenarios [32–37]: (i) highly locally dispersive system, and (ii) large channel spacing. However, XPM should be taken into account for optical transmission systems deploying NZDSF fiber where local dispersion values are small. Thus, SPM is usually the dominant nonlinear effect for systems employing transmission fiber with high local dispersions, e.g., SSMF and DCF. The effect of SPM is normally coupled with the nonlinear phase shift ϕ_{NL} defined as [3]

$$\phi_{NL} = \int_0^L \gamma P(z)\,dz = \gamma L_{eff} P$$

$$\gamma = \omega_c \bar{n}_2 / (A_{eff} c) \tag{4.113}$$

$$L_{eff} = \left(1 - e^{-\alpha L}\right) / \alpha$$

where ω_c is the lightwave carrier, L_{eff} is the effective transmission length and α is the fiber attenuation factor which normally has a value of 0.17 to 0.2 dB/km in the 1550 nm spectral window. The temporal variation of the nonlinear phase ϕ_{NL} results in the generation of new spectral components far apart from the lightwave carrier ω_c, indicating the broadening of the signal spectrum. This spectral broadening $\delta\omega$ can be obtained from the time dependence of the nonlinear phase shift as follows:

$$\delta\omega = -\frac{\partial\phi_{NL}}{\partial T} = -\gamma\frac{\partial P}{\partial T}L_{eff} \qquad (4.114)$$

Equation 4.114 indicates that $\delta\omega$ is proportional to the time derivative of the average signal power P. Additionally, the generation of new spectral components occur mainly at the rising and falling edges of optical pulses, i.e., the amount of generated chirps are larger for an increased steepness of the pulse edges.

The wave propagation equation can be represented as

$$\frac{\partial A(z,t)}{\partial z} + \frac{\alpha}{2}A(z,t) + \beta_1\frac{\partial A(z,t)}{\partial t} + \frac{j}{2}\beta_2\frac{\partial^2 A(z,t)}{\partial t^2} - \frac{1}{6}\beta_3\frac{\partial^3 A(z,t)}{\partial t^3}$$

$$= -j\gamma|A(z,t)|^2 A(z,t) - \frac{1}{\omega_0}\frac{\delta}{\delta t}\left(|A|^2 A\right) - T_R A\frac{\delta\left(|A|^2\right)}{\delta t} \qquad (4.115)$$

in which we have ignored the pure delay factor involving β_1. The last term in the right-hand side represents the Raman scattering effects.

4.8.2 Modulation Instability

The mutual effect between the nonlinear dispersion effects and the nonlinear effects can lead to the modulation of the lightwave pulses and thus unstable states of the optical pulses. This phenomenon is usually called the modulation instability, normally observed in soliton lasers.

The gain spectrum of the modulation instability is shown in Figure 4.32 [23].

4.8.3 Effects of Mode Hopping

Up to now we have assumed that the source center emission wavelength was unaffected by the modulation. In fact when a short current pulse is applied to a semiconductor laser, its center emission wavelength may hop from one mode to its neighbor a longer wavelength. In the case where a multi-longitudinal mode laser is used this hopping effect is negligible, however it is very significant for a single longitudinal laser.

4.9 Advanced Optical Fibers: Dispersion-Shifted, Flattened and Compensated Optical Fibers

At the beginning of the 1980s, there was great interest in reducing the total dispersion $[M(\lambda) + D(\lambda)]$ of SMF at 1550 nm where the loss is lowest for silica fiber. There were two

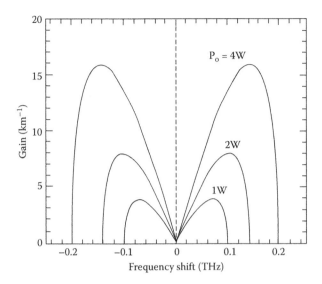

FIGURE 4.32
Spectrum of the optical gain due to modulation instability at three different average power level in an optical fiber with $\beta_2 = 20$ ps^2/km and $\gamma = 2$ W/km.

significant trends, one was to reduce the linewidth and to stabilize the laser center wavelength and the other was to reduce the dispersion at this wavelength. The fibers designed for long-haul transmission systems usually exhibit a near zero dispersion at a certain spectral window. These are dispersion shifted fibers, that is at this wavelength we prefer to have the total dispersion $= [M(\lambda) + D(\lambda)] \sim 0$. The material dispersion factor $M(\lambda)$ is natural and slightly affected by variation of doping material and concentration. However $D(\lambda)$ can be tailored, by designing appropriate refractive index profiles and geometrical structure, to balance the material dispersion effects. Note that the dispersion factors due to material and waveguide take algebraic values, thus they can be designed to take opposite values to cancel each other.

Advanced optical fiber design techniques can offer the design of dispersion flatten fibers where the dispersion factor is flat over the wavelength range from 1300 to 1600 nm by tailoring the refractive index profile of the core of optical fibers in such distribution as the W-profile, the segmented profile and multilayer core structure etc.

Question: What is the principal phenomenon for an optical fiber so that the dispersion characteristic is flattened over the wavelength range 1300–1550 nm?

Another type of optical fiber which would be required for compensating the dispersion effect of optical signal after transmission over a length of optical fiber is the dispersion compensated fiber, whose dispersion factor is many times larger than that of the standard communication fiber with an opposite sign. This can be designed by setting the total dispersion to the required compensated dispersion and thus the waveguide dispersion can be found over the required operating range. Thence optical fiber structures can be tailored.

4.10 Numerical Solution: Split Step Fourier Method

4.10.1 Symmetrical Split Step Fourier Method (SSFM)

The evolution of slow varying complex envelopes $A(z, t)$ of optical pulses along an SMF is governed by the NLSE [3]:

$$\frac{\partial A(z,t)}{\partial z} + \frac{\alpha}{2} A(z,t) + \beta_1 \frac{\partial A(z,t)}{\partial t} + \frac{j}{2}\beta_2 \frac{\partial^2 A(z,t)}{\partial t^2} - \frac{1}{6}\beta_3 \frac{\partial^3 A(z,t)}{\partial t^3}$$

$$= -j\gamma |A(z,t)|^2 A(z,t)$$

(4.116)

where z is the spatial longitudinal coordinate, α accounts for fiber attenuation, β_1 indicates DGD, β_2 and β_3 represent second and third order dispersion factors of fiber CD, and γ is the nonlinear coefficient. In a single channel transmission, Equation 4.116 includes the following effects: fiber attenuation, fiber CD and PMD, dispersion slope and SPM nonlinearity. Fluctuation of optical intensity caused by the Gordon-Mollenauer effect [38] is also included in this equation.

The solution of NLSE and hence the modeling of pulse propagation along an SMF is solved numerically by using SSFM [3]. In SSFM, fiber length is divided into a large number of small segments δz. In practice, fiber dispersion and nonlinearity are mutually interactive at any distance along the fiber. However, these mutual effects are small within δz and thus effects of fiber dispersion and fiber nonlinearity over δz are assumed to be statistically independent of each other. As a result, SSFM can separately define two operators: i) the linear operator that involves fiber attenuation and fiber dispersion effects, and ii) the non-linearity operator that takes into account fiber nonlinearities. These linear and nonlinear operators are formulated as follow:

$$\hat{D} = -\frac{j\beta_2}{2}\frac{\partial^2}{\partial T^2} + \frac{\beta_3}{6}\frac{\partial^3}{\partial T^3} - \frac{\alpha}{2}$$

$$\hat{N} = j\gamma |A|^2$$

(4.117)

where $j = \sqrt{-1}$; A replaces $A(z, t)$ for simpler notation and $T = t - z/v_g$ is the reference time frame moving at the group velocity. Equation 4.117 can be rewritten in a shorter form, given by:

$$\frac{\partial A}{\partial z} = (\hat{D} + \hat{N})A$$

(4.118)

and the complex amplitudes of optical pulses propagating from z to $z + \delta z$ are calculated using the following approximation

$$A(z+h,T) \approx \exp(h\hat{D})\exp(h\hat{N})A(z,T)$$

(4.119)

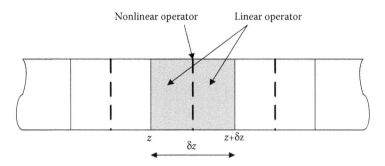

FIGURE 4.33
Schematic illustration of symmetric SSFM.

Equation 4.119 is accurate to the second order of the step size δz [3]. The accuracy of SSFM can be improved by including the effect of fiber nonlinearity in the middle of the segment rather than at the segment boundary (see Figure 4.33). This modified SSFM is known as the symmetric SSFM.

Equation 4.120 can now be modified as:

$$A(z+\delta z, T) \approx \exp\left(\frac{\delta z}{2}\hat{D}\right)\exp\left(\int_{z}^{z+\delta z}\hat{N}(z')dz'\right)\exp\left(\frac{\delta z}{2}\hat{D}\right)A(z,T) \qquad (4.120)$$

This method is accurate to the third order of the step size δz. In symmetric SSFM, the optical pulse propagates along a fiber segment δz in two stages. Firstly, the optical pulse propagates through the linear operator that has a step of $\delta z/2$ in which the fiber attenuation and dispersion effects are taken into account. Then, the fiber nonlinearity is calculated in the middle of the segment. After that, the pulse propagates through the second half of the linear operator. The process continues repetitively in consecutive segments of size δz until the end of the fiber. It should be highlighted that the linear operator is computed in frequency domain while the nonlinear operator is calculated in the time domain.

4.10.2 MATLAB® Program and MATLAB Simulink Models of the SSFM

A Matlab program is given below. This program performs the propagation of the optical signals along optical fiber transmission distance as shown in Figure 4.34. This program must be included in the folder storing the MATLAB Simulink model. In this folder an initialization program (Appendix 5) must be also included to set the data and parameters required for the Simulink model and subroutines. Furthermore the SSMF including Raman gain amplification effects is given in Appendix 4.

4.10.2.1 MATLAB Program

```
function output = ssprop_matlabfunction_modified(input)
nt = input(1);
u0 = input(2:nt+1);
dt = input(nt+2);
```

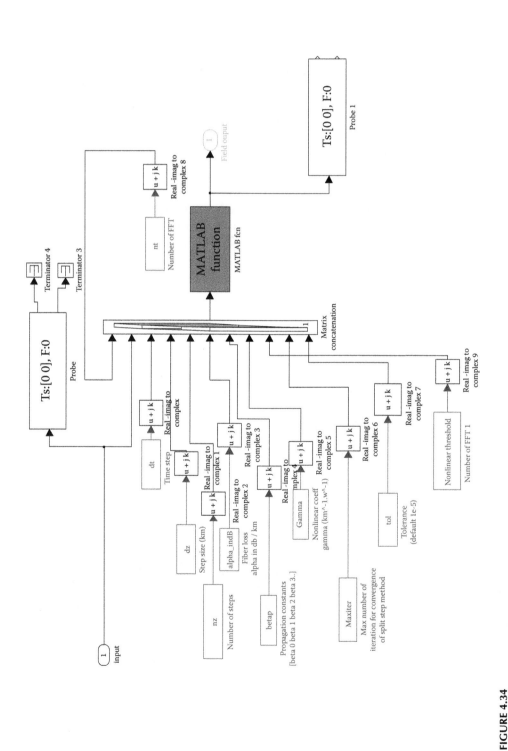

FIGURE 4.34

MATLAB Simulink model of the SSMF, MATLAB function includes the Matlab program, other inputs are data to pass to the Matlab program SSMF, usually defined in Initialization file (given in Appendix 4).

```
dz = input(nt+3);
nz = input(nt+4);
alpha_indB = input(nt+5);
betap = input(nt+6:nt+9);
gamma = input(nt+10);
P_non_thres = input(nt+11)
maxiter = input(nt+12);
tol = input(nt+13);
tic;
%tmp = cputime;
% This section solves the NLSE for pulse propagation in an optical fiber
using the SSF method
% The following effects are included: group velocity dispersion
% (GVD),higher order dispersion, loss, and self-phase modulation (gamma).
%
% USAGE
%
% u1 = ssprop(u0,dt,dz,nz,alpha,betap,gamma);
% u1 = ssprop(u0,dt,dz,nz,alpha,betap,gamma,maxiter);
% u1 = ssprop(u0,dt,dz,nz,alpha,betap,gamma,maxiter,tol);
%
% INPUT
%
% u0 - starting field amplitude (vector)
% dt - time step - [in ps]
% dz - propagation stepsize - [in km]
% nz - number of steps to take, ie, ztotal = dz*nz
% alpha - power loss coefficient [in dB/km], need to convert to linear to
% %have P=P0*exp(-alpha*z)
% betap - dispersion polynomial coefs, [beta_0 ... beta_m] [in ps^(m-1)/
km]
% gamma - nonlinearity coefficient [in (km^-1.W^-1)]
% maxiter - max number of iterations (default = 4)
% tol - convergence tolerance (default = 1e-5)
%% OUTPUT
%% u1 - field at the output
%
% Convert alpha_indB to alpha in linear domain
%---------------
alpha = log(10)*alpha_indB/10; % alpha (1/km)
%---------------

ntt = length(u0);

w = 2*pi*[(0:ntt/2-1),(-ntt/2:-1)]'/(dt*nt);
%w = 2*pi*[(ntt/2:ntt-1),(1:ntt/2)]'/(dt*ntt);

clear halfstep

halfstep = -alpha/2;
for ii = 0:length(betap)-1;
halfstep = halfstep - j*betap(ii+1)*(w.^ii)/factorial(ii);
end
```

```
clear LinearOperator
% Linear Operator in Split Step method
LinearOperator = halfstep;
% pause
halfstep = exp(halfstep*dz/2);


u1 = u0;
ufft = fft(u0);

% Nonlinear operator will be added if the peak power is greater than the
% Nonlinear threshold
iz = 0;
while (iz < nz) & (max((abs(u1).^2 + abs(u0).^2)) > P_non_thres)
iz = iz+1;

uhalf = ifft(halfstep.*ufft);

for ii = 1:maxiter,
uv = uhalf .* exp(-j*gamma*(abs(u1).^2 + abs(u0).^2)*dz/2);
ufft = halfstep.*fft(uv);
uv = ifft(ufft);

%fprintf('You are using SSFM\n');

if (max(uv-u1)/max(u1) < tol)
u1 = uv;
break;
else
u1 = uv;
end
end
if (ii == maxiter)
warning(sprintf('Failed to converge to %f in %d iterations',...
tol,maxiter));
end

u0 = u1;

end

if (iz < nz) & (max((abs(u1).^2 + abs(u0).^2)) < P_non_thres)

% u1 = u1.*rectwin(ntt);
ufft == fft(u1);
ufft = ufft.*exp(LinearOperator*(nz-iz)*dz);
u1 = ifft(ufft);
%fprintf('Implementing Linear Transfer Function of the Fiber
Propagation');
end
toc;
output = u1;
```

4.10.2.2 MATLAB Simulink Model

The MATLAB program (subroutine) is incorporated in the MALAB Simulink model for signal propagation. Under the mask of the block Matlab Function [ssprop_matlab-function_modified(input)] is the inclusion of the Matlab program subroutine as shown in Figure 4.35.

4.10.2.3 Modeling of Polarization Mode Dispersion

First-order PMD can be implemented by modeling the optical fiber as two separate paths representing the propagation of two PSPs. The symmetrical SSFM can be implemented on each polarized transmission path and then their outputs are superimposed to form the output optical field of the propagated signals. The transfer function to represent the first-order PMD is given by

$$H(f) = H^{+}(f) + H^{-}(f) \tag{4.121}$$

where

$$H^{+}(f) = \sqrt{k} \exp\left[j2\pi f\left(-\frac{\Delta\tau}{2} \right) \right] \tag{4.122}$$

and

$$H^{-}(f) = \sqrt{k} \exp\left[j2\pi f\left(-\frac{\Delta\tau}{2} \right) \right] \tag{4.123}$$

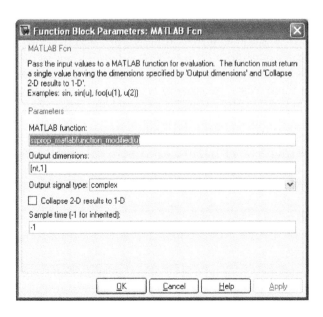

FIGURE 4.35
Screen of MATLAB Simulink model under mask of the Matlab function of the SSMF algorithm.

in which k is the power splitting ratio; $k = 1/2$ when a 3-dB or 50:50 optical coupler/splitter is used, $\Delta\tau$ is the instantaneous DGD value following a Maxwell distribution (refer to Equation 4.93) [24,25].

4.10.2.4 Optimization of Symmetrical SSFM

4.10.2.4.1 Optimization of Computational Time

A huge amount of time is spent in symmetric SSFM for fast Fourier transform (FFT) and the inverse FFT (IFFT) operations, in particular when fiber nonlinear effects are involved. In practice, when optical pulses propagate towards the end of a fiber span, the pulse intensity has been greatly attenuated due to the fiber attenuation. As a result, fiber nonlinear effects are getting negligible for the rest of that fiber span and hence, the transmission is operating in a linear domain in this range. In this research, a technique to configure symmetric SSFM is proposed in order to reduce the computational time. If the peak power of an optical pulse is lower than the nonlinear threshold of the transmission fiber, for example around −4 dBm, symmetrical SSFM is switched to a linear mode operation. This linear mode involves only fiber dispersions and fiber attenuation and its low-pass equivalent transfer function for the optical fiber is:

$$H(\varpi) = \exp\left\{-j\left[(1/2)\beta_2\varpi^2 + (1/6)\beta_3\varpi^3\right]\right\} \tag{4.124}$$

If β_3 is not considered in this fiber transfer function, which is normally the case due to its negligible effects on 40 Gb/s and lower bit rate transmission systems, the above transfer function has a parabolic phase profile [26,27].

4.10.2.4.2 Mitigation of Windowing Effect and Waveform Discontinuity

In symmetric SSFM, mathematical operations of FFT and IFFT play very significant roles. However, due to a finite window length required for FFT and IFFT operations, these operations normally introduce overshooting at two boundary regions of the FFT window, commonly known as the windowing effect of FFT. In addition, since the FFT operation is a block-based process, there exists the issue of waveform discontinuity, i.e., the right-most sample of the current output block does not start at the same position of the left-most sample of the previous output block. The windowing effect and the waveform discontinuity problems are resolved with the following solutions (see Figure 4.36).

The actual window length for FFT/IFFT operations consists of two blocks of samples (2N sample length). The output, however, is a truncated version with the length of one block (N samples) and output samples are taken in the middle of the two input blocks. The next FFT window overlaps the previous one by one block of N sample.

4.10.3 Remarks

The attenuation and dispersion of optical signals transmitted through silica optical fibers are described. Attenuation can be reduced by using the optical wavelength in the longer wavelength range. For example for silica fiber the preferred wavelength is at 1.55 μm. However natural forces are not kind to us and a dispersion factor of about 18 ps/(nm · km) generates pulse broadening for signal transmitted at this wavelength in a circular fiber.

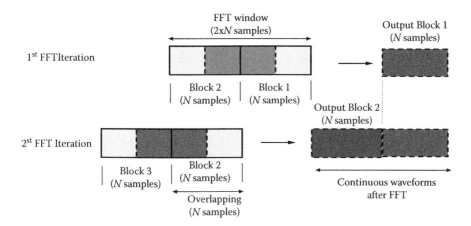

FIGURE 4.36
Proposed solution for mitigating windowing effect and waveform discontinuity caused by FFT/IFFT operations.

Longer wavelength carriers can be used in the mid-IR range of about 2.5 to 5 μm. At this wavelength range different kind of glasses must be used, such as chalcogenite type or fluoride type. Another technique presently used to compensate for the dispersion effect is to reduce the linewitdth of the lasers or by equalizing techniques such as spectrum inversion at the transmitter ends or at the center of the transmission length by optical filtering at the receiving end. Alternatively the optical fiber can be tailored to achieve dispersion shifted or flatten characteristics.

4.11 Appendix: MATLAB Program for the Design of Optical Fibers

```
%------------OPTICAL COMMUNICATIONS LABORATORY 1---------------%
% Copyright CRC Press
% clear after each run.

c=2.997925e8;

G1=0.711040;
G2=0.408218;
G3=0.704048;

lambda1=0.064270e-6;
lambda2=0.129408e-6;
lambda3=9.425478e-6;

a=4.1e-6;
delta=0.0025;

start=input('Enter lambda start point (nm) ---: ');
finish=input('Enter lambda end point (nm) -----: ');
resolution=input('Enter lambda resolution (nm) ----: ');
disp('');
```

```
lambda=start*1e-9;
lambdavector(1,1)=lambda;

for(i=1:(((finish-start)/resolution)+1))

 n1squared=1+((G1*power(lambda,2))/(power(lambda,2)-power(lambda1,2)))+((
G2*power(lambda,2))/(power(lambda,2)-power(lambda2,2)))+((G3*power(lam
bda,2))/(power(lambda,2)-power(lambda3,2)));
 n1=sqrt(n1squared);
 n1vector(1,i)=n1;

 n2=n1*(1+delta);
 n2vector(1,i)=n2;

 V=(2*pi/lambda)*a*n1*sqrt(2*delta);
 Vvector(1,i)=V;

 dy1dx=(-2*G1*power(lambda1,2)*lambda)/(power(power(lambda,2)-
power(lambda1,2),2));
 dy2dx=(-2*G2*power(lambda2,2)*lambda)/(power(power(lambda,2)-
power(lambda2,2),2));
 dy3dx=(-2*G3*power(lambda3,2)*lambda)/(power(power(lambda,2)-
power(lambda3,2),2));

 d2y1dx2=(2*G1*power(lambda1,2)*(3*power(lambda,2)+power(lambda1,2)))/
(power(power(lambda,2)-power(lambda1,2),3));
 d2y2dx2=(2*G2*power(lambda2,2)*(3*power(lambda,2)+power(lambda2,2)))/
(power(power(lambda,2)-power(lambda2,2),3));
 d2y3dx2=(2*G3*power(lambda3,2)*(3*power(lambda,2)+power(lambda3,2)))/
(power(power(lambda,2)-power(lambda3,2),3));

 d2ndx2=0.5*(((d2y1dx2+d2y2dx2+d2y3dx2)*power(n1,2)-
0.5*(power(dy1dx+dy2dx+dy3dx,2)))/power(n1,3));

 M=(-d2ndx2/c)*lambda;
 Mvector(1,i)=M; %row vector

 Dw=(-n2*delta)/c*(0.080+0.549*power(2.834-V,2))*(1/lambda);
 Dwvector(1,i)=Dw;

 if(i < (((finish-start)/resolution)+1))
 lambdavector(1,i+1)=lambdavector(1,i)+(resolution*1e-9);
 lambda=lambdavector(1,i+1);
 end
end

plot(lambdavector,Mvector,lambdavector,Dwvector,lambdavector,
Mvector+Dwvector);
grid;
```

%**Program Listings for Design of Standard Single Mode Fiber**
```
% totdisp_SMF.m
%
```

```
% MatLab script for calculating of total dispersion for
% Non-Zero Dispersion Shifted Fiber. The script plots
% the material dispersion, the waveguide dispersion, and
% the total dispersion for the designed fiber.
%
% Optical Fiber Design
%


lambda = [1.1:0.01:1.700]*1e-6;

G1=0.7028554;     %Sellmeier's coefficients for germanium
G2=0.4146307;     %doped silica (concentration B in table)
G3=0.8974540;
lambda1=0.0727723e-6;  %Wavelengths for germanium doped silica
lambda2=0.1143085e-6;
lambda3=9.896161e-6;

c = 299792458;    %Speed of light
pi = 3.1415926;   %Greek letter pi
a = 4.1e-6;   %Core radius
delta = 0.003;    %Greek letter delta (ref. index difference %between core
                  and cladding)

% Calculating the refractive index
% -----------------------------------

npow2oflambda = 1 + (G1.*lambda.^2./(lambda.^2.-lambda1*lambda1)) ...
 + (G2.*lambda.^2./(lambda.^2.-lambda2*lambda2)) ...
 + (G3.*lambda.^2./(lambda.^2.-lambda3*lambda3));

noflambda=sqrt(npow2oflambda);

pointer=find(lambda==1.550e-6);
n1=noflambda(pointer)      %Refractive index in the core

% Calculating the material dispersion
% -----------------------------------

t1 = diff(noflambda);
t2 = diff(lambda);
t3 = t1./t2;

t4 = diff(t3);
t5 = diff(lambda);
t5 = adjmat(t5);

lambda=adjmat(lambda);
lambda=adjmat(lambda);

%Material dispersion
Matdisp = - (lambda./c) .* (t4./t5);

% Converting to ps/nm.km
```

```matlab
Matdisp = Matdisp.*1e6;
figure(1)
clf
hold

xlabel('nm')
ylabel('ps/nm.km')
title('Standard Single Mode Fiber')
plot(lambda, Matdisp, '.-')
grid on

% Calculating waveguide dispersion
% -------------------------------- '

V = (2 * pi * a * n1 * sqrt(2 * delta)) ./(lambda);

Dlambda1 = - (n1 * delta)./(c * lambda);
%plot(lambda,Dlambda2)

Dlambda2 = 0.080 + 0.549 * (2.834 - V).^2;
%plot(lambda,Dlambda2)

Dlambda = Dlambda1 .* Dlambda2;

% Converting to ps/nm.km
Dlambda = Dlambda.*1e6;
plot(lambda,Dlambda, '-')

% Calculating total dispersion
% --------------------------------

TotDisp = Matdisp + Dlambda;
plot(lambda, TotDisp, ':')

legend('Material Dispersion', 'Waveguide Dispersion', 'Total Dispersion',
0)

% Finding the dispersion at 1460, 1550 and 1625 nm
% -------------------------------------------------

pointer=find(lambda==1.460e-6);
Disp1460=TotDisp(pointer)
pointer=find(lambda==1.550e-6);
Disp1550=TotDisp(pointer)
pointer=find(lambda==1.6250e-6);
Disp1625=TotDisp(pointer)
% refind_SMF.m
%
% MatLab script for calculating of possible values for
% refractive index in the core, n1, and cladding, n2,
% its corresponding relative refractive index.
%
```

```
% PROJECT DESIGN: Optical Fiber Design
%

lambda = [1.1:0.01:1.700]*1e-6;

n1 = [1.0487:0.001:1.8587];  %Refractive index of the core
n2 = [1.0435:0.001:1.8535];  %Refractive index of the cladding

% Calculating the refractive index for the index profile
% -------------------------------------------------------

delta = (n1 - n2) ./ n1;
deltap = delta * 100;

plot(n1, deltap)
grid on

xlabel('n1')
ylabel('Delta - Relative Refractive Index(%)')
title('Refractive Index')
```

% Program Listings of The Design of Nonzero Dispersion-Shifted Fiber

```
% totdisp_NZDSF.m
%
% MatLab script for calculating of total dispersion for
% Non-Zero Dispersion Shifted Fiber. The script plots
% the material dispersion, the waveguide dispersion and
% the total dispersion of the designed fiber.
%
% PROJECT 1: Optical Fiber Design
%

lambda = [1.1:0.001:1.700]*1e-6;

G1=0.7028554;    %Sellmeier's coefficients for germanium
G2=0.4146307;    %doped silica (concentration B in table)
G3=0.8974540;

lambda1=0.0727723e-6;  %Wavelengths for germanium doped silica
lambda2=0.1143085e-6;
lambda3=9.896161e-6;

c = 299792458;   %Speed of light
pi = 3.1415926;  %Greek letter pi
a = 2.4e-6;      %Core radius
delta = 0.0043;  %Greek letter delta (ref. index difference %between
                 core and cladding)

% Calculating the refractive index
% --------------------------------

npow2oflambda = 1 + (G1.*lambda.^2./(lambda.^2.-lambda1*lambda1)) ...
  + (G2.*lambda.^2./(lambda.^2.-lambda2*lambda2)) ...
  + (G3.*lambda.^2./(lambda.^2.-lambda3*lambda3));
```

```
noflambda=sqrt(npow2oflambda);

pointer=find(lambda==1.550e-6);
n1=noflambda(pointer)    %Refractive index in the core @1550nm

% Calculating the material dispersion
% ---------------------------------

t1 = diff(noflambda);
t2 = diff(lambda);
t3 = t1./t2;

t4 = diff(t3);
t5 = diff(lambda);
t5 = adjmat(t5);

lambda=adjmat(lambda);
lambda=adjmat(lambda);

%Material dispersion
Matdisp = - (lambda./c) .* (t4./t5);

% Converting to ps/nm.km
Matdisp = Matdisp.*1e6;
figure(1)
clf
hold

xlabel('nm')
ylabel('ps/nm.km')
title('Non-Zero Dispersion Shifted Fiber')
plot(lambda, Matdisp, '.')
grid on

% Calculating waveguide dispersion
% ------------------------------

V = (2*pi*a*n1*sqrt(2*delta))./(lambda);

pointer=find(lambda==1.550e-6);
V1550=V(pointer)

Dlambda1 = - (n1 * delta)./(c * lambda);
Dlambda2 = 0.080 + 0.549 * (2.834 - V).^2;
Dlambda = Dlambda1 .* Dlambda2;

% Converting to ps/nm.km
Dlambda = Dlambda.*1e6;
plot(lambda,Dlambda, '-')

% Calculating total dispersion
% ------------------------------
```

```
TotDisp = Matdisp + Dlambda;
plot(lambda, TotDisp, '+')

legend('Material Dispersion', 'Waveguide Dispersion', 'Total Dispersion', 0)

% Finding the dispersion at 1460, 1550 and 1625 nm
% ------------------------------------------------

pointer=find(lambda==1.460e-6);
Disp1460=TotDisp(pointer)

pointer=find(lambda==1.550e-6);
Disp1550=TotDisp(pointer)

pointer=find(lambda==1.6250e-6);
Disp1625=TotDisp(pointer)

% refine_NZDSF.m
%
% MatLab script for calculating of possible values for
% refractive index in the core, n1, and cladding, n2,
% its corresponding relative refractive index.
%

lambda = [1.1:0.01:1.700]*1e-6;

n1 = [1.0487:0.001:1.8587];   %Refractive index of the core
n2 = [1.0324:0.001:1.8424];   %Refractive index of the cladding

% Calculating the refractive index for the index profile
% -------------------------------------------------------

delta = (n1 - n2) ./ n1;
deltap = delta * 100;

plot(n1, deltap)
grid on

xlabel('n1')
ylabel('Delta - Relative Refractive Index(%)')
title('Refractive Index')
```

4.12 Program Listings of the Split Step Fourier Method with Self Phase Modulation and Raman Gain Distribution

```
function output = ssprop_matlabfunction_raman(input)

nt = input(1);
u0 = input(2:nt+1);
```

```
dt = input(nt+2);
dz = input(nt+3);
nz = input(nt+4);
alpha_indB = input(nt+5);
betap = input(nt+6:nt+9);
gamma = input(nt+10);
P_non_thres = input(nt+11);
maxiter = input(nt+12);
tol = input(nt+13);
%Ld = input(nt+14);
%Aeff = input(nt+15);
%Leff = input(nt+16);

tic;
%tmp = cputime;


%-------------------------------------------------------------
% Original author: Thomas E. Murphy (tem@alum.mit.edu)
% Adapted and modified by Thanh Liem Huynh (thanh.huynh@eng.monash.edu.
au)
%-------------------------------------------------------------
% This function ssolves the nonlinear Schrodinger equation for
% pulse propagation in an optical fiber using the split-step
% Fourier method described in:
%
% Agrawal, Govind. Nonlinear Fiber Optics, 2nd ed. Academic
% Press, 1995, Chapter 2
%
% The following effects are included in the model: group velocity
% dispersion (GVD), higher order dispersion, loss, and self-phase
% modulation (gamma). Raman gain as a distributed amplification
%
% USAGE
%
% u1 = ssprop(u0,dt,dz,nz,alpha,betap,gamma);
% u1 = ssprop(u0,dt,dz,nz,alpha,betap,gamma,maxiter);
% u1 = ssprop(u0,dt,dz,nz,alpha,betap,gamma,maxiter,tol);
%
% INPUT
%
% u0 - starting field amplitude (vector)
% dt - time step - [in ps]
% dz - propagation stepsize - [in km]
% nz - number of steps to take, ie, ztotal = dz*nz
% alpha - power loss coefficient [in dB/km], need to convert to linear to
have P=P0*exp(-alpha*z)
% betap - dispersion polynomial coefs, [beta_0 ... beta_m] [in ps^(m-1)/
km]
% gamma - nonlinearity coefficient [in (km^-1.W^-1)]
% maxiter - max number of iterations (default = 4)
% tol - convergence tolerance (default = 1e-5)
%
% OUTPUT
%
```

```
% u1 - field at the output
%---------------
% Convert alpha_indB to alpha in linear domain
%---------------
alpha = 1e-3*log(10)*alpha_indB/10; % alpha (1/km) - see Agrawal p57
%---------------
%P_non_thres = 0.0000005;

ntt = length(u0);
w = 2*pi*[(0:ntt/2-1),(-ntt/2:-1)]'/(dt*nt);
%t = ((1:nt)'-(nt+1)/2)*dt;

gain = numerical_gain_hybrid(dz,nz);

for array_counter = 2:nz+1
 grad_gain(1) = gain(1)/dz;
 grad_gain(array_counter) = (gain(array_counter)-gain(array_counter-1))/
dz;
end
gain_lin = log(10)*grad_gain/(10*2);

clear halfstep
 halfstep = -alpha/2;
 for ii = 0:length(betap)-1;
 halfstep = halfstep - j*betap(ii+1)*(w.^ii)/factorial(ii);
 end

 square_mat = repmat(halfstep, 1, nz+1);
 square_mat2 = repmat(gain_lin, ntt, 1);
 size(square_mat);
 size(square_mat2);
 total = square_mat + square_mat2;

clear LinearOperator
 % Linear Operator in Split Step method
 LinearOperator = halfstep;
 halfstep = exp(total*dz/2);

u1 = u0;
ufft = fft(u0);

% Nonlinear operator will be added if the peak power is greater than the
% Nonlinear threshold
iz = 0;
while (iz < nz) && (max((gamma*abs(u1).^2 + gamma*abs(u0).^2)) > P_non_
thres)
 iz = iz+1;

 uhalf = ifft(halfstep(:,iz).*ufft);

 for ii = 1:maxiter,
 uv = uhalf .* exp((-j*(gamma)*abs(u1).^2 + (gamma)*abs(u0).^2)*dz/2);
 ufft = halfstep(:,iz).*fft(uv);
 uv = ifft(ufft);
```

```
if (max(uv-u1)/max(u1) < tol)
u1 = uv;
break;
else
u1 = uv;
end


end
% fprintf('You are using SSFM\n');
if (ii == maxiter)

fprintf('Failed to converge to %f in %d iterations',tol,maxiter);
end

 u0 = u1;

end

if (iz < nz) && (max((gamma*abs(u1).^2 + gamma*abs(u0).^2)) < P_non_
thres)

% u1 = u1.*rectwin(ntt);
 ufft = fft(u1);
 ufft = ufft.*exp(LinearOperator*(nz-iz)*dz);
 u1 = ifft(ufft);

 %fprintf('Implementing Linear Transfer Function of the Fiber
Propagation');
end

%toc;

output = u1;
```

4.13 Program Listings of an Initialization File (Linked with Split Step Fourier Method of Section 4.12)

```
% This file initialization file - declaring all parameters and data
required for Simulink model and Split Step Fourier

clear all
close all

% CONSTANTS

c = 299792458; % speed of light (m/s)

% NUMERICAL PARAMETERS
```

```
numbitspersymbol = 1
P0 = 0.003; % peak power (W)
FWHM = 25 % pulse width FWHM (ps)
%halfwidth = FWHM/1.6651 % for Gaussian pulse
halfwidth = FWHM % for square pulse

bitrate = 1/halfwidth; % THz
baudrate = bitrate/numbitspersymbol;
signalbandwidth = baudrate;

%%%%%%%%%%%%%%%%%%%%%%%%%%%%%%%%%%%
% for DPSK
Vpi=5;
halfVpi = Vpi/2;
twoVpi=Vpi*2;

% nt = 2^8; % number of points in FFT
PRBSlength = 2^5;

% Make sure : FFT time window (=nt*dt) = PRBSlength * FWHM...
% FFTlength nt = PRBSlength/block * numbersamples/bit = PRBSlength *
(FWHM/dt)
% num_samplesperbit = FWHM/dt should be about 8 - 16 samples/bit
num_samplesperbit = 32; % should be 2^n
dt = FWHM/num_samplesperbit ; % sampling time(ps); % time step (ps)
nt = PRBSlength*num_samplesperbit; % FFT length

% nt = 2^9;
% nt =num_samplesperbit;

dz = 0.2; % distance stepsize (km)
nz = 500;

%melbourne to gippsland
%170km two spans
nz_MelbToGipps = 500;

%undersea link
%290km
nz_Raman = 250;
nz_undersea = 950;
nz_DCF = 145;

%George Town to Hobart
nz_GtownToHobart = 500;

% number of z-steps
maxiter = 10; % max # of iterations
tol = 1e-5; % error tolerance

% OPTICAL PARAMETERS

nonlinearthreshold = 0.010; % 10mW -- % Nonlinear Threshold Peak Power
```

```
lambda = 1550; % wavelength (nm)
optical_carrier = c/(lambda*1e-9);
%dBperkm = 0.2; % loss (dB/km)
alpha_indB = 0.17; % loss (dB/km)
D = 18.5; % GVD (ps/nm.km); if anomalous dispersion(for compensation),D
is negative
beta3 = 0.06; % GVD slope (ps^3/km)


ng = 1.46; % group index
n2 = 2.6e-20; % nonlinear index (m^2/W)
Aeff = 76; % effective area (um^2)

% CALCULATED QUANTITIES

T = nt*dt; % FFT window size (ps) -Agrawal: should be about 10-20 times
of the pulse width
alpha_loss = log(10)*alpha_indB/10; % alpha (1/km)
beta2 = -1000*D*lambda^2/(2*pi*c); % beta2 (ps^2/km);

%------------------------------------------------------------
% beta 3 can be calculated from the Slope Dispersion (S) as follows:]
% Slope Dispersion
% S = 0.092; % ps/(nm^2.km)
% beta31 = (S - (4*pi*c./lambda.^3))./(2*pi*c./lambda.^2)
%------------------------------------------------------------
gamma = 2e24*pi*n2/(lambda*Aeff); % nonlinearity coef (km^-1.W^-1)
t = ((1:nt)'-(nt+1)/2)*dt; % vector of t values (ps)
t1 = [(-nt/2+1:0)]'*dt; % vector of t values (ps)
t2 = [(1:nt/2)]'*dt; % vector of t values (ps)

w = 2*pi*[(0:nt/2-1),(-nt/2:-1)]'/T; % vector of w values (rad/ps)
v = 1000*[(0:nt/2-1),(-nt/2:-1)]'/T; % vector of v values (GHz)
vs = fftshift(v); % swap halves for plotting
v_tmp = 1000*[(-nt/2:nt/2-1)]'/T;

% STARTING FIELD

% P0 = 0.001 % peak power (W)
% FWHM = 20 % pulse width FWHM (ps)
%halfwidth = FWHM/1.6651 % for Gaussian pulse

%For square wave input, the FWHM = Half Width
%halfwidth = FWHM;

L = nz*dz

Lnl = 1/(P0*gamma) % nonlinear length (km)
Ld = halfwidth^2/abs(beta2) % dispersion length (km)
N = sqrt(abs(Ld./Lnl)) % governing the which one is dominating:
dispersion or Non-linearities
ratio_LandLd = L/Ld % if L << Ld --> NO Dispersion Effect
ratio_LandLnl = L/Lnl % if L << Lnl --> NO Nonlinear Effect
```

```
% Monitor the broadening of the pulse with relative the Dispersion Length
% Calculate the expected pulsewidth of the output pulse
% Eq 3.2.10 in Agrawal "Nonlinear Fiber Optics" 2001 pp67
FWHM_new = FWHM*sqrt(1 + (L/Ld)^2)

% N<<1 --> GVD ; N >>1 ---> SPM
Leff = (1 - exp(-alpha_loss*L))/alpha_loss
expected_normPout = exp(-alpha_loss*2*L)
NlnPhaseshiftmax = gamma*P0*Leff

betap = [0 0 beta2 beta3]';

% Constants for ASE of EDFA
% PSD of ASE: N(at carrier freq) = 2*h*fc*nsp*(G-1) with nsp = Noise
% Figure/2 (assume saturated gain)
%*************** Standdard Constant ****************************
h = 6.626068e-34; %Plank's Constant
%*******************************************
```

4.14 Problems

1. a. Give a brief account of the optical transmission loss of silica fibers for optical communications systems as a function of the operating wavelength in the C-band, L-band, and S-band wavelength regions.

   b. The refractive index of the cladding silica can be approximated as $n(\lambda)) = c_1 + c_2\lambda^2 + c_3\lambda^{-2}$. Show that the material dispersion factor is zero at the wavelength given by $\lambda^4 = -\dfrac{3c_3}{c_2}$. The coefficients c_1, c_2, and c_3 for pure and GeO$_2$-doped silica fiber are:

Coefficients	Pure Silica	7.9% GeO$_2$-doped Silica
c_1	1.45084	1.46286
c_2 in μm^{-2}	– 0.00334	– 0.00331
c_3 in μm^2	0.00292	0.00320

   Find the zero dispersion wavelength of the material dispersion curve of the GeO$_2$-doped silica fiber. Hence, estimate the material dispersion factor for silica fiber at 1550 nm wavelength.

   c. Give an expression of the waveguide dispersion factor as a function of b and V where b is the normalized propagation constant and V is the normalized frequency. Design a non-zero dispersion shifted single mode optical fiber at 1550 nm where its total dispersion factor is +2.0 ps/nm/km at this wavelength. It is recommended that the following parameters should be specified for the designed fiber: the fiber diameter, the relative refractive index difference, the cutoff wavelength, and the total dispersion at the desired wavelength of operation.

2. An optical fiber with a step index profile, a core diameter of 62.5 micrometers, and a numerical aperture of 0.2 at a wavelength of 1300 nm is used for signal distribution and transmission in a local area network.

   a. What is the *V*-parameter of this optical fiber?

   b. How many guided modes would it support. Can you comment on this number regarding the velocities of lightwaves?

   c. Select a cladding diameter. Give reasons for your selection.

   d. Find the maximum acceptance angle of this fiber. Estimate the coupling loss of a laser source with a uniform radiation cone of 30 degrees.

3. An optical fiber has the following parameters: index profile = step-like; core diameter = 9.0 µm; numerical aperture NA = 0.11; and a cladding refractive index = 1.48.

   a. Find the normalized frequency of the fiber at 1550 nm wavelength.

   b. Is the fiber operating in the single mode or multimode region at 1550 nm? If it is in the single mode region, estimate its mode field diameter and its spot size. Sketch its field and intensity distribution across the fiber cross section.

   c. Find the cutoff wavelength of the fiber. If lightwaves of a smaller wavelength than this cutoff wavelength are launched into the fiber, is the fiber still operating in the single mode region?

4. A single mode step index optical fiber has the following parameters: core diameter = 8.0 µm; cladding diameter = 0.125 mm; core refractive index = 1.460; and relative index difference = 0.2% at 1550 nm.

   a. Confirm that the fiber can be operating in the single mode region at 1550 nm wavelength.

   b. Find the fiber cutoff wavelength.

   c. What is the fiber mode field diameter if it is operating at 1550 nm wavelength.

5. For the optical fiber in Problem 2.2B, if the refractive index profile is parabolic ($\alpha = 2$) or triangular ($\alpha = 1$) with the above numerical aperture at the central axis, repeat (a), (b), and (c).

6. Single mode optical fibers produced by Corning (and Optical Waveguides Pty. Ltd., Noble Park, Victoria, Australia) have typical characteristics as per technical data sheet attached at the end of this chapter.

   a. Using the fiber's physical characteristics and technical data on its numerical aperture, confirm the fiber functional characteristics such as the cutoff wavelength range, and so on.

   b. If this fiber is launched with an 850 nm laser, how many modes would it support? Sketch the mode fields for LP_{01} and LP_{11} modes.

   c. If lightwaves at 1300 nm travel over 10 km of this fiber, calculate the travel time of these waves.

   d. Estimate the fiber mode field diameter at 1300 nm wavelength.

   e. If the same spot size of (d) is required for the fiber to operate at 1550 nm, can you advise the manufacturer on any change of the fiber physical parameters?

7. a. The optical fiber in Problem 2(c) is used in an optical fiber transmission system with a laser source operating at 1310 nm and with an output power of 1.0 mW. The fiber length is 50 km. The optical receiver can detect an average optical power of 0.1 µW. Is it possible to detect the optical power at the end of the fiber length?

b. Referring to the technical data of the standard optical fiber, estimate the spreading of optical pulse after transmitting through the 50 km length of fiber if the source has an optical line width of 2.0 nm.

8. A step index optical fiber used for an optical communications system operating at 1310 nm has a core radius of 25 μm, and refractive indices in the core and cladding regions of 1.460 and 1.4550 respectively.

   a. What is the numerical aperture of the fiber?

   b. Estimate the number of guided modes.

9. a. Show that for a graded index fiber having a core refractive index

   $$n^2(r) = n_2^2\left[1+2\Delta s\left(\frac{r}{a}\right)\right]$$ with $s(r/a) = 1 - (r/a)^\alpha$ the acceptance angle $\alpha(r)$ is given by

   $$\sin\alpha(r) = \left[n^2(r) - n_2^2\right]^{1/2}.$$

   b. If the optical fiber has a parabolic profile shape, show that

   $$\sin\alpha(r) = NA\sqrt{1-\left(\frac{r}{a}\right)^{1/2}}, \quad \text{where } NA = n_2(2\Delta)^{1/2}(1+\Delta).$$

   c. A parabolic graded index silica optical fiber has a cladding refractive index of 1.460 and a relative index difference at the core axis of 1%. Find the maximum acceptance angle at the core axis of the fiber. Plot $\sin\alpha(r)$ as a function of r. What is the acceptance angle of the fiber at the core and cladding interface? Comment on the launch of a laser source into this fiber.

10. a. For a single mode optical fiber having a graded index central dip – that is, $s(r/a)$ $= 1 - (1 - r/a)^\alpha$ the ESI parameters of V, and the radius are given by:

    $$\frac{V_e}{V} = \left[1 - \frac{2\gamma}{(\alpha+1)(\alpha+2)}\right]^{1/2} \quad \text{and} \quad \frac{a_e}{a} = \frac{(\alpha+1)(\alpha+2)(\alpha+3)-6\gamma}{(\alpha+1)(\alpha+2)(\alpha+3)-2\gamma},$$

    where $V = ka(2\Delta)^{1/2}$.

    b. The fiber has a physical core radius of 8.0 μm, a maximum relative index difference of 0.3%, and a cladding refractive index of 1.460. Find its ESI parameters for the normalized frequency and radius at 1550 nm wavelength. Also, find its ESI cutoff wavelength, and its mode field diameter at this wavelength.

11. What are the wavelength ranges of infrared light, ultraviolet light, and far infrared light? What are the approximate wavelengths of the colors in the color band of resistors? Do they correspond with the colors of the rainbow?

12. A GeO$_2$-doped silica-based optical fiber has the following parameters: (i) step-index-profile, (ii) refractive index difference at the core of 0.5%, and (iii) core diameter of 9.0 micrometers.

    a. Calculate the refractive index of the fiber core and cladding at wavelengths of 1.310 and 1.55 micrometers respectively.

    b. What is the estimated loss of this fiber at the above wavelengths?

c. What are the *V*-parameters of the fiber at these wavelengths?

d. What are the material dispersion and waveguide dispersion factors at these wavelengths? What are the total dispersion factors?

e. This fiber is to be used in optical systems of bit rates of 2.2 Gb/sec. What is the maximum fiber length that the signal can be transmitted without suffering the allowable signal degradation?

The material dispersion factor is given as:

$$M(\lambda) = -\frac{\lambda}{c}\frac{d^2 n}{d\lambda^2}$$

with the refractive index as a function of wavelength given as.

$$n^2(\lambda) = 1 + \sum_k \frac{G_k \lambda^2}{(\lambda^2 - \lambda_k^2)}$$

Thus, with different doping concentration of the impurities, one can measure the Sellmeier coefficients and the refractive index, thence the dispersion factor. The waveguide dispersion factor is given as:

$$D(\lambda) = -\frac{n_2(\lambda)\Delta}{c\lambda} V \frac{d^2(Vb)}{dV^2}.$$

The factor $V\frac{d^2(Vb)}{dV^2}$ can be approximated for single mode optical fiber as:

$$V\frac{d^2(Vb)}{dV^2} \cong 0.080 + 0.549(2.834 - V)^2.$$

Therefore, it is now a matter of substituting the parameters into these equations in order to determine the materials and waveguide dispersion curves which are similar to those given in Figure 4.6.

13. a. Give a brief account of the pros and cons for optical fiber communications systems operating at 810 nm, 1300 nm, and 1550 nm wavelength regions.

b. Why does silica optical fiber become very lossy at a 1400 nm wavelength region?

c. What are the typical optical fiber losses at the above wavelength regions? Give typical cable losses.

14. a. Show that the material dispersion factor is zero at the wavelength given by

$$\lambda^4 = -\frac{3c_3}{c_2}.$$

b. The coefficients c_1, c_2, and c_3 for pure and GeO_2-doped silica fiber are:

Coefficients c's	Pure Silica	7.9% GeO$_2$-doped Silica
c_1	1.45084	1.46286
c_2 in μm^{-2}	-0.00334	-0.00331
c_3 in μm^2	0.00292	0.00320

Find the zero dispersion wavelengths due to material of these fibers.

c. Derive an expression for the group delay per km unit length. Plot this group delay versus wavelength for Part (b).

d. Find the transit time difference of lightwaves propagating through the fiber emitted by light sources centered at 810 nm and 1550 nm with a line width of 10 nm.

15. Using the approximate expression for the normalized propagation constant b as a function of V, derive the group velocity delay due to waveguide, and hence the dispersion factor due to waveguide as a function of V.

16. Using the data of the SMF optical fiber manufactured by Corning, calculate:

a. The material dispersion factor, and

b. The waveguide dispersion factor

at 1330 nm and 1550 nm, respectively.

17. Design dispersion-shifted single mode optical fibers with the dispersion-shifted zero at 1550 nm. Each group of students (three to four) are requested to select a particular group of doped material with different Sellmeier coefficients for the design.

18. a. Sketch the geometrical structure and the refractive index profile of circular optical fibers where the core diameter is 10 micron and the outside diameter is 125 micron. The refractive index is constant in the core of 1.48 at 1550 nm wavelength in free space, and cladding is 1.477. Make sure that you have the coordinate system defined for the transverse plane and the propagation direction.

b. What is the frequency of the optical wave at 1550 nm? What is the color of the lightwave at this wavelength?

c. What is the phase velocity of the lightwaves through the glass material whose refractive index is the same as that of the core material? What is the wavelength of the lightwave inside the core along the propagation direction of the wave which is not necessarily the z-direction?

d. What is the condition for the lightwave to be weakly guided along a fiber?

19. a. State the expression of the V-parameter. Comment on the dependence of the V-parameter with respect to the core radius, the relative refractive index difference, and the operating wavelength of the lightwave.

b. State the condition for the V-parameter so that it supports only one fundamental mode LP$_{01}$.

c. A silica fiber consists of a core region of Ge:doped silica with a refractive index of 1.488 at 1550 nm, and a cladding region of pure silica of 1.482 at the same wavelength. Find the relative refractive index difference of the fiber. Calculate the V-parameter of the fiber if the core diameter is 8 micron and the operating wavelength is 1550 nm. Does the fiber operate in the single mode region or multimode region?

   d. What is the *V*-parameter of the fiber if it is excited with lightwaves of 633 nm wavelength? How many modes does the fiber support?

20. a. Write down the wave equation of lightwaves guided in a circular optical fiber whose coordinate system must be cylindrical. You may assume that the solutions of the wave equation follows a Bessel solution as the oscillating waves inside the core, and exponential decay in the cladding. The phase propagation along the z-direction and time dependent can be easily found.

   b. Apply the boundary condition to obtain the eigenvalue equation of the wave propagation. State the condition for the *V*-parameter for the optical fiber to operate in the single mode region.

21. Refer to the technical specification of the Corning fiber SMF-28.

   a. Extract the values of the relative refractive index difference, the core diameter, and the cladding diameter.

   b. Estimate the *V*-parameter of this fiber and confirm that the fiber is operating in the single mode region at 1550 nm wavelength.

   c. Estimate the mode spot size or the mode field diameter of this fiber using the theoretical formulae and check against the value given in the specification.

   d. Sketch the field distribution in the transverse plane of the fiber.

   e. Find the attenuation curve as a function of wavelength. Give reasons why the attenuation is high in a number of spectral regions.

   f. A lightwave source with an optical power of +10 dBm at 1550 nm wavelength is launched into 100 km of the SMF-28 optical fiber. What is the power (in dBm and mW) of the lightwaves at the output of this fiber strand?

22. This problem relates to the attenuation and dispersion characteristics of optical fibers

   a. Write down the approximate dispersion factor due to materials of the standard single mode optical fiber. Estimate this factor at a wavelength of 1552 nm.

   b. Write down the expression of the waveguide dispersion factor as a function of the *V*-parameter and the normalized propagation constant *b* of the single mode optical fiber. Use the approximation of *b* with respect to *V*; obtain an expression for the dispersion factor due to the guiding of the fundamental mode LP_{01}.

   c. If the total of the single mode fiber is to be designed to be less than 5 ps/nm/km, estimate the value of *V*. Calculate the radius of the fiber if a relative refractive index difference of 0.003 can be fabricated.

   d. What is the cutoff wavelength and the mode spot size of the fiber designed in (c)?

See solution in Problem Set 4-Subset 2.

23. a. Sketch the attenuation of silica-based glass fiber as a function of wavelength. Indicate the theoretical approximation and relevant scattering and/or absorption phenomena in these fibers.

   b. What are the attenuation factors at the wavelength regions 1300 nm, 1380 nm, and 1530–1565 nm? Specify these factors in both log and linear scales.

   c. A lightwave source with an optical power of +10 dBm (i.e., 10 mW) of 1550 nm wavelength is launched into 100 km of optical fibers of (a) and (b). What is the power of the lightwaves at the output of this fiber strand?

d. The total dispersion factor of the fiber is +17 ps/nm/km. The lightwave source is now modulated with 40 Gb/sec ON-OFF keying pulse sequence.

   i. Sketch the signal power spectrum—refer to materials of ECE2011—signal processing. Then estimate the 3dB bandwidth (in nm) of the signal bands.

   ii. The group of waves in (i) is propagating through the fiber; what is the broadening of the pulse sequence?

   iii. Show how you could compensate for this broadening effect of the optical signals after 100 km fiber propagation.

24. Optical fibers

e. Write down the approximate dispersion factor due to materials of the G.655 Corning LEAF fiber. Estimate this factor at a wavelength of 1552 nm.

f. Write down the expression of the waveguide dispersion factor as a function of the V-parameter and the normalized propagation constant b of the single mode optical fiber. Use the approximation of b with respect to V; obtain an expression for the dispersion factor due to the guiding of the fundamental mode LP_{01}.

g. If the total dispersion factor of the single mode fiber is to be designed to be less than 5 ps/nm/km, estimate the value of V. Calculate the radius of the fiber if a relative refractive index difference of 0.003 can be fabricated.

h. What is the cutoff wavelength and the mode spot size of the fiber designed in (c)?

25. e. Sketch the attenuation of silica-based glass fiber of the Corning Fiber SMF-28e as a function of wavelength. Indicate the theoretical approximation and relevant scattering and/or absorption phenomena in these fibers.

f. What are the attenuation factors at the wavelength regions 1300 nm, 1380 nm, and 1530–1565 nm? Specify these factors in both log and linear scales.

g. A lightwave source with an optical power of +10 dBm (i.e., 10 mW) of 1550 nm wavelength is launched into 100 km of optical fibers of (a) and (b). What is the power of the lightwaves at the output of this fiber strand?

h. The total dispersion factor of the fiber is +17 ps/nm/km. The lightwave source is now modulated with 40 Gb/s ON-OFF keying pulse sequence.

   i. Sketch the signal power spectrum—refer to materials of ECE2011—signal processing. Then estimate the 3 dB bandwidth (in nm) of the signal bands.

   ii. The group of waves in (i) is propagating through the fiber; what is the broadening of the pulse sequence?

   iii. Show how you could compensate this broadening effect of the optical signals after 100 km fiber propagation.

26. A single mode optical fiber is to be designed for operation in the 1550 nm region. The following operational parameters are required:

- Single mode in the 1300 nm to 1620 nm wavelength.
- Total chromatic dispersion factor of 5.0 ps/nm/km, and within 10% variation in the 1520 nm to 1565 nm wavelength.

   a. Obtain an expression of the waveguide dispersion factor as a function of b and V where b is the normalized propagation constant and V is the normalized frequency.

   b. Design the single mode optical fiber at the operating wavelength of 1550 nm where its total dispersion factor is +5.0 ps/nm/km.

It is recommended that the following parameters should be specified for the designed fiber: the fiber diameter, the relative refractive index difference, the cut-off wavelength, and the total dispersion at the desired wavelength of operation.

   The following steps can be used for the design:

     i. Estimate the material dispersion factor at 1550 nm and the waveguide dispersion required for the operating wavelength.

     ii. Use the waveguide dispersion factor to find the required *V*-parameter at the operating wavelength.

     iii. Select either a value of the fiber radius or a value for the relative refractive index difference. Determine the other fiber parameters.

     iv. Estimate the fiber spot size and the cutoff wavelength of the designed fiber.

   The refractive index of pure silica at 1550 nm is 1.480.

Some Questions

1. *Why it is possible to assume that there is no free charge in an optical fiber in applying the Maxwell's equations?*

2. *What are the superstrate, substrate, and core regions of a planar optical waveguide? Can you interpret these regions with those of a circular optical fiber?*

3. *Sketch a symmetric planar optical waveguide and give some typical values of refractive indices for all regions.*

4. *Define and explain the physical meanings of the parameters u, v, and V for guided modes and radiating modes of a generalized planar symmetric optical waveguide.*

5. *In the analysis of planar optical waveguides we did not attempt to solve for the eigenvalues of the propagation of the guided modes. Can you suggest a technique for obtaining these eigenvalues for a symmetric planar optical waveguide?*

6. *Can you suggest a structure for an asymmetric planar optical waveguide? What are the advantages and disadvantages of an asymmetric planar waveguide as compared to that of a planar guide?*

7. *Can you comment on the polarization of the guided mode of a single mode optical fiber? Does the polarization stay fixed in one or two directions?*

8. *Does the term "single mode" really mean "only one mode" in a single mode optical fiber? When we say "single mode fiber," do we mean single mode at all wavelength regions?*

9. *Why do telecommunications engineers prefer single mode optical fibers rather than multi-mode fibers for long-haul telecommunications systems?*

10. *Indicate the three "optical windows" where optical communications systems have been deployed. Comment on the two "highest windows"?*

11. *Comment on the future development of telecommunications networks in the 1300 nm and 1550 nm windows.*

12. *What are the frequencies corresponding to the operating wavelengths 1300 nm and 1550 nm? For two optical lightwaves having wavelengths of 1550 nm and 1551 nm, estimate the useable operating bandwidth between them.*

13. If the channel spacing between lightwaves is 1.0 nm and the difference in attenuation coefficients between lightwaves channels is 0.2dB/km, how many channels can we use in the 1300 nm and 1550 nm windows? Use your estimate in the previous question to estimate the useable bandwidth for information transmission.

14. Can you suggest some techniques to minimize the pulse-broadening effects in signal transmission in telecommunication systems?

15. What is the definition of group velocity? What is group delay?

16. What does chromatic dispersion mean? What are the material and waveguide dispersion factors?

17. What are the differences between the dispersion mechanisms of lightwaves propagating in single and multimode circular optical fibers? Explain why telecommunications engineers prefer single mode fiber to the multimode type for long-haul optical communications systems?

18. What are chromatic dispersion effects due to materials and waveguides? What do we really mean by negative dispersion? Do the transmitted pulses compress in a "totally negative dispersion" fiber? Can you make use of the negative and positive dispersion factors of chromatic dispersion to design dispersion-zero optical fibers?

19. Are you aware of optical amplifiers? What are they? How do they work and transform the design and operation of optical communications systems and networks?

20. If engineers wish to establish a telecommunication system between satellites, would they be able to use optical fiber communications systems?

21. Would we be able to employ optical fiber communications systems and networks for intercontinental communications?

22. Are you aware of the number of optical fiber manufacturers in America? How many manufacturers are there in America or Europe or China?

23. Would you be able to estimate the total length of optical fiber cables that have been installed in US telecommunications networks?

24. How extensive is the optical fiber communications networks between the West and East Coasts of America? Where are the gateways from America to the rest of the world?

References

1. G.P. Agrawal, "Fiber Optic Communications Systems," 3rd Ed., J. Wiley 2002.
2. R.H. Stolen, E.P. Ippen, and A.R. Tynes, *Appl. Phys. Let.*, 20, 62, 1972.
3. E.P. Ippen and R. H. Stolen, *Appl. Physics Lett.*, 21, 539, 1972.
4. R.G. Smith, *Appl. Optics*, 11, 2489, 1972.
5. G.P. Agrawal, Fiber optic communications systems, 60. J. Wiley, 2nd Ed., NY 2002.
6. G.P. Agrawal, "Fiber optic communications systems," 67. 3rd Ed., J Wiley, NY 2002.
7. J.P. Gordon and H. Kogelnik, "PMD fundamentals: Polarization mode dispersion in optical fibers," *PNAS*, 97(9), pp. 4541–4550, April 2000.
8. Corning. Inc., "An Introduction to the Fundamentals of PMD in Fibers," *White Paper*, July 2006.
9. A. Galtarossa and L. Palmieri, "Relationship between pulse broadening due to polarization mode dispersion and differential group delay in long single mode fiber," *Electronics Letters*, 34(5), March 1998.

10. J.M. Fini and H.A. Haus, "Accumulation of polarization-mode dispersion in cascades of compensated optical fibers," *IEEE Photonics Tech. Lett.*, 13(2), 124–126, February 2001.
11. Corning. Inc., "An Introduction to the Fundamentals of PMD in Fibers," *White Paper*, July 2006
12. A. Carena, V. Curri, R. Gaudino, P. Poggiolini, and S. Benedetto, "A time-domain optical transmission system simulation package accounting for nonlinear and polarization-related effects in fiber," *IEEE J. Sel. Areas in Comm.*, 15(4), 751–765, 1997.
13. S.A. Jacobs, J.J. Refi, and R.E. Fangmann, "Statistical estimation of PMD coefficients for system design," *Elect. Lett.*, 33(7), 619–621, March 1997.
14. G.P. Agrawal, "Fiber optic communication systems," Academic Press, N.Y., 2002.
15. A.F. Elrefaie, R.E. Wagner, D.A. Atlas, and D.G. Daut, "Chromatic Dispersion Limitations in Coherent Lightwave Transmission Systems," *IEEE J. Lightw. Tech.*, 6(6), 704–709, 1998.
16. Jau Tang, "The Channel Capacity of a Multispan DWDM System Employing Dispersive Nonlinear Optical Fibers and an Ideal Coherent Optical Receiver," *IEEE J. Lightw. Tech.*, 20(7), 1095–1101, 2002.
17. Bo Xu and Maïté Brandt-Pearce, "Comparison of FWM- and XPM-Induced Crosstalk Using the Volterra Series Transfer Function Method," *IEEE J. Lightw. Tech.*, 21(1), 40–54, 2003.
18. J. Tang, "The Shannon Channel Capacity of Dispersion-Free Nonlinear Optical Fiber Transmission," *IEEE J. Lightw. Tech.*, 19(8), 1104–1109, 2001.
19. J. Tang, "A Comparison Study of the Shannon Channel Capacity of Various Nonlinear Optical Fibers," *IEEE J. Lightw. Tech.*, 24(5), 2070–2075, 2006.
20. J.G. Proakis, *Digital Communications,* 4th ed., 185–213. New York: McGraw-Hill, 2001.
21. L.N. Binh, "Digital Optical Communications," CRC Press, Florida USA, 2009.
22. L.N. Binh and N.D. Nguyen, "Generation of High-order Multi-bound-Solitons and Propagation in Optical Fibers," Optics Communications, 282, 2394–2406, 2009.
23. G.P. Agrawal, "Nonlinear Fiber Optics," 3rd Ed., Academic Press, San Diego, CA, USA 2001.
24. A.F. Elrefaie and R.E. Wagner, "Chromatic dispersion limitations for FSK and DPSK systems with direct detection receivers," *IEEE Photonics Tech. Lett.*, 3(1), 71–73, 1991.
25. A.F. Elrefaie, R.E. Wagner, D.A. Atlas, and A.D. Daut, "Chromatic dispersion limitation in coherent lightwave systems," *IEEE J. Lightw. Tech.*, 6(5), 704–710, 1988.
26. A.F. Elrefaie and R.E. Wagner, "Chromatic dispersion limitations for FSK and DPSK systems with direct detection receivers," *IEEE Photonics Tech. Lett.*, 3(1), 71–73, 1991.
27. A.F. Elrefaie, R.E. Wagner, D.A. Atlas, and A.D. Daut, "Chromatic dispersion limitation in coherent lightwave systems," *IEEE J. Lightw. Tech.*, 6(5), 704–710, 1988.

5

Design of Single-Mode Optical Fiber Waveguides

5.1 Introduction

Since the prediction of the feasibility of optical fiber communications by Kao and Hockham [1], numerous stages of development have been recorded, especially in the last quarter of the twentieth century. As the bit rate has been pushed higher and higher towards 100 Gb/s in the second decade of the twenty-first century, the demand of low and flattening dispersion becomes intense. Each stage is characterized by new breakthroughs. A historical perspective covering the initial phase of development of optical fibers was presented by Li [2]. Then the design and fabrication of single mode optical fibers (SMFs) were described by Ainslie and Day [3] and Khoe and Lydtin [4]. Modern advanced optical fibers must also be considered to offer distributed amplification as well as low dispersion and low loss [5].

The design of optical fibers has been pushed towards low loss low dispersion and optimized for Raman distributed amplification gain for ultra-high bit rate optical information transmission. This is the figure of merit for digital pulse transmission in long-haul optically amplified communication systems.

The fist experiments concentrated on the reduction of the material impurities in the fiber so that it would be cost-competitive with coaxial cables. In the 1970s the targeted 20 dB/km barrier was broken [6]. This led to the prediction of the ultimate lower limit on attenuation by Keck, Maurer, and Schultz [7]. However, the first such fiber fabricated was single mode and problems with splicing led to the intensive development of larger core multimode fibers from 1972 to 1979. During the same period the operating wavelength was also shifted from the 1300 nm region to 1550 nm window.

The year 1979 was marked with the design of SMFs and the establishment of multimode optical fibers and practical limits on transmission systems [8]. The main limitations are caused by the extreme sensitivity of the fiber bandwidth to the profile shape [9] and the excess loss due to the high level of doping required [2]. These limitations cause the practical bit rate-distance product to fall well below the theoretical limit of $1.4/\Delta^2$ (Mb/s)-km [10]. Thus it became apparent that the maximum achievable bit rate-distance product could only be achieved if single mode fibers were used. This is due to the low doping levels required and the absence of unpredictable inter-modal interferences.

The push to a longer wavelength spectral region to take advantage of low loss and large bandwidth-distance product of the single mode fiber is due to two main factors: firstly the Rayleigh scattering loss is low, about 0.19 dB/km in the 1550 nm spectral region; secondly, the major dispersive components in a single mode fiber are material and waveguide dispersion. In addition, the polarization mode dispersion due to the random non-uniformity of the fiber core and stress-induced refractive indices would also influence the transmission performance at ultra-high speed, e.g., greater than 10 Gb/s.

Since the invention of inline optical amplifiers, especially the Er:doped fiber amplifiers (EDFA) the fiber loss and insertion losses of optical modulators was no longer the major obstacle and the dispersion of SMF became the principal issue. The design of SMF has been structured towards low and flattening in the 1550 nm region. There must be a small amount of dispersion in order to avoid the four wave mixing effects, at the same time offering optimum performance on the distributed Raman amplification, especially when undersea transmission is required.

This chapter thus describes an extensive design strategy for the advanced optical fibers for advanced modulation format long-haul transmission. We first present a unified formulation of the optical fiber waveguide problems, thence followed by a simplified design of single mode fibers, thence a case study on the design of modern optical fibers having dispersion flattening properties.

5.2 Unified Formulation of Optical Fiber Waveguide Problems

Inspired by Snitzer [11] it can be proved that the vector analysis could be replaced by its highly accurate 0th order solution provided that the index difference between the core and the cladding is small [12]. This led to tremendous simplification of the equations describing the waveguide. This simplified scalar analysis was applied to the design of practical waveguide by Gloge [13]. Gloge classified these fibers as weakly guiding and invented the term linearly polarized (LP) mode designation that is well known today. These modes are compatible with the transverse electromagnetic modes (TEM) of lasers or microwave waveguides [14].

The use of SMFs in optical communications is now firmly established [15]. However early works on fiber design were hampered by the non-existence of analytical solutions to profiles other than the simple step index waveguide. Thus numerical methods of solutions were developed [16]. At about the same time the Wentzel–Kramers–Brillouin (WKB) method was applied to the analysis of multimode waveguides [17]. The latter, however, is insufficiently accurate of single mode waveguides due to the large extent of the evanescent field.

There are two approaches to the early numerical approaches which were adopted. Dil and Blok [16] integrate the full vector set of equations based on the early theoretical works [18]. This requires that extensive computational and numerical methods of solving the simplified scalar wave equation have been tried. The errors due to the use of the scalar wave equation in weakly guiding fibers have been categorized and found to be negligible for a mode which is excited at a frequency sufficiently far from the cutoff region [19]. Corrections to the scalar results have been developed by Tjaden [20] and generalized by Snyder and Love [21]. Thence these corrections to the 0th order solution were applied to fibers with multiple index profile structures designed for dispersion-flattened applications [22].

Most analysis carried out on the propagation characteristics of SMFs with simple graded profiles have been based on the 0th order scalar wave equation [23]. In the next section a method for the first order [24] is presented that allows the first order scalar wave equation be derived from the full vector solutions originally developed by Kurtz and Streifer [18] and extended by Okoshi [25]. Thence a correction due to polarization to the propagation constant is made directly from this equation. These corrections can be easily incorporated in numerical analysis schemes.

The design of fibers with complicated refractive index (RI) profiles is facilitated if we have access to the waveguide component of the chromatic dispersion. This is given in the next section. The total dispersion is calculated from the non-normalized wave equation giving no direct knowledge of the contribution of each component of the total dispersion [26–28].

5.2.1 First Order Scalar Wave Equation

From Maxwell's equation given in Appendix B of Chapter 2 and a scalar approximation, it is well known that the sinusoidal steady state cylindrical field components of the full vector modes (hybrid) satisfying Maxwell's equations in a source free, loss less, dielectric and non-magnetic medium can be derived from the longitudinal field components E_z and H_z. This is because the radial and azimuthal field components are coupled [29]. Thus we can write, omitting the mode characteristics factor $e^{j(\omega t - \beta z - v\theta)}$ common to all the field components

$$
\frac{d^2 E_z}{dr^2} + \frac{1}{r} \frac{dE_z}{dr} + \left[k_0^2 n^2(r) - \beta^2 - \frac{v^2}{r^2} \right] E_z
$$
$$
= \left\{ \frac{2\beta^2}{k_0^2 n^2(r) - \beta^2} \frac{1}{n(r)} \frac{dn(r)}{dr} \right\} \frac{dE_z}{dr} - j \frac{2\beta\mu_0\omega v}{k_0^2 n^2(r) - \beta^2} \frac{1}{r \cdot n(r)} \frac{dn}{dr} H_z \tag{5.1}
$$

and

$$
\frac{d^2 H_z}{dr^2} + \frac{1}{r} \frac{dH_z}{dr} + \left[k_0^2 n^2(r) - \beta^2 - \frac{v^2}{r^2} \right] H_z
$$
$$
= \left\{ \frac{2k_0^2 n^2(r)}{k_0^2 n^2(r) - \beta^2} \frac{1}{n(r)} \frac{dn(r)}{dr} \right\} \frac{dH_z}{dr} + j \frac{2k_0^2 \beta v}{k_0^2 n^2(r) - \beta^2} \frac{n(r)}{r} \frac{dn}{dr} E_z \tag{5.2}
$$

where $n(r)$ is the radial RI distribution function; β is the axial propagation constant; v is the azimuthal mode number; k_0 is the free space wave number; ω is the angular frequency; μ_0 is the magnetic permeability. The geometry is shown in Figure 5.1.

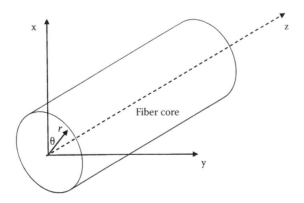

FIGURE 5.1
Geometrical structure of a circular optical fiber waveguide.

Following Okamoto and Okoshi [30] we introduce the radial functions $\phi(R)$; $\varphi(R)$ such that

$$E_z = \frac{k_0^2 n^2 \phi(R)}{\beta}$$

$$H_z = -\frac{j}{\omega\mu_0} k_0^2 n^2 \phi(R)$$

(5.3)

where n_0 denotes the maximum index at the center of the core and $R = r/a$. To facilitate further simplification and the extraction of well-known parameters we introduce the normalized waveguide parameters:

$$V = ak_0\sqrt{n_0^2 - n_{cl}^2} = ak_0 n_{cl}\sqrt{2\Delta}$$

$$b = \frac{\left(\dfrac{\beta}{k_0}\right)^2 - n_{cl}^2}{n_0^2 - n_{cl}^2}$$

$$S(R) = \frac{n^2(r) - n_{cl}^2}{n_0^2 - n_{cl}^2}$$

(5.4)

with

$$\Delta = \frac{n_0^2 - n_{cl}^2}{2n_{cl}^2}$$

and

$$n^2(r) = \begin{cases} n_{cl}^2\left[1 + 2\Delta S(R)\right]; & R \le 1 \\ n_{cl}^2; & R > 1 \end{cases}$$

The definition here for the relative RI is slightly different as compared to a number of published works. No significant effects on the results are observed.

   The application of Equation 5.4 gives, after some tedious algebra, the normalized coupled equations

$$\left[S(R) - b\right]\frac{1}{R}\frac{d}{dR}\left\{\frac{1}{S(R) - b}\right\}R\frac{d\phi}{dR} + \left[V^2\left(S(R) - b\right) - \frac{v^2}{r^2}\right]\phi$$

$$+\frac{v\left(S(R) - b\right)}{R}\frac{d}{dR}\left[\frac{1}{S(R) - b}\right]\phi - 2\Delta\frac{dS(R)}{dR}\left[\frac{d\phi}{dR} + \frac{v}{R}\phi\right] = 0$$

(5.5)

and

$$\left[S(R) - b\right]\cdot\frac{1}{R}\frac{d}{dR}\left\{\frac{1}{1 - b - S(R)}\right\}R\frac{d\phi}{dR} + \left[V^2\left(S(R) - b\right) - \frac{v^2}{r^2}\right]\phi$$

$$+\frac{v\left(S(R) - b\right)}{R}\frac{d}{dR}\left[\frac{1}{S(R) - b}\right]\phi = 0$$

(5.6)

where we have made the approximation $\Delta \ll 1$ in arriving at the coupled equations, Equations 5.5 and 5.6. Since we are interested in the modes with $v \neq 0$, these equations cannot be decoupled without introducing some new techniques. The simplified analysis arising from setting $v = 0$ leads to the transverse electronic (TE) and transverse magnetic (TM) modes of the waveguide, respectively [11]. Note that Equation 5.5 contains the extra terms involving Δ. In Okoshi analysis [25] this term was eliminated. The fact that $\Delta \neq 0$ and $(dS(R))/dR \neq 0$ led to the hybrid nature of the vectorial modes. These conclusions have been arrived at by an independent numerical study by Morishita, Kondoh, and Kumagai [19]. These are known collectively as ∇n^2 gradient terms [21]. However Kokubun and Iga [24] found that only a tiny fraction of ∇n^2 participates in the coupling. Following Ref. [11] we define the mode classification parameter given by

$$P = \frac{j\omega\mu_0}{\beta}\frac{H_z}{E_z} = \frac{\Phi}{\phi} = \frac{G_1 - G_2}{G_1 + G_2} \tag{5.7}$$

with

$$G_1(R) = \frac{\phi(R) + \Phi(R)}{2}$$

and

$$G_2(R) = \frac{\phi(R) - \Phi(R)}{2}$$

As the first step towards the decoupling of the wave equations, if we substitute Equation 5.7 into 5.5 and obtain the sum and difference of the result we end up with the auxiliary set of equations

$$[S(R)-b]\frac{1}{R}\frac{d}{dR}\left\{\frac{1}{S(R)-b}\right\}R\frac{dG_1}{dR} + \left[V^2(S(R)-b)-\frac{v^2}{R^2}\right]G_1$$

$$+\frac{v(S(R)-b)}{R}\frac{d}{dR}\left[\frac{1}{S(R)-b}\right]G_1 - \frac{dS(R)}{dR}\left[\frac{d(G_1-G_2)}{dR}+\frac{v}{R}(G_1-G_2)\right] = 0 \tag{5.8}$$

and

$$[S(R)-b]\cdot\frac{1}{R}\frac{d}{dR}\left\{\frac{1}{1-b-S(R)}\right\}R\frac{dG_2}{dR} + \left[V^2(S(R)-b)-\frac{v^2}{R^2}\right]G_2$$

$$+\frac{v(S(R)-b)}{R}\frac{d}{dR}\left[\frac{1}{S(R)-b}\right]G_2 - \Delta\frac{dS(R)}{dR}\left[\frac{d(G_1+G_2)}{dR}+\frac{v}{R}(G_1-G_2)\right] = 0 \tag{5.9}$$

According to the mode classification scheme of Snitzer [11] in a step index profile fiber $P \to -1$ for the HE mode and $P \to +1$ for the EH modes. If we assume that this is approximately correct for profiles other than a simple step, we now have:

$$G_1 \approx 0; G_2 \neq 0; \text{ HE modes}$$
$$G_2 \approx 0; G_1 \neq 0; \text{ EH modes} \tag{5.10}$$

This simplification decouples Equations 5.8 and 5.9. We can further simplify by defining

$$\varphi(R) = \frac{1}{2a\Delta(S(R)-b)}\left[\frac{dG_i}{dR} \pm \frac{v}{R}G_i\right]; \quad i = 1,2 \tag{5.11}$$

The plus and minus signs correspond to the modes HE and EH, respectively. The first order scalar wave equation can be obtained. After some considerable algebra, the first order scalar wave equation is finally obtained as for the HE modes (taking only the plus sign)

$$\frac{1}{R}\frac{d}{dR}\left[R\frac{d\varphi}{dR}\right] + \left[V^2(S(R)-b) - \frac{l^2}{R^2}\right]\varphi - \Delta\frac{dS(R)}{dR}\left[\frac{d\varphi}{dR} - \frac{l}{R}\varphi\right] = 0 \tag{5.12}$$

where $l = v - 1$ is the equivalent LP-designated azimuthal mode number. Equation 5.12 reduces to the 0th order scalar wave equation if we set $\Delta \to 0$. The later has been derived by Gambling and Matsumura [31] following a somewhat similar approach.

5.2.2 Eigenvalue Equation

In the cladding $dS/dR \to 0$ and Equation 5.12 reduces to the 0th order scalar wave equation. Thus we write

$$\varphi(R) = AK_l(WR); \quad R > 1 \tag{5.13}$$

where

$$W = V\sqrt{b}$$

and K_l is the modified Bessel function of the second kind of order l and A is a constant. The eigenvalue equation is obtained by matching the field components at the core/cladding interface. The exact boundary conditions require the continuity of all six electromagnetic field components of the hybrid mode stated as

$$E_z(a^+) = E_z(a^-)$$

$$E_\theta(a^+) = E_\theta(a^-)$$

$$E_r(a^+) = \frac{n^2(a^-)}{n_{cl}^2}E_r(a^-) \tag{5.14}$$

$$H_z(a^+) = H_z(a^-)$$

$$H_\theta(a^+) = H_\theta(a^-)$$

$$H_r(a^+) = H_r(a^-)$$

These boundary conditions can be simplified by setting $n^2(a^-) = n_{cl}^2$ in which case we can only require the logarithmic derivative of φ to be continuous at $R = 1$. However, to include the first order corrections in the eigenvalue equation we like to point out that Equation 5.14 can be written in terms of the continuity of circularly polarized components

$$\begin{bmatrix} F_z^\pm \\ E^\pm \end{bmatrix}_{r=a^+} = \begin{bmatrix} 1 & 0 \\ 0 & \dfrac{n_{cl}^2 + n^2(a^-)}{2n_{cl}^2} \end{bmatrix} \begin{bmatrix} F_z^\pm \\ E^\pm \end{bmatrix}_{r=a^-}$$

with

$$E^\pm = E_r \pm jE_\theta; \quad H^\pm = H_r \pm jH_\theta \tag{5.15}$$

and with

$$F_z^\pm = jE_z \pm j\frac{\omega\mu_0}{\beta} H_z$$

where the radial and azimuthal field components can be expressed in term of $\varphi(R)$ via Maxwell's equation and Equations 5.3, 5.7 and 5.11 reduce to the required eigenvalue equation

$$\frac{\varphi'(1)}{\varphi(1)} = l - \left[1 - \Delta S(1^-)\right] W \frac{K_{l+1}(W)}{K_l(W)} \tag{5.16}$$

Thus, for profiles having smooth core-cladding transition, $S(1^{-1}) = 0$ and Equation 5.16 reduces to the eigenvalue equation given in Ref. [31]. In all cases the product of Δ and $S(1^{-1})$ should be exceedingly small. Therefore we use only the simplified form of Equation 5.16.

5.2.3 Polarization Correction to b

To show that Equation 5.12 is indeed correct, we rewrite it in the form, for the fundamental mode ($l = 0$)

$$\frac{1}{R}\frac{d}{dR}\left[R\frac{d\varphi}{dR}\right] + \left[V^2\left(S(R)-b\right) - \frac{l^2}{R^2}\right]\varphi = \Delta\frac{dS(R)}{dR}\left[\frac{d\varphi}{dR}\right] = 0 \tag{5.17}$$

Now multiplying Equation 5.17 by φR and integrating from $R = 0$ to infinitive gives

$$\frac{1}{R}\left[\frac{d\varphi}{dR}\right]_0^\infty - \left[bV^2\right]\int_0^\infty \varphi(R)\,R\,dR + V^2\int_0^\infty S(R)\varphi(R)\,dR$$

$$= \Delta\int_0^\infty \frac{dS(R)}{dR}\left[\frac{d\varphi}{dR}\right]R\,dR \tag{5.18}$$

It can be easily shown using large asymptotic expansion of $K_l(WR)$ in Equation 5.13 that for large R, $\lim\limits_{R\to\infty} R\dfrac{d\varphi}{dR} \to 0$. Thus Equation 5.18 becomes, after dividing through by $\int_0^\infty \varphi(R)\,dR$

$$b_1 = b_0 - \frac{\Delta\displaystyle\int_0^\infty \frac{dS(R)}{dR}\left[\frac{d\varphi}{dR}\right]R\,dR}{\displaystyle\int_0^\infty \varphi(R)\,R\,dR}$$

with
$$b_0 = \frac{\int_0^1 S(R)\ (R)dR}{\int_0^\infty (R)R\,dR} \tag{5.19}$$

where b_1 and b_0 are the 0th and first order approximations of the normalized propagation constant, respectively. Thus Equation 5.19 is the normalized propagation constant resulting from the solution of the 0th order scalar wave equation.

5.2.4 Waveguide Characteristics Parameters

From the point of view of the design of optical waveguide, the two most significant factors affecting its performance as a transmission medium are waveguide dispersion and bending losses. The degree of pulse broadening is extremely sensitive to the exact matching of the zero dispersion wavelength and the center wavelength [32]. The received pulse width is directly related to the transmission data rate in bits per second [33]. For silica-based single mode fibers designed for dispersion shifting the dominant influence on the design comes from the waveguide component of total dispersion. This is the only component that can be significantly altered by suitable modification of the RI profile shape.

The bending losses consist of two components. The pure bend loss is due to radiation losses of the mode induced by a single bend. It is an indication of the mode confinement capability of the waveguide as measured by the mode size. A more critical component of bending losses is that, due to the microbending induced along the axis of the fiber, the guided mode loses its power to the higher order leaky modes. For conventional non-segmented core fibers, microbending loss is predictable and calculated from Petermann's simplified spot size formula [34]. Its magnitude is proportional to large powers of the spot size [35]. Previously, it was confirmed experimentally that to eliminate micobending loss at the operating wavelength, it is sufficient to keep the mode field diameter below 10 μm [36]. However Petermann's spot size simplified formula is only valid for near Gaussian fundamental modal fields. This has been confirmed numerically by Francois and Vassalo [37]. It has now been established that the fundamental modal field distribution of the dispersion of dispersion-flattened fibers deviates significantly from Gaussian dispersion [38]. This led to the development of lower and upper limits on the microbending loss based on the spectral behavior of three spot sizes [39]. These three spot sizes, two of which are measurable, are required for the accurate characterization of microbending losses. The two measurable spot sizes are related to the waveguide dispersion and splice losses [40,41].

In this section we present simplified formulae for calculating the spot sizes and waveguide dispersion over a range of frequencies. We include a unified treatment of theoretical development which suggests that the amount of waveguide dispersion that can be designed by profile shaping may be constrained by the requirement of low bending loss.

5.2.4.1 Chromatic Fiber Dispersion

There are three main formulae used for the calculation of total dispersion from the normalized wave equation. The first attempt to break up total dispersion into its waveguide and material dispersion components is due to Gloge [13]. It was improved by Gambling, Matsumura, and Ragdale [42] and later improved by Francois [43] to include fibers with more than one dopant. The last formula was developed for the design of fluorine-doped depressed-cladding fibers. However preliminary calculations performed by Ruhl [44] suggest that it is only required for fibers with severely depressed F-doped claddings. The majority of fiber designs have been arrived at by calculating the total dispersion directly from the non-normalized wave equation [45–47]. However for better understanding of the influence of profile shaping on dispersion it is advantageous to follow the approach by Gambling et al. Gloge's simplified formula is still being used but significant discrepancies have been reported [48]. Here we employ Gloge's formula in the same form as Gambling's more complete description to allow us to assess its applicability.

In the process of normalization we have made an implicit assumption that the profile shape function is independent of wavelength. Early work on optimization of the α-profile for designing high bandwidth fibers indicated that this assumption is well founded.

The total dispersion coefficient of the fiber can be expressed in picoseconds per kilometer length and the spectral width of the source or the modulated signal. In modern optical communication the source line width is considered very narrow compared with the bandwidth of the modulation which dominates the broadening of the lightwave carrier. The dispersion is given by

$$\frac{d\tau}{d\lambda} = -\frac{\lambda}{c} \frac{d^2 n_e}{d\lambda^2} \tag{5.20}$$

with

$$n_e = \frac{\beta}{k_0}$$

n_e is the effective RI of the guided mode LP_{01} along the z-axis of the fiber and τ is the measurable group delay. For the very small value of the relative RI difference, it can be expressed as (see also from Equation 5.4)

$$\Delta \approx \frac{n_0 - n_{cl}}{n_{cl}} \tag{5.21}$$

Thus the simplified mapping relation becomes

$$n_e \simeq n_{cl} \left[1 + b\Delta \right] \tag{5.22}$$

This equation can be put into the form related to the normalized frequency V-parameter as

$$n_e \simeq n_{cl} + \frac{bV^2}{8\pi^{2a^2}} \frac{\lambda^2}{n_{cl}} \tag{5.23}$$

Thus the differential equation (Equation 5.20) can be rewritten as

$$\frac{d^2 n_e}{d\lambda^2} = \frac{d^2 n_{cl}}{d\lambda^2} + \frac{1}{8\pi^{2a^2}} \left[\begin{array}{l} bV^2 \frac{d^2}{d\lambda^2} \left(\frac{\lambda^2}{n_{cl}} \right) + 2\frac{dV}{d\lambda} \frac{d}{d\lambda} \left(\frac{\lambda^2}{n_{cl}} \right) \frac{d(bV^2)}{dV} \\ + \left(\frac{\lambda^2}{n_{cl}} \right) \left\{ \left(\frac{dV}{d\lambda} \right)^2 \frac{d^2(bV^2)}{dV^2} + \left(\frac{d^2 V}{d\lambda^2} \right) \frac{d(bV^2)}{dV} \right\} \end{array} \right] \tag{5.24}$$

Expanding and rearranging the term we end up with the dispersion composite terms

$$\frac{d\tau}{d\lambda} = S_{cmd} + S_{wd} + S_{cpd}$$

with

$$S_{cmd} = -\frac{\lambda}{c}\left[A(V)\frac{d^2 n_0}{d\lambda^2}\right] + \left[1 - A(V)\frac{d^2 n_{cl}}{d\lambda^2}\right]; \quad A(V) = \frac{1}{2}\left[b + \frac{d(bV)}{dV}\right]$$

$$S_{wd} = -\frac{n_{cl}\Delta}{\lambda c}V\frac{d^2\left(bV^2\right)}{dV^2} \tag{5.25}$$

$$S_{cpd} = \frac{n_{cl}}{c}D(V)\frac{d\Delta}{d\lambda}; \quad D(V) = \frac{d^2\left(bV^2\right)}{dV^2} - \frac{d(bV)}{dV} - b$$

These are composite material, waveguide and composite profile dispersion, respectively. In step index fiber $A(V)$ denotes the fraction of modal power in the core. Thus it is a measure of the distribution of power between the core and the cladding. The profile dispersion term takes into account the fact that even though the profile shape is independent of wavelength, its height is not. This is due to the slight difference in the dispersive properties of the core from that of the cladding caused by the introduction of a dopant.

The dominant term in Equation 5.25 is the normalized waveguide dispersion coefficient $V(d^2(bV^2)/dV^2)$. The significance of this equation is that the fiber design is simplified by realizing that only the waveguide dispersion component is affected to a significant extent by profile shaping. The other two components are functions of doping density. To keep the losses low, this factor is often constrained. Thus, the overall design process can be concentrated on the profile shape design alone. Direct numerical integration of the normalized wave equation yields the dispersion characteristics (b-V curves) from which waveguide dispersion can be obtained by differentiation.

Gloge's dispersion formula is given by [13]

$$\frac{d\tau}{d\lambda} = -\frac{1}{\lambda c}k_0\frac{dN_{cl}}{dk_0} + k_0\frac{d(N_0 - N_{cl})}{dk_0}\frac{d(bV)}{dV} + \frac{\left(N_0^2 - N_{cl}^2\right)}{n_0 + n_{cl}}V\frac{d^2\left(bV^2\right)}{dV^2}$$

$$N_{cl} = \frac{d(k_0 n_{cl})}{dk_0}; \quad N_0 = \frac{d(k_0 n_0)}{dk_0} \tag{5.26}$$

with N_{cl}; N_0 are the material group indices of the cladding and the core, respectively. Equation 5.26 was derived from the exact mapping relation given in Equation 5.4 of the normalized propagation constant. The approximated relation of this normalized quantity is

$$\frac{n_0 - n_{cl}}{n_{cl}} = \frac{N_0 - N_{cl}}{N_{cl}} \ll 1 \tag{5.27}$$

Equation 5.27 is equivalent to the approximation

$$\Delta \ll 1$$

$$\frac{\lambda}{n_{cl}} \frac{dn_0}{d\lambda}; \frac{\lambda}{n_{cl}} \frac{dn_{cl}}{d\lambda} \ll 1 \tag{5.28}$$

which are the same approximations used to derive the simplified form of Equation 5.25. This may cause some discrepancies with the equations provided by Gloge and Gambling's dispersion formulae. Expanding the terms involving $N_0; N_{cl}$ in Equation 5.27 and using approximations in Equation 5.28 we can rewrite that

$$\frac{d\tau}{d\lambda} = -\frac{1}{\lambda c} \left[\frac{d(bV)}{dV} \frac{d^2 n_0}{d\lambda^2} \right] + \left[1 - \frac{d(bV)}{dV} \frac{d^2 n_{cl}}{d\lambda^2} \right]$$

$$- \frac{n_{cl}\Delta}{\lambda c} V \frac{d^2 (bV^2)}{dV^2} + \frac{n_{cl}}{c} V \frac{d^2 (bV^2)}{dV^2} \frac{d\Delta}{d\lambda} \tag{5.29}$$

This dispersion relation is remarkably similar to Equation 5.25 provided that we force

$$b = \frac{d(Vb)}{dV} \to 1 \tag{5.30}$$

5.2.4.2 Spot Size

5.2.4.2.1 Definition
Three definitions of the spot size are given as [34]

$$r_0^2 = \frac{2 \int_0^\infty \varphi^2(R) R^3 dR}{\int_0^\infty \varphi^2(R) R dR} \tag{5.31}$$

r_0 is required for the calculation of microbending loss [34] and splice loss due to small angular offset. It is a measurable spot size. A more accurate definition

$$\bar{r}_0^2 = \frac{2 \int_0^\infty \varphi^2(R) R dR}{\int_0^\infty \left[\varphi^2(R) \right] dR} \tag{5.32}$$

$\bar{r}_0$ places a lower bound on the microbending loss and is required for the splice loss due to small offset. Pask [41] proves that it is directly related to the measurable root mean square (RMS) width of the far field mode spot size. Thus it can be used as a defined mode spot size obtained from experimental measurement.

The third spot size definition places an upper limit to the microbending loss calculated from the RMS spot size [39]. It is defined as

$$r_\infty^2 = \frac{2}{a^2 k_0 n_0 (\beta - k_0 n_{cl})} \tag{5.33}$$

No experimental techniques have been established to measure this mode spot size but the bending loss can be measured and related to the spot size.

5.2.4.2.2 Relation to Normalized Parameters

It can be shown that [49] the spot size is related to the propagation constant as

$$V \frac{d^2(Vb)}{dV^2} = 4 \frac{d}{dV}\left(\frac{1}{V\bar{r}_0^2}\right)$$

(5.34)

This is important between the spectral variation of the spot size and the normalized waveguide dispersion parameter. This equation can be rewritten as

$$\bar{r}_0^2 = \frac{4}{V^2\left[\dfrac{d(Vb)}{dV} - b\right]}$$

(5.35)

The equation can then be recast into the form

$$\bar{r}_0^2 = \frac{4}{V^2\Omega_0[\eta_0 - b]}$$

$$\Omega_0 = \int S(R)RdR \equiv \text{profile volume}$$

(5.36)

$$\eta_0 = \text{power fraction in the core}$$

$$\eta_0 = \frac{1}{2\Omega_0}\frac{d(bV)}{dV} + \left[2 - \frac{1}{2\Omega_0}\right]\frac{b}{2}$$

Equation 5.36 is an exact mathematical statement of Chang's explanation of waveguide dispersion as being due to the change in the modal power distribution with wavelength [50]

The spot size definition in Equation 5.33 can thus be written as (using $\Delta \ll 1$ and Equation 5.22)

$$\bar{r}_\infty^2 = \frac{4}{V^2 b}$$

(5.37)

Thus we can observe that

$$\bar{r}^2 \leq \bar{r}_0^2 \leq \bar{r}_\infty^2$$

(5.38)

Using the results given in Equations 5.35 and 5.37 we obtain a simple estimate of the cutoff for LP_{11} mode as

$$b = \frac{1}{2}\frac{d(Vb)}{dV}\bigg|_{V=V_c}$$

(5.39)

This relation suggests that for a fiber operating near the cutoff region of LP$_{11}$ mode the fractional power is given by (using also Hussey's relation [51])

$$\eta_0 \approx \left[1 + \frac{1}{4\Omega_0}\right] b \tag{5.40}$$

It is noted here that this relation may not be valid at $\Omega_0 = 0$ which can happen for the case of depressed cladding fiber.

5.2.4.2.3 Close Form Formula for Spot Size

Using the scalar wave equation for the step index fiber the spot size for single mode fiber can be written as

$$r_0^2 = \frac{4}{3}\left[\frac{1}{2} + \frac{1}{v^2} + \frac{1}{u^2} + \frac{J_0(u)}{uJ_1(u)}\right] \tag{5.41}$$

At far field the spot size of Equation 5.41 can be written as

$$\overline{r_0}^2 = \frac{2}{bV^2}\left[\frac{J_1(u)}{J_0(u)}\right]^2 \tag{5.42}$$

5.2.4.3 Fiber Extinct Loss Formulae

For completeness the expressions for microbending loss are included in this section on Ge-doped silica fiber.

5.2.4.3.1 Microbending Loss

The improved loss formulae developed by Petermann and Kuhne [39], given as

$$\alpha_m = \frac{A}{4}(k_0 n_0 r_0)^2 \left[\frac{k_0 n_0 r^2(p)}{2}\right]^{2p} \text{ dB/km} \tag{5.43}$$

where A is the fiber constant, p is an integer derived from the statistical distribution of the microbends and

$$\overline{r}^2 \le r_0 \overline{r} \le r^2(p) \le r_\infty^2; 0 \le p \le \infty \tag{5.44}$$

Thus Equation 5.43 is valid on the lower and upper limit of the mode spot size for arbitrary statistical distribution of the microbends.

5.2.4.3.2 Macrobending Loss

Several authors have presented the loss formulae due to pure bending [31,52,53]. For a weakly guiding fiber these formulae are identical and given as

$$\alpha_b = 10\log_{10}\left\{\frac{a}{8}\left(\frac{a\pi r_\infty^3}{R_c}\right)^{1/2}\right\}S(V,b)e^{-\frac{8\Delta\,bR_c}{3a\,r_\infty}}\,\text{dB/km} \tag{5.45}$$

where

$$S(V,b) = \frac{A^2}{\displaystyle\int_0^\infty \varphi^2(R)dR}\;; $$

a = core radius; R_c is the bend radius and A refers to the amplitude of the field at the core/cladding interface. Note that for fibers designated with a fixed Δ, the dominant loss factor is independent of details of the profile shape. Thus the bend loss is dominated by the value of b of the mode. In practice a critical bend loss is measured at a critical bend radius R_{cc} of 1.0 m loop test. In numerical calculation it is convenient to set this critical bend loss at 10 dB/m.

5.2.4.3.3 Excess Rayleigh Loss
The excess Rayleigh loss is estimated by

$$\alpha_e = \frac{\displaystyle\int_0^\infty \alpha(R)\varphi^2 RdR}{\displaystyle\int_0^\infty \varphi^2(R)RdR} \tag{5.46}$$

where $\alpha(R)$ is the profile grading loss factor given by

$$\alpha(R) = CS(R) \tag{5.47}$$

with C as a constant. Thus

$$\alpha_e = C\frac{\displaystyle\int_0^\infty S(R)\varphi^2 RdR}{\displaystyle\int_0^\infty \varphi^2(R)RdR} \tag{5.48}$$

However the integral expression has been expressed in terms of the far field measurable RMS spot size. Thus, the elimination of the quotient from Equation 5.48 gives the simple excess loss formula

$$\alpha_e = C\left(b + \frac{2}{V^2 r^2}\right) \tag{5.49}$$

For $V \to \infty$ the power is totally confined to the core and we have $\alpha_e = C$. Thus the constant C can be measured from the loss of multimode fiber.

5.2.4.4 Generalized Mode Cutoffs

The mode cutoffs of the next higher modes define the range of SMF operation. Thus the correct identification and calculation of this parameter is an essential part of any fiber design facility. In fibers with complicated multiple index structures, the second mode cut-off can belong to the LP_{02}; LP_{11} modes. This phenomenon is due to mode coupling effects caused by the presence of a secondary waveguide in the cladding. At cutoff $b \equiv 0$ and $V = V_C$. If we denote the field at cutoff as $\varphi_C(R)$ then

$$V_C = \frac{\int_0^\infty \left[\left(\varphi_C'(R) \right)^2 + l^2 \frac{\varphi_C^2(R)}{R^2} \right] R dR}{\int_0^1 S(R) \varphi_C^2(R) R dR} \tag{5.50}$$

where the dash denotes the derivative with respect to R. In the cladding $(S(R) = 0)$ the cut-off is simply given as

$$\varphi_C(R) = \frac{A}{R^l}; \quad R \geq 1 \tag{5.51}$$

with A as a constant. Thus the cutoff V-parameter is then given as

$$V_C = \frac{A^2 l + \int_0^1 \left[\left(\varphi_C'(R) \right)^2 + l^2 \frac{\varphi_C^2(R)}{2} \right] R dR}{\int_0^1 S(R) \varphi_C^2(R) R dR} \tag{5.52}$$

The LP_{11} mode cutoff is the smallest value of V_C calculated by setting $l = 1$ where as LP_{01}; LP_{02} mode cutoffs correspond to the lowest two values of V_C, respectively, of Equation 5.52 with $l = 0$. The LP_{01} mode can have a non-zero cutoff frequency in profiles with depressed claddings such that $\Omega_0 < 0$, the profile volume defined above.

5.3 Simplified Approach to the Design of Single-Mode Optical Fibers

5.3.1 Introductory Remarks

As mentioned above we can decompose the dispersion parameter into its composite material, waveguide and profile dispersion components. Of these three components we notice that only the waveguide dispersion term is significantly affected by the waveguide profile shaping at a fixed doping density. Thus, the process of designing dispersion-shifted fibers by altering the shape of the RI profile is greatly simplified.

Dispersion shifted to the spectral region near but not in the 1550 nm window is very important for fibers employed in ultra-dense multi-wavelength optical communications so that there would be no four wave mixing nonlinear effects and some minute dispersion factor.

In the direct approach total dispersion is calculated directly from the un-normalized wave equation [54]. Although it has already been established that certain features of the profile are needed for dispersion shifting and flattening, the mechanism responsible for it has only been quantitatively explained [55]. This approach only allows one to distinguish between classes of fibers. Within each class more explicit empirical knowledge is required. The direct approach cannot yield the answer because the effect of the fine profile features on the waveguide component of dispersion is transparent to the designer.

The problem of recognizing features of the profile which affect the waveguide dispersion was identified by Stewart [56]. The approximate mathematical procedure is based on earlier work on the profile moment method. No further works along these lines have been reported. The inherent limitation of the moment method is that the use of only two moments means that the reference profile has to be carefully chosen to ensure good accuracy. Waveguide dispersion is very sensitive to profile shape. Even the improved three-moment equivalent step index (ESI) developed by Hussey and Pask [57] fail to give accurate estimates of the waveguide dispersion [58,59].

This followed earlier reports of the inaccuracies of Stewart's original two-moment ESI procedure [60,61]. However, the moment ESI approach is a useful tool for fiber characterization and design optimization provided the fiber profile does not deviate significantly from the reference profile.

The problem of modifying profile shape to satisfy a certain waveguide dispersion characteristic can be simplified by splitting the core into a few segments. Within each segment, up to two parameters need to be changed to alter the dispersion characteristics. These are the segment relative height and width. Subject to manufacturing constraints and loss considerations, the entire fiber design exercise can be implemented with reduced computational effort. We introduce a systematic scheme of profile segmentation and classification that is directly related to the waveguide dispersive properties of the fiber. This improved approach streamlines the entire design exercise and reduces trial and error. Although the systematic core segmentation philosophy has been reported [62] and the important link between the waveguide dispersion-modified fibers established [61] no general empirical relationships have been established. Most fiber designs have been arrived at by using the numerical technique of the un-normalized wave equation [46,54,63].

We integrate this simplified approach together with simplified numerical analysis algorithm to study features of the profile shapes which affect the waveguide dispersion significantly. This knowledge, compiled as a set of empirical rules, is intended to assist fiber designers carrying out a complete design in minimum time. Our aim is to identify the particular profile shape that is most likely to satisfy a certain waveguide dispersion characteristic. Optimization of profile parameters can be carried out by synthesis techniques. The next section presents a technique that identifies the new profiles which flatten the waveguide dispersion characteristics. This leads to a dispersion zero and is also loss sensitive to change in the core radius. This is incorporated into an existing dispersion-shifted fiber to obtain improved overall performance.

5.3.2 Classification Scheme for Single-Mode Optical Fibers

It is possible to classify different types of fibers according to the magnitude and slope of the normalized waveguide at the operating point. In term of normalized frequency V, this is usually located in the range $0.5 < V/V_C < 1.0$ where V_C is the mode cutoff of the modes referred to LP_{11} or LP_{02}. Four classes of fibers can be categorized. These are listed in the following.

5.3.2.1 Fiber with Small Waveguide Dispersion

These are first generation single mode fibers which operate with zero dispersion at 1300 nm. Since this is very close to the zero dispersion due to materials of pure silica (1270 nm) only a very small waveguide dispersion is required to shift the zero to the low loss window of 1550 nm. For a step index profile only two parameters, a and Δ, can be varied to position λ_0 at 1300 nm. A large-Δ small core design is undesirable due to the doping induced excess Rayleigh losses. Such fibers are also susceptible to excessive drawing-induced losses. In addition, the zero dispersion wavelength is shifted to the wavelength of 1350 nm due to the higher Ge doping level and larger waveguide dispersion. Doping a fiber with Fe increases the zero material dispersion wavelengths [64]. This is too close to the OH– peak in the fiber attenuation characteristics. Thus a small Δ and large core radius is desirable. However, too small a relative RI difference would induce higher bending losses. A more flexible design is the depressed cladding step index fiber which allows λ_0 shifted close to the 1300 nm spectral region while at the same time allowing a core radius and the selection of Δ that provides tight mode confinement and low loss.

Modern fiber manufacturing has eliminated to OH– peak [65] thus this type of dispersion shift fibers in the 1300 nm region may find significant application in fiber to the premises (FTP) with upstream modulation using lightwave in the 1300 nm spectral region. The ultra-high bit rate reaching 100 Gb/s may find this type of fiber very useful.

5.3.2.2 Fibers with Large Uniform Waveguide Dispersion

In theory any profile shape can be designed to yield large waveguide dispersion for dispersion shifting. Large waveguide dispersion is needed at longer wavelengths due to the increase of the material dispersion. Uniform waveguide dispersion at the operating region is advantageous because the allowable tolerance on the core radius is related to the slope of the waveguide dispersion curve. The key to the design problem is how to obtain large uniform waveguide dispersion while at the same time maintaining tight mode confinement. To avoid bending losses, it is necessary to design a fiber with a large cutoff wavelength, λ_C. In our terminology this is equivalent to keeping the operating wavelength such that $V/V_C \approx 1$ since $\lambda \approx \lambda_C(V/V_C)$.

The first fully dispersion-shifted low loss single mode fiber was designed by White [36] and fabricated by Ainslie et al. [3]. This follows earlier problems with excess loss associated with the fabrication of fibers with small core, large Δ and step profile gradient [66]. However this triangular core fiber has sufficient waveguide dispersion at $V/V_C \approx 1$>. Thus its operating point is to be positioned at $V/V_C \approx 0.55$ to take advantage of large waveguide dispersion, resulting in the cutoff at about 850 nm. This fiber has been found to be bending loss sensitive.

Two earlier approaches have been taken to shift the cutoff wavelength towards the 1550 nm but <1300 nm to avoid closing the lower wavelength region. A depressed cladding has been found to shift the cutoff wavelength to 1100 nm [67] while maintaining the zero dispersion wavelength at 1540 nm>, however, a more elegant solution is to introduce a ring around the core [61] the effect of the ring is to shift the point of maximum waveguide dispersion towards $V/V_C \approx 1$>. Thus this option not only increases the cutoff wavelength but also ensures that the operating wavelength sits on the flat portion of the waveguide dispersion versus V/V_C curve. Since V is proportional to the core radius ($V = (2\pi)/(\lambda_0)\, an_{cl}\sqrt{2}$) the uniform waveguide dispersion of the ringed profile implies that the shifted wavelength is tolerable to the changes in the core radius. Modern types of dispersion-shifted fibers are the enhanced ring profile [54,68], Gaussian profile [69], and the convex profile [25,70]. Calculations by these works have shown that these profiles are insensitive to bending losses and dispersion-shifting has been achieved without imposing serious constraints on the reproducibility of the shifted wavelength.

5.3.2.3 Fibers with Very Large Steep Waveguide Dispersion

One of the inherent constraints of tailoring the dispersion to balance material dispersion is that waveguide dispersion design is eventually restricted by the dispersion characteristics of the material forming the fiber core. For Ge-doped silica-core fibers, the material dispersion curve in the 1270 nm and 1800 nm low loss window is steeply increasing. Thus, broadband fibers designed to minimize total dispersion over the same spectral range are required to possess a negative steeply-decreasing waveguide dispersion curve. This implies that the total dispersion curve will be very sensitive to changes in the core radius.

The first attempt to design dispersion-flattened fibers for broadband applications, especially in dense multi-wavelength optical communications and to avoid nonlinear four wave mixing effects is due to Okamoto [71]. However attempts to widen the low dispersion window of these W-fibers failed to produce a practical fiber. Detailed analysis by Francois [43] showed that for W-profile fiber with non-zero LP_{01} mode cutoffs, their dispersion characteristics are very sensitive to geometrical parameters variations. Such fibers with non-zero LP_{01} mode cutoff frequencies are just the type of fibers needed to reduce dispersion over a sufficiently wide spectral range. Furthermore, the LP_{01} modes in these structures are inherently leaky [32]. Thus, the total overall performance of these fibers is poor compared to conventional fibers.

Cohen, Mamel, and Lang [32] concluded that these W-profile fibers cannot maintain low losses simultaneously with zero-dispersion crossings near 1300 nm and 1550 nm. They first suggested that a practical method of reducing the fundamental mode leakage is to re-trap the lightwaves by enclosing the original profile by coupling effects between the inner core and the outer claddings. However due to mode coupling effects between the inner and the outer claddings, the upper limit to single mode operation can be limited by the LP_{02} mode cutoff [72]. Truly single mode dispersion-flattened fibers with dispersion <1.5 ps/nm/km over the spectral range 1330 nm to 1550 nm window have been designed and fabricated [73]. Triple-clad fibers with fluorine-doped claddings and pure SI core have an advantage over the quadruply-clad Ge-doped fibers. F-doping shifts the zero material wavelength below the 1270 nm window [26]. Mass production of these fibers has been reported [74].

5.3.2.4 Fiber with Ultra-Large Waveguide Dispersion

This group of single mode fibers is designed for nonlinear applications where the sign and the slope of the total dispersion are important. For fibers designed for soliton propagation [75,76] a large negative dispersion factor is required to maintain the stationary of the transmitted pulse whose width is <1 ps. A fiber profile with a very deep depressed inner cladding has been designed and fabricated for soliton transmission [77].

5.3.3 Practical Limit of Single-Mode Optical Fiber Design

Due to numerous parameters involved in the design of single mode fibers with multiple index structures, certain design rules have to be observed. This reduces the overall design effort to a managerial level. The design rules are dictated by several factors. These are derived from limitations of the tests, certain empirical rules pertaining to propagation properties of the fiber and constraints relationship that exist between optical operational parameters.

For fibers designed to operate in the 1550 nm window the intrinsic loss falls below 0.2 dB/km. Thus other mechanisms of fiber loss due to structural perturbations caused by external handling become significant.

Firstly, to keep excess Rayleigh loss to a minimum, Ge-doping in the core is kept below 1% [78]. If F-doped depressed claddings are required then a combination of deep cladding depression and low core doping level results in a small value of the normalized propagation constant b. This can lead to leaky mode losses. Furthermore, narrow and deep depressed claddings are difficult to fabricate. Fluorine diffuses from regions of low index to adjacent layers.

An additional complication caused by profiles with deep depressed claddings is that the small value of normalized propagation causes the mode spot size to increase substantially. This combination results in large macrobending losses. Furthermore, a large microbending loss is characteristic of fiber having too large a spot size [34]. For a dispersion-shifted fiber, the mode field distribution is intrinsically non-Gaussian. Its microbending loss excess exceeds Petermann's predicted loss [39]. On the other hand a large spot size is required to reduce splicing losses at joints. Again, from practical experiences a spot size of 5 µm is a good size for SMFs. However for dispersion-shifted and dispersion-flattened fibers the field at 1550 nm deviates significantly from the Gaussian profile [38]. Thus the standard method of spot size characterization becomes unreliable. Several new spot size definitions are available.

5.3.4 Fiber Design Methodology

In the design we structure the program into segments by modules each performing a specific task so that the parameters can be tuned without any complications. The original idea of processing is based on the reduction of total processing time on geometrical design rule checking for very large scale integration (VLSI) circuit.

In the analysis of dispersion characteristics of single mode fibers described above we can find some means for decomposing the total dispersion parameters into its constituents. Thus we may be able to trace the behavior of the dispersion as each parameter is varied.

The complete calculation of total dispersion at an arbitrary wavelength can be carried out once the Ge-doping level of Δ is specified. Further simplification can be made if we fix the value of the relative RI around a nominal and practical range. All state-of-the-art fiber manufacturing processes indicate that $\Delta < 1\%$ to achieve low loss single mode fibers for communication fibers with low dispersion, dispersion shifted and broadband applications. We thus restrict our design exercise to this nominal value of Δ. Note that a variation of the relative RI would require material engineers to alter the composition of the doping concentration of the core materials.

Material dispersion of several composite core refractive indices at different doping levels indicates very small change between them. Thus the normalized waveguide dispersion component is the single dominating factor. Thus the process of dispersion shifting can be implemented. At a fixed value of Δ and a specified zero-dispersion wavelength to be shifted, the waveguide dispersion and thence the factor $V(d^2(Vb)/dV^2)$ can be found. One can scan the factor $V(d^2(Vb)/dV^2)$ versus V/V_C until the correct amount of dispersion can be found. For certain profile shapes, this search may fail since the restricted range is set at $0.55 < V/V_C < 1$. Then one has to change profile shapes either to increase the level of waveguide dispersion or shift the maximum value of dispersion to the values of V/V_C close to unity.

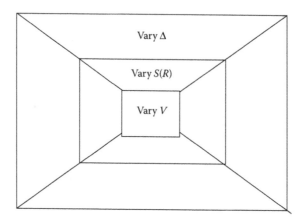

FIGURE 5.2
Systematic approach to single mode fiber design: a hierarchical scheme.

If this too fails to produce a solution, the value of Δ may be not be suitable, thus a change to maximum value of Δ, i.e., maximum doping level in the core, remains as the last option. A hierarchical approach to the design of single mode fiber is summarized in Figure 5.2.

5.3.5 Design Parameters and Equations

5.3.5.1 Group Velocity Dispersion (GVD)

The total dispersion factor, D (in ps/(km-nm)) of an SMF is given by:

$$D = -\left(\frac{2\pi c}{\lambda^2}\right)\beta_2 \equiv D_M + D_W \tag{5.53}$$

where β_2 is the well known GVD parameter which is the second order derivative of β, the propagation constant of the guided mode. D_M and D_W are the material and waveguide dispersion parameters, respectively. Although these factors are well known we believe that a brief outline of their meaning and expressions is essential to present our new design algorithm for multiple-clad dispersion-flattening fibers.

The third order or dispersion slope, essential for dense wavelength division multiplexed (WDM) transmission should be taken into account and can be obtained from the higher order derivatives of the propagation constant as

$$\beta_3 = \frac{d\beta_2}{d\omega} = \frac{d^3\beta}{d\omega^3} \tag{5.54}$$

The material dispersion in an optical fiber is due to the wavelength dependence of the RI of the core and cladding. The RI $n(\lambda)$ is approximated by the well known Sellmeier's equation

$$n^2(\lambda) = 1 + \sum_{j=1}^{M} \frac{B_j \lambda^2}{(\lambda^2 - \lambda_j^2)} \tag{5.55}$$

TABLE 5.1

Sellmeier's Coefficients for Silica-based Material and Doping Concentration for the Design

Type	Doping Conc.	SiO$_2$	B$_1$	B$_2$	B$_3$	λ$_1$	λ$_2$	λ$_3$
A (1)	0%	100%	0.696 1663	0.407 9426	0.897 4794	0.068 4043	0.116 2414	9.896 161
B (2)	3.1%	96.9%	0.702 8554	0.414 6307	0.897 4540	0.072 7723	0.114 3085	9.896 161
C (3)	5.8%	94.2%	0.708 8876	0.420 6803	0.895 6551	0.060 9053	0.125 4514	9.896 162
D (4)	7.9%	92.1%	0.713 6824	0.425 4807	0.896 4226	0.061 7167	0.127 0814	9.896 161
E (5)	0%	pure	0.696 750	0.408 218	0.890 815	0.069 066	0.115 662	9.900 559
F (6)	13.5%	86.5%	0.711 040	0.408 218	0.704 048	0.064 270	0.129 408	9.425 478
G (7)	9.1%	90.9%	0.695 790	0.452 497	0.712 513	0.061 568	0.119 921	8.656 641
H (8)	13.3%	86.7%	0.690 618	0.401 996	0.898 817	0.061 900	0.123 662	9.098 960
I (9)	1%	99%	0.691 116	0.399 166	0.890 423	0.068 227	0.116 460	9.993 707
J (10)	48.7%	51.3%	0.796 468	0.497 614	0.358 924	0.094 359	0.093 386	5.999 652

where λ_j indicates the ith resonance wavelength and B_j is its corresponding oscillator strength. n stands for n_1 or n_2 for core or cladding regions. These constants are tabulated in Table 5.1 for several material types. The first three Sellmeier terms, B_1, B_2 and B_3, are normally used.

The first, second and third order derivatives of Equation 5.55 with respect to wavelength can be easily obtained using symbolic manipulation as listed in Appendix A. They are then used to find the material dispersion factor, D_M which can be obtained by

$$D_M = -\frac{\lambda}{c}\left(\frac{d^2 n(\lambda)}{d\lambda^2}\right) \tag{5.56}$$

where c is the velocity of light in a vacuum. For pure silica and over the spectral range of 1.25–1.66 µm, D_M can be approximated by an empirical relation [79]

$$D_M \approx 122\left(1 - \frac{\lambda_{ZD}}{\lambda}\right) \tag{5.57}$$

where λ_{ZD} is the zero material dispersion wavelength. For instance, $\lambda_{ZD} = 1.276$ µm is only for pure silica. λ_{ZD} can vary in the range 1.27–1.29 µm for optical fibers whose core and cladding are doped to vary the RI.

The waveguide dispersion D_W can be approximated as [80]:

$$D_W = -\left(\frac{n_1 - n_2}{\lambda_c}\right)V\frac{d^2\left(Vb\right)}{dV^2} \tag{5.58}$$

where b is the normalized propagation constant which would be defined for different regions across the entire area of the fiber cross section. The normalized waveguide dispersion parameter $V(d^2(Vb)/dV^2)$ depends strongly on the normalized frequency parameter V. It plays a central role in the waveguide dispersion effect.

Segmented-layer-index profiles can be used for flattening and lowering the fiber dispersion factor over the two low loss spectral windows of 1300 nm and 1550 nm. It is apparent that dispersion flattening in the wavelength region of interest can be achieved only if a layer with lower RI than that of the cladding is introduced close to the core, i.e., a depressed cladding [83].

5.3.5.2 Dispersion Slope

The dispersion slope can be defined as

$$S = \frac{dD}{d\lambda} \, \text{ps/nm}^2\text{km} \tag{5.59}$$

or alternatively as [12,83]

$$S = \left(2\pi c/\lambda^2\right)\beta_3 + \left(4\pi c/\lambda\right)\beta_2 \tag{5.60}$$

where β_2 and β_3 are the higher order derivatives of the propagation constant. Their expressions are listed in Appendix B.

5.3.6 Triple-Clad Profile

In this section the relationship between the fiber structural geometry, the index profile, its total dispersion and the mode spot size are examined, especially for multiple-clad index profile fibers. Figure 5.3 shows the RI profile distribution of a triple-clad optical fiber. Ten different fiber material types can be used for the core and/or cladding regions. The maximum dispersion not higher than 3 ps/(nm-km) is set as an example in this paper, over the wavelength range of 1300–1580 nm. These materials and their Sellmeier coefficients are tabulated in Table 5.1. They are numerically coded as Type 1 to Type 10 or alphabetically corresponding from A to H.

The waveguide dispersion factor plays an important role in "*shaping*" the total dispersion curve. As there are three different layers in the cladding regions (see Figure 5.1), it is expected the waveguide dispersion factors for the three cladding regions, namely D_M, D_W and D_{TOT} constitute the total dispersion based on the principle of superposition. Hence, the effects of each structural parameter are designed to satisfy specific fiber dispersion properties.

5.3.6.1 Profile Construction

The non-normalized and normalized RI profiles of a triple-clad step index fiber are shown schematically in Figure 5.1a and b, respectively, a_i—the ith outer radius, n_i—the RI of the ith layer and n—the RI of the uniform cladding. The RI of the ith layer relative to that of the uniform cladding is thus given by [81,82]:

$$\Delta_i = \frac{n_i^2 - n_{cl}^2}{2n_{cl}^2} \approx \frac{n_i - n_{cl}}{n_{cl}} \tag{5.61}$$

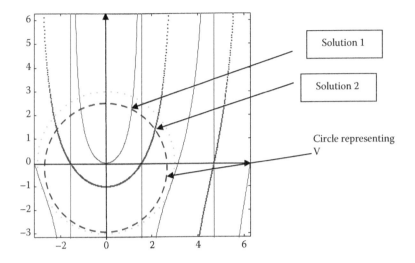

FIGURE 2.22
Graphical solution of Equations 2.64, 2.72, or 2.73: _____ $v = u \tan u$, _ _ _ _ $v = -u/\tan u$, and -------.$V^2 = u^2 + v^2$.

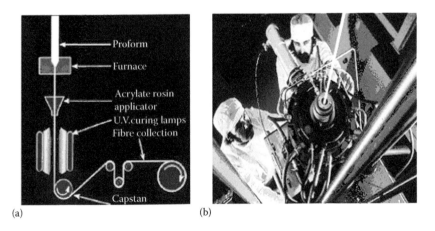

FIGURE 4.9
(a) Schematic of fiber drawing machine. (b) Picture of fiber microwave furnace and diameter monitoring and feedback control.

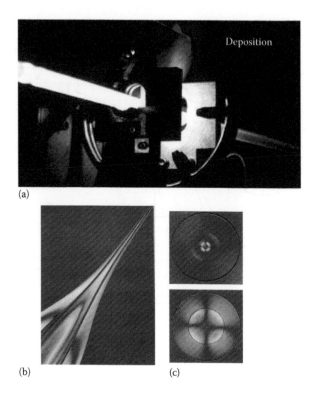

(a)

(b)   (c)

FIGURE 4.10
(a) Schematic of a fiber deposition and fabrication of a fiber perform, deposition of core material and collapsing.
(b) Fiber perform, before drawing into fiber stands. (c) Cross section of a fiber perform with refractive index
profile exactly the same as the fiber index profile of SM (upper) and multimode (lower) types.

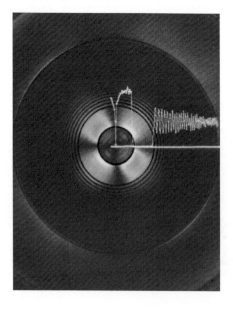

FIGURE 4.11
Real index profile across an SM fiber perform. Note: non-step like profile—so why modeled as step index
structure?

FIGURE 4.12
Installation of fiber cables. (a) Installation of fiber cable by hanging. (b) Installation of fiber cable by ploughing. (c) Installation of undersea fiber cable. (d) Splicing two optical fibers.

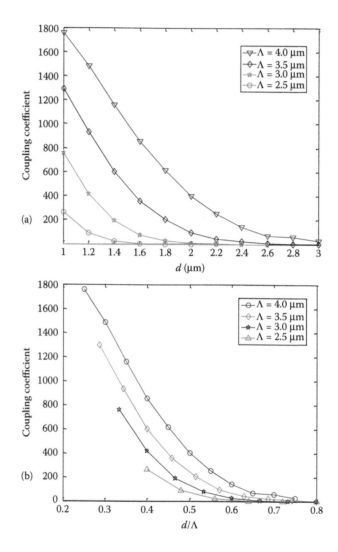

FIGURE 7.31

(a) Coupling coefficient vs. hole diameter, Λ varies from 2.5 to 4.0 μm. (b) Coupling coefficient with the hole-pitch size as a parameter, Λ varies from 2.5 to 4.0 μm.

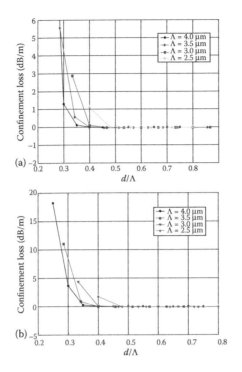

FIGURE 7.33
(a) Losses for structure D = 2 × pitch. (b) Losses for structure D = 4 × pitch.

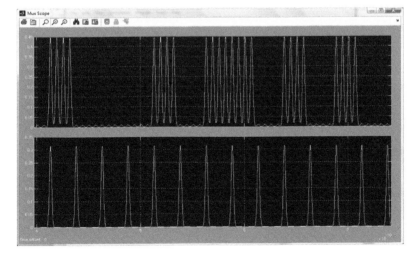

FIGURE 8.37
Upper curve depicts multiplexed pulse sequence; lower traces displays demultiplexed sequence.

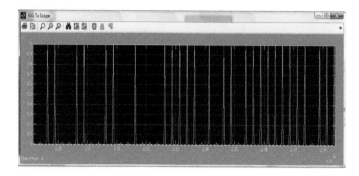

FIGURE 8.38
Triple product pulse received at the output of the nonlinear optical waveguide.

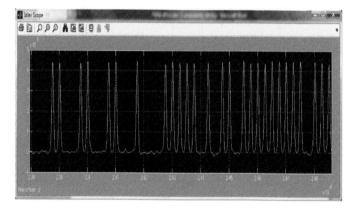

FIGURE 8.39
Decoded pulse sequence from product of Figure 8.38.

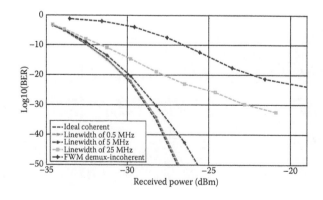

FIGURE 8.52
Back to back performance of incoherent receiver and coherent receiver with different phase noise levels of pulsed laser.

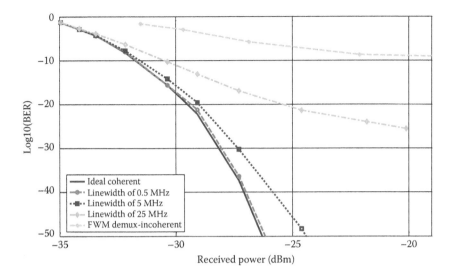

FIGURE 8.53
Transmission performance of incoherent receiver and coherent receiver with different phase noise levels of pulsed laser.

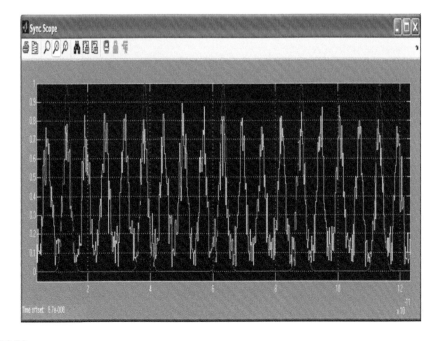

FIGURE 8.54
Synchronization between the control and the OTDM pulses.

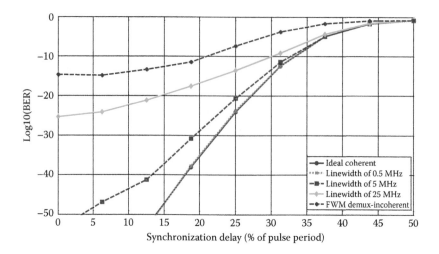

FIGURE 8.57
BER curves versus the synchronization delay.

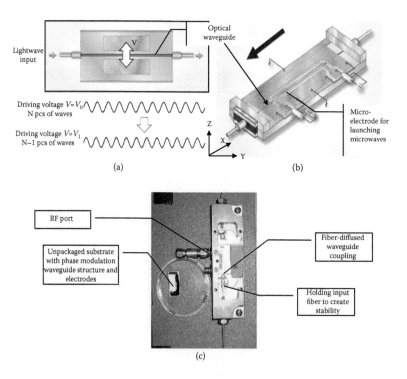

FIGURE 9.1
EO-PM in an integrated modulator using LiNbO$_3$. Electrode impedance matching is not shown. (a) Schematic diagram; (b) integrated optic structure; and (c) photograph of a packaged modulator.

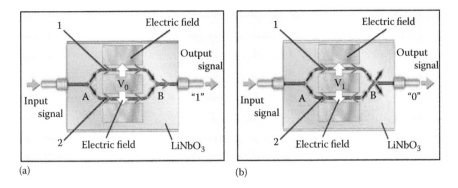

(a)   (b)

FIGURE 9.2
Intensity modulation using interferometric principles in guide wave structures in LiNbO$_3$: (a) ON—constructive interference mode; (b) destructive interference mode—OFF. Optical guided wave paths 1 and 2. Electric
field is established across the optical waveguide.

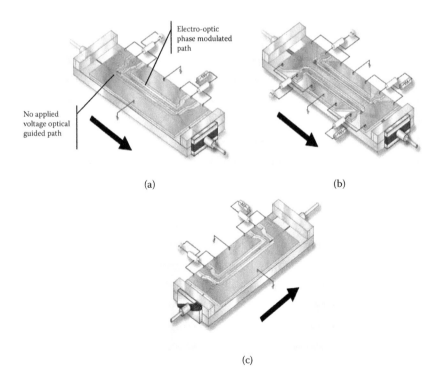

(a)   (b)

(c)

FIGURE 9.8
Intensity modulators using LiNbO$_3$: (a) single drive electrode (b) dual electrode structure; and (c) EO polarization scrambler using LiNbO$_3$.

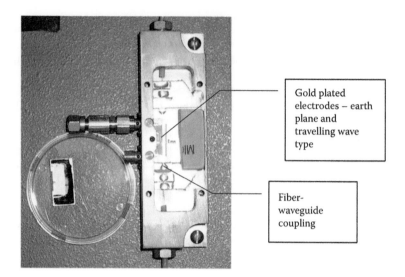

FIGURE 9.26
Fabricated and packaged optical modulator: the substrate and fabricated electrode shown on the left side of the photograph and the packaged 26 GHz modulator with fiber pigtail shown on the right side.

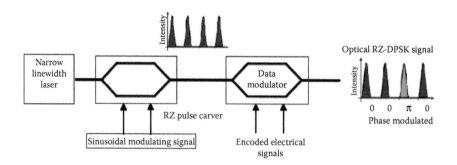

FIGURE 9.44
Block diagram of RZ-DPSK photonics transmitter. Phase modulated shown by the intensity of the shaded region under the pulses.

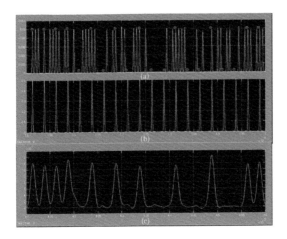

FIGURE 10.14
Time traces of: (a) the 160 Gb/s OTDM signal, (b) the control signal, and (c) the 40 Gb/s demultiplexed signal.

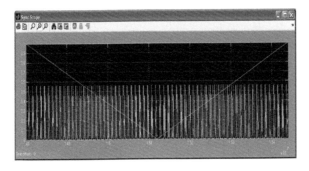

FIGURE 10.17
The variation in time domain of the time delay (cyan), the original signal (violet) and the delayed signal (yellow).

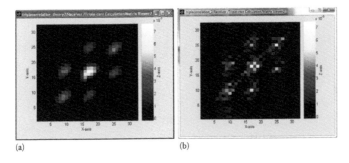

(a)   (b)

FIGURE 10.20
Triple correlation of the dual-pulse signal based on: (a) theoretical estimation, and (b) FWM in NL waveguide.

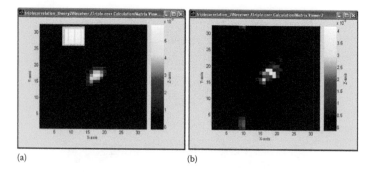

(a)   (b)

FIGURE 10.21
Triple correlation of the single-pulse signal based on: (a) theoretical estimation, and (b) FWM in NL waveguide.
(Inset: the single-pulse pattern.)

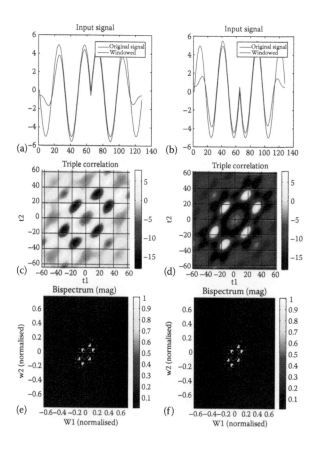

FIGURE 10.23
Input waveform with phase changes at the transitions (a and b), triple correlation and bispectrum (c–h) of both
phase and amplitude.

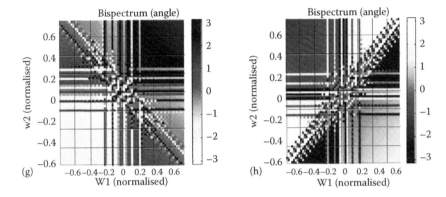

(g)

(h)

FIGURE 10.23 (Continued)

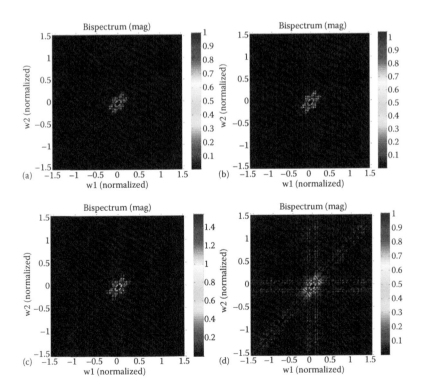

(a)

(b)

(c)

(d)

FIGURE 10.24

Effect of Gaussian noise on the bispectrum (a and c) amplitude distribution in two dim (b and d) phase spectral distribution.

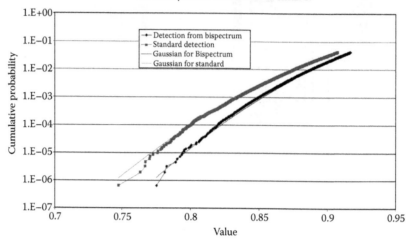

Cumulative probability distribution, with 10% filtered Gaussian noise for conventional detection and for detection from the bispectrum. 750000 symbols. Normalised to mean 1.

FIGURE 10.25
Error estimation version detection level of the HOS processor.

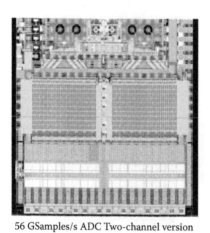

56 GSamples/s ADC Two-channel version
using CHAIS architecture
(a)

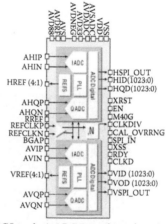

56 GSamples/s ADC Four-channel version

(b)

FIGURE 10.26
Plane view of the Fujitsu ADC operating at 56 GSamples/s: (a) integrated view, and (b) operation schematic.

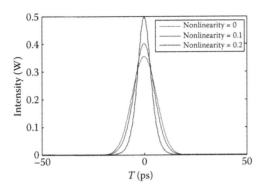

FIGURE 10.29
Pulses with different cavity nonlinearity values.

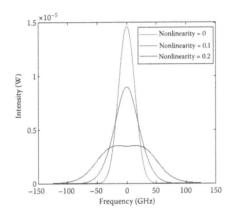

FIGURE 10.31
Pulse spectra with different cavity nonlinearity values.

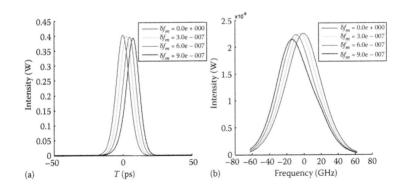

FIGURE 10.33
Steady-state pulses and spectra of an actively mode-locked fiber laser having cavity nonlinearity of 0.1 with different detuning values.

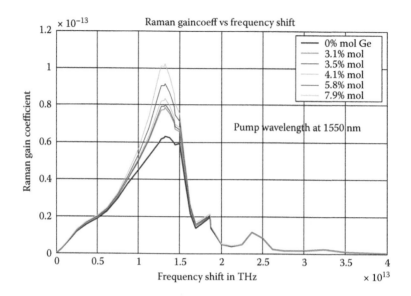

FIGURE 10.58
Spectrum of Raman gain coefficient with various Germanium doping concentrations.

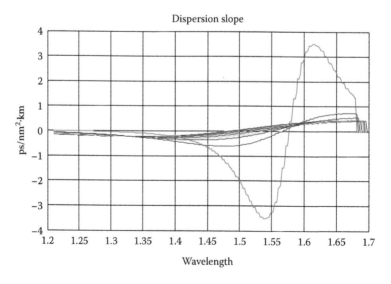

FIGURE 10.65
Dispersion slope of profile of Table 10.7.

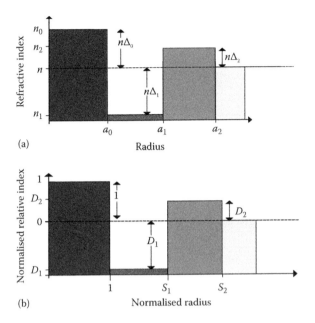

FIGURE 5.3
RI profile of a triple-clad fiber: (a) non-normalized index profile, and (b) normalized profile.

The normalized outer radius is also defined as

$$S_i = \frac{a_i}{a_0} \tag{5.62}$$

The normalized relative index of the ith layer is then given by

$$D_i = \frac{\Delta_i}{\Delta_0} \tag{5.63}$$

with $S_0 = 1$ and $D_0 = 1$. It is convenient to express the degrees of freedom in terms of the structural parameters a_0, S_1, S_2, n_0, D_1, D_2 and n.

A uniform cladding of pure silica (material type A) is normally chosen for analysis and the Sellmeier expansion is used to calculate the RI and its derivatives with wavelength. Ten different silica-based material types of different doping concentrations of dopants, as tabulated in Table 5.1, are used in the analysis as the core and/or cladding materials of the triple-clad profile for tailoring the total dispersion factor. A maximum total dispersion limit, not larger than 3 ps/(nm-km) set in this work, over 1300–1580 nm window. This low and non-zero dispersion ceiling allows sufficiently wide length-bandwidth product for ultra-high bit rate transmission, compatible with that of the common non-zero dispersion-shifted fibers (ITU G655 type) and ease of manufacturing.

A MATLAB design platform is developed for multiple-clad dispersion-flattened and compensating single mode fibers. The V-dependent parameter representing D_W are numerically determined. Seven fiber parameters that affect the performance of the triple-clad fibers are: core radius (a_0), first cladding radius (a_1), second cladding radius (a_2), core index (n_0), first cladding index (n_1), second cladding index (n_2) and outer cladding index (n) are used as the degree of freedom in the designs.

5.3.6.2 Waveguide Guiding Parameters of Triple-Clad Profile Fiber

The transverse propagation constants of the guided lightwaves u/a and v/a in the core and cladding regions, respectively, are given for the core and the first and second cladding layers (corresponding to subscripts 0, 1 and 2, respectively) of the triple-clad index profile fibers as

$$u_0 = a_0\sqrt{k^2 n_0^2 - \beta_0^2}\,;\ u_1 = a_1\sqrt{k^2 n_1^2 - \beta_1^2}\,;\ u_2 = a_2\sqrt{k^2 n_2^2 - \beta_2^2}\,; \tag{5.64}$$

$$v_0 = a_0\sqrt{\beta_0^2 - k^2 n^2}\,;\ v_1 = a_1\sqrt{\beta_1^2 - k^2 n^2}\,;\ v_2 = a_2\sqrt{\beta_2^2 - k^2 n^2}\,; \tag{5.65}$$

where β_0, β_1 and β_3 are the propagation constants of the guided waves in the core, the first and second cladding layers are given by

$$\beta_0 = \sqrt{k^2\left(b_0\left(n_0^2 - n^2\right) + n^2\right)}\,;\ \beta_1 = \sqrt{k^2\left(b_1\left(n_1^2 - n^2\right) + n^2\right)}\,;$$

$$\beta_2 = \sqrt{k^2\left(b_2\left(n_2^2 - n^2\right) + n^2\right)} \tag{5.66}$$

Hence the normalized frequencies for all layers can be expressed as

$$V_0 = a_0 k\sqrt{n_0^2 - n^2}\,;\quad V_1 = a_1 k\sqrt{\left|n_1^2 - n^2\right|}\,;\quad V_2 = a_2 k\sqrt{n_2^2 - n^2} \tag{5.67}$$

and the effective V of the multiple-clad fiber can be defined as

$$V_{eff} = k a_0 \sqrt{2n\left(\left(n_0 - n_1\right) + \left(n_1 - n_2\right) + \left(n_2 - n\right)\right)} \tag{5.68}$$

The spot size r_0 can be found analytically as [12]

$$r_0 = \sqrt{\frac{a_0^2}{\ln V_{eff}^2}} \tag{5.69}$$

The spot size can be approximated so that not only the minimum dispersion is achieved but also the requirement for maximum effective area is satisfied.

The radial modal intensity distribution $I(r)$ is given by

$$I(r) \cong \exp\left[-\frac{1}{2}\left(\frac{r}{r_0}\right)^2\right] \tag{5.70}$$

The waveguide dispersion factors of different fiber layers can be obtained as:

$$D_{w0} = -\left(\frac{n_0 - n_1}{\lambda c}\right) V_0 \frac{d^2\left(V_0 b\right)}{dV_0^2} \tag{5.71a}$$

$$D_{W1} = -\left(\frac{n_1 - n_2}{\lambda c}\right) V_1 \frac{d^2 (V_1 b)}{dV_1^2} \tag{5.71b}$$

$$D_{W2} = -\left(\frac{n_2 - n}{\lambda c}\right) V_2 \frac{d^2 (V_2 b)}{dV_2^2} \tag{5.71c}$$

where the generic normalized waveguide dispersion coefficient is $V(d^2(Vb)/dV^2)$. For tri-ple-clad fibers we have the following three normalized waveguide dispersion parameters

$$\frac{V_0 d^2 (V_0 b)}{dV_0^2} = 2\left(\frac{u_0}{V_0}\right)^2 \times \left\{ K_0 (1 - 2K_0) + \frac{2}{v_0} \left(v_0^2 + u_0^2 K_0\right) \sqrt{K_0} \left(K_0 + \frac{1}{v_0}\sqrt{K_0} - 1\right) \right\} \tag{5.72a}$$

$$\frac{V_1 d^2 (V_1 b)}{dV_1^2} = 2\left(\frac{u_1}{V_1}\right)^2 \times \left\{ K_1 (1 - 2K_1) + \frac{2}{v_1} \left(v_1^2 + u_1^2 K_1\right) \sqrt{K_1} \left(K_1 + \frac{1}{v_1}\sqrt{K_1} - 1\right) \right\} \tag{5.72b}$$

$$\frac{V_2 d^2 (V_2 b)}{dV_2^2} = 2\left(\frac{u_2}{V_2}\right)^2 \times \left\{ K_2 (1 - 2K_2) + \frac{2}{v_2} \left(v_2^2 + u_2^2 K_2\right) \sqrt{K_2} \left(K_2 + \frac{1}{v_2}\sqrt{K_2} - 1\right) \right\} \tag{5.72c}$$

with

$$K_0 = \frac{BESSELK_1(v_0)}{BESSELK_0(v_0)}; \ K_1 = \frac{BESSELK_1(v_1)}{BESSELK_0(v_1)}; \ K_2 = \frac{BESSELK_1(v_2)}{BESSELK_0(v_2)} \tag{5.73}$$

where $BESSELK_i$ is the ith order modified Bessel function. Finally, using the superposition position the total dispersion of multiple-clad fibers is obtained as:

$$D_{TOT} = D_M + D_{W0} + D_{W1} + D_{W2} \tag{5.74}$$

5.4 Dispersion Flattening and Compensating

This section considers the design of dispersion-flattened and compensating fibers for 40 Gb/s DWDM channels over the C, L and S bands. Generic observation and approximation of the waveguide dispersion are given. They are then used in the designs for dispersion flattening and compensating.

The design of a dispersion flattened fibers (DFF) was first considered in the late 1970s [85]. In Ref. [85], a comprehensive investigation describes the doubly clad fiber design for DFF, with in-depth calculations of the cutoff properties and index profile influence. Practical results presented in this reference are compared in this work. In Ref. [85], the Bessel functions as solutions of the wave equations are used for the parameter $V(d^2(Vb)/dV^2)$.

Dispersion compensating fiber (DCF) have been analyzed since the early 1970s. However at that time the repeaterless distance was limited by the fiber loss until the advent of optical amplifiers in the 1990s, its development has attracted much attention. In Ref. [22], extensive mathematical concepts of DCF were described. Although the RI profile is not an ideal step index, the superposition principles have been used in the design of DCFs. In Ref. [20], a figure of merit defined as the ratio of dispersion to attenuation has been proposed to investigate the interplay between the fiber dispersion, attenuation and its effective area to evaluate the performance of a dispersion-managed optical transmission system. In Ref. [21], the concept of the figure of merit is further extended. An optimal design for DCF and its fabrication are also given in Ref. [5].

We define a dispersion slope compensation ratio (DSCR) of a DCF, termed as the "Kappa" factor as

$$DSCR = \frac{S_{DCF}}{D_{DCF}} \Big/ \frac{S_{DFF}}{D_{DFF}} \tag{5.75}$$

where S_{DCF}, S_{DFF}, D_{DCF} and D_{DFF} are the dispersion slopes and factors of the dispersion compensating and dispersion-flattened fibers, respectively. Similarly, a term "Kappa" [18] is used instead for DSCR which is specified by a number of fiber manufacturers for DCF modules.

This section describes detailed designs of both DFF and DCF with multiple-clad index profiles using materials listed in Table 5.1. Optimum profiles, material types and their effects on the fiber dispersion properties are considered. Fibers with the least dispersion ripple over the two low loss spectral windows are given for optimum ultra-wideband, ultra-long reach and ultra-high capacity DWDM optical communication systems.

5.4.1 Approximation of Waveguide Dispersion Parameter Curves

Figure 5.4 shows the complete shape of the waveguide parameters as given in Equation 5.20 and the approximated curve for its right half where the *V*-parameter range is fallen into the design range of SMFs. Figure 5.5 shows three $V(d^2(Vb)/dV^2)$ exact curves as represented by Figures 5.4 through 5.6. Neither simple approximation nor exact representation of $(Vd^2(Vb)/dV^2)$ curves has been found in published literatures. Most published works [15,16] and the

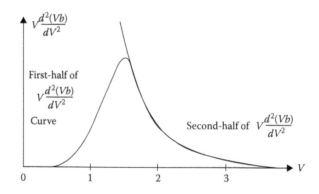

FIGURE 5.4
$V(d^2(Vb)/dV^2)$ complete curve, dotted line is obtained from Equations 5.72a–d.

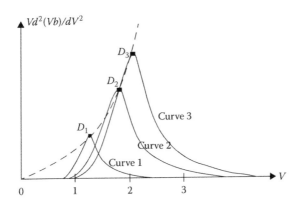

FIGURE 5.5
Family curves of $((Vd^2(Vb)/dV^2))$.

developed algorithm in Ref. [22] are far too complicated for the design process, particularly at the design inception stage. Therefore, we propose in this section an *algorithm* to predict the behavior of $(Vd^2(Vb)/dV^2)$ in the following steps:

Step 1: Approximate the right half of the exact $V(d^2(Vb)/dV^2)$ curves as given in Figures 5.4 through 5.6, by a simple analytical expression with an error of <2% for $1 < V < 3$ [14]. Inspecting a number of a family of $V(d^2(Vb)/dV^2)$ curves [10,19,22] shown in Figure 5.5, the peaks follow a simple cubic equation given by

$$\left(V\frac{d^2(Vb)}{dV^2}\right) = V^3 \tag{5.76}$$

Thus, the intercepting point $D(Dx, Dy)$ in Figure 5.5 is the corresponding peak of one of the curves. A correction factor, mainly the constant of the cubic equation can be used to modify, if required. Higher order polynomial is unnecessary in this case.

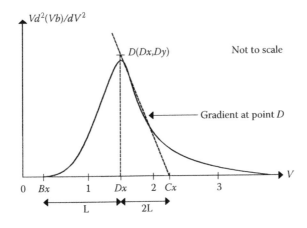

FIGURE 5.6
The intersection point D and the approximated $((Vd^2(Vb)/dV^2))$ curve.

Step 2: Referring to the point $D(Dx, Dy)$ of Figure 5.6, we could find the gradient at that point and hence the point Cx given by

$$Cx = Dx - \frac{Dy}{\left(\dfrac{d}{dV} \left(\dfrac{Vd^2(Vb)}{dV^2} \right) \right)_{at(Dx,Dy)}} \tag{5.77}$$

Step 3: Having found the location of $C(Cx, 0)$ we could predict point $B(Bx, 0)$, that is the cutoff point of $V(d^2(Vb)/dV^2)$ curve as

$$Bx = Dx - \frac{1}{2}(Cx - Dx) \tag{5.78}$$

Step 4: Now considering the curve from point B to point D, additional points are to be introduced to obtain a desired shape. Likewise, a few points are selected to represent the curve in the right-half of the $V(d^2(Vb)/dV^2)$ curve. We found that ten points to represent the $(Vd^2(Vb)/dV^2)$ curve would be adequate. Having obtained these significant points, the task is interpolating them to form a smooth curve. *Spline interpolation* method has been adopted for this purpose.

Step 5: As we are using the $(Vd^2(Vb)/dV^2)$ curve to find the waveguide dispersions defined in Equation 5.72a–c, the curves as shown in Figure 5.5 can be represented by a general mathematical expression as a function of wavelength. Thus, a polynomial of the 9th order has been chosen for this purpose (see Equation 5.81).

For single mode, the normalized frequency is given by,

$$V_i = 2.405 \left(\frac{\lambda_{ci}}{\lambda} \right) \tag{5.79}$$

where λ_{ci} is the cutoff wavelength of the *i*th region (Figure 5.7).

Hence, a 9th order of $V(d^2(Vb)/dV^2)$ polynomial can be approximated by

$$V\frac{d^2(Vb)}{dV^2} = P_0 + P_1 \left(\frac{2.405\lambda_c}{\lambda} \right) + P_2 \left(\frac{2.405\lambda_c}{\lambda} \right)^2 + P_3 \left(\frac{2.405\lambda_c}{\lambda} \right)^3 + \cdots + P_9 \left(\frac{2.405\lambda_c}{\lambda} \right)^9 \tag{5.80}$$

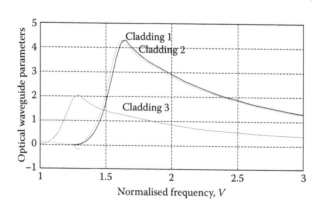

FIGURE 5.7
The optical waveguide parameter curves for optical fibers with three cladding types.

or

$$V\frac{d^2(Vb)}{dV^2} = \sum_{m=0}^{9} P_m\left(\frac{2.405\lambda_c}{\lambda}\right)^m \tag{5.81}$$

where $P_0, P_1, P_2, \ldots, P_9$ are the polynomial constants obtained by *polyfiting of MATLAB®* [23]. The above steps allow seven degrees of freedom for fiber designs. By analyzing the effect of each parameter we would be able to predict the changes of the dispersion factor and hence identify the main factors that would play a principal role for tailoring the dispersion.

As mentioned above, material types A, B, C, D, E, F, G, I and J can be alternatively used as either core or cladding materials to satisfy the dispersion-flattening limit. Material type H has the total dispersion of about 9 ps/(nm-km) in the range 1300–1580 nm. Analyzing all the modeled total dispersion curves it can be observed that the fiber with material type J gives the optimum dispersion flattening. A total dispersion <3 ps/(nm-km) from 1280–1620 nm can be achieved. There is a maximum of three zero dispersion points. The first and second zero dispersion points are approximately located in the two spectral windows at 1300 nm and 1550 nm.

5.4.2 Effect of Core and Cladding Radius on the Total Dispersion

Figures 5.8 through 5.11 show the variations of core/cladding radii with respect to the total dispersion factor. We observe the following effects of the variation of the core/cladding radii on the dispersion factor over the entire 1300–1600 nm spectral region

1. As the core radius, a_0, is increased the total dispersion curve is shifted upwards, at the same time the second zero dispersion point is shifted to a higher wavelength region. Meanwhile, the first zero dispersion point remains unchanged and the third zero dispersion point gradually shifts to a lower wavelength region. By analyzing the behavior of the dispersion curves of several simulated results, we

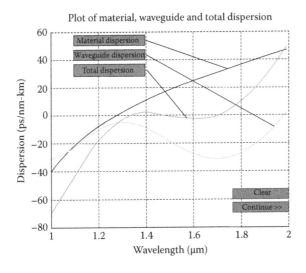

FIGURE 5.8
Material, waveguide and total dispersion of "type A" as the core material of a triple-clad profile as a parameter.

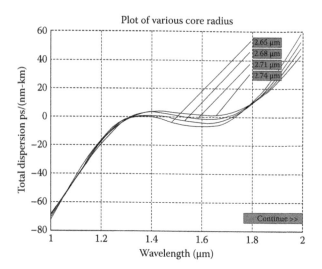

FIGURE 5.9
Total dispersion factor versus wavelength with the core radius of the triple-clad optical fiber as a parameter.

obtain the maximum sensitivity of changes in a_0 to total dispersion is about 88.88 ps/(nm-km-µm).

2. As the first cladding radius, a_1, is decreased the total dispersion curve shifts upwards. The maximum sensitivity due to incremental change in a_1 to total dispersion is about 64.68 ps/(nm-km-µm).

3. As the second cladding radius a_2 is decreased, the total dispersion curve shifts upwards, but the first zero dispersion point shifts to a lower wavelength region,

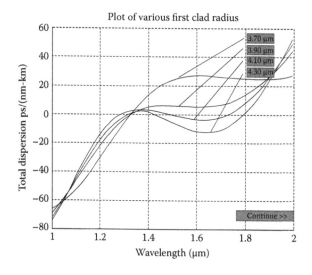

FIGURE 5.10
Total dispersion factor versus wavelength with the first clad radius of the triple-clad optical fiber as a parameter.

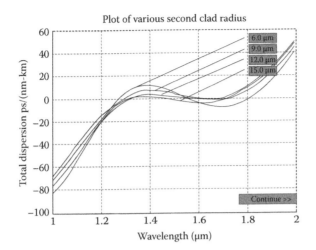

FIGURE 5.11
Total dispersion factor versus wavelength with the second clad radius of the triple-clad optical fiber.

a maximum sensitivity due to incremental changes of a_2 to total dispersion in the fiber windows region of 0.99 ps/(nm-km-μm) is obtained.

Therefore it can be concluded that the change in core radius, a_0 is very sensitive to the total dispersion (i.e., 88.88 unit dispersion per μm) as compared to that of the outer radius of the second layer a_2. Thus for the triple-clad fiber the selection of a_0 is very critical to achieve a specific non-zero dispersion factor. The sensitivity of each core radius is compared with respect to that of the second cladding layer and tabulated in Table 5.2 that reconfirms that the most sensitive factor is the central core radius. Thus the manufacturing tolerance of the fiber core radius must be controlled accurately as compared to those of the cladding layers.

5.4.3 Effects of Refractive Indices of the Cladding Layers on the Total Dispersion Parameter

Figures 5.12 through 5.15 show various curves representing the ratio of the core and cladding refractive indices versus the total dispersion parameter. The following effects of each core/cladding radius on the total dispersion factor can be noted

1. As the core RI, n_0, is reduced, the total dispersion curve shifts upwards. A maximum sensitivity of incremental changes in n_0 to fiber total dispersion is 50,000 ps/(nm-km) per unit RI is obtained.
2. When the first cladding RI, n_1, is increased the total dispersion curve shifts upwards but the first zero dispersion point shifts to a higher wavelength region. A

TABLE 5.2

Normalized Sensitivity Comparison of Core, First and Second Cladding Radius

	a_0	a_1	a_2
Normalized Sensitivity	89.7	65.3	1

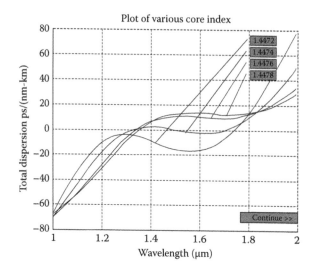

FIGURE 5.12
Total dispersion factor versus wavelength with the core RI of the triple-clad optical fiber as a parameter.

maximum sensitivity of incremental changes in n_1 to total dispersion is 59,200 ps/(nm-km) per unit RI is obtained.

3. As the second cladding RI, n_2, is decreased the total dispersion curve shifts upwards but the first zero dispersion point shifts to a lower wavelength region. Meanwhile, the second zero dispersion point remains almost unchanged. A maximum sensitivity of incremental change in n_1 to total dispersion of 6666 ps/(nm-km) per unit RI is obtained.

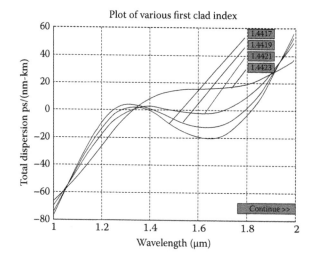

FIGURE 5.13
Dispersion factor versus wavelength with the first cladding RI of the triple-clad profile as a parameter.

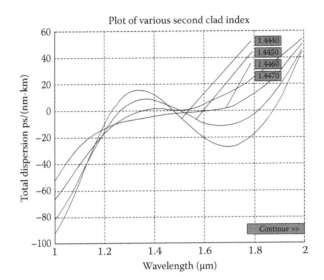

FIGURE 5.14
Total dispersion factor as a function of the second cladding RI of the triple-clad profile.

4. As the outer cladding RI, n, is decreased the total dispersion curve shifts upwards but the first and second zero dispersion points are shifted to a higher wavelength region. We obtain a maximum sensitivity of changes in n_1 to fiber total dispersion of about 142,850 ps/(nm-km) per unit RI.

We thus conclude that changes in outer cladding RI, n is sensitive to the total dispersion (i.e., 142,850 unit dispersion per unit RI) compared to the n_2. Hence selecting n is very

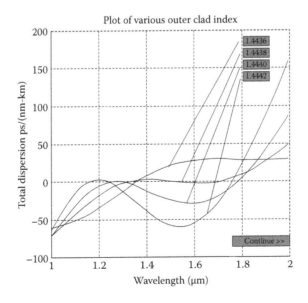

FIGURE 5.15
Total dispersion parameter varies with variation of the outer cladding RI of the triple-clad profile.

TABLE 5.3

Normalized Sensitivity Comparison of Core, First,
Second and the RI of the Outer Most Cladding Region

	n_0	n_1	n_2	n
Normalized Sensitivity	7.50	8.88	1.00	21.43

critical in the design of triple-clad step index optical fibers. The normalized sensitivity of the refractive indices of the core and the cladding layers with respect to the RI of the third cladding layer n_2 is tabulated in Table 5.3.

This layer is chosen for normalization due to its closeness to the outer most cladding layer. It shows clearly that the outer most cladding layer is the most sensitive.

5.4.4 Effect of Doping Concentration on the Total Dispersion

Figure 5.14 indicates that increasing doping concentration would shift the total dispersion curve down slightly. The change is minute. An estimated change of 0.5 unit dispersion per unit concentration is measured. Thus the doping concentration in the core region does not play a major role in the flattening of the total dispersion curve.

Hence, the doping concentration would thus now be the last factor to be considered for the design of triple-clad step index fibers. This factor contributes significantly to the fiber attenuation.

5.5 Design Algorithm

5.5.1 Design Algorithm for DFF

Step 1: Initially the material dispersion is calculated based on the materials tabled in Table 5.1. Figure 5.15 shows typical spectral distribution of the material, waveguide and total dispersion.

Step 2: The criteria for the design of a low non-zero dispersion fiber is to compensate for the material dispersion by the waveguide dispersion, that means the group delays of the red and blue shifts are in opposite directions with each other. The waveguide dispersion can be calculated for various cladding layers, which would "balance" the material dispersion over 1.3–1.7 μm spectral range. The waveguide dispersion can be calculated with an accurate estimation of the parameter $V(d^2(Vb)/dV^2)$ so that it would "cancel" the material dispersion over a specific spectral range as illustrated in Figure 5.17.

Step 3: The waveguide dispersion is superimposed over the material dispersion. Non-zero and low total dispersion for the range of 1.3–1.7 μm can be achieved. The total dispersion curve is shown in Figure 5.17.

5.5.2 Design Algorithm for DCF

The algorithm developed for designing DCF is similar to that of DFF. The principal difference is the condition that the DCF is a section of a fiber span deployed in the optically

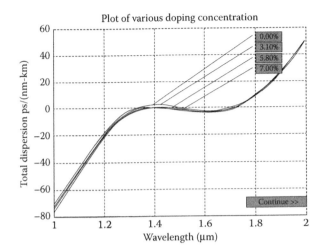

FIGURE 5.16
Total dispersion factor varies with variation of doping concentration of the triple-clad profile.

amplified link. It must compensate the GVD of various spectral components. The group velocities of red-shift signals are different from blue-shift ones. When modulated signals travel through a long stretch of DFF fiber, they lag or lead each other in distance due to various velocity of propagation. The DCF compensates for this anomaly. The compensating condition for a transmission span is constrained by

$$|D_{DFF} \, L_{DFF}| = |D_{DCF} \, L_{DCF}| \tag{5.82}$$

The index is varied so that the fiber dispersion becomes very negative and satisfies Equation 5.82 as shown in Figures 5.16 and 5.17.

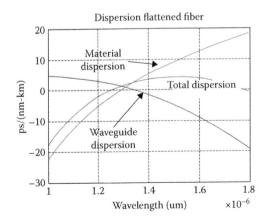

FIGURE 5.17
Spectral variations of the material dispersion, waveguide dispersion and total dispersion of a fiber with profile: core type = material B (2), core radius = 3.8625 μm, index difference = 0.98%.

5.6 Design Cases

The designed dispersion spectral characteristics can be matched with any specific dispersion characteristics for both the DFF and DCF. This section presents two case studies of typical designs of the dispersion-flattened and compensated fibers. It is imperative to justify the results with the values obtained for all the parameters used such as the number of guided modes, the core and cladding layer radii, relative index difference, spot size etc. The selection of the parameters can be inserted in windows provided in Figure 5.18. This allows fine adjustment of the parameters. The results are presented in Figures 5.22 and 5.23 with the windows numbered from left to right and top to bottom for the design cases presented in the next sections.

5.6.1 Design Case 1

In this case the core material is of type B and results are obtained for a dispersion compensated span of DFF and DCF of Figure 5.19 for transmission of at least 40 Gb/s DWDM channels over several hundred kilometers (Figure 5.20).

The obtained results are illustrated in the windows in Figure 5.21 as follows:

- Window 1: indicates the fiber dispersion factor. A non-zero DFF (<3ps/(nm · km)) is designed over 1300–1600 nm region. This window is used as the guiding window in the design example as follows:
- Window 2: dispersion factor of the DCF is obtained with a unity Kappa value.
- Window 3: the dispersion slopes of both DFF and DCF are plotted where the upper and lower lines indicate the dispersion slopes of DFF and DCF, respectively.
- Window 4: shows the plot of "$V(d^2(Vb)/dV^2$" versus V. The lower and upper lines indicate the characteristics of $V(d^2(Vb)/dV^2)$ versus V_{12} and V_{13}, respectively.
- Window 5: shows fiber RIP. The upper most line shows the RIP of core varying from 1.4525 at 1.3 µm to 1.4475 at 1.6 µm. The RI of first cladding (middle line)

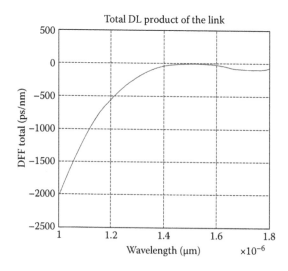

FIGURE 5.18
Dispersion × length product over ultra-wide spectral range.

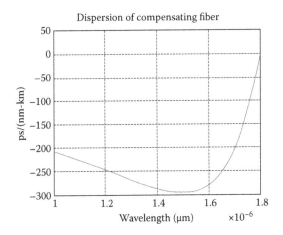

FIGURE 5.19
Dispersion factor of the DCF with a profile of: core type = aterial G (7), radius = 1.59938 μm, index difference = 1.57% and DCF length = 1.39241 km.

varies from 1.4425 at 1.3 μm to 1.4375 at 1.6 μm. Finally RIPs (lowest lines) of outer
cladding vary from 1.448 at 1.3 μm to 1.4475 at 1.6 μm.

- Window 6: shows spectral distribution of the spot size of the DFF. Designed spot
  size is compatible with the standard SMFs ITU G.652 or G.655.
- Window 7: shows the rise time budget. Theoretical value (green line) is taken as
  0.5/40 ns when non-return-to-zero (NRZ) is used.
- Window 8: gives the Kappa value which would reach unity.
- Window 9: gives the dispersion-length of the dispersion-managed link which is
  close to zero.

5.6.2 Design Case 2

In this case the core is of type E material for the DFF and DCF as shown in Figure 5.22.
Figure 5.23 illustrates the design outputs as follows

- Window 1: shows a finite non-zero dispersion (< a few ps/nm-km) DFF over the
  spectral range 1300–1600 nm.
- Window 2: view of the DCF total dispersion factor of the DCF with a Kappa value
  approaching unity.
- Window 3: shows the dispersion slopes of both the DFF (upper) and DCF (lower).
- Window 4: shows $V(d^2(Vb)/dV^2)$ versus V_{12} (lower) and V_{13} (upper).

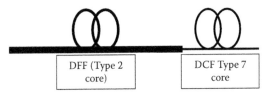

FIGURE 5.20
A dispersion compensated fiber transmission span.

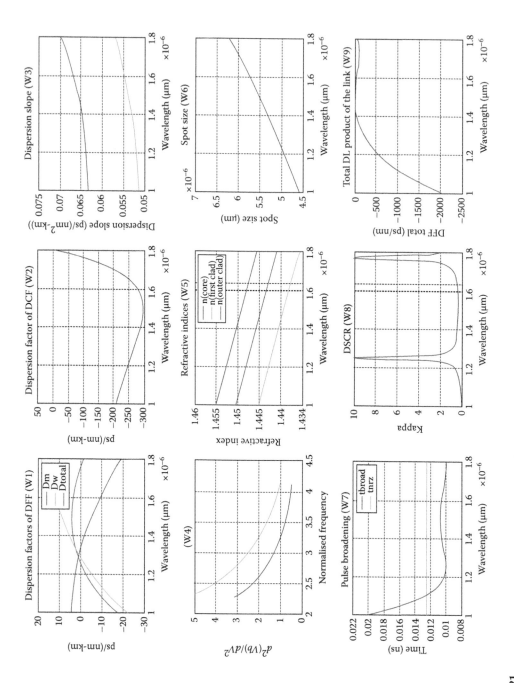

FIGURE 5.21
Design windows of DCF and DFF using material Type 2 and 7 as the central core materials, respectively.

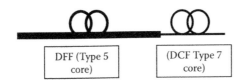

FIGURE 5.22
A dispersion compensated fiber transmission span.

- Window 5: shows the refractive index profile (RIP) across core (upper) varying from 1.4475 at 1.3 µm to 1.444 at 1.6 µm; RIP of first cladding (lowest) varying from 1.445 at 1.3 µm to 1.4425 at 1.6 µm and outer cladding (middle) varying from 1.4474 at 1.3 µm to 1.4435 at 1.6 µm. Note the relative index difference is minute, in order of < 1%.
- Window 6: shows the spectral property of the DFF spot size which is similar to that obtained in Case 1.
- Window 7: gives the rise time budget of the fiber using NRZ transmission format.
- Window 8: shows Kappa approaching unity value.
- Window 9: gives the DL product of the dispersion-managed link, which is very close to zero.

5.6.3 Design Summary

From the design cases it can be concluded that material types J, B and E satisfy the constraint of low non-zero dispersion (<3 ps/(nm-km)) over a wide spectral range (i.e., 1280–1620 nm). The total dispersion factor in this region is almost uniform with very low dispersion ripple (± 0.01 ps/(nm-km)). The waveguide dispersion $V(d^2(Vb)/dV^2)$ curves can be approximated for the design of multiple-clad fibers. The extreme sensitivity (88.88 unit dispersion per unit µm) of fiber total dispersion to incremental change of the core radius is a unique property of DFFs and DCFs. The changes of the outer cladding RI are critical (i.e., 142,850 unit dispersion per unit RI). Doping concentration has little effect on the total dispersion factor with only 0.5 unit dispersion per unit concentration.

The total dispersion factor is very sensitive to changes in core radius, a_0 and outer cladding RI, n. Consequently the contributions of these two parameters should be considered first before varying the other five parameters. Analyzing the effects of each of the seven geometrical and index profile parameters and the contribution of material dispersion due to different types of a set of materials as core or cladding to the waveguide dispersion factor, one could design the triple-clad profile to satisfy any dispersion flattening and compensating factors over an ultra-wide spectral range as illustrated in the two design cases. Furthermore the RI difference is quite small and therefore practical for manufacturing of low loss fibers as well as for satisfying the weakly guiding mechanism of the LP mode usually required for long-haul optical transmission systems.

5.7 Concluding Remarks

The demands for low flattening and compensating dispersion factors over the wideband of operational wavelength 1300–1550 nm (and even wider) can be satisfied by fibers with

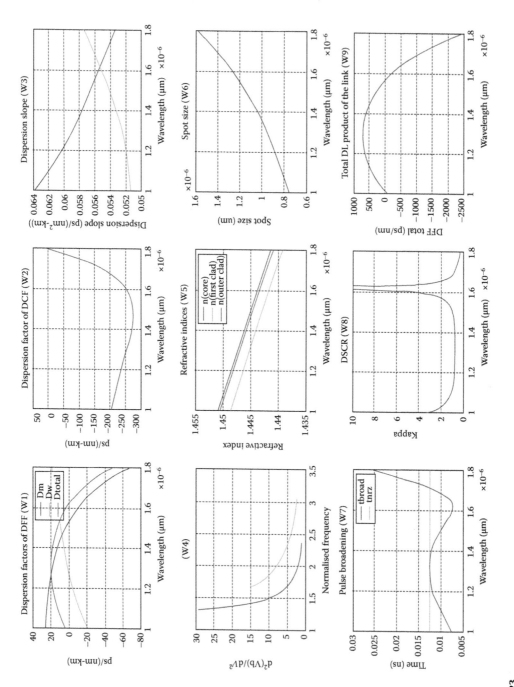

FIGURE 5.23
Design windows for DFF and DCF with core material Type 5 and 7 respectively.

triple-clad index profiles. Several stringent conditions on the seven geometrical and index profile parameters of multiple-clad optical fibers are described. A simplified approach and a sequence of design steps are given. We have also presented efficient design algorithms based on the approximation and fitting the saddle points of the curves that represent the waveguide dispersion factor. Surprisingly these points follow a simple cubic function. A reasonably accurate prediction of the waveguide dispersion is then obtained allowing a best fit curve for tailoring the total dispersion factor over an ultra-wide spectral region. This has been demonstrated in the two case studies using different core materials.

Varying the seven degrees of freedom based on the geometrical and index parameters the effects of core and cladding radii, refractive indices and doping concentrations on the total dispersion are analyzed and the design sensitivities are drawn. The developed method presented in this paper is also valid and applicable to other geometrical and index profiles.

The sensitivity of DFFs and DCFs to the polarization dispersion is of much interest and will be reported in the near future. The design guidelines presented in this chapter would be applicable to the design and manufacturing of broadband DFF and DCFs for ultra-long reach and ultra-wideband DWDM optical communications transmission systems.

The demands of long-haul optical transmission systems also require the employment of Raman distributed and lumped amplification. Currently we are investigating the design of DCF and DFF for optimum Raman gain coefficients and the manipulation of the profiles of these fibers presented in this paper will be utilized.

5.8 Problems

1. Mini-project: Design of Single Mode Optical Fibers

   Project Objectives:

   Design

   To design the geometrical and index profile of silica optical fibers to meet certain dispersion properties as required. Furthermore, a set of fiber performance with dispersion as the main factor must be investigated as a function of the fiber core radius and the relative refractive index difference.

   System Applications

   Your designed fiber must be incorporated with a dispersion-compensated fiber—in this case the standard single mode optical fiber (SMF) whose specifications are given in the lectures. If the total length of the transmission is 10,000 km, specify the length of your designed fiber and the SMF so that the average fiber dispersion is 0.01 ps/nm/km. It can be assumed that the spacing of optical amplifiers is 100 km.

   Design Specifications

   A number of types of single mode optical fibers must be designed for optical communications systems in long and short-haul transmissions. Table 5.4 shows the characteristics of the required fibers.

TABLE 5.4

Index Distribution Profile and Requirements for Dispersion of the Fibers to be Designed

Fib. No.	Optical Fibers Profile	Material Type Core	Systems Requirement	Other Requirements
		See Tables Below for the Sellmeier Constants		Maximum Dispersion in the Wavelength Range 1510 to 1590 nm in ps/nm/km
1	Triangular	A-J	Dispersion-shifted wavelength at 1550 nm	<1.5
2	Parabolic	A-J	Dispersion-shifted wavelength at 1520 nm	<1.5
3	Triangular	A-J	Dispersion-shifted wavelength at 1560 nm	<2
4	Parabolic	A-J	Dispersion-shifted wavelength at 1530 nm	<3
5	Triangular	A-J	Dispersion -compensated wavelength at 1550 nm	>0.5
6	Parabolic	A-J	Dispersion-compensated wavelength at 1540 nm	>1.0
9	Triple-clad	A-J	Dispersion-flattened over 1500 to 1590 nm	not more than 0.2 ps/nm/km
10	Triangular in core and segmented in cladding	A-J	Dispersion-flattened over 1500 to 1590 nm	not more than 0.2 ps/nm/km

Tables 5.5 and 5.6 give the Sellmeier coefficients of different materials and doping concentrations so that different refractive indices can be achieved.

Mini-Project: Design of Optical Fibers

Practical limits for the fiber core radius and relative index difference must be taken into account. A set of curves must be obtained with the core radius or the relative index difference as a parameter. Make sure that the material dispersion factors are correctly modeled.

TABLE 5.5

Sellmeier Coefficients for Several Optical Fiber Silica-Based Materials with Germanium Doped in the Core Region. Wavelength in μm and G_k in μm^{-2}

Sellmeier Constants	Germanium Concentration, C % Mole			
Types	A	B	C	D
	0	3.1%	5.8%	7.9%
G_1	0.6961663	0.7028554	0.7088876	0.7136824
G_2	0.4079426	0.4146307	0.4206803	0.4254807
G_3	0.8974794	0.8974540	0.8956551	0.8964226
λ_1	0.6840432	0.0727723	0.0609053	0.0617167
λ_2	0.1162414	0.1143085	0.1254514	0.1270814
λ_3	9.896161	9.896161	9.896162	9.896161

Source: After Y. Ishida, M. Kojima, T. Kobayashi and Y. Sugakawa, "Optical Fiber for Communication", US Patent No.4,114,981, 1978.

TABLE 5.6

Sellmeier Coefficients for Several Optical Fiber Silica-based Materials with Germanium Doped in the Core Region Wavelength in m and G_k in μm^{-2}

Sellmeier Constants	Concentration Composition			
Types	E	F	G	H
	Quenched SiO_2	13.5 GeO_2:86.5 SiO_2	9.1 P_2O_5:90.0 SiO_2	13.3B_2O_3:86.7SiO_2
G_1	0.696750	0.711040	0.695790	0.690618
G_2	0.408218	0.408218	0.452497	0.401996
G_3	0.890815	0.704048	0.712513	0.898817
λ_1	0.069066	0.064270	0.061568	0.061900
λ_2	0.115662	0.129408	0.119921	0.123662
λ_3	9.900559	9.425478	8.656641	9.098960
Sellmeier Constants	Concentration Composition			
Types	I	J	K	L
	1.0 F :99.0 SiO_2	16.2 Na_2O :32.5 B_2O_3 : 50.6 SiO_2		
G_1	0.691116	0.796468		
G_2	0.399166	0.497614		
G_3	0.890423	0.358924		
λ_1	0.068227	0.094359		
λ_2	0.116460	0.093386		
λ_3	9.993707	5.999652		

Source: After J.W. Fleming et al., "Low-Loss Single-Mode Fibers Prepared by Plasma-Enhanced MCVD," *Electronics Letters*, vol. 17, 1978, pp. 326–332.

Software Environment

The preferred package is MATLAB, or MATLAB 7.0 and above for Windows.

Assessment of Design Assignments

A major design assignment on optical fibers for communication systems counts for 10% of the total marks allocated for the optical systems part. Twenty percent will be awarded to design groups selecting index profile types 9 or 10.

The design assignment is specified for different group of students. The maximum number of each group is two. A larger number group can only be accepted in exceptional circumstances, and in such cases the complexity of the design assignment is increased accordingly.

Appendix A: Derivatives of the RI with Respect to Wavelength

The first derivative is given by

$$\frac{dn(\lambda)}{d\lambda} = \frac{\displaystyle\sum_{j=1}^{3}\left[\frac{2\lambda B_j}{\lambda^2 - \lambda_j^2}\left(1 - \frac{\lambda^2 B_j}{\left(\lambda^2 - \lambda_j^2\right)}\right)\right]}{2\sqrt{1 + \displaystyle\sum_{j=1}^{3}\frac{\lambda^2 B_j}{\left(\lambda^2 - \lambda_j^2\right)}}} \tag{5.83}$$

The second derivative is then obtained as

$$\frac{d^2 n(\lambda)}{d\lambda^2} = \frac{\displaystyle\sum_{j=1}^{3}\frac{2 B_j}{\lambda^2 - \lambda_j^2}\left(1 - \frac{5\lambda^2}{\left(\lambda^2 - \lambda_j^2\right)} + \frac{4\lambda^4}{\left(\lambda^2 - \lambda_j^2\right)^2}\right)}{2\sqrt{1 + \displaystyle\sum_{j=1}^{3}\frac{\lambda^2 B_j}{\left(\lambda^2 - \lambda_j^2\right)}}}$$

$$- \frac{\displaystyle\sum_{j=1}^{3}\frac{2\lambda B_j}{\lambda^2 - \lambda_j^2}\left(1 - \frac{\lambda^2}{\left(\lambda^2 - \lambda_j^2\right)}\right)}{4\left(1 + \displaystyle\sum_{j=1}^{3}\frac{\lambda^2 B_j}{\left(\lambda^2 - \lambda_j^2\right)}\right)^{3/2}} \tag{5.84}$$

and the third derivative is then followed by

$$\frac{d^3 n(\lambda)}{d\lambda^3} = \frac{3\displaystyle\sum_{j=1}^{3}\frac{2\lambda B_j}{\lambda^2 - \lambda_j^2}\left(1 - \frac{\lambda^2}{\left(\lambda^2 - \lambda_j^2\right)}\right)}{8\left(1 + \displaystyle\sum_{j=1}^{3}\frac{\lambda^2 B_j}{\left(\lambda^2 - \lambda_j^2\right)}\right)^{5/2}} + \frac{\displaystyle\sum_{j=1}^{3}\frac{24\lambda B_j}{\lambda^2 - \lambda_j^2}\left(-1 + \frac{3\lambda^2}{\left(\lambda^2 - \lambda_j^2\right)} - \frac{2\lambda^4}{\left(\lambda^2 - \lambda_j^2\right)^2}\right)}{2\sqrt{1 + \displaystyle\sum_{j=1}^{3}\frac{\lambda^2 B_j}{\left(\lambda^2 - \lambda_j^2\right)}}}$$

$$- \frac{3\displaystyle\sum_{j=1}^{3}\frac{2 B_j}{\lambda^2 - \lambda_j^2}\left(1 - \frac{5\lambda^2}{\left(\lambda^2 - \lambda_j^2\right)} + \frac{4\lambda^4}{\left(\lambda^2 - \lambda_j^2\right)^2}\right)\displaystyle\sum_{j=1}^{3}\frac{2\lambda B_j}{\lambda^2 - \lambda_j^2}\left(1 - \frac{\lambda^2}{\left(\lambda^2 - \lambda_j^2\right)}\right)}{4\left(1 + \displaystyle\sum_{j=1}^{3}\frac{\lambda^2 B_j}{\left(\lambda^2 - \lambda_j^2\right)}\right)^{3/2}} \tag{5.85}$$

Appendix B: Higher Order Derivatives of the Propagation Constant

$$\beta_2 = \lambda^2 \frac{(1-\Delta)}{2\pi c^2} \left(\lambda \left(1+b(v)\delta\right) \frac{d^2 n}{d\lambda^2} + 2\lambda\Delta \frac{dn}{d\lambda} \frac{db(v)}{d\lambda} + \Delta\lambda n \frac{d^2 b}{d\lambda^2} \right) \tag{5.86}$$

and

$$\beta_3 = \frac{\left(\lambda^2 (1-\Delta)\right)}{4\pi^2 c^3} 3\lambda^2 \left(1+b(v)\right)\Delta \frac{d^2 n}{d\lambda^2} + 3\Delta\lambda^3 \frac{d^2 n}{d\lambda^2} \frac{db(v)}{d\lambda} + \dots$$

$$+\left(1+b(v)\right)\Delta\lambda^3 \frac{d^3 n}{d\lambda^3} + 6\lambda^2\Delta \frac{dn}{d\lambda} \frac{db(v)}{d\lambda} + \dots \tag{5.87}$$

$$\dots 3\Delta\lambda^3 \frac{dn}{d\lambda} \frac{db(v)}{d\lambda^2} + 3\lambda^2\Delta n \frac{d^2 b(v)}{d\lambda^2} + \lambda^3 n \frac{d^3 b(v)}{d\lambda^3}$$

with

$$\frac{db(v)}{d\lambda} = 1.9920 \left(\frac{2\pi}{\lambda}\right) a_1 \sqrt{2\Delta} \left(1.1428 - \frac{0.9960}{V_{12}}\right) \frac{\left(\frac{dn}{d\lambda}\right)}{V_{12}^2} \tag{5.88}$$

$$\frac{d^2 b(v)}{d\lambda} = 1.9920 \left(2\pi / \lambda\right) a_1 \sqrt{2\Delta} \left(\left(0.9960 / V_{12}^4\right) \frac{dn}{d\lambda} \left(2\pi / \lambda\right) a_1 \sqrt{2\Delta} \left(\frac{dn}{d\lambda}\right)^2 + \dots \right.$$

$$\left. \dots \left(1.1428 - 0.9960 / V_{12}\right) \left(\frac{d^2 n}{d\lambda^2} / V_{12}^2 - \left(2\pi / \lambda\right) a 1 \sqrt{2\Delta} \left(\frac{dn}{d\lambda}\right)^2 / V_{12}^3 \right) \right) \tag{5.89}$$

$$\frac{d^3 b(v)}{d\lambda^3} = 1.984 \left(\left(2\pi / \lambda\right) a 1 \sqrt{2\Delta} \right)^2$$

$$\times \left(2 \frac{dn}{d\lambda} \frac{d^2 n}{d\lambda^2} / V_{12}^4 - 4 \left(\left(2\pi / \lambda\right) a 1 \sqrt{2\Delta} \right) \left(\frac{dn}{d\lambda}\right)^3 / V_{12}^5 \right) \tag{5.90}$$

$$\frac{dn}{d\lambda} = -(\lambda / n) n_a \tag{5.91}$$

with $n_a = b(i)\lambda(i)^2 / \left(\lambda^2 - \lambda(i)^2\right)^2$.

$$\frac{d^2 n}{d\lambda^2} = \left(A / \left(2\sqrt{(1+B)}\right)\right) - \left(c / 4\sqrt{(1+B)}\right) \tag{5.92}$$

$$\frac{d^3n}{d\lambda^3} = \left(D/\left(8^{1.5}\sqrt{(1+E)}\right)\right) + \left(F/\left(2\sqrt{(1+E)}\right)\right) - 3GH/\left(4(1+E)^{1.5}\right) \tag{5.93}$$

in which the coefficients A–H are given by

$$D = 2\lambda b(i)\left(1 - \lambda^2/\lambda_1\right)/\lambda_1 \text{ with } \lambda_1 = \lambda^2 - \lambda(i)^2 \tag{5.94a}$$

$$E = \lambda^2 b(i)/\lambda_1; \quad F = \left(24\lambda b(i)\left(-1 + 3\lambda^2/\lambda_1\right) - 2\lambda^4/\lambda_1^2\right)/\lambda_1 \tag{5.94b}$$

$$G = \left(2b(i)\left(1 - 5\lambda^2/\lambda_1 + 4\lambda^4/\lambda_1^2\right)\right)/\lambda_1 \tag{5.94c}$$

$$H = \left(2\lambda b(i)\left(1 - \lambda^2/\lambda_1\right)\right)/\lambda_1 \tag{5.94d}$$

$$A = \left(2b(i)\left(1 - 5\lambda^2/\lambda_1 + 4\lambda^4/\lambda_1^2\right)\right)/\lambda_1; \quad B = \left(\lambda^2 b(i)\right)/\lambda_1 \tag{5.94e}$$

$$C = \left(2\lambda b(i)\left(1 - \lambda^2/\lambda_1\right)\right)/\lambda_1 \tag{5.94f}$$

MATLAB Program for Design of Single-Mode Optical Fibers

```
%DCF
function Dffiber2()
%Dffibers designed to be a callback function for dff3. It uses the values
from the controls

plot_theoretical=0;

dcfradius_slider_text=findobj('Tag','dcfradius_slider_text');
dcfradius_slider=findobj('Tag','dcfradius_slider');
dcfdelta_slider_text=findobj('Tag','dcfdelta_slider_text');
dcfdelta_slider=findobj('Tag','dcfdelta_slider');
dcflength_slider_text=findobj('Tag','dcflength_slider_text');
dcflength_slider=findobj('Tag','dcflength_slider');
ttype1_button=findobj('Tag','ttype1_button');
ttype2_button=findobj('Tag','ttype2_button');
ttype3_button=findobj('Tag','ttype3_button');
ttype4_button=findobj('Tag','ttype4_button');
ttype5_button=findobj('Tag','ttype5_button');
ttype6_button=findobj('Tag','ttype6_button');
ttype7_button=findobj('Tag','ttype7_button');
ttype8_button=findobj('Tag','ttype8_button');
ttype9_button=findobj('Tag','ttype9_button');
ttype10_button=findobj('Tag','ttype10_button');
dff31_fig=findobj('Tag','dff31_fig');


ttype1_on=get(ttype1_button,'value'); %is it Type 1?
ttype2_on=get(ttype2_button,'value');%is it Type 2?
ttype3_on=get(ttype3_button,'value');%is it Type 3?
ttype4_on=get(ttype4_button,'value'); %is it Type 3?
ttype5_on=get(ttype5_button,'value'); %is it Type 3?
ttype6_on=get(ttype6_button,'value'); %is it Type 3?
ttype7_on=get(ttype7_button,'value'); %is it Type 3?
ttype8_on=get(ttype8_button,'value'); %is it Type 3?
ttype9_on=get(ttype9_button,'value'); %is it Type 3?
ttype10_on=get(ttype10_button,'value'); %is it Type 3?


%Calculate Waveguide Dispersion
 bs(1)=0.6961663;
 bs(2)=0.4079426;
 bs(3)=0.89474794;
 lamds(1)=0.0684043*1e-6;
 lamds(2)=0.1162414*1e-6;
 lamds(3)=9.896161*1e-6;
 b2(1)=0.7028554;
 b2(2)=0.4146307;
 b2(3)=0.8974540;
 lamd2(1)=0.0727723*1e-6;
 lamd2(2)=0.11430854*1e-6;
```

```
lamd2(3)=9.896161*1e-6;
b3(1)=0.7088876;
b3(2)=0.4206803;
b3(3)=0.8956551;
lamd3(1)=0.0609053*1e-6;
lamd3(2)=0.1254514*1e-6;
lamd3(3)=9.9896162*1e-6;
b4(1)=0.7136824;
b4(2)=0.4254807;
b4(3)=0.8964226;
lamd4(1)=0.0617167*1e-6;
lamd4(2)=0.1270814*1e-6;
lamd4(3)=9.896161*1e-6;
b5(1)=0.696750;
b5(2)=0.408218;
b5(3)=0.890815;
lamd5(1)=0.069066*1e-6;
lamd5(2)=0.115662*1e-6;
lamd5(3)=9.900559*1e-6;
b6(1)=0.711040;
b6(2)=0.408218;
b6(3)=0.704048;
lamd6(1)=0.064270*1e-6;
lamd6(2)=0.129408*1e-6;
lamd6(3)=9.424578*1e-6;
b7(1)=0.695790;
b7(2)=0.452497;
b7(3)=0.712513;
lamd7(1)=0.061568*1e-6;
lamd7(2)=0.119921*1e-6;
lamd7(3)=8.656641*1e-6;
b8(1)=0.690618;
b8(2)=0.401996;
b8(3)=0.898817;
lamd8(1)=0.061900*1e-6;
lamd8(2)=0.123662*1e-6;
lamd8(3)=9.098960*1e-6;
b9(1)=0.691116;
b9(2)=0.399166;
b9(3)=0.890423;
lamd9(1)=0.068227*1e-6;
lamd9(2)=0.116460*1e-6;
lamd9(3)=9.993707*1e-6;
b10(1)=0.796468;
b10(2)=0.497614;
b10(3)=0.358924;
lamd10(1)=0.094359*1e-6;
lamd10(2)=0.093386*1e-6;
lamd10(3)=5.999652*1e-6;
rad=get(dcfradius_slider,'value');
a=(rad)*1e-6;
delta=get(dcfdelta_slider,'value');
del=(delta)*1;
L=get(dcflength_slider,'value');
```

```
c=2.997925e8;
a1=1.405*a;
lamdax=1.0e-6;
global dcfdff1;
global dcfdff;
global dcftbroad;
for lamda=1.0e-6:.01e-6:1.8e-6
%Calculate Material Dispersion and Dispersion Slope
    n111=0;
    n112=0;
    n113=0;
    na=0;
    if(ttype1_on==1)
        for i=1:3
            lamda1=lamda^2-(lamds(i))^2;
            n11=bs(i)*lamda^2/lamda1;
            n111=n11+n111;
            n12=(bs(i)*lamds(i)^2)*(3*lamda^2+lamds(i)^2)/(lamda1^3);
            n112=n112+n12;
            n13=(bs(i)*lamda*lamds(i)^2/lamda1)^2;
            n113=n113+n13;
            na=bs(i)*lamds(i)^2/lamda1^2;
        end
      for i=1:3
          lamda1=lamda^2-(lamds(i))^2;
          A=(2*bs(i)*(1-(5*lamda^2)/lamda1+(4*lamda^4)/
            lamda1^2))/lamda1;
          B=(lamda^2*bs(i))/lamda1;
          C=(2*lamda*bs(i)*(1-lamda^2/lamda1))/lamda1;
      end
      for i=1:3
          lamda1=lamda^2-(lamds(i))^2;
          D=(2*lamda*bs(i)*(1-lamda^2/lamda1))/lamda1;
          E=(lamda^2*bs(i))/lamda1;
          F=(24*lamda*bs(i)*(-1+3*lamda^2/lamda1-2*lamda^4/
            lamda1^2))/lamda1;
          G=(2*bs(i)*(1-5*lamda^2/lamda1+4*lamda^4/lamda1^2))/
            lamda1;
          H=(2*lamda*bs(i)*(1-lamda^2/lamda1))/lamda1;
      end
    end
    if(ttype2_on==1)
        for i=1:3
            lamda1=lamda^2-(lamd2(i))^2;
            n11=b2(i)*lamda^2/lamda1;
            n111=n11+n111;
            n12=(b2(i)*lamd2(i)^2)*(3*lamda^2+lamd2(i)^2)/
              (lamda1^3);
            n112=n112+n12;
            n13=(b2(i)*lamda*lamd2(i)^2/lamda1)^2;
            n113=n113+n13;
            na=b2(i)*lamd2(i)^2/lamda1^2;
        end
      for i=1:3
```

```
    lamda1=lamda^2-(lamd2(i))^2;
    A=(2*b2(i)*(1-(5*lamda^2)/lamda1+(4*lamda^4)/
      lamda1^2))/lamda1;
    B=(lamda^2*b2(i))/lamda1;
    C=(2*lamda*b2(i)*(1-lamda^2/lamda1))/lamda1;
end
for i=1:3
    lamda1=lamda^2-(lamd2(i))^2;
    D=(2*lamda*b2(i)*(1-lamda^2/lamda1))/lamda1;
    E=(lamda^2*b2(i))/lamda1;
    F=(24*lamda*b2(i)*(-1+3*lamda^2/lamda1-2*lamda^4/
      lamda1^2))/lamda1;
    G=(2*b2(i)*(1-5*lamda^2/lamda1+4*lamda^4/lamda1^2))/
      lamda1;
    H=(2*lamda*b2(i)*(1-lamda^2/lamda1))/lamda1;
end
end
if(ttype3_on==1)
    for i=1:3
        lamda1=lamda^2-(lamd3(i))^2;
        n11=b3(i)*lamda^2/lamda1;
        n111=n11+n111;
        n12=(b3(i)*lamd2(i)^2)*(3*lamda^2+lamd3(i)^2)/
          (lamda1^3);
        n112=n112+n12;
        n13=(b3(i)*lamda*lamd3(i)^2/lamda1)^2;
        n113=n113+n13;
        na=b3(i)*lamd3(i)^2/lamda1^2;
    end
for i=1:3
    lamda1=lamda^2-(lamd3(i))^2;
    A=(2*b3(i)*(1-(5*lamda^2)/lamda1+(4*lamda^4)/
      lamda1^2))/lamda1;
    B=(lamda^2*b3(i))/lamda1;
    C=(2*lamda*b3(i)*(1-lamda^2/lamda1))/lamda1;
end
for i=1:3
    lamda1=lamda^2-(lamd3(i))^2;
    D=(2*lamda*b3(i)*(1-lamda^2/lamda1))/lamda1;
    E=(lamda^2*b3(i))/lamda1;
    F=(24*lamda*b3(i)*(-1+3*lamda^2/lamda1-2*lamda^4/
      lamda1^2))/lamda1;
    G=(2*b3(i)*(1-5*lamda^2/lamda1+4*lamda^4/lamda1^2))/
      lamda1;
    H=(2*lamda*b3(i)*(1-lamda^2/lamda1))/lamda1;
end
end
if(ttype4_on==1)
    for i=1:3
        lamda1=lamda^2-(lamd4(i))^2;
        n11=b4(i)*lamda^2/lamda1;
        n111=n11+n111;
        n12=(b4(i)*lamd4(i)^2)*(3*lamda^2+lamd4(i)^2)/
          (lamda1^3);
```

```
                n112=n112+n12;
                n13=(b4(i)*lamda*lamd4(i)^2/lamda1)^2;
                n113=n113+n13;
                na=b4(i)*lamd4(i)^2/lamda1^2;
         end
    for i=1:3
         lamda1=lamda^2-(lamd4(i))^2;
         A=(2*b4(i)*(1-(5*lamda^2)/lamda1+(4*lamda^4)/
           lamda1^2))/lamda1;
         B=(lamda^2*b4(i))/lamda1;
         C=(2*lamda*b4(i)*(1-lamda^2/lamda1))/lamda1;
    end
    for i=1:3
         lamda1=lamda^2-(lamd4(i))^2;
         D=(2*lamda*b4(i)*(1-lamda^2/lamda1))/lamda1;
         E=(lamda^2*b4(i))/lamda1;
         F=(24*lamda*b4(i)*(-1+3*lamda^2/lamda1-2*lamda^4/
           lamda1^2))/lamda1;
         G=(2*b4(i)*(1-5*lamda^2/lamda1+4*lamda^4/lamda1^2))/
           lamda1;
         H=(2*lamda*b4(i)*(1-lamda^2/lamda1))/lamda1;
    end
    end
    if(ttype5_on==1)
         for i=1:3
             lamda1=lamda^2-(lamd5(i))^2;
             n11=b5(i)*lamda^2/lamda1;
             n111=n11+n111;
             n12=(b5(i)*lamd5(i)^2)*(3*lamda^2+lamd5(i)^2)/
                (lamda1^3);
             n112=n112+n12;
             n13=(b5(i)*lamda*lamd5(i)^2/lamda1)^2;
             n113=n113+n13;
             na=b5(i)*lamd5(i)^2/lamda1^2;
         end
       for i=1:3
           lamda1=lamda^2-(lamd5(i))^2;
           A=(2*b5(i)*(1-(5*lamda^2)/lamda1+(4*lamda^4)/
             lamda1^2))/lamda1;
           B=(lamda^2*b5(i))/lamda1;
           C=(2*lamda*b5(i)*(1-lamda^2/lamda1))/lamda1;
    end
    for i=1:3
         lamda1=lamda^2-(lamd5(i))^2;
         D=(2*lamda*b5(i)*(1-lamda^2/lamda1))/lamda1;
         E=(lamda^2*b5(i))/lamda1;
         F=(24*lamda*b5(i)*(-1+3*lamda^2/lamda1-2*lamda^4/
           lamda1^2))/lamda1;
         G=(2*b5(i)*(1-5*lamda^2/lamda1+4*lamda^4/lamda1^2))/
           lamda1;
         H=(2*lamda*b5(i)*(1-lamda^2/lamda1))/lamda1;
      end
    end
    if(ttype6_on==1)
```

```
for i=1:3
    lamda1=lamda^2-(lamd6(i))^2;
    n11=b6(i)*lamda^2/lamda1;
    n111=n11+n111;
    n12=(b6(i)*lamd6(i)^2)*(3*lamda^2+lamd6(i)^2)/
        (lamda1^3);
    n112=n112+n12;
    n13=(b6(i)*lamda*lamd6(i)^2/lamda1)^2;
    n113=n113+n13;
    na=b6(i)*lamd6(i)^2/lamda1^2;
end
for i=1:3
    lamda1=lamda^2-(lamd6(i))^2;
    A=(2*b6(i)*(1-(5*lamda^2)/lamda1+(4*lamda^4)/
        lamda1^2))/lamda1;
    B=(lamda^2*b6(i))/lamda1;
    C=(2*lamda*b6(i)*(1-lamda^2/lamda1))/lamda1;
end
for i=1:3
    lamda1=lamda^2-(lamd6(i))^2;
    D=(2*lamda*b6(i)*(1-lamda^2/lamda1))/lamda1;
    E=(lamda^2*b6(i))/lamda1;
    F=(24*lamda*b6(i)*(-1+3*lamda^2/lamda1-2*lamda^4/
        lamda1^2))/lamda1;
    G=(2*b6(i)*(1-5*lamda^2/lamda1+4*lamda^4/lamda1^2))/
        lamda1;
    H=(2*lamda*b6(i)*(1-lamda^2/lamda1))/lamda1;
end
end
if(ttype7_on==1)
    for i=1:3

        lamda1=lamda^2-(lamd7(i))^2;
        n11=b7(i)*lamda^2/lamda1;
        n111=n11+n111;
        n12=(b7(i)*lamd7(i)^2)*(3*lamda^2+lamd7(i)^2)/
            (lamda1^3);
        n112=n112+n12;
        n13=(b7(i)*lamda*lamd7(i)^2/lamda1)^2;
        n113=n113+n13;
        na=b7(i)*lamd7(i)^2/lamda1^2;
    end
for i=1:3
    lamda1=lamda^2-(lamd7(i))^2;
    A=(2*b7(i)*(1-(5*lamda^2)/lamda1+(4*lamda^4)/
        lamda1^2))/lamda1;
    B=(lamda^2*b7(i))/lamda1;
    C=(2*lamda*b7(i)*(1-lamda^2/lamda1))/lamda1;
end
for i=1:3
    lamda1=lamda^2-(lamd7(i))^2;
    D=(2*lamda*b7(i)*(1-lamda^2/lamda1))/lamda1;
    E=(lamda^2*b7(i))/lamda1;
    F=(24*lamda*b7(i)*(-1+3*lamda^2/lamda1-2*lamda^4/
```

```
            lamda1^2))/lamda1;
      G=(2*b7(i)*(1-5*lamda^2/lamda1+4*lamda^4/lamda1^2))/
         lamda1;
      H=(2*lamda*b7(i)*(1-lamda^2/lamda1))/lamda1;
  end
  end
  if(ttype8_on==1)
      for i=1:3
            lamda1=lamda^2-(lamd8(i))^2;
            n11=b8(i)*lamda^2/lamda1;
            n111=n11+n111;
            n12=(b8(i)*lamd8(i)^2)*(3*lamda^2+lamd8(i)^2)/
               (lamda1^3);
            n112=n112+n12;
            n13=(b8(i)*lamda*lamd8(i)^2/lamda1)^2;
            n113=n113+n13;
            na=b8(i)*lamd8(i)^2/lamda1^2;
      end
    for i=1:3
      lamda1=lamda^2-(lamd8(i))^2;
      A=(2*b8(i)*(1-(5*lamda^2)/lamda1+(4*lamda^4)/
         lamda1^2))/lamda1;
      B=(lamda^2*b8(i))/lamda1;
      C=(2*lamda*b8(i)*(1-lamda^2/lamda1))/lamda1;
    end
  for i=1:3
      lamda1=lamda^2-(lamd8(i))^2;
      D=(2*lamda*b8(i)*(1-lamda^2/lamda1))/lamda1;
      E=(lamda^2*b8(i))/lamda1;
      F=(24*lamda*b8(i)*(-1+3*lamda^2/lamda1-2*lamda^4/
         lamda1^2))/lamda1;
      G=(2*b8(i)*(1-5*lamda^2/lamda1+4*lamda^4/lamda1^2))/
         lamda1;
      H=(2*lamda*b8(i)*(1-lamda^2/lamda1))/lamda1;
   end
  end
  if(ttype9_on==1)
      for i=1:3
            lamda1=lamda^2-(lamd9(i))^2;
            n11=b9(i)*lamda^2/lamda1;
            n111=n11+n111;
            n12=(b9(i)*lamd9(i)^2)*(3*lamda^2+lamd9(i)^2)/
               (lamda1^3);
            n112=n112+n12;
            n13=(b9(i)*lamda*lamd9(i)^2/lamda1)^2;
            n113=n113+n13;
            na=b9(i)*lamd9(i)^2/lamda1^2;
      end
  for i=1:3
      lamda1=lamda^2-(lamd9(i))^2;
      A=(2*b9(i)*(1-(5*lamda^2)/lamda1+(4*lamda^4)/
         lamda1^2))/lamda1;
      B=(lamda^2*b9(i))/lamda1;
      C=(2*lamda*b9(i)*(1-lamda^2/lamda1))/lamda1;
```

```
    end
    for i=1:3
        lamda1=lamda^2-(lamd9(i))^2;
        D=(2*lamda*b9(i)*(1-lamda^2/lamda1))/lamda1;
        E=(lamda^2*b9(i))/lamda1;
        F=(24*lamda*b9(i)*(-1+3*lamda^2/lamda1-2*lamda^4/
          lamda1^2))/lamda1;
        G=(2*b9(i)*(1-5*lamda^2/lamda1+4*lamda^4/lamda1^2))/
          lamda1;
        H=(2*lamda*b9(i)*(1-lamda^2/lamda1))/lamda1;
    end
    end
    if(ttype10_on==1)
        for i=1:3
            lamda1=lamda^2-(lamd10(i))^2;
            n11=b10(i)*lamda^2/lamda1;
            n111=n11+n111;
            n12=(b10(i)*lamd10(i)^2)*(3*lamda^2+lamd10(i)^2)/
              (lamda1^3);
            n112=n112+n12;
            n13=(b10(i)*lamda*lamd10(i)^2/lamda1)^2;
            n113=n113+n13;
            na=b10(i)*lamd10(i)^2/lamda1^2;
        end
      for i=1:3
          lamda1=lamda^2-(lamd10(i))^2;
          A=(2*b10(i)*(1-(5*lamda^2)/lamda1+(4*lamda^4)/
            lamda1^2))/lamda1;
          B=(lamda^2*b10(i))/lamda1;
          C=(2*lamda*b10(i)*(1-lamda^2/lamda1))/lamda1;
      end
      for i=1:3
          lamda1=lamda^2-(lamd10(i))^2;
          D=(2*lamda*b10(i)*(1-lamda^2/lamda1))/lamda1;
          E=(lamda^2*b10(i))/lamda1;
          F=(24*lamda*b10(i)*(-1+3*lamda^2/lamda1-2*lamda^4/
            lamda1^2))/lamda1;
          G=(2*b10(i)*(1-5*lamda^2/lamda1+4*lamda^4/lamda1^2))/
            lamda1;
          H=(2*lamda*b10(i)*(1-lamda^2/lamda1))/lamda1;
      end
    end
    dnlamda2=(A/(2*sqrt(1+B)))-(C/(4*(1+B)^1.5));
    dnlamda3=(D/(8*(1+E)^2.5))+(F/(2*sqrt(1+E)))-(3*G*H/
      (4*(1+E)^1.5));
    ncore=sqrt(1+n111);
    ddn=(n112/ncore)-(n112/(ncore^3));
    dnlamda1=-(lamda/ncore)*na;
    neff=ncore-(lamda*dnlamda1);
    dmat=-(lamda/c)*ddn*1e6;
%Calculation of Waveguide Dispersion
%First Cladding of fiber
    nclad=ncore*(1-del);
%Outer cladding of pure SILICA
```

```
nclad21=0;
dnn2=0;
for i=1:3
    lamda1=lamda^2-(lamds(i))^2;
    nclad2=bs(i)*lamda^2/lamda1;
    nclad21=nclad2+nclad21;
    dnn=lamds(i)^2*bs(i)^2/(lamda1^2);
    dnn2=dnn2+dnn;
end
nclad2=sqrt(1+nclad21);
dn=-(lamda/nclad2)*dnn2;
N=nclad2-lamda*dn;
delta12=(ncore-nclad);
delta23=(nclad-nclad2);
delta=delta23/delta12;
deltaa=(delta12+delta23)/nclad2;
k=2*pi/lamda;
v12=a*k*sqrt(ncore^2-nclad^2);
v13=a1*k*sqrt(abs(nclad^2-nclad2^2));
spotsize=a1*(0.65+1.619*(v12)^(-1.5)+2.879*(v12)^(-6));
b(1)=(1.1428-0.996/v12)^2;
b(2)=(1.1428-0.996/v13)^2;
beta(1)=sqrt(k^2*(b(1)*(ncore^2-nclad2^2)+nclad2^2));
beta(2)=sqrt(k^2*(b(2)*(nclad^2-nclad2^2)+nclad2^2));
u(1)=a*sqrt(k^2*ncore^2-beta(1)^2);
u(2)=a1*sqrt(k^2*nclad^2-beta(2)^2);
v(1)=a*sqrt(beta(1)^2-k^2*nclad2^2);
v(2)=a1*sqrt(beta(2)^2-k^2*nclad2^2);
k0=besselk(1,v(1))/(besselk(0,v(1)));%Bessel Function
k1=besselk(1,v(2))/(besselk(0,v(2)));%Bessel Function
dv=2*(u(1)/v12)^2*(k0*(1-2*k0)+(2/v(1))*(v(1)^2+(u(1)^2)*k0)
  *sqrt(k0)*(k0+sqrt(k0)/v(1)-1));
dv1=abs(2*(u(2)/v13)^2*(k1*(1-2*k1)+(2/v(2))*(v(2)^2+(u(2)
  ^2)*k1)*sqrt(k1)*(k1+sqrt(k1)/v(2)-1)));
bv=(1.1428-0.9960/v12)^2;
AA=(2*pi/(lamda))*a1*sqrt(2*del);
dblamda1=1.9920*AA*(1.1428-0.9960/v12)*(dnlamda1/v12^2);
dblamda2=1.9920*AA*((0.9960/v12^4)*AA*(dnlamda1)
  ^2+(1.1428-0.9960/v12)*(dnlamda2/v12^2-
  2*AA*(dnlamda1)^2/v12^3));
dblamda3=1.984*AA^2*(2*dnlamda1*dnlamda2/v12^4-
  *AA*dnlamda1^3/v12^5)+...
     1.984*AA^2*(dnlamda1*(dnlamda2/v12^2-2*AA*dnlamda1^2/
       v12^3)/v12^2)+...
     1.9920*AA*(1.1428-0.9960/v12)*(dnlamda3/v12^2-
     6*AA*dnlamda1*dnlamda2/v12^3+6*AA*dnlamda1^3/v12^4);
beta2=(lamda^2*(1-del)/(2*pi*c^2))*(lamda*(1+bv*delta)*dnl
  amda2+2*lamda*del*dnlamda1*dblamda1+del*lamda*ncore dblamda2);
beta3=-(lamda^2*(1-del))/(4*pi^2*c^3)*(3*lamda^2*(1+bv*del)
  *dnlamda2+3*del*lamda^3*dnlamda2*dblamda1+...
     (1+bv*del)*lamda^3*dnlamda3+6*lamda^2*del*dnlamda1*dblamda1+3*de
     l*lamda^3*dnlamda1*dblamda2+3*lamda^2*del*ncore*dblamda2+lamda^3
     *ncore*dblamda3);
S=((2*pi*c/lamda^2)^2*beta3+(4*pi*c/lamda^3)*beta2)*1e-3;;
```

```
      dw0=-((ncore-nclad)/(lamda*c))*v12*dv;
      dw1=-((nclad-nclad2)/(lamda*c))*v13*dv1;
      dwtotal=(dw0+dw1)*1e6;
      dcfdff=dmat+dwtotal;
      BW=100;
      global dcfS;
      S=((2*pi*c/lamda^2)^2*beta3+(4*pi*c/lamda^3)*beta2)*1e-3;;
      dcfS=S;
      dcftbroad=dcfdff*L*0.01*1e-3;
      global dcfdffl;
      dcfdffl=dcfdff*L;
      if lamda==1.0e-6
          V1=v12;
          V2=v13;
          dvv=dv;
          dv11=dv1;
          dwtotal1=dwtotal;
          S1=S;
          lamdax=lamda;
          dmaty=dmat;
          dcfdff1=dcfdff;
          dcftbroad1=dcftbroad;
          dcfdffl1=dcfdffl;
      end
      V1=[V1 v12];
      V2=[V2 v13];
      dvv=[dvv dv];
      dv11=[dv11 dv1];
      dwtotal1=[dwtotal1 dwtotal];
      dcfdff1=[dcfdff1 dcfdff];
      dmaty=[dmaty dmat];
      lamdax=[lamdax lamda];
      S1=[S1 S];
      dcfdffl1=[dcfdffl1 dcfdffl];
end
global dcfS1;
dcfS1=S1;



function Dffiber1()
%Dffibers designed to be a callback function for dff31. It uses the
values from the controls

plot_theoretical=0;

%get the handles
radius_slider_text=findobj('Tag','radius_slider_text');
radius_slider=findobj('Tag','radius_slider');
delta_slider_text=findobj('Tag','delta_slider_text');
delta_slider=findobj('Tag','delta_slider');
length_slider_text=findobj('Tag','length_slider_text');
length_slider=findobj('Tag','length_slider');
type1_button=findobj('Tag','type1_button');
```

```
type2_button=findobj('Tag','type2_button');
type3_button=findobj('Tag','type3_button');
type4_button=findobj('Tag','type4_button');
type5_button=findobj('Tag','type5_button');
type6_button=findobj('Tag','type6_button');
type7_button=findobj('Tag','type7_button');
type8_button=findobj('Tag','type8_button');
type9_button=findobj('Tag','type9_button');
type10_button=findobj('Tag','type10_button');
dcfradius_slider_text=findobj('Tag','dcfradius_slider_text');
dcfradius_slider=findobj('Tag','dcfradius_slider');
dcfdelta_slider_text=findobj('Tag','dcfdelta_slider_text');
dcfdelta_slider=findobj('Tag','dcfdelta_slider');
ttype1_button=findobj('Tag','ttype1_button');
ttype2_button=findobj('Tag','ttype2_button');
ttype3_button=findobj('Tag','ttype3_button');
ttype4_button=findobj('Tag','ttype4_button');
ttype5_button=findobj('Tag','ttype5_button');
ttype6_button=findobj('Tag','ttype6_button');
ttype7_button=findobj('Tag','ttype7_button');
ttype8_button=findobj('Tag','ttype8_button');
ttype9_button=findobj('Tag','ttype9_button');
ttype10_button=findobj('Tag','ttype10_button');
dff31_fig=findobj('Tag','dff31_fig');
dffiber2;
if(isempty(dff31_fig)) %is dff3 running?
  return;
end


%*********** setup figure windows
%test to se if the plot figure already exists
h=findobj('Tag','Dffiber1_fig');
if(isempty(h))
   scrsz=get(0,'Screensize');
   h=figure('position',[10 0 0.9*scrsz(3) 0.9*scrsz(4)],'name','Dispersion
   Flattened Fiber','Tag','Dffiber1_fig');
else
   figure(h); %make the plot figure the current figure for
     drawing in
   clg; %clear the graphs
end


type1_on=get(type1_button,'value'); %is it Type 1?
type2_on=get(type2_button,'value');%is it Type 2?
type3_on=get(type3_button,'value');%is it Type 3?
type4_on=get(type4_button,'value'); %is it Type 3?
type5_on=get(type5_button,'value'); %is it Type 3?
type6_on=get(type6_button,'value'); %is it Type 3?
type7_on=get(type7_button,'value'); %is it Type 3?
type8_on=get(type8_button,'value'); %is it Type 3?
type9_on=get(type9_button,'value'); %is it Type 3?
type10_on=get(type10_button,'value'); %is it Type 3?
ttype1_on=get(ttype1_button,'value'); %is it Type 1?
ttype2_on=get(ttype2_button,'value');%is it Type 2?
```

```
ttype3_on=get(ttype3_button,'value');%is it Type 3?
ttype4_on=get(ttype4_button,'value'); %is it Type 3?
ttype5_on=get(ttype5_button,'value'); %is it Type 3?
ttype6_on=get(ttype6_button,'value'); %is it Type 3?
ttype7_on=get(ttype7_button,'value'); %is it Type 3?
ttype8_on=get(ttype8_button,'value'); %is it Type 3?
ttype9_on=get(ttype9_button,'value'); %is it Type 3?
ttype10_on=get(ttype10_button,'value'); %is it Type 3?

%**************** setup simulation parameters
%Calculate Waveguide Dispersion
global dcfdff1;
global dcfdff;
global dcfdff11;
 bs(1)=0.6961663;
 bs(2)=0.4079426;
 bs(3)=0.89474794;
 lamds(1)=0.0684043*1e-6;
 lamds(2)=0.1162414*1e-6;
 lamds(3)=9.896161*1e-6;
 b2(1)=0.7028554;
 b2(2)=0.4146307;
 b2(3)=0.8974540;
 lamd2(1)=0.0727723*1e-6;
 lamd2(2)=0.11430854*1e-6;
 lamd2(3)=9.896161*1e-6;
 b3(1)=0.7088876;
 b3(2)=0.4206803;
 b3(3)=0.8956551;
 lamd3(1)=0.0609053*1e-6;
 lamd3(2)=0.1254514*1e-6;
 lamd3(3)=9.9896162*1e-6;
 b4(1)=0.7136824;
 b4(2)=0.4254807;
 b4(3)=0.8964226;
 lamd4(1)=0.0617167*1e-6;
 lamd4(2)=0.1270814*1e-6;
 lamd4(3)=9.896161*1e-6;
 b5(1)=0.696750;
 b5(2)=0.408218;
 b5(3)=0.890815;
 lamd5(1)=0.069066*1e-6;
 lamd5(2)=0.115662*1e-6;
 lamd5(3)=9.900559*1e-6;
 b6(1)=0.711040;
 b6(2)=0.408218;
 b6(3)=0.704048;
 lamd6(1)=0.064270*1e-6;
 lamd6(2)=0.129408*1e-6;
 lamd6(3)=9.424578*1e-6;
 b7(1)=0.695790;
 b7(2)=0.452497;
 b7(3)=0.712513;
 lamd7(1)=0.061568*1e-6;
```

```
lamd7(2)=0.119921*1e-6;
lamd7(3)=8.656641*1e-6;
b8(1)=0.690618;
b8(2)=0.401996;
b8(3)=0.898817;
lamd8(1)=0.061900*1e-6;
lamd8(2)=0.123662*1e-6;
lamd8(3)=9.098960*1e-6;
b9(1)=0.691116;
b9(2)=0.399166;
b9(3)=0.890423;
lamd9(1)=0.068227*1e-6;
lamd9(2)=0.116460*1e-6;
lamd9(3)=9.993707*1e-6;
b10(1)=0.796468;
b10(2)=0.497614;
b10(3)=0.358924;
lamd10(1)=0.094359*1e-6;
lamd10(2)=0.093386*1e-6;
lamd10(3)=5.999652*1e-6;
rad=get(radius_slider,'value');
a=(rad)*1e-6;
delta=get(delta_slider,'value');
L=get(length_slider,'value');
del=(delta)*1;
c=2.997925e8;
a1=1.405*a;
lamdax=1.0e-6;
R=40;
for lamda=1.0e-6:.01e-6:1.8e-6
%Calculate Material Dispersion and Dispersion Slope
    n111=0;
    n112=0;
    n113=0;
    na=0;
    if(type1_on==1)
        for i=1:3
            lamda1=lamda^2-(lamds(i))^2;
            n11=bs(i)*lamda^2/lamda1;
            n111=n11+n111;
            n12=(bs(i)*lamds(i)^2)*(3*lamda^2+lamds(i)^2)/
               (lamda1^3);
            n112=n112+n12;
            n13=(bs(i)*lamda*lamds(i)^2/lamda1)^2;
            n113=n113+n13;
            na=bs(i)*lamds(i)^2/lamda1^2;
         end
     for i=1:3
        lamda1=lamda^2-(lamds(i))^2;
        A=(2*bs(i)*(1-(5*lamda^2)/lamda1+(4*lamda^4)/
          lamda1^2))/lamda1;
        B=(lamda^2*bs(i))/lamda1;
        C=(2*lamda*bs(i)*(1-lamda^2/lamda1))/lamda1;
    end
```

```
for i=1:3
    lamda1=lamda^2-(lamds(i))^2;
    D=(2*lamda*bs(i)*(1-lamda^2/lamda1))/lamda1;
    E=(lamda^2*bs(i))/lamda1;
    F=(24*lamda*bs(i)*(-1+3*lamda^2/lamda1-2*lamda^4/
       lamda1^2))/lamda1;
    G=(2*bs(i)*(1-5*lamda^2/lamda1+4*lamda^4/lamda1^2))/
       lamda1;
    H=(2*lamda*bs(i)*(1-lamda^2/lamda1))/lamda1;
end
end
if(type2_on==1)
    for i=1:3
        lamda1=lamda^2-(lamd2(i))^2;
        n11=b2(i)*lamda^2/lamda1;
        n111=n11+n111;
        n12=(b2(i)*lamd2(i)^2)*(3*lamda^2+lamd2(i)^2)/
           (lamda1^3);
        n112=n112+n12;
        n13=(b2(i)*lamda*lamd2(i)^2/lamda1)^2;
        n113=n113+n13;
        na=b2(i)*lamd2(i)^2/lamda1^2;
    end
for i=1:3
    lamda1=lamda^2-(lamd2(i))^2;
    A=(2*b2(i)*(1-(5*lamda^2)/lamda1+(4*lamda^4)/
       lamda1^2))/lamda1;
    B=(lamda^2*b2(i))/lamda1;
    C=(2*lamda*b2(i)*(1-lamda^2/lamda1))/lamda1;
end
for i=1:3
    lamda1=lamda^2-(lamd2(i))^2;
    D=(2*lamda*b2(i)*(1-lamda^2/lamda1))/lamda1;
    E=(lamda^2*b2(i))/lamda1;
    F=(24*lamda*b2(i)*(-1+3*lamda^2/lamda1-2*lamda^4/
       lamda1^2))/lamda1;
    G=(2*b2(i)*(1-5*lamda^2/lamda1+4*lamda^4/lamda1^2))/
       lamda1;
    H=(2*lamda*b2(i)*(1-lamda^2/lamda1))/lamda1;
end
end
if(type3_on==1)
    for i=1:3
        lamda1=lamda^2-(lamd3(i))^2;
        n11=b3(i)*lamda^2/lamda1;
        n111=n11+n111;
        n12=(b3(i)*lamd2(i)^2)*(3*lamda^2+lamd3(i)^2)/
           (lamda1^3);
        n112=n112+n12;
        n13=(b3(i)*lamda*lamd3(i)^2/lamda1)^2;
        n113=n113+n13;
        na=b3(i)*lamd3(i)^2/lamda1^2;
    end
for i=1:3
```

```
        lamda1=lamda^2-(lamd3(i))^2;
        A=(2*b3(i)*(1-(5*lamda^2)/lamda1+(4*lamda^4)/
          lamda1^2))/lamda1;
        B=(lamda^2*b3(i))/lamda1;
        C=(2*lamda*b3(i)*(1-lamda^2/lamda1))/lamda1;
    end
    for i=1:3
        lamda1=lamda^2-(lamd3(i))^2;
        D=(2*lamda*b3(i)*(1-lamda^2/lamda1))/lamda1;
        E=(lamda^2*b3(i))/lamda1;
        F=(24*lamda*b3(i)*(-1+3*lamda^2/lamda1-2*lamda^4/
          lamda1^2))/lamda1;
        G=(2*b3(i)*(1-5*lamda^2/lamda1+4*lamda^4/lamda1^2))/
          lamda1;
        H=(2*lamda*b3(i)*(1-lamda^2/lamda1))/lamda1;
    end
    end
    if(type4_on==1)
        for i=1:3
            lamda1=lamda^2-(lamd4(i))^2;
            n11=b4(i)*lamda^2/lamda1;
            n111=n11+n111;
            n12=(b4(i)*lamd4(i)^2)*(3*lamda^2+lamd4(i)^2)/
              (lamda1^3);
            n112=n112+n12;
            n13=(b4(i)*lamda*lamd4(i)^2/lamda1)^2;
            n113=n113+n13;
            na=b4(i)*lamd4(i)^2/lamda1^2;
        end
    for i=1:3
        lamda1=lamda^2-(lamd4(i))^2;
        A=(2*b4(i)*(1-(5*lamda^2)/lamda1+(4*lamda^4)/
          lamda1^2))/lamda1;
        B=(lamda^2*b4(i))/lamda1;
        C=(2*lamda*b4(i)*(1-lamda^2/lamda1))/lamda1;
    end
    for i=1:3
        lamda1=lamda^2-(lamd4(i))^2;
        D=(2*lamda*b4(i)*(1-lamda^2/lamda1))/lamda1;
        E=(lamda^2*b4(i))/lamda1;
        F=(24*lamda*b4(i)*(-1+3*lamda^2/lamda1-2*lamda^4/
          lamda1^2))/lamda1;
        G=(2*b4(i)*(1-5*lamda^2/lamda1+4*lamda^4/lamda1^2))/
          lamda1;
        H=(2*lamda*b4(i)*(1-lamda^2/lamda1))/lamda1;
    end
    end
    if(type5_on==1)
        for i=1:3
            lamda1=lamda^2-(lamd5(i))^2;
            n11=b5(i)*lamda^2/lamda1;
            n111=n11+n111;
            n12=(b5(i)*lamd5(i)^2)*(3*lamda^2+lamd5(i)^2)/
              (lamda1^3);
```

```
                n112=n112+n12;
                n13=(b5(i)*lamda*lamd5(i)^2/lamda1)^2;
                n113=n113+n13;
                na=b5(i)*lamd5(i)^2/lamda1^2;
            end
        for i=1:3
            lamda1=lamda^2-(lamd5(i))^2;
            A=(2*b5(i)*(1-(5*lamda^2)/lamda1+(4*lamda^4)/
              lamda1^2))/lamda1;
            B=(lamda^2*b5(i))/lamda1;
            C=(2*lamda*b5(i)*(1-lamda^2/lamda1))/lamda1;
        end
        for i=1:3
            lamda1=lamda^2-(lamd5(i))^2;
            D=(2*lamda*b5(i)*(1-lamda^2/lamda1))/lamda1;
            E=(lamda^2*b5(i))/lamda1;
            F=(24*lamda*b5(i)*(-1+3*lamda^2/lamda1-2*lamda^4/
              lamda1^2))/lamda1;
            G=(2*b5(i)*(1-5*lamda^2/lamda1+4*lamda^4/lamda1^2))/
              lamda1;
            H=(2*lamda*b5(i)*(1-lamda^2/lamda1))/lamda1;
        end
    end
    if(type6_on==1)
        for i=1:3
                lamda1=lamda^2-(lamd6(i))^2;
                n11=b6(i)*lamda^2/lamda1;
                n111=n11+n111;
                n12=(b6(i)*lamd6(i)^2)*(3*lamda^2+lamd6(i)^2)/
                   (lamda1^3);
                n112=n112+n12;
                n13=(b6(i)*lamda*lamd6(i)^2/lamda1)^2;
                n113=n113+n13;
                na=b6(i)*lamd6(i)^2/lamda1^2;
            end
        for i=1:3
            lamda1=lamda^2-(lamd6(i))^2;
            A=(2*b6(i)*(1-(5*lamda^2)/lamda1+(4*lamda^4)/
              lamda1^2))/lamda1;
            B=(lamda^2*b6(i))/lamda1;
            C=(2*lamda*b6(i)*(1-lamda^2/lamda1))/lamda1;
        end
        for i=1:3
            lamda1=lamda^2-(lamd6(i))^2;
            D=(2*lamda*b6(i)*(1-lamda^2/lamda1))/lamda1;
            E=(lamda^2*b6(i))/lamda1;
            F=(24*lamda*b6(i)*(-1+3*lamda^2/lamda1-2*lamda^4/
              lamda1^2))/lamda1;
            G=(2*b6(i)*(1-5*lamda^2/lamda1+4*lamda^4/lamda1^2))/
              lamda1;
            H=(2*lamda*b6(i)*(1-lamda^2/lamda1))/lamda1;
        end
    end
    if(type7_on==1)
```

```
    for i=1:3
        lamda1=lamda^2-(lamd7(i))^2;
        n11=b7(i)*lamda^2/lamda1;
        n111=n11+n111;
        n12=(b7(i)*lamd7(i)^2)*(3*lamda^2+lamd7(i)^2)/
           (lamda1^3);
        n112=n112+n12;
        n13=(b7(i)*lamda*lamd7(i)^2/lamda1)^2;
        n113=n113+n13;
        na=b7(i)*lamd7(i)^2/lamda1^2;
    end
for i=1:3
    lamda1=lamda^2-(lamd7(i))^2;
    A=(2*b7(i)*(1-(5*lamda^2)/lamda1+(4*lamda^4)/
      lamda1^2))/lamda1;
    B=(lamda^2*b7(i))/lamda1;
    C=(2*lamda*b7(i)*(1-lamda^2/lamda1))/lamda1;
end
for i=1:3
    lamda1=lamda^2-(lamd7(i))^2;
    D=(2*lamda*b7(i)*(1-lamda^2/lamda1))/lamda1;
    E=(lamda^2*b7(i))/lamda1;
    F=(24*lamda*b7(i)*(-1+3*lamda^2/lamda1-2*lamda^4/
      lamda1^2))/lamda1;
    G=(2*b7(i)*(1-5*lamda^2/lamda1+4*lamda^4/lamda1^2))/
      lamda1;
    H=(2*lamda*b7(i)*(1-lamda^2/lamda1))/lamda1;
end
end
if(type8_on==1)
    for i=1:3
        lamda1=lamda^2-(lamd8(i))^2;
        n11=b8(i)*lamda^2/lamda1;
        n111=n11+n111;
        n12=(b8(i)*lamd8(i)^2)*(3*lamda^2+lamd8(i)^2)/
           (lamda1^3);
        n112=n112+n12;
        n13=(b8(i)*lamda*lamd8(i)^2/lamda1)^2;
        n113=n113+n13;
        na=b8(i)*lamd8(i)^2/lamda1^2;
    end
 for i=1:3
    lamda1=lamda^2-(lamd8(i))^2;
    A=(2*b8(i)*(1-(5*lamda^2)/lamda1+(4*lamda^4)/
      lamda1^2))/lamda1;
    B=(lamda^2*b8(i))/lamda1;
    C=(2*lamda*b8(i)*(1-lamda^2/lamda1))/lamda1;
end
for i=1:3
    lamda1=lamda^2-(lamd8(i))^2;
    D=(2*lamda*b8(i)*(1-lamda^2/lamda1))/lamda1;
    E=(lamda^2*b8(i))/lamda1;
    F=(24*lamda*b8(i)*(-1+3*lamda^2/lamda1-2*lamda^4/
      lamda1^2))/lamda1;
    G=(2*b8(i)*(1-5*lamda^2/lamda1+4*lamda^4/lamda1^2))/
```

```
            lamda1;
        H=(2*lamda*b8(i)*(1-lamda^2/lamda1))/lamda1;
    end
end
if(type9_on==1)
    for i=1:3
        lamda1=lamda^2-(lamd9(i))^2;
        n11=b9(i)*lamda^2/lamda1;
        n111=n11+n111;
        n12=(b9(i)*lamd9(i)^2)*(3*lamda^2+lamd9(i)^2)/
            (lamda1^3);
        n112=n112+n12;
        n13=(b9(i)*lamda*lamd9(i)^2/lamda1)^2;
        n113=n113+n13;
        na=b9(i)*lamd9(i)^2/lamda1^2;
    end
    for i=1:3
        lamda1=lamda^2-(lamd9(i))^2;
        A=(2*b9(i)*(1-(5*lamda^2)/lamda1+(4*lamda^4)/
            lamda1^2))/lamda1;
        B=(lamda^2*b9(i))/lamda1;
        C=(2*lamda*b9(i)*(1-lamda^2/lamda1))/lamda1;
    end
    for i=1:3
        lamda1=lamda^2-(lamd9(i))^2;
        D=(2*lamda*b9(i)*(1-lamda^2/lamda1))/lamda1;
        E=(lamda^2*b9(i))/lamda1;
        F=(24*lamda*b9(i)*(-1+3*lamda^2/lamda1-2*lamda^4/
            lamda1^2))/lamda1;
        G=(2*b9(i)*(1-5*lamda^2/lamda1+4*lamda^4/lamda1^2))/
            lamda1;
        H=(2*lamda*b9(i)*(1-lamda^2/lamda1))/lamda1;
    end
end
if(type10_on==1)
    for i=1:3
        lamda1=lamda^2-(lamd10(i))^2;
        n11=b10(i)*lamda^2/lamda1;
        n111=n11+n111;
        n12=(b10(i)*lamd10(i)^2)*(3*lamda^2+lamd10(i)^2)/
            (lamda1^3);
        n112=n112+n12;
        n13=(b10(i)*lamda*lamd10(i)^2/lamda1)^2;
        n113=n113+n13;
        na=b10(i)*lamd10(i)^2/lamda1^2;
    end
    for i=1:3
        lamda1=lamda^2-(lamd10(i))^2;
        A=(2*b10(i)*(1-(5*lamda^2)/lamda1+(4*lamda^4)/
            lamda1^2))/lamda1;
        B=(lamda^2*b10(i))/lamda1;
        C=(2*lamda*b10(i)*(1-lamda^2/lamda1))/lamda1;
    end
    for i=1:3
```

```
        lamda1=lamda^2-(lamd10(i))^2;
        D=(2*lamda*b10(i)*(1-lamda^2/lamda1))/lamda1;
        E=(lamda^2*b10(i))/lamda1;
        F=(24*lamda*b10(i)*(-1+3*lamda^2/lamda1-2*lamda^4/
          lamda1^2))/lamda1;
        G=(2*b10(i)*(1-5*lamda^2/lamda1+4*lamda^4/lamda1^2))/
          lamda1;
        H=(2*lamda*b10(i)*(1-lamda^2/lamda1))/lamda1;
    end
  end
  dnlamda2=(A/(2*sqrt(1+B)))-(C/(4*(1+B)^1.5));
  dnlamda3=(D/(8*(1+E)^2.5))+(F/(2*sqrt(1+E)))-(3*G*H/
    (4*(1+E)^1.5));
  ncore=sqrt(1+n111);
  ddn=(n112/ncore)-(n112/(ncore^3));
  dnlamda1=-(lamda/ncore)*na;
  neff=ncore-(lamda*dnlamda1);
  dmat=-(lamda/c)*ddn*1e6;
%Calculation of Waveguide Dispersion
%Outer cladding of pure SILICA
  nclad21=0;
  dnn2=0;
  for i=1:3
      lamda1=lamda^2-(lamds(i))^2;
      nclad2=bs(i)*lamda^2/lamda1;
      nclad21=nclad2+nclad21;
      dnn=lamds(i)^2*bs(i)^2/(lamda1^2);
      dnn2=dnn2+dnn;
  end
  nclad2=sqrt(1+nclad21);
%First Cladding of fiber
  nclad=ncore-del;
  dn=-(lamda/nclad2)*dnn2;
  N=nclad2-lamda*dn;
  delta12=(ncore-nclad);
  delta23=(nclad-nclad2);
  delta=delta23/delta12;
  deltaa=(delta12+delta23)/nclad2;
  k=2*pi/lamda;
  v12=a*k*sqrt(ncore^2-nclad^2);
  v13=a1*k*sqrt(abs(nclad^2-nclad2^2));
  spotsize=a1*(0.65+1.619*(v12)^(-1.5)+2.879*(v12)^(-6));
  b(1)=(1.1428-0.996/v12)^2;
  b(2)=(1.1428-0.996/v13)^2;
  beta(1)=sqrt(k^2*(b(1)*(ncore^2-nclad2^2)+nclad2^2));
  beta(2)=sqrt(k^2*(b(2)*(nclad^2-nclad2^2)+nclad2^2));
  u(1)=a*sqrt(k^2*ncore^2-beta(1)^2);
  u(2)=a1*sqrt(k^2*nclad^2-beta(2)^2);
  v(1)=a*sqrt(beta(1)^2-k^2*nclad2^2);
  v(2)=a1*sqrt(beta(2)^2-k^2*nclad2^2);
  k0=besselk(1,v(1))/(besselk(0,v(1)));%Bessel Function
  k1=besselk(1,v(2))/(besselk(0,v(2)));%Bessel Function
  dv=2*(u(1)/v12)^2*(k0*(1-2*k0)+(2/v(1))*(v(1)^2+(u(1)^2)
    *k0)*sqrt(k0)*(k0+sqrt(k0)/v(1)-1));
```

```
dv1=abs(2*(u(2)/v13)^2*(k1*(1-2*k1)+(2/v(2))*(v(2)^2+(u(2)
   ^2)*k1)*sqrt(k1)*(k1+sqrt(k1)/v(2)-1)));
bv=(1.1428-0.9960/v12)^2;
AA=(2*pi/(lamda))*a1*sqrt(2*del);
dblamda1=1.9920*AA*(1.1428-0.9960/v12)*(dnlamda1/v12^2);
dblamda2=1.9920*AA*((0.9960/v12^4)*AA*(dnla
   mda1)^2+(1.1428-0.9960/v12)*(dnlamda2/v12^2-
   2*AA*(dnlamda1)^2/v12^3));
dblamda3=1.984*AA^2*(2*dnlamda1*dnlamda2/v12^4-
   4*AA*dnlamda1^3/v12^5)+...
       1.984*AA^2*(dnlamda1*(dnlamda2/v12^2-2*AA*dnlamda1^2/
          v12^3)/v12^2)+...
       1.9920*AA*(1.1428-0.9960/v12)*(dnlamda3/v12^2-6
       *AA*dnlamda1*dnlamda2/v12^3+6*AA*dnlamda1^3/v12^4);
beta2=(lamda^2*(1-del)/(2*pi*c^2))*(lamda*(1+bv*delta)*dnlam
   da2+2*lamda*del*dnlamda1*dblamda1+del*lamda*ncore*dblamda2);
beta3=-(lamda^2*(1-del))/(4*pi^2*c^3)*(3*lamda^2*(1+bv*del)
   *dnlamda2+3*del*lamda^3*dnlamda2*dblamda1+...
       (1+bv*del)*lamda^3*dnlamda3+6*lamda^2*del*dnlamda1
       *dblamda1+3*del*lamda^3*dnlamda1*dblamda2+3*lamda^2
       *del*ncore*dblamda2+lamda^3*ncore*dblamda3);
S=((2*pi*c/lamda^2)^2*beta3+(4*pi*c/lamda^3)*beta2)*1e-3;;
dw0=-((ncore-nclad)/(lamda*c))*v12*dv;
dw1=-((nclad-nclad2)/(lamda*c))*v13*dv1;
dwtotal=(dw0+dw1)*1e6;
dff=dmat+dwtotal;
spotsize=a1*(0.65+1.619*(v12)^(-1.5)+2.879*(v12)^(-6));
BW=100;
S=((2*pi*c/lamda^2)^2*beta3+(4*pi*c/lamda^3)*beta2)*1e-3;;
tnrz=0.50/40;%average of 0.35/40 and 0.75/40
dpmd=1;
tbroad=sqrt((dff*L*0.01*1e-3)^2+(dpmd*sqrt(L)*1e-3)^2);
dffl=dff*L;%DL product of DFF
neff=ncore-(lamda*dnlamda1);
b1=b(1);
modes=0.5*(pi*2*a/lamda)^2*(ncore^2-nclad^2);
global dcfS;
dscr=abs((S/dff)*(dcfdff/dcfS));
if lamda==1.0e-6
    bb=b1;
    dscr1=dscr;
    V1=v12;
    V2=v13;
    dvv=dv;
    dv11=dv1;
    dwtotal1=dwtotal;
    S1=S;
    lamdax=lamda;
    dmaty=dmat;
    dff1=dff;
    tbroad1=tbroad;
    tnrz1=tnrz;
    dffl1=dffl;
    nclad22=nclad2;
```

```
              ncore1=ncore;
              nclad1=nclad;
              modes1=modes;
              spotsize1=spotsize;
       end
       bb=[bb b1];
       dscr1=[dscr1 dscr];
       dwtotal1=[dwtotal1 dwtotal];
       dff1=[dff1 dff];
       dmaty=[dmaty dmat];
       lamdax=[lamdax lamda];
       tbroad1=[tbroad1 tbroad];
       tnrz1=[tnrz1 tnrz];
       S1=[S1 S];
       dff11=[dff11 dff1];
       spotsize1=[spotsize1 spotsize];
       modes1=[modes1 modes];
       nclad22=[nclad22 nclad2];
       ncore1=[ncore1 ncore];
       nclad1=[nclad1 nclad];
       V1=[V1 v12];
       V2=[V2 v13];
       dvv=[dvv dv];
       dv11=[dv11 dv1];
end
global dcfS1;
          dtdff1=dff1+dcfdff1;
          dtdff11=dff11+dcfdff11;
          subplot(3,3,1);
          plot(lamdax,dwtotal1,lamdax,dmaty,lamdax,dff1);
          title('Dispersion Factors of DFF (W1)');
          ylabel('ps/(nm-Km)');
          legend('Dm','Dw','Dtotal')
          xlabel('wavelength (µm)');
          grid on;
          subplot(3,3,2);
          plot(lamdax,dcfdff1);
          title('Dispersion Factor of DCF (W2)');
          ylabel('ps/(nm-Km)');
          xlabel('wavelength (µm)');
          grid on;
          subplot(3,3,3);
          plot(lamdax,S1,lamdax,dcfS1);
          title('Dispersion Slope (W3)');
          xlabel('wavelength(µm)');
          ylabel('Dispersion Slope (ps/(nm^2-Km))');
          grid on;
          subplot(3,3,4);
          plot(V1,dvv,V2,dv11);
          title(' (W4)');
          xlabel('Normalised Frequency');
          ylabel('d^2(Vb)/dv^2');
          grid on;
          subplot(3,3,5);
```

```
      plot(lamdax,ncore1,lamdax,nclad1,lamdax,nclad22);
      title('Refractive Indices (W5)');
      ylabel('Refractive Index');
      legend('n(core)','n(first clad)','n(outer clad)')
      xlabel('wavelength(μm)');
      grid on;
      subplot(3,3,6);
      plot(lamdax,spotsize1);
      title('Spot Size (W6)');
      ylabel('spotsize (um)');
      xlabel('Wavelength (μm)');
      grid on;
      subplot(3,3,7);
      plot(lamdax,tbroad1,lamdax,tnrz1);
      title('Pulse Broadening (W7)');
      xlabel('Wavelength (μm)');
      legend('tbroad','tnrz');
      ylabel('TIME (ns)');
      grid on;
      subplot(3,3,8);
      plot(lamdax,dscr1);
      title('DSCR (W8)');
      xlabel('wavelength (μm)');
      ylabel('KAPPA');
      grid on;
      axis([1.0e-6 1.8e-6 0 10]);
      subplot(3,3,9);
      plot(lamdax,dtdff11);
      title('Total DL product of the link (W9)');
      ylabel('DFF TOTAL (ps/nm)');
      xlabel('wavelength (μm)');
      grid on;


function dff31(action)
if (nargin<1)
    action='initialize';
end;


if strcmp(action,'initialize')
  h=figure('position',[10 0 650 500],'name','Core Type','Tag','dff31_
  fig');
  %size is a 4-element vector [left, bottom, width,height]


  %setup Type buttons
  Type1_button=uicontrol('style','radiobutton','Tag','type1_
button','string','CoreType 1',...
   'Callback','dff31(''type1'')','Position',[10,450,100,20]);
  Type2_button=uicontrol('style','radiobutton','Tag','type2_
button','string','CoreType 2',...
   'Callback','dff31(''type2'')','Position',[10,400,100,20],
```

```
'value',1);
  Type3_button=uicontrol('style','radiobutton','Tag','type3_
button','string','CoreType 3',...
  'Callback','dff31(''type3'')','Position',[10,350,100,20]);
  Type4_button=uicontrol('style','radiobutton','Tag','type4_
button','string','CoreType 4',...
  'Callback','dff31(''type4'')','Position',[10,300,100,20]);
  Type5_button=uicontrol('style','radiobutton','Tag','type5_
button','string','CoreType 5',...
  'Callback','dff31("type5")','Position',[10,250,100,20]);
  Type6_button=uicontrol('style','radiobutton','Tag','type6_
button','string','CoreType 6',...
  'Callback','dff31("type6")','Position',[10,200,100,20]);
  Type7_button=uicontrol('style','radiobutton','Tag','type7_
button','string','CoreType 7',...
  'Callback','dff31("type7")','Position',[10,150,100,20]);
  Type8_button=uicontrol('style','radiobutton','Tag','type8_
button','string','CoreType 8',...
  'Callback','dff31("type8")','Position',[10,100,100,20]);
  Type9_button=uicontrol('style','radiobutton','Tag','type9_
button','string','CoreType 9',...
  'Callback','dff31("type9")','Position',[10,50,100,20]);
  Type10_button=uicontrol('style','radiobutton','Tag',
'type10_button','string','CoreType 10',...
  'Callback','dff31("type10")','Position',[10,0,100,20]);


  TType1_button=uicontrol('style','radiobutton','Tag',
'ttype1_button','string','DCFCoreType 1',...
  'Callback','dff31("ttype1")','Position',[300,450,100,20]);
  TType2_button=uicontrol('style','radiobutton','Tag', 'ttype2_
button','string','DCFCoreType 2',...
  'Callback','dff31("ttype2")','Position',[300,400,100,20],
'value',1);
  TType3_button=uicontrol('style','radiobutton','Tag', 'ttype3_
button','string','DCFCoreType 3',...
  'Callback','dff31("ttype3")','Position',[300,350,100,20]);
  TType4_button=uicontrol('style','radiobutton','Tag','ttype4_
button','string','DCFCoreType 4',...
  'Callback','dff31("ttype4")','Position',[300,300,100,20]);
  TType5_button=uicontrol('style','radiobutton','Tag', 'ttype5_
button','string','DCFCoreType 5',...
  'Callback','dff31("ttype5")','Position',[300,250,100,20]);
  TType6_button=uicontrol('style','radiobutton','Tag', 'ttype6_
button','string','DCFCoreType 6',...
  'Callback','dff31("ttype6")','Position',[300,200,100,20]);
  TType7_button=uicontrol('style','radiobutton','Tag', 'ttype7_
button','string','DCFCoreType 7',...
  'Callback','dff31("ttype7")','Position',[300,150,100,20]);
  TType8_button=uicontrol('style','radiobutton','Tag', 'ttype8_
button','string','DCFCoreType 8',...
  'Callback','dff31("ttype8")','Position',[300,100,100,20]);
  TType9_button=uicontrol('style','radiobutton','Tag', 'ttype9_
button','string','DCFCoreType 9',...
```

```
  'Callback','dff31("ttype9")','Position',[300,50,100,20]);
  TType10_button=uicontrol('style','radiobutton','Tag', 'ttype10_
button','string','DCFCoreType 10',...
  'Callback','dff31(''ttype10'')','Position',[300,0,100,20]);

%setup radius slider
  radius_slider=uicontrol('style','slider','Tag','radius_slider','Min',0.250
0,'Max',6.5000,...
  'Sliderstep',[0.0001 .0006],'String','Bit width','Callback','dff31(''ra
dius_slider'')','Position',[150,300,150,20], 'value',0.25000);

  uicontrol('style','text','string','Radius (in um)','Posit
ion',[150,350,150,20]);
  radius_slider_text=uicontrol('style','text','Tag','radius_slider_text',
'Position',[150,325,150,20]);
  set(radius_slider_text,'string',get(radius_slider,'value'));

%setup delta slider
  delta_slider=uicontrol('style','slider','Tag','delta_slider','Min',0.00
01,'Max',0.015,...
  'Sliderstep',[0.0001 0.0006],'String','delta','Callback',
'dff31(''delta_slider'')','Position',[150,100,150,20], 'value',0.0001);

  uicontrol('style','text','string','index diff','Posit
ion',[150,150,150,20]);
  delta_slider_text=uicontrol('style','text','Tag', 'delta_slider_text','
Position',[150,125,150,20]);
  set(delta_slider_text,'string',get(delta_slider,'value'));
%set up length slider
  length_slider=uicontrol('style','slider','Tag','length_
slider','Min',1.0,'Max',200,...
  'Sliderstep',[0.0001 0.0006],'String','delta','Callback','dff31(''len
gth_slider'')','Position',[150,20,150,20],'value',1.0);

  uicontrol('style','text','string','LENGTH','Position',
[150,50,150,20]);
  length_slider_text=uicontrol('style','text','Tag', 'length_slider_text'
,'Position',[150,75,150,20]);
  set(length_slider_text,'string',get(length_slider,'value'));

  dcflength_slider=uicontrol('style','slider','Tag', 'dcflength_
slider','Min',0.1,'Max',50,...
  'Sliderstep',[0.0001 0.0006],'String','delta','Callback',
'dff31(''dcflength_slider'')','Position',[410,20,150,20], 'value',0.1);

  uicontrol('style','text','string','DCFLENGTH','Position',
[410,50,150,20]);
  dcflength_slider_text=uicontrol('style','text','Tag', 'dcflength_
slider_text','Position',[410,75,150,20]);
  set(dcflength_slider_text,'string',get(dcflength_slider,'value'));

%setup radius slider
```

```
  dcfradius_slider=uicontrol('style','slider','Tag', 'dcfradius_slider','
Min',0.2500,'Max',6.5000,...
  'Sliderstep',[0.0001 .0006],'String','Bit width','Callback','dff31(''dc
fradius_slider'')','Position',[410,300,150,20],'value',0.25000);

  uicontrol('style','text','string','Radius (in um)','Posit
ion',[410,350,150,20]);
  dcfradius_slider_text=uicontrol('style','text','Tag','dcfradius_slider_
text','Position',[410,325,150,20]);
  set(dcfradius_slider_text,'string',get(dcfradius_slider, 'value'));

  %setup delta slider

  dcfdelta_slider=uicontrol('style','slider','Tag', 'dcfdelta_slider','Mi
n',0.0001,'Max',0.025,...
  'Sliderstep',[0.0001 0.0006],'String','delta','Callback',
  'dff31(''dcfdelta_slider'')','Position',[410,100,150,20],
  'value',0.0001);

  uicontrol('style','text','string','index diff','Posit
ion',[410,150,150,20]);
  dcfdelta_slider_text=uicontrol('style','text','Tag','dcfdelta_slider_
text','Position',[410,125,150,20]);
  set(dcfdelta_slider_text,'string',get(dcfdelta_slider,'value'));

  Dffiber1;    %make it go once to get the display set

%******************end of initialisation phase***************************
**
%***************** start of callback phases ******************
  else  %there is an input argument so must be in the feedback phase
h=findobj('Tag','dff31_fig'); %test to see if has been initialised.
  if(isempty(h)) %if it hasn't, then initialise it
    dff31;
    return; %just initialise then finish
  end
  radius_slider=findobj('Tag','radius_slider');
  radius_slider_text=findobj('Tag','radius_slider_text');
  radius_slider_label=findobj('Tag','radius_slider_label');
  delta_slider=findobj('Tag','delta_slider');
  delta_slider_text=findobj('Tag','delta_slider_text');
  delta_slider_label=findobj('Tag','delta_slider_label');
  length_slider=findobj('Tag','length_slider');
  length_slider_text=findobj('Tag','length_slider_text');
  length_slider_label=findobj('Tag','length_slider_label');
  dcflength_slider=findobj('Tag','dcflength_slider');
  dcflength_slider_text=findobj('Tag','dcflength_slider_
    text');
  dcflength_slider_label=findobj('Tag','dcflength_slider_
    label');
  type1_button=findobj('Tag','type1_button');
  type2_button=findobj('Tag','type2_button');
  type3_button=findobj('Tag','type3_button');
  type4_button=findobj('Tag','type4_button');
```

```
type5_button=findobj('Tag','type5_button');
type6_button=findobj('Tag','type6_button');
type7_button=findobj('Tag','type7_button');
type8_button=findobj('Tag','type8_button');
type9_button=findobj('Tag','type9_button');
type10_button=findobj('Tag','type10_button');

type1_on=get(type1_button,'value');
type2_on=get(type2_button,'value');
type3_on=get(type3_button,'value');
type4_on=get(type4_button,'value');
type5_on=get(type5_button,'value');
type6_on=get(type6_button,'value');
type7_on=get(type7_button,'value');
type8_on=get(type8_button,'value');
type9_on=get(type9_button,'value');
type10_on=get(type10_button,'value');

dcfradius_slider=findobj('Tag','dcfradius_slider');
dcfradius_slider_text=findobj('Tag','dcfradius_slider_
    text');
dcfradius_slider_label=findobj('Tag','dcfradius_slider_
    label');
dcfdelta_slider=findobj('Tag','dcfdelta_slider');
dcfdelta_slider_text=findobj('Tag','dcfdelta_slider_text');
dcfdelta_slider_label=findobj('Tag','dcfdelta_slider_
    label');
ttype1_button=findobj('Tag','ttype1_button');
ttype2_button=findobj('Tag','ttype2_button');
ttype3_button=findobj('Tag','ttype3_button');
ttype4_button=findobj('Tag','ttype4_button');
ttype5_button=findobj('Tag','ttype5_button');
ttype6_button=findobj('Tag','ttype6_button');
ttype7_button=findobj('Tag','ttype7_button');
ttype8_button=findobj('Tag','ttype8_button');
ttype9_button=findobj('Tag','ttype9_button');
ttype10_button=findobj('Tag','ttype10_button');

ttype1_on=get(ttype1_button,'value');
ttype2_on=get(ttype2_button,'value');
ttype3_on=get(ttype3_button,'value');
ttype4_on=get(ttype4_button,'value');
ttype5_on=get(ttype5_button,'value');
ttype6_on=get(ttype6_button,'value');
ttype7_on=get(ttype7_button,'value');
ttype8_on=get(ttype8_button,'value');
ttype9_on=get(ttype9_button,'value');
ttype10_on=get(ttype10_button,'value');

if(type1_on==0 & type3_on==0 & type4_on==0 & type5_on==0 & type6_on==0
& type7_on==0 & type8_on==0 & type9_on==0 & type10_on==0)
    set(type2_button,'value',1); %make sure at least one button is on
end;
```

```
if strcmp(action,'type1')
  set(type2_button,'value',0);
  set(type3_button,'value',0);
  set(type4_button,'value',0);
  set(type5_button,'value',0);
  set(type6_button,'value',0);
  set(type7_button,'value',0);
  set(type8_button,'value',0);
  set(type9_button,'value',0);
  set(type10_button,'value',0);

elseif strcmp(action,'type2')
  set(type1_button,'value',0);
  set(type3_button,'value',0);
  set(type4_button,'value',0);
  set(type5_button,'value',0);
  set(type6_button,'value',0);
  set(type7_button,'value',0);
  set(type8_button,'value',0);
  set(type9_button,'value',0);
  set(type10_button,'value',0);

elseif strcmp(action,'type3')
  set(type1_button,'value',0);
  set(type2_button,'value',0);
  set(type4_button,'value',0);
  set(type5_button,'value',0);
  set(type6_button,'value',0);
  set(type7_button,'value',0);
  set(type8_button,'value',0);
  set(type9_button,'value',0);
  set(type10_button,'value',0);

elseif strcmp(action,'type4')
  set(type1_button,'value',0);
  set(type2_button,'value',0);
  set(type3_button,'value',0);
  set(type5_button,'value',0);
  set(type6_button,'value',0);
  set(type7_button,'value',0);
  set(type8_button,'value',0);
  set(type9_button,'value',0);
  set(type10_button,'value',0);

elseif strcmp(action,'type5')
  set(type1_button,'value',0);
  set(type2_button,'value',0);
  set(type3_button,'value',0);
  set(type4_button,'value',0);
  set(type6_button,'value',0);
  set(type7_button,'value',0);
  set(type8_button,'value',0);
  set(type9_button,'value',0);
```

```
          set(type10_button,'value',0);

elseif strcmp(action,'type6')
     set(type1_button,'value',0);
     set(type3_button,'value',0);
     set(type4_button,'value',0);
     set(type5_button,'value',0);
     set(type2_button,'value',0);
     set(type7_button,'value',0);
     set(type8_button,'value',0);
     set(type9_button,'value',0);
     set(type10_button,'value',0);

elseif strcmp(action,'type7')
     set(type1_button,'value',0);
     set(type3_button,'value',0);
     set(type4_button,'value',0);
     set(type5_button,'value',0);
     set(type6_button,'value',0);
     set(type2_button,'value',0);
     set(type8_button,'value',0);
     set(type9_button,'value',0);
     set(type10_button,'value',0);

elseif strcmp(action,'type8')
     set(type1_button,'value',0);
     set(type3_button,'value',0);
     set(type4_button,'value',0);
     set(type5_button,'value',0);
     set(type6_button,'value',0);
     set(type7_button,'value',0);
     set(type2_button,'value',0);
     set(type9_button,'value',0);
     set(type10_button,'value',0);

elseif strcmp(action,'type9')
     set(type1_button,'value',0);
     set(type3_button,'value',0);
     set(type4_button,'value',0);
     set(type5_button,'value',0);
     set(type6_button,'value',0);
     set(type7_button,'value',0);
     set(type8_button,'value',0);
     set(type2_button,'value',0);
     set(type10_button,'value',0);

elseif strcmp(action,'type10')
     set(type1_button,'value',0);
     set(type3_button,'value',0);
     set(type4_button,'value',0);
     set(type5_button,'value',0);
     set(type6_button,'value',0);
     set(type7_button,'value',0);
```

```
    set(type8_button,'value',0);
    set(type9_button,'value',0);
    set(type2_button,'value',0);

  elseif strcmp(action,'radius_slider')
    set(radius_slider_text,'String',get (radius_slider,'value'));

  elseif strcmp(action,'delta_slider')
    set(delta_slider_text,'String',get(delta_slider,'value'));

  elseif strcmp(action,'length_slider')
    set(length_slider_text,'String',get (length_slider,'value'));
  end; %end of fiding out which control had been activated

%FOR DCF

  if(ttype1_on==0 & ttype3_on==0 & ttype4_on==0 & ttype5_on==0 & ttype6_
on==0 & ttype7_on==0 & ttype8_on==0 & ttype9_on==0 & ttype10_on==0)
      set(ttype2_button,'value',1); %make sure at least one button is on
  end;

  if strcmp(action,'ttype1')
    set(ttype2_button,'value',0);
    set(ttype3_button,'value',0);
    set(ttype4_button,'value',0);
    set(ttype5_button,'value',0);
    set(ttype6_button,'value',0);
    set(ttype7_button,'value',0);
    set(ttype8_button,'value',0);
    set(ttype9_button,'value',0);
    set(ttype10_button,'value',0);

  elseif strcmp(action,'ttype2')
    set(ttype1_button,'value',0);
    set(ttype3_button,'value',0);
    set(ttype4_button,'value',0);
    set(ttype5_button,'value',0);
    set(ttype6_button,'value',0);
    set(ttype7_button,'value',0);
    set(ttype8_button,'value',0);
    set(ttype9_button,'value',0);
    set(ttype10_button,'value',0);

  elseif strcmp(action,'ttype3')
    set(ttype1_button,'value',0);
    set(ttype2_button,'value',0);
    set(ttype4_button,'value',0);
    set(ttype5_button,'value',0);
    set(ttype6_button,'value',0);
    set(ttype7_button,'value',0);
    set(ttype8_button,'value',0);
    set(ttype9_button,'value',0);
    set(ttype10_button,'value',0);
```

```
elseif strcmp(action, 'ttype4')
  set(ttype1_button, 'value',0);
  set(ttype2_button, 'value',0);
  set(ttype3_button, 'value',0);
  set(ttype5_button, 'value',0);
  set(ttype6_button, 'value',0);
  set(ttype7_button, 'value',0);
  set(ttype8_button, 'value',0);
  set(ttype9_button, 'value',0);
  set(ttype10_button, 'value',0);

elseif strcmp(action, 'ttype5')
  set(ttype1_button, 'value',0);
  set(ttype2_button, 'value',0);
  set(ttype3_button, 'value',0);
  set(ttype4_button, 'value',0);
  set(ttype6_button, 'value',0);
  set(ttype7_button, 'value',0);
  set(ttype8_button, 'value',0);
  set(ttype9_button, 'value',0);
  set(ttype10_button, 'value',0);

elseif strcmp(action, 'ttype6')
  set(ttype1_button, 'value',0);
  set(ttype3_button, 'value',0);
  set(ttype4_button, 'value',0);
  set(ttype5_button, 'value',0);
  set(ttype2_button, 'value',0);
  set(ttype7_button, 'value',0);
  set(ttype8_button, 'value',0);
  set(ttype9_button, 'value',0);
  set(ttype10_button, 'value',0);

elseif strcmp(action, 'ttype7')
  set(ttype1_button, 'value',0);
  set(ttype3_button, 'value',0);
  set(ttype4_button, 'value',0);
  set(ttype5_button, 'value',0);
  set(ttype6_button, 'value',0);
  set(ttype2_button, 'value',0);
  set(ttype8_button, 'value',0);
  set(ttype9_button, 'value',0);
  set(ttype10_button, 'value',0);

elseif strcmp(action, 'ttype8')
  set(ttype1_button, 'value',0);
  set(ttype3_button, 'value',0);
  set(ttype4_button, 'value',0);
  set(ttype5_button, 'value',0);
  set(ttype6_button, 'value',0);
  set(ttype7_button, 'value',0);
  set(ttype2_button, 'value',0);
  set(ttype9_button, 'value',0);
  set(ttype10_button, 'value',0);
```

```
elseif strcmp(action,'ttype9')
  set(ttype1_button,'value',0);
  set(ttype3_button,'value',0);
  set(ttype4_button,'value',0);
  set(ttype5_button,'value',0);
  set(ttype6_button,'value',0);
  set(ttype7_button,'value',0);
  set(ttype8_button,'value',0);
  set(ttype2_button,'value',0);
  set(ttype10_button,'value',0);

elseif strcmp(action,'ttype10')
  set(ttype1_button,'value',0);
  set(ttype3_button,'value',0);
  set(ttype4_button,'value',0);
  set(ttype5_button,'value',0);
  set(ttype6_button,'value',0);
  set(ttype7_button,'value',0);
  set(ttype8_button,'value',0);
  set(ttype9_button,'value',0);
  set(ttype2_button,'value',0);

elseif strcmp(action,'dcfradius_slider')
  set(dcfradius_slider_text,'String',get(dcfradius_slider,'value'));

elseif strcmp(action,'dcfdelta_slider')
  set(dcfdelta_slider_text,'String',get(dcfdelta_slider,'value'));

elseif strcmp(action,'dcflength_slider')
  set(dcflength_slider_text,'String',get(dcflength_slider,'value'));
end; %end of fiding out which control had been activated

Dffiber1;%make it go with the changes caused by buttons and sliders

end; % end of all
```

References

1. K.C. Kao and G. Hockham, "Dielectric fiber surface waveguides for optical frequencies," *Proc. IEE* 113, 1151–1158, 1966.
2. T. Li, "Advanced in optical fiber communications: An historical perspective," *IEEE J. Sle. Areas Comm.*, SAC-1, 356–372, 1983.
3. B.J. Ainslie et al., "The design and fabrication of monomode optical fibers by MOCVD," *IEEE J. Quant. Elect.*, QE-18, 514–523, 1982.
4. G.D. Khoe and H. Lydtin, "European fibers and passive components: Status and trend," *IEEE J. Select. Areas Comm.*, SAC-4, 457–471, 1986.
5. N. Carlie, L. Petit, and K. Richardson, " Engineering of glasses for advanced optical fiber Applications," *J. Eng. Fibers and Fabrics*, 4(4), 21–29,2009, http://www.jeffjournal.org.
6. P.F. Kapron, D. Keck, and R.D. Maurer, "Radiation losses in glass optical waveguides," *Appl. Phys. Lett.*, 17, 423–425, 1970.

7. D.B. Keck, R.D. Maurer, and P.C. Shultz, "On the ultimate lower limit of attenuation in glass optical waveguides," *Appl. Phys Lett.*, 22, 307–309, 1973.

8. D.B. Keck and R. Bouillie, "Measurements on high-bandwidth optical waveguides," *Opt. Comm.*, 25, 43–48, 1978.

9. D. Marcuse and H.M. Presby, "Effects of profile deformations on fiber bandwidth," *Appl. Opt.*, 18, 2758–2763, 1979.

10. Marcatilli, E.A.J., "Modal dispersion in optical fibers with arbitrary numerical aperture and profile dispersion," *Bell Syst. Tech. J.*, 56, 49–63, 1977.

11. Snitzer, E., "Cylindrical dielectric waveguide modes," *J. Opt. Soc. Am.*, 51, 491–498, 1961.

12. A.W. Snyder, "Understanding monomode optical fibers," *Proc. IEEE.*, 69, 6–13, 1981.
    (b) A.W. Snyder, "Asymptotic expressions for eigen functions and eigenvalues of a dielectric or optical waveguide," *IEEE Trans. Microw. Th. Tech.*, MTT_17, 1130–1138, 1969.

13. D. Gloge, "Weakly guiding fibers," *Appl. Opt.*, 10, 2442–2445, 1971.

14. D. Gloge, "Propagation effects in optical fibers," *IEEE Trans. Microw. Th. Tech.*, MTT_23, 106–120, 1975.

15. J.E. Midwinter, "Current status of optical communications Technology," *IEEE J. Lightw. Tech.*, LT-3, 927–930, 1985.

16. J.G. Dil and H. Blok, "Propagation of electromagnetic surface waves in a radially inhomogeneous optical waveguides," *Opto-electronics*, 5, 415–428, 1973.

17. D. Gloge and E.A.J. Martillini, "Multimode theory of graded core fibers," *Bell Syst. Tech. J.*, 52, 1563–1578, 1973.

18. C.N. Kurtz and W. Streifer, "Guided waves in homogeneous focusing media: Part I: Formulation, solution for quadratic inhomogeneity," *IEEE Trans. Microw. Th. Tech.*, MTT_17, 11–15, 1969.

19. K. Morishita, Y. Kondo, and N. Kumagai, "On the accuracy of scalar approximation technique in optical fiber analysis," *IEEE Trans. Microw. Th. Tech.*, MTT_28, 33–36, 1980.

20. D.L.A. Tjaden, "First-order connection to weak guidance approximation in fiber optics," *Phillips J. Res.*, 33, 103–112, 1978.

21. A.W. Snyder and J. Love, *"Optical Waveguide Theory,"* Chapman and Hall, London, 1983.

22. P.K. Bachman et al., " PCVD DFSM fibers: Performance, limitations, design optimization," *Proc. IOOC-ECOC 1985*, 197–200, Venice, Italy, 1–4th October, 1985.

23. W.A. Gambling and E. Matsumura, "Cot-off frequency in radially inhomogeneous single mode fiber," *Elect. Lett.*, 13, 139–140, 1978.

24. Y. Kokubun and K. Iga, "Formulae for TE01 cut-off in optical fibers with arbitrary index profile," *J. Opt. Soc. Am.*, 70, 36–40, 1980.

25. T. Okoshi, *"Optical Fibers (Chapter 5),"* Academic Press, NY, 1982.

26. U.C. Paek, G.E. Peterson, and A. Carnevale, "Effects of depressed cladding on the transmission characteristics of single mode optical fibers with graded index profiles," *Appl. Opt.*, 21, 3430–3436, 1985.

27. K. Okamoto, "Comparison of calculated and measured impulse responses of optical fibers," *Appl. Opt.*, 18, 2199–2206, 1979.

28. R.W. Davies, D. Davidson, and M.P. Singh, "Single mode optical fiber with arbitrary refractive-index profile: propagation solution by Numerov method," *IEEE J. Lightw. Tech.*, LT-3, 619–627, 1985.

29. T. Tanaka and Y. Suematsu, "An exact analysis of cylindrical fiber with index distribution by matrix method and its application to focusing fiber," *Trans. IECE*, E59, 1–8, 1976.

30. K. Okamoto and T. Okoshi, "Vectorial wave analysis of inhomogeneous optical fibers using finite element method," *IEEE Trans. Microw. Th. Tech.*, MTT-26, 109–114, 1978.

31. W.A. Gambling and H. Matsumura, "Propagation in radially-inhomogeneous single mode fiber," *Opt. Quant. Elect.*, 10, 31–40, 1978.

32. L.G. Cohen, W.L. Mammel, and S.L. Lang, "Low loss quadruply clad single-mode lightguides with dispersion below 2 ps/nm/km over the 1.28 μm to 1.65 μm wavelength range," *Elect. Lett.*, 18, 1023–1024, 1982.

33. S.D. Personick, "Receiver design for digital fiber optic communication systems," *Bell Syst. Tech. J.*, 52, 843–874, 1973.

34. K. Petermann, "Fundamental mode microbending loss in graded-index and W-profile fibers," *Opt. Quant. Elect.*, 9, 167–175, 1977.

35. S. Hornung, N.J. Doran, and R. Allen, "Monomode fiber microbending loss measurements and their interpretation," *Opt. Quant. Elect.*, 14, 359–362, 1982.

36. K.I. White, "Design parameters for dispersion-shifted triangular profile fibers," *Elect. Lett.*, 18, 725–727, 1982.

37. P.L. Francois and C. Vassalo, "Comparison between pseudomode and radiation mode methods for deriving microbending losses," *Elect. Lett.*, 22, 261–262, 1986.

38. F. Welling, "Non-Gaussian intensity distribution and dispersion compensation in single-mode fibers," *Elect. Lett.*, 21, 811–812, 1985.

39. K. Petermann and R. Kuhne, "Upper and lower limits for the microbending loss in arbitrary single mode fibers," *IEEE J. Lightw. Tech.*, LT-4, 2–7, 1985.

40. K. Petermann, "Constraints for fundamental mode-size for broadband dispersion-compensated single mode fibers," *Elect. Lett.*, 19, 712–714, 1983.

41. C. Pask, "Physical interpretation of Peterman's strange spot size for single mode fibers," *Elect. Lett.*, 20, 144–145, 1984.

42. W.A. Gambling, H. Matsumura, and C.M. Ragdale, "Wave propagation in a single mode fiber with dip in the refractive index," *Opt. Quant. Elect.*, 10, 301–309, 1979.

43. P.L. Francois, "Tolerance requirements for dispersion free single mode fiber design: influence of geometrical parameters, dopant diffusion and axial dip," *IEEE J. Quant. Elect.*, QE-18, 1490–1499, 1982.

44. F.F. Ruhl, "Computer analysis of transmission properties of single mode fibers," Research report 7798, *Tel. Aust. Research Lab.*, 1–41, 1985.

45. C. Pask and R.A. Sammut, "Experimental characterisation of graded-index single-mode fibres," *Electron. Lett.*, 16(9), 1980.

46. K. Okamoto and T. Miya, "Zero total dispersion in single mode optical fibers over an extended frequency range," *Radio Sc.*, 17, 31–36, 1982.

47. D.M. Cooper et al., "Optical processing technique for spot size measurements in single mode fibers," *Elect. Lett.*, 21, 56–57, 1985.

48. (a) R.W. Davies and D. Sahm, "Correlation of zero-dispersion wavelength with mode confinement parameters in single mode fibers: analysis of simple step, triangular core, and dispersion shifted models," *IEEE J. Lightw. Tech.*, LT-4, 1393–1401, 1985.
(b) D.W. Davies, D. Davidson, and M.P. Singh, "Single mode optical fibers with arbitrary index propfile: propagation solution by the Numerov method," *IEE Proc. Opto-electronics*, 5, 415–428, 1973.

49. S.V. Chung, "Simplified analysis and design of single mode optical waveguides," PhD Dissertation, Monash University, Australia, 1986.

50. C.T. Chang, "Minimum dispersion at 1550 nm for single mode step-index fibers," *Elect. Lett.*, 15, 765–767, 1979.

51. C.D. Hussey, "Field to dispersion relationship in single mode optical fibers," *Elect. Lett.*, 20, 1051–1052, 1984.

52. D.C. Chang and E.F. Kuester, "Resonance characteristics of an rectangular microstrip antenna," in Proc. Workshop Printed [67] Circuit Antenna Tech., New Mexico State Univ., Las Cruces, 28, 1–18, 1979.

53. J. Sakai and T. Kimura, "Bending Loss of propagation modes in arbitrary index profile optical fibers," *Appl. Opt.*, 17, 14–99, 1978.

54. D.M. Cooper et al., "Dispersion-shifted single-mode optical fibers," *British Telecom. Tech. J.*, 3, 52–57, 1985.

55. J.B. Jeunhomme, *Single Mode Fibre Optics, Principles and Applications*, 2nd Ed., Marcel Dekker Pub., 1990.

56. W.J. Stewart, "Simplified parameter-based analysis of single-mode optical guides," *Elect. Lett.*, 16, 380–382, 1980.

57. (a) C.D. Hussey and C. Pask, "Single mode fibers in a few moments," *Elect. Lett.*, 14, 359–362, 1981.

(b) C.D. Hussey and C. Pask, "Theory of the profile moments description of single-mode fibers," *Proc. IEE* (Part H), 129, 123–132, 1982.

58. P.J. Samson, "Usage-based comparison of ESI techniques," *IEEE J. Lightw. Tech.*, LT-3, 165–175, 1985.

59. F. Martiner and C.D. Hussey, "Enhanced ESI for prediction of waveguide dispersion single-mode optical fibers," *Elect. Lett.*, 20, 1019–1021, 1984.

60. B.P. Nelson and J.V. Wright, "Problems in the use of ESI parameters in specifying monomode fibers," *British Telecom. Tech. J.*, 2, 81–85, 1984.

61. V.A. Bhagavatula, M.S. Spotz, and D.E. Quinn, "Uniform waveguide dispersion segmented core designs for dispersion shifted single-mode fibers," OFC 1984, Paper MA-2, New Orleans, LA, 1984, 1985.

62. P.L. Francois, "Propagation mechanism in quadruply-clad fibers: Mode coupling, dispersion and pure bend losses," *Elect. Lett.*, 19, 885–886, 1983.

63. U.C. Pas, G.E. Peterson, and A. Carnevale, "Dispersionless single mode lightguides with index profiles," *Bell Syst. Tech. J.*, 60, 583–589, 1981.

64. U.C. Paek, G.E. Peterson, and A. Carnevale, "Parametric effects on the bandwidth of a single-mode fiber with experimental verification," *Appl. Opt.*, 21, 704–709, 1982.

65. Corning Inc., "Corning SMF-28e+ fiber with next NextCor Technology: Product information" Issued September 2007, Corning, NY, USA, 2007.

66. B.J. Ainslie et al., "Interplay of design parameters and fabrication conditions on the performance of monomode fibers made by MCVD," *IEEE J. Quant. Elect.*, QE-17, 854–857, 1981.

67. H.T. Shang et al., "Dispersion-shifted depressed-clad triangular profile single mode fibers," *Elect. Lett.*, 21, 201–202, 1985.

68. L.N. Binh, T.L. Huynh, K.Y. Chin, and D. Sharma, "Design of dispersion flattened and compensating fibers for dispersion-managed optical communications systems," *Int. J. Wireless and Opt. Comm.*, 2(1), 63–82, 2004.

69. R. Yamauchi et al., "Design and performance of Gaussian profile dispersion-shifted fibers manufactured by VAD process," *IEEE J. Lightw. Tech.*, LT-4, 997–1004, 1986.

70. N. Kuwaki et al., "Dispersion shifted convex index single-mode fibers," *Elect. Lett.*, 21, 1186–1187, 1985.

71. K. Okamoto, "Comparison of calculated and measured impulse responses of optical fibers," *Appl. Opt.*, 15, 729–731, 1979.

72. P. Francois, "Tolerance requirements for dispersion free single-mode fiber design: Influence of geometrical parameters, dopant diffusion, and axial dip," *Trans. Microwave Theory Tech.*, MTT-30, 1478–1487, 1982.

73. P.L. Francois, J.F. Bayon, and F. Alard, "Design of mono-mode quadruply-clad fibers," *Elect. Lett.*, 22, 261–262, 1986.

74. W. Lieber et al., "Three-step index strictly single mode only F-doped silica fibers for broadband low dispersion," *IEEE J. Lightw. Tech.*, LT-4, 715–719, 1986.

75. A. Hasegawa and Y. Kodama, "Signal transmission by optical solitons in mono-mode fibers," *Proc. IEEE*, 69, 1145–1150, 1981.

76. L.N. Binh, "Digital optical communications," CRC Press, 2007.

77. B.P. Nelson, B.J. Ainslie, and D.M. Cooper, "Design and manufacture of fibers with specific dispersion properties," *Elect. Lett.*, 21, 274–276, 1985.

78. T.D. Croft, J.E. Ritter, and V.A. Bhagavatula, "Low losss diepsrsion-shifted single-mode fiber manufactured by the OVD process," *IEEE J. Lightw. Tech.*, LT-3, 931–934, 1985.

79. G.P. Agrawal, *Fibre-optic communications systems*, John Wiley & Sons, 2001.

80. L.N. Binh, and S.V. Chung, "A generalized approach to single-mode dispersion-modified optical fibre design," *Opt. Eng.*, 1996.

81. Y. Li and C.D. Hussey, "Triple-clad single-mode fibres for dispersion flattening," *Opt. Eng.*, 33, 3999–4005, 1994.

82. Y. Li, C.D. Hussey, and T.A. Birks, "Triple-clad single-mode fibres for dispersion shifting," *IEEE J. Lightw. Tech.*, LT-11, 1812–1819, 1993.

83. M. Hirano, A. Tada, T. Kato, M. Onishi, Y. Makio, and M. Nishimura, "Dispersion Compensating Fibre over 140 nm-Bandwidth," *Proc. 27th ECOC, Th.M.1.4*, 494–495.

84. A. Monerie, "Propagation in Doubly Clad Single Mode Optical Fibres," *IEEE J. Quant. Elect.*, QE-18, 535–542, April 1982.

85. (a) L.N. Binh, "Design guidelines for ultra-broadband dispersion flatten optical fibres with segmented core index profile," Dept. Elect. Comp. Syst. Eng, Monash Uni. Tech.Rept.No.ECSE14 - 2003, http://www.ds.eng.monash.edu.au/techrep/reports/post2003index.html;
    (b) F. Forghieri, R.W. Tkach, A.R. Chraplyvy, A.M Vengsarkar, and AT&T Bell Laboratories, "Dispersion Compensating Fibre: Is there a figure of merit" in *OFC'96, Tech. Digest*, 255–256.
    (c) G.E. Berkey and M.R. Sozanki, "Negative Slope Dispersion Compensating Fibres," *Sc. and Tech. Div., Corning*, WM14-1-WM14-3.

86. Y. Ishida, M. Kojima, T. Kobayashi and Y. Sugakawa, "Optical Fiber for Communication", US Patent No.4,114,981, 1978.

87. J.W. Fleming et al., "Low-Loss Single-Mode Fibers Prepared by Plasma-Enhanced MCVD," Electronics Letters, vol. 17, 1978, pp. 326–332.

6

Scalar Coupled-Mode Analysis

6.1 Introduction

In this chapter, a brief introduction to the waveguide coupler configurations is given in Section 6.2. In Section 6.3, a set of new parameters, called power parameters, are first defined and applied in the reformulation of the simplified (first-order) coupled-mode theory (CMT) to demonstrate simple and systematic analytical solutions for both symmetric and asymmetric two-mode guided-wave couplers. The composite, asymmetric fiber-slab couplers are dealt with separately in Section 6.4 through simplified first-order scalar coupled-mode and compound-mode equations, including a generalized, asymmetric index profile and corresponding sets of new coupling coefficients. The effects of the fiber-bend curvature and the distributed coupling are investigated in Section 6.5. In Section 6.6 the algorithm and computer programs are introduced and applied in the numerical solutions of the above scalar coupled-mode and compound-mode equations for the fiber-slab couplers. The numerical results of the computerized calculation and the discussion follow in Section 6.7.

Note that the validity of common scalar approximations of weak guidance and weak coupling is generally assumed whenever a scalar formulation is used. It is therefore sufficient to carry out the analysis using the scalar formulation of CMT, instead of rather more complicated vectorial calculations.

Simulation programs for asymmetric couplers are given in Appendix 4.

6.2 Coupler Configurations

6.2.1 Overview

Optical guided-wave couplers involve coupling or interaction of identical or nonidentical guided modes propagating along the closely spaced or coupling region. Figure 6.1 illustrates various configurations of optical guided-wave couplers.

6.2.1.1 Two-Mode Couplers

Two-mode couplers refer to those guided-wave couplers that involve coupling between two modes only. The two guided modes may be identical (similar) or nonidentical (dissimilar) and the couplers may be symmetric or asymmetric, depending on the constituent waveguides and coupler configurations.

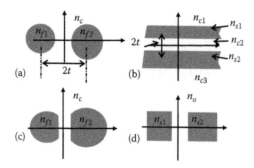

FIGURE 6.1

Schematic cross-sectional view of various directional couplers (DC): (a) fiber DC; (b) slab DC; (c) D-fiber DC; (d) channel waveguide DC.

6.2.1.2 Fiber-Slab Couplers

A fiber-slab coupler is a composite waveguide structure that allows coupling between an optical fiber and a generally asymmetric slab (thin-film or planar) waveguide.

6.2.1.3 Grating-Assisted Couplers

It is well known that small periodic perturbations on the index profiles of coupled waveguides may facilitate (or reduce) the coupling, especially for nonidentical, phase-mismatched waveguide modes [1,2]. Figure 6.2 shows a typical configuration of a slab-slab coupler with a grating structure.

6.2.2 Configurations

Optical waveguide couplers may be categorized in several ways so that they can be studied systematically. To be consistent in terminology throughout this book, the coupler configurations are classified according to their mode configurations, e.g., the number and symmetry of the coupled modes. Most of the present analysis is devoted to the novel, asymmetric fiber-slab couplers, which can be regarded as a class of multimode couplers,

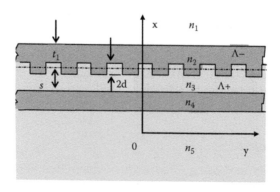

FIGURE 6.2

Schematic cross-sectional view of a grating-assisted slab directional coupler.

whilst a systematic reformulation in the power parameters and analytical solutions of the scalar CMT is obtained for the conventional two-mode couplers.

6.2.3 Two-Mode Couplers

The two-mode couplers may be further classified into those with two identical or non-identical modes, depending on the identity of the coupled modes, and therefore treated separately due to their different coupling properties.

In practice, the configuration of two identical modes may represent couplers composed of two single-mode, identical fibers, slab, or channel waveguides, whilst the configuration of two nonidentical modes may represent those made of single-mode, nonidentical waveguides such as fiber-fiber, slab-slab, and channel-channel couplers (e.g., with different core sizes or different optical constants), or composite, fiber-channel waveguides.

6.2.4 Multimode Couplers

Multimode couplers basically cover a broad range of coupler configurations that cannot be considered a two-mode system as defined above. Practical examples may include identical and nonidentical fiber-fiber, slab-slab, and channel-channel couplers, as illustrated in Figure 6.3, with at least one of the guides supporting two or more modes or a composite waveguide with more than two coupled modes in a composite fiber-slab, fiber-channel, or slab-channel structure [2].

6.2.5 Fiber-Slab Couplers

The composite fiber-slab couplers represent an interesting and special case and are singled out in this work for special treatment. This is mainly because

1. They represent a novel, hybrid, and practical coupler structure that has aroused interest for its distinctive, asymmetric coupling features and increasing practical applications in modern fiber-optics and integrated-optics technologies [3–5].
2. They do not fall into the above seemingly exhaustive categories of waveguide couplers. Even if the fiber and the slab each guide a single mode, a coupler base on both involves coupling between the fiber mode and a continuum of that slab

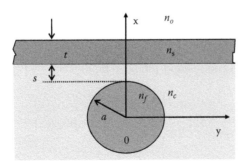

FIGURE 6.3
Cross-sectional view of a fiber-slab waveguide coupler.

mode propagating in all the possible directions within the plane of the slab, namely multimodes [6].

Figure 6.3 shows the cross-section geometry, the structural and optical constants for the fiber-slab coupler system, and analysis.

The fundamental mode confined in an optical fiber is, in fact, two-dimensional in nature, whilst that in an optical slab waveguide can be considered one-dimensional (i.e., bound to the slab only). In other words, the propagation of light launched initially into the fiber and then coupled into the slab waveguide is not restricted to the direction of the fiber-mode propagation. As illustrated later in this chapter, this unique mode configuration will allow a complex multimode coupling which represents couplings of an infinite number of modes. That is, the coupling involves the fundamental mode of the fiber and that of the slab, but with a span of a continuous radiation spectrum in the plane of the slab (i.e., the y-z plane in Figure 6.3). To mathematically facilitate the analysis, the continuum of the slab mode may be transformed by common techniques to a discrete set of so-called "transverse modes" having the same transverse mode (field) profiles but propagating in different directions (along the z-axis) in the plane of the slab. Therefore, it seems justified to treat the composite fiber-slab waveguide coupler as a special case of multimode couplers, rather than a two-mode coupler.

6.3 Two-Mode Couplers

This section documents a new treatment and a reformulation of CMT in terms of so-called "power parameters." It also presents the application of the new formulation of CMT in obtaining the power coupling and redistribution characteristics of both the identical and nonidentical two-mode couplers.

6.3.1 Coupled-Mode Equations

For general couplers of two modes, the total fields can be expressed as a linear superposition of the individual guided modes such as

$$\underline{E} = a\underline{E}_a + b\underline{E}_b \tag{6.1}$$

and the following coupled-mode equations can be derived by [7]

$$a'(z) = -j(\beta_a + Q_a)a(z) - jK_{ab}b(z)$$
$$b'(z) = -j(\beta_b + Q_b)b(z) - jK_{ba}a(z) \tag{6.2}$$

where the prime represents differentiation with respect to z, a and b represent the modal amplitudes, $\{\beta_a, \beta_b\}$ are the corresponding modal propagation constants, and $\{Q_a, Q_b\}$ and $\{K_{ab}, K_{ba}\}$ are the self- and cross-coupling coefficients, respectively.

The solution of the coupling coefficients depends on the details of the coupler configuration, the mode identity, and the appropriate formulations adopted. Generally speaking, the above coupled-mode equations and the reformulation derived below are not restricted to conventional (first-order) CMT. This is clear when the review of the different formulations of the CMT in Chapter 2 is recalled, such as the first-order, full scalar, and vector CMT. Another example will follow in Chapter 4 where the simplified and full scalar CMT for a composite fiber-slab waveguide coupler are compared.

6.3.2 Power Parameters

To facilitate an analytical solution, we introduce the following power parameters in relation to the z-dependent mode- and cross-power terms:

$$P_a(z) = a^*(z)a(z)$$
$$P_b(z) = b^*(z)b(z)$$
$$P_r(z) = \text{Re}\left[a^*(z)b(z)\right]$$
$$P_i(z) = \text{Im}\left[a^*(z)b(z)\right]$$

(6.3)

where a and b generally represent the amplitudes of the individual guided modes, and Re and Im indicate the real and imaginary components, respectively. However, a and b may also represent the compound-mode amplitudes of the whole coupler system, if so desired (Snyder and Love, 1992 [7]). Note that the above four parameters are all real numbers and only three of them are independent parameters.

To transform the solutions of the above parameters back to their original modal amplitudes as they are represented in Equation 6.2, we have

$$|a| = (P_a)^{1/2}$$
$$|b| = (P_b)^{1/2}$$
$$a^*b = P_r + jP_i$$
$$ab^* = P_r - jP_i$$

(6.4)

where the z dependence of all variables is implied for brevity, and this convention applies in the following, unless otherwise stated.

In order to gain some insight into the physical meaning of the power parameters, the complex modal amplitudes a and b can be expressed in terms of their real amplitudes ($|a|$ and $|b|$) and their phases (φ_a and φ_b), respectively. That is

$$a \equiv |a|e^{j\varphi_a}$$
$$b \equiv |b|e^{j\varphi_b}$$
$$\Phi \equiv \varphi_b - \varphi_a$$

(6.5)

so that Equation 6.3 becomes

$$P_r = (P_a P_b)^{1/2}\cos\Phi$$
$$P_i = (P_a P_b)^{1/2}\sin\Phi$$

(6.6)

and conversely

$$P_a P_b = P_r^2 + P_i^2$$
$$\Phi = \tan^{-1}\left(P_i/P_r\right)$$

(6.7)

where Φ represents the phase difference (i.e., phase mismatch) between the modes a and b.

From the above definitions and equations, it can be seen that P_a and P_b of power-orthogonal couplers simply represent the z-dependent modal powers, respectively, whilst the phase-mismatch can be determined by the ratio of P_i to P_r.

General initial operation conditions expressed as $\{a(0), b(0), a'(0), b'(0)\}$ can be passed on to $\{P_a(0), P_b(0), P_r(0), P_i(0)\}$ and $\{P'_a(0), P'_b(0), P'_r(0), P'_i(0)\}$ by using Equation 6.3.

6.3.3 Symmetric Two-Mode Coupler

Couplers consisting of two identical waveguides are symmetric in both waveguide configurations and power distributions. Alternatively, they may be viewed as a special case of couplers consisting of two nonidentical waveguides, as will be dealt with in the next section. The term "identical" simply means that in the structure of the coupler there are no differences between the two guides in terms of waveguide geometry and optical configurations.

6.3.3.1 Coupled-Mode Equations

For symmetric coupler of two identical modes, the scalar coupled-mode Equation 6.2a and b are reduced to

$$a' = -j(\beta + Q)a - jKb$$

(6.8a)

$$b' = -j(\beta + Q)b - jKa$$

(6.8b)

In the power parameters, they become

$$P'_a = 2KP_i$$

(6.9a)

$$P'_b = -2KP_i$$

(6.9b)

$$P'_r = 0$$

(6.9c)

$$P'_i = K(P_b - P_a)$$

(6.9d)

From the above coupled-mode equations in power parameters, the following general observations are immediately in order

1. From Equation 6.9a and b, we have a constant of motion, i.e., a z-invariant:

$$P = P_a(z) + P_b(z) = P_a(0) + P_b(0)$$

(6.10)

This is an indication that the total power is conserved at a value P, as determined by the total power of the coupling system at any point along the light propagation. Typically, this is given by the initial launching conditions of the coupler.

2. Equations 3.6a and 3.9c show that P_r is also a z-invariant:

$$P_r = P_a(z)P_b(z)\cos\Phi(z) = P_a(0)P_b(0)\cos\Phi(0) \tag{6.11}$$

3. Zero power-coupling condition can be expressed as: $P_a' = P_b' = 0$. From Equations 9.9a and 6.9b, $P_i \equiv 0$ and then, using Equation 6.9d, we have

$$P_a = P_b \tag{6.12}$$

Note that the above observations are clear from the reformulated coupled-mode equations plus the definition and property of the power variables and they are unrelated to the details or initial launching conditions of the coupler. The new coupled-mode equations are simply expressed in terms of the power parameters, easy to solve and yet quite rich in the physical information about the operation conditions of the two-mode coupler.

In addition, if the initial conditions of the coupled modes are known, the corresponding analytical solutions of the above simple coupled-mode equations can be obtained with ease. In this way the coupler properties such as power distributions and coupling can be fully revealed, as demonstrated below.

6.3.3.2 Analytical Solutions

1. Given the general initial conditions $|a_v(0)| = a_{v0}$, $\Phi(0) = \Phi_0$, and $a_v'(0) = 0$ ($v = a$, b), the corresponding power parameters, and their first-order derivatives, can be expressed as follows:

$$P_v(0) = P_{v0} \ (v = a, b, r, i) \tag{6.13}$$

$$P_v'(0) = 0 \ (v = a, b, r, i) \tag{6.14}$$

From Equations 6.9 and 6.10, and the above initial conditions, the following differential equations can be obtained:

$$P_a'' + 4K^2 P_a = 2K^2 P \tag{6.15}$$

with solution

$$P_a(z) = P/2 + (P_{a0} - P/2)\cos(2Kz) \tag{6.16a}$$

$$P_b(z) = P/2 + (P_{b0} - P/2)\cos(2Kz) \tag{6.16b}$$

The above are the general solutions describing the power distribution in each of the two single-mode waveguides. Solutions corresponding to a specific set of initial conditions can be obtained from this general solution by applying the initial conditions of the coupler. As an illustration, the above power distribution is displayed in Figure 6.4 (assuming a normalized total power, i.e., $P = 1$).

It can be seen that the power in each mode oscillates (due to the light propagation as well as coupling) sinusoidally along the propagation direction and they swap with each other after a period of $\pi/(2K)$. In fact, given the beat length L of a coupler supporting two identical modes

$$L = 2\pi/(\beta_+ - \beta_-), \tag{6.17}$$

where β_+ and β_- represent the propagation constants of the symmetric and anti-symmetric compound modes, respectively, of the coupler, we have

$$\pi/(\pi_+ - \pi_-) = 2K \tag{6.18}$$

This is a well-known relation for symmetric two-mode couplers [7]. Interestingly, the total power exchanged between the two guides is also dependent on the initial launching conditions and can be determined by the power mismatch between the initial powers launched into the two modes, i.e., $|P_{b0} - P_{a0}|$. In addition, the coupler of two identical modes is a symmetric coupler, namely, the above-observed power distribution for one of the modes can be swapped with the other, if the initial launching conditions are also swapped.

Finally, as demonstrated by the straight dot-line, i.e., Equation 6.4 in Figure 6.3, when the initial modal powers are equal, i.e., $P_{b0} = P_{a0}$, there is no coupling

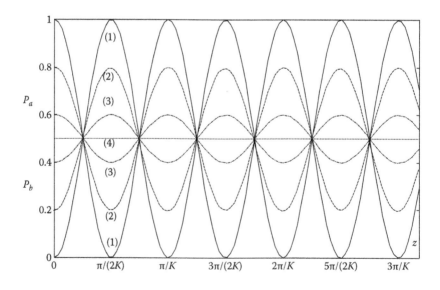

FIGURE 6.4
Diagram of power distribution of P_a and P_b, with the total power $P_a + P_b = 1$, $P_a(z) = (1/2)[1 + \delta\cos(2Kz)]$, $P_b(z) = (1/2)[1 - \delta\cos(2Kz)]$, with $\delta \equiv P_{a0} - P_{b0}$. (1) Solid line: $P_{a0} = 1$ and $P_{b0} = 0$, $\delta = 1$; (2) dashed line: $P_{a0} = 0.8$ and $P_{b0} = 0.2$, $\delta = 0.6$; (3) dash-dot line: $P_{a0} = 0.6$ and $P_{b0} = 0.4$, $\delta = 0.2$; (4) dotted line: $P_{a0} = 0.5$ and $P_{b0} = 0.5$, $\delta = 0$.

between the two modes at all. This has confirmed the result of Equation 3.12. That is, $|b|^2 = |a|^2 = 1/2$ or $b = \pm a$. Utilizing Equation 6.5, we have

$$|b|e^{-j\Phi z} = \pm|a| \tag{6.19}$$

or

$$\Phi = 0 \text{ for } b = a \text{ and } \Phi = \pi \text{ for } b = -a \tag{6.20}$$

This agrees with the fact that, for equal input powers into the two modes (waveguides), the coupler is able to support a compound mode (eigenmode). This mode is symmetric when the phases of the two modes are synchronous and antisymmetric when they are out of phase, with the following compound-mode propagation constants, respectively

$$\beta_e = \beta + Q \pm K \left(+ \text{ for symmetric mode, } - \text{ for antisymmetric mode}\right) \tag{6.21}$$

2. Alternatively, without deriving the general solution in the above, the coupled-mode equations can also be solved for any specific initial conditions. Typically, with a constant power P launched into one of the two modes only, say, mode a, the initial conditions become

$$P_a(0) = |a(0)|^2 = P \tag{6.22a}$$

$$P_b(0) = |b(0)|^2 = 0 \tag{6.22b}$$

$$P_a'(0) = P_b'(0) = 0 \tag{6.23}$$

and the coupled-mode Equation 6.9 can be solved accordingly.

In power parameters, the additional initial conditions can be expressed as

$$P_r(0) = P_i(0) = 0 \tag{6.24}$$

$$P_r'(0) = 0, \; P_i'(0) = -KP \tag{6.25}$$

The set of coupled-mode Equation 6.9 leads to

$$P_a'(z) + P_b'(z) = 0 \tag{6.26}$$

$$P_a(z) + P_b(z) = P \tag{6.27}$$

The above result reveals the fact that the total guided power is indeed conserved in the structure of the coupler. Then, using the above initial conditions, we have

$$\left[P_a(P - P_a)\right]^{-1/2} dP_a = \pm 2Kdz \tag{6.28}$$

and the solution is

$$P_a(z) = (P/2)\left[1 + \cos(2Kz)\right] \tag{6.29}$$

Using Equation 6.28, the power conservation law, we have

$$P_b(z) = (P/2)[1 - \cos(2Kz)]$$

(6.30)

The above solutions are clearly the same as those of the general solution Equation 6.15 when the same set of initial conditions Equation 6.20 is also applied.

6.3.4 Asymmetric Two-Mode Coupler

Couplers consisting of two nonidentical modes are asymmetric couplers. The term "nonidentical" means that in the structure of the coupler there are some differences between the two guides in terms of waveguide geometry and/or optical configurations.

6.3.4.1 Coupled-Mode Equations

The mode coupling of the asymmetric two-mode coupler can be described by the standard coupled-mode Equation 6.2a and b and, in power parameters as defined in the above section, they become

$$P_a'(z) = 2K_{ab}P_i$$

(6.31a)

$$P_b'(z) = -2K_{ba}P_i$$

(6.31b)

$$P_r'(z) = [(\beta_b - \beta_a) + (Q_b - Q_a)]P_i$$

(6.31c)

$$P_i'(z) = -K_{ba}P_a + K_{ab}P_b - [(\beta_b - \beta_a) + (Q_b - Q_a)]P_r$$

(6.31d)

A feature of two coupled nonidentical modes distinctive from that of two identical modes can be observed by a comparison with Equation 6.9a and b. The addition of the right side of Equation 6.31a and b is not zero because in general $K_{ab} \neq K_{ba}$ for the two nonidentical coupled modes. This means that the coupler system is generally not power-orthogonal, i.e., the power cannot be conserved between the modal powers of the two modes only and the nontrivial cross-power terms are necessary to satisfy the conservation of energy in the coupling system.

In fact, from Equations 3.31a–c, the following two interesting z-invariants can be found.

$$P = P_a(z) + P_b(z) + 2P_r(z)(K_{ba} - K_{ab})/(\beta_b + Q_b - \beta_a - Q_a)$$

(6.32)

$$[P_a(z) - P_a(0)]/K_{ab} + [P_b(z) - P_b(0)]/K_{ba} = 0$$

(6.33)

Instead of Equations 3.10 and 3.31 should be interpreted as the law of the power conservation for the two coupled, nonidentical modes. The mode nonorthogonality due to the butt-coupling of nonidentical modes will be dealt with in more detail in Chapter 4 where the full, nonorthogonal formulation of CMT will be derived. In addition, Equation 6.33 can be utilized to determined the power distribution between $P_a(z)$ and $P_b(z)$. Figure 6.5 illustrates the simple, linear relationship between the two variables $P_a(z)$ and $P_b(z)$ with initial values of $P_a(0)$ and $P_b(0)$ respectively.

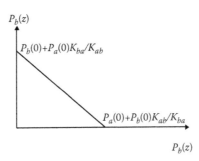

FIGURE 6.5
Illustration of relation between $P_a(z)$ and $P_b(z)$.

Figure 6.5 may be used to assist the analysis or design of the asymmetric two-mode coupler. For example, given any value of P_ν ($\nu = a, b$) at an arbitrary propagation distance z, the value of P_μ ($\mu = a, b$ but $\mu \neq \nu$) can be calculated from Equation 6.33 or estimated utilizing Figure 6.5.

6.3.4.2 Analytical Solutions

1. Assume the initial conditions of Equations 6.13 and 6.14. From Equation 6.31a and b, we have

$$P_i(0) = 0 \tag{6.34}$$

and, by definition

$$\Phi(0) = \tan^{-1}\left(P_i(0)/P_r(0)\right) = 0 \tag{6.35}$$

$$P_r(0) = P_{r0} = \left(P_{a0}P_{b0}\right)^{1/2} \tag{6.36}$$

$$P_r'(0) = 0 \tag{6.37}$$

$$P_i'(0) = P_{id} = -K_{ba}P_{a0} + K_{ab}P_{b0} + \left(\beta_a - \beta_b + Q_a - Q_b\right)\left(P_{a0}P_{b0}\right)^{1/2} \tag{6.38}$$

To solve the coupled-mode Equations 6.31a–d, we derive the following first-order differential equation:

$$P_i'(z)dP_i'(z) = -BP_i(z)dP_i(z) \tag{6.39}$$

where

$$B = \left[4K_{ab}K_{ba} + \left(\beta_b + Q_b - \beta_a - Q_a\right)^2\right] > 0 \tag{6.40}$$

This leads to another first-order differential equation of P_i

$$dP_i / \left(P_{id}^2 - BP_i^2 \right)^{1/2} = \pm dz \tag{6.41}$$

and its solution leads to

$$P_a(z) = P_a(0) + 2\left[\cos\left(B^{1/2}z\right) - 1\right]|P_{id}|K_{ab}/B \tag{6.42a}$$

$$P_b(z) = P_b(0) + 2\left[1 - \cos\left(B^{1/2}z\right)\right]|P_{id}|K_{ab}/B \tag{6.42b}$$

where K_{ab} is the cross-coupling coefficient, and P_{id} and B are constants defined by Equations 3.38 and 3.40, respectively.

The above power distributions are sinusoidal. The initial power-values can be determined from the initial launching conditions of the coupler. The amplitudes of the power oscillation are not only dependent on the constants of initial power distribution, but also on those of the cross-coupling and the phase-mismatched self-coupling. The beat length (or frequency) of the power distribution is therefore a function of both the cross-coupling and the phase-mismatched self-coupling constants.

As with the case of two identical coupled modes, the zero-mode coupling condition (i.e., $P_v(z) \equiv P_v(0)$, $v = a, b$) is equivalent to setting $P_a' = P_b' = 0$. From Equation 6.31a and b, we have $P_i \equiv 0$ (i.e., $P_{id} = 0$) and, utilizing Equation 3.38, we derive the following launching condition for the zero-coupling situation

$$P_{b0} = P_{a0}\left\{\beta_b - \beta_a + Q_b - Q_a + \left[\left(\beta_b - \beta_a + Q_b - Q_a\right)^2 + 4K_{ab}K_{ba}\right]^{1/2}\right\}^2 / \left(4K_{ab}^2\right) \tag{6.43a}$$

or, equivalently

$$b_0 = a_0 \left\{ (\beta_b - \beta_a) + (Q_b - Q_a) \pm \left[\left[(\beta_b - \beta_a) + (Q_b - Q_a) \right]^2 + 4K_{ab}K_{ba} \right]^{1/2} \right\} / (2K_{ab}) \tag{6.43b}$$

It can be verified that Equation 6.43 is exactly the analytical solution (eigenvectors) of the compound-mode equations of the general, asymmetric two-mode coupler, whilst the compound-mode propagation constant (i.e., eigenvalue) is

$$\beta_e = (1/2)\left\{ \beta_b + \beta_a + Q_b + Q_a \pm \left[\left(\beta_b - \beta_a + Q_b - Q_a \right)^2 + 4K_{ab}K_{ba} \right]^{1/2} \right\} \tag{6.44}$$

When $\beta_b \to \beta_a$, $Q_b \to Q_a$ and $K_{ab} \to K_{ba}$, we have $P_{b0} \to P_{a0}$, and Equation 6.43 is reduced to Equation 6.12, the zero-coupling initial condition and the compound-mode solutions for the symmetric two-mode couplers.

2. Applying the typical initial conditions Equations 6.22 and 6.23, we have the following initial conditions for the power parameters

$$P_a(0) = P \tag{6.45a}$$

$$P_b(0) = P_r(0) = P_i(0) = 0 \tag{6.45b}$$

$$P_a'(0) = P_b'(0) = P_r'(0) = 0 \tag{6.45c}$$

$$P_i'(0) = P_{id} = -K_{ba}P \tag{6.45d}$$

Following the similar procedures of part A of Section 6.3.4, the coupled-mode Equations 6.45a–d can be solved with some specific initial conditions such as the above. Here, the set of general solutions Equation 6.42 obtained above can be directly reduced to the following, by applying the above initial conditions

$$P_a(z) = P\left\{1 + F\left[\cos\left(B^{1/2}z\right) - 1\right]\right\} \tag{6.46a}$$

$$P_b(z) = PF\left\{1 - \left[\cos\left(B^{1/2}z\right)\right]\right\}K_{ba}/K_{ab} \tag{6.46b}$$

where P is the power launched initially into the guide a, K_{ab}, and K_{ba} are the cross-coupling coefficients, B is given by Equation 6.40, and F is defined by

$$F = 2K_{ab}K_{ba}/B = (1/2)\left\{1 + \left[(\beta_b - \beta_a + Q_b - Q_a)/(4K_{ab}K_{ba})^2\right]^2\right\}^{-1} \tag{6.47}$$

It is straightforward to show that the coupled-mode solutions, as given above by Equations 6.40, 6.46, and 6.47, are a more general form of those from other formulations (e.g., [8, 9]).

Assuming a normalized launching power (i.e., $P = 1$), the power distributions along the z-axis (the light-propagation direction) are shown in Figure 6.6.

Unlike the symmetric power coupling displayed in Figure 6.4 for the symmetric two-mode coupler, Figure 6.6 shows an asymmetric power coupling and an incomplete power transfer for asymmetric two-mode couplers. In fact, for an asymmetric coupler, the following is generally valid [10, 11]

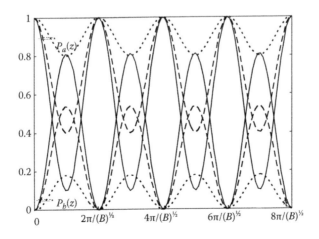

FIGURE 6.6
Power distributions along the propagation direction. $K_{ba}/K_{ab} = 0.9$, $P_a(z) = 1 + F[\cos(B^{1/2}z) - 1]$, $P_b(z) = 0.9F\{1 - \cos(B^{1/2}z)]\}$; (1) solid-line: $F = 0.45$; (2) dash-line: $F = 0.3$; (3) dash-dot-line: $F = 0.1$.

$$\beta_b \neq \beta_a, \, Q_b \neq Q_a \text{ and } K_{ba} \neq K_{ab} \tag{6.48}$$

$$B > 4K_{ab}K_{ba} \tag{6.49}$$

From Equation 6.48, it can be seen that the asymmetry of the power coupling and distribution is governed by the ratio of K_{ba}/K_{ab}, whilst the amplitudes of the power distribution or the completeness of the power transfer is given by the ratio of F.

For coupler of two identical modes, $K_{ab} = K_{ab} = K$, $F = 0.5$, and $B = 4K^2$, the asymmetric power distributions described by Equation 6.46 reduce to the symmetric one of Equations 6.29 and 6.30.

6.4 Fiber-Slab Couplers

This section presents the general analysis of scalar CMT for a composite fiber-slab coupler in which the optical fiber is coupled with an *asymmetric* slab waveguide. Basically, it generalizes the model for the coupler of a single-mode fiber and a *symmetric* slab waveguide [6]. This generalization allows the effect of asymmetric slab geometry on the lightwave coupling between the two waveguides to be revealed. Next, the scalar coupled-mode and compound-mode equations of the composite fiber-slab waveguide are determined and, finally, the results and discussion from the numerical calculations of the simplified version of the coupled-mode equations are presented.

6.4.1 Coupled-Mode Equations

The composite coupler configuration of the fiber and asymmetric slab waveguide and the relevant geometrical and optical parameters are illustrated in Figure 6.3. The fiber and the slab waveguide are unperturbed or uncoupled if they are remote from each other. If the fiber and the slab are brought into sufficiently close proximity, the evanescent mode fields of the two waveguides overlap and the two waveguides are perturbed or coupled. Unlike conventional single- or multimode directional fiber couplers, the fiber modes are not orthogonal to the slab modes. Each of the slab modes with propagation constant β_s is free to travel in all possible directions in the plane of the slab, i.e., in the (x, y) plane, and forms a continuum of guided modes with β_s as the maximum z-projection of the propagation constants. That is, the fiber-slab guided-wave coupler involves coupling between infinitely many modes. As a result, the reformulated approach successfully undertaken for the two-mode couplers cannot be adopted here and a numerical method must be used to solve the set of differential equations of the large number of coupled modes.

To facilitate the numerical calculation, the above continuum of the slab modes is transformed into a set of discrete guided modes $\{S_n\}$ by introducing into the y-component of the slab-mode expression a parameter D [6], with $2D$ representing the distance between the two perfectly conducting planes, theoretically imposed so that the continuum of the slab mode is discretized into a series of so-called transverse modes with a boundary condition of $S_n = 0$ at $y = \pm D$. This set of discrete slab modes is characterized by $\{\beta_{sn}\}$, a set of propagation constants as the projections of β_s on the z axis. Figure 6.7 illustrates the geometric (vectorial) relationship between β_s, $\{\beta_{sn}\}$, and $\sigma_n = \pi \, (n+1/2)/D$, as given by

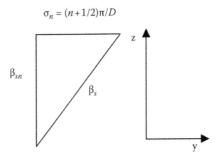

FIGURE 6.7
Geometric relations between the propagation constant β_s of the slab mode and the axial component $\{\beta_{sn}\}$.

$$\beta_{sn}^2 = \beta_s^2 - \sigma_n^2 \text{ for } n = 0, 1, 2, \ldots \tag{6.50}$$

The common scalar approximations of a weak guidance of the fiber and slab wave-guides and weak coupling between these two waveguides is assumed throughout this section. Weak guidance requires that all the refractive indices involved are close to each other whilst the weak coupling implies that effects of the fiber and slab field overlap are negligible. Under the above approximations, the transverse electric field in the perturbed or coupled fiber and slab waveguides can be well represented by the superposition of linearly polarized fiber and slab scalar modes. The waveguiding and coupling behavior of the structure can be described by recourse to the scalar wave equation for the transverse electric field E of the fiber-slab system, with the index profile of the fiber-slab coupler defined by

$$n^2(x, y) = n_f^2(x, y) + n_s^2(x, y) - n_c^2 \tag{6.51}$$

where $n_f(x, y)$ and $n_s(x, y)$ are the index profiles for the unperturbed fiber and slab wave-guide, respectively

$$n_f(x, y) = \begin{cases} n_f & \text{for } x^2 + y^2 \leq a^2 \\ n_c & \text{for } x^2 + y^2 > a^2 \end{cases} \tag{6.52a}$$

$$n_s(x, y) = \begin{cases} n_c & \text{for } x < a+s \\ n_s & \text{for } a+s \leq x \leq a+s+t \\ n_o & \text{for } x > a+s+t \end{cases} \tag{6.52b}$$

For simplicity, it is also assumed that only the dominant, linearly polarized LP_{01} mode [12] is first launched and then excited in the optical fiber, i.e., $a_n = 0$ for $n > 0$. The transverse electric field in the coupled fiber and slab waveguides can be well represented by the superposition of the guided modes of the fiber and the slab, i.e., $\{F_0, S_n\}$:

$$E(x, y, z) = a_0(z) F_0(x, y) + \sum_n b_n(z) S_n(x, y) \tag{6.53}$$

where $\{a_0(z), b_n(z)\}$ $(n = 0, 1, 2,...)$ is a set of z-dependent expansion (excitation) coefficients (modal amplitudes), whilst the summation is carried out over the set of discretized transverse slab modes.

In the coordinate system shown in Figure 6.3, the dominant LP_{01} mode (the transverse electric field) of the isolated fiber is given by

$$F_0 = N_f \begin{cases} \dfrac{J_0(k_f r)}{J_0(k_f a)} & \text{for } r \le a \\[3mm] \dfrac{K_0(\gamma_f r)}{K_0(\gamma_f a)} & \text{for } r > a \end{cases} \tag{6.54}$$

where J_0 and J_1 are the Bessel functions of the first kind, K_0 is the modified Bessel functions of the second kind, and

$$N_f = \frac{\gamma_f J_0(k_f a)}{\sqrt{\pi} V_f J_1(k_f a)}$$

is the normalization constant of the fiber mode (see Appendix 4). The parameter a denotes the fiber radius, $k_f^2 = n_f^2 k^2 - \beta_{f0}^2$ and $\gamma_f^2 = \beta_{f0}^2 - n_c^2 k^2$ are constants in which $k = 2\pi/\lambda$ (λ is the free-space light wavelength), β_{f0} is the propagation constant obtained from the LP_{01} mode dispersion equation (see [14], Chapter 2, Equations 2.5 and 2.6), and n_f and n_c are the refractive indices of the fiber core and cladding respectively. $V_f = ka(n_f^2 - n_c^2)^{1/2}$ is a dimensionless waveguide parameter related to k_f and γ_f via

$$V_f^2 = a^2 \left(k_f^2 + \gamma_f^2 \right) \tag{6.55}$$

In the same coordinate system (i.e., Figure 6.3), the nth guided transverse slab mode is obtained in the form

$$S_n(x, y) = N_s \cos(s_n y) \begin{cases} V_{so} \exp[\gamma_c (x - h)] / V_{sc} & x < a + s \\[2mm] \{\cos[k_s(x - h - t)] - (\gamma_o/k_s)\sin[k_s(x - h - t)]\} & \text{for } a + s \le x \le a + s + t \\[2mm] \exp[-\gamma_o(x - h - t)] & x > a + s + t \end{cases} \tag{6.56}$$

where

$$n = 0, 1, 2, ...,$$
$$k = 2\pi/\lambda, \ k_s^2 = n_s^2 k^2 - \beta_s^2,$$
$$\gamma_c^2 = \beta_s^2 - n_c^2 k^2, \ \gamma_o^2 = \beta_s^2 - n_o^2 k^2,$$
$$V_{sc} = kt \left(n_s^2 - n_c^2 \right)^{1/2}/2, \ V_{so} = kt \left(n_s^2 - n_o^2 \right)^{1/2}/2,$$
$$N_s = \left[\frac{2\gamma_c\gamma_o}{(\gamma_o + \gamma_c + \gamma_c\gamma_o t)D} \right]^{1/2}$$

is the normalization constant and σ_n is defined in Equation 6.50. The parameter t denotes the slab thickness, s the minimum distance between the surface of the fiber core and that of the polished flat (see Figure 6.3), β_s is the propagation constant of the slab mode, and n_s and n_o are the refractive indices of the slab and the overlay cladding, respectively. V_{sc} and V_{so} are dimensionless waveguide parameters related to k_s, γ_c, and γ_o via

$$V_{sc}^2 = t^2 \left(k_s^2 + \gamma_c^2 \right)/4 \tag{6.57a}$$

$$V_{so}^2 = t^2 \left(k_s^2 + \gamma_o^2 \right)/4 \tag{6.57b}$$

The propagation constant β_s of the slab guided mode of the mth order can be obtained from the well-known dispersion equation of the asymmetric slab waveguide

$$k_s t - \tan^{-1}\left(\gamma_c/k_s \right) - \tan^{-1}\left(\lambda_0/k_s \right) = m\pi \quad \left(m = 0, 1, 2, \ldots \right) \tag{6.58}$$

Coupled-mode equations are obtained when the field expansion Equation 6.53 is substituted into the wave equation of the total field, and then the resultant equation is multiplied successively with F_0 and S_n and integrated over the transverse (x, y) plane whilst making use of the wave equations and orthogonal relations of the fiber and slab modes. Using the scalar approximations, i.e., assuming a slowly varying envelope of the modal amplitudes and weakly coupled fiber and slab modes [13], both the second-order derivatives and the field overlap integrals may be neglected [6] and the following first-order coupled-mode equations are derived

$$a_0' = -j\left(\beta_{f00} + Q_{f00} \right)a_0 - j\sum_n K_{f0n}b_n \tag{6.59a}$$

$$b_m' = -j\sum_n \left(\delta_{mn}\beta_{sn} + Q_{smn} \right) b_n - jK_{sm0}a_0 \tag{6.59b}$$

where $a_0' = da_0/dz$, $b_m' = db_m/dz$, m, $n = 0, 1, 2, \ldots$, δ_{mn} is the Kronecker delta function, and $\{Q_{f00}, Q_{smn}\}$ and $\{K_{f0n}, K_{sm0}\}$ are the self- and cross-coupling coefficients, respectively.

6.4.2 Compound-Mode Equations

Conventional compound-mode equations can be obtained either from the nonorthogonal compound-mode equations by applying the scalar approximations, or straight from the above scalar coupled-mode equations by introducing a transform from the guided modes of the fiber and the slab to the compound modes of the fiber-slab coupler

$$A_0 = a_0 e^{-jbz} \tag{6.60a}$$

$$B_n = b_n e^{-jbz} \tag{6.60b}$$

The set of compound-mode equations corresponding to the set of coupled-mode Equation 6.59a and b thus becomes

$$Q_{f00}A_0 + \sum_n K_{f0n}B_n = (\beta - \beta_{f0})A_0 \tag{6.61a}$$

$$K_{sm0}A_0 + \sum_n Q_{smn}B_n = (\beta - \beta_{sm})B_m \quad \text{for } m = 0, 1, 2, \ldots \tag{6.61b}$$

where β is the eigenvalue and $\{A_0, B_n \ (n = 0, 1, 2,\ldots)\}$, the eigenvectors, are to be solved.

6.4.3 Coupling Coefficients

The coupling coefficients in the coupled-mode and compound-mode equations, i.e., Equations 6.59 and 6.61, respectively, are defined as follows

$$Q_{f00} = \frac{k^2}{2\beta_{f0}} \int_{A_\infty} \left[n^2(x, y) - n_f^2(x, y) \right] F_0 F_0 dA \tag{6.62a}$$

$$Q_{smn} = \frac{k^2}{2\beta_{sm}} \int_{A_\infty} \left[n^2(x, y) - n_s^2(x, y) \right] S_m S_n dA \tag{6.62b}$$

$$K_{f0n} = \frac{k^2}{2\beta_{f0}} \int_{A_\infty} \left[n^2(x, y) - n_s^2(x, y) \right] F_0 S_n dA \tag{6.63a}$$

$$K_{sm0} = \frac{k^2}{2\beta_{sm}} \int_{A_\infty} \left[n^2(x, y) - n_f^2(x, y) \right] S_m F_0 dA \tag{6.63b}$$

where $m, n = 0, 1, 2,\ldots$, $dA = dxdy$, and A_∞ indicates integration over the infinite cross-section, i.e., the entire transverse (x, y) plane, of the coupler system. From the refractive index profiles defined in Equations 6.51 and 6.52, the coupling coefficients defined above can be categorized in two ways. $\{Q_{f00}, Q_{smn}\}$ are self-coupling coefficients that represent the coupling among the fiber or slab modes due to the presence of the other, $\{K_{f0n}, K_{sm0}\}$ are cross-coupling coefficients that couple the slab modes to the fiber mode or vice versa.

The above integrals are all solved exactly except Q_{f00}, for which a large-argument asymptotic approximation of the modified Bessel function K_0 has to be used. The closed-form expressions of the coupling coefficients are given below

$$Q_{f00} = \frac{\pi^2 N_f^2 V_{sc}^2}{\sqrt{2}\beta_{f0}\gamma_f^2 t^2 K_0^2(\gamma_f a)}$$

$$\times \left[\text{erf}\left(\sqrt{2\gamma_f}(a+s+t)\right) - \text{erf}\left(\sqrt{2\gamma_f}(a+s)\right) + A_s \text{erfc}\left(\sqrt{2\gamma_f}(a+s+t)\right) \right] \tag{6.64a}$$

$$Q_{smn} = \frac{\pi N_s^2 k_s^2 V_f^2 t^2 e^{-2\gamma_c(a+s)}}{8\beta_{sm}V_{sc}^2}\left[\frac{I_1\left(\sqrt{4\gamma_c^2-(\sigma_m+\sigma_n)^2}\,a\right)}{\sqrt{4\gamma_c^2-(\sigma_m+\sigma_n)^2}\,a}+\frac{I_1\left(\sqrt{4\gamma_c^2-(\sigma_m-\sigma_n)^2}\,a\right)}{\sqrt{4\gamma_c^2-(\sigma_m-\sigma_n)^2}\,a}\right]$$

$$(6.64b)$$

$$K_{f0n} = \frac{\pi N_f N_s V_f^2 k_s t e^{-\gamma_c(a+s)}}{2\beta_{f0}aV_{sc}J_0\left(k_f a\right)\left(k_f^2+\gamma_c^2-\sigma_n^2\right)}$$

$$\times\left[\sqrt{\gamma_c^2-\sigma_n^2}J_0\left(k_f a\right)I_1\left(\sqrt{\gamma_c^2-\sigma_n^2}\,a\right)+k_f J_1\left(k_f a\right)I_0\left(\sqrt{\gamma_c^2-\sigma_n^2}\,a\right)\right]$$

$$(6.65a)$$

$$K_{sn0} = \frac{\pi N_f N_s V_{sc} k_s e^{-\sqrt{\gamma_f^2+\sigma_n^2}(a+s)}}{\beta_{sn}K_0\left(\gamma_f a\right)t\sqrt{\gamma_f^2+\sigma_n^2}}$$

$$\times\left\{\frac{\sqrt{\gamma_f^2+\sigma_n^2}+\gamma_c-B_s\left(\sqrt{\gamma_f^2+\sigma_n^2}-\gamma_0\right)e^{-\sqrt{\gamma_f^2+\sigma_n^2}t}}{k_s^2+\gamma_f^2+\sigma_n^2}+\frac{A_sB_se^{-\sqrt{\gamma_f^2+\sigma_n^2}t}}{\gamma_0+\sqrt{\gamma_f^2+\sigma_n^2}}\right\}$$

$$(6.65b)$$

where $V_f = ka\sqrt{n_f^2-n_c^2}$, $V_{sc} = (kt/2)\sqrt{n_s^2-n_c^2}$, $V_{so} = (kt/2)\sqrt{n_s^2-n_c^2}$, $A_s = (n_0^2-n_c^2)/(n_s^2-n_c^2)$, $B_s = V_{sc}/V_{so}$, erf(x) is the error function, erfc(x) = 1− erf(x), and I_0 and I_1 are the modified Bessel functions of the first kind.

In the above expressions, the new parameters A_s and B_s are introduced to quantify the geometric asymmetry of the index profile of the slab waveguide. For the case involving a symmetric slab waveguide, we have $A_s = 0$ and $B_s = 1$.

6.4.4 Attenuation Coefficients

Assuming an exponential decay for a_0, i.e., $a_0 = A_0\exp[\alpha z/2]$, negligible contributions from the self-coupling of Q_{smn} terms to the fiber-mode loss, and a slowly varying amplitude a_0 in comparison to the beat wavelength $\Lambda = 2\pi/|\beta_{f0}-\beta_{sn}|$ of the cross-coupling, the following approximate expression [6] for the power-loss coefficient α of the fiber mode will be used in calculation

$$\alpha = 2j\sum K_{f0n}K_{sn0}\left\{1-e^{j(\beta_{f0}-\beta_{sn})z}\right\}/(\beta_{f0}-\beta_{sn})$$

$$(6.66)$$

where β_{f0} and $\{\beta_{sn}\}$ ($n = 0,1,2,...$) are the propagation constants of the fiber and slab transverse modes, respectively, and $\{K_{f0n}\}$ and $\{K_{sn0}\}$ ($n = 0,1,2,...$) are the cross-coupling coefficients defined in Equation 6.65.

In addition, the above expression has a simplified analytical result by using the phase match (i.e., $\beta_{sm} = \beta_{f0}$) at certain mode-order $n = m$ and following relation by using Equation 6.50 and setting $v \equiv m + \mu$

$$\beta_{f0}-\beta_{sn} = \beta_{sm}-\beta_{sn} \approx \sigma_v\rho_\mu/\beta_{f0}$$

$$(6.67)$$

where $r_m = m\pi/D(\mu \equiv v; m = 0,1,2,...)$.

Finally, the analytical result of the fiber power-loss coefficient α becomes

$$\alpha = K_{f0m}K_{sm0}\beta_{f0}D/\sigma_m \tag{6.68}$$

where $m = v$ evaluated at $\beta_{sv} = \beta_{f0}$ and $\sigma_m = (m + 1/2)\pi/D$.

In the calculation, both Equqtions 6.66 and 6.68 would be used so that consistency and accuracy are retained.

6.5 Fiber Bending

In this analysis, the fiber bending, i.e., the axially distributed coupling in a composite coupler composed of a single-mode fiber half-block with an asymmetric slab overlay, is also considered by the scalar CMT. The dependence of the distributed light coupling, power distribution, and transfer on the asymmetry of the slab overlay is described by the gradually varied distance between the coupled fiber and slab modes, i.e., the distributed coupling coefficients. Numerical results will be presented in Section 6.6, along (and in comparison) with the special case of a straight fiber.

6.5.1 Fiber Bend Expression

The coupler configuration of a polished fiber half-block with a dielectric slab overlay is illustrated by Figure 6.8. Note that, whilst the parameter s_0 represents the minimum distance of spacing between the surface of the fiber core and that of the overlay, the local distance s at any position z along the fiber length may be described by

$$s(z) \approx s_0 + z^2/(2R)\dashv \tag{6.69}$$

where R is the curvature radius of the fiber.

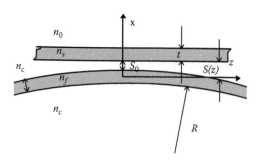

FIGURE 6.8

Longitudinal views of the geometry of the composite fiber-slab coupler and the essential optical waveguide parameters. n_μ ($\mu = 0, s, c, f, s$) represents the refractive index in the different cross-sections of the coupler structure.

6.5.2 Effects on Coupling

Assuming a large value of R, its effects on the mode coupling can be investigated by considering the fiber bend as a succession of a large number of small sections of straight fibers that are parallel but at a gradually varying distance s to the slab (Chapter 2, Section 2.4.2.2B). Given below is a different version of the above scalar CMT with the coupling coefficients made z-dependent by using Equation 6.69.

In scalar approximation, the total transverse electric field E of the fiber-slab structure can be represented by a superposition of the set of fiber and slab modes $\{F_0, S_n \ (n = 0,1,2,...)\}$

$$E(x, y, z) = a_0(z) F_0(x, y) e^{jb_{f0}z} + S_n(x, y) e^{-jb_{sn}z} \tag{6.70}$$

where $a_0(z)$ and $\{b_n(z)\}$ are the z-dependent modal amplitudes. In comparison with modal expansion Equation 6.53 and that using Equation 6.60, we have

$$a_0 = a_0 e^{j\beta_{f0}z} \tag{6.71a}$$

$$b_n = b_n e^{j\beta_{sn}z} \tag{6.71b}$$

$$a_0 = A_0 e^{-j(\beta-\beta_{f0})z} \tag{6.72a}$$

$$b_n = B_n e^{-j(\beta-\beta_{sn})z} \tag{6.72b}$$

The coupled-mode equations for the new modal amplitudes can be derived as

$$a_0'(z) = -j\left\{ S_{f00}(z)a_0(z) + \sum_n C_{f0n}(z)b_n e^{j(\beta_{f0}-\beta_{sn})z} \right\} \Big/ (2\beta_{f0}) \tag{6.73a}$$

$$b_m'(z) = -j\left\{ \sum_n S_{smn}(z)b_n e^{j(\beta_{sm}-\beta_{sn})z} + C_{sm0}(z)e^{j(\beta_{sm}-\beta_{f0})z}a_0(z) \right\} \Big/ (2\beta_{sm}) \tag{6.73b}$$

where the self- and the cross-coupling coefficients $\{S_{f00}, S_{smn}\}$ and $\{C_{f0n}, C_{sm0}\}$ are defined in relation to Equations 6.62 and 6.63, respectively. In fact, from Equations 6.59 and 6.73, we have

$$S_{f00}(z) \equiv 2\beta_{f0}Q_{f00}\big|_{s=s(z)} \tag{6.74a}$$

$$S_{smn}(z) \equiv 2\beta_{sm}Q_{smn}\big|_{s=s(z)} \tag{6.74b}$$

$$C_{f0n}(z) \equiv 2\beta_{f0}K_{f0n}\big|_{s=s(z)} \tag{6.75a}$$

$$C_{sm0}(z) \equiv 2\beta_{sm}K_{sm0}\big|_{s=s(z)} \tag{6.75b}$$

where the parameter s in Equations 6.62 and 6.63 is replaced by the z-dependent function Equation 6.69. The above-defined coupling coefficients, along with those of Equations 6.62 and 6.63, are employed in the full (nonorthogonal) formulation of scalar CMT.

6.6 Numerical Calculations

Numerical results are presented to allow comparison with the previous work of [6] on a coupler of a single-mode fiber and a symmetric slab waveguide so that, when the slab waveguide is asymmetric, the effect of the asymmetry on the coupling can be explicitly displayed or, when the slab waveguide becomes symmetric, the results can be reduced back to those of Ref. [6].

6.6.1 Optical and Structural Parameters

6.6.1.1 Uniform Fiber-Slab Couplers

All the numerical calculations have been carried out using common, practical values of the parameters, i.e., a light wavelength $\lambda = 1.3$ μm, fiber radius $a = 2.5$ μm, slab thickness $t = 3$ μm, refractive index of the fiber cladding $n_c = 1.46$, and the distance between the fiber and the slab $s = 0.5$ μm (see Figure 6.3). The refractive index of the slab cladding (superstrate) is designated as n_0 to account for the asymmetric index profile of the slab. Table 6.1 shows the values of the fiber parameters, i.e., n_f, Δ_f, V_f, and β_{f0}, and Table 6.2 shows those of parameters of the slab waveguide, i.e., n_o, Δ_{sc}, V_{s0}, and β_s. Obviously the cases with $n_o = 1.46$ ($n_o = n_c$) correspond to the symmetric case presented in Figures 2a through 5a of Ref. [6]. The results calculated using the coupled-mode Equations 6.59 or 6.73 and compound-mode Equation 6.61 reduce to those of Marcuse [6] when n_o is set to n_c or, equivalently, $A_s = 0$.

The common form of the profile height parameter and the waveguide parameter, defined as $\Delta_f = (n_f^2 - n_c^2)/(2n_f^2)$, $V_f = ka\sqrt{n_f^2 - n_c^2}$ for the optical fiber and $\Delta_{sc} = (n_s^2 - n_c^2)/(2n_s^2)$, $V_{so} = (kt/2)\sqrt{n_s^2 - n_o^2}$ for the asymmetric slab guide, respectively, were adopted here instead of $\Delta_f = (n_f^2 - n_o^2)/(2n_o^2)$, $V_f = kn_f a\sqrt{(n_f^2 - n_o^2)/n_o^2}$ for the fiber and $\Delta_f = (n_s^2 - n_o^2)/(2n_o^2)$, $V_{s0} = (kn_s t/2)\sqrt{(n_s^2 - n_0^2)/n_0^2}$ for the slab, as used in Ref. [6], where the slab waveguide is symmetric, i.e., $n_o = n_c$.

The values of all the refractive indices involved are close to each other. The propagation constant of the fiber LP_{01} mode β_{f0} was obtained as the solution of the eigenvalue equation for weakly guiding fibers. The refractive index of the slab was set as $n_s = 1.4745$. It is also assumed that only the LP_{01} mode is launched initially through the fiber at $z = 0$, i.e., $a_0(0) = 1$ and $b_m(0) = 0(m = 0,1,2,...)$ so that the behavior of the mode-coupling can be demonstrated over the wide range of waveguide parameters without allowing higher-order modes to complicate the matter.

TABLE 6.1

Waveguide Parameters of the Optical Fiber, as Used in Figure 6.9a–d

Figure 6.9	n_f	$\Delta_f = ((n_f^2 - n_c^2)/(2n_f^2))$	$V_f = ka\sqrt{n_f^2 - n_c^2}$	β_{f0}
Figure 6.9a	1.4817	0.0145	3.0529	7.1247
Figure 6.9b	1.4756	0.0105	2.5858	7.0987
Figure 6.9c	1.4745	0.0098	2.4925	7.0942
Figure 6.9d	1.4709	0.0074	2.1597	7.0801

TABLE 6.2

Waveguide Parameters of the Slab, as Used for Different Curves

Curve	n_0	$\Delta_{sc} = \left(n_s^2 - n_c^2\right)/\left(2n_s^2\right)$	$V_{so} = (kt/2)\sqrt{n_s^2 - n_o^2}$	β_s
(1)	1.40	0.0492	3.3550	7.0932
(2)	1.46	0.0098	1.4955	7.1005
(3)	1.47	0.0030	0.8345	7.1090

From Table 6.3, it appears that all the cases, except the last one, may allow a second fiber mode to propagate. Indeed, if $2.405 < V_f < 3.832$, both LP_{01} and LP_{11} modes can propagate in the fiber. However, because the phase or propagation constants of the LP_{11} mode or higher-order modes are not matched with those of the LP_{01} mode and the slab modes excited through the dominant phase-matched coupling, negligible power will be built up in the higher-order modes and, thus, the higher-order modes do not come into play [6].

As mentioned in Section 6.4.1, the discretization of the continuum spectrum of the transverse slab modes involves imposing a pair of hypothetical walls placed at a large distance $y = \pm D$ from the origin of co-ordinates, where D is a constant, so that the transverse slab mode S_n ($n = 0, 1, 2,...$) approaches zero when y reaches $\pm D$ [6]. Figure 6.9 shows that any propagation constant of the set of transverse slab modes, traveling in the (y, z) plane but at certain angles θ to the z axis (see Figure 6.3), can be denoted as β_{sn} ($n = 0, 1, 2,...$), which is related to the propagation constant β_s of the slab mode traveling along the z axis, where $\beta_{sn}^2 = \beta_s^2 - \sigma_n^2$ and $\sigma_n = \pi(n+1/2)/D$ ($n = 0, 1, 2,...$). The greater the value of D, the greater the total number of discredited transverse slab modes (i.e., the more the allowed values for n to simulate the continuum) and thus the higher the accuracy with the cost of a longer computing time. A typical value of $D = 500$ μm is used here to minimize the computing time without sacrificing the effective accuracy. In fact, the typical value of D is selected by comparing all the solutions of using $D > 500$ μm with those of using $D = 500$ μm until the difference between the results becomes invisible or negligible.

6.6.1.2 Couplers with Bend Fibers

Numerical calculation is also carried out for the practical, composite fiber-slab coupler with a slab overlay on the flat cladding surface of a side-polished or D-shaped fiber with curvature. In this case, the distance (separation) s between the fiber and the slab is varied (distributed), i.e., z-dependent, as shown in Figure 6.8. As a result, all the coupling coefficients, as analytically solved in Equations 6.64 and 6.65, become z-dependent, and,

TABLE 6.3

The Ridge-mode Propagation Constants of the Composite Waveguide System, as Used in Figures 6.8a–d, 6.9a–d, 6.10a and b

Figure	Curve 1	Curve 2	Curve 3
Figure 6.10a and Figure 6.11a	7.125735	7.125755	7.125680
Figure 6.10b and Figure 6.11b	7.102339	7.104438	7.109725
Figure 6.10c and Figure 6.11c	7.099693	7.103230	7.109608
Figure 6.10d and Figure 6.11d	7.096633	7.102521	7.109758
Figure 6.12a and Figure 6.12b	7.094045	7.100725	7.109033

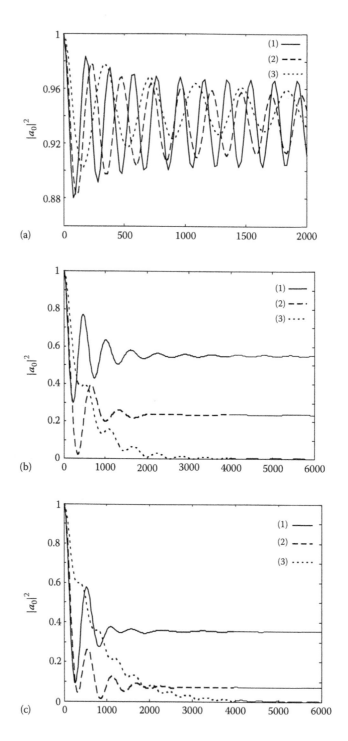

FIGURE 6.9

(a) Diagram of $|a_0(z)|^2$. $n_f = 1.4817$. (1) $n_0 = 1.40$; (2) 1.46; (3) 1.47. (b) Diagram of $|a_0(z)|^2$. $n_f = 1.4756$. (1) $n_0 = 1.40$; (2) 1.46; (3) 1.47. (c) Diagram of $|a_0(z)|^2$. $n_f = 1.4745$. (1) $n_0 = 1.40$; (2) 1.46; (3) 1.47. (d) Diagram of $|a_0(z)|^2$. $n_f = 1.4709$. (1) $n_0 = 1.40$; (2) 1.46; (3) 1.47. (e) Fiber-mode loss to the slab, i.e., the exponential curves fit to curves (1), (2), and (3) of (d), respectively.

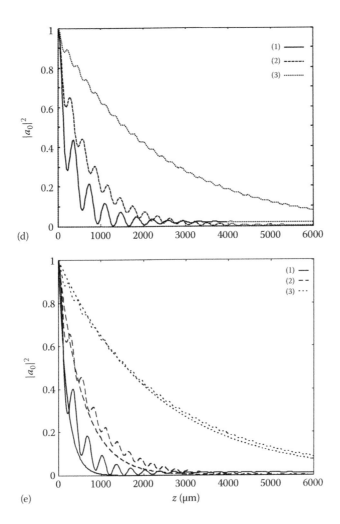

(d)

(e)

FIGURE 6.9 (Continued)

in the numerical calculation, the parameter s in Equations 6.64 and 6.65 is replaced by $s(z)$, as given by Equation 6.69. Typical values of the minimum distance (separation) between the fiber and the slab $s_0 = 0.5$ μm (the same as that used by [6]) and of the curvature radius of the fiber $R = 50$ cm are used in the calculation, except for cases otherwise specified.

6.7 Results and Discussion

The simplified version of the scalar coupled-mode Equations 6.59 and 6.73 (with constant coupling coefficients for uniform couplers) can be employed for numerical calculation. Figure 6.9a–e illustrate the z-dependent absolute square magnitude $|a_0|^2$ of the fiber mode; the results and discussion are presented in Section 6.7.1.

Section 6.7.2 presents the properties of the dominant compound mode (ridge modes). The set of simplified versions of the compound-mode Equation 6.61, is calculated numerically (see Section 6.6.2). Table 6.3 shows the ridge-mode propagation constants of the composite fiber-slab waveguide structure.

The field distribution of the ridge modes is shown in Figure 6.10a–d, i.e., the compound modes with the largest propagation constants, as a function of x in the plane of $y = 0$ (see Figure 6.3). Figure 6.11a–d display the corresponding fields as a function of y in the plane of $x = 4.5$ (µm), i.e., at the center of the slab.

From Figures 6.9 to 6.10, all the diagrams numbered (a)–(d) have the refractive index of the fiber core n_f varied around that of the slab ($n_s = 1.4745$), i.e., $n_f > n_s$, $n_f \geq n_s$, $n_f = n_s$, and $n_f < n_s$ respectively, so that the effect of relative values of n_f and n_s can be determined. Moreover, within each diagram of (a)–(d) the value of n_0 is also varied around that of ($n_c = 1.46$), resulting in the three curves numbered (1), (2), and (3) with $n_0 < n_c$, $n_0 = n_c$, and $n_0 > n_c$ respectively, so that the effect of relative values of n_0 and n_c is shown in Figure 6.9.

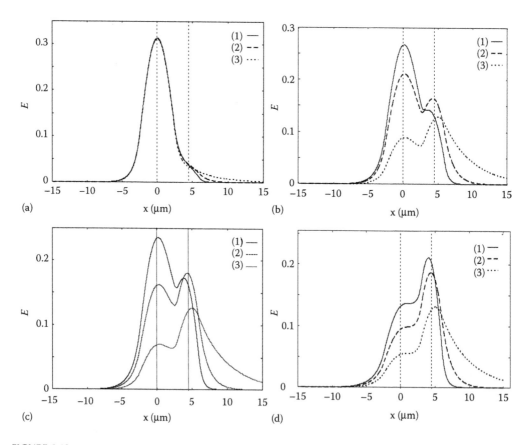

FIGURE 6.10
(a) $E(x)$ of the dominant compound-mode of the fiber-slab coupler in the plane of $y = 0$; $n_f = 1.4817$; $n_0 = 1.40$, 1.46, and 1.47 in curves (1), (2), and (3), respectively. (b) $E(x)$ of the dominant compound-mode of the fiber-slab coupler in the plane of $y = 0$; $n_f = 1.4756$; $n_0 = 1.40$, 1.46, and 1.47 in curves (1), (2), and (3), respectively. (c) $E(x)$ of the dominant compound-mode of the fiber-slab coupler in the plane of $y = 0$; $n_f = 1.4745$; $n_0 = 1.40$, 1.46, and 1.47 in curves (1), (2), and (3), respectively. (d) $E(x)$ of the dominant compound-mode of the fiber-slab coupler in the plane of $y = 0$; $n_f = 1.4709$; $n_0 = 1.40$, 1.46, and 1.47 in curves (1), (2), and (3), respectively.

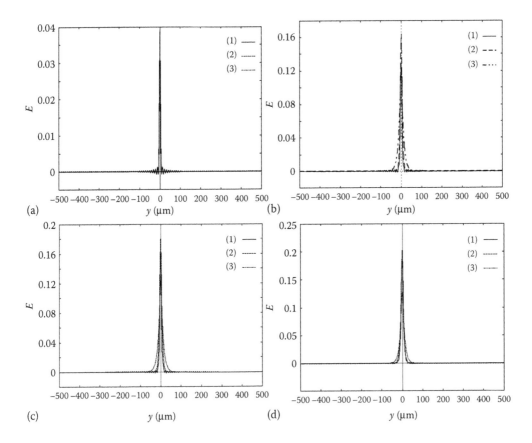

FIGURE 6.11
(a) Field $E(y)$ of the ridge mode of the fiber-slab coupler in the plane of $x = 4.5$ μm. $n_f = 1.4817$; (1) $n_o = 1.40$, (2) $n_o = 1.46$, (3) $n_o = 1.47$. (b) Field $E(y)$ of the ridge mode of the fiber-slab coupler in the plane of $x = 4.5$ μm. $n_f = 1.4756$; (1) $n_o = 1.40$, (2) $n_o = 1.46$, (3) $n_o = 1.47$. (c) Field E_y of the ridge mode of the fiber-slab coupler in the plane of $x = 4.5$ μm. $n_f = 1.4745$; (1) $n_o = 1.40$, (2) $n_o = 1.46$, (3) $n_o = 1.47$. (d) Field E_y of the ridge mode of the fiber-slab coupler in the plane of $x = 4.5$ μm. $n_f = 1.4709$; (1) $n_o = 1.40$, (2) $n_o = 1.46$, (3) $n_o = 1.47$.

6.7.1 Characteristics of Mode Coupling

Figure 6.9a indicates that, when $n_f > n_s$, only a fraction of the launched power in the fiber fluctuates between the fiber and the slab, and the light power remains largely in the fiber. From Table 6.1, $\beta_{f0} > \beta_s$, and thus, in the absence of coupling, the fiber mode is not in phase synchronism with any of the transverse slab modes, so that very little power is coupled from the fiber to the slab waveguide. It can also be observed that the greater the value of n_o, the longer the oscillation period, but the lower the oscillation amplitude. The former can be explained by the increasing beat wavelength caused by the decreasing value of β_s (see Table 6.2) and thus that of the phase mismatch, which is proportional to the inverse of the phase mismatch [6, 7], whilst the latter is due to the decreasing coupling strength as the field of the slab modes is shifted further from that of the fiber with an increasing value of n_o.

New features are displayed in Figure 6.9b as the value of n_f gets close to (but is still greater than) that of n_s. Lowering the value n_f, and that of β_{f0}, generally causes the power oscillation to decay. The rate of decay of the power in fiber is increased with the increasing value of n_o and β_s, and thus power is transferred from the fiber to the slab due to reduced phase mismatch. This therefore results in strengthened coupling between the fiber and slab modes. If $n_o \leq n_c$ ($A_s \leq 0$), the power oscillates with a high strength and dies out. If $n_o > n_c$ ($A_s > 0$) the fiber power decays significantly and radiates out to the slab. The amount of power coupled out of the fiber is strongly dependent on the value of n_o, or the asymmetry of the slab waveguide. When $n_o - n_c = 0.01$, the absolute square of the fiber mode amplitude $|a_0|^2$ reduces to zero beyond a certain propagation distance (typically 2000 μm), whilst it only reduces to about 80% of the initial value of $|a_0|^2$ at $z = 0$ in the symmetric case ($n_o = n_c$).

In the case when the value of n_f equals that of n_s, similar features of the fiber-mode power beating and decaying are illustrated in Figure 6.9c. In comparison with Figure 6.9b, it can be seen that if $n_o \leq n_c$ ($A_s \leq 0$) more power is coupled out from the fiber and if $n_o > n_c$ ($A_s > 0$) the fiber-mode attenuation coefficient becomes smaller due to the similarity of the values of β_{f0} and β_s (the phase synchronism). As with Figure 6.9b, the amount of power coupled out of the fiber is also strongly dependent on the value of n_o or the asymmetry of the slab waveguide. For example, when $n_o - n_c = -0.06$ the value of $|a_0|^2$ at $z \geq 2000$ μm is more than four times greater than that of the symmetric case when $n_o = n_c$.

Finally, when the value of n_f is less than that of n_s, Figure 6.9d shows that the fiber mode decays in such a way that the smaller the value of n_o, the greater the attenuation, as expected by the resultant stronger coupling, and the more the power goes to excite the continuum of modes that are not ridge modes, i.e., modes bound to the slab but unbound in the vicinity of the fiber. This is demonstrated in Figure 6.9e, showing an exponential curve fit by $|a_0|^2 = e^{-\alpha z}$ to each of the curves in Figure 6.9d, where α is the power-attenuation coefficient (given by Equations 6.66 and 6.68) of the fiber due to radiation loss to the slab.

6.7.2 Characteristics of Ridge Modes

Figure 6.10a illustrates that, when $n_f > n_s$, most of the compound-mode field remains in the vicinity of the fiber. However, there are some deformations of the symmetric mode of an isolated fiber due to the presence of the slab. The higher the value of n_o, the more distinct the deformation appears. This can be explained by the fact that the increased value of n_o has an effect of increasing the field value of the slab in the outside cladding given by Equation 6.56. As a linear summation of the fiber and transverse slab guided modes, the compound mode of the fiber-slab coupler system will accordingly show the increased mode profile in the same region.

As the value of n_f becomes close to, but still greater than, that of n_s, as shown in Figure 6.10b, there are extra peaks of the field around the position of the slab. Increasing n_o causes more deformation of the pure fiber mode in such a way that the higher the value of n_o, the lower the field in the fiber becomes and the field in the slab increases in height relative to that of the fiber.

Similar field shifts are illustrated by Figure 6.10c when $n_o = n_s$. Finally, when $n_f < n_s$, Figure 6.10d shows that large fractions of the mode power are now transferred to the slab.

The above observations are in agreement with the above coupled-mode analysis, as shown in Figure 6.9a–d, and can be explained by the increased coupling strength of the fiber and slab guided-modes and the amount of power transferred from the fiber into the slab because of the decreasing value of refractive index of the fiber core n_f relative to that of the slab n_s or, equivalently, the shifting to the phase synchronism. Figure 6.10a–d also display the effect of the value of n_o or A_s on the compound mode in that the higher the value of n_o or A_s, the greater the distance away the field extends into the slab as well as outside the cladding and, as a result, the field amplitude in the fiber decreases. This may be explained by the shift of the peak areas of the slab mode-fields away from that of the fiber due to the increasing value of the refractive index n_o outside the slab guiding layer.

Figure 6.11a–d indicate that these dominant compound-modes are indeed ridge modes because the fields at the center of the slab are concentrated in the vicinity of the fiber. Except in Figure 6.11a, where the field is little affected by varying the value of n_o, as also shown in Figure 6.10a, increasing the value of n_o generally causes the field to extend further away from the vicinity of the fiber. It should also be noted that the peak values of the ridge-mode fields are dependent not only on the ratio of n_f/n_s, but also on that of n_o/n_c.

For some optical parameters, the fiber-slab coupler may support a second ridge mode. This is demonstrated here by the compound modes with the second largest propagation constants when $n_f > n_s$. As an example, the field distribution of the second ridge-mode as functions of x at $y = 0$ (see Figure 6.3) is shown in Figure 6.12a whilst that as functions of y at $x = 0$ is shown in Figure 6.12b. Note that the larger peak of the mode field in Figure 6.12a is shifted to the region of the slab, with the sign reversed, in comparison to those in Figure 6.10a, and some energy remains in the fiber. Increasing value of n_o causes the field to extend much further from the fiber whilst maintaining a lower peak in the slab. Figure 6.12b shows the confinement of the field around the center of the fiber, even though most of the light energy is concentrated in the plane of the slab.

As shown in Table 6.3, the propagation constants $\{\beta_2\}$ of the second ridge-modes are lower in value than those $\{\beta_1\}$ of the first ridge-modes and the beat wavelengths $\{\Lambda\}$ are calculated using $\Lambda = 2\pi/|\beta_1 - \beta_2|$. $\Lambda = 198, 251$, and $377\ \mu m$ are found for $n_o = 1.40, 1.46$, and 1.47 ($n_c = 1.46$), respectively, and these results show very good agreement with the oscillation periods observed in Figure 6.9a.

As illustrated by Figure 6.13a and b, other higher-order compound modes of the fiber-slab coupler have a much lower profile along the x-direction than the ridge modes and are shown as basically different harmonics along the y-direction, spreading over to the boundaries where the field vanishes. These field distributions are obviously non-ridge modes. By contrast, the first two (dominant) compound modes are shown to be ridge modes. Interestingly, the energy of the first is concentrated in the fiber and the second in the slab, similar to the field distributions of the isolated (unperturbed) fiber and slab modes, respectively. This observation is in agreement with that of [14, 15] on the compound-mode properties of simple, asymmetric two-mode slab-slab couplers, namely, the two compound-modes of the coupler are nearly the same as the two nonsynchronous slab guided-modes.

6.7.3 Effects of Other Waveguide Parameters

As demonstrated above, the characteristics of the mode coupling and power transfer between the fiber and the slab guided-modes are generally described in terms of their mode synchronism, i.e., match of the values of the propagation constants (phase-match) of the fiber and slab guided-modes. Ultimately, it is the waveguide parameters that govern the mode synchronism and, along with the coupler parameters, determine the coupling

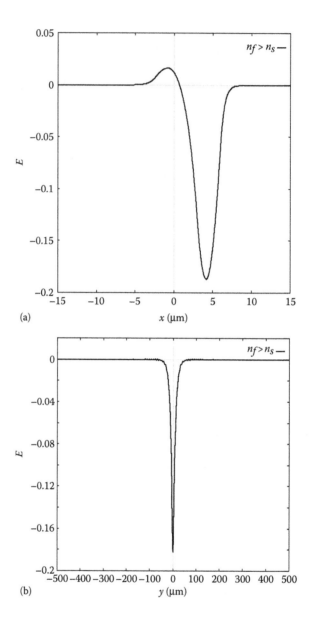

FIGURE 6.12
Field E of the second ridge-mode for $n_f = 1.4817$ in the plane of (a) $y = 0$; (b) $x = 0$. $n_o = 1.40$, 1.46, and 1.47 in curves (1), (2), and (3), respectively.

features of the guided-wave coupler. Given below are some specific examples that can demonstrate the dependence of the power coupling behavior on a particular waveguide or coupler parameter.

6.7.3.1 Effect of Light Wavelength

Figure 6.14a and b display the following effects of the light wavelength on the power coupling.

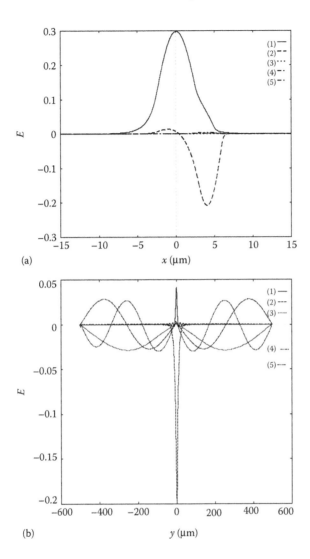

FIGURE 6.13
Field E of the first five compound-modes, i.e., curves (1), (2), (3), (4), and (5), respectively, for $n_f = 1.4756$ and $n_o = 1.0$ in the plane of (a) $y = 0$; (b) $x = 0$.

1. From the figure captions, it can be seen that the fiber and slab propagation constants, and the difference between them, are dependent on the light wavelength. Generally speaking, the greater the value of the light wavelength, the smaller the value of the propagation constants and the difference between them, and vice versa.

2. As shown in Figure 6.14, the change of the wavelength has not only tuned the magnitude, but also the beat length of the power exchange between the fiber and slab modes. The greater the value of the light wavelength becomes, the greater the power-coupling magnitude and the shorter the beat length, and vice versa.

3. Figure 6.14b represents cases ($n_f = n_s$) that are generally closer to an exact phasematch than those of Figure 6.14b ($n_f > n_s$). As a result, it can be observed that the

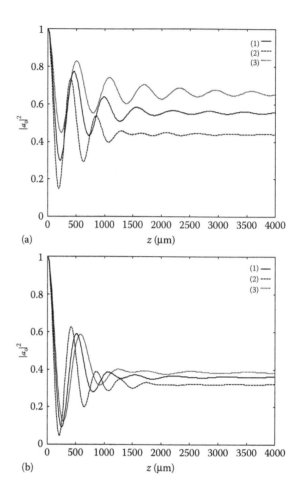

FIGURE 6.14
(a) Effect of light wavelength on $|a_o(z)|^2$. $n_f = 1.4756$, $n_s = 1.4745$, $n_o = 1.40$. (1) $\lambda = 1.3$, $\beta_{f0} = 7.0989$, $\beta_s = 7.0932$; (2) $\lambda = 1.55$, $\beta_{f0} = 5.9470$, $\beta_s = 5.9427$; (3) $\lambda = 1.15$, $\beta_{f0} = 8.0306$, $\beta_s = 8.0239$. (b) Effect of light wavelength on $|a_o(z)|^2$. $n_f = 1.4745$, $n_s = 1.4745$, $n_o = 1.40$. (1) $\lambda = 1.3$, $\beta_{f0} = 0.0944$, $\beta_s = 7.0932$; (2) $\lambda = 1.55$, $\beta_{f0} = 5.9436$, $\beta_s = 5.9427$; (3) $\lambda = 1.2$, $\beta_{f0} = 7.6891$, $\beta_s = 7.6878$.

effect of the light wavelength is obviously greater in Figure 6.14a, where the coupled modes are less synchronized than in Figure 6.14b, where the coupled modes are nearly phase-matched.

6.7.3.2 Effect of Guide-Layer Size

By analogy with the above observations and with the other parameters fixed, the variation of the thickness of the slab t also has an effect of tuning the mode synchronism and therefore coupling between the fiber and slab modes. Table 6.4 shows the propagation constants and Figure 6.15a–c illustrate the effects of varying the slab guide thickness.

When the value of t is reduced from 3 μm to 2 μm, along with the variation of the superstrate refractive index of the slab, the following can be observed

TABLE 6.4

Values and the Differences of the Calculated Propagation Constants in Figure 6.15

Figure 6.15	(1)	(2)	(3)	(4)
(a) $n_0 = 1.40$	$\beta_{f0} = 7.0944$	$\beta_{f0} = 7.0944$	$\beta_{f0} = 7.0802$	$\beta_{f0} = 7.0802$
(b) $n_0 = 1.46$	$\beta_s = 7.0749$	$\beta_s = 7.0932$	$\beta_s = 7.0749$	$\beta_s = 7.0932$
(c) $n_0 = 1.47$	$\beta_{f0}-\beta_s = 0.0195$	$\beta_{f0}-\beta_s = 0.0012$	$\beta_{f0}-\beta_s = 0.0053$	$\beta_{f0}-\beta_s = -0.013$

1. Similar to the effects of λ, variation of the value of t changes the magnitude and beat length of the mode coupling, the decay of the launching power in the fiber, and the amount of the power coupled from the fiber to the slab. When $\beta_{f0} > \beta_s$, a thinner slab has a smaller propagation constant and hence the coupled modes are further away from a possible phase match. This results in less power coupled from the fiber to the slab for $t = 2$ (μm) than for $t = 3$ (μm).

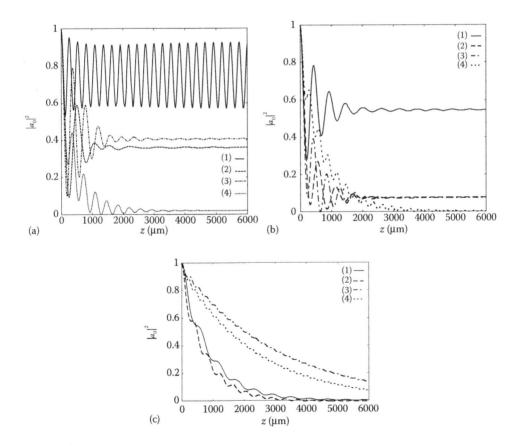

FIGURE 6.15

(a) Diagram of $|a_0(z)|^2$. $n_0 = 1.4$, $\beta_{f0} = 7.0944$ for (1) and (2); $n_f = 1.4709$, $\beta_{f0} = 7.0802$ for (3) and (4); $t = 2$ (μm), $\beta_s = 7.0749$ for (1) and (3), $t = 3$ (μm), $\beta_s = 7.0932$ for (2) and (4). (b) Diagram of $|a_0(z)|^2$. $n_0 = 1.46$; $\beta_{f0} = 7.0944$ for (1) and (2); $n_f = 1.4709$, $\beta_{f0} = 7.0802$ for (3) and (4); $t = 2$ (μm), $\beta_s = 7.0749$ for (1) and (3), $t = 3$ (μm), $\beta_s = 7.0932$ for (2) and (4). (c) Diagram of $|a_0(z)|^2$. $n_0 = 1.47$; $n_f = 1.4745$, $\beta_{f0} = 7.0944$ for (1) and (2); $n_f = 1.4709$, $\beta_{f0} = 7.0802$ for (3) and (4); $t = 2$ (μm), $\beta_s = 7.0749$ for (1) and (3), $t = 3$ (μm), $\beta_s = 7.0932$ for (2) and (4).

2. The combined effects with the variation of n_o are even more interesting when $n_o \leq n_c$. The variation of the amount of power exchange caused by the change of t is quite dramatic (up to about 40% of the total power launched) and the amount of power coupled from the fiber to the slab is generally in the reversed order of the coupler's closeness to mode synchronism (see Table 6.4 and Figure 6.15a and b). However, when $n_o < n_c$, the power coupling results in largely an exponential decay of the power in the fiber. There is much less variation of the power exchange compared to the case of $n_o < n_c$ and greater attenuation for cases with greater values of n_f and β_{f0}.

6.7.3.3 Effect of the Refractive Index of the Cladding

Because the cladding refractive index of the fiber and that of the slab substrate are the same ($n_c = 1.46$), only the effect of the superstrate index of refraction n_o (i.e., the outside slab cladding in Figure 6.3) needs to be considered. Figure 6.16a–c display the levels of power in the fiber mode during the coupling in relation to the value of n_o and the varied degree of mode synchronism. This clearly demonstrates that, with all the other parameters fixed, increasing the value of n_o generally increases the amount of power transferred from the fiber to the slab.

6.7.4 Distributed Coupling

The constituent waveguide parameters are the same as those used for the uniform fiber-slab coupler given above. However, the light is now launched initially through the fiber at $z = z_0 < 0$, i.e., $a_0 (z_0) = 1$ and $b_m (z_0) = 0$ ($n = 0, 1, 2,...$), where the coupling is trivial or vanishing. Calculated from the distributed (i.e., z-dependent) coupled-mode Equation 6.73a and b, Figures 6.17 and 6.18 reveal some new power-coupling features of the polished fiber half-block with an asymmetric slab overlay.

6.7.4.1 Fixing n_o Each Time while Varying n_f, with Respect to n_s

First, the effect of the fiber curvature and the resultant distributed coupling can be clearly observed, when compared with Figure 6.9a–d for the uniform fiber-slab coupler. There appears to be an effective coupling (interaction) length during the light propagation ($-z_0 \leq z \leq z_0$, $z_0 \approx 2000$ μm) with onset and offset points ($z = -z_0$ and $z = z_0$, respectively), equally distant from the center of the coupler ($z = 0$). The variation of the coupling strength along the light propagation direction is also obvious. These can be explained by the symmetric nature of the z-dependent parameter s, the local distance between the fiber and slab waveguides, which is related to the fiber curvature by Equation 6.69. In addition, all the coupling coefficients defined in Equations 6.64 and 6.65, via Equations 6.74 and 6.75, are now exponentially dependent on s as well as z^2, utilizing Equation 3.69, and this results in the distributed coupling strength along the z direction.

When $n_o = 1.40$, Figure 6.17a shows the smaller value of n_f around that of n_s causes more power to be coupled out of the fiber into the slab. Complete power recovery is observed when $n_f > n_s$; power oscillation occurs when some power is tapped into the slab from the fiber when $n_f \approx n_s$ and the complete power transfer is possible, as demonstrated by the curve (4), where there is a phase match between the fiber and slab modes.

When $n_s = 1.4745$ and $n_o = 1.46$ (i.e., when the slab waveguide is symmetric), Figure 6.17b shows that much more power (over 80%) is transferred from the fiber to the slab than

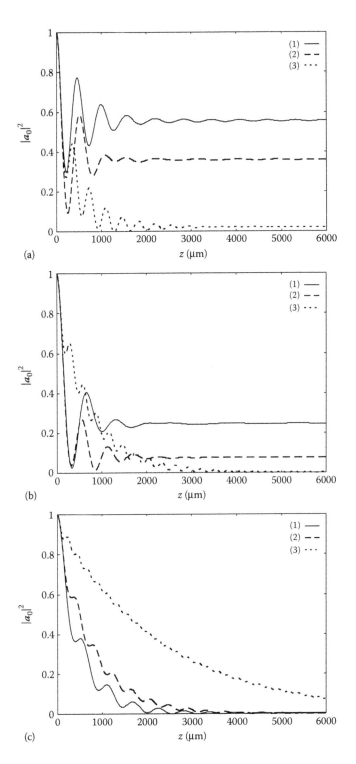

FIGURE 6.16

Diagram of $|a_0(z)|^2$. (a) $n_o = 1.40$; (b) $n_o = 1.46$; (c) $n_o = 1.47$. $t = 3$ (μm); (1) $n_f = 1.4756$; (2) $n_f = 1.4745 (= n_s)$; (3) $n_f = 1.4709$.

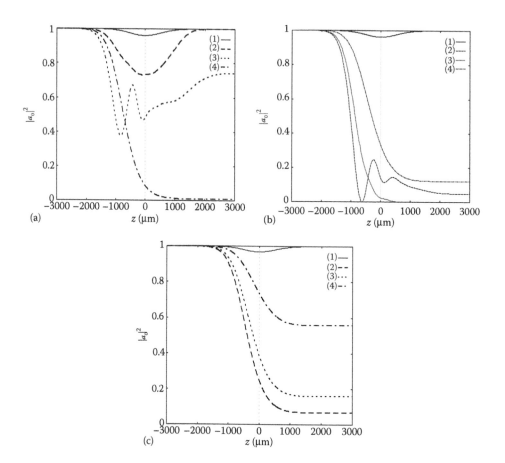

FIGURE 6.17
(a) Diagram of $|a_0(z)|^2$. $n_o = 1.40$, $t = 3$ (µm), $R = 50$ cm; (1) $n_f = 1.4817$; (2) $n_f = 1.4756$; (3) $n_f = 1.4745$ $(= n_s)$; (4) $n_f = 1.4709$. (b) Diagram of $|a_0(z)|^2$. $n_o = 1.46$, $t = 3$ (µm), $R = 50$ cm; (1) $n_f = 1.4817$; (2) $n_f = 1.4756$; (3) $n_f = 1.4745$ $(= n_s)$; (4) $n_f = 1.4709$. (c) Diagram of $|a_0(z)|^2$. $n_o = 1.47$, $t = 3$ (µm), $R = 50$ cm; (1) $n_f = 1.4817$; (2) $n_f = 1.4756$; (3) $n_f = 1.4745$ $(= n_s)$; (4) $n_f = 1.4709$.

the above case, except for the complete power recovery of the first curve when $n_f = 1.4817$. Nearly complete power transfer is possible other than for curve (3) this time.

An interesting feature involves comparison of curve (3) in Figure 6.17a and curve (2) in Figure 6.17b. The rate of the power attenuation (i.e., the slope of the curve) through coupling is so large that the power oscillates in between the fiber and slab. This generally results in a reduced amount of the power eventually being coupled out of the fiber and a great amount of power being restored in the fiber, as demonstrated by Figure 6.17a.

Figure 6.17c shows a similar curve (1) but three smooth curves (2), (3), and (4) as the power decays in the fiber. Now $n_s = 1.47$, the case of the most power transfer is curve (2) when $n_f = 1.4756$.

Overall, the above diagrams have revealed some more interesting and sophisticated features due to the distributed coupling than the features of the uniform coupling primarily dominated by the mode (phase) synchronism and symmetry.

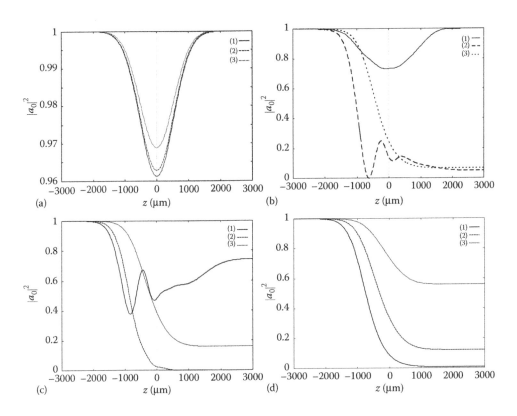

FIGURE 6.18

(a) Diagram of $|a_o(z)|^2$. $n_f = 1.4817$, $t = 3$ (µm), $R = 50$ cm; (1) $n_o = 1.40$; (2) $n_o = 1.46$; (3) $n_o = 1.47$. (b) Diagram of $|a_o(z)|^2$. $n_f = 1.4756$, $t = 3$ (µm), $R = 50$ cm; (1) $n_o = 1.40$; (2) $n_o = 1.46$; (3) $n_o = 1.47$. (c) Diagram of $|a_o(z)|^2$. $n_f = 1.4745$ ($= n_s$), $t = 3$ (µm), $R = 50$ cm; (1) $n_o = 1.40$; (2) $n_o = 1.46$; (3) $n_o = 1.47$. (d) Diagram of $|a_o(z)|^2$. $n_f = 1.4709$, $t = 3$ (µm), $R = 50$ cm; (1) $n_o = 1.40$; (2) $n_o = 1.46$; (3) $n_o = 1.47$.

6.7.4.2 *Fixing n_f Each Time while Varying n_o*

Rearranging the curves in Figure 6.17, Figure 6.18a indicates that, when $n_f > n_s$, the launched power remains in the fiber initially until the light reaches a certain propagation distance (here, $-z_0 \approx 2000$ µm) from the symmetry center (i.e., $z = 0$) of the fiber half-block. A very weak coupling appears at $z = -z_0$ with a small decay of the fiber power as light propagates towards $z = 0$. The power of the fiber is then gradually recovered after the light passes through $z = 0$. The curves of the power decay and recovery look symmetric about $z = 0$. This distributed coupling is in sharp contrast with Figure 6.9a where the fiber is straight and the light power fluctuates between the fiber and the slab during the propagation but remains mainly in the fiber. From Table 6.1, $\beta_{f0} > \beta_s$, and it is then not possible for the fiber mode to be phase-matched with any of the transverse slab modes. Very little power is thus coupled from the fiber to the slab waveguide. In addition, the effect of n_o can also be observed such that the greater the value of n_o, the weaker the coupling strength as the field of the slab modes is made to shift further (along x direction in Figure 6.1) from that of the fiber with an increasing value of n_o, the overlay refractive-index on the other side of the slab from the fiber.

As the value of n_f gets close to (but is still greater than) that of n_s, Figure 6.18a and b show that when the fiber is curved, coupling is distributed. Whether or not light is coupled from the fiber to the slab waveguide is strongly dependent on the value of n_o. If $n_o \leq n_c$ ($A_s \leq 0$), the fiber power decay due to light propagating into the coupling region is small and can be fully recovered after the light is guided past $z = 0$. If $n_o > n_c$ ($A_s > 0$), the fiber power decay is so large that when the light propagates into the coupling region it cannot be recovered after the light is guided through the structure at $z = 0$. That is, the light is almost completely coupled from the fiber to the slab waveguide. In Figure 6.7b, when the fiber is straight, lowering the value n_f and that of β_{f0}, generally causes power oscillation or decay, depending on the values of n_o, and β_s. If $n_o \leq n_c$ ($A_s \leq 0$), the power oscillates with a high strength and dies out with only a fraction of the power coupled into the slab waveguide. If $n_o > n_c$ ($A_s > 0$), the fiber power also decays significantly and radiates out to the slab.

In the case when the value of n_f equals that of n_s, distributed coupling and interesting features of the fiber-mode power decay can be observed in Figure 6.18c for the case of a curved fiber in comparison with the power beating and decaying in Figure 6.9c for a straight fiber. The lower the value of n_o, the higher the rate of the decaying power. If the fiber is curved, the lower value of n_o and the sharper attenuation result in much more of the power being coupled from the fiber to the slab. However, if the fiber is straight, the lower value of n_o and sharper attenuation cause the power to oscillate at a higher strength and then die out with less power coupled from the fiber at the end.

Finally, when the value of n_f is less than that of n_s, Figure 6.18d displays distributed coupling, whilst both Figures 6.9d and 6.18d show that the fiber mode decays in such a way that the smaller the value of n_o, the greater the power attenuation, as expected due to the resultant stronger coupling, i.e., more power goes to excite the continuum of modes in the slab waveguide.

6.8 Concluding Remarks

General conclusions can be drawn from the above coupled-mode analysis according to the coupler configurations so that the relevant and distinctive features of that particular type of coupler in analysis can be emphasized in contrast to each other.

6.8.1 Symmetric and Asymmetric Two-Mode Coupling Systems

Reformulating the standard coupled-mode equations in new variables such as power parameters has been shown to be a fruitful process in that, generally speaking, the new coupled-mode equations derived are more physically revealing, mathematically simpler, and thus easier to solve.

1. The power conservation law is expressed in one of the new coupled-mode equations for power orthogonal couplers composed of two identical modes and is readily drawn from the new formulation for general, nonorthogonal couplers composed of two nonidentical modes.

2. Reformulation itself is not only a systematic step further in obtaining the possible analytic solutions of the original coupled-mode equations the new formulation replaces, but also in obtaining some vital information about the operational features of the two-mode coupling systems, e.g., in the form of constants of motion (i.e., conserved quantities along the light propagation and coupling).

3. It has been demonstrated that couplers of two identical modes are indeed power orthogonal and symmetric. That is, the power is conserved and the exchange of power is complete between the two coupled modes. By contrast, couplers of two nonidentical modes have been shown to be power nonorthogonal and asymmetric in general. The power may not be well conserved between the two coupled modes and an extra part of the so-called cross-power term generally comes into play to satisfy the power conservation.

6.8.2 Uniform Fiber-Slab Coupling Systems

Generally asymmetric, composite fiber-slab couplers have been studied and it was found that the characteristics of light coupling and mode propagation depend on a combination of relative values of n_f and n_s as well as those of n_o and n_c, i.e., asymmetry of the slab waveguide index profile, as quantified by the parameter A_s. The properties of the ridge modes support the results obtained from the coupled-mode theory and the discussion presented above of the characteristics of the fiber and slab mode-coupling, especially the features of the power oscillations and beatings.

When n_s is lower than but not very close to n_f, the light is well confined in the fiber core and is only slightly affected by the asymmetry because of the phase mismatch. However, as soon as n_s is close to, equal to, or higher than n_f (the two waveguides are close to phase synchronism), the asymmetry of the slab cladding becomes very important. The mode beating, decay, and the amount of the power transferred can be readily tuned by a small variation in the value of n_o.

This analysis and the features of the composite fiber-slab coupling system may find applications, for example, in the practical design and selection of the fiber, slab waveguide, and cladding materials before fabrication of the coupler. Such a process is important and necessary for the couplers to achieve the light coupling and power transfer performance desired by using an optimized combination of the material and structural parameters.

6.8.3 Distributed Fiber-Slab Coupling Systems

An optical fiber half-block with a dielectric overlay has been studied theoretically and it was found that the characteristics of light coupling depend on the curvature of the fiber, the relative values of n_f and n_s, as well as those of n_o and n_c, i.e., asymmetry of the slab waveguide index profile, as quantified by the parameter A_s.

During the light propagation through the composite structure, the distributed coupling of light between the curved fiber and the slab waveguide is clearly demonstrated in comparison with the coupling between a straight fiber and a slab waveguide. The effective coupling region of the composite fiber-slab waveguide coupler can be varied by changing the curvature of the fiber or the minimum distance between the fiber and the slab waveguides.

When n_s is lower than but not very close to n_f, the light is well confined in the fiber core and is not substantially affected by the asymmetry due to the phase mismatch. However, as soon as n_s is close to, equal to, or higher than n_f (the two waveguides are close to phase synchronism), both the curvature of the fiber and the asymmetry of the slab cladding refractive-index come strongly into play. The mode decay and the amount of the power transferred from the fiber to the slab is quite different between the case of a straight and that of a curved fiber, and can be readily tuned by a small variation in the value of n_o.

This analysis and the features of the composite fiber-slab waveguide coupling system composed of a fiber half-block with a slab overlay should find applications in many practical situations. These may include the selection of the fiber and its curvature, slab waveguide and cladding materials, design of the coupler structure, and prediction of the coupling behaviors before fabrication. Once again, this simple analysis should help the design process so that the couplers may achieve the light coupling and power transfer performance desired by use of an optimal combination of both the material and structural parameters.

6.9 Problems

1. Consider the coupler consisting of a circular waveguide and a planar optical waveguide as shown in Figure 6.3.

    a. Assuming that the fiber is the standard single fiber (SSMF) as given in Chapters 4 and 5, determine the minimum distance required for the coupling of guided mode launched in the fiber to the slab waveguide whose refractive index difference is the same as that of the SSMF. You can assume that the refractive index difference between the guiding area and the cladding region of the planar guide is the same as that of the fiber core and cladding.

    b. What would happen if the planar waveguide became a channel waveguide, as in Chapter 3?

2. Refer to Figure 6.4 for the coupling of a two-mode coupler. Find the distance along the propagation direction in terms of the coupling coefficient κ so that a complete power coupling occurs. Similarly for 50% power and 0% power coupled from one guide to the other.

3. Refer to Figure 6.5 for the coupling of an asymmetric two-mode coupler. Find the distance along the propagation direction in terms of the cross-coupling coefficient B so that a complete power coupling occurs. Similarly for 50% power and 0% power coupled from one guide to the other.

4. Write down the dispersion equation of an asymmetric planar waveguide. Make sure that you declare all parameters of the equation.

References

1. A. Yariv, "Coupled-mode theory for guided-wave optics," *IEEE J. Quantum Electron.*, QE-9, 919–933, 1993.
2. T. Tamir, "*Integrated Optics*," Springer-Verlag, Berlin, 1975.
3. B. Lamouroux, P. Morel, P. Prade, and Y. Vinet, "Evanescent-field coupling between a monomode fiber and a high-index medium of limited thickness," *J. Opt. Soc. Am. A.*, 2(5), 759–764, 1985.
4. D. Marcuse, Personal communications in relation to errors in the paper of Marcuse, 1989.
5. W. Johnstone, S. Murray, G. Thursby, M. Gill, A. McDonach, D. Moodie, and B. Culshaw, "Fiber optical modulators using active multimode waveguide overlays," *Electron. Lett.*, 27, 894–896, 1991.

6. D. Marcuse, "Investigation of coupling between a fiber and an infinite slab," *IEEE J. Lightwave Technol.*, 7, 122–130, 1989.
7. A.W. Snyder and J.D. Love, "*Optical Waveguide Theory*," Chapman, London 1983.
8. D. Marcuse, "*Light Transmission Optics*", 1st Ed., Van Nostrand Reinhold, New York, 1972.
9. A.W. Snyder, "Coupled-mode theory for optical fibers," *J. Opt. Society Am.*, 62, 1267–1277, 1972.
10. E.A.J. Marcatili, "Improved coupled-mode equations for dielectric guides," *IEEE J. Quantum Electron.*, QE-22, 988–993, 1986.
11. E.A.J. Marcatili, L.L. Buhl, and R.C. Alferness, "Experimental verification of the improved coupled-mode equations," *Appl. Phys. Lett.*, 49, 1692–1693, 1986.
12. D. Gloge, "Weakly guiding fibers," *Appl. Opt.*, 10, 2252–2258, 1971.
13. A.W. Snyder and A. Ankiewicz, "Optical fiber couplers-optimum solution for unequal cores," *IEEE J. Lightwave Technol.*, 6, 463–474, 1988.
14. D. Marcuse, "Directional couplers made of non-identical asymmetric slabs. Part I : Synchronous couplers," *IEEE J. Lightwave Technol.*, LT-5(1), 113–118, 1987.
15. D. Marcuse, "Directional couplers made of non-identical asymmetric slabs. Part II : Grating-assisted couplers," *IEEE J. Lightwave Technol.*, LT-5(2), 268–273, 1987.

7

Full Coupled-Mode Theory

7.1 Full Coupled-Mode Analysis

7.1.1 Introduction

In this chapter, a full, scalar, coupled-mode analysis is presented for both the two-mode and the fiber-slab guided-wave couplers. It generalizes the simplified analysis of Chapter 6 and, when the weakly guiding and coupling conditions are valid, should reduce to that described in Chapter 6. An effort has been made to keep the format of the full formulations similar to that in Chapter 6 so that most of the analytical results derived there can be utilized, except for the fact that the full term is now retained throughout the analysis.

The next section, subsection 7.1.2, reformulates the full coupled-mode theory (CMT) in the power parameters for the two-mode coupler, whilst subsections 7.1.3 and 7.1.4 present the analytical analysis and numerical calculations, respectively, for the fiber-slab couplers. All are carried out in comparison with the corresponding results from the simplified analysis described in Chapter 6.

Section 7.2 gives the analytical approach to scalar coupled-mode theory with vectorial corrections. Thence Section 7.3 illustrates the application of the CMT with vectorial correction to the evaluation of a grating-assisted fiber-slab optical waveguide coupling.

7.1.2 Two-Mode Couplers

As demonstrated in Chapter 6, the power-orthogonal two-mode couplers can be treated as a special case of general, power, nonorthogonal, two-mode couplers. In this section, only the reformulated full coupled-mode equations and their analytical solutions for the asymmetric two-mode couplers are presented. Coupled-mode equations and analytical solutions of the symmetric two-mode coupler are readily derivable from the asymmetric ones by simply considering two identical modes and thus using symmetric mode parameters and coupling coefficients in the general formulation.

7.1.2.1 Full Coupled-Mode Equations

For weakly guiding the coupling of the two modes of modal amplitudes a and b, the following general coupled-mode equations can be derived (e.g. [1]):

$$a'(z) = -j(\beta_a + Q_a)a(z) - jK_{ab}b(z) \tag{7.1a}$$

$$b'(z) = -j(\beta_b + Q_b)b(z) - jK_{ba}a(z) \tag{7.1b}$$

where the prime represents differentiation with respect to z, a and b represent the modal amplitudes (i.e., expansion coefficients) of the two modes, respectively, $\{\beta_a, \beta_b\}$ are the corresponding mode propagation constants, and $\{Q_a, Q_b\}$ and $\{K_{ab}, K_{ba}\}$ are now the full self- and cross-coupling coefficients, respectively, which are related to the first-order coupling coefficients $\{Q_a, Q_b\}$ and $\{K_{ab}, K_{ba}\}$, as presented in Chapter 6, by

$$Q_a = \left(Q_a - P_{ab}K_{ba}\right)/\left(1 - P_{ab}^2\right) = Q_a - P_{ab}K_{ba} \tag{7.2a}$$

$$Q_b = \left(Q_b - P_{ab}K_{ab}\right)/\left(1 - P_{ab}^2\right) = Q_b - P_{ab}K_{ab} \tag{7.2b}$$

$$K_{ab} = \left(K_{ab} - P_{ab}Q_b\right)/\left(1 - P_{ab}^2\right) = K_{ab} - P_{ab}Q_b \tag{7.3a}$$

$$K_{ba} = \left(K_{ba} - P_{ab}Q_a\right)/\left(1 - P_{ab}^2\right) = K_{ba} - P_{ab}Q_a \tag{7.3b}$$

In terms of the power variables, as introduced in Section 6.3 of Chapter 6, Equation 7.1a and b become

$$P_a'(z) = 2K_{ab}P_i \tag{7.4a}$$

$$P_b'(z) = -2K_{ba}P_i \tag{7.4b}$$

$$P_r'(z) = \left[(\beta_b - \beta_a) + (Q_b - Q_a)\right]P_i \tag{7.4c}$$

$$P_i'(z) = -K_{ba}P_a + K_{ab}P_b - \left[(\beta_b - \beta_a) + (Q_b - Q_a)\right]P_r \tag{7.4d}$$

7.1.2.2 Analytical Solutions

From the above full coupled-mode equations in power parameters, the following constants of motion can be found:

$$P = P_a(z) + P_b(z) + 2P_r(z)(K_{ba} - K_{ab})/(\beta_b + Q_b - \beta_a - Q_a) \tag{7.5a}$$

$$\Gamma = P_a(z)/K_{ab} + P_b(z)/K_{ba} = P_a(0)/K_{ab} + P_b(0)/K_{ba} \tag{7.5b}$$

In comparison to Equation 6.32 of Chapter 6, Equation 7.5a is now the generalized law of the power conservation for the two coupled nonidentical modes. From the definition of the Poynting's vector and then the z-propagating total power in the coupler of two nonidentical modes, we have

$$P = P_a(z) + P_b(z) + 2P_{ab}P_r(z) \tag{7.6}$$

$$K_{ba} - K_{ab} = P_{ab}(\beta_b + Q_b - \beta_a - Q_a) \tag{7.7}$$

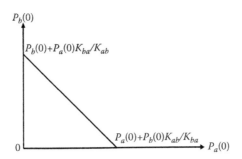

FIGURE 7.1
Illustration of relation between $P_a(z)$ and $P_b(z)$.

The above relationship has been verified by other full formulations of CMT, e.g., the vector CMT of [2,3]. Furthermore, Equation 7.5b can be utilized to determined the power distribution in between $P_a(z)$ and $P_b(z)$. Figure 7.1 illustrates a simple, straight-line relation between the two variables $P_a(z)$ and $P_b(z)$ with initial values of $P_a(0)$ and $P_b(0)$ respectively.

The below simple diagram is similar to that of Figure 6.3 (Chapter 6), except for the expressions of cross-coupling coefficients that now contain the higher-order terms. As before, this can be used to aid in full coupled-mode analysis or design of the asymmetric two-mode coupler. First, some general initial conditions such as those given by Equations 6.13 and 6.14 of Chapter 6 are assumed. From Equation 7.4a and b:

$$P_i(0) = 0 \tag{7.8}$$

$$\Phi(0) = \tan^{-1}\left(P_i(0)/P_r(0)\right) = 0 \tag{7.9}$$

$$P_r(0) = P_{r0} = \left(P_{a0}P_{b0}\right)^{1/2} \tag{7.10}$$

$$P_r'(0) = 0 \tag{7.11}$$

$$P_i'(0) = P_{id} = -K_{ba}P_{a0} + K_{ab}P_{b0} + \left(\beta_a - \beta_b + Q_a - Q_b\right)\left(P_{a0}P_{b0}\right)^{1/2} \tag{7.12}$$

Similar to the derivation of Equation 6.39 of Chapter 6, we arrive at the following first-order differential equation:

$$P_i'(z)\mathrm{d}P_i'(z) = -BP_i(z)\mathrm{d}P_i(z) \tag{7.13}$$

where

$$B = \left[4K_{ab}K_{ba} + \left(\beta_b + Q_b - \beta_a - Q_a\right)^2\right] > 0 \tag{7.14}$$

The solution of Equation 7.14 can be expressed as

$$P_a(z) = P_a(0) + 2\left[\cos\left(B^{1/2}z\right) - 1\right]|P_{id}|K_{ab}/B \tag{7.15a}$$

$$P_b(z) = P_b(0) + 2\left[1 - \cos\left(B^{1/2}z\right)\right]|P_{id}|K_{ab}/B \tag{7.15b}$$

where K_{ab} is the cross-coupling coefficient, and P_{id} and B are constants defined by Equations 7.12 and 7.14, respectively.

The above power distributions of a full CMT are still sinusoidal, but all the coupling coefficients now contain the higher-order terms. The initial values can be determined by the initial conditions of the coupler. The amplitudes of the power oscillation are not only dependent on the constants of initial power distribution, but also on those of the generalized cross- and self-coupling. The beat length (or frequency) of the power distribution is also a function of the generalized cross- and self-coupling constants.

As in the case of the first-order CMT of two identical coupled modes, the zero-mode coupling condition (i.e., $P_v(z) \equiv P_v(0)$, $v = a,b$) is equivalent to setting $P_a' = P_b' = 0$. From Equation 7.4a and b, we have $P_i \equiv 0$ (i.e., $P_{id} = 0$) and, utilizing Equation 7.12, the following launching condition for the zero coupling situation can be derived:

$$P_{b0} = P_{a0}\left\{\beta_b - \beta_a + Q_b - Q_a + \left[(\beta_b - \beta_a + Q_b - Q_a)^2 + 4K_{ab}K_{ba}\right]^{1/2}\right\}^2 \Big/ (4K_{ab}^2) \tag{7.16}$$

$$b_0 = a_0\left\{(\beta_b - \beta_a) + (Q_b - Q_a) \pm \left[(\beta_b - \beta_a) + (Q_b - Q_a)\right]^2 + 4K_{ab}K_{ba}\right\}^{1/2}\Big/ (2K_{ab}) \tag{7.17}$$

It can be verified that Equations 7.16 or 7.17 is exactly the analytical solution (i.e., eigen-vectors) of the full compound-mode equations of the general, asymmetric two-mode coupler, whilst the compound-mode propagation constant (i.e., eigenvalue) is given by

$$\beta_e = (1/2)\left\{\beta_b + \beta_a + Q_b + Q_a \pm \left[(\beta_b - \beta_a + Q_b - Q_a)^2 + 4K_{ab}K_{ba}\right]^{1/2}\right\} \tag{7.18}$$

When $\beta_b \to \beta_a$, $Q_b \to Q_a$ and $K_{ab} \to K_{ba}$, we have $P_{b0} \to P_{a0}$, which is the zero-coupling initial condition and the full compound-mode solutions for the symmetric two-mode couplers.

Secondly, applying the typical initial launching conditions, Equations 6.21 through 6.23 of Chapter 6, leads to the following initial conditions for the power parameters:

$$P_a(0) = P \tag{7.19a}$$

$$P_b(0) = P_r(0) = P_i(0) = 0 \tag{7.19b}$$

$$P_a'(0) = P_b'(0) = P_r'(0) = 0 \tag{7.19c}$$

$$P_i'(0) = P_{id} = -K_{ba}P \tag{7.19d}$$

A set of general solutions obtained above can be directly reduced to the following, by applying the above initial conditions:

$$P_a(z) = P\left\{1 + F\left[\cos\left(B^{1/2}z\right) - 1\right]\right\} \tag{7.20a}$$

$$P_b(z) = PF\left\{1 - \cos\left(B^{1/2}z\right)\right\}K_{ba}/K_{ab} \tag{7.20b}$$

where P is the total power launched initially into the guide a, K_{ab} and K_{ba} are the full cross-coupling coefficients, B is given by Equation 7.14 and F is defined by

$$F = 2K_{ab}K_{ba}/B = (1/2)\left\{1 + \left[\left(\beta_b - \beta_a + Q_b - Q_a\right)/\left(4K_{ab}K_{ba}\right)^{1/2}\right]^2\right\}^{-1} \tag{7.21}$$

As demonstrated by Equations 6.46 and 6.47 of Chapter 6, unless $(\beta_b - \beta_a + Q_b - Q_a)^2 \ll 4K_{ab}K_{ba}$, no substantial power exchange would occur. Making use of Equation 7.7, we have

$$F = (1/2)\left\{1 + \left[\left(K_{ba}/K_{ab}\right)^{1/2} - \left(K_{ab}/K_{ba}\right)^{1/2}\right]^2 \Big/ \left(4P_{ab}^2\right)\right\}^{-1} \tag{7.22}$$

where P_{ab} is the field-overlap integral (i.e., butt-coupling coefficient) of the two modes.

Assuming a normalized launching power (i.e., $P = 1$), the power distributions along the light-propagation direction are as shown in Figure 7.2. It can be observed from Equation 7.20 that the asymmetry of the power coupling and distribution is governed by the ratio of K_{ba}/K_{ab}, whilst the fractions of the power and distribution or the completeness of the power transfer is given by the ratio of F, which is related to P_{ab} and K_{ba}/K_{ab} by Equation 7.23. A special case is a coupler structured by two identical modes, where $K_{ab} = K_{ab} = K$, $F = 0.5$,

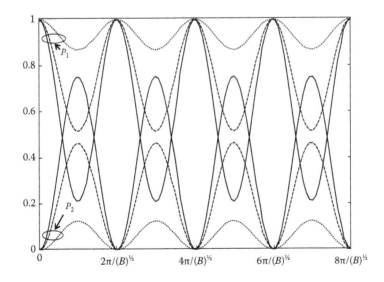

FIGURE 7.2
Power distributions along the propagation direction $K_{ba}/K_{ab} = 0.95$. (1) Solid line: $P_{ab} = 0.5$; (2) dashed line: $P_{ab} = 0.25$; (2) dotted line: $P_{ab} = 0.1$.

and $B = 4K^2$, and the above asymmetric power distribution described by Equation 7.20 is reduced to a symmetric one.

7.1.3 Fiber-Slab Couplers

In this section, the scalar, the full coupled-mode, and the compound-mode equations of the fiber-slab coupler are derived. This is followed by presentation of the analytical solutions of the full coupling coefficients. Finally, the conditions of power conservation are discussed and critically compared with the vector, full coupled-mode formulation.

7.1.3.1 Full Coupled-Mode Equations

As described above for the two-mode couplers, the nonorthogonal CMT is also termed as a full CMT because its formulation retains the full terms, i.e., the first and the higher-order components. The procedures for deriving the first-order (or simplified) formulation of the scalar CMT are therefore applicable here except for the treatment of those higher-order terms, such as the field overlapping and the polarization effects. Conversely, a full formulation of CMT should be reduced to the first-order one whenever the scalar and other approximations are valid. Recall that two types of field expressions can be used to derive the scalar coupled-mode equations, namely

$$\varphi = \sum \left\{ \tilde{a}_n(z) F_n(x, y) e^{-j\beta_{fn}z} + \tilde{b}_{an}(z) S_{NP}(x, y) e^{-j\beta_{San}z} \right\} \tag{7.23}$$

$$\varphi = \sum \left\{ \tilde{a}_n(z) F_n(x, y) e^{-j\beta_{fn}z} + \tilde{b}_n(z) S_{NP}(x, y) e^{-j\beta_{San}z} \right\} \tag{7.24}$$

and the numerical results should be the same in terms of the absolute square of the modal amplitudes because of the following modal transforms:

$$a_n = \tilde{a}_n e^{-j\beta_{fn}z} \tag{7.25a}$$

$$b_n = \tilde{b}_n e^{-j\beta_{Sn}z} \tag{7.25b}$$

7.1.3.1.1 Derivation using Field Expression

First the field expansion Equation 7.23 is substituted into the scalar wave Equation 7.24, and the resultant equation multiplied successively with the unperturbed modes F_m and S_m, and integrated over the transverse (x, y) plane whilst making use of the modal equations and the orthogonality relations for the fiber and slab modes. As the result, we have

$$a''_m + \sum_n P_{fmn} b''_n + \beta^2_{fm} a_m + \sum_n S_{fmn} a_n + \sum_n \left(C_{fmn} + \beta^2_{sn} P_{fmn} \right) b_n = 0 \tag{7.26a}$$

$$b''_m + \sum_n P_{smn} a''_n + \beta^2_{sm} b_m + \sum_n S_{smn} b_n + \sum_n \left(C_{fmn} + \beta^2_{fn} P_{smn} \right) a_n = 0 \tag{7.26b}$$

where $a_m'' = d^2 a_m/dz^2$, $b_m'' = d^2 b_m/dz^2$, and $m = 0,1,2,...$; the new self-coupling coefficients $\{S_{fmn}, S_{smn}\}$ and the cross-coupling coefficients $\{C_{fmn}, C_{smn}\}$ are related to the first-order coupling coefficients defined in Chapter 6 by

$$S_{fmn} = 2\beta_{fm}Q_{fmn} \tag{7.27a}$$

$$S_{smn} = 2\beta_{sm}Q_{smn} \tag{7.27b}$$

$$C_{fmn} = 2\beta_{fm}K_{fmn} \tag{7.28a}$$

$$C_{smn} = 2\beta_{sm}K_{smn} \tag{7.28b}$$

where $\{Q_{fmn}, Q_{smn}\}$ and $\{K_{fmn}, K_{smn}\}$ are the self- and cross-coupling coefficients, respectively, as defined in Equations 6.62 and 6.63 (Chapter 6) and analytically solved in Equations 6.64 and 6.65 (Chapter 6). $\{P_{fmn}, P_{smn}\}$ are butt-coupling coefficients (i.e., field-overlap integrals) due to the nonorthogonality of the mode fields $\{F_m, S_n\}$ and are defined as

$$P_{fmn} = \int_{A_\infty} F_m S_n dA \tag{7.29a}$$

$$P_{smn} = \int_{A_\infty} S_m F_n dA \tag{7.29b}$$

Note that not all the coupling coefficients are independent. We derive

$$P_{fmn} = P_{snm} \tag{7.30}$$

and utilizing the two-dimensional Green's theorem and the fact that the guided-mode fields of the fiber and the slab waveguide decrease exponentially far from the fiber core and guiding layer of the slab waveguide, the following relationship holds:

$$C_{fmn} - C_{snm} = P_{fmn}\left(\beta_{fm}^2 - \beta_{sn}^2\right) \tag{7.31}$$

Within the limit of the scalar approximation, the above analytical relationship is generally applicable in composite two-guide coupler systems, e.g., optical fiber couplers with unequal cores [4]. Note that it generalizes to the full terms of the well-known equation

$$K_{fmn} - K_{snm} = P_{fmn}\left(\beta_{fm} - \beta_{sn}\right) \tag{7.32}$$

which holds when the value of β_{fm} is close to that of β_{sn}

For the sake of simplicity, a single mode is initially launched in the fiber, i.e., $a_n = 0$ for $n > 0$. With some manipulation, the following coupled-mode equations can be derived from Equation 7.27:

$$a_0'' + \left(S_{f00} + \beta_{f0}^2\right)a_0 + \sum_n C_{f0n}b_n = 0 \tag{7.33a}$$

$$b_0'' + \left(C_{sm0}a_0 + \beta_{f0}^2\right)a_0 + \sum_n \left(S_{smn} + \delta_{mn}\beta_{sn}^2\right)b_n = 0 \tag{7.33b}$$

where a_0 is the modal amplitude of the single, linearly polarized fiber mode commonly noted as the LP_{01} mode [5], whilst b_m is that of the guided slab mode of the mth transverse order. $\{\beta_{f0}, \beta_{sm}\}$ are the corresponding propagation constants, respectively.

The new, full coupling coefficients introduced above are related to those defined in (Equations 7.27 through 7.29) by

$$S_{f00} = S_{f00} - \sum_n P_{f0n}C_{sn0} \tag{7.34a}$$

$$S_{smn} = S_{smn} - P_{sm0}C_{f0n} \tag{7.34b}$$

with C_{f0n} and C_{sn0} defined as

$$C_{f0n} = C_{f0n} - \sum_l P_{f01}S_{sln} \tag{7.35a}$$

$$C_{sm0} = C_{sm0} - P_{sm0}S_{f00} \tag{7.35b}$$

and $A \equiv A/(1 - \sum_n P_{f0n}P_{sn0})$

7.1.3.1.2 Derivation using Field Expression

Using the field expression Equation 7.25 and the coupling coefficients defined above, and following the similar procedures outlined above, the following coupled-mode equations can be obtained:

$$\tilde{a}_0'' - j2\beta_{f0}\tilde{a}_0' + S_{f00}\tilde{a}_0 + \sum_n C_{f0n}b_n e^{j(\beta_{f0} - \beta_{sn})z} = 0 \tag{7.36a}$$

$$\tilde{b}_m'' - j2\beta_{sm}\tilde{b}_m' + C_{sm0}\tilde{a}_0 e^{j(\beta_{sm} - \beta_{f0})z} + \sum_n S_{smn}\tilde{b}_n e^{j(\beta_{sm} - \beta_{sn})z} = 0 \tag{7.36b}$$

where, once again, a single mode excitation for the fiber, i.e., $a_n = 0$ for ... $n > 0$ is assumed. Obviously, Equation 7.36 is equivalent to Equation 7.33 via the transformation of Equation 7.25.

7.1.3.1.3 Relation to Simplified Coupled-Mode Equations

The above full coupled-mode equations generalize the standard nonorthogonal ones, and lead to a set of coupled-mode equations that only involves the first-order derivatives. To facilitate a comparison with the standard nonorthogonal formulations of scalar or vector CMT, simplifications can be made in the scalar approximation limit. In addition, the analytical and numerical results can be studied in comparison with those obtained in Chapter 6, where both the second-order derivatives and the field overlap integrals are neglected. This way the effects of the modal nonorthogonality (overlap) can be revealed by contrast.

Firstly, a slowly varying envelope approximation is assumed, that is

$$|\tilde{a}_0''| \ll 2\beta_{f0}|\tilde{a}_0'| \tag{7.37a}$$

$$|\tilde{b}_m''| \ll 2\beta_{sm}|\tilde{b}_m'| \tag{7.37b}$$

The coupled-mode Equations 7.34 and 7.37 become, respectively

$$a_0' = -j\left(\beta_{f00} + Q_{f00}\right)a_0 - j\sum_n K_{f0n}b_n \tag{7.38a}$$

$$b_m' = -j\sum_n \left(\delta_{mn}\beta_{sn} + Q_{smn}\right)b_n - jK_{sm0}a_0 \tag{7.38b}$$

$$\tilde{a}_0' = -jQ_{f00} - j\sum_n K_{f0n}\tilde{b}_n e^{j(\beta_{f0}-\beta_{sn})z} \tag{7.39a}$$

$$\tilde{b}_m' = -j\sum_n Q_{smn}\tilde{b}_n e^{j(\beta_{sm}-\beta_{sn})z} - j\tilde{a}_0 K_{sm0} e^{j(\beta_{sm}-\beta_{f0})z} \tag{7.39b}$$

where

$$Q_{f00} = S_{f00}/(2\beta_{f0}) \tag{7.40a}$$

$$Q_{smn} = S_{smn}/(2\beta_{sm}) \tag{7.40b}$$

$$K_{f0n} = C_{f0n}/(2\beta_{f0}) \tag{7.41a}$$

$$K_{sm0} = C_{sm0}/(2\beta_{sm}) \tag{7.41b}$$

Secondly, further approximations can be made by neglecting the effect of the field overlap of the fiber and slab modes. This is valid if the fiber and slab waveguides are considered electromagnetically well separated or very weakly coupled. The set of coupling coefficients $\{Q_{f00}, Q_{smn}, K_{f0n}, K_{sm0}\}$ are reduced simply to $\{Q_{f00}, Q_{smn}, K_{f0n}, K_{sm0}\}$ and the set of coupled-mode Equations 7.39 and 7.40 now appear in the following common first-order results:

$$a_0' = -j\left(\beta_{f00} + Q_{f00}\right)a_0 - j\sum_n K_{f0n}b_n \tag{7.42a}$$

$$b_m' = -j \sum_n (\delta_{mn} b_{sn} + Q_{smn}) b_n - jK_{sm0} a_0 \tag{7.42b}$$

$$\tilde{a}_0' = -jQ_{f00} \tilde{a}_0 - j \sum_n K_{f0n} \tilde{b}_n e^{j(\beta_{f0} - \beta_{sn})z} \tag{7.43a}$$

$$\tilde{b}_m' = -j \sum_n Q_{smn} \tilde{b}_n e^{j(\beta_{sm} - \beta_{sn})z} - jK_{sm0} \tilde{a}_0 e^{j(\beta_{sm} - \beta_{f0})z} \tag{7.43b}$$

The full coupling coefficients introduced in the above are all analytically solved. The solutions of the integrals of the first-order coupling coefficients presented in Chapter 6 can be used and the butt-coupling coefficients are solved in closed form as given in Appendix 6.

7.1.4 Full Compound-Mode Equations

The full, scalar compound-mode equations of the composite fiber-slab waveguide coupler can be derived using the following field expression:

$$\varphi = \sum \left[A_n F_n (x, y) + B_n S_n (x, y) \right] e^{-j\beta z} \tag{7.44}$$

where $\{A_n, B_n\}$ are the expansion constants and β is the compound-mode propagation constant of the coupler. Alternatively, the compound-mode equations can be obtained utilizing the coupled-mode derivations and the following transformations:

$$a_n (z) = A_n e^{-j\beta z} \tag{7.45a}$$

$$b_n (z) = B_n e^{-j\beta z} \tag{7.45b}$$

$$\tilde{a}_n = A_n e^{j(\beta_{fn} - \beta)z} \tag{7.46a}$$

$$\tilde{b}_n = B_n e^{j(\beta_{sn} - \beta)z} \tag{7.46b}$$

As a result, we obtain the full, scalar compound-mode equations as

$$\left(S_{f00} + b_{f0}{}^2 \right) A_0 + \sum_n C_{f0n} B_n = \beta^2 A_0 \tag{7.47a}$$

$$C_{sm0} A_0 + \sum_n \left(S_{smn} + \delta_{mn} \beta_{sn}{}^2 \right) B_n = \beta^2 B_m \tag{7.47b}$$

which correspond to the set of full, scalar coupled-mode Equations 7.34 or 7.37. Similarly, we have

$$\left(\beta_{f00} + Q_{f00}\right)A_0 + \sum_n K_{f0n}B_n = \beta A_0 \tag{7.48a}$$

$$\sum_n \left(\delta_{mn}\beta_{sn} + Q_{smn}\right)B_n + K_{sm0}A_0 = \beta B_m \tag{7.48b}$$

corresponding to the standard nonorthogonal coupled-mode equation of 7.39. Equation 7.49 can be obtained, for example, by using the assumption of slowly varying envelope,

$$\tilde{a}_0'' - j2\beta_{f0}\tilde{a}_0' = \left(\beta_{f0}^2 - \beta^2\right)\tilde{a}_0 \tag{7.49a}$$

$$\tilde{b}_m'' - j2\beta_{sm}\tilde{b}_m' = \left(\beta_{sm}^2 - \beta^2\right)\tilde{b}_m \text{ for } m = 0, 1, 2, \ldots \tag{7.49b}$$

thus the approximation in Equation 7.38 leads to

$$\beta_{f0}^2 - \beta^2 \approx 2\beta_{f0}\left(\beta_{f0} - \beta\right) \tag{7.50a}$$

$$\beta_{sm}^2 - \beta^2 \approx 2\beta_{sm}\left(\beta_{sm} - \beta\right) \tag{7.50b}$$

Further approximations are possible, e.g., using the weak-coupling conditions, the set of compound-mode Equation 7.49 can be further reduced to

$$\left(\beta_{f00} + Q_{f00}\right)A_0 + \sum_n K_{f0n}B_n = \beta A_0 \tag{7.51a}$$

$$\sum_n \left(\delta_{mn}\beta_{sn} + Q_{smn}\right)B_n + K_{sm0}A_0 = \beta B_m \tag{7.51b}$$

which is identical to the first-order compound-mode equation as given in Chapter 6.

7.1.5 Power Conservation

In this section, vector formulations of power conservation are derived in relation to the full, scalar CMT formulated above. The scalar results are reconciled with those of the vector ones so that the specific conditions attached for the full, scalar coupled-mode formulation, if any, can be found.

7.1.5.1 Power Conservation Law

The total power conservation during the light propagation along the z-axis of the coupler can be expressed as

$$\frac{\partial P}{\partial z} = \frac{1}{4}\frac{\partial}{\partial z}\iint_{A_\infty} \left(\underline{E} \times \underline{H}^* + \underline{E}^* \times \underline{H}\right) \cdot \hat{z}dA = 0 \tag{7.52}$$

7.1.5.2 Full Scalar Coupled-Mode Expression

Using the scalar-mode expansion described in Chapter 6 (Equation 6.53), we have

$$P = \frac{1}{2k}\sqrt{\frac{\varepsilon_0}{\mu_0}}\left[\beta_{f0}|a_0|^2 + \sum_n \beta_{sn}|b_n|^2 + \sum_n \frac{1}{2}\left(\beta_{f0}+\beta_{sn}\right)P_{f0n}\left(a_0^*b_n + a_0 b_n^*\right)\right] \tag{7.53}$$

$$= P_a + P_b + P_{ab} \tag{7.54}$$

and, utilizing the full coupled-mode equations, e.g., Equation 7.39,

$$P_a' = -\frac{1}{2k}\sqrt{\frac{\varepsilon_0}{\mu_0}}\sum_n C_{f0n}\mathrm{Re}\left(ja_0 {}^* b_n\right) \tag{7.55a}$$

$$P_b' = \frac{1}{2k}\sqrt{\frac{\varepsilon_0}{\mu_0}}\left\{\sum_n C_{sn0}\mathrm{Re}\left(ja_0 {}^* b_n\right) - \sum_{m,n} S_{smn}\mathrm{Re}\left(jb_m {}^* b_n\right)\right\} \tag{7.55b}$$

$$P_{ab}' = \frac{1}{2k}\sqrt{\frac{\varepsilon_0}{\mu_0}}\times\left\{\sum_n\left[\left(\beta_{f0}+\beta_{sn}\right)P_{f0n}\left(\beta_{f0}+Q_{f00}\right)-\sum_m\left(\beta_{f0}+\beta_{sm}\right)P_{f0m}\left(\delta_{mn}\beta_{sn}+Q_{smn}\right)\right]\right.$$

$$\left.\times\mathrm{Re}\left(ja_0 {}^* b_n\right) + \sum_{m,n}Q_{f0m}\left(\beta_{f0}+\beta_{sn}\right)P_{f0n}\mathrm{Re}\left(jb_m {}^* b_n\right)\right\} \tag{7.56}$$

Note that, in deriving Equation 7.56, the coefficient transformation of Equations 7.40, 7.41, and the orthogonal relations $\int_{A_\infty}F_m S_n = d_{mn}(m,n=0,1,2,\ldots)$ are used.

Power conservation requires $P'=0$ and this leads to the following equations:

$$C_{f0n} - C_{sn0} = P_{f0n}\left(\beta_{f0}{}^2 - \beta_{sn}{}^2\right) + \left(\beta_{f0}+\beta_{sn}\right)P_{f0n}Q_{f00}$$

$$-\sum_m\left(\beta_{f0}+\beta_{sm}\right)P_{f0m}Q_{smn} \tag{7.57}$$

$$S_{smn} = K_{f0m}\left(\beta_{f0}+\beta_{sn}\right)P_{f0n} \tag{7.58}$$

When coupling is weak, i.e., P_{f0n} is small, Equation 7.58 and the numerical calculations have shown that $Q_{f00}\ll\beta_{f0}$, $Q_{smn}\ll\beta_{sm}$, $C_{f0n}\to C_{f0n}$ and $C_{sn0}\to C_{sn0}$, so that Equation 7.57 is reduced to Equation 7.31.

7.1.6 Numerical Results and Discussion

Numerical results are calculated to allow comparison with the first-order results presented in Chapter 6 and visualization of the effect of the asymmetry of the slab waveguide on the coupling.

TABLE 7.1

Waveguide Parameters of the Optical Fiber

n_f	β_{f0}	k_f	γ_f	$V_f = ka\,(n_f^2 - n_c^2)^{1/2}$
1.4817	7.12500	0.7213	0.9856	3.101
1.4756	7.09886	0.6857	0.7744	2.617
1.4745	7.09436	0.6771	0.7319	2.520
1.4709	7.08019	0.641	0.5787	2.177

7.1.6.1 Parameters and Computer Programs

For consistency and comparison purposes, the optical and structural parameters used in the numerical solution of the full coupled-mode and compound-mode equations are the same as those in Section 3.61, except for where stated otherwise. The initial conditions at $z = 0$ are $a_0(0) = 1$...and ...$b_m(0) = 0$ $(m = 0,1,2,...)$.

The computer programs in Sections 6.2 of Chapter 6 have been upgraded with all the analytical solutions of the full coupling coefficients.

Tables 7.1 and 7.2 display values of some of the important fiber and slab parameters and constants used in the following comparative analysis. Other slab constants include $n_s = 1.4745$ and $n_c = 1.46$ (see Chapter 6).

In Table 7.1, $\gamma_f = (\beta_{f0}^2 - k^2 n_c^2)^{1/2}$, β_{f0}, and ...k_f are calculated using a scalar coupled mode approach, whilst in Table 7.2, k_s is formed using full vectorial analysis, $\beta_s = (k^2 n_s^2 - k_s^2)^{1/2}$ and $\gamma_s = (\gamma_s^2 - k^2 n_0^2)^{1/2}$. Note that the phase mismatch between the fiber and slab modes is dependent not only on the relative values of the refractive indices of the fiber ($\propto n_f$) with respect to those of the slab ($\propto n_s$) but also on the asymmetry of the slab waveguide, as indicated by n_0 in Table 7.2 (or n_0/n_c), or A_s, a parameter representing the slab asymmetry as defined in Section 6.3 of Chapter 6.

7.1.6.2 Effects of Higher-Order Terms

7.1.6.2.1 Distribution of Butt-Coupling Coefficient

It is found through the numerical examples of the above-derived analytical expressions that the butt-coupling coefficient P_{f0n} is very small compared to unity. This indicates that the coupling between the fiber- and the slab-guided modes is weak. Interestingly, upon examination, the value of P_{f0n} is much larger than the self- and cross-coupling coefficients, as Figure 7.3a illustrates. As a result, it can be seen that the distribution of the other coupling coefficients against the transverse slab-mode order n has been suppressed due to the large scale for displaying the butt-coupling coefficients. The values of the butt-coupling coefficients are monotonically decreasing with the increasing transverse order of the slab mode or, equivalently, angle θ, at which the particular transverse slab mode propagates away from the z axis (in the $\{y, z\}$ plane of the slab).

TABLE 7.2

Waveguide Parameters of the Slab Waveguide

n_0	β_s	k_s	γ_0	$V_{s0} = (kt/2)(n_s^2 - n_0^2)^{1/2}$
1.40	7.07487	0.8570	2.06595	2.2366
1.46	7.08829	0.7377	0.6706	0.9970
1.47	7.10515	0.5523	0.06713	0.5563

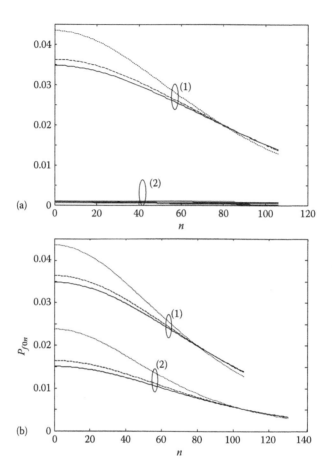

FIGURE 7.3

(a) A comparison of values of the butt-coupling coefficients with those of the self- and cross-coupling coefficients. $n_0 = 1.46$. Solid lines: $n_f = 1.4756$; dashed lines: $n_f = 1.4745$; and dotted line: $n_f = 1.4709$. (1) Dependence of butt-coupling coefficient on n, the transverse slab mode order; (2) values of the self- and cross-coupling coefficients. (b) Dependence of P_{f0n} on the transverse slab mode order n. Solid lines: $n_f = 1.4767$; dashed lines: 1.4736; dotted lines: 1.4689. (1) $n_0 = 1.46$; (2) $n_0 = 1.47$.

Figure 7.3a and b show that the values of the butt-coupling coefficients are dependent on the relative phase-difference between the fiber and slab modes, i.e., on both the values of n_f and n_s, when other constants are fixed. Generally speaking, the lower the value of n_f is with respect to that of n_s, the larger the value of P_{f0n}. For $n_f = 1.4709$ in Figure 7.3a and $n_f = 1.4689$ in Figure 7.3b, there is a phase match between the fiber mode and one of the slab transverse modes, i.e., $\beta_{sn}|_{n=53} = \beta_{f0} = 7.08019$, and the value of the field-overlap integral is well above those of solid and dashed lines where there are no such phase matches.

In addition, Figure 7.3b illustrates the effect of the asymmetry of the slab waveguide on the butt-coupling coefficients. The values of P_{f0n} of $n_0 = 1.47$ are generally smaller than those corresponding to the value of $n_0 = 1.46$. This can be explained by the fact that, when the value of n_0 is increased with respect to that of n_c (=1.46), the field distribution of the slab mode is shifted away from the peak of the fiber mode and this clearly reduces the field overlap, i.e., the butt-coupling coefficients.

7.1.6.2.2 Effects on Other Coupling Coefficients

From the definition (see Equations 7.36 and 7.42), K_{fon} and K_{f0n} are the first-order and the full cross-coupling coefficients, respectively, representing slab-to-fiber mode -coupling. Their n distribution and dependence on the values of n_f are displayed in Figure 7.4a.

The hidden n distributions of line-group (2) in Figure 7.3a, suppressed due to scaling, are now clearly revealed. It can be observed that both K_{f0n} and K_{fon} are decreased when n_f is decreased or when n is increased, indicating less power transferred from the slab modes to the fiber mode. The dependence on n_f may be explained by the shift of the fiber mode to a phase match with the slab modes (thus more fiber-to-slab mode-coupling) due to the decreasing n_f, whilst the n distribution is due to the power radiation within the slab at different angles with respect to the z axis (i.e., different values of n) and thus n-distributed coupling strength. In addition, we have $K_{fon} < K_{f0n}$ generally. That is, the slab-to-fiber mode-coupling is slightly reduced by the butt-coupling (see Equation 7.36a).

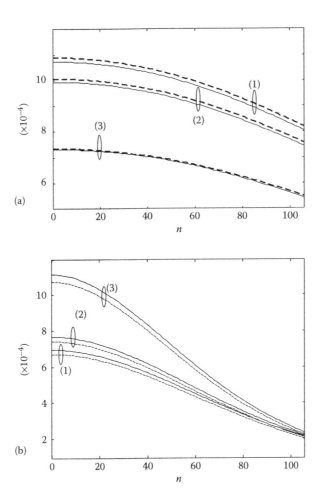

FIGURE 7.4
(a) Solid lines: C_{fon}; dashed lines: K_{fon}. $n_0 = 1.46$. (1) $n_f = 1.47674$; (2) $n_f = 1.47357$; (3) $n_f = 1.46890$. (b) Solid line: C_{sn0}; dashed line: K_{sn0}. $n_0 = 1.46$. (1) $n_f = 1.4756$; (2) $n_f = 1.4745$; (3) $n_f = 1.4709$. (c) Solid grid: C_{smn}; dashed grid: Q_{smn}. $n_f = 1.47674$, $n_0 = 1.46 = n_c$. (d) Solid grid: C_{smn}; dashed grid: Q_{smn}. $n_f = 1.47674$, $n_0 = 1.47 > n_c$.

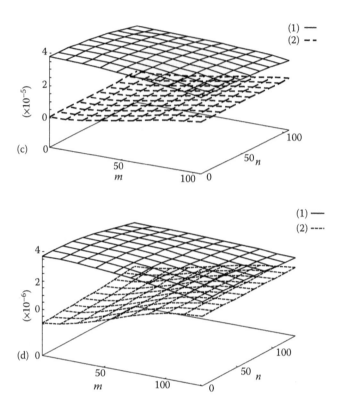

FIGURE 7.4 (Continued)

Similarly, K_{sn1} and K_{sn0} are the first-order and the full-term cross-coupling coefficients, respectively, representing the fiber-to-slab mode coupling. Their n distribution and dependence on the values of n_f are displayed in Figure 7.4b.

Contrary to what is observed in Figure 7.4a, both coefficients K_{sn1} and K_{sn0} are increased when n_f is decreased. This may be explained as described above in terms of the fiber-to-slab phase shift, as this allows more power to be coupled from the fiber to the slab. The same explanation provided above applies to the n distribution of K_{sn1} and K_{sn0} in Figure 7.4b. Interestingly, it appears that the second term in the right-hand side of Equation 7.36b has a sign reversal, which explains why we have $K_{sn0} > K_{sn1}$. In fact this is possible according to Equations 7.35a and 7.41a, i.e., with a negative, full, fiber, self-coupling coefficient Q_{f00}, and has been verified by the numerical calculation.

The two-dimensional $\{m, n\}$-distributions of C_{smn} and Q_{smn} (defined by Equations 7.35b and 7.41b) are shown in Figure 7.4c and d. The higher-order effects are quite different from those on the cross-coupling coefficients discussed so far, because $\{C_{smn}, Q_{smn}\}$ represent the self-coupling among the slab modes due to the presence of the fiber. First, they are much smaller than the cross-coupling coefficients and the fiber-mode self-coupling coefficients. Secondly, the higher-order corrections are greater for the lower transverse orders (i.e., towards $m, n = 0$) of the slab modes than for those of the higher orders. This may be explained by the fact that the butt-coupling of the fiber mode (i.e., the higher-order effects) is stronger with the slab transverse-modes propagating in the vicinity of the fiber (i.e., with smaller $\{m, n\}$ values) than with those radiating away from the fiber axis (i.e., with larger $\{m, n\}$ values). In addition, when the value of n_0 is increased with respect to that of n_c, the

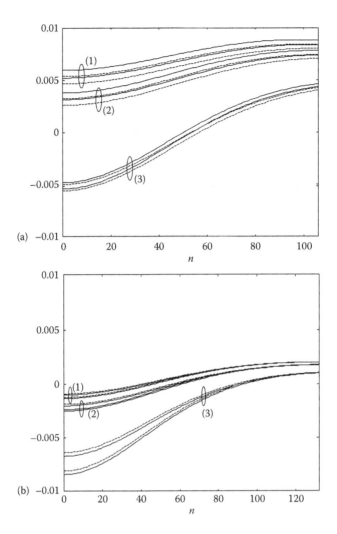

FIGURE 7.5
Power conservation. A comparison of the simplified and full CMT. (a) $n_0 = n_c = 1.46$; (b) $n_0 = 1.47 > n_c$. (1) $n_f = 1.4756$; (2) $n_f = 1.4745$; (3) $n_f = 1.4709$.

values of C_{smn} and Q_{smn} are decreased due to the reduced influence of the fiber core on the self-coupling of the slab modes.

7.1.6.2.3 Power Conservation Expressions

Numerical calculations are carried out to verify the validity of the analytical power-conservation formulations and the results are illustrated in Figure 7.5a and b, where N is the number of iterations.

Note that, within each line group, i.e., (1), (2), or (3), the upper solid and dashed lines represent the values of the left-hand side (LHS) of Equations 7.31 and 7.56, respectively, whilst the lower solid and dashed lines represent those of the corresponding right-hand side (RHS).

The n distribution of Equations 7.31 and 7.56 demonstrates that the LHS is generally not zero, because the fiber and the slab are very dissimilar waveguides. In addition, the values of the LHS of Equations 7.31 and 7.56 are generally in good agreement with the corresponding RHS and so are the values between Equations 7.31 and 7.56, i.e., the first-order and corresponding full-term expressions.

7.1.6.3 Characteristics of Mode Coupling

The numerical examples in this section are calculated with the same coupler parameters as in Section 6.1 of Chapter 6, except that $t = 2$ μm. The results of the full CMT are solutions of (42) or (43).

First, Figure 7.6a displays some interesting higher-order effects as the value of n_f is slightly higher than that of n_s, i.e., $n_f = 1.4756$, $n_s = 1.4745$. When $n_0 = n_c$, i.e., when the slab waveguide is symmetric, the effect of the higher-order terms on the power-beat amplitude and period is clear. The power beating dies out much more quickly than the first-order result, whilst the period of the oscillation does not change much. When $n_0 > n_c$, when the power in the fiber decay greatly, the effect of the higher-order terms is also greatly reduced. From Tables 4.1 and 4.2, we have $\beta_{f0} = 7.09886$, $\beta_s = 7.08829$ for line group (1) and $\beta_s = 7.10515$ for (2). The power decay and complete transfer from the fiber to the slab in line group (2) can be explained by the following phase-match condition we found: $\beta_{sn}|_{n=47} = \beta_{f0} = 7.09886$. This resonance condition leads to resonant coupling (Snyder and Love, 1983) and, as a result, a substantial power transfer from the fiber to the slab. Note that this is made possible with the variation of a single parameter, n_0 ($n_0 = 1.47$ in this case), which represents asymmetry of the slab waveguide. For line group (1), $n_0 = 1.46$, $\beta_{f0} > \beta_s$, namely, the value of the fiber-mode propagation constant is above those of the slab transverse modes (a band of continuum) and it is therefore not possible to have a resonance condition as described above.

Secondly, when the value of n_f is lowered to just below that of n_s, similar higher-order effects can be observed in Figure 7.6b except that more power is coupled out of the fiber (see line group (1)). This is attributed to the fact that the value of n_f is getting close to that of n_s, and, as a result, the composite fiber-slab coupler is approaching a phase-match condition between the fiber and slab guided-modes (e.g., $\beta_{f0} = \beta_{sm}$, $0 < m < N$). By comparing with the line groups (2) in Figure 7.6a, it can be seen that the rate of fiber-mode power decay is reduced in Figure 7.6b.

Finally, when the value of n_f is even lower than that of n_s, Figure 7.6c demonstrates that higher-order effects appear in the power decay curve (i.e., line group (2)) as well. When the slab is symmetric, the coupling strength is increased and much more power is coupled out of the fiber into the slab than in the two examples described above. When $n_0 > n_c$, the rate of the power decay in the fiber is further reduced, indicating an even lower power attenuation coefficient.

7.1.6.4 Characteristics of Ridge Modes

The numerical examples in this section are calculated with the same coupler parameters as in Section 3.6.1. The results of the full CMT are solutions of Equation 7.48, whilst those of the first-order CMT are solutions described in Chapter 6 Equation 6.61.

First, when n_f is slightly greater than n_s, the effect of the higher-order terms begins to show up, as demonstrated in Figure 7.7a. There are obvious shoulders of the field around the position of the slab. Increasing n_0 causes more deformation of the pure fiber mode in that the higher the value of n_0, the lower the field in the fiber becomes and the shoulder in the slab increases in height relative to that of the fiber. Similar field shifts are illustrated by Figure 7.7b when n_f equals n_s. Finally, when $n_f < n_s$, Figure 7.7c shows that large fractions of the mode power are transferred to the slab waveguide.

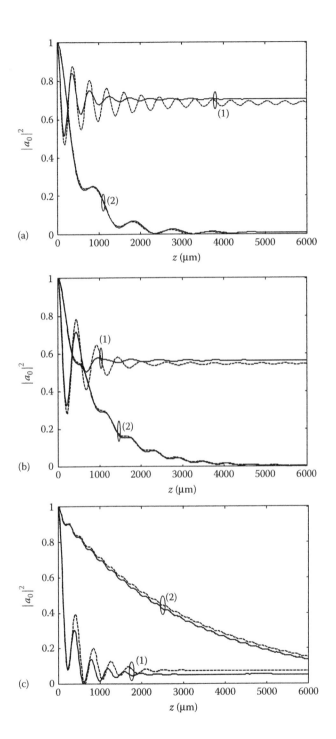

FIGURE 7.6
(a) Diagram of $|a_0(z)|^2$. $n_f = 1.4767 > n_s$. Solid lines: full CMT; dashed lines: first-order CMT. (1) $n_0 = 1.46 = n_c$; (2) $n_0 = 1.47 > n_c$. (b) Diagram of $|a_0(z)|^2$. $n_f = 1.4736 < n_s$. Solid lines: full CMT; dashed lines: first-order CMT. (1) $n_0 = 1.46 = n_c$; (2) $n_0 = 1.47 > n_c$. (c) Diagram of $|a_0(z)|^2$. $n_f = 1.4689 < n_s$. Solid lines: full CMT; dashed lines: first-order CMT. (1) $n_0 = 1.46 = n_c$; (2) $n_0 = 1.47 > n_c$.

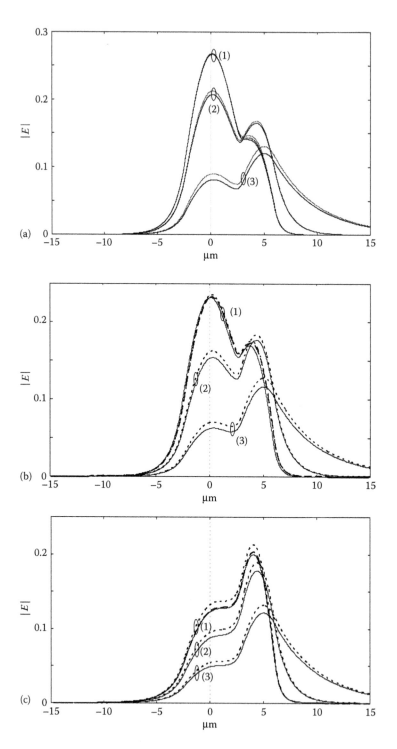

FIGURE 7.7

(a) x distribution of compound-mode field. $n_f = 1.4756$. Solid lines: results from (39); dashed lines: results from (38); and dotted lines: first-order results. (1) $n_0 = 1.40$; (2) $n_0 = 1.46$; (3) $n_0 = 1.47$. (b) As (a) except that $n_f = 1.4745$. (c) As (a) except that $n_f = 1.4709$.

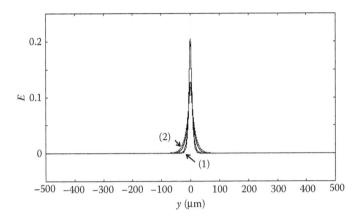

FIGURE 7.8
y distribution of the typical dominant compound-modes. Solid lines: from the first-order CMT; dashed lines: from the full CMT. (1) $n_0 = 1.40$; (2) $n_0 = 1.47$.

The above observations confirm the coupled-mode analysis as shown in Figure 7.6a–c. The increasing coupling strength of the fiber and slab guided-modes and amount of power transfer from the fiber into the Figure 7.7 slab waveguide can be explained by the decreasing value of refractive index of the fiber core n_f relative to that of the slab n_s and the resultant phase synchronism between the fiber and slab guided-modes. Figure 7.7a–c also display the effect of the value of n_0 or A_s on the compound modes in that the higher the value of n_0 or A_s, the greater the distance away the field extends into the slab as well as outside the cladding and, as a result, the field amplitude in the fiber decreases. This may be explained by the shift of the peak areas of the slab mode-fields away from that of the fiber due to the increasing value of the refractive index n_0 outside the slab guiding layer.

Figure 7.8 shows the typical y distribution of the dominant compound-modes. They are indeed ridge modes, because the fields at the center of the slab are concentrated in the vicinity of the fiber. Increasing the value of n_0 generally causes the field to slightly extend away from the vicinity of the fiber. Note that the peak values of the ridge-mode fields are dependent not only on the relative values of n_f and n_s, but also on those of n_0 and n_c. This is because both the cores (with n_f and n_s) and the claddings (with n_c and n_0) contribute to the individual fiber and slab guided-modes, and therefore the compound modes of the composite fiber-slab system.

For some optical parameters, the composite fiber-slab waveguide may support a second ridge mode. This is demonstrated here by the compound mode with the second largest propagation constants when $n_f > n_s$. The field distributions of the second ridge-modes as functions of x at $y = 0$ are shown in Figure 7.9a whilst those as functions of y at $x = 0$ are shown in Figure 7.9b. Note that the peaks of the mode fields in Figure 7.9a are shifted to the region of the slab, whilst some energy remains in the fiber. The higher-order effects are quite obvious, reducing the x-distribution peak value by more than 50%. Figure 7.9b shows the confinement of the fields around the center of the fiber, even though most of the light energy is concentrated in the plane of the slab. Once again, the higher-order effect is clear, with the y distribution peak reduced more than 50% but with the field pattern broadened. The higher-order effects for this second ridge mode and other compound modes, larger than previous examples, may contribute to the larger higher-order corrections seen in the power-beating curves than in the power decay curves in Figure 7.6.

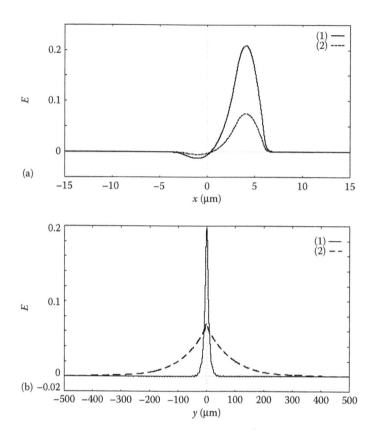

FIGURE 7.9
(a) x distributions of the compound modes of second largest propagation constants. $n_f = 1.4817$, $n_0 = 1$. (1) First-order CMT; (2) full CMT. (b) y distributions of the compound modes of second largest propagation constants. $n_f = 1.4817$, $n_0 = 1$. (1) First-order CMT; (2) full CMT.

7.1.7 Concluding Remarks

The full coupled-mode and compound-mode analysis presented in this chapter generalizes the previous first-order CMT and provides a check-up on the higher-order effects such as the butt-coupling effects neglected in the first-order CMT. In this section, general conclusions are drawn from the above full coupled-mode and compound-mode analysis of both the general two-mode and composite fiber-slab couplers.

7.1.7.1 Full CMT of Two-Mode Coupling Systems

1. The full coupled-mode equations can be reformulated into the format of the corresponding simplified ones, except with the coupling coefficients updated to the full-term version. The analytical solutions can be obtained utilizing the power-parameter reformulation and the previous results described in Chapter 6, Section 6.3.

2. The power conservation law is generalized to include the higher-order terms, with a format that is easily comparable with the simplified results obtained in Chapter 6.

3. Reformulation in the power parameters is also straightforward from the full coupled-mode equations, resulting in the full-term constants of motion and power diagram that reveal vital information about the operation of the two-mode coupler, such as power redistribution and conservation.

4. All the formulations of the full CMT can be simplified, utilizing the valid scalar approximations such as weak-coupling conditions, and finally reduced to the corresponding first-order ones. The full generalization of the previous simplified formulation and results of CMT is thus achieved.

7.1.7.2 Full CMT of Fiber-Slab Coupling Systems

1. Generally asymmetric, composite fiber-slab couplers have been studied in depth using the full CMT, which generalizes previous first-order results. However, most of the features of light coupling and mode propagation are similar to those observed in Chapter 6. That is, most of the higher-order corrections are minor, indicating a general validity of the scalar approximations with the optical and structural parameters adopted.

2. Generalization of the first-order formulation to a full-term one has been made, including the coupled- and compound-mode equations, analytical expression of all the coupling coefficients, and the power conservation law. As a result, the full coupled- and compound-mode equations remain in the simple format of the first-order ones as described in Chapter 3 whilst all the first-order coupling coefficients are replaced by the corresponding full-term expressions.

3. Extensive numerical calculations have been carried out, in comparison with the previous first-order results, covering distributions of all the coupling coefficients, power conservation, coupling features, and compound-mode field. Higher-order effects are generally minor and they appear more evident in power beating cases ($\beta_{f0} > \beta_s$ generally) than in power decay cases ($\beta_{f0} \leq \beta_s$ generally).

4. The above full CMT may find applications in design, analysis, and optimization of coupler systems consisting of generally asymmetric and hybrid waveguides which are weakly guided but more closely coupled than the first-order cases.

7.2 Scalar CMT with Vectorial Corrections

7.2.1 Introduction

The previous section has laid down the foundations of both full and simplified versions of the scalar coupled-mode and compound-mode theory, applicable to waveguide coupler systems such as symmetric, asymmetric two-mode couplers and composite fiber-slab couplers. In this chapter, the analytically derived vector corrections are included as additional coupling coefficients in numerical calculations of scalar CMT of the composite fiber-slab waveguides. The slab waveguide is generally asymmetric, with moderate light-wave guiding and coupling assumed, and the polarization effects are to be addressed. It is found that the modifications to light propagation and coupling coefficients determined previously by first-order scalar CMT are dependent mainly on the strength of coupled fiber and slab modes, the thickness of the guiding layers, and the ratio of refractive indices of guiding to cladding layer (asymmetry of the slab waveguide). The vector corrections observed are not significant compared to the previous scalar CMT results, as the optical constants defined appear to satisfy the approximations necessary for the scalar analysis.

In this section we derive the analytical vector corrections to the scalar CMT of fiber-slab couplers in order to gain some insights into vector effects on light guiding and coupling. Comparisons with the previous results obtained in Chapter 6 from the simplified scalar CMT, as well as some new examples of the scalar CMT, are given so that the vector corrections due to polarization effects of the dissimilar waveguide couplers can be determined.

In Section 2.2, the equations of scalar CMT with vector corrections and the additional coupling coefficients due to the polarization effects are analytically derived for the composite fiber-slab waveguide. In Section 2.3, the vector corrections on the scalar coupling coefficients are investigated in detail. Finally, the results and discussion from the numerical calculations using analytical formulations are presented.

7.2.2 Formulations for Fiber-Slab Couplers

In Chapter 6, transverse electric (TE) modes have been assumed for the slab modes in analytical formulation and numerical calculations, owing to the inability of the pure scalar CMT to deal with the polarization effect (e.g., from the transverse magnetic, TM modes). To deliberately derive a formulation of CMT with vector correction for the polarization effect in the composite fiber-slab couplers, we first present the analytical expression of the fiber-slab coupled-mode equations.

7.2.2.1 Field Expression and Index Profile

Figure 6.1 of Chapter 6 shows the cross-section view of the composite fiber-slab coupler under investigation, and the key notations of structural and optical constants used in the analysis. The fiber and the asymmetric slab are assumed to be weakly guiding and in sufficiently close proximity to have some overlap between the evanescent fields of the two waveguides, whilst not strongly coupled. From CMT it is well known that power exchange takes places in a directional coupler only between TE-TE or TM-TM modes [6].

Following the notation and procedures given in Chapter 3, the total transverse electric field E_t in the coupled fiber-slab waveguides is represented by the superposition of the linearly polarized (i.e., x- or y-polarized) fiber and slab scalar modes (F_0, $\{S_n\}$) [7] and the scalar wave equation of E is replaced by the vector wave equations of E_t [8–10], leading to

$$E(x, y, z) = a_0(z)F_0(x, y) + \sum_{n}^{N} b_n(z)S_n(x, y) \tag{7.59}$$

which generally satisfies the following vector wave equation:

$$\nabla_t^2 E_t + \frac{\partial^2 E_t}{\partial z^2} + k^2 n^2 (x, y) E_t = -\nabla_t \left(E_t \cdot \nabla_t \ln n^2 \right) \tag{7.60}$$

$$\begin{cases} \nabla_t^2 E_x + \dfrac{\partial^2 E_x}{\partial z^2} + k^2 n^2 (x, y) E_x = -\dfrac{\partial}{\partial x}\left(E_x \dfrac{\partial \ln n^2}{\partial x} \right) - \dfrac{\partial}{\partial x}\left(E_y \dfrac{\partial \ln n^2}{\partial y} \right) \\[4mm] \nabla_t^2 E_y + \dfrac{\partial^2 E_y}{\partial z^2} + k^2 n^2 (x, y) E_y = -\dfrac{\partial}{\partial y}\left(E_x \dfrac{\partial \ln n^2}{\partial x} \right) - \dfrac{\partial}{\partial y}\left(E_y \dfrac{\partial \ln n^2}{\partial y} \right) \end{cases} \tag{7.61}$$

where $\nabla_t = \hat{x}(\partial/\partial x) + \hat{y}(\partial/\partial y)$, $\nabla_t^2 = (\partial^2/\partial x^2) + (\partial^2/\partial y^2)$ and $n(x, y)$ defines the refractive index profile throughout the transverse (x, y) plane such that

$$n^2 (x, y) = n_f^2 (x, y) + n_s^2 (x, y) - n_c^2 \tag{7.62}$$

where

$$S(x) = \begin{cases} 0 \ (x \le 0) \\ 1 \ (x > 0) \end{cases} \tag{7.63}$$

is a step function.

7.2.2.2 Coupled-Mode Equations

The scalar coupled-mode equations with vectorial corrections are obtained when the vector field expansion Equation 7.59 is substituted into the vector wave Equation 7.60, whilst the resultant equation is dot-multiplied successively with F_0 and S_n and integrated over the entire transverse (x, y) plane, utilizing the scalar wave equations and orthogonal relations for the scalar fiber and slab modes $\{F_0, S_n\}(n = 0, 1, 2, \ldots)$ [7]. As the scalar approximations are equivalent to the assumption of a slowly varying envelope [11] of the mode amplitudes and weakly coupled fiber-slab waveguides, the second-order derivatives and the field overlap integrals may be neglected [7] and the following simplified scalar coupled-mode and compound-mode equations are derived:

$$\begin{cases} a_0' = -i\left(\beta_{f0} + Q_{f00} + V_{f00}\right) a_0 - i \sum_n \left(K_{f0n} + V_{f0n} \right) b_n \\[4mm] b_m' = -i \sum_n \left(\delta_{mn}\beta_{sn} + Q_{smn} + V_{smn} \right) b_n - i \left(K_{sm0} + V_{sm0} \right) a_0 \end{cases} \tag{7.64}$$

$$\begin{cases} \beta A_0 = \left(\beta_{f0} + Q_{f00} + V_{f00}\right) A + \sum_n \left(K_{f0n} + V_{f0n} \right) B_n \\[4mm] \beta B_m = \sum_n \left(\delta_{mn}\beta_{sn} + Q_{smn} + V_{smn} \right) B_n + \left(K_{sm0} + V_{sm0} \right) A_0 \end{cases} \tag{7.65}$$

where $a_0' = \dfrac{da_0}{dz}$, $b_m' = \dfrac{db_m}{dz}$, $m, n = 0, 1, 2, \ldots$, δ_{mn} is the Kronecker delta function, and β_{f0}, $\{\beta_{sn}\}$, and β are the propagation constants of the LP_{01} mode of the fiber, the set of transverse modes of the slab, and the compound-mode of the fiber-slab system, respectively.

$\{Q_{f00}, Q_{smn}, K_{f0n}, K_{sm0}\}$ are the same coupling coefficients of the scalar CMT as they appear in Chapter 3, whilst $\{V_{f00}, V_{smn}, V_{fs0n}, V_{sm0}\}$ represent the new, additional coupling coefficients due to the vector corrections and are given by

$$V_{f00} = -\int_{A_\infty} (\nabla_t \cdot F_0 e_t)(F_0 e_t \cdot \nabla_t \ln n^2) dA \tag{7.66a}$$

$$V_{smn} = -\int_{A_\infty} (\nabla_t \cdot S_m e_t)(S_n e_t \cdot \nabla_t \ln n^2) dA \tag{7.66b}$$

$$V_{fs0n} = -\int_{A_\infty} (\nabla_t \cdot F_0 e_t)(S_n e_t \cdot \nabla_t \ln n^2) dA \tag{7.67a}$$

$$V_{sfm0} = -\int_{A_\infty} (\nabla_t \cdot S_m e_t)(F_0 e_t \cdot \nabla_t \ln n^2) dA \tag{7.67b}$$

In the above equations, $m, n = 0, 1, 2, \ldots, dA = dx \cdot dy$, and A_∞ indicates integration over the entire transverse (x, y) plane, i.e., the infinite cross-section of the fiber-slab waveguide system. $\{V_{f00}, V_{sm0}\}$ can represent the vector corrections to the self-coupling coefficients $\{Q_{f00}, Q_{smn}\}$, whilst $\{V_{fs0n}, V_{sfm0}\}$ represent the vector corrections to the cross-coupling coefficients $\{K_{f0n}, K_{sm0}\}$.

7.2.2.3 Vector-Correcting Coupling Coefficients

For y-polarized modes (e.g., TE in the slab), there is no vector correction due to the index discontinuity of the slab boundaries according to Equations 7.4 or 7.7, although there may exist some (minor) vector corrections due to the fiber. Indeed, we have

$$\frac{\partial}{\partial y} \ln n^2(x, y) = \frac{y}{\sqrt{x^2 + y^2}} \delta\left(\sqrt{x^2 + y^2} - a\right) \ln\left(\frac{n_c}{n_f}\right)^2 \tag{7.68}$$

where $\delta(x) = (dS(x)/dx)$ is the standard δ function, which indicates that the only possible vector corrections come from the index discontinuity of the fiber. If the fiber is truly weakly guiding [1], no major vector corrections are observed.

For x-polarized modes (e.g., TM in the slab), however, there may exist vectorial corrections to the scalar mode-coupling due to the index discontinuity at both the fiber and slab waveguide boundaries. This arises as seen from consideration

$$\frac{\partial}{\partial y} \ln n^2(x, y) = \frac{y}{\sqrt{x^2 + y^2}} \delta\left(\sqrt{x^2 + y^2} - a\right) \ln\left(\frac{n_c}{n_f}\right)^2$$

$$+ \delta(x - a - s) \ln\left(\frac{n_s}{n_c}\right)^2 + \delta(x - a - s - t) \ln\left(\frac{n_o}{n_s}\right)^2 \tag{7.69}$$

In this case, the additional coupling coefficients introduced as vector corrections can be simplified as follows:

$$V_{f00} = -\frac{1}{2\beta_{f0}} \int_{A\infty} \left(\frac{\partial F_0}{\partial x} \frac{\partial \ln n^2}{\partial x} \right) F_0 dA \tag{7.70a}$$

$$V_{smn} = -\frac{1}{2\beta_{sm}} \int_{A\infty} \left(\frac{\partial S_m}{\partial x} \frac{\partial \ln n^2}{\partial x} \right) S_n dA \tag{7.70b}$$

$$V_{fs0n} = -\frac{1}{2\beta_{f0}} \int_{A\infty} \left(\frac{\partial F_0}{\partial x} \frac{\partial \ln n^2}{\partial x} \right) S_n dA \tag{7.71a}$$

$$V_{sfm0} = -\frac{1}{2\beta_{sm}} \int_{A\infty} \left(\frac{\partial S_m}{\partial x} \frac{\partial \ln n^2}{\partial x} \right) F_0 dA \tag{7.71b}$$

For x-polarized modes, the additional coupling coefficients $\{V_{f00}, V_{smn}, V_{fs0n}, V_{sfm0}\}$ are solved exactly, except for V_{f00}, for which an asymptotic approximation for large arguments had to be made. The final results are expressed in a short series of exponential integrals, the closed-form solutions of which are presented in Appendix 5. Note again that the parameter $A_s = (n_0^2 - n_c^2)/(n_s^2 - n_c^2)$ represents an asymmetry factor of index profile of the slab waveguide.

7.2.3 Numerical Results and Discussion

Numerical results are firstly obtained using the parameters in Tables 6.1 and 6.2 of Chapter 6, the same as those given in the references, with the slab thickness $t = 3$ μm and the refractive index of the slab $n_s = 1.4745$. In this way, the dependence or otherwise of the polarization effects on the asymmetry factor or the optical constants of the waveguides can be comparatively observed, as can their influence on light propagation and coupling. Table 7.1 shows the values of the fiber parameters, i.e., n_f, $\Delta_f = (n_f^2 - n_c^2)/(2n_f^2)$, $V_f = ka\sqrt{n_f^2 - n_c^2}$, and β_{f0}; and Table 7.2 shows those of the parameters of the slab waveguide, i.e., n_0, $\Delta_{sc} = (n_s^2 - n_c^2)/(2n_s^2)$ $V_{so} = \frac{kt}{2}\sqrt{n_s^2 - n_c^2}$ and β_s. The slab mode may propagate on any direction along the (y, z) plane at an angle to the waveguide axis z. Other general default values are light wavelength $\lambda = 1.3$ μm, fiber radius $a = 2.5$ μm, refractive index of the fiber cladding $n_c = 1.46$, $D = 500$ μm, and the distance between the fiber and the slab $s = 0.5$ μm (see Figure 6.1 of Chapter 6).

The propagation constant of the LP_{01} mode, β_{f0}, was obtained as the solution of the eigenvalue equation for weakly guiding fibers and only the LP_{01} mode is launched initially into the fiber at $z = 0$, i.e., $a_0(0) = 1$ and $b_m(0) = 0$ ($m = 0, 1, 2,...$). For simplicity, only the z-dependent, fiber-mode, absolute-square magnitude $|a_0|^2$ is plotted.

7.2.3.1 Effects on Mode Coupling

Results of numerical calculations using Equation 7.64 and the above optical constants show that when $n_0 = 1.4–1.47$, no significant vector corrections or modifications to the previous

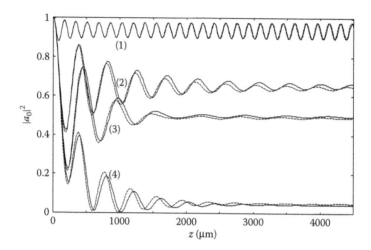

FIGURE 7.10

Propagation of $|a_0|^2(z)$. Solid lines: results with vector corrections; dashed lines: results from simplified scalar CMT. $n_0 = 1.00$. (1) $n_f = 1.4817$; (2) $n_f = 1.4756$; (3) $n_f = 1.4745$; (4) $n_f = 1.4709$.

results by scalar CMT are observed. The results of scalar CMT with vector corrections are in very good agreement with those of scalar CMT (see Chapter 6), as expected since the optical constants used satisfy the initial, scalar approximations.

However, if air ($n_0 = 1.0$) is used as the superstrate cladding of the slab waveguide, as is the case in some practical devices (such as those with the slab or thin-film guiding layer exposed to air), there are observable (but minor) vector corrections, as shown in Figure 7.10. These corrections add to the first-order results of some amplitude and phase modulations, which are dependent on the fiber-slab mode synchronism. As the refractive index of the fiber (n_f) core gets close to that of the slab (n_s), the coupling becomes stronger (see Chapter 6) and the vector corrections also become more evident. This may be explained by the increased level of discontinuity of the refractive indices at the outside boundary of the slab waveguide. Given a value of n_s, the lower the value of n_0, the greater the discontinuity at the boundary of n_s to n_0. Here $n_s = 1.4745$, the ratio of the refractive indices n_s/n_0 is increased to 1.4745 when $n_0 = 1.00$, whilst in the analysis of simplified scalar CMT (see Chapter 6) the ration $n_s/n_0 = 1.0532$ when $n_0 = 1.40$, $n_s/n_0 = 1.0099$ when $n_0 = 1.46$, and $n_s/n_0 = 1.0031$ when $n_0 = 1.47$.

7.2.3.2 Effect of Slab Thickness

The effect of the slab-guide thickness on the vectorial corrections can also be determined. Figure 7.11 illustrates a comparison made for the absolute square of the fiber-mode expansion coefficient, i.e., $|a_0|^2$ (z) when $t = 3.0$ μm and 2.0 μm.

For simplicity, the line groups (1) and (2) in Figure 7.11a–c are the same as the line groups (1) and (2), respectively, in Figure 7.10, where $t = 3.0$ μm. For the line group (1) in Figure 7.11, all the structural and optical constants remain the same except for the value of the slab thickness t, i.e., $t = 2.0$ μm, and the vectorial corrections now become more obvious, demonstrating that the stronger coupling between the fiber and the slab guided-modes leads to greater vectorial corrections. In other words, the lower the value of n_f (when close to n_s), the greater the fraction of power coupled from the fiber into the slab waveguide and the greater the magnitude

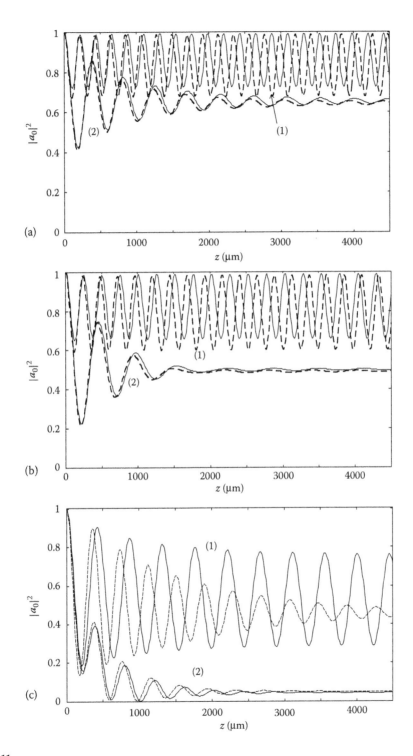

FIGURE 7.11
Propagation of $|a_0|^2(z)$. Solid lines: results with vector corrections; dashed lines: results from simplified scalar CMT. (a) $n_f = 1.4756$ or $n_f/n_s = 1.0008$; (b) $n_f = 1.4745$ or $n_f/n_s = 1.0$; (c) $n_f = 1.4709$ or $n_f/n_s = 0.9976$. (1) $t = 2$ μm; (2) $t = 3$ μm. $n_0 = 1.00$.

of the vectorial corrections. Similarly, the thinner the slab thickness, the greater the vectorial corrections. Thus vectorial corrections seem to affect the fiber-slab power couplings in two ways: by changing the amplitudes and oscillation periods of the modal powers. When $n_f \geq n_s$ and $n_f = n_s$, the amplitudes of the power coupling are steadily reduced with propagation. When $n_f < n_s$, the amplitude of the power oscillation is surprisingly well maintained in the z-propagation direction, as shown by the solid line of group (1) in Figure 7.11c, rather than damped, as shown by the dashed line of the same group in Figure 7.11c. The effect on the power-propagation phases appears to be consistent throughout Figure 7.11a–c: the periods of the mode-power oscillations are gradually increased with the light propagation along the z-axis and stronger coupling leading to larger phase shifts.

When $n_o = 1.40$ (with the ratio of n_0/n_c, or the parameter A_s defined in Section 6.4 of Chapter 6, representing the degree of asymmetry of the slab waveguide) and $n_s = 1.4745$ (with the ratio of $n_s/n_0 = 1.0532$ representing the strength of guidance), there are no noticeable vector corrections when $n_f > n_s$ and $n_f \gtrsim n_s$ (i.e., $n_f = 1.4817$ and 1.4756 respectively), but there appear to be some visible effects when $n_f = n_s$ (i.e., $n_f = 1.4745$) and even more so when $n_f < n_s$ ($n_f = 1.4709$), as shown in Figure 7.12. There are no major vector corrections in the cases where $n_0 = 1.46$ ($n_s/n_0 = 1.0099$) and $n_0 = 1.47$ ($n_s/n_0 = 1.0031$). The onset of the vector corrections can be seen as additional couplings appear to be dependent on the levels of the index discontinuity at the boundaries between the guiding cores and claddings (i.e., the ratio of n_f/n_s, n_s/n_c, and n_s/n_0), as given by Equation 7.69.

7.2.3.3 Effects on Coupling Coefficients

To examine the vectorial corrections in more detail, we have plotted the additional cross-coupling coefficients against those of scalar CMT (Chapter 6, Section 6.4.3). Figure 7.13 shows that the slab-to-fiber coupling coefficient $(K_{f0n} + V_{fs0n})$ is slightly higher in value than that of K_{f0n}, whilst Figure 7.14 reveals that there appear greater changes for fiber-to-slab coupling coefficient K_{sm0} (in the corresponding scale) due to vectorial corrections than for K_{f0n}. In fact, the stronger the coupling (e.g., the better the fiber and slab modes are synchronized), the greater the difference between the values of K_{sm0} and $(K_{sm0} + V_{sfm0})$. The comparison between the values of the fiber-mode self-coupling coefficient $(Q_{f00} + V_{f00})$ and

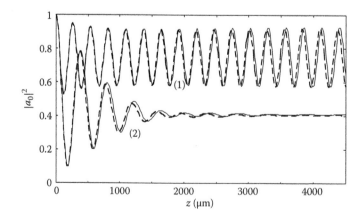

FIGURE 7.12
Propagation of $|a_0|^2(z)$. Solid lines: results with vector corrections; dashed lines: results from simplified scalar CMT. $n_0 = 1.40$; $t = 2$ μm. (1) $n_f = 1.4745$; (2) $n_f = 1.4709$.

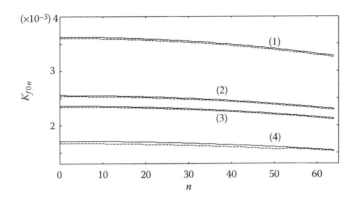

FIGURE 7.13
$(K_{f0n} + V_{fs0n})$ against K_{f0n}. $n_0 = 1.0$, $t = 2.0$ μm. Solid lines: vector-corrected; dashed lines: scalar CMT. (1) $n_f = 1.4817$; (2) $n_f = 1.4756$; (3) $n_f = 1.4745$; (4) $n_f = 1.4709$.

those of Q_{f00}, as well as between the slab-mode self-coupling coefficient $(Q_{smn} + V_{smn})$ and Q_{smn} were calculated but are not shown here. There are only very small vectorial corrections to Q_{smn} whilst Q_{f00} is a distinct number, rather than one-dimensional arrays such as $\{K_{f0n}, V_{fs0n}, K_{sm0}, V_{sfm0}\}$ or two-dimensional arrays such as $\{Q_{smn}, V_{smn}\}$, where $m, n = 0, 1, 2,...$ refer to the order of the transverse slab modes.

7.2.3.4 Effects on Compound Modes

The compound-mode Equation 7.65 are used to investigate the vectorial effects on the compound modes, especially the ridge modes (see Section 7.2).

Figure 7.15 illustrates typical comparisons of the propagation constants of the compound modes between the scalar CMT and the vector-corrected formulations when $n_0 = 1.0$ and $t = 2.0$ μm. Once again, only minor vectorial corrections are found for most of the compound modes. Interestingly, in the scale of Figure 7.15, the propagation constants of most of the compound modes form a smoothly curved baseline, whilst only those of the first and/or the second compound-mode are the exceptions. These first and second compound-modes

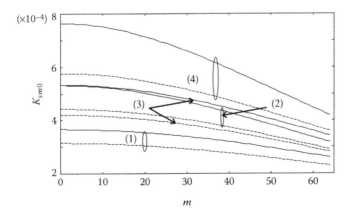

FIGURE 7.14
$(K_{sm0} + V_{sfm0})$ against K_{sm0}. $n_0 = 1.0$, $t = 2.0$ μm. Solid lines: vector-corrected; dashed lines: scalar CMT. (1) $n_f = 1.4817$; (2) $n_f = 1.4756$; (3) $n_f = 1.4745$; (4) $n_f = 1.4709$.

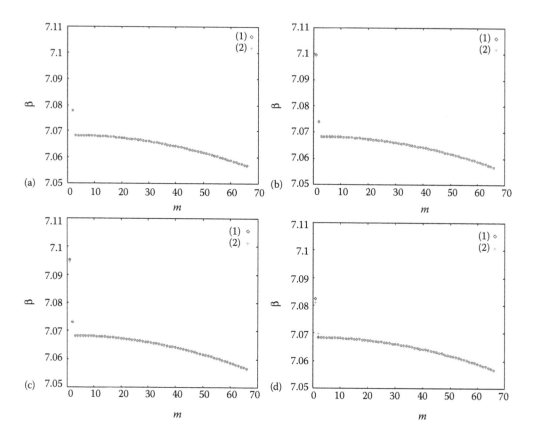

FIGURE 7.15

$\beta(m)$ distribution with the transverse slab mode-order m. $n_0 = 1.0$, $t = 2.0$ μm. (a) $n_f = 1.4817$; (b) $n_f = 1.4756$; (c) $n_f = 1.4745$; (d) $n_f = 1.4709$. (1) Results of the scalar CMT; (2) results with the vector corrections.

correspond to the ridge modes (see Section 6.7.2 of Chapter 6), which have propagation constants well above those of other higher-order, compound modes. This behavior may be explained by the observation that the gap between the propagation constants of the first and second ridge-modes, or the relative values of the compound-mode propagation constants in general, seem to be governed by the gap between those of the fiber- and slab-modes, or by the measure of nonsynchronization between the two individual guided-modes in general. For example, $n_f = 1.4756$, $n_s = 1.4745$, $\beta(1) \approx 7.0996$, $\beta(2) \approx 7.0739$, $\beta(3) \approx 7.0683,...$, whilst $\beta_{f0} \approx 7.0989$, $\beta_s \approx 7.0715$, $\beta_s(1) \approx 7.0683,...$; the propagation constants of the first and second ridge-modes are indeed quite close in value to those of the unperturbed fundamental guided-modes of the fiber and slab, respectively. In addition, the propagation constants of the higher compound-modes (i.e., $\beta(3)$, $\beta(4),...$, which form the baseline) are found to be very close in value to the axial components of propagation constants of those slab-modes propagating in a direction along the (y, z) plane at an angle to the waveguide axis z.

7.2.4 Concluding Remarks

The scalar CMT with vectorial corrections has been applied to fiber-slab couplers and the results compared to those of scalar CMT alone. As expected intuitively, it was found that the scalar CMT does produce reliable results concerning the light coupling of the composite structure. This is true even for the x-polarized field (i.e., TM in the slab), provided the index discontinuity is small and the coupling is moderate.

When the level of index discontinuity is greater (e.g., n_s/n_0 and/or n_f/n_c are increased) or the size of the guide cores (e.g., t) changes, some vectorial corrections may occur and the additional couplings from these corrections appear to depend on synchronism between the fiber and slab modes. In the case of a weakly guiding fiber coupled with a slab with $n_s/n_0 = 1.4745$ and if n_f is much greater than n_s, the light is well confined in the fiber and little correction is required because of the phase mismatch and only minor influence from the slab boundaries. However, as soon as n_f is close to, equal to, or lower than n_s (the fiber and slab modes are close to phase synchronism), the index discontinuity and the asymmetry of the slab become more important. The calculations of mode beating period, decay, and the amount of power transferred can be greatly changed by using vectorial corrections.

The scalar CMT with vectorial corrections is very useful in scalar CMT analysis of coupled waveguides whenever the level of index discontinuity of guided layers to the cladding layers is of concern, or polarization birefringence or coupling is of interest. As shown above, it can also be used to check on the validity of the scalar CMT to ensure the scalar approximation is met. Wherever necessary, vectorial corrections should be included to improve the accuracy of, and confidence in, the analysis and its associated assumptions.

7.3 Grating-Assisted Fiber-Slab Couplers

7.3.1 Introduction

This section presents a coupled-mode analysis of general, grating-assisted fiber-slab couplers. First, the analytical expressions of the scalar coupled-mode equations and the additional coupling coefficients are derived based on the scalar, simplified coupled-mode formulations presented in Chapter 3. Secondly, the analytical solutions of the additional (i.e., grating-assisted) coupling coefficients are obtained and numerical algorithms for the solutions of the new coupled-mode equations are described. Finally, the numerical calculations and discussion are carried out to reveal the grating-assisted coupling features.

7.3.2 Analytical Formulation

This section deals with the grating-assisted coupling in a general, asymmetric, and composite fiber-slab waveguide. For simplicity, scalar CMT will be formulated and the necessary additional coupling coefficients will be analytically derived. This is followed by a few numerical examples to demonstrate the effects of the grating structure on the mode (power) coupling between the guided fiber and slab modes.

7.3.2.1 Coupled-Mode Equations

A scalar formulation of CMT is adopted in this chapter. The scalar approach is validated by the previous results given in Chapter 6, Section 6.1 and references therein, and by the correctness of the scalar approximation, owing to the range of the structural and optical constants of the fiber-slab coupler system.

Scalar coupled-mode equations with grating-assisted couplings can be derived straightforwardly by using similar procedures of derivation to those outlined in previous chapters. The effect of the grating perturbation can be expressed as an axial perturbation on the index profile of the fiber-slab coupler:

$$n_g^2(x, y, z) = n^2(x, y) + \Delta n^2(x, y, z) \tag{7.72}$$

with the following condition as a small perturbation:

$$\Delta n(x, y, z) \ll n(x, y) \tag{7.73}$$

In the above expressions, n_g represents the index profile of the whole fiber-slab coupler with grating, whilst n represents that of the whole coupler without grating. Assuming a grating of rectangular shape, Figure 7.16 illustrates the boundary (solid line) of the grating and that (dashed line) of the unperturbed coupler without the grating.

Note that the different sections of the periodic grating structure are marked by the signs + and −, in which both the index perturbation $\Delta n^2(x, y)$ and the fields are different (Hong and Huang, 1995). In addition, Λ_+ and Λ_- are the corresponding periods of the grating, and $2d$ stands for the depth of the grating and K_g ($=2\pi/\lambda$) the equivalent wave number. Table 7.3 shows the changes of refractive indices within the grating structure, i.e., for areas defined by $a + s - d \leq x \leq a + s + d$:

After some algebra, the following scalar coupled-mode equations are obtained:

$$a'_{0\pm} = -j\left[\beta_{f00} + Q_{f00} + Q^{(g)}_{f00\pm}\right]a_0 - j\sum_n\left[K_{f0n} + K^{(g)}_{f0n\pm}\right]b_n \tag{7.74a}$$

$$b'_{m\pm} = -j\sum_n\left[\delta_{mn}\beta_{sn} + Q_{smn} + Q^{(g)}_{smn\pm}\right]b_{n\pm} - j\left[K_{sm0} + K^{(g)}_{sm0\pm}\right]a_{0\pm} \tag{7.74b}$$

where $\{Q^{(g)}_{f00\pm}, Q^{(g)}_{smn\pm}\}$ and $\{K^{(g)}_{f0n\pm}, K^{(g)}_{sm0\pm}\}$ are the additional self- and cross-coupling coefficients due to the presence of the grating perturbation and defined by, respectively,

$$Q^{(g)}_{f00\pm} = \frac{k^2}{2\beta_{f0}}\int_{A_\infty}\Delta n^2(x, y, z)F_{0\pm}^2\,dA \tag{7.75a}$$

$$Q^{(g)}_{smn\pm} = \frac{k^2}{2\beta_{sm}}\int_{A_\infty}\Delta n^2(x, y, z)S_{m\pm}S_{n\pm}\,dA \tag{7.75b}$$

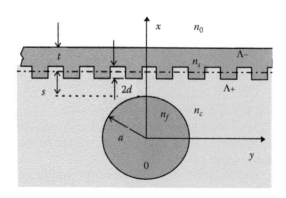

FIGURE 7.16
Schematic, cross-sectional view of the grating-assisted fiber-slab coupler.

TABLE 7.3

Change of Refractive Index and Electric Field in Different Grating Sections

$n_g(x, y, z), e(x, y, z)$	+ Sections without grating	− Sections without grating	+ Sections with grating	− Sections with grating
$a + s - d \leq x \leq a + s$	n_c, e_c	n_c, e_c	n_s, e_s	n_c, e_c
$a + s < x \leq a + s + d$	n_s, e_s	n_s, e_s	n_s, e_s	n_c, e_c

$$K^{(g)}_{f0n\pm} = \frac{k^2}{2\beta_{f0}} \int_{A_\infty} \Delta n^2(x, y, z) F_{0\pm} S_{n\pm} \, dA \tag{7.76a}$$

$$K^{(g)}_{sm0\pm} = \frac{k^2}{2\beta_{sm}} \int_{A_\infty} \Delta n^2(x, y, z) S_{m\pm} F_{0\pm} \, dA \tag{7.76b}$$

7.3.2.2 Additional Coupling Coefficients

Analytical solutions of the above-defined additional coupling coefficients are dependent on the expression of $\Delta n^2(x, y, z)$, i.e., the cross-sectional locations as well as the longitudinal profiles of the grating perturbation. The index modulation by the grating structure (Figure 7.16) can be approximately expressed as (e.g., [12])

$$\Delta n^2(x, y, z) \propto \cos(K_g z) \tag{7.77}$$

and, from Table 7.3, the expressions of the additional grating-induced coupling coefficients can be derived as follows:

$$Q^{(g)}_{f00\pm} = \frac{k^2}{2\beta_{f0}} \left(n_c^2 - n_s^2\right) \cos(K_g z) \int_{a+s-d}^{a+s+d} dx \int_{-\infty}^{\infty} dy F_0^2(x, y) \tag{7.78a}$$

$$Q^{(g)}_{smn\pm} = \frac{k^2}{2\beta_{sm}} \left(n_c^2 - n_s^2\right) \cos(K_g z) \int_{a+s-d}^{a+s+d} dx \int_{-\infty}^{\infty} dy S_{m\pm} S_{n\pm} \tag{7.78b}$$

$$K^{(g)}_{f0n\pm} = \frac{k^2}{2\beta_{f0}} \left(n_c^2 - n_s^2\right) \cos(K_g z) \int_{a+s-d}^{a+s+d} dx \int_{-\infty}^{\infty} dy F_0 S_{n\pm} \tag{7.79a}$$

$$K^{(g)}_{sm0\pm} = \frac{k^2}{2\beta_{sm}} \left(n_c^2 - n_s^2\right) \cos(K_g z) \int_{a+s-d}^{a+s+d} dx \int_{-\infty}^{\infty} dy S_{m\pm} F_0 \tag{7.79b}$$

The above integrals are all analytically solved and the closed-form expressions of these additional coupling coefficients can be found in Appendix 6. The effects of grating on the mode (power) coupling can now be investigated by numerical calculations using the new set of scalar coupled-mode equations given by Equation 7.74. Note, however, that the coupled-mode equations with the subindex + and − have to be used periodically in the grating sections Λ_+ and Λ_-, respectively, during the light propagation.

7.3.3 Numerical Results and Discussion

Numerical calculations are carried out in comparison with the results of the simplified scalar CMT for fiber-slab couplers without grating structure (see Chapter 6).

7.3.3.1 Effects on Mode Coupling

All the numerical calculations have been carried out using the practical conditions described in Chapter 6 (e.g., see Section 6.1), i.e., at a light wavelength $\lambda = 1.3$ μm, with fiber radius $a = 2.5$ μm, slab thickness $t = 3$ μm, refractive index of the fiber cladding $n_c = 1.46$, refractive index of the slab guide $n_s = 1.4745$, and the distance between the fiber and the slab $s = 0.5$ μm.

Numerical examples are first calculated when the refractive index of the fiber core n_f is varied against that of the slab guide n_s so that the modulation of the grating with different coupling conditions can be directly observed. The initial values for the grating are: (1) the grating depth $2d = 1.0$ μm; (2) the grating wave-number $K_g = |\beta_{f0} - \beta_{s0}|$.

Figure 7.17 demonstrates the interesting effects of grating on the coupling when the index of refraction $n_f = 1.4817$, 1.4756, and 1.4709, respectively, whilst the slab-guide refractive index $n_s = 1.4745$. The perturbation of the grating on the asymmetric fiber-slab guided-wave coupler obviously changes both the power-beating lengths and the coupling strength.

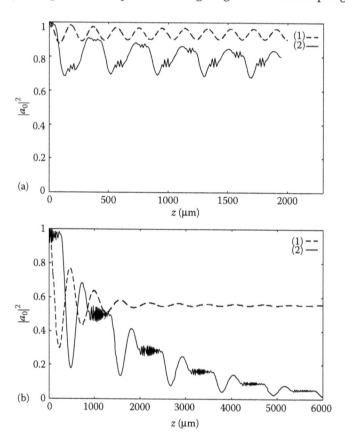

FIGURE 7.17
Diagram of $|a_0|^2$. (a) $n_f = 1.4817$; (b) $n_f = 1.4756$. $t = 3$ μm, $n_0 = 1.40$. (1) Scalar CMT without grating; (2) scalar CMT with grating.

Figure 7.17a shows that when $n_f > n_s$, the grating has approximately doubled the power-beating length whilst strengthening the coupling between the fiber and slab modes. Note that the periodic index modulation of the fiber-slab coupler structure also affects the power coupling, resulting in intermittent zigzag-shaped modulations on the mode-power $|a_0|^2$.

The effect of the grating-assisted coupling is more evident in the next example, when $n_f = 1.4756$, as shown in Figure 7.17b. According to the scalar CMT, the coupler without grating is not phase matched (although close) between the fiber and the slab modes and, as a result, more than 50% of the launched power remains in the fiber. However, for the coupler with a grating, a phase match is introduced via the value of K_g of the grating and the coupling is therefore assisted to achieve a complete power transfer from the fiber to the slab after a certain distance of propagation, in this case some 6000 μm (Figure 7.18).

Finally, when $n_f < n_s$, there is a phase match between the fiber mode and one of the transverse slab modes even in the absence of the grating structure. It can be seen that the additional phase match between the fiber mode and the lowest-order slab transverse slab mode does not "assist" the coupling much at all. Instead, it appears that the grating manifests its effect mainly through the phase modulation, i.e., altering the values of beat length, whilst slightly tuning the power decay in the fiber and the level of power transfer from the fiber to the slab.

7.3.3.2 Effects of Grating Parameters

The two grating parameters have been varied one at a time to see what effects they have on the power coupling.

7.3.3.2.1 Effects of Grating Period

Firstly, comparison is made when the depth of the grating is fixed whilst the period of the grating is set as $pd_1 = 2\pi / |\beta_{f0} - \beta_{s0}|$ and $pd_2 = 2\pi / |\beta_{f0} - \beta_{s10}|$, respectively. That is, the wave number of the grating, K_g, is set to compensate the phase mismatch between the fundamental fiber and transverse slab mode (i.e., $|\beta_{f0} - \beta_{s0}|$), and that between the fiber mode and the 11th slab transverse mode (i.e., $|\beta_{f0} - \beta_{s10}|$). This illustrates the effect of a small variation of the grating period.

It can be seen from Figure 7.18 that when $n_f > n_s$, line group (1) shows that the grating-assisted couplings have similar patterns of the power modulation, with about 10% less

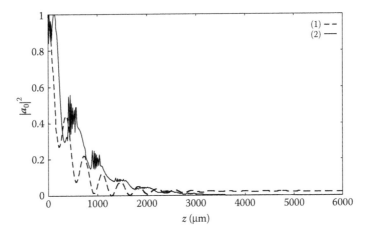

FIGURE 7.18
$n_f = 1.4709$, $t = 3$ μm, $n_o = 1.47$. (1) Scalar CMT without grating; (2) scalar CMT with grating.

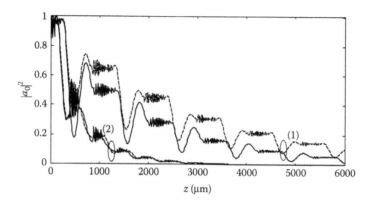

FIGURE 7.19

Effect of grating period. $d = 0.05$ μm, $n_0 = 1.40$. Solid lines: $K_g = |\beta_f - \beta_{s0}|$; dashed lines: $K_g = |\beta_f - \beta_{s10}|$. (1) $n_f = 1.4756 > n_s = 1.4745$; (2) $n_f = 1.4709 < n_s = 1.4745$.

power coupled from the fiber to the slab when the grating period $(2\pi/K_g)$ is decreased from pd_1 to pd_2. When $n_f < n_s$, there is already a phase match between the fundamental fiber mode and one of the transverse slab modes, and line group (2) reveals that the grating-introduced phase-compensation does not alter the coupling properties as much as it does in the case of line group (1) (Figure 7.19).

<u>$n_0 = 1.46$ and 1.47</u> No visible effects of the grating period on the power coupling can be observed in this case. That is, the diagrams of fiber mode-power $|a_0(z)|^2$ appear indistinguishable from those coupled waveguide systems with $n_0 = 1.46$ and 1.47.

Secondly, comparison is made when the depth of the grating is fixed whilst the period of the grating is set to $2\pi/|\beta_{f0} - \beta_{s0}|$ and 160 (μm) (much smaller than $2\pi/|\beta_{f0} - \beta_{s0}|$). This is to show the effect of a large variation of the grating period.

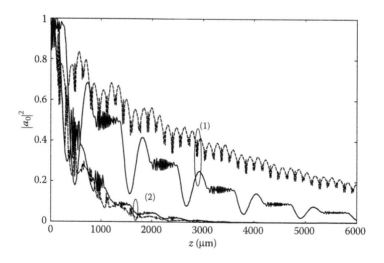

FIGURE 7.20

Effect of grating period on $|a_0(z)|^2$. $d = 0.05$ μm, $n_0 = 1.40$. Solid lines: $K_g = |\beta_f - \beta_{s0}|$; dashed lines: $K_g = 2\pi/160$. (1) $n_f = 1.4756 > n_s = 1.4745$; (2) $n_f = 1.4709 < n_s = 1.4745$.

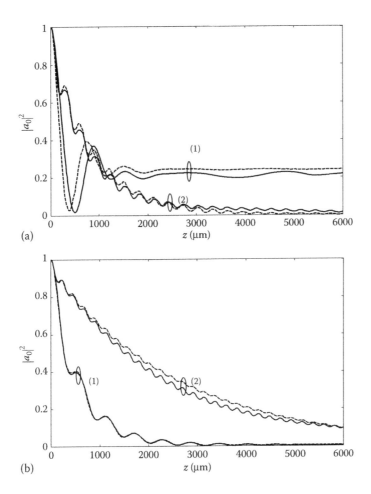

FIGURE 7.21
Effect of grating period on $|a_0(z)|^2$. $d = 0.05$ μm; (a) $n_0 = 1.46$; (b) $n_0 = 1.47$. Solid lines: $K_g = |\beta_f - \beta_{s0}|$; dashed lines: $K_g = 2\pi/160$. (1) $n_f = 1.4756 > n_s = 1.4745$; (2) $n_f = 1.4709 < n_s = 1.4745$.

$\underline{n_0 = 1.40}$ The modulation on power coupling due to the large variation of the grating period is clearly observed in Figure 7.20. The dashed lines indicate the reduction of both the grating period in modulations of the power coupling and the amount of power being transferred from the fiber to the slab. Similar to Figure 7.19, the grating effect appears stronger when there is no exact phase match for the coupler in the absence of the grating structure.

$\underline{n_0 = 1.46 \text{ and } 1.47}$ Both Figure 7.21a and b show that, in these cases, the effects of the grating-period variation are very small in comparison to the case when $n_0 = 1.40$. This may be explained by the fact that, in the absence of the grating structure, the coupler system is already in, or close to, a phase match between the fiber mode and the transverse slab modes. As a result, the natural coupling (i.e., coupling without the aid of the grating) becomes dominant and the mechanism of grating-assisted coupling is suppressed.

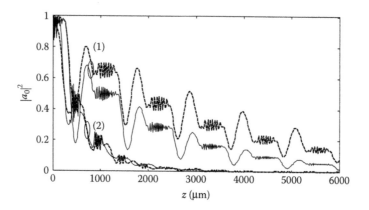

FIGURE 7.22

Effect of grating depth. $K_g = |\beta_f - \beta_{s0}|$. Solid lines: $d = 0.05\,\mu m$; dashed lines: $d = 0.025\,\mu m$. (1) $n_f = 1.4756 > n_s = 1.4745$; (2) $n_f = 1.4709 < n_s = 1.4745$.

7.3.3.2.2 Effects of Grating Depth

Comparison is made when the period of the grating is fixed, whilst the depth is varied from $d = 0.05$ μm to $d = 0.025$ μm. Figure 7.22 displays changes of power modulations either when $n_f > n_s$ (i.e., no fiber-slab phase synchronism) or $n_f < n_s$ (i.e., possible fiber-slab phase synchronism). Similar to Figure 7.19, it demonstrates that the effect of the grating is reduced when there is a phase match between the fiber and slab modes in the absence of the grating (see Figure 7.21, line group (2)).

7.3.4 Conclusions

Grating-assisted coupling in a waveguide coupler is analyzed utilizing the approved scalar CMT whilst treating the effect of grating as a periodic perturbation on the index profile of the whole coupler structure. Accordingly, the coupled-mode equations for fiber-slab coupler with grating structure can be obtained as follows:

The coupled-mode equations for the fiber-slab coupler with grating can be written in a similar way as those without grating, but with the coupler index profile replaced by the new one (overlaid by a periodic index modulation) in each of the analytical expressions of coupling coefficients defined in Chapter 6 (Equations 6.62 and 6.63). As a result, each of the new coupling coefficients will be those solved as in Chapter 3, plus an additional grating-assisted coupling coefficient solved as shown above.

The numerical calculations as a case study have shown interesting coupling features that are quite pertinent to the name of the coupler itself: i.e., 'grating-assisted' fiber-slab coupler. When the parameters of the coupler are such that there is no phase synchronism between the fiber and the slab modes, the grating comes into play and it assists coupling as seen in both the resultant amplitude and phase modulation. However, when there is a phase synchronism between the two sets of guided modes, the effects of the grating structure are greatly reduced, i.e., the grating in this case does not assist the coupling very much, if at all.

The above coupling features add to the fiber-slab coupler another unique dimension of amplitude and phase modulation, and thus increase the flexibility of system design to achieve the desired level of power transfer from the fiber to the slab waveguide. This may find practical applications in design or analysis of general asymmetric or hybrid wave-guide couplers.

7.4 Analysis of Nonlinear Waveguide Couplers

In this section the coupled-mode analysis of optical nonlinear (third-order) directional couplers (NLDC) is presented. Section 7.4.1 describes a first-order and generalized, full coupled-mode formulation in power parameters for both power-orthogonal and power-nonorthogonal two-mode NLDC, respectively, whilst Section 7.4.2 demonstrates the solutions of a nonlinear fiber-slab coupler, utilizing a simplified, scalar CMT formulation.

7.4.1 Nonlinear Two-Mode Couplers

This section deals with nonlinear light-wave coupling between two guided modes by con-sidering the third-order (Kerr-like) nonlinear refractive indices of the waveguide materials.

Previous, important nonlinear coupled-mode formulations have been introduced in Chapter 6 and Chapters 8 and 9. The solutions of first-order coupled-mode equations (CME) for identical, power-orthogonal two-mode NLDC with Kerr-like nonlinearity can be expressed in standard elliptic integrals, whereas those for nonidentical modes can-not, due to the mode asymmetry [13]. The full, nonlinear CME can be solved with the aid of the superposition of two compound-mode amplitudes, a linear combination of the linear waveguide-mode amplitudes. The power orthogonality can be enforced upon the formulation. The full CME can be later improved by [14] to replace the linear individual guided modes with the nonlinear ones. As a result, all the coupling coefficients in their CME become power dependent. However, after introducing the Stokes parameters in an attempt to analytically solve the CME, the power-nonorthogonality can be totally ignored because only the case with a negligible butt-coupling coefficient is considered.

The analysis is based on the power (i.e., guided-mode and cross-mode power) param-eters. As an example, they are first applied to the simple, power-orthogonal two-mode NLDC and the law of power conservation and redistribution derived. Then a generalized, full coupled-mode formulation is proposed for power-nonorthogonal two-mode NLDC and, in the power parameters, the analytical solutions of the NLDC obtained. In particu-lar, as demonstrated in Chapters 6 and 7, the formulations in the power parameters are self-contained in that the total power is conserved and the full, power-nonorthogonal for-mulation generalizes the previously published power-orthogonal results (e.g., for NLDC: [13,14]).

7.4.1.1 Power Parameters

In the analysis, the following four power parameters can be employed:

$$P_a(z) = a^*(z)a(z) \tag{7.80a}$$

$$P_b(z) = b^*(z)b(z) \tag{7.80b}$$

$$P_r(z) = \text{Re}\left[a^*(z)b(z)\right]$$ (7.80c)

$$P_i(z) = \text{Im}\left[a^*(z)b(z)\right]$$ (7.80d)

In comparison with the standard Stokes parameters, as defined by (Equation 2.88) in Chapter 2 [15], we have the following correspondence:

$$P_a \Leftrightarrow (s_0 + s_1)/2$$ (7.81a)

$$P_b \Leftrightarrow (s_0 - s_1)/2$$ (7.81b)

$$P_r \qquad s_2/2$$ (7.81c)

$$P_i \qquad s_3/2$$ (7.81d)

In the power parameters, the formulation of the mode-power distribution (i.e., $P_a(z)$ or $P_b(z)$) and power conservation etc. are more explicit than that in the Stokes parameters.

7.4.1.2 Simplified CMT

The scalar, first-order, coupled-mode equations for two generally nonidentical modes can be derived as

$$a' = -i\left(Q_a + Q_{aa}|a|^2 + 2Q_{ba}|b|^2\right)a - iK_{ab}b$$ (7.82a)

$$a' = -i\left(Q_b + Q_{bb}|b|^2 + 2Q_{ab}|a|^2\right)b - iK_{ba}a$$ (7.82b)

In power parameters, they become

$$P_a' = 2K_{ab}P_i$$ (7.83a)

$$P_b' = -2K_{ba}P_i$$ (7.83b)

$$P_r' = \left[Q_b - Q_a + (2Q_{ab} - Q_{aa})P_a - (2Q_{ba} - Q_{bb})P_b\right]P_i$$ (7.83c)

$$P_i' = K_{ab}P_b - K_{ba}P_a + \left[Q_a - Q_b + (2Q_{ba} - Q_{bb})P_b - (2Q_{ab} - Q_{aa})P_a\right]P_r$$ (7.83d)

If the law of power conservation is expressed as

$$P' = P_a' + P_b' + P_{ab}' = 0$$ (7.84)

we have, from Equation 7.83a and b:

$$P_{ab}' = 2(K_{ba} - K_{ab})P_i \qquad (7.85)$$

and the following constant of motion Γ:

$$\Gamma = K_{ba}P_a + K_{ab}P_b \qquad (7.86)$$

Note that the linear limit of the nonlinear coupler of two nonidentical modes, with the nonlinear perturbation on the index profile and thus additional coupling coefficients, is included. Similarly, Figure 7.23 displays a simple relation between P_a and P_b.

7.4.1.3 Generalized Full CMT

In this section, a generalized, full scalar CMT is proposed for symmetric two-mode NLDC with Kerr nonlinearity. Firstly, the coupled-mode equations are derived, with all the coupling coefficients defined, and reformulated in the power parameters. Then the constants of motion (i.e., the z-invariants along the z-axis) are derived, including the power conservation law, and analytical solutions are attempted and described.

7.4.1.3.1 Coupled-Mode Equations

Without losing generality, the new, power-nonorthogonal coupled-mode equations are derived for a symmetric two-mode NLDC consisting of two asymmetric slab waveguides. Application of the resultant formulation to other symmetric, nonlinear two-mode coupler configurations are straightforward.

Figure 7.24a shows the cross-sectional view of a third-order, symmetric, two-mode NLDC, whilst Figure 7.24b illustrates the constituent asymmetric waveguides a and b, respectively. The parameter t represents the thickness of the slab guides, s the distance between the two slab guides, and n_1 and n_2 the refractive indices of the outside claddings and slab guides, respectively. The inside cladding with a refractive index of $n_3 + n_{3L}I$ is optical nonlinear, where n_3 is the linear component (as n_1 and n_2) and $n_{3L}I$ is the nonlinear component with a third-order nonlinear coefficient n_{3L} and local light intensity of I.

The influence of the nonlinearity of the waveguide materials on the coupled modes in a waveguide coupler can be described by the nonlinear polarization (e.g., [14]) related to

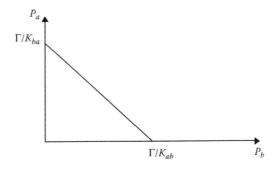

FIGURE 7.23
Illustration of relation between $P_a(z) = |a(z)|^2$ and $P_b(z) = |b(z)|^2$.

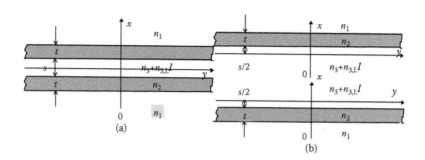

FIGURE 7.24

A cross-sectional schematic of the third-order nonlinear two-mode couplers: (a) the coupler; (b) the constituent waveguides.

the index change or, equivalently, by a perturbation to the refractive-index profiles. The perturbed index profiles of the two-mode coupler, waveguide a and b, i.e., $n(x)$, $n_a(x)$, and $n_b(x)$, can be expressed as a superposition of a linear (low power, with a subindex of L) and Kerr-like nonlinear (high power, with a subindex of $3NL$) refractive indices, respectively:

$$n = n_L + n_{3NL,I} \tag{7.87a}$$

$$n_\nu = n_{L\nu} + n_{3NL,I\nu} \; (\nu = a, b) \tag{7.87b}$$

$$n_L = \begin{cases} n_1 \\ n_3 \\ n_2 \end{cases} \text{and} \quad n_{3NL,I} = \begin{cases} 0 & \text{for} & |x| > s+t \\ n_{3,I}I & \text{for} & s \le |x| \le s+t \\ 0 & \text{for} & |x| \le s \end{cases} \tag{7.88a}$$

$$n_{La} = \begin{cases} n_1 \\ n_2 \\ n_3 \end{cases} \text{and} \quad n_{3NL,Ia} = \begin{cases} 0 & \text{for} & x > s+t \\ 0 & \text{for} & s \le x \le s+t \\ n_{3,I_a}I_a & \text{for} & x < s \end{cases} \tag{7.88b}$$

$$n_{Lb} = \begin{cases} n_3 \\ n_2 \\ n_1 \end{cases} \text{and} \quad n_{3NL,Ib} = \begin{cases} n_{3,I_b}I_b & \text{for} & x > -s \\ 0 & \text{for} & -s-t \le x \le -s \\ 0 & \text{for} & x < s \end{cases} \tag{7.88c}$$

where I is the local light intensity in W/m² and $\{n_{3,I\mu}\}$ ($\mu = I, I_a, I_b$) is a set of third-order non-linear coefficients of the coupler system. Alternatively, in terms of the electric fields, we have

$$n^2 = n_L^2 + n_{3NL,E} \tag{7.89a}$$

$$n_\nu^2 = n_{L\nu}^2 + n_{3NL,E\nu} \tag{7.89b}$$

$$n_{3NL,E} = \begin{cases} 0 & \text{for} & |x| > s+t \\ n_{3,E}\,|E|^2 & \text{for} & s \le |x| \le s+t \\ 0 & \text{for} & |x| \le s \end{cases} \tag{7.90a}$$

$$n_{3NL,Ea} = \begin{cases} 0 & \text{for} \quad x > s+t \\ 0 & \text{for} \quad s \leq x \leq s+t \\ n_{3,E_a} \mid E_a \mid^2 & \text{for} \quad x < s \end{cases} \tag{7.90b}$$

$$n_{3NL,Eb} = \begin{cases} n_{3,E_b} \mid E_b \mid^2 & \text{for} \quad x > -s \\ 0 & \text{for} \quad -s-t \leq x \leq -s \\ 0 & \text{for} \quad x < s \end{cases} \tag{7.90c}$$

where the nonlinear coefficients $n_{3,E}$ and $n_{3,Ei}$ ($i = 1, 2$) are related to $n_{3,I}$ and $n_{3,Iv}$ respectively, by

$$n_{3,E} = c\varepsilon_0 n_L^2 \dot{n}_{3,I} \tag{7.91a}$$

$$n_{3,En} = c\varepsilon_0 n_{Lv}^2 n_{3,Iv} \quad (v = a, b) \tag{7.91b}$$

Assuming TE wave propagation and utilizing the vector reciprocity principle (derived from the Maxwell equations, see Appendix 2), the following full, Kerr-like nonlinear and power-conserved coupled-mode equations are derived:

$$a'(z) + P_{ab}b'(z) = -jQ_{az}a(z) - jK_zb(z) \tag{7.92a}$$

$$b'(z) + P_{ab}a'(z) = -jQ_{bz}b(z) - jK_za(z) \tag{7.92b}$$

with the prime sign ′ representing the first-order derivative with respect to z, and Q_{1z}, Q_{2z} and K_z equivalent to

$$Q_{az} = Q + (K_c - K_s)|b(z)|^2 + K_t\left[a(z)b(z)^* + a(z)^* b(z)\right] \tag{7.93a}$$

$$Q_{bz} = Q + (K_c - K_s)|a(z)|^2 + K_t\left[a(z)b(z)^* + a(z)^* b(z)\right] \tag{7.93b}$$

$$K_z = K + K_c\left[a(z)b(z)^* + a(z)^* a(z)\right] \tag{7.93c}$$

where Q, K, and P_{ab} are the new power-dependent self-, cross-, and butt-coupling coefficients, whilst K_s, K_c, and K_t are the self-, cross-, phase-modulation, and nonlinear coupling coefficients, as defined in the following:

$$Q = \frac{\varepsilon_0\omega}{4P_0}\int_{A_\infty}\left(n^2 - n_v^2\right)|\underline{E}_v|^2 \, dA \, (v = a \text{ or } b) \tag{7.94a}$$

$$K = \frac{\varepsilon_0\omega}{4P_0}\int_{A_\infty}(n^2 - n_\mu^2)\underline{E}_\mu\underline{E}_v^* \, dA \, (\mu, \eta = a, b \text{ but } \mu \neq v) \tag{7.94b}$$

$$K_s = \frac{\varepsilon_0\omega}{4P_0}\int_{A_\infty} n_{3NL,E}|\underline{E}_n|^4 \, dA \quad (v = a \text{ or } b) \tag{7.95a}$$

$$K_c = \frac{\varepsilon_0 \omega}{4P_0} \int_{A_\infty} n_{3NL,E} \, |\mathbf{E}_\mu|^2 |\mathbf{E}_v|^2 \, dA \qquad (\mu, v = a, b \text{ but } \mu \neq v) \tag{7.95b}$$

$$K_t = \frac{\varepsilon_0 \omega}{4P_0} \int_{A_\infty} n_{3NL,E} \, |\mathbf{E}_\mu|^2 \, \underset{\sim}{\mathbf{E}}_\mu \underset{\sim}{\mathbf{E}}_v{}^* \, dA \qquad (\mu, v = a, b \text{ but } \mu \neq v) \tag{7.95c}$$

$$P_{ab} = \frac{1}{4} P_0 \int_{A_\infty} \left(\underset{\sim}{\mathbf{E}}_\mu \times \underset{\sim}{\mathbf{H}}_v{}^* + \underset{\sim}{\mathbf{E}}_v{}^* \times \underset{\sim}{\mathbf{H}}_\mu \right) . \hat{z} . dA \qquad (\mu, v = a, b \text{ but } \mu \neq v) \tag{7.96a}$$

$$P_0 = \frac{1}{4} \int_{A_\infty} \left(\underset{\sim}{\mathbf{E}}_v \times \underset{\sim}{\mathbf{H}}_v{}^* + \underset{\sim}{\mathbf{E}}_v{}^* \times \underset{\sim}{\mathbf{H}}_v \right) . \hat{z} dA \qquad (v = a \text{ or } b) \tag{7.96b}$$

where P_0 is the power of the vth guided-mode and $\hat{z}$ is the unit vector along the z axis.

Alternatively, Equation 7.92 can be transformed into the following set of differential equations:

$$a'(z) = -j \left(Q_{az} - P_{ab} K_z \right) a(z) - j \left(K_z - P_{ab} Q_{bz} \right) b(z) \tag{7.97a}$$

$$b'(z) = -j \left(Q_{bz} - P_{ab} K_z \right) b(z) - j \left(K_z - P_{ab} Q_{az} \right) a(z) \tag{7.97b}$$

where the underline sign "_" represents an operation of having the variable or constant divided by $(1-P_{ab})$. For example, we define

$$\underline{A} \equiv A / (1 - P_{ab}) \tag{7.98}$$

With the above-defined coupling coefficients, the coupled-mode Equation 7.97a and b can be reformulated in power parameters as follows:

$$P_a' = \left(Q_{ba} + 2K_{ct} P_r + P_{ab} K_{sc} P_a \right) P_i \tag{7.99a}$$

$$P_b' = -\left(Q_{ba} + 2K_{ct} P_r + P_{ab} K_{sc} P_b \right) P_i \tag{7.99b}$$

$$P_r' = -K_{sc} \left(P_a - P_b \right) P_i \tag{7.99c}$$

$$P_i' = \left(-Q_{ba} + 2K_{st} P_r \right) \left(P_a - P_b \right) \tag{7.99d}$$

where the following constants are defined in relation to the above coupling coefficients:

$$K_{sc} = K_s - K_c \tag{7.100a}$$

$$Q_{ba} = Q_{bz} - P_{ab} Q_{az} \tag{7.100b}$$

$$K_{ct} = K_c - P_{ab} K_t \tag{7.100c}$$

$$K_{st} = K_{sc} - 2K_{ct} = K_s - 3K_c + 2P_{ab} K_t \tag{7.100d}$$

7.4.1.3.2 Constants of Motion

From the above reformulated coupled-mode Equations 7.99a–d, a few constants of motion can be readily derived. They include, as was demonstrated in Chapters 3 and 4 for the linear two-mode couplers, the law of power conservation (in the new formulation), and other equations in the power parameters, which are very helpful in achieving possible analytical solutions to the coupled-mode equations.

7.4.1.3.2.1 Power Conservation   From Equations 7.99a–c, the following equation in relation to the conservation of the total guided power can be readily derived:

$$P_a' + P_b' = -2P_{ab}P_r' \tag{7.101a}$$

$$\Gamma_1 = P_a + P_b + 2P_{ab}P_r = P \tag{7.101b}$$

where Γ_1 is a constant of motion and P represents the total power guided along the waveguide (coupler) axis. In fact, in comparison with the standard definition of the total guided power in the Poynting vectors, it can be found that P_a and P_b represent the guided power in the modes a and b, respectively, whilst P_r represents the cross-power term that accounts for the mode (power) nonorthogonality of the coupler system.

7.4.1.3.2.2 Other Constants of Motions   From Equation 7.99, the following constants of motion can also be found:

$$\Gamma_2 = (P_a - P_b)^2 - \left[1 - 4(2Q_{21}/K_{sc} + P_{ab})P_r - 4(2K_{ct}/K_{sc} - P_{ab}^2)P_r^2\right] \tag{7.102a}$$

$$\Gamma_3 = P_i^2 - \left[2(Q_{ba}/K_{sc})P_r - (K_{st}/K_{sc})P_r^2\right] \tag{7.102b}$$

This is one of the advantages of the reformulated coupled-mode Equations 7.99a–d in power parameters, compared to those formulated in modal amplitudes Equations 7.92a–b or 7.97a–b. Firstly, these constants of motion are z-invariant, namely, they remain constant during the light propagation and coupling in the coupler system. In other words, during the operation of the coupler, these special constants of motion, which are functions of the power parameters, may completely reveal and describe the operation conditions of the whole coupler system. Secondly, these constants of motion can be determined by the initial launching conditions of the coupler and can therefore be used to search for the trajectories of the motion or to conduct stability analysis [16]. Finally, these special relations among the power parameters are very useful in the search for possible analytical solutions in terms of the mode or cross-mode powers.

7.4.1.3.3 Analytical Solutions

In this section, an analytical solution is presented for the generalized and optical nonlinear coupled-mode Equations 7.99a–d and the results are expressed in terms of the power parameters.

7.4.1.3.3.1 Initial Conditions   Without losing generality, the practical one-guide launching conditions of a two-mode coupler are used in the following analysis, namely

$$a(0) = 1 \text{ and } b(0) = 0 \tag{7.103}$$

Then the initial conditions for the power parameters become, by their definition Equation 7.80:

$$P_a(0) = 1 \tag{7.104a}$$

$$P_b(0) = P_r(0) = P_i(0) = 0 \tag{7.104b}$$

and, utilizing the coupled-mode Equations 7.99a–d, we arrive at

$$P_a'(0) = P_b'(0) = P_r'(0) = 0 \tag{7.105a}$$

$$P_i'(0) = -Q_{ba} \tag{7.105b}$$

where Q_{ba} is defined in Equation 7.100b.

7.4.1.3.3.2 Analytical Solutions From Equations 7.99 to 7.102, the following relations can be derived:

$$(P_a - P_b) = \left[-(\eta/\zeta)(2P_{r-} - \alpha_1)(2P_r - \alpha_2) \right]^{1/2} \tag{7.106a}$$

$$P_i = \left[\gamma P_r (\beta_2 - 2P_r) \right]^{1/2} \tag{7.106b}$$

in which the following constants are used:

$$\zeta = \left(4Q_{ba}/K_{sc} + 2P_{ab} \right)^{-1} \tag{7.107a}$$

$$\eta = \left(2K_{ct}/K_{sc} - P_{ab}^2 \right)\left(4Q_{ba}/K_{sc} + 2P_{ab} \right)^{-1} \tag{7.107b}$$

$$\alpha_1 = -\left[1 + (1 + 4\zeta\eta)^{1/2} \right]/(2\eta) \cong -1/\eta \tag{7.107c}$$

$$\alpha_2 = -\left[1 - (1 + 4\zeta\eta)^{1/2} \right]/(2\eta) \cong \zeta \tag{7.107d}$$

$$\beta_2 = 4Q_{ba}/K_{st} = \left\{ 1 - P_{ab}\left[2(\zeta - \eta) + P_{ab} \right]/\gamma \right\}/(\zeta - \eta) \cong (1 - 2P_{ab}\zeta)/(\zeta - \eta) \tag{7.107e}$$

$$\gamma = K_{st}/K_{sc} = (\zeta - \eta)/\zeta + P_{ab}^2 \cong (\zeta - \eta)/\zeta \tag{7.107f}$$

where, in Equations 7.107c–f, the condition of $P_{ab}^2 \ll 1$ is used, which is equal to assuming that the two modes are not strongly coupled.

From definitions of Equations 7.94 and 7.95, $K_c \ll K$ and $K_c \ll K_s$ are generally valid, which leads to the following approximations:

$$|\zeta| \gg |\eta|, \; |\eta| \ll 1 \text{ and } \zeta\eta \ll 1 \tag{7.108a}$$

$$|\alpha_1| \gg |\alpha_2| \text{ and } |\alpha_1| \gg |\beta_2| \tag{7.108b}$$

Note that if the effects of P_{ab} are totally neglected, the above full coupled-mode equations, coupling coefficients, and constants, i.e., those of Equations 7.92 through 7.107 containing the underlined constants, are reduced to those defined by [14]. For example, when $P_{ab} \ll 1$, Equations 7.107a–e, i.e., $\underline{\zeta} \to \zeta$, $\underline{\eta} \to \eta$, $\underline{\alpha}_1 \to \alpha_1$, $\underline{\alpha}_2 \to \alpha_2$, $\underline{\beta}_2 \to \beta_2$, respectively. Therefore, the formulation generalizes the previous ones described by [13] and [14].

From Equations 7.99c, 7.96a and b, we have

$$2P_r' = K_{sc} \left[(\eta\gamma/\zeta) 2P_r \left(2P_{r-} - \alpha_1 \right) \left(2P_r - \alpha_2 \right) \left(2P_r - \beta_2 \right) \right]^{1/2} \tag{7.109a}$$

$$\int_0^{2P_r(z)} \left[2P_r \left(2P_{r-} - \alpha_1 \right) \left(2P_r - \alpha_2 \right) \left(2P_r - \beta_2 \right) \right]^{-1/2} \delta(2P_r) = K_{sc} \left(\eta\gamma/\zeta \right)^{1/2} z \tag{7.109b}$$

Utilizing the last equation of Equation 7.107f, we have

$$K_{sc} \left(\eta\gamma/\zeta \right)^{1/2} z \cong 4Q_{ba} \left[\eta(\zeta - \eta) \right]^{1/2} / \left(1 - 2P_{ab}\zeta \right) \tag{7.110}$$

Without the loss of generality, self-focusing Kerr nonlinearity is assumed, that is

$$n_{3,E} \ngtr 0, \ \alpha_1 < 0, \ \alpha_2 \ngtr 0, \ \text{and} \ n_{3,E} \ngtr 0, \ \beta_2 > 0 \tag{7.111}$$

where $n_{3,E}$ is the Kerr nonlinear coefficient defined by Equation 7.91, and α_1, α_2, and β_2 are defined by Equations 7.107c–e, respectively.

The solution of the elliptic integral in Equation 7.109 can be first expressed in the odd and even Jacobian elliptic functions of sn and cn (see Appendix 7; [14]), respectively, and, utilizing the approximations of Equation 7.108, it leads to

$$P_r = \begin{cases} \dfrac{P}{2} \left[1 + \mathrm{cn}\left(\kappa_+ z \mid m \right) \right] & m \le 1 \quad (\alpha_2 \le \beta_2) \\[3mm] \dfrac{P}{2} \left[1 + \mathrm{dn}\left(\kappa_- z \mid m^{-1} \right) \right] & m > 1 \quad (\alpha_2 > \beta_2) \end{cases} \tag{7.112}$$

where $\{\kappa_+, \kappa_-\}$ and m are constants, and the modulus of the elliptic functions $\{\mathrm{cn}, \mathrm{dn}\}$, respectively, are defined as follows:

$$\kappa_+ = 2Q_{ba} \left(1 + 4\zeta\eta \right)^{1/4} / \left(1 - 2P_{ab}\zeta \right) \tag{7.113a}$$

$$\kappa_+ = 2Q_{ba} \left[\zeta(\zeta - \eta) \right]^{1/2} / \left(1 - 2P_{ab}\zeta \right) \tag{7.113b}$$

$$m = \alpha_2 \left(\beta_2 - \alpha_1 \right) / \left[\beta_2 \left(\alpha_2 - \alpha_1 \right) \right] \cong \alpha_2/\beta_2 \cong \zeta(\zeta - \eta) / \left(1 - 2P_{ab}\zeta \right) \tag{7.114}$$

For the last two equations of Equation 7.114, the conditions of Equation 7.110 and $P_{ab}^2 \ll 1$ are used. From the periodic properties of the elliptic functions of cn and dn, the half-beat length of the nonlinear coupler are given by

$$L_c = \begin{cases} \left(1 - 2P_{ab}\zeta \right) k(m) \Big/ \left[Q_2 \left(4\underline{\zeta}\underline{\eta} + 1 \right)^{1/4} \right] & m \le 1 \quad (\alpha_2 \le \beta_2) \\[3mm] \left(1 - 2P_{ab}\zeta \right) k\left(m^{-1} \right) \Big/ \left[2Q_2 \left(\underline{\zeta}(\underline{\zeta} - \underline{\eta}) \right)^{1/2} \right] & m \le 1 \quad (\alpha_2 > \beta_2) \end{cases} \tag{7.115}$$

Clearly, the analytical solutions Equations 7.112 and 7.115 generalize those of [14], in that the field-overlap effects (i.e., butt-coupling coefficients and power-nonorthogonality) are accounted for in the coupled-mode equations and solutions without more complicated formulation. The special case of weak coupling with negligible field-overlap integrals can be derived from the formulation and solution by applying the approximation $P_{ab} \ll 1$ and the above results are reduced to those of [14].

7.4.2 Nonlinear Fiber-Slab Couplers

The reformulation technique in the power parameters demonstrated for the optical two-mode coupler systems is clearly not suitable for the optical composite fiber-slab guided-wave couplers. This is because, as described in Chapter 3, the fiber-slab mode-coupling involves multimodes and, therefore, an analytical solution to a set of a large number of differential equations seems intractable. In this section, weak guidance, and weak linear and nonlinear couplings are assumed. The effect of the nonlinear coupling is expressed through the index modulation (perturbation) and thus the induced additional nonlinear coupling coefficients based on the linear, coupled fiber-slab guided modes.

In Section 7.4.3.1, a set of scalar, first-order, coupled-mode equations is derived for a typical composite fiber-slab coupler configuration. The additional coupling coefficients are derived analytically in Section 7.4.3.2. Finally, a numerical example is given in Section 7.4.3.3 to demonstrate the Kerr-like nonlinear effects on the power coupling.

7.4.2.1 Simplified Scalar CMT

As with the treatment of the grating-assisted fiber-slab coupler in Chapter 6, the fiber-slab coupler with third-order nonlinear index distribution can be considered as a linear coupler with a nonlinear perturbation on its index profile. The fiber-slab coupler configuration is the same as shown in Chapter 6. The refractive indices of the coupler $n(x, y)$ (i.e., over the whole coupler cross-section A_∞) and of its different cross-sections $\{nv(x, y)\}$ (i.e., over the cross-sectional areas of A_o, A_s, A_c and A_f, with $A_\infty \equiv A_o + A_s + A_c + A_f$) can be expressed as

$$n^2 = n_L{}^2 + n_2 |E|^2 \; (x, y \in A_\infty) \tag{7.116a}$$

$$n_v{}^2 = n_{Lv}{}^2 + n_{2v} |E_v|^2 \; (x, y \in A_v) \tag{7.116b}$$

$$n_2 \equiv \{n_{2o}, n_{2s}, n_{2c}, n_{2f}\} (x, y \in A_\infty) \tag{7.117}$$

where $v = \{o, s, c f\}$ represents the overlay of the slab (A_o), the slab (A_s), the fiber cladding (A_c), and core (A_f), and the first and second terms correspond to the linear and nonlinear index profiles, respectively. $\{E, n_2\}$ and $\{E_v, n_{2v}\}$ represent the electric field and the third-order refractive coefficient of the coupler and each cross-section, respectively.

Under scalar approximations, i.e., assuming weak guidance and coupling of the fiber and slab guided-modes, the following simplified, first-order, coupled-mode equations can be derived using the procedures as described in Section 6.4.1 and the above nonlinear index perturbation (7.116):

$$a_0' = -i\left(\beta_{f0} + Q_{f00} + I_{f0}|a_0|^2 + \sum_{m,n}^{N} I_{fsmn}b_m b_n^* \right)a_0 - i\sum_n^N K_{f0n}b_n \tag{7.118a}$$

$$b'_m = -i \sum_n \left(\delta_{mn}\beta_{sn} + Q_{smn} + \sum_{p,l} I_{mnpl} b_p b_l^* + I_{sfmn} |a_0|^2 \right) b_n - iK_{smo}a_0 \qquad (7.118b)$$

where a_0 and $\{b_m\}$ are the fiber and slab modal amplitudes in the total-field expansion, the prime sign' represents the first derivative with respect to the propagation distance z, $\{\beta_{f0}, Q_{f00}, K_{f0n}\}$ and $\{\beta_{sn}, Q_{smn}, K_{smo}\}$ are the propagation constants, self- and cross-coupling coefficients of the fiber and slab, respectively, whilst $\{I_{f0}, I_{mnpl}\}$, $\{I_{fsmn}, I_{sfmn}\}$ are the additional fiber and slab coupling coefficients corresponding to the self- and cross-phase modulations, respectively, defined as follows:

$$I_{f0} = \frac{k^2}{2\beta_{f0}} \int_{A_\infty} n_2 F_0^4 dA \qquad (7.119a)$$

$$I_{mnpl} = \frac{k^2}{2\beta_s} \int_{A_\infty} n_2 S_m S_n S_p S_l dA \qquad (7.119b)$$

$$I_{fsmn} = \frac{k^2}{2\beta_{f0}} \int_{A_\infty} n_2 F_0^2 S_m S_n dA \qquad (7.120a)$$

$$I_{sfmn} = \frac{k^2}{2\beta_s} \int_{A_\infty} n_2 F_0^2 S_m S_n dA \qquad (7.120b)$$

where $k = 2\pi/\lambda$ is the free-space wave number, n_2 is the nonlinear coefficient as defined in Equation 7.116, and F_0 and $\{S_n\}$ are the fundamental fiber mode and the slab mode of transverse order n, respectively. Note that n_2 is generally a cross-sectional distribution of nonlinear coefficients of the cores and claddings, i.e., $n_2 = \{n_{2o}, n_{2s}, n_{2c}, n_{2f}\}$, and, therefore, it should remain as a part of the integrand as in Eaquations 7.107 and 7.108, rather than be taken out of the integral.

In Equation 7.118, the third and fourth terms in the right-hand side are the dominant additional nonlinear terms representing the self- and cross-phase modulation effects, respectively, whilst the nonlinear modulation on the coupling coefficients and nonlinear coupling at high intensity are assumed negligible [13].

7.4.2.2 Coupling Coefficients

For simplicity, it is assumed that only the overlay of the slab is Kerr-like nonlinear. In order to obtain an analytical solution of the integrals of the above-defined, nonlinear coupling coefficients, further approximations can be made. Firstly, from the definitions Equations 7.119 and 7.120, the cross-phase modulation terms can be ignored in comparison with the self-phase modulation terms, because the field-overlap terms in Equation 7.120 (of evanescent fields) between the two guides are generally smaller than the mode-power terms in Equation 7.119. That is, the fourth terms on the right-hand side of Equation 7.118 are small compared to the third terms. Secondly, from the above-defined nonlinear index distribution and the notation of Equation 7.117, we have $n_2 = \{n_{2o}, 0, 0, 0\}$. Therefore, the integral of Equation 7.119a is carried out in the cross-section area of the slab-overlay cladding only (i.e., $\int_{A_\infty} \rightarrow \int_{A_o}$) and, because the integrand is then composed of very weak evanescent field (i.e., the tails of the field extending into the slab overlay) of the fiber mode, the integral I_{f0} can be neglected in comparison with the propagation constant β_{f0} and the self-coupling coefficient Q_{f00}, i.e., the first and second terms on the right-hand side of Equation 7.118a.

The nonlinear self-modulation coefficient I_{mnpl} is thus exactly solved as follows

(1) For the above nonlinear-index distribution, Equation 7.119b becomes

$$I_{mnpl} = \frac{k^2}{2} n_{2o} \int_{A_s} S_m S_n S_p S_l dA \tag{7.121}$$

(2) For the slab overlay, the scalar mode field is given below,

$$S_n \big|_{A_o} = \frac{N_s k_s t}{2V_{so}} e^{-\gamma_o [x - (a+s+t)]} \cos(\sigma_n y) \tag{7.122}$$

where the constants are defined and described in Section 6.1 of Chapter 6.

(3) Substituting Equations 7.122 into 7.121, the exact solution of I_{mnpl} can be expressed as

$$I_{mnpl} = \frac{\alpha_o D k^2}{\gamma_o} \left(\frac{N_s k_s t}{4V_{so}} \right)^4$$

$$\times \left\{ \delta_{(m+n-p-l),0} + \delta_{(m+n+p-l+1),0} + \delta_{(m+n-p+l+1),0} \right.$$

$$\left. + \delta_{(m-n+p+l+1),0} + \delta_{(m-n-p-l-1),0} + \delta_{(m-n+p-l),0} + \delta_{(m-n-p+l),0} \right\} \tag{7.123}$$

7.4.2.3 Power Tuning Effects

Numerical calculations have been carried out using the practical parameters as described in Section 6.1 of Chapter 6, i.e., at a light wavelength $\lambda = 1.3$ μm, with fiber radius $a = 2.5$ μm, slab thickness $t = 3$ μm, refractive index of the fiber cladding $n_c = 1.46$, $n_s = 1.4745$, and the distance between the fiber and the slab $s = 0.5$ μm (see Figure 7.3a). The typical value of the self-focusing, nonlinear, refractive index (see Equation 7.89 $n_{3,lo} = 10^{-9}$ m²/W [14] is used and, from Equation 7.93, $n_{2o} = c\varepsilon_o n_o^2 n_{3,lo}$ can be obtained, where n_o is the slab-overlay index of refraction, whilst c and ε_o are the free-space light speed and dielectric constant, respectively.

The Fortran program $a_o\_b_n$.listed in Section 3.6.2 of Chapter 3 is modified to include the additional nonlinear self-phase modulation coefficient solved in Equation 7.123 and is used to obtain the following numerical results with initial conditions of $a_0(0) = 1$ and $b_n = 0$ ($n = 0, 1, 2,...$). That is, initially, light is launched (with unit power) into the fiber to excite its fundamental mode.

Figure 7.25 shows that, when the core index of refraction of the fiber is greater than that of the slab, i.e., $n_f = 1.4817 > n_s = 1.4745$ (the fiber mode is out of phase with those of the slab transverse modes according to Section 6.1 of Chapter 6), the nonlinear effect is mainly displayed in the form of phase shift. The launched power largely remains in the fiber whilst a small amount couples back and forth between the fiber and the slab with an extended beat length (as a result of the nonlinear phase modulation of the guided modes).

When the fiber-core index of refraction is reduced to $n_f = 1.4756$ (greater than $n_s = 1.4745$, although the coupler is getting close to a phase match), Figure 7.26 demonstrates the interesting nonlinearity-assisted coupling, which transfers up to a possible 100% of the total power from the fiber to the slab. However, it appears that the exact amount of power

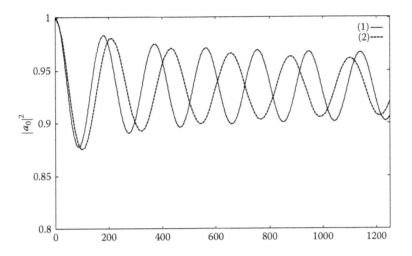

FIGURE 7.25
Propagation of $|a_0(z)|^2$. $n_f = 1.4817$, $n_s = 1.4745$, $n_c = 1.40$. (1) Scalar CMT without nonlinearity; (2) scalar CMT with nonlinearity.

transferred depends on the interaction (i.e., total coupling) length of the coupler due to the slow oscillation of the power between the two guides.

When the core refractive indices become equal, i.e., $n_f = n_s = 1.4745$, the coupler is almost exactly phase-matched (i.e., $\beta_{f0} \cong \beta_s$; see Tables 3.1 and 3.2 of Chapter 3) and Figure 7.27 further demonstrates the nonlinearity-assisted coupling, with a nearly complete power transfer in this case. The above phenomenon of optically enhanced coupling may be explained by the phase-detuning effect [13] due to the nonlinear phase modulation. The phase modulation is optically induced, initially starting in the fiber (the launching guide), and moves into the slab as the light is gradually coupled from the fiber to the slab. Here the slab self-phase modulation Equation 7.121 is the contributing factor to the enhanced, phase-matched coupling, which drives the coupler to the state of a full power transfer (in analogy to the crossed state of a two-mode coupler).

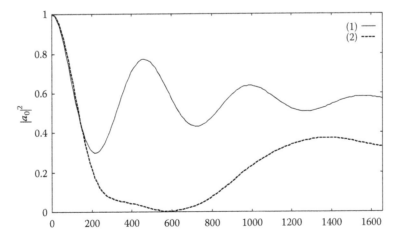

FIGURE 7.26
Propagation of $|a_0(z)|^2$. $n_f = 1.4756$, $n_s = 1.4745$, $n_c = 1.40$. (1) Scalar CMT without nonlinearity; (2) scalar CMT with nonlinearity.

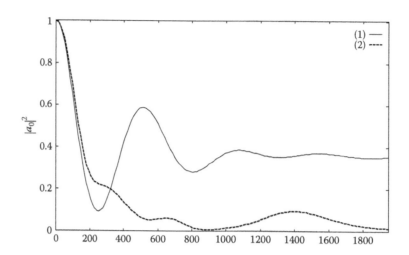

FIGURE 7.27

Propagation of $|a_0(z)|^2$. $n_f = 1.4745$, $n_s = 1.4745$, $n_c = 1.40$. (1) Scalar CMT without nonlinearity; (2) scalar CMT with nonlinearity.

Finally, when the fiber-core refractive index becomes smaller than that of the slab, i.e., $n_f = 1.4709 < n_s = 1.4745$, the fiber mode is phase-matched with one of the slab transverse modes (i.e., in the absence of the nonlinear effect). As a result, the nonlinearity does not seem to significantly assist the coupling in terms of the power transfer, if at all. Instead, it alters the power-beating properties, such as the beat length and the amplitudes of the power oscillation, through the nonlinear phase self-modulation (Figure 7.28).

7.4.3 Concluding Remarks

7.4.3.1 Nonlinear Two-Mode Couplers

A symmetric, nonlinear, two-mode coupler was analyzed utilizing the scalar CMT whilst treating the effect of Kerr-like nonlinearity as a perturbation on the index-profile of the

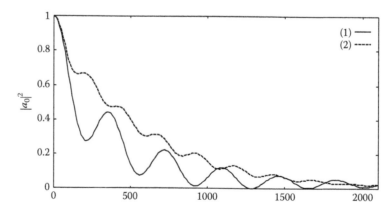

FIGURE 7.28

Propagation of $|a_0(z)|^2$. $n_f = 1.4709$, $n_s = 1.4745$, $n_c = 1.40$. (1) Scalar CMT without nonlinearity; (2) scalar CMT with nonlinearity.

whole coupler structure. The most general cases (i.e., including the nonlinear guided-modes and power-nonorthogonality) are considered and a full, power-conserved and nonorthogonal coupled-mode equations in the power parameters proposed and analytically solved. Without losing simplicity (e.g., the format of the simplified analysis), the new formulations and solutions generalize the previous results [13,14] with all the coupling coefficients and constants updated to include the mode-nonorthogonality (i.e., the filed-overlap effects) ignored previously.

This generalization technique may find immediate application in the design and analysis of an optical nonlinear guided two-mode coupler system, especially for the cases of weak guidance and weak to moderate couplings.

7.4.3.2 Nonlinear Fiber-Slab Couplers

A simplified, scalar coupled-mode analysis is presented to demonstrate the essential nonlinear effects on the optical composite fiber-slab guided-wave couplers.

A simple, numerical example is given to demonstrate some interesting nonlinear coupling features that are quite similar to those of the grating-assisted coupling in a fiber-slab coupler described in Chapter 6. When the parameters of the coupler are such that there is no phase synchronism between the fiber and the slab modes, the nonlinear phase-modulation comes into play and it assists coupling, resulting in up to a total power transfer from the fiber to the slab. However, when there is a phase synchronism between the two sets of guided modes, the effects of the Kerr-like nonlinearity seem greatly reduced and are displayed as some minor changes to the small power beating on the power-decay curve.

Based on the above nonlinearity-assisted coupling feature, we propose an all-optical, in-line fiber to slab coupler device that is easy to fabricate, considering the matured polished-fiber or D-fiber technique. This composite fiber-slab guided-wave system may find immediate applications in the areas of in-line fiber coupler and sensor technology. It may be designed to achieve the desired level of power transfer from the fiber (in-line) to the slab waveguide through an optimized selection of the materials and launched power.

7.5 Coupling in Dual-Core Microstructure Fibers

7.5.1 Introduction

As briefly described in Chapter 4, MOF is a new class of optical fiber that has emerged in recent years [17]. It is formed with an array of air holes running along its length. MOFs are fascinating because of their various novel properties, including endlessly single-mode operation [18], large and scalable dispersion and nonlinearity, surprising phenomenon of a short wavelength bend loss edge, etc. The holes within the fiber act as cladding and light can be guided using two kinds of mechanisms. One is due to the average index difference between the silica core and the periodic holey cladding, which is similar to a conventional total internal reflection mechanism; the second guide light uses the photonic bandgap (PBG) effects [19]. In this section, we briefly describe index-guiding MOFs and the defects which form the two virtual cores of a coupling system of a dual core MOF (DC-MOF).

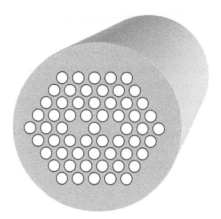

FIGURE 7.29
Dual-core photonic crystal fiber.

The fiber is fabricated by drawing down a stack of hundreds of silica capillaries, with a solid silica cane embedded in the structure to form the fiber core. For MOF to be used in a communication system, some basic fiber components based on MOF should be available. One of the main components is a fiber coupler. In normal MOFs, there is only one defect at the center and light is guided along this higher index defect. By incorporating several solid canes, MOF with several cores will be fabricated without any additional effort. Multiple cores can be coupled or uncoupled depending on the properties of the fiber. Such fibers can find applications in bending sensing, bidirectional coupling, and interferometry. If we introduce two defects, the MOF structure will have dual cores and a MOF coupler structure is formed, as shown in Figure 7.29.

In this paper, coupling characteristics of DC-MOFs are investigated by using the effective index model and the refractive index is calculated using the beam propagation method (BPM). BPM is essentially a particular approach for approximating the exact wave equation for monochromatic waves and solving the resulting equations numerically. The scalar field assumption allows the wave equation to be written in the form of the well-known Helmholtz equation (refer to Equation 7.124) for monochromatic waves. This method is conceptually straightforward and efficient. It can automatically include the effects of guided and radiating modes, mode coupling and conversion, polarization, and nonlinearity.

The wave equations of the E and H polarized waves are given by

$$\begin{cases} \nabla_{\perp}^{2}E + k_0^{2}\left(\varepsilon_r - n_{eff}^{2}\right)E = 0 \\ \nabla_{\perp}^{2}H + k_0^{2}\left(\varepsilon_r - n_{eff}^{2}\right)H = 0 \end{cases} \tag{7.124}$$

7.5.2 Coupling Characteristics

We consider dual-core photonic crystal fiber structure, as shown in Figure 7.30a, where Λ is the hole-to-hole lattice distance, d is the hole diameter, and the background silica index is assumed to be 1.45. The separation between the centers of two cores, S, is 2Λ. The effective index method is used to get the equivalent step index of the dual-core MOF. We firstly find out the fundamental space-filling mode [20], which propagates in the infinite photonic crystal cladding. Then the effective index of the cladding acts as cladding index of the fiber, shown in Figure 7.30b, and pure silica acts as core, thus making an equivalent step index fiber. (R is the equivalent core radius.) Figure 7.30c and d show the transverse mode profiles of DC-MOF

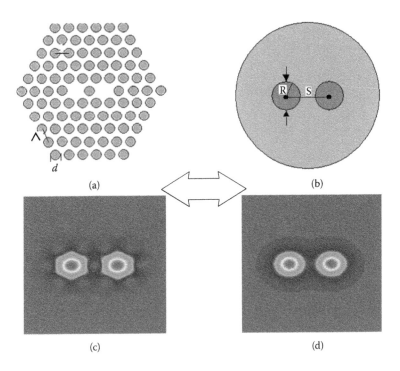

FIGURE 7.30
Effective index model: cross-section of (a) the dual-core MOF, (b) the equivalent dual-core fiber; transverse mode profile of (c) the dual-core MOF; (d) the equivalent dual-core fiber.

as well as the equivalent dual-core fiber. The effective index of the equivalent two-core fiber is set to 1.440978 (same as DC-MOF), by tuning the equivalent core radius R.

The coupling coefficient C_{pq}^{js} is a measure of the overlap of the pth mode in core j and the qth mode in core s, and generally expressed as described in the above sections:

$$C_{pq}^{js} = \frac{\omega}{2} \int_{A^{(s)}} (\varepsilon^{(s)} - \varepsilon) E_p^{(j)} \cdot E_q^{(s)} dA \tag{7.125}$$

where both fields are at frequency ω, $(\varepsilon^{(s)} - \varepsilon)$ is the difference between the dielectric constants of the core s and its cladding, and $E_p^{(j)}(x, y)$ is the transverse electric field profile of mode p in core j without the existence of core s. In general, the coupling coefficients between the two cores are not equal. For identical cores, we have $C_{12} = C_{21}$. For single-mode two-core fibers in which the two cores are identical, Equation 7.125 can be simplified [1]:

$$C_{12} = C_{21} = C = \sqrt{2\Delta} \frac{u^2}{\rho V^3} \frac{K_0\left(\dfrac{wd}{\rho}\right)}{K_1^2(w)} \tag{7.126}$$

where d is the core-to-core separation between the two cores, ρ is the radius of the cores, V is the normalized frequency, which is given by $V = k_0 \rho n_{co} \sqrt{2\Delta}$, and Δ is the refractive index difference between the cores and the cladding, which is defined as $\Delta = (n_{co}^2 - n_{cl}^2) / 2n_{co}^2$, with n_{co} and n_{cl} the refractive index of the core and the cladding. The parameter u is a solution of the characteristic equation (see also Chapter 4)

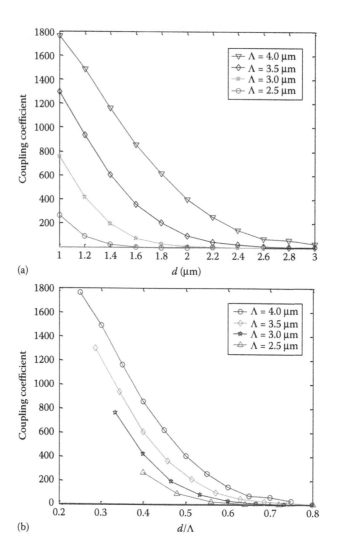

FIGURE 7.31

(See Color Insert) (a) Coupling coefficient vs. hole diameter, Λ varies from 2.5 to 4.0 μm. (b) Coupling coefficient with the hole-pitch size as a parameter, Λ varies from 2.5 to 4.0 μm.

$$uJ_1(w)/J_0(w) = wK_1(w)/K_0(w) \tag{7.127}$$

where $w = \sqrt{(k_0\rho_i)^2(n_{coi}^2 - n_{cl}^2) - u^2}$. J_0 and J_1 are zero- and first-order Bessel functions of the first kind, and K_0 and K_1 are zero- and first-order modified Bessel functions of the second kind, respectively.

The coupling coefficients can then be calculated from the coupling Equation 7.125. Figure 7.31 shows the hole size dependence of coupling coefficient with Λ = 4.0, 3.5, 3.0, 2.5 μm. By increasing the pitch size (Λ), or decreasing the hole size, d, which is equivalent to increasing the core radius, the coupling coefficients are increased. Within the range of $d = 1$–2 μm, coupling coefficient plots for different Λ are clearly separated, which has a potential application in sensor industry. For the $d/\Lambda > 0.4$, the coupling coefficient is less than 10^3.

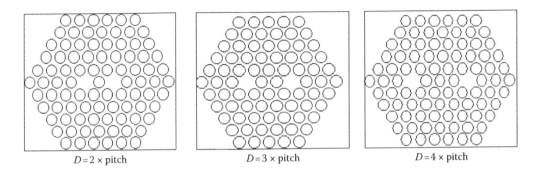

FIGURE 7.32
Dual-core MOF structures of different pitches.

In the fabrication process of MOFs, collapse of air holes normally happened, but light will still be confined in the desired solid core if the structure is within the low coupling range.

7.5.3 Dual-Core MOF Design without Loss

Recently, most of the MOF couplers have been fabricated by the fused biconical method [21]. A pair of twisted MOFs have been elongated while they were heated by a hydrogen flame. So far the best result with a five-stack MOF was 33/67. By reducing various losses and finding proper fabrication conditions, one might expect to develop a practical MOF coupler. We are preparing apparatus for a proper experiment. Reducing confinement loss by proper structure design for DC-MOF is the focus of this part. By setting the mode solver to be correlation, complex effective index can be achieved, and thus the confinement losses can be calculated as follows [22,23]:

$$\text{Confinement loss } (\text{dB/m}) = \frac{20 \times 10^6}{\ln 10} \frac{2\pi}{\lambda [\mu m]} \text{Im}(n_{eff}) \tag{7.128}$$

Three structures of three DC-MOFs are depicted in Figure 7.32. By increasing the core-to-core distance from 2Λ to 4Λ, we calculate the loss for three different structures according to Equation 7.128, and the results are listed in Table 7.4 and plotted in Figure 7.33. By analyzing the results, we can conclude that $d/\Lambda = 0.4$–0.6 is the optimized structure

TABLE 7.4

Losses of Different Dual-Core MOF Structures

d/Λ	L1 (dB/m)	L2 (dB/m)	L3 (dB/m)
0.25	2.900100	16.721000	11.097000
0.30	0.135650	2.334100	0.966660
0.35	0.005879	0.228960	0.052615
0.40	0.000184	0.014530	0.002457
0.45	0.000142	0.000950	0.000135
0.50	0.000079	0.000117	0.000162
0.55	0.000184	0.000026	0.000417
0.60	0.000174	0.000011	0.000213

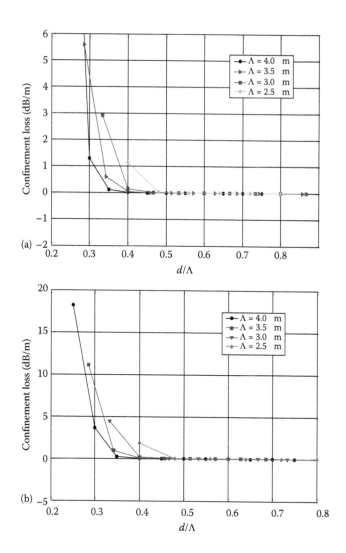

FIGURE 7.33
(See Color Insert) (a) Losses for structure $D = 2 \times$ pitch. (b) Losses for structure $D = 4 \times$ pitch.

for dual-core MOF with low loss. During the fabrication of MOF, collapse of air holes happens often, but if the structure is within the low coupling range, light will still be confined in the desired solid core. Transmission is high through each core despite many unintentional defects in the cladding, indicating that the guidance is determined by the holes near the core.

7.5.4 Remarks

This section illustrates numerically the coupling coefficient in a DC-MOF by using the effective index model. The coupling characteristics are investigated for potential application in sensing and multiplexing. The relationship between dual-core structure and confinement loss is analyzed. Optimized structure ($d/\Lambda = 0.4$) with low loss and promising for coupling is derived from the simulation results.

An optical fiber coupler is a device that distributes the light beam guided along a fiber into two or more branches. It can be used as a power divider, a wavelength-division multiplexer (WDM), and an optical switch. As the applications of photonic crystal fibers are increased, MOF couplers will be necessary in optical communications. Optimized DC-MOF structure for coupling is derived in this paper. Structures that are promising for light confinement when multiple defects are presented in MOF are also indicated.

7.6 Problems

1. Write down the coupling equations between two waveguides using full coupled-mode theory relating to the evolution of the amplitudes of the field guided in the waveguides and their corresponding mode propagation constants and their full self- and cross-coupling coefficients.

2. Refer to the power distribution along the propagation direction of the two couplers (Figure 7.2). What is the coupling ratio of the cross coupling between the waveguides to achieve such power coupling and distribution? Determine the locations along the propagation direction for 50% cross coupling.

References

1. A.W. Snyder, "Coupled-mode theory for optical fibers," *J. Opt. Society Am.*, 62, 1267–1277, 1972.
2. S.L. Chuang, "A coupled-mode formulation by reciprocity and a variational principle," *J. Lightwave Technol.*, LT-5, 5–15, 1987.
3. S.L. Chuang, "Application of the strongly coupled-mode theory to integrated optical devices," *IEEE J. Quantum Electron.*, QE-23, 499–509, 1987.
4. A.W. Snyder and A. Ankiewicz, "Fiber couplers composed of unequal cores," *Electronics Lett.*, 22, 1237–1238, 1986.
5. D. Gloge, "Weakly Guiding Fibers," *Appl. Opt.*, 10, 2252–2258, 1971.
6. A. Yariv, "Optical Electronics," 3rd Ed., Wiley, NY, 1985.
7. D. Marcuse, "Investigation of coupling between a fiber and an infinite slab," *IEEE J. Lightwave Technol.*, LT-7, 122–130, 1989.
8. A.W. Snyder and J.D. Love, "Optical Waveguide Theory," Chapman and Hall, London, 1983.
9. W.P. Huang, S.T. Chu, and S.K. Chaudhuri, "A scalar coupled-mode theory with vector correction," *IEEE J. Quantum* Electron., 28, 184–193, 1992.
10. W-P. Huang, "Coupled-mode theory for optical waveguides: an overview," *Opt. Soc. Am. A*, 11(3), 963–983, 1994.
11. A.W. Snyder and A. Ankiewicz, "Optical Fiber Couplers-Optimum Solution for Unequal Cores," *IEEE J. Lightwave Technol.*, 6, 463–474, 1988.
12. G. Grieffel, M. Itzkovitch, and A. Hardy, "Coupled-mode formulation for directional couplers with longitudinal perturbations," *IEEE J. Lightwave Technol.*, 27, 985–994, 1991.
13. S.M. Jensen, "The nonlinear coherent coupler," *IEEE J. Quantum Electron.*, QE-18, 1580–1583, 1982.
14. X.J. Meng and N. Okamoto, "Improved Coupled-Mode Theory for Nonlinear Directional Couplers," *IEEE J. Quantum Electron.*, 27, 1175–81, 1991.
15. Shu Zheng, "Optical Composite Guided-Wave Coupler Systems (Chapter 2)," PhD Dissertation, Monash University, 1999.

16. B. Daino, G. Gregori, and S. Wabnitz, "Stability analysis of nonlinear coherent coupling," *J. Appl. Phys.*, 58(12), 4513–4514, 1985.

17. A. Birks, P.J. Roberts, P.St.J. Russell, D.M. Atkin, and T.J. Shepherd, Full 2-D photonic band gaps in silica/air structures. *Eletron. Lett.*, 31, 1941, 1995.

18. J.C. Knight, T.A. Birks, and P.St.J. Russel, "All-silica single-mode operational fiber with photonic crystal cladding," *Opt. Lett.*, 21, 1547–1549, 1996.

19. T.A. Birks, P.J. Robert, and P.St.J. Russell, "Full 2-D photonic bandgaps in silica/air structures," *Electron. Lett.*, 31, 1941–1942, 1997.

20. T.A. Birks, J.C. Knight, and P.St.J. Russel, "Endlessly single-mode photonic crystal fiber," *Opt. Lett.*, 22, 961–963, 1997.

21. B.H. Lee, J.B. Eom, J. Kim, D.S. Moon, and U. Paek, "Photonic Crystal Fiber coupler," *Opt. Lett.*, 27, 812–814, 2002.

22. V. Finazzi, T.M. Monro, and D.J. Richardson, "Smaller-core silica holey fibers: Nonlinearity and confinement loss trade-offs," *J. Opt. Soc. Amer.*, 20, 1427–1436, 2003.

23. A. Bjarklev, J. Broeng, and A.S. Bjarklev, *Photonic Crystal Fibres*, Kluwer Academic Publishers, Boston, 2003.

8

Nonlinear Optical Waveguides: Switching, Parametric Conversion and Systems Applications

In this chapter, the formulation using an extended operator of the full vector electric field wave equations of loss less dielectric media is developed for a nonlinear guided wave problem. This formulation is based on the method of moments and solved by the Garlerkin finite element method (FEM) using first-order moments. Any spurious modes occurring are suppressed by a penalty function method. The solving system includes normalization procedures, acceleration techniques and a nonlinear solution method. All optical switching is observed and analyzed.

Parametric conversion of ultra-high speed optical signals and superimposed by pump beam is described and analyzed for application in the demultiplexing of optical time division multiplexed (OTDM) at 320 Gb/s bit rate in which the signals are then detected coherently and incoherently for differentially quadrature phase shift keying (DQPSK) modulated data sequences.

8.1 Introduction

In previous chapters the guiding of lightwaves in structures whose refractive index is linear and non-dependent on the power of the guided waves was discussed. However in practice there are cases when the total average power or the peak power of lightwave pulses or sequence of pulses is high coupled with some materials of intensity-dependent permittivity, or refractive index, thence the wave equation describing the behavior of the guided waves becomes nonlinear. Thus we have to deal with the problem of nonlinear wave equations with a view to obtaining realistic solutions and seeking applications of such phenomenon in photonic systems such as switching, memory, etc.

This chapter deals with nonlinear optical waveguides in planar or channel structures and provides a robust procedure for systematic investigations of these types of guided wave structures. It starts with the formulation of the full vectorial electromagnetic (EM) wave equation with an assumption that the nonlinearity is weak and may be highly nonlinear but not strongly nonlinear, that is mostly satisfied for practical cases.

Then the operator equation can be approximated by the methods of moment and solved by a nonlinear FEM. An extended operator may be used in order to overcome the restrictions on admissible functions imposed by the discontinuity and boundary conditions. High order meshes may be used to further reduce the uncertainty of the numerical solutions which are unavoidable for solving the nonlinear guided wave problems.

A scalar approximation procedure is developed for weakly guiding wave structures. A qualitative comparison of this formulation and the full vectorial representation/

computation is given. The stability of nonlinear modes is critical to the application of non-linear modal methods. Robust procedures dedicated to scalar nonlinear wave propagation are presented.

We illustrate the application of nonlinear planar and channel optical waveguides in the processing of lightwave modulated signals as a pre-processing photonic circuit for ultra-high speed optical communications systems. Several important features of the analysis are presented in the modeling of the nonlinear guided wave problems including an automatic mesh generation taking into account the boundary conditions and the concentration of expected optical fields, smooth and bandwidth reduction to minimize the size of the matrix elements are also considered, finite element solver is also given and demonstrated to be very effective for the power and high nonlinear operating regions of the problem.

This paper is thus organized as follows: Section 8.2 introduces the formulation of the wave equations and the operators for the nonlinear guided wave problems; Section 8.3 introduces the FEM in association with meshing of the problems and numerical examples to illustrate the effectiveness of the problem solving methods.

8.2 Formulation of Electromagnetic Wave Equations for Nonlinear Optical Waveguides

8.2.1 Introductory Remarks

Nonlinear optical waveguides can be formulated in a linear way, provided that an iteration procedure is incorporated to handle the nonlinearity, within an approximated linear region as the nonlinearity of several known materials is very small. In Chapters 6 and 7 we presented that a nonlinear optical waveguide can be represented by a homogenous wave equation and its solution can be formulated in either vectorial form or in scalar form as a weakly-guidance approximation.

Among vectorial formulations favored by many the $\vec{H}$-vector version [1–4] are due to its convenience in handling the boundary discontinuity of the permittivity between the core and cladding regions because the $\vec{H}$-field components are automatically satisfied as compared with the $\vec{E}$-field representation [5,6]. With full vectorial formulation using both $\vec{E}$ and $\vec{H}$, not only the discontinuity problems but more complex computations are required. If the transverse $(\vec{E}_t, \vec{H}_t)$ and longitudinal $(\vec{E}_z, \vec{H}_z)$ formulations are used then it is valid only for some restricted anisotropy of the media.

Although optical waveguides generally are open boundary problems, the closed boundary approximation can be made or adopted for most applications in the field of guided wave optics with modal analysis since only guided modes are of interest.

The efficiency of the approximate solutions can be associated with numerical techniques such as the FEM with larger area elements at the region near the open boundaries. The constitutive relations between the $(\vec{E}_z, \vec{H}_z)$ vectors are rather complicated in nonlinear media, and depend on the detailed origin of the nonlinearity and the operating frequency [7]. Nonlinear dielectric susceptibilities are related to the microscopic structure of the media and can be properly evaluated only with full quantum mechanical calculations.

In the next section the nonlinear wave evaluations and constitutive relations are presented so that numerical solutions can be formed in order to obtain the physical mechanism of the switching or guiding in these nonlinear guided wave devices.

8.2.2 Nonlinear Wave Equations and Constitutive Relations

The waveguide structures under consideration are uniform along the propagation direction z and have arbitrary index profiles in the x-y plane. The dielectric media under study are source free and non-dissipative. The EM wave equation is therefore of homogeneous form and the permittivity and permeability dyadics are Hermitian. When the dispersion of the media is relatively small and the spectrum of the laser beam is narrow enough, thence time harmonic analysis applies.

Due to the duality of the $(\vec{E}_z, \vec{H}_z)$ fields, the homogeneous and time harmonic nonlinear wave equation has the form for both electric and magnetic fields. It can be written in the general form with the phase factor $e^{j\omega t - \beta z}$ being implied:

$$\nabla \times \left(\hat{p}^{-1} \cdot \left(\nabla \times \underline{V}\right)\right) - k_0^2 \hat{q} \underline{V} = 0; \ k_0^2 \neq 0 \tag{8.1}$$

Defined in $\Omega \subset R^2$ with homogeneous boundary and interface conditions

$$\underline{n} \times \underline{V} = 0 \quad \text{on } C_1$$

$$\underline{n} \times \left(\hat{p}^{-1} \cdot \left(\nabla \times \underline{V}\right)\right) = 0 \quad \text{on } C_2$$

$$\left(\underline{n} \times \underline{V}\right)_{diff} = 0 \quad \text{on } \Sigma \tag{8.2}$$

$$\underline{n} \times \left(\hat{p}^{-1} \cdot \left(\nabla \times \underline{V}\right)\right)_{diff} = 0 \quad \text{on } \Sigma$$

where the subscript *diff* denotes the difference between quantities on the interior side and on the exterior side of Σ, and

$$\nabla \triangleq \vec{a}_x \frac{\delta}{\delta x} + \vec{a}_y \frac{\delta}{\delta y} + \vec{a}_z \frac{\delta}{\delta z};$$

$R^2 \triangleq$	two dimensional Euclidean space;
$\bar{\Omega} \triangleq$	a connected and closed set in R^2;
$\Omega \triangleq \bar{\Omega}$	strictly excluding all boundaries and media interfaces—an open set;
$\omega:$	angular frequency;
$\beta:$	the propagation constant of the guided wave;

$k_0 = \dfrac{\omega}{c} = \dfrac{\omega}{\sqrt{\mu_0 \varepsilon_0}}$: free space wave number;

$C_1:$	the Dirichlet boundary;
$C_2:$	the Neumann boundary;
$\Sigma:$	the interface between different media;
$\vec{n}:$	the outward unit vector normal to Σ or $C_1 \cup C_2$;

TABLE 8.1

Mapping of Variable for the $\vec{E}$ and $\vec{H}$ Formulations

Formulation	$\hat{p}$	$\hat{q}$	$\vec{V}$
E-vector	$\hat{\mu}_r$	$\hat{\varepsilon}_r$	$\vec{E}$
H-vector	$\hat{\varepsilon}_r$	$\hat{\mu}_r$	$\vec{H}$

The mapping of $\hat{p}, \hat{q}, \vec{V}$ to the relative permeability permittivity and the vectorial fields $\vec{E}, \vec{H}$ is given in Table 8.1.

The nonlinear permittivity dyadic can be expressed in terms of a linear term plus a nonlinear term (via the third-order susceptibility) as:

$$\hat{\varepsilon}_r = \hat{\varepsilon}_r^l + \delta\hat{\varepsilon}_r^{NL} \tag{8.3}$$

The nonlinear term $\hat{\varepsilon}_r^{NL}$ as a function of $\vec{E}$ is restricted to be monotonic and such that

$$\left\|\hat{\varepsilon}_r^{NL}\right\| \ll \left\|\hat{\varepsilon}_r^l\right\| \quad \forall \vec{E} \tag{8.4}$$

Thus a simple scheme of iteration can be applied. This equation also implies that the nonlinearity can reach saturation level. This saturation effect is essential when the self focusing action is stronger than the diffraction effect that commonly exists in physical systems.

8.2.3 Extended Operator and Penalty Function Method

Let L be the differential operator defined by

$$Lv \triangleq \vec{\nabla}\times\left(\hat{\mu}_r^{-1}\cdot\left(\vec{\nabla}\times v\right)\right) - \hat{\varepsilon}_r \cdot v \tag{8.5}$$

And the $\langle .,\,.\rangle$ be the inner product in a complex Hilbert space H defined as

$$\langle a, b \rangle \triangleq \int_\Omega a^* \cdot b\, d\bar{x}d\bar{y} \tag{8.6}$$

The star (*) denotes the complex conjugate

$$\vec{\nabla} \triangleq \frac{1}{k_0}\nabla; \quad \bar{x} = k_0 x; \quad \bar{y} = k_0 y \tag{8.7}$$

Now, $\forall w, v \in H$ application of the general method of moments to Equation 8.5 with w as the testing function yields

$$\langle w, Lv \rangle \triangleq \int_{\Omega} w^* \cdot Lv d\bar{x} d\bar{y}$$

$$= \int_{\Omega} w^* \cdot \left[\bar{\nabla} \times \left(\hat{\mu}_r^{-1} \cdot (\bar{\nabla} \times v) \right) \right] d\bar{x} d\bar{y} - \int_{\Omega} w^* \cdot \hat{\varepsilon}_r \cdot v d\bar{x} d\bar{y}$$

$$= \int_{\Omega} \left[(\bar{\nabla} \times w)^* \cdot \hat{\mu}_r^{-1} \cdot (\bar{\nabla} \times v) \right] d\bar{x} d\bar{y} - \int_{\Omega} w^* \cdot \hat{\varepsilon}_r \cdot v d\bar{x} d\bar{y}$$

$$\quad - \int_{\Sigma_-^+ + C} (n \times w)^* \cdot \left(\hat{\mu}_r^{-1} \cdot (\bar{\nabla} \times v) \right) d\bar{s}$$

$$= \int_{\Omega} \left[\bar{\nabla} \times \left(\cdot \hat{\mu}_r^{-1} \cdot (\bar{\nabla} \times w) \right)^* \cdot v \right] d\bar{x} d\bar{y} - \int_{\Omega} (\hat{\varepsilon}_r \cdot w)^* \cdot v d\bar{x} d\bar{y} \qquad (8.8)$$

$$\quad - \int_{\Sigma_-^+ + C} (n \times w)^* \cdot \left(\hat{\mu}_r^{-1} \cdot (\bar{\nabla} \times v) \right) d\bar{s}$$

$$= \langle L^a w, v \rangle - \int_{\Sigma_-^+ + C} (n \times w)^* \cdot \left(\hat{\mu}_r^{-1} \cdot (\bar{\nabla} \times v) \right) d\bar{s}$$

$$\quad + \int_{\Sigma_-^+ + C} \left(\hat{\mu}_r^{-1} \cdot (\bar{\nabla} \times w) \right)^* \times (n \times v) \right) d\bar{s}$$

with $$\sum_-^+ = \sum^+ \sum_-$$

denotes the union over both the internal (+) and external sides of the dielectric interfaces, $\bar{S} = k_0 S$ the normalized element, n is the normal direction to the dielectric interface Σ or the boundary $C = C_1 \cup C_2$, $\otimes$ denotes the Hermitian conjugate and L^a the operator adjoint to L.

The "volume rules" of L^a and L are identical on Ω provided that $\hat{\mu}_r; \hat{\varepsilon}_r$ are Hermitian dyadics. With the restrictions of a uniaxial property of the permittivity and permeability given as

$$\hat{q} = \begin{bmatrix} q_{xx} & q_{xy} & q_{xz} \\ q_{yx} & q_{yy} & q_{yz} \\ -jq_{xz} & -jq_{yz} & q_{zz} \end{bmatrix}$$

$$\varepsilon_r^{NL} = \begin{bmatrix} \varepsilon_1^{NL} & 0 & 0 \\ 0 & \varepsilon_2^{NL} & 0 \\ 0 & 0 & \varepsilon_3^{NL} \end{bmatrix} \qquad (8.9)$$

Thence the operator L can be extended from formally self adjoint to self adjoint as follows: The integral terms along the dielectric interface in Equation 8.8 are of the form

$$\oint_{\Sigma_-^+} a \cdot b\, d\bar{s} = \oint_{\Sigma} \left(a_{diff} \cdot b_{av} + b_{diff} \cdot a_{av} \right) d\bar{s}$$

(8.10)

where $\quad \{\cdot\}_{diff} = \{\cdot\}_{int} - \{\cdot\}_{ext} \quad \{\cdot\}_{av} = \dfrac{1}{2}\left(\{\cdot\}_{int} - \{\cdot\}_{ext} \right)$

and Σ denotes just a single integration over the dielectric interface itself using an outward unit normal vector.

The extended operator is defined by sharing the integrals between the original and adjoint operator

$$\langle w, L_e \bar{a} \rangle_e \triangleq \langle w, Lv \rangle$$

$$- \oint_{\Sigma} \left[w_{av}^* \left(n \times \left(\hat{\mu}_r^{\otimes -1} \cdot \left(\bar{\nabla} \times w \right) \right)_{diff} \right) + \hat{\mu}_r^{\otimes -1} \cdot \left(\bar{\nabla} \times w \right) \times \left(n \times v \right)_{diff} \cdot \right] d\bar{s}$$

$$- \int_{C_1} \left(\hat{\mu}_r^{\otimes -1} \cdot \left(\bar{\nabla} \times w \right)^* \times \left(n \times v \right) \right) d\bar{s} - \int_{C_1} w^* \cdot \left(n \times \hat{\mu}_r^{\otimes -1} \cdot \left(\bar{\nabla} \times w \right)^* \times \left(n \times v \right) \right) d\bar{s}$$

$$= \int_{\Omega} \left[\left(\bar{\nabla} \times w \right)^* \cdot \hat{\mu}_r^{-1} \cdot \left(\bar{\nabla} \times v \right) \right] d\bar{x}d\bar{y} - \int_{\Omega} w^* \cdot \hat{\varepsilon}_r \cdot v\, d\bar{x}d\bar{y}$$

$$- \oint_{\Sigma} \left(\hat{\mu}_r^{\otimes -1} \cdot \left(\bar{\nabla} \times w \right)_{av}^* \times \left(n \times v \right)_{diff} + \left(\left(n \times w \right)_{diff} \cdot \left(\hat{\mu}_r^{-1} \cdot \left(\bar{\nabla} \times w \right) \right)_{av} \right) \right) d\bar{s}$$

$$- \int_{C_1} \left[\left(\hat{\mu}_r^{\otimes -1} \cdot \left(\bar{\nabla} \times w \right)^* \times \left(n \times v \right) \right) + \left(n \times w \right)^* \cdot \hat{\mu}_r^{-1} \cdot \left(\bar{\nabla} \times v \right) \right] d\bar{s}$$

$$= \langle L^a w, v \rangle - \oint_{\Sigma} \left(n \times \hat{\mu}_r^{\otimes -1} \cdot \left(\bar{\nabla} \times w \right)_{diff}^* \cdot v_{av} + \left(\left(n \times w \right)_{diff}^* \cdot \left(\hat{\mu}_r^{-1} \cdot \left(\bar{\nabla} \times v \right) \right)_{av} \right) \right) d\bar{s}$$

$$- \int_{C_1} \left[\left(n \times w \right)^* \cdot \left(\hat{\mu}_r^{-1} \cdot \left(\bar{\nabla} \times v \right) \right)^* \right] d\bar{s} - \int_{C_1} \left[n \times \left(\hat{\mu}_r^{-1} \cdot \left(\bar{\nabla} \times w \right) \right)^* \cdot v \right] d\bar{s}$$

(8.11)

$$\triangleq \langle L_e^a v, w \rangle_e$$

It can be shown that the rules of L_e^a and L_e are identical in Ω as well as on Σ and C_1, C_2. Hence the extended operator is self adjoint in spite of the discontinuity and boundary conditions. A projection solution proceeds in this extended domain in the same manner as in the original domain, but the expansion and testing function need not satisfy all the discontinuity and boundary conditions.

8.2.4 Eigenvalues and Methods of Moments

The cross section of the waveguide can be meshed into a number of first-order triangular elements and the electric field within each element can be expended with nodal variables as coefficients

$$E_e = \sum_i E_i v_i; \qquad i = 3l + k \qquad l = 0,1,2; \qquad k = 1,2,3 \tag{8.12}$$

where l and k are constructed in the form

$$v_i \triangleq \begin{cases} L_i u_k & k = 1,2 \\ jL_i u_k & k = 3 \end{cases} \qquad j = \sqrt{-1} \tag{8.13}$$

with the local coordinates defined as

$$\begin{bmatrix} L_0 \\ L_1 \\ L_2 \end{bmatrix} = \begin{bmatrix} x_0 & x_1 & x_2 \\ y_0 & y_1 & y_2 \\ 1 & 1 & 1 \end{bmatrix} \begin{bmatrix} x \\ y \\ 1 \end{bmatrix} \tag{8.14}$$

where (x_1, y_1) are the Cartesian coordinates of the vertex of the triangle. In analyzing nonlinear optical waveguides, the dispersion relation, β versus P, is essential with k_0 and waveguide geometrical dimensions.

Applying the Garlerkin FEM to Equation 8.11 yields the following generalized eigenvalue equation

$$[S = pU]x - Tx = 0 \tag{8.15}$$

with the eigenvalue $\bar{\beta}$ embedded in S and U, and x is an $n \times 1$ vector of modal variables and S, U, T are $n \times n$ real symmetric matrices which are assembled from element matrices:

$$[S_e]_{ij} \triangleq \int_\Omega \left[\left(\bar{\nabla} \times v_i \right)^* \cdot \hat{\mu}_r^{-1} \cdot \left(\bar{\nabla} \times v_j \right) \right] d\bar{x} d\bar{y}$$

$$- \oint_{\Sigma_e} \left(\hat{\mu}_r^{\otimes -1} \cdot \left(\bar{\nabla} \times v_i \right)_{av}^* \cdot \left(n \times v_j \right)_{diff} \right. \tag{8.16}$$

$$+ \left(\left(n \times v_i \right)_{diff} \cdot \left(\hat{\mu}_r^{-1} \cdot \left(\bar{\nabla} \times v_i \right) \right)_{av} \right) d\bar{s}$$

$$[U_e]_{ij} \triangleq \int_{\Omega_e} \left[\bar{\nabla} \cdot \left(\varepsilon_r^{\otimes 1} \cdot v_i \right) \right]^* \left[\bar{\nabla} \cdot \left(\varepsilon_r^1 \cdot v_j \right) \right] d\bar{x} d\bar{y}$$

$$- \oint_{\Sigma_e} \left[\left(\bar{\nabla} \cdot \left(\hat{\varepsilon}_r^{\otimes 1} \cdot v_i \right) \right)_{av} \right) \cdot \left(n \cdot \hat{\varepsilon}_r \cdot v_j \right)_{diff} \right]$$

$$+ \left(\left(n \cdot \hat{\varepsilon}_r^{\otimes} \cdot v_i \right)_{diff}^* \left(\left(\bar{\nabla} \cdot \left(\hat{\varepsilon}_r \cdot v_i \right) \right)_{av} \right) d\bar{s} \tag{8.17}$$

$$- \int_{C_{2e}} \left[\left(\bar{\nabla} \cdot \left(\hat{\varepsilon}_r^{\otimes 1} \cdot v_i \right)^* \right) \cdot \left(n \cdot \hat{\varepsilon}_r \cdot v_j \right) \right]$$

$$+ \left(\left(n \cdot \hat{\varepsilon}_r^{\otimes} \cdot v_i \right)^* \left(\bar{\nabla} \cdot \left(\hat{\varepsilon}_r \cdot v_i \right) \right) \right) d\bar{s}$$

and

$$[T_e]_{ij} = \int_{\Omega_e} v_i^* \cdot \hat{\varepsilon}_r \cdot v_j \, d\bar{x} d\bar{y} .$$

(8.18)

Here the Ω_e denotes the domain of the element; $\Sigma_e = \Omega_e \cap \Sigma$; $C_{ie} = \Omega_e \cap C_i$ $i = 1, 2$.

Again the line integral in terms of Equations 8.16 through 8.18 vanish in cases where all the discontinuity and boundary conditions can be forced explicitly.

The guided power can be evaluated by integrating the x-component of the Poynting vector over the cross section.

$$P = \frac{1}{2Z_0 \varepsilon_n k_0^2} \left[-\text{Im} \left\{ \int_\Omega \left(E^* \times \left(\bar{\nabla} \times \bar{E} \right) \cdot e_x d\bar{x} d\bar{y} \right) \right\} \right]$$

(8.19)

where e_x is the unit vector, Z_0 is the intrinsic impedance of vacuum. The normalized power is then given as

$$\bar{P} = 2Z_0 \varepsilon_n k_0^2 P$$

(8.20)

Equation 8.15 can then be written in the form of a global matrix equation in terms of the actual eigenvalue $\bar{\beta} = \beta/k_0$ as:

$$\left[\bar{\beta}^2 A + \bar{\beta} B + C \right] \underset{\sim}{x} = 0$$

(8.21)

This is a quadratic eigenvalue equation and is solvable with some complicated numerical method. Alternatively it can be converted into a standard form

$$\left[R(\bar{\beta}) \right] \underset{\sim}{x} = \lambda T \underset{\sim}{x}$$

(8.22)

$$\text{where} \qquad R \triangleq R + pU$$

Thence solved iteratively by modifying $\bar{\beta}$ till $\lambda \to 1$. The following procedure can be employed to solve Equation 8.21: (a) specifying the required parameters as input data and estimate $[R]$ & $[T]$ in the linear region by assuming no nonlinearity; (b) solving the linear eigenvalue problem for λ, x; (c) modifying $\bar{\beta}$, $\bar{\varepsilon}_r$ to obtain updated $[R]$ & $[T]$; (d) repeating (b) to obtain modified λ, x; (e) going back to (c) if the eigenpair do not satisfy within the desired criterion.

For optical waveguide problems, usually only one or two of the lowest modes are required. With iteration procedures for nonlinear optical waveguides, the eigenvalues can only be found one by one. Furthermore, due to the banded property of the matrix of Equation 8.22, it has been found that successive over relaxation (SOR) method combined with the Rayleigh quotient (RQ) is extremely fast for finding the fundamental mode in steps (b) and (d). As for higher order modes, the bisection and inverse iteration method has proven to be necessary. Naturally caution in the exercise must be taken in interpreting the "modes" in a nonlinear situation.

However due to the large size of the band matrix for high accurate mode eigenvalue or propagation constant, the Newton acceleration method is appropriate. Also the vectorial version of Aitken's δ^2-method has been found not to give convergence while the over relaxation method has been found to offer a 20% reduction of computational time.

8.2.5 Solution Methods for Nonlinear Generalized Eigenvalue Problems

With reference to a computing system of limited computational resources, the anisotropy of the guided medium is limited to the case such that the global matrices are real and symmetric. As discussed previously, the efficiency of an algorithm depends on the complexity of the problem. Further examination of the problem is needed prior to selecting the solution method. Table 8.2 lists the general properties of the linearized algebraic eigenvalue equation. For a nonlinear problem eigenpairs have to be found individually whenever a nonlinear interaction scheme is employed. Furthermore at each nonlinear iteration step, only one eigenpair has a physical meaning. Thus one must exercise care in computing the physical modes as the mode index of a physical mode might be interchanged with that of an unphysical solution during the nonlinear iteration, especially for higher order modes.

Taking into account the properties of the nonlinear problems, we think that the SOR [8] and vector iteration (VI) [9,10] methods are the two most effective methods for the computation of the fundamental mode.

The methods selected for solving the linear generalized eigenvalue problems are described in the following:

8.2.5.1 Successive over Relaxation and Rayleigh Quotient

The SOR method is one of the most successful linear iterative methods for sparse matrix computations. In most textbooks the SOR method computes the deterministic problem

$$[A][x] = [b] \tag{8.23}$$

The iteration is implemented in the form

$$[D]x_{i+1} = \omega_r \left([C_L] x_{i+1} + [C_U] x_i + b \right) - (1 - \omega_r)[D] x_i \tag{8.24}$$

TABLE 8.2

General Properties of Linearized Algebraic Eigenvalue Problems

Property	Value	Remarks
Number of degree of freedom	~1000	Approx. 1000 unknowns corresponding to about 330 nodes
Symmetry	Symmetric	Matrices and unknowns are real
Definiteness	Positive definite	Both R and T matrices
Sparsity	Balanced	The semi-bandwidth is around 8% of NDF when NDF~1000 for first-order elements and is about 15% for second-order elements
Eigenvalue and/or eigenvector required	Both	Only one or two lowest modes
Computational resources	Both are limited	CPU time more crucial than storage

with a starting vector x_0, where ω_r is the real number known as the relaxation factor or relaxation parameter, and $[D]$, $[-C_L]$ and $[-C_U]$ are, respectively, the diagonal, strictly lower triangular, and strictly upper triangular parts of A.

If $[A]$ is symmetric and positive (or negative) definite, SOR methods converge to the solution for any fixed value of ω_r in the range $0 < \omega_r < 2$. Only over relaxation ($\omega_r > 1$) is of importance for fast convergence, after which the method is named. When $\omega_r = 1$ then the SOR reduces to the Gauss-Siedel method.

The SOR method can be directly adapted to the generalized real symmetric eigenvalue problem, to solve for the eigenpair corresponding to the lowest eigenvalue [11] with the relationship given as

$$[K]\underset{\sim}{x} = \lambda[M]\underset{\sim}{x} \tag{8.25}$$

In this connection, we can rewrite as

$$[A] = [K] - \lambda[M] \qquad \text{and} \qquad \bar{b} = 0 \tag{8.26}$$

After each step of iteration the eigenvalue λ is estimated by the RQ [12] so that

$$\lambda \to \lambda_i = \frac{x_i^T [K] x_i}{x_i^T [M] x_i} \tag{8.27}$$

which is used for the next step of iteration until some convergence criterion is met; and the eigenvector $\bar{x}$ is normalized with respect to the quadratic form for $[M]$ so that

$$\bar{x} \to \bar{x}_i = \frac{\bar{x}_i}{\left(\bar{x}_i^T [M] \bar{x}_i \right)^{1/2}}$$

after each step of iteration to avoid overflow on a computer of limited word length. The SOR-RQ described here has a global convergence to the eigenpair corresponding to the smallest eigenvalue. The convergence of the SOR method is guaranteed only if $[A]$ is real symmetric and definite. $\{A\}$ becomes singular when the estimated eigenvalue approaches its exact value. Thus it might be a real danger of divergence of the scheme. Furthermore in using the SOR it is quite important to select a good value of ω_r so that the convergence rate can be accelerated.

8.2.5.2 *Vector Iteration*

The other alternative method for solving the eigenvalue equation for the fundamental mode is the power iteration method [13] which can be classified as forward iteration or reverse iteration in which the former converges to the largest eigenvalue and to the smallest by the latter. The eigenvector $\bar{x}_i$ of the lowest eigenvalue can be obtained by solving

$$[K]\underset{\sim}{x}_{i+1} = [M]\underset{\sim}{x}_i \tag{8.28}$$

and the lowest eigenvalue is estimated by the RQ as described above. Thence the approximated eigenvector is substituted in Equation 8.26 for the next iteration to obtain the eigenvalue. This can be resorted to the use of the Sturn sequence property as given in Ref. [10].

The principal minor $p_r(\bar{\lambda}) \triangleq \det([K]_r - \lambda[M]_r); r = 0,1,......,q$ of the symmetric matrix $[K] - \bar{\lambda}[M]$ with $[M]$ being positive definite, form a Sturn sequence provided that the corresponding matrix equation has no degenerate eigenvalues, where $p_0(\bar{\lambda}) \equiv 1$; $[K]_r$ and $[M]_r$ denote the principal rth order submatrices of $[K]$ and $[M]$, respectively. Consequently the number of eigenvalues strictly greater than $\bar{\lambda}$ is equal to the number of agreements in sign of this sequence provided that $p_r(\bar{\lambda})$ is taken to have the opposite sign to that of $p_{r-1}(\bar{\lambda})$ if $p_r(\bar{\lambda}) = 0$ [12]; or equivalently, the number of eigenvalues strictly less than $\bar{\lambda}$ is equal to the number of changes in sign of this sequence provided that $p_r(\bar{\lambda})$ is taken to have the same sign as that of $p_{r-1}(\bar{\lambda})$ if $p_0(\bar{\lambda}) \equiv 0$. By using this property the kth eigenvalue can be isolated by repeated bisection and counting the number of sign changes/agreements of the sequence.

In this chapter the principal mirrors are evaluated by Gaussian elimination with partial pivoting to take advantage of the banded structure. With partial pivoting, it is found that the Sturn sequence count may give misleading results whenever one or more of the pivots used are too small, which usually happens as the test eigenvalue is brought close to one of the exact eigenvalues and the separation between that eigenvalue and its neighbors is relatively small. These situations can be detected easily by comparing the number of sign changes/agreements when using the test eigenvalue with those obtained when using its upper and lower bounds.

8.2.5.3 Posteri Error Estimate

For nonlinear problems the error-bound approach is not as tractable as for linear problems, thus one has to adopt an alternative approach to estimate the residuals upon obtaining a solution. For the general eigenvalue problem, we introduce a residual vector $\tilde{r}$ defined by

$$\tilde{r} \triangleq [K]\tilde{x} - \tilde{\lambda}[M]\tilde{x} \tag{8.29}$$

where $(\tilde{\lambda}, x^\sim)$ is an approximate eigenpair. Then the length of the residual may be evaluated by using the norm associated with an appropriate inner product:

$$\|\tilde{r}\| \triangleq < \tilde{r}, \tilde{r} >_q^{1/2} = \left(\langle K\tilde{x}, K\tilde{x} \rangle_q + \tilde{\lambda}^2 \langle M\tilde{x}, M\tilde{x} \rangle_q - 2\tilde{\lambda} \langle K\tilde{x}, K\tilde{x} \rangle_q \right)^{1/2} \tag{8.30}$$

In which the real symmetric property of $[K]$ and $[M]$ has been utilized.

The above error estimate does not make much sense as its magnitude depends on the normalization of $\tilde{x}$. To establish a useful estimate we can introduce a scalar quantity, the r, relative residual as

$$r_r = \frac{\|\tilde{r}\|}{\left(\langle K\tilde{x}, K\tilde{x} \rangle_q + \tilde{\lambda}^2 \langle M\tilde{x}, M\tilde{x} \rangle_q - 2\tilde{\lambda} \langle K\tilde{x}, K\tilde{x} \rangle_q \right)^{1/2}}$$

$$= \left[1 - \frac{2\langle K\tilde{x}, K\tilde{x} \rangle_q}{\left(\langle K\tilde{x}, K\tilde{x} \rangle_q + \tilde{\lambda}^2 \langle M\tilde{x}, M\tilde{x} \rangle_q - 2\tilde{\lambda} \langle K\tilde{x}, K\tilde{x} \rangle_q \right)^{1/2}} \right]^{1/2} \tag{8.31}$$

It is apparent that this relative residual belongs to the set [0, 1] and is independent of the scaling of $\tilde{x}$ and r, vanishes as the appropriate eigenpair approaches the exact one. Until now we have not defined the inner product $<.,.>$ in R_q though the error estimate has been established. To be specific, one can choose

$$\langle x, y \rangle \triangleq x^T y \qquad \forall\, x, y \in R^q \tag{8.32}$$

In order to compare the results obtained here with the exact ones of a classical linear example, one has to establish another error estimate. Let $\tilde{v}$ be an approximate mode computed by the method described herewith and let the corresponding exact mode be denoted by $\hat{v}$ with the sign being consistent with that of $\tilde{v}$ so that $< \tilde{v}, \tilde{v} \geq 0$.

Following a procedure similar to the steps described above, one can define a scalar quantity $e_r(\tilde{v}, \tilde{v})$ as a relative error

$$e_r\left(v, \hat{v}\right) \triangleq \left(1 - \langle v, \hat{v} \rangle\right)^{1/2} \tag{8.33}$$

subject to the normalization

$$\langle \tilde{v}, \tilde{v} \rangle \triangleq \langle \hat{v}, \hat{v} \rangle \equiv 1 \tag{8.34}$$

8.2.5.4 Nonlinear Acceleration Techniques

The solution procedures for nonlinear algebraic equations are iterative in nature and thus can be accelerated. Either the eigenvalue or the eigenvector can be accelerated but not both to avoid falling into the divergence. For the eigevalue the best known acceleration technique is Aitken's δ^2-method [14] for scalar quantity. Each component of an eigenvector is a scalar quantity and therefore may be accelerated by this method.

Let $\tilde{x}_k(k = 1, 2, 3 \ldots)$ be an eigenvector obtained at the kth step of the nonlinear iteration procedure after stage b and assume that it is normalized so

$$\|x_k\|_\infty = 1.0 \tag{8.35}$$

The eigenvector is denoted by $\bar{x}_k$ after the acceleration and renormalization. Then the acceleration technique is described by

$$y_k = \bar{x}_{k-1} + A_{cc}\left(x_k - \bar{x}_{k-1}\right)$$

and

$$\tag{8.36}$$

$$\bar{x}_{k-1} = \frac{y_k}{\|y_k\|_\infty}$$

with A_{cc} is the nonlinear acceleration factor.

Thus given the guided power P_g, the eigenvector for the nonlinear problem $\hat{x}_k$ can be expressed as

$$\hat{x}_k = \sqrt{\frac{P_g}{P(\bar{x}_k)}} x_k \tag{8.37}$$

with $P(\bar{x}_k)$ as the guided power determined by $\bar{x}_k$. A fixed acceleration factor is naturally not optimal. However with a conservative value of about 1.3, the method can be shown to be beneficial for the examples given in this chapter. A saving of computing resources of about 20% is achieved using this acceleration procedure.

8.3 Numerical Examples of Nonlinear Optical Waveguides

8.3.1 Waveguides of Non-Saturation Nonlinear Permittivity

8.3.1.1 *Embedded Channel*

The structures of the optical waveguide examined in this section include a channel waveguide of rectangular geometry surrounded by a cladding region as shown in Figure 8.1.
   Consider the case of a linear core bounded by self focusing nonlinear cladding. The nonlinear perturbation term of the relative permittivity dyadic is assumed to be uniaxial as follows:

$$\hat{\varepsilon}_r^n = \begin{bmatrix} \varepsilon_1^n & & \\ & \varepsilon_2^n & \\ & & \varepsilon_3^n \end{bmatrix} \tag{8.38}$$

For a Kerr nonlinear medium then

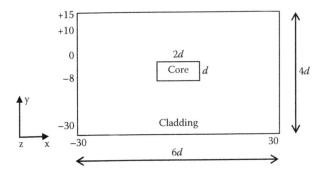

FIGURE 8.1
Geometry and profile of nonlinear optical channel waveguide: (a) structure and coordinate system, (b) mesh for finite element.

$$\varepsilon_i^n = \gamma \left(|E_i|^2 + b \sum_{\substack{j=1 \\ j \neq 1}}^{3} |E_j|^2 \right); \qquad i = 1, 2, 3. \tag{8.39}$$

where $\gamma [m^2/V^2]$ is the nonlinear coefficient defined as

$$\gamma = c_0 \varepsilon_0 \varepsilon_r^l n_2 \tag{8.40}$$

is the term of the usual nonlinear optical coefficient n_2, c_0 is the velocity of light in vacuum, ε_0; ε_r^l are the permittivity of vacuum and relative permittivity in the linear region. The coefficient b depends on the origin of the nonlinear mechanism

$$b = \begin{cases} 1 & \text{electrorestriction} \\[2mm] \dfrac{1}{3} & \text{electronic distortion} \\[2mm] -\dfrac{1}{2} & \text{molecular orientation} \end{cases} \tag{8.41}$$

The other parameters are chosen as follows: the free space wavelength is of 550 nm, $n^2 = 1.0 \times 10^{-9} \, m^2/W$; $\beta d = 10$ and

$$\varepsilon_r^l = \begin{cases} 1.55^2 & \text{core} \\ 1.52^2 & \text{cladding} \end{cases} \tag{8.42}$$

The structure in Figure 8.1 is meshed into 116 second-order elements with 256 nodes as shown in Figure 8.2. The maximum node index difference between any two related nodes is 37 after node renumbering. The fundamental mode of the waveguide under investigation corresponds to either H_{11}^x or E_{11}^y mode in the linear operating region whose contour of amplitude is plotted in Figure 8.3 for components in the x, y and z directions.

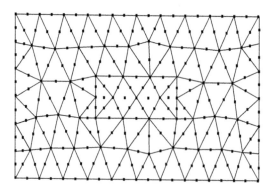

FIGURE 8.2
Mesh network of the channel waveguide, core and cladding regions, 265 nodes and 116 second-order elements.

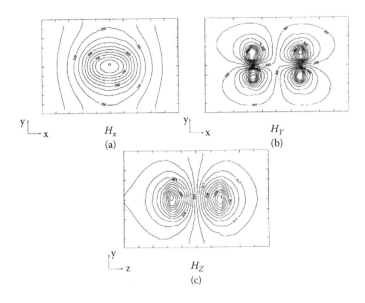

FIGURE 8.3
Magnetic field distribution of the fundamental mode H_{11}^x or E_{11}^y guided in the rectangular waveguide in the linear operating region (a) H_x (b) H_y (c) H_z. Horizontal and vertical axes: linear scale with the origin at the center, contour scale arbitrary with relative scaling.

As the structure is normalized with respect to the propagation constant β, the eigenvalue λ in the generalized eigenvalue equation now represents $(k_0/\beta)^2$. The power dispersion curve can be found by the conventional way by specifying the guided power P and calculating β/k_0 which is defined as the effective refractive index, n_{eff} of the guided mode, that means that the guide mode along the propagation direction z would see the guided medium by a slowing factor of n_{eff}. The computed power dispersion curve for an isotropic medium is shown in Figure 8.3 for $b = 1$. This dispersion curve is contrasted with those obtained in Ref. [15] in which the H-formulation was also employed. This shows agreement of the method described here (Figure 8.4).

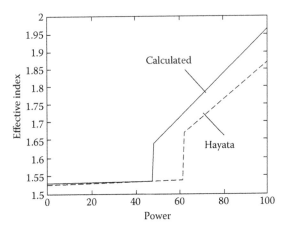

FIGURE 8.4
Power dispersion curve of the fundamental nonlinear mode: β/k_0 (effective index) versus power P [μW]. ---- Yahata [16] results and ___ calculated using method described.

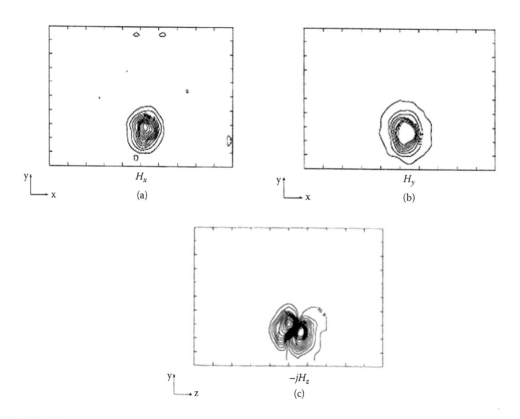

FIGURE 8.5

Magnetic field components of the fundamental nonlinear mode excited at $P = 80\ \mu W$ (a) H_x, (b) H_y and (c) $-jH_z$. Horizontal and vertical axes: linear scale with the origin at the center, contour scale arbitrary with relative scaling.

The field components of the guided mode in the nonlinear region are shown in Figure 8.5 at an exited power of 80 µW.

Under nonlinear operation the convergence may not be obtained with the mesh of Figure 8.2, thus a refined mesh is used with 176 second-order elements and 389 nodes as shown in Figure 8.6. The power dispersion curve corresponding to the original and

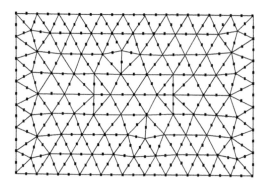

FIGURE 8.6

Refined finite element mesh.

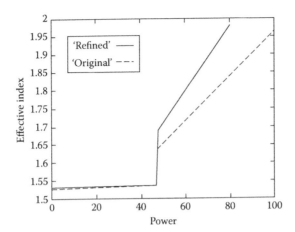

FIGURE 8.7
Power dispersion curve of the fundamental nonlinear mode: β/k_0 (effective index) versus power P [μW]. ---- original results and — refined mesh.

the refined meshes are plotted in Figure 8.7. Better accuracy is obtained with refined and denser mesh. The convergence of the mode with the change of the excitation power is shown in Figure 8.8. The maximum value of the effective index, i.e., related to the eigenvalue, is obtained when the iteration step reaches the 23rd node. The instability or switching of the guided mode is understandable as obtained in Figures 8.7 and 8.8. There would also be bistability as demonstrated later in this chapter.

Figure 8.9 shows the self focusing magnetic field distribution at $P = 80$ μW using a finer mesh size. The mode fields are more confined as compared with the coarser mesh show in Figure 8.5.

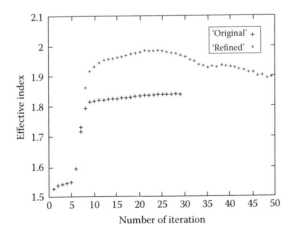

FIGURE 8.8
Divergence behavior of the effective index with respect to the number of iteration procedure at 80 μW for meshes of 116 elements (original) and 176 elements (refined of second-order).

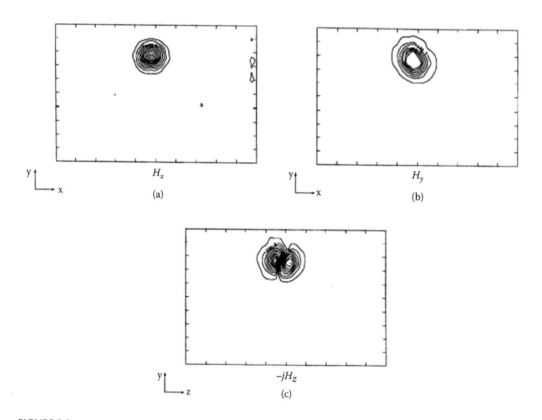

FIGURE 8.9
Magnetic field components of the fundamental nonlinear mode excited at P = 80 μW using refined mesh: (a) H_x, (b) H_y and (c) $-jH_z$. Horizontal and vertical axes: linear scale with the origin at the center, contour scale arbitrary with relative scaling.

8.3.1.2 Overlay Nonlinear Film and Linear Embedded Channel

Figure 8.10a shows the geometry of a nonlinear optical channel waveguide, consisting of an ion-exchanged channel (linear) in a glass substrate and a nonlinear overlay. The relative permittivity is given as:

$$\hat{\varepsilon}_r = \begin{cases} \hat{I} & |x| \le 30; 10 < y \le 15 \\[2mm] \left(1.53^2 + \dfrac{f(E)}{1+f(E)}\right)\hat{I} & |x| \le 30; 0 < y \le 10 \\[2mm] \left(1.54 + 0.045\,\mathrm{erfc}\left(-y/8\right)^2 \hat{I}\right) & |x| \le 8; y < 0 \\[2mm] 1.54^2\hat{I} & 8 < |x| \le 30; y < 0 \end{cases}$$

(8.43)

where $\hat{I}$ is the identity dyadic, and $f(E) = \Sigma_{i=1}{}^3 |E_i|^2 = ||E||_2{}^2$ ($b = 1$, $\alpha = 1$). The boundary conditions are assumed to represent perfect conductors. The whole structure is meshed into 424 first-order elements with 237 nodal points, as shown in Figure 8.10b. The boundary conditions are enforced explicitly by the technique described in Silvester and Ferrari [17]

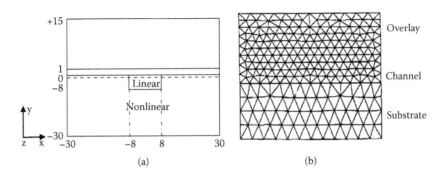

FIGURE 8.10
Nonlinear optical channel waveguide: (a) structure and coordinate system; (b) mesh for finite element analysis.

on a node-by-node basis. The discontinuity condition at dielectric interfaces is enclosed by matching the normal component of the electric flux density at the nodes of the interfaces. In the inhomogeneous region (within the channel and near its side walls) $\hat{\varepsilon}_r$ is linearly interpolated in terms of its nodal values on an element-by-element basis, which makes the numerical integration necessary to compute $[R]$ and $[T]$.

The calculated dispersion curve is plotted in Figure 8.11 which shows the effective index $\bar{\beta}$ is strongly dependent on the total power transmitted P_n. The all-optical switching phenomenon is clearly shown in Figure 8.12 where the fraction of the switched power is defined as the power in the $y > 0$ region divided by the total power. It is apparent that pulse amplification besides optical switching can be achieved. The field distribution in the regions of consideration is shown in Figure 8.13a and b for the case of low and high input power, respectively. It shows clearly the switching of the guide optical field in the channel to the upper region.

In conclusion, this example shows the effectiveness of the E-vector finite element analysis to obtain the solution of the nonlinear guided wave problems. The algorithm adopted here includes normalization, acceleration techniques for reaching the eigenvalue solutions and a nonlinear iteration scheme.

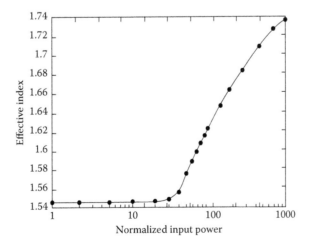

FIGURE 8.11
Dispersion curve of the nonlinear optical waveguide.

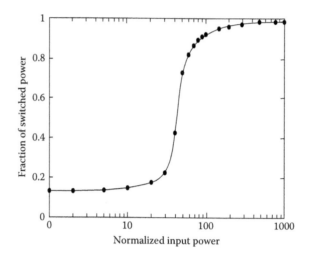

FIGURE 8.12
All optical switching characteristics of curve from the channel optical waveguide region to upper region.

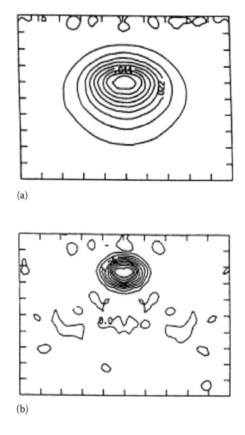

(a)

(b)

FIGURE 8.13
Intensity distribution of the electric field $\|E\|_2$ at (a) low power ($P_a = 1$) (b) high power ($P_a = 1000$). Horizontal and vertical axes: linear scale with the origin at the center, contour scale arbitrary with relative scaling.

8.3.1.3 Waveguides of Nonlinear Permittivity with Saturation

Broadly, there are two principal saturation models given as

$$\varepsilon_r^n = \frac{\gamma |E|^2}{1 + \alpha \gamma |E|^2}$$

and   (8.44)

$$\varepsilon_r^n = \frac{1}{\alpha} \left[1 - e^{\alpha \gamma |E|^2} \right]$$

where α is the saturation parameter, and γ is the optical nonlinear coefficient. Without the loss generality, we adopt the former model of saturation that normally represents the saturation of a two-level electronic system in materials far from resonance. By analogy with the scalar model the relative permittivity can be written related to the vectorial E-field $\underset{\sim}{E}$ as

$$\varepsilon_i^n = \frac{\gamma |E|^2}{1 + \alpha \gamma |\underset{\sim}{E}|^2} ; \quad i = 1, 2, 3$$ (8.45)

This model is now used to replace the Kerr nonlinearity, for the isotropic case $b = 1$, Equation 8.39 becomes

$$\varepsilon_i^n = \frac{\gamma \|\underset{\sim}{E}\|_2^2}{1 + \alpha \gamma \|\underset{\sim}{E}\|_2^2} ; \quad i = 1, 2, 3$$ (8.46)

The dispersion characteristics of the medium with saturated permittivity are shown in Figure 8.14 and typical field distributions of the *H*-field vector are shown in Figure 8.15 for coarse mesh and for finer mesh in Figure 8.16. We observe that: (a) the saturation effect makes the field pattern mire diffused; (b) the intense focusing action enables the guided wave to be self channeling and therefore more independent of the core region; and (c) the

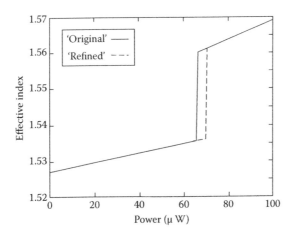

FIGURE 8.14
Power dispersion curve of the fundamental nonlinear mode with saturated permittivity: β/k_0 (effective index) versus power P [µW]. ___ coarse mesh and ___ refined mesh.

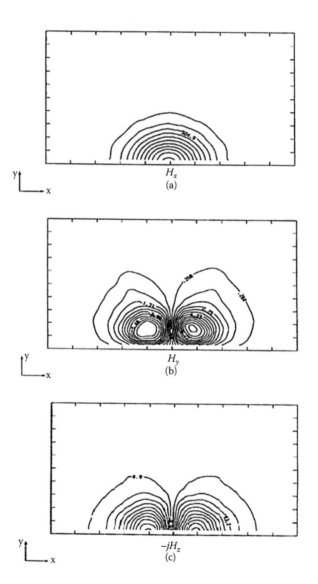

FIGURE 8.15
Magnetic field components of the fundamental nonlinear mode excited at P = 80 μW under saturation of permittivity using coarse mesh: (a) H_x, (b) H_y and (c) $-jH_z$. Horizontal and vertical axes: linear scale with the origin at the center, contour scale arbitrary with relative scaling.

order of this "half-pattern" mode, which makes use of the electric wall as a symmetry plane, is lower than that of the corresponding "full-pattern" mode, if it exists, for the given closed structure with fixed power; therefore the mode obtained here is consistent with the solution method for finding the lowest mode.

Finally we can observe that the symmetric structure can support asymmetric mode when strong self focusing occurs. The reason is quite simple. The nonlinear materials, symmetry at low powers does not mean that symmetry remains at very high powers as the field itself can alter the distribution of the permittivity in the nonlinear region. Consequently the so-called even (or symmetric) or odd (anti-symmetric) modes can no longer exist physically when the power exceeds some certain critical level.

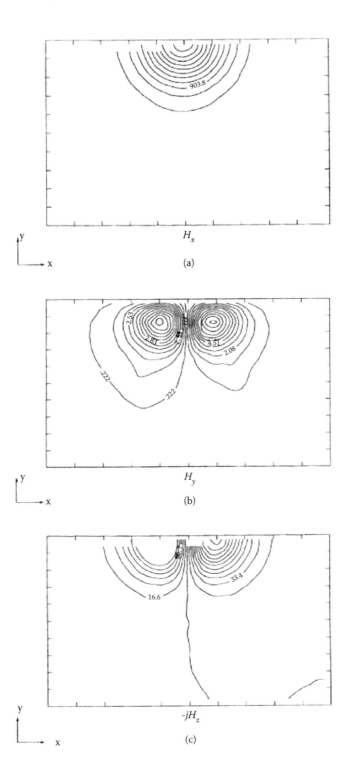

FIGURE 8.16

Magnetic field components of the fundamental nonlinear mode excited at P = 80 μW excitation power under saturation of permittivity using refined mesh: (a) H_x, (b) H_y and (c) $-jH_z$. Horizontal and vertical axes: linear scale with the origin at the center, contour scale arbitrary with relative scaling.

We can thus make some concluding remarks that: (i) in simulating the nonlinear effects in quasi-3D structures, it is essential to incorporate saturation into the nonlinear permittivity model for both physical and mathematical reasons; (ii) symmetric structures can support asymmetric modes in nonlinear optical waveguiding systems. Thus it is not recommended that the use made for the symmetry of a structure unless the field pattern being sought is known to be symmetric *a priori*.

8.3.1.4 Bistability Phenomena in Nonlinear Optical Waveguide

When the optical power reaches above a certain threshold, the field distribution for the saturation nonlinear permittivity model does not represent any useful physical solution as they confine themselves to the artificial boundary. We thus consider the waveguide structures shown in Figure 8.17 consisting of a linear channel waveguide and loaded with a nonlinear strip. The substrate is also linear. The relative permittivity dyadic is taken as

$$
\hat{\varepsilon}_r = \begin{cases}
\hat{I} & |\bar{x}| \le 30 & 15 < \bar{y} \le 30 \\
8 < |\bar{x}| \le 30 & 15 < \bar{y} \le 30 \\
\left(1.52^2 + \dfrac{|E|_2^2}{1+\alpha|E|_2^2}\right)\hat{I} & |\bar{x}| \le 8 & 0 < \bar{y} \le 15 \\
1.55^2\hat{I} & |\bar{x}| \le 8 & -12 < \bar{y} \le 0 \\
1.55^2\hat{I} & 8 < |\bar{x}| \le 30 & -12 < \bar{y} \le 0 \\
|\bar{x}| \le 30 & -30 < \bar{y} \le -12 &
\end{cases}
\tag{8.47}
$$

where $\hat{I}$ is the identity dyadic, and the boundary conditions are assumed to represent a perfect conductor, and the saturation power α is taken to be 1.0 without loss of generality.

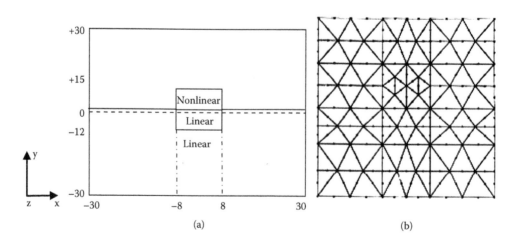

(a)   (b)

FIGURE 8.17
Nonlinear strip loaded channel optical waveguide: (a) structure and coordinate system, (b) mesh network for finite element analysis (126 second-order element, 281 nodes).

The whole structure is meshed into 126 second-order elements and 281 nodal points as depicted in Figure 8.17b. After node numbering, the maximum node index difference between any related nodes is 49. The power dispersion curve computed by the E formulation is shown in Figure 8.18a and a jump is observed which suggests that there may be two possible values of the effective refractive indices, thus the evaluation of the algorithm is to be reformulated to take account of this possibility.

For the saturating nonlinear isotropic permittivity model, we can introduce the dimensionless field variables defined by

$$\tilde{E} \triangleq \sqrt{\alpha} E$$

$$\tilde{H} \triangleq Z_0 \sqrt{\alpha} H$$

(8.48)

Consequently the relation between $\hat{E}$ and $\hat{H}$ is governed by

$$\tilde{H} \triangleq j \frac{\rho}{k_0} \hat{\mu}_r^{-1} \nabla \times \tilde{E}$$

$$\tilde{E} \triangleq -j \frac{\rho}{k_0} \mu \hat{\varepsilon}_r^{-1} \nabla \times \tilde{H}$$

(8.49)

Thence the saturation model becomes

$$\varepsilon_i'' = \left(1.52^2 + \frac{\|E\|_2^2}{1 + \alpha \|E\|_2^2} \right) \qquad i = 1, 2, 3$$

(8.50)

With the field variables normalized, the corresponding guided power P is also normalized that is given by

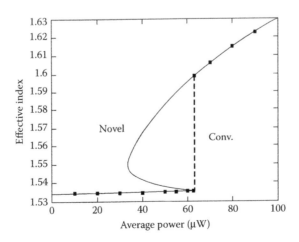

FIGURE 8.18
Dispersion curve of the nonlinear strip loaded waveguide. Effective index (β/k) versus optical power (mW) for conventional algorithm and new algorithm.

$$\tilde{P} \triangleq \frac{k_0}{\rho} \Im \left\{ -\int_\Omega \tilde{V}^* \times \left(\hat{p}^{-1} \cdot \left(\bar{\nabla} \times \tilde{V} \right) \right) \cdot a_z d\bar{x} d\bar{y} \right\}$$

(8.51)

$$\tilde{V} \equiv \tilde{E}, \tilde{H}$$

Now if we introduce $\hat{V} = A_m V'$ with A_m as a real coefficient and subject to

$$\frac{k_0}{\rho} \Im \left\{ -\int_\Omega \tilde{V}'^* \times \left(\hat{p}^{-1} \cdot \left(\bar{\nabla} \times \tilde{V}' \right) \right) \cdot a_z d\bar{x} d\bar{y} \right\} \equiv 1$$

(8.52)

so that $\tilde{P} = A_m^2$ then (A_m, V') may be solved as the eigenpair by specifying the effective index β/k_0. For simplicity the surface terms resulting from the extended operator formalism will be ignored for the time being. The partial differential equation corresponding to the final penalty formulation (see Section 8.2) can be shown to be

$$\bar{\nabla} \times \left(\hat{p}^{-1} \cdot \left(\bar{\nabla} \times V \right) \right) - p\hat{q} \bar{\nabla} \bar{\nabla} \cdot \left(\hat{q} \cdot V \right) = \lambda \hat{q} \cdot V$$

(8.53)

$$\text{with} \qquad \lambda = \left(\frac{k_0}{\rho} \right)^2$$

This equation is valid for both $\tilde{V}, V'$. Now if using the E-formulation then the guided power being the eigenvalue for a given effective index of the following equation

$$\bar{\nabla} \times \left(\hat{\mu}_r^{-1} \cdot \left(\bar{\nabla} \times E'_{i+1} \right) \right) - p\hat{\varepsilon}_{r(i)} \bar{\nabla} \bar{\nabla} \cdot \left(\hat{\varepsilon}_{r(i)} . E'_{i+1} \right) = A_{m(i+1)}^2 \left\{ \delta \frac{\hat{\varepsilon}_{r(i)}}{A_{m(i)}^2} \right\} \cdot E'_{i+1}$$

(8.54)

where i is the iteration index, δ defined as

$$\hat{\varepsilon}_{r(i)} \equiv \hat{\varepsilon}_r^l + \delta \hat{\varepsilon}_{r(i)}^n$$

(8.55)

$$\hat{\varepsilon}_{r(i)}^n \equiv \hat{\varepsilon}_r^n \left(A_{m(i)} E'_{(i)} \right)$$

Thus we can compute the eigenvalue and then the value A_m or this as the eigenvalue. Note that this A_m^2 is directly related to the normalized guide power. The power dispersion curve is obtained as given in Figure 8.18 and a bistability is observed. The following reasons can be offered for this bistability

- When the guide power is relatively low, the refractive index induced by the nonlinearity is not high enough for the nonlinear region to support a mode; and therefore the stable modal field must be mainly distributed in the linear region.
- When the guide power becomes relatively high, the modal field penetrating into the nonlinear region would guide the total field into self focusing. Therefore the fundamental mode can no longer be confined in the linear region but propagates in the nonlinear region.

- Between these extreme cases, there exists a range of guided powers within which the modal field of the fundamental mode can propagate either in the linear region or in the nonlinear region, depending on the initial condition. The modal field of the fundamental mode, whether in the linear region or in the nonlinear region, is expected to be stable.

- Now we may ask whether the portion of the lower branch of the power dispersion curve within this range of power can be termed a "higher order mode" as the effective index is lower than that of the upper branch. Also "mode" in the nonlinear situation is loosely defined. No, there is only one extreme in the field pattern, and this field pattern has the same polarization as the one on the same branch but corresponding to a lower power level. More importantly, when the modal field propagates in the linear region for power within this range, the corresponding portion of the upper branch of the power dispersion curve does not actually exist, and the modal field confined in the linear region can never sense such a virtual branch due to a kind of "quantum well", which makes both branches coexist and represent the fundamental mode.

- The middle portion of the power dispersion curve of negative slope is expected to be unstable. It serves as a transient bridge joining the two stable branches.

The field distribution of the component E_x of the guided mode in the lower, middle and upper branches of the bistable curve are shown in Figure 8.19 at the launched power in the region where nonlinear switching and bistability occur.

8.4 Nonlinear Optical Waveguide for Optical Transmission Systems

8.4.1 Introduction

In this section we describe the uses of nonlinear optical waveguides for optical transmission systems. Three typical applications are given. Firstly, the parametric amplification is analyzed in which amplification of optical signals can be achieved. Furthermore phase conjugation can also be obtained via this parametric amplification. This is further described in Chapter 10.

The second use of the nonlinear waveguide is the generation of the triple correlation product, which multiplies in the optical domain, the original pulse and two delayed versions of such signal or pulse. This triple product acts as an optical signal pre-processor to increase the sensitivity of optical receivers which are termed as the bispectrum receivers. Thus this section gives a simulation of the optical integrated circuit (IC) consisting of a linear signal processing part and a nonlinear optical waveguide with four wave mixing (FWM) phenomena to obtain the triple product of the original and delayed beams.

Recently, ultrafast optical signal processing has emerged as a promising technology for optical communication networks. An advanced optical network requires a variety of signal processing functions including optical regeneration, wavelength conversion, optical switching and signal monitoring. An attractive way to realize this optical signal processing in transparent and high speed mode is to exploit the third-order nonlinearity in optical waveguides due to the ultrafast response time of optical nonlinearities (several femtoseconds).

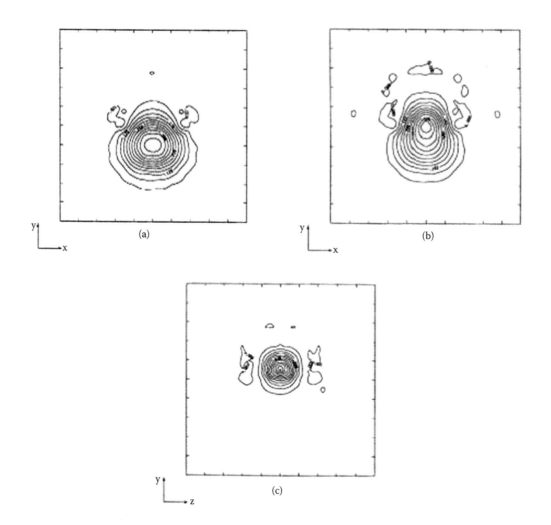

FIGURE 8.19
E_x of the guided mode in: (a) lower branch; (b) middle branch; and (c) upper branch.

The popular optical waveguide in optical communication systems is optical fiber. The nonlinearity of conventional fiber is often very small to prevent the degradation of the transmission signal from nonlinear effects. However much higher nonlinearity is required in optical processing, therefore highly nonlinear fibers (HNLF) are used. The nonlinear coefficient of HNLF is ten times higher than that of conventional fibers. Apart from optical fibers, there is another class of optical waveguides, which are fabricated on flat substrates, such as slab waveguides, channel waveguides and rib waveguides. Because of their geometries, these waveguides are called planar waveguides. A planar waveguide can confine a high intensity of light within an area comparable to the wavelength of light, over a short distance (several centimeters). Hence they are compatible with optical integrated circuits which produce very compact photonic devices for signal processing.

In this section we review the mathematical representation of third-order nonlinearity and some basic propagation equations which are used to model the nonlinear interactions in optical waveguides. The model is developed in MATLAB then integrated into an optical transmission Simulink platform which allows us to investigate optical signal processing in waveguides similar to that in practical systems. Several system applications using the third-order nonlinearity have been demonstrated to validate our model.

8.4.2 Third-Order Nonlinearity and Propagation Equation

Propagation in optical waveguides is generally governed by Maxwell's equations. From Maxwell's equations, the equation of EM wave propagation in nonlinear waveguide in the time-spatial domain in vector form [18] is:

$$\nabla^2 \vec{E} - \frac{1}{c^2}\frac{\partial^2 \vec{E}}{\partial t^2} = \mu_0 \left(\frac{\partial^2 \overrightarrow{P_L}}{\partial t^2} + \frac{\partial^2 \overrightarrow{P_{NL}}}{\partial t^2} \right) \tag{8.56}$$

where E is the EM field of the pulse, P_L, P_{NL} are the linear and nonlinear components, respectively, of the polarization vector which are defined by:

$$
\begin{aligned}
P_L(r,t) &= \varepsilon_0 \int_{-\infty}^{\infty} \chi_{xx}^1(t-t')E(r,t')e^{j\omega t'}\,dt' \\
&= \frac{\varepsilon_0}{2\pi}\int_{-\infty}^{\infty}\chi_{xx}^1(\omega)E(r,\omega-\omega_0)e^{j(\omega-\omega_0)t'}\,d\omega \\
&= \varepsilon_0\chi^{(3)}\,E(r,t)E(r,t)E(r,t)
\end{aligned}
\tag{8.57}
$$

where $\chi^{(3)}$ is the third-order susceptibility with the following assumptions: (i) only one mode propagating along z is present in the waveguide; (ii) the material is perfectly transparent and the wavelength, as well as harmonics of the wavelength of the applied field, are far away from any material resonances; (iii) the variations of complex amplitude, both in real and phase amplitude, and its derivatives in propagation distance are sufficiently small with respect to the carrier; and (iv) nonlinearity can be seen as a small disturbance of the linear behavior.

These assumptions are valid in most cases in practice. Thus the electrical field $\vec{E}$ can be written as $\vec{E}(\vec{r},t) = \frac{1}{2}\hat{x}\{F(x,y)A(z,t)\exp[i(\beta_0 z - \omega_0 t)]+c\cdot c\}$ where $A(z,t)$ is the slowly varying complex envelope. After some algebra, using methods of separating variables, the following equation for pulse evolution in optical waveguide is

$$
\begin{aligned}
\frac{\partial A(z,t)}{\partial z} + \frac{1}{2}\sum_{n=0}^{\infty}\frac{j^n\alpha_n}{n!}\frac{\partial^n A(z,t)}{\partial t^n} - j\sum_{n=0}^{\infty}\frac{j^n\beta_n}{n!}\frac{\partial^n A(z,t)}{\partial t^n} \\
= j\sum_{n=0}^{\infty}\frac{j^n\gamma_n}{n!}\frac{\partial^n}{\partial t^n}\left(1+\frac{j}{\omega_0}\frac{\partial}{\partial t}\right)A(z,t)\int_{-\infty}^{\infty}g(t-t')\left|A(z,(t-t'))\right|^2 dt'
\end{aligned}
\tag{8.58}
$$

where the effects of propagation constant β, the loss α and the nonlinear term γ around ω_0 are Taylor expanded as:

$$\beta(\omega) = \sum_{n=0}^{\infty} \frac{\beta_n}{n!} \Omega^n$$

$$\alpha_{eff}(\omega) = \sum_{n=0}^{\infty} \frac{\alpha_n}{n!} \Omega^n \tag{8.59}$$

$$\gamma(\omega) = \sum_{n=0}^{\infty} \frac{\beta_n}{n!} \Omega^n$$

$g(t)$ is the nonlinear response function including the electronic and nuclear contributions. With the above assumptions, $\gamma(\omega)$ and $\alpha(\omega)$ are constant over the pulse spectrum, Equation 8.58 can be rewritten as:

$$\frac{\partial A(z,t)}{\partial z} + \frac{\alpha}{2} A(z,t) - j \sum_{n=0}^{\infty} \frac{j^n \beta_n}{n!} \frac{\partial^n A(z,t)}{\partial t^n}$$

$$= j\gamma \left(1 + \frac{j}{\omega_0} \frac{\partial}{\partial t}\right) A(z,t) \int_{-\infty}^{\infty} g(t-t') \left| A(z(t-t')) \right|^2 dt' \tag{8.60}$$

For the pulses wide enough to contain many optical cycles, Equation 8.51 can be simplified as

$$\frac{\partial A}{\partial z} + \frac{\alpha}{2} A + \frac{i\beta_2}{2} \frac{\partial^2 A}{\partial \tau^2} - \frac{\beta_3}{6} \frac{\partial^3 A}{\partial \tau^3} = i\gamma \left[|A|^2 A + \frac{i}{\omega_0} \frac{\partial (|A|^2 A)}{\partial \tau} - T_R A \frac{\partial (|A|^2)}{\partial \tau} \right] \tag{8.61}$$

where a frame of reference moving with the pulse at the group velocity v_g is used by making the transformation $\tau = t - z/v_g \equiv t - \beta_1 z$, and A is the total complex envelope of propagation waves, α, β_k are linear loss and dispersion coefficients, respectively, $\gamma = \omega_0 n_2 / cA_{eff}$ is the nonlinear coefficient and the first moment of the nonlinear response function is defined as

$$T_R \equiv \int_0^{\infty} tg(t') dt' \tag{8.62}$$

Thus this is the basic propagation equation, called the nonlinear Schrodinger equation (NLSE), for studying third-order nonlinearity in optical waveguides. However Equation 8.61 is commonly used for modeling the pulse propagation in optical fiber, it is still limited in studying ultra-short pulse propagation in highly nonlinear waveguides, thus we can write

$$\frac{\partial A}{\partial z} + \frac{\alpha}{2} A + \frac{i\beta_2}{2} \frac{\partial^2 A}{\partial \tau^2} - \frac{\beta_3}{6} \frac{\partial^3 A}{\partial \tau^3}$$

$$= i\gamma \left(1 + \frac{i}{\omega_0} \frac{\partial}{\partial \tau}\right) \times \left((1 - f_R) \left[|A|^2 + \frac{1}{3} e^{-i2\omega_0 \tau} A^2\right] A + f_R g(z, \tau, A)\right) \tag{8.63}$$

In the process of FWM we have

$$A = \sum_i A_i e^{\left[j(k_i z - \omega_i \tau)\right]}$$

thus

$$\frac{\partial A_n}{\partial z} + \frac{\alpha}{2} A_n + \frac{i\beta_{2n}}{2} \frac{\partial^2 A_n}{\partial \tau^2} - \frac{\beta_{3n}}{6} \frac{\partial^3 A_n}{\partial \tau^3}$$

$$= i\gamma A_n \left[|A_n|^2 + 2 \sum_{k \neq n} |A_k|^2 \right] + i\gamma \sum_{\substack{n=k+l-m \\ k,l \neq m}} A_k A_l A_m^* \exp(-i\Delta k z) \tag{8.64}$$

$$- \sum_{k=1}^{n-1} \frac{\omega_n}{\omega_k} \frac{g_{nk}}{2} |A_k|^2 + \sum_{k=n+1}^{N} \frac{\omega_n}{\omega_k} \frac{g_{nk}}{2} |A_k|^2$$

with $\Delta k = k_k + k_l - k_m - k_n$.

8.4.3 Simulation Model

From the propagation equations mentioned above, nonlinear interactions can be simulated by two approaches. The first approach is to use Equation 8.53 in which the total complex amplitude is the summation of individual complex amplitudes with different carrier frequencies. The second approach is to use the coupled equation system. In the second approach, interactions between different waves are clearly represented by terms in each coupled equation, the simulation required is more complex than the first one. Hence we have focused on the first to build the model of propagation in highly nonlinear waveguides. This option also offers an easy possibility of integration into the Simulink model

$$A = \sum_i A_i e^{\left[j(\omega_i - \omega_0)\tau\right]}. \tag{8.65}$$

8.4.3.1 Parametric Amplification

One of important applications of the $\chi^{(3)}$ nonlinearity is parametric amplification. The optical parametric amplifiers (OPA) offer a wide gain bandwidth, high differential gain and optional wavelength conversion and operation at any wavelength [19]. These important properties of OPA are obtained because the parametric gain process does not rely on energy transitions between energy states, but it is based on highly efficient FWM in which two photons at one or two pump wavelengths with arbitrary phases will interact with a signal photon. A fourth photon, the idler, will be formed with a phase such that the phase difference between the pump photons and the signal and idler photon satisfies a phase matching condition. As the Kerr effect, similarly to the Raman process, relies on nonlinear interactions in the fiber, the intrinsic gain response for an OPA is in the same order as for the Raman amplifier. Hence the amplifier can operate in a very high saturated power mode and allows for ultrafast all-optical signal processing. A scheme of the fiber-based parametric amplifier is shown in Figure 8.20.

(a)

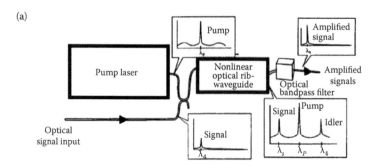

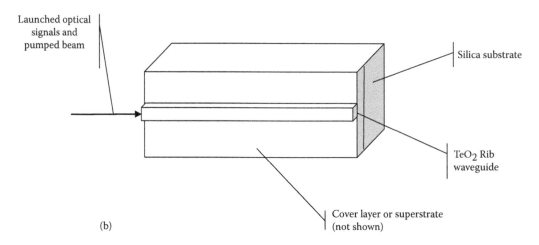

(b)

FIGURE 8.20
General scheme of phase-insensitive fiber-based OPA (extracted from Ref. [18]): (a) using nonlinear effects in rib-optical waveguide (fiber); (b) nonlinear optical rib-waveguide structure with high nonlinear coefficient to shorten the interaction length, hence the length of waveguide.

From an EM point of view, the interaction of three stationary co-polarized waves angular frequencies ω_p, ω_i, ω_s can be characterized by the slowly varying electric field with complex amplitudes A_p, A_i, A_s, respectively. Using the basic propagation (Equation 8.64) or from Equation 8.63, we can derive from the three coupled equations for complex field amplitude of the three waves as follows:

$$\frac{\partial A_p}{\partial z} = -\frac{\alpha}{2}A_p + i\gamma A_p\left[\left|A_p\right|^2 + 2\left(\left|A_s\right|^2 + \left|A_i\right|^2\right)\right] + i2\gamma A_s A_i A_p^* \exp(-i\Delta kz) \qquad (8.66)$$

$$\frac{\partial A_s}{\partial z} = -\frac{\alpha}{2}A_s + i\gamma A_s\left[\left|A_s\right|^2 + 2\left(\left|A_p\right|^2 + \left|A_i\right|^2\right)\right] + i\gamma A_p^2 A_i^* \exp(-i\Delta kz) \qquad (8.67)$$

$$\frac{\partial A_i}{\partial z} = -\frac{\alpha}{2}A_i + i\gamma A_i\left[\left|A_i\right|^2 + 2\left(\left|A_s\right|^2 + \left|A_p\right|^2\right)\right] + i\gamma A_p^2 A_s^* \exp(-i\Delta kz) \qquad (8.68)$$

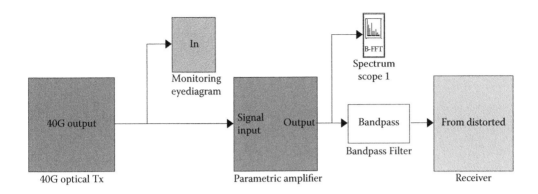

FIGURE 8.21
MATLAB Simulink set up of the parametric amplification embedded in the photonic transmitter and receiver.

Here the dispersion effects have been neglected. The last term of these equations is responsible for the energy transfer between the interacting waves.

The parametric amplification is implemented by a MATLAB program and incorporated into a Simulink block as shown in Figure 8.21, including the transmitter and receiver whose structures are shown in Figures 8.22 through 8.24. Using a return-to-zero pulse

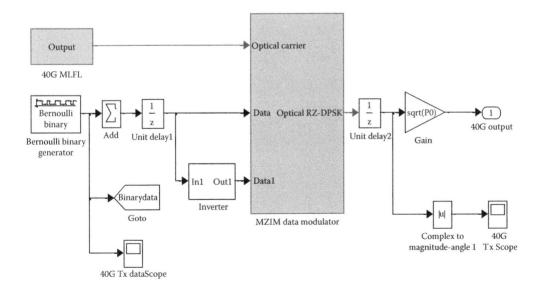

FIGURE 8.22
Structure of 40 Gb/s photonic transmitter (see also Chapter 9). MLFL = mode-locked fiber laser.

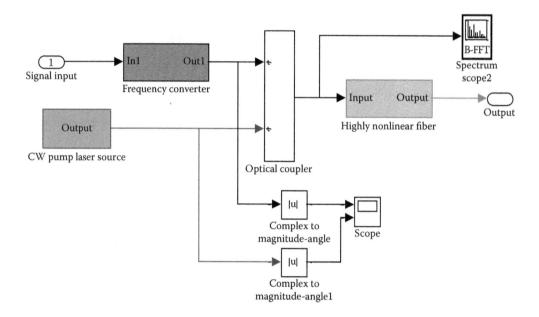

FIGURE 8.23
Simulink model of a parametric amplifier as an optical pre-amplifier. The wavelengths of the laser pump source and the signals (wavelength converted via the frequency converter) are separated by a spectral distance to achieve parametric amplification.

shaping modulated input signal of 40 Gb/s, the important simulation parameters are tabulated in Table 8.3.

Figure 8.25 shows the signals before and after the amplifier in time domain. The graph indicates the amplitude fluctuation of the amplified signal as a noisy source from the wave mixing process. Their corresponding spectra are shown in Figure 8.26a and b.

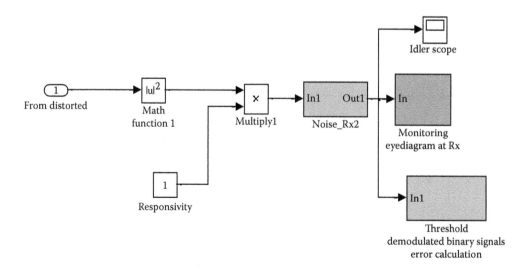

FIGURE 8.24
Simulink model of the optical receiving sub-system including noises and monitoring the receiver performance.

TABLE 8.3

Simulation Parameters

HNLF	Wavelengths	Power
$D = 0.02$ ps/(nm.km),	$\lambda_c = 1559$ nm	$P_s^{peak} = 1$ mW
$S = 0.03$ ps/(nm2.km)	$\lambda_s = 1548$ nm	$Ppump = 1$ W
$A_{eff} = 12$ um²	$\lambda_p = 1560.07$ nm	
$\gamma = 0.01345$ W⁻¹/m,	$\Delta\lambda_{BPF} = 0.64$ nm	
$\alpha = 0.5$ dB/km, L = 500 m		

8.4.3.2 Demultiplexing of the Optical Time Division Multiplexed Signal

Innovations of the OTDM enable the feasibility of the Tbaud transmission for terabyte per second Ethernet applications which meet the increasing demand in optical networks. One of the key ultrafast OTDM technologies is optical switching/demultiplexing of low bit rate individual channels from an ultra-high rate OTDM signal. The demultiplexing of a 1.28 Tbaud OTDM signal using FWM in tens of meters of HNLF has been reported. The working principle of the FWM-based demultiplexing is described in Figure 8.27. However the advances in nonlinear waveguide fabrication make the integrated all-optical solutions feasible and more advantageous. The promising optical materials for large nonlinear waveguide have been demonstrated recently, such as As_2S_3 and TeO_2. These nonlinear waveguides offer a lot of advantages such as no free-carrier absorption, they are stable at room temperature, there is no requirement of quasi-phase matching and the possibility of dispersion engineering.

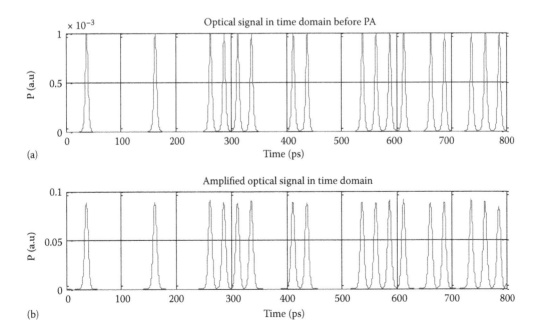

(a)

(b)

FIGURE 8.25
Time traces of the optical signal: (a) before, and (b) after the parametric amplifier.

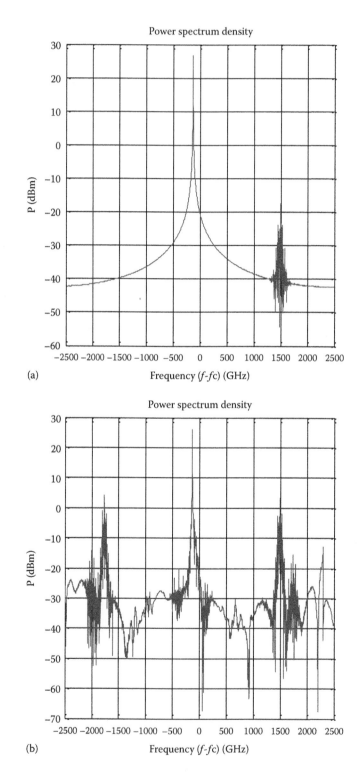

FIGURE 8.26
Corresponding spectra of the optical signal: (a) before, and (b) after the parametric amplifier.

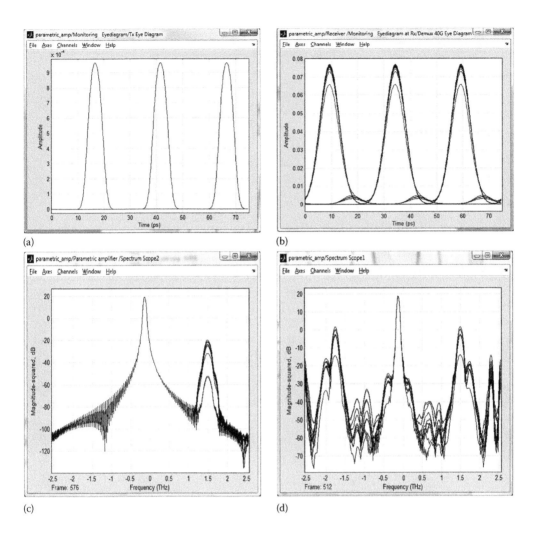

(a)   (b)

(c)   (d)

FIGURE 8.27
Spectrum of signal and pump wavelength: (a) pump and signals; (b) eyediagram of amplified signals; (c) spectrum of pump and signals; (d) parametric amplified spectrum.

The demultiplexing OTDM signal is simulated to demonstrate the validity of our model in different cases including signal propagation in nonlinear waveguide. The simulation is achieved by FWM-based demultiplexing of 40 Gb/s signal from the aggregate 160 Gb/s OTDM signal using the highly nonlinear waveguide. We note that the model presented in this section is to demonstrate the correctness of the FWM process in a nonlinear optical waveguide by the split step method processing of the NLSE. The important parameters in Table 8.4 are used in the simulation.

Figure 8.28 shows the spectra of input and output of the waveguide, respectively. The output spectrum shows the idler signal generated by the FWM then filtered by a bandpass filter (BPF) of 1 nm. Similar to the parametric amplification example, a Simulink model of OTDM demultiplexing is constructed as shown in Figure 8.29. The parametric gain characteristic is shown in Figure 8.43 with the parameters specified in Table 8.3. Figure 8.30 shows the MATLAB Simulink mode.

TABLE 8.4

Parameters of Nonlinear Waveguide

Nonlinear Waveguide	Wavelengths	Power
D = 28 ps/(nm.km),	λ_c = 1559 nm	$P_s{}^{peak}$ = 1 mW
S = 0.03 ps/(nm².km)	λ_s = 1548 nm	P_{pump} = 1 W
A_{eff} = 1 um²	λ_p = 1560.07 nm	
γ = 10.9 W⁻¹/m,	$\Delta\lambda_{BPF}$ = 0.64 nm	
α = 0.5 dB/cm, L = 7 cm		

8.4.3.3 Triple Correlation Simulation Model

One of the possible applications exploiting the $\chi^{(3)}$ nonlinearity of the nonlinear waveguide materials is implementation of the generation of the triple correlation product in optical domain. A MATLAB Simulink model set up for generating the triple product is shown in Figure 8.30a and the summation of three delayed signals and the original are structured in Figure 8.30b. The optical transmitter is also included. An optical BPF is placed at the output of the triple product to filter the FWM generated signals that means to extract the FWM signal spectrum. This is essential in order to reduce the noises contributed by the unfiltered original signal. Note that the two delayed signals are frequency shifted by wavelength converters in order to satisfy the phase matching condition. The spectrum of the combined signals is shown in Figure 8.31. The output of the nonlinear waveguide is monitored by a spectrum analyzer whose spectrum is shown in Figure 8.32. The FWM process is modeled by using the nonlinear coupled equations, Equations 8.66 through 8.68. The nonlinear coefficients are taken as those of As₂S₃ material. The FWM signals are then filtered by an optical BPF whose spectrum is expected to represent that of the triple product as shown in Figure 8.33.

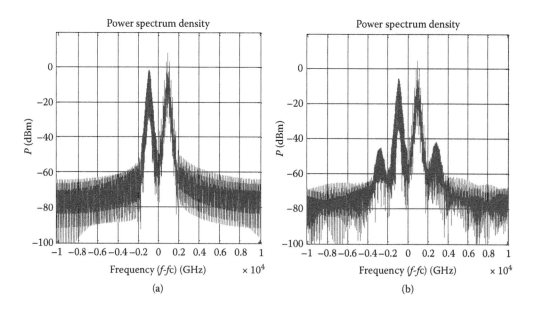

FIGURE 8.28
Spectra of optical signals at the input (a) and output (b) of a nonlinear optical waveguide.

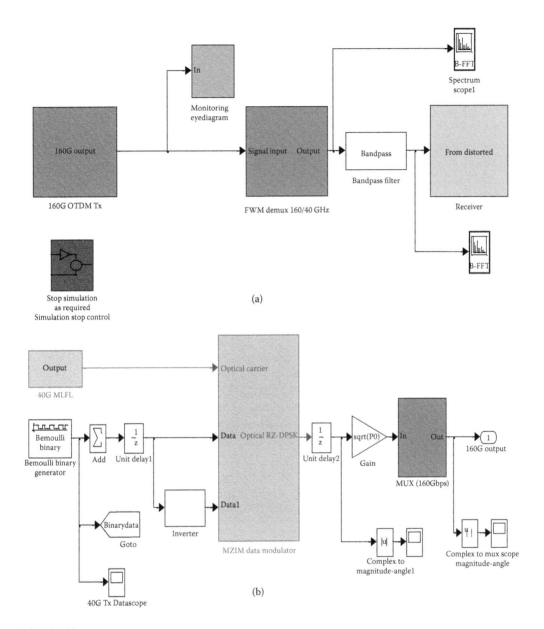

FIGURE 8.29
MATLAB Simulink model of parametric amplification of pulse propagating through nonlinear optical wave-guide: (a) general model of system for parametric amplification and FWM of the triple correlation generator; (b) optical transmitter for generation of short pulse to emulate signal input signals; (c) time domain multiplexed delayed signals at output port to be fed to the nonlinear optical waveguide; (d) short pulse pump sources and interaction in nonlinear optical waveguide—Simulink model.

The temporal signals at different stages of the triple product generator are shown in Figure 8.34. The upper trace depicts the summation of the original and the two delayed signals, while the middle trace is that of the pre-filtered FWM signals and the lower trace is the triple product signals at the output of the optical BPF. This triple product signal in the time domain will be fed to a bispectrum analyzer to determine the signals and noises.

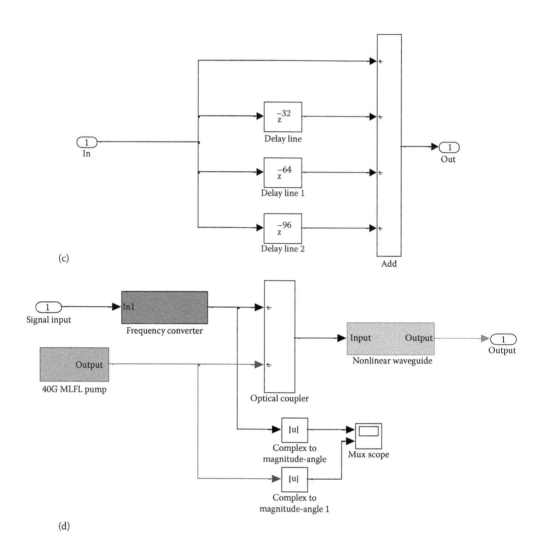

(c)

(d)

FIGURE 8.29 (*Continued*)

8.4.3.4 Concluding Remarks

This section has demonstrated the simulation for the generation of triple product in the optical domain. The main component is the nonlinear optical waveguide whose nonlinear coefficient is about 1000 times larger than that of silica fiber, through which the FWM is employed with the wavelength of the channels/signals shifted to appropriate positions to take advantage of the phase matching, thence the conversion of these signals to a triple product in the time domain.

Furthermore we have also demonstrated the demultiplexing of OTDM signals by pumping very short pulse sequence into the nonlinear waveguide to convert the sequence to a shifted wavelength optical region and filtering out the desired temporal data sequence. This is implemented in order to justify the correctness of our modeling of the nonlinear waveguide and the FWM the spectra of the signals at the input and output ports of the nonlinear waveguide after the FWM effect are shown in Figures 8.35a and 8.35b respectively.

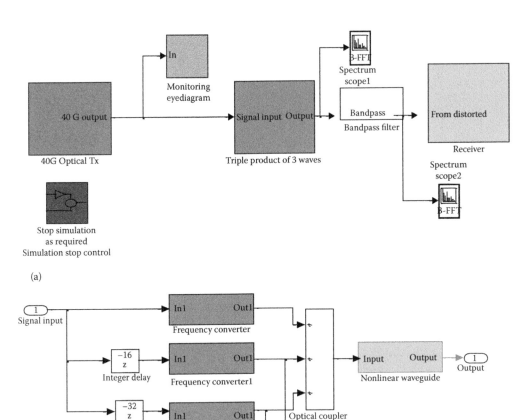

FIGURE 8.30
MATLAB Simulink model set-up for generation of triple product of the signals: (a) generic model; (b) three beam, signal and two delayed copied signals.

8.5 Demultiplexing 320 Gb/s Optical Time Division Multiplexed-Differential Quadrature Phase Shift Keying Signals Using Parametric Conversion in Nonlinear Optical Waveguides

In this section we demonstrate a system application of the use of the parametric conversion in the demultiplexing and demodulation of 320 Gb/s OTDM-DQPSK signals using either "conventional", i.e., demultiplexing then detection, or coherent detections with the local oscillator (LO) operating in pulsed sequence. A continuous-wave LO can be replaced by a short-pulse laser source in a homodyne coherent receiver enabling us to simultaneously demultiplex and detect the multiplexed channels of the ultra-high speed OTDM phase modulated signals. Simulated results also indicate at least 5 dB improvement of the receiver sensitivity over the conventional technique by the pulsed LO coherent receiver.

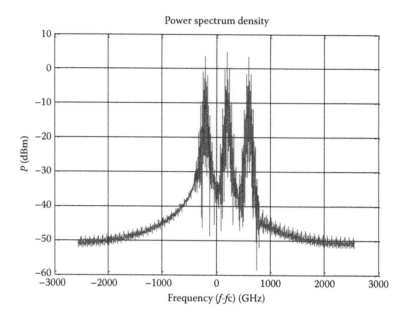

FIGURE 8.31

Spectrum of the original and two delayed copied signals—f_c is center frequency—reference frequency is normally set at 1550 nm for the central data channel of dense wavelength division multiplexed optical systems.

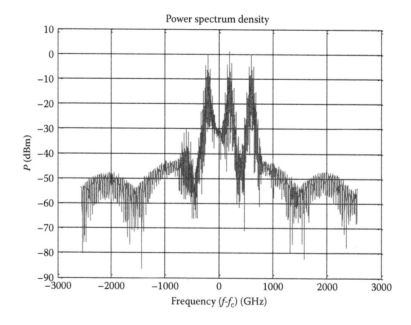

FIGURE 8.32

Spectrum of the FWM generated signals of the three delayed signals.

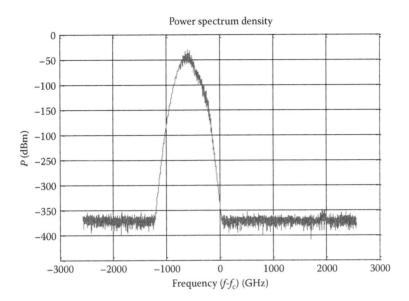

FIGURE 8.33
Spectrum of the filtered FWM signals of the three delayed signals, which is the triple correlated product at the output of the nonlinear waveguide.

Figures 8.36a and 8.36b shows the pump signal sequence to be feed to the optical transmitter and the received eye diagram at the output data sequence at the receiver. Then Figure 8.37 depicts multiplexed pulse sequence (upper curve) and the traces (lower waveform) displays the demultiplexed sequence. Figure 8.38 depicts the time domain waveform of the triple product pulse received at the output of the nonlinear optical waveguide. Figure 8.39 splays the decoded pulse sequence from product of (See Color Insert) Upper curve depicts multiplexed pulse sequence; lower traces displays demultiplexed sequence. Figure 8.40 displays the simulated spectrum of pump source and signals before FWM generation under parametric amplification. Figure 8.41 shows the spectrum of original signals and FWM generated signals and the pump beam after the parametric amplification process in the guided nonlinear medium. Figure 8.42 shows the spectrum of filtered signals after the nonlinear FWM effects in nonlinear optical waveguides. Then Figure 8.43 shows the simulated gain for a fiber OPA with parameters as given in Table 8.3. Finally Figure 8.44 shows the recovered amplitude modulated sequence in the time domain.

8.5.1 Introduction

The demand of high capacity in communication networks has been dramatically increasing year by year. The ultra-high data rate channels toward the terabyte per second regime for Ethernet application is required in future optical networks. This demand poses several technical challenges in the transmission physical layer. One of feasible technologies for terabyte per second Ethernet is the OTDM. Moreover the combination of OTDM with multilevel modulation formats such as DQPSK or quadrature amplitude modulation (QAM) easily enables the implementation of single channel several terabyte per second transmissions [20]. The optical transmitter can be implemented without much difficulty. The

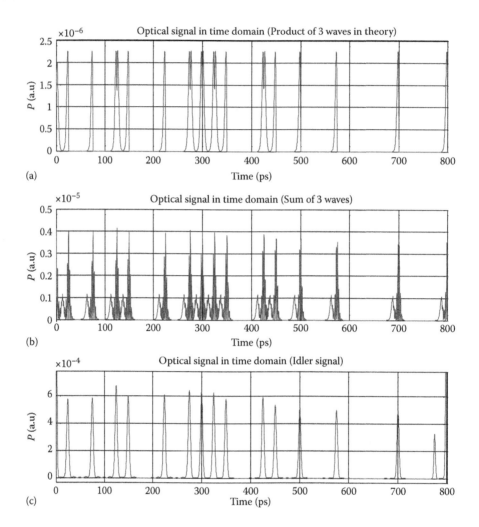

FIGURE 8.34
Waveforms: (a) MATLAB theoretical triple product signals upper sequence; (b) filtered sequence; (c) triple product sequence in time domain at the output of the nonlinear optical waveguide and the optical BPF.

most challenging issues of this advanced OTDM system rests on the receiving process that requires the realization of two principal functions: demultiplexing and demodulation.

In a "conventional" OTDM receiver, two functions are separately executed. Initially, the OTDM signal sequence is demultiplexed to lower rate tributaries which are then demodulated by incoherent receivers to detect the data stream [21]. On the other hand, OTDM receivers based on homodyne coherent detection have also been recently proposed [20,22]. When the LO is a short-pulsed optical sequence, both functions can be simultaneously performed. Thus this coherent receiver much simplifies the OTDM receiving sub-system. Furthermore the digital signal processors (DSP) can also be integrated in the coherent receiver to improve its sensitivity and possibly to mitigate the transmission impairments. However, a comprehensive comparison of the performance of these OTDM receiver structures has not been conducted.

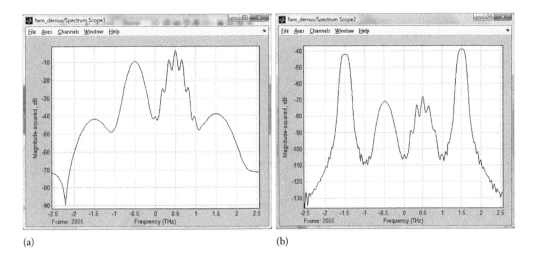

(a)   (b)

FIGURE 8.35
Spectra of signal input: (a) and at output, (b) of the nonlinear optical waveguide.

In this section we demonstrate the implementation of both receiver structures based on the Simulink modeling platform developed for simulation of optical communication systems [23] and evaluate the performance of these receivers for OTDM sequences under DQPSK modulation formats. The paper is organized as follows: Section 8.5.2 gives us some basic principles of OTDM-DQPSK system as well as demultiplexing and demodulation functions in both incoherent and coherent receivers. Simulation models based on Simulink platforms are introduced in Section 8.5.3. Section 8.5.4 shows the simulation results to compare the performance of both receivers which indicates a remarkable feature of the coherent OTDM receiver. Finally, the conclusions are given in Section 8.6.

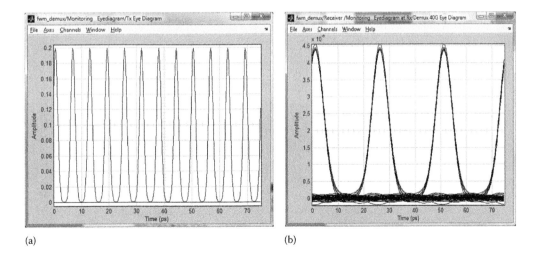

(a)   (b)

FIGURE 8.36
(a) Pump signal sequence to be feed to the optical transmitter. (b) Received eyediagram of the output data sequence at the receiver.

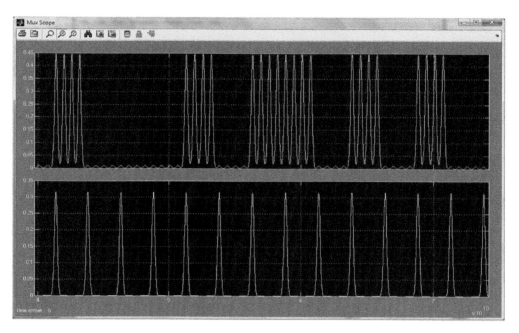

FIGURE 8.37
(See Color Insert) Upper curve depicts multiplexed pulse sequence; lower traces displays demultiplexed sequence.

8.5.2 Operational Principles

Figure 8.45 shows a typical configuration of an OTDM transmission system using advanced modulation format. At the transmitter, a mode-locked laser (MLL) is used to generate the ultra-short pulses before passing through the DQPSK modulator. The DQPSK signals at the output of tributary transmitters proceed to the optical time multiplexer which is structured by accurate optical delay lines to combine all tributaries into an ultra-high speed OTDM signal. The transmission link is characterized by the M number of fiber spans. Each span consists of a standard single mode fiber (SSMF), a

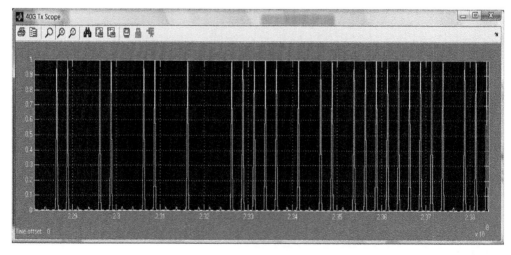

FIGURE 8.38
(See Color Insert) Triple product pulse received at the output of the nonlinear optical waveguide.

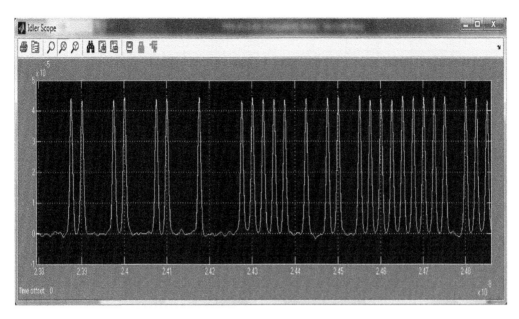

FIGURE 8.39
(See Color Insert) Decoded pulse sequence from product of Figure 8.38.

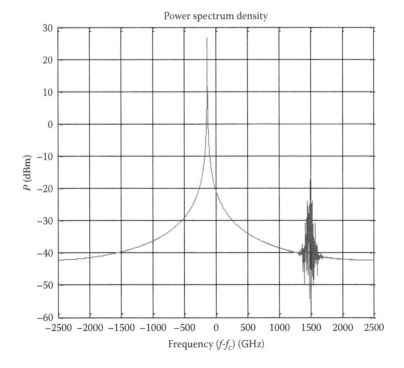

FIGURE 8.40
Spectrum of pump source and signals before FWM generation with parametric amplification.

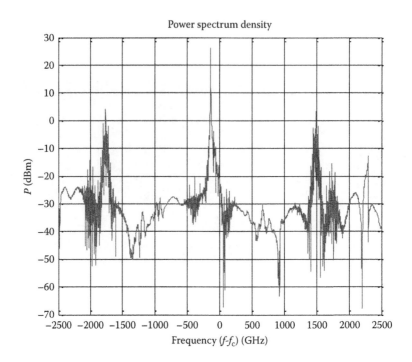

FIGURE 8.41
Spectrum of original signals and FWM generated signals and the pump beam after the parametric amplification.

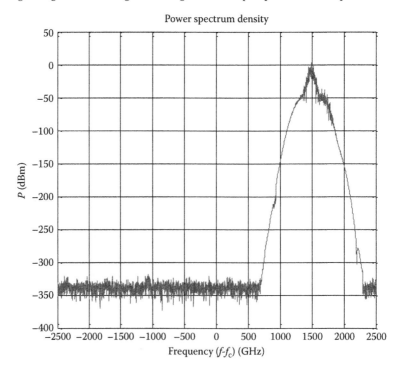

FIGURE 8.42
Spectrum of filtered signals after the nonlinear FWM effects in nonlinear optical waveguides.

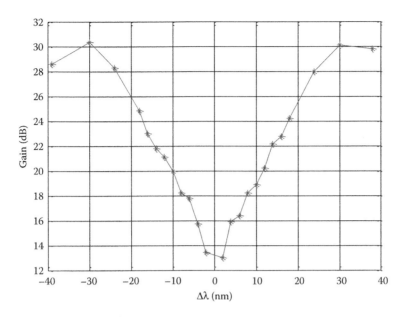

FIGURE 8.43
Simulated gain for a fiber OPA with parameters shown in Table 8.3.

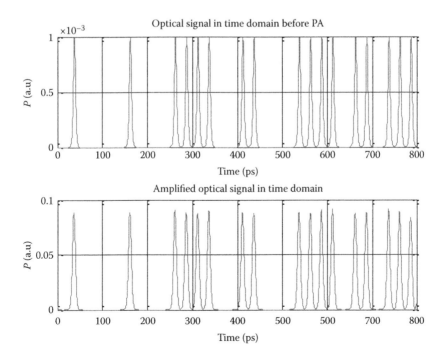

FIGURE 8.44
Recovered amplitude modulated sequence.

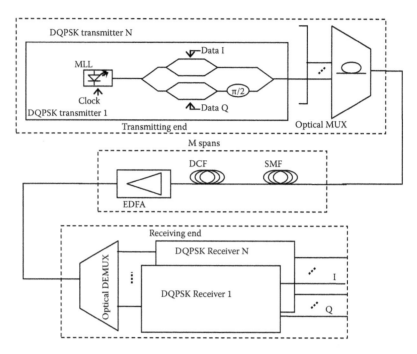

FIGURE 8.45
A typical configuration of a DQPSK-OTDM transmission system.

dispersion compensating fiber (DCF) and Er:doped fiber amplifiers (EDFAs) which fully equalize the total loss of the whole span. The receiving end implements two important functions as mentioned above, the demultiplexing of the OTDM signal and the demodulation of DQPSK states. The structure of an OTDM-DQPSK receiver can be classified into two types: the conventional receiver (demultiplexer and incoherent receiver) and the coherent receiver.

8.5.2.1 Conventional Demultiplexing Technique

8.5.2.1.1 Demultiplexing

Most demultiplexing techniques are based on the nonlinearity in optical fibers/waveguides such as HNLF [24], semiconductor optical amplifier (SOA) [25,26] and optical planar waveguides [27]. These techniques rely on the exploitation of the nonlinear effects in the nonlinear waveguide such as cross-phase modulation (XPM) or FWM to enable switching at ultra-high speed. The most popular structure of the OTDM demultiplexer is the nonlinear optical loop mirror (NOLM) using HNLF. However its stability is still a serious obstacle. Recently, nonlinear waveguides have emerged as a promising device for ultra-high speed photonic processing [28,29]. This device is very compact and the FWM is exploited demultiplexing as shown in Figure 8.45a. The control pulses generated from an MLL at tributary rate are pumped and co-propagated with the OTDM signal through the nonlinear waveguide. The mixing process between the control pulses and the OTDM signal during propagation through the nonlinear waveguide converts the desired tributary channel to a new idler wavelength. Then the

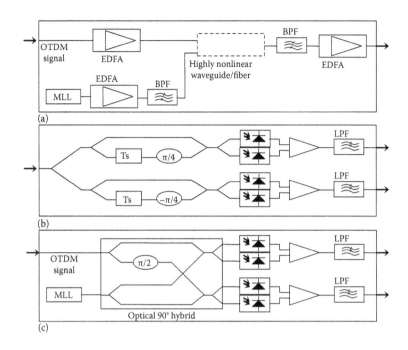

FIGURE 8.46
Configurations of: (a) nonlinear effect-based OTDM demultiplexer; (b) the incoherent DQPSK receiver; and (c) the coherent demultipexing-demodulation OTDM-DQPSK receiver.

demultiplexed signal at the idler wavelength is extracted by a BPF before going to a DQPSK demodulator.

8.5.2.1.2 *Demodulation of Differential Quadrature Phase Shift Keying Signal*

A DQPSK signal can be incoherently demodulated by two Mach-Zehnder delay interferometers (MZDI) followed balanced detector as shown in Figure 8.46b. The inphase and quadrature (I&Q) components are detected by the conversion of the differential phase modulation into intensity modulation through the interference in MZDIs [30]. In this receiver, the different length and the phase shift of the interferometer must be exactly tuned to avoid a high penalty in performance.

8.5.2.2 **Optical Coherent Demultiplexing and Demodulation**

Instead of the incoherent receiver, a homodyne coherent optical receiver with demultiplexing function has been proposed [21]. In this scheme, an ultra-short pulse laser is used as a LO to mix with the incoming OTDM signal. The I&Q components of only one tributary are directly demodulated by synchronizing the LO pulses with time slots of this tributary in the OTDM signal. Structure of a coherent demux-DQPSK receiver is shown in Figure 8.46c. Thus, the demultiplexing and demodulation functions are simultaneously implemented by using the coherent receiver with a pulsed LO source. This receiver offers a range of important advantages such as a simplification of the OTDM system structure, more flexible in processing the tributaries and a remarkable reduction of pump power. Furthermore, the digital signal processing (DSP) can be easily applied to compensate deteriorative effects on the signal.

8.5.3 Simulation Models

Simulation models of the OTDM-DQPSK system are developed from the Simulink modeling platform [21]. Figure 8.48 shows the diagrams and key simulation blocks of the whole system using FWM-based demultiplexing and incoherent receiver (Figure 8.49a) and using coherent receiver with demultiplexing function (Figure 8.49b). Simulink blocks of basic components in the system such as EDFA, transmission fibers, optical modulators and detectors are described in detail in Ref. [31]. In this section, some important parameters of the main blocks and a brief description of receiving models developed for the OTDM-DQPSK system are given.

8.5.3.1 Optical Time Division Multiplexed-Differential Quadrature Phase Shift Keying Transmitter

The description of the RZ-DQPSK transmitter model on the Simulink platform can be found in Table 8.5. However an MLL instead of a pulse carver is used in our setup to generate ultra-short pulses for the OTDM signal. Then the outputs of tributary transmitters are time-division multiplexed to generate the OTDM singal of 160 Gsymbols/s in our setup (Figures 8.47 and 8.48).

8.5.3.2 Fiber Link

Based on the configuration in Figure 8.47, the link in simulation is developed with ten spans. Blocks such as optical fibers and EDFA in each span are explained in detail in Ref. [31]. Propagation of the optical signal in fibers can be modeled by the NLSE in a nonlinear scheme or a transfer function in a linear scheme. EDFA block with amplified spontaneous emission (ASE) noise is modeled as a black box. The important parameters of the span are summarized in Table 8.1.

8.5.3.3 Demultiplexer and Receiver

Figure 8.49a shows the Simulink diagram of the FWM-based demultiplexer. The FWM process in the nonlinear waveguide to implement the demultiplexing function is also

TABLE 8.5

Important Parameters of Main Simulation Blocks in the OTDM-DQPSK System

---

OTDM-DQPSK Transmitter

MLL: $P_0 = 1$ mW, $T_p = 2.5$ ps, $f_m = 40$ GHz
Modulation format: DQPSK; OTDM multiplexer: 4×40 Gsymbols/s

Fiber transmission link

SMF: $L_{SMF} = 80$ km, $D_{SMF} = 17$ ps/nm/km, $\alpha = 0.2$ dB/km
DCF: $L_{DCF} = 20$ km, $D_{DCF} = -68$ ps/nm/km, $\alpha = 0.2$ dB/km
EDFA: Gain = 20 dB, NF = 5dB; Number of spans: 10

FWM demultiplexer

Pumped control: $P_p = 500$ mW, $T_p = 2.5$ps, $f_m = 40$GHz, $\lambda_p = 1556.55$ nm
Input signal: $P_s = 35$ mW, $\lambda_s = 1548.51$ nm
Waveguide: Lw = 7cm, Dw = 28 ps/km/nm, $\alpha = 0.5$ dB/cm, $\gamma = 10^4 \cdot 1/$W/km

Coherent receiver

MLL: $P0 = 12.5$ mW ($P_{av} \approx 1$ mW), $T_p = 2.5$ ps, $f_m = 40$ GHz
Balanced Rx: $B_e = 28$GHz, $i_{eq} = 20$ pA/Hz$^{1/2}$, $i_d = 10$ nA.

---

320 Gbits/s OTDM-DQPSK system with FWM base demux and incoherent demodulation

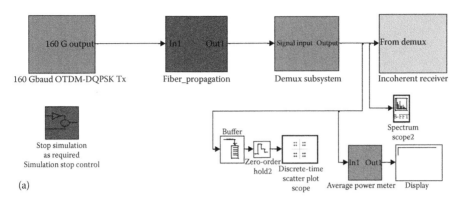

(a)

320 Gbits/s OTDM-DQPSK system with coherent demux and demodulation

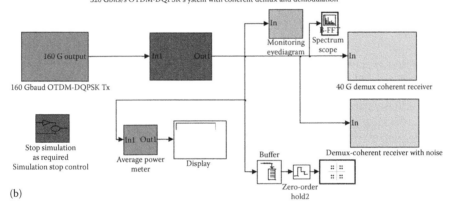

(b)

FIGURE 8.47
Simulink models of the OTDM-DQPSK transmission system using: (a) the conventional receiver; and (b) the coherent receiver.

modeled by NLSE 0. In the block, the complex fields of the signal and the pump are frequency shifted to the corresponding wavelengths (see Table 8.4) prior to the launching into the nonlinear waveguide. Optical Gaussian BPFs in the model are used to reduce the ASE noise and to extract the demultiplexed signal (the idler signal). Then the demultiplexed signal is incoherently demodulated by the MZDI receivers as described in Ref. [22].

A homodyne coherent receiver with demultiplexing function is developed as shown in Figure 8.49b. An MLL to generate short pulses similar to that in the transmitter is used as the LO. The phase noise of the laser is also included in the model and described as a complex Gaussian process with a variance $\sigma_\phi^2 = 4\pi\Delta fT$, where Δf is the linewidth of the longitudinal modes of pulsed laser, T is the symbol period. Mixing of the OTDM signal with the LO pulses takes place in the optical 90° hybrid followed by a couple of balanced detectors to detect directly I- and Q-components of the desired channel. We note that the peak power of control pulses in the coherent receiver is much lower than that in the conventional receiver.

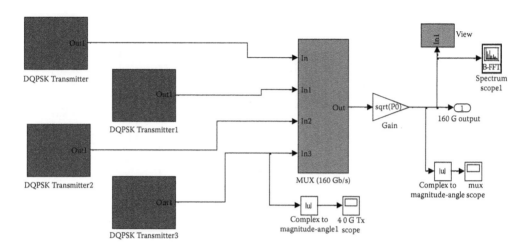

FIGURE 8.48
The Simulink model of the OTDM-DQPSK transmitter.

8.5.3.4 Performance of Optical Time Division Multiplexed-Differential Quadrature Phase Shift Keying Receivers: A Comparison

Figure 8.50a shows the spectrum of the signals after propagating through the nonlinear waveguide. Then the demultiplexed signal is extracted by the BPF as shown in Figure 8.50b, and d show the eyediagrams of the OTDM signal and the demultiplexed signal, respectively. After demultiplexing, the signal is demodulated by MZDI receivers, the signals of the tributaries are directly demultiplexed and demodulated in the coherent scheme. Figure 8.51 shows the constellations and the eyediagrams of the demultiplexed and demodulated tributary signal of the incoherent receiver and coherent receiver with laser linewidth of 5 MHz. Due to the nonlinear effect during propagation in the waveguide, the phase states of the DQPSK signal before the incoherentre receiver are rotated in the phase plane as shown in Figure 8.51a. In contrast, the phase states in the coherent receiver are only affected by the phase noise of LO laser besides the ASE noise from EDFAs.

Figure 8.52 shows the back to back performance of two OTDM-DQPSK receivers. The coherent receiver with different laser linewidths offers much better performance than the incoherent receiver. In the conventional receiver the use of many EDFAs in the demultiplexing process also reduces the signal to noise ratio which degrades the performance of the receiver. Furthermore, the nonlinear propagation in the waveguide may also enhance the phase noise which causes a highin the BER curve of the incoherent scheme. Figure 8.53 shows the performance after transmission through the link of 1000 km (10 fully dispersion-compensated spans). While the incoherent scheme shows a strong degradation of the performance after transmission, the coherent receiver still remains good bit error rate (BER) performance.

8.5.4 Influence of Synchronization

Synchronization plays a key factor in the ultra-high OTDM system. To ensure maximum efficiency of demultiplexing, the control pulses must be synchronized with the time slots of the demultiplexed tributary in the OTDM signal as shown in Figure 8.50. Figure 8.52

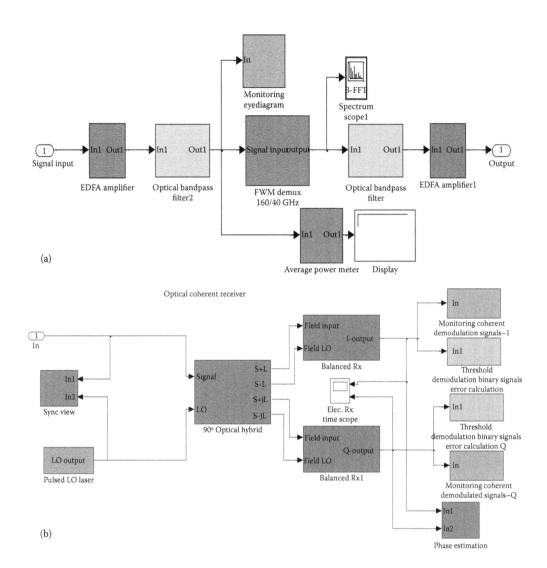

FIGURE 8.49
The Simulink models of: (a) FWM-based demultiplexer; and (b) coherent receiver with pulsed LO laser.

shows the back to back performance of incoherent receiver and coherent receiver with different phase noise levels of pulsed laser. On other hand, the performance of demultiplexing is degraded by any time mismatche between the control and OTDM pulses as displayed in Figures 8.54 and 8.55.

In the FWM-based demultiplexer, the efficiency of demultiplexing is proportional to the power of the idler signal, then the currents from the receiver depends on the square of pumped power of the control pulses $i_{RX} \propto P_p^2 P_s$, where P_p is the pump power, P_s is the power of the OTDM signal. While the currents from the coherent receiver are proportional to the square root of LO power: $i_{RX} \propto \sqrt{P_{LO}P_s}$, where P_{LO} is the power of LO. Therefore the signal level in the incoherent receiver is more sensitive to the mismatch of the synchronization than that in the coherent receiver. Figure 8.56 shows the eyediagrams of demodulated signals in the incoherent and coherent receivers at different synchronization delays. They

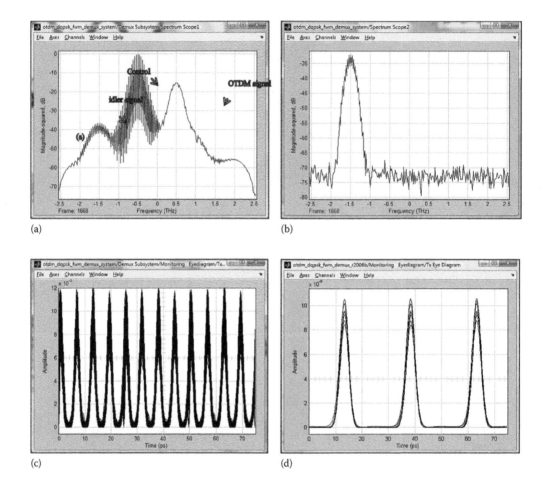

FIGURE 8.50
Spectra at the outputs: (a) of nonlinear waveguide; (b) of optical BPF; the eyediagrams of (c) the OTDM signal; and (d) demultiplexed signal.

also indicate that the eye opening of the signals in the coherent receiver is better than that in the incoherent receiver.

Figure 8.57 depicts the BER curves versus the synchronization delay in terms of the pulse period of both types of OTDM receivers. The results indicate a larger delay tolerance of the coherent receiver which facilitates the requirements of the OTDM-DQPSK receiver structure.

8.6 Concluding Remarks

In conclusion, this example shows the effectiveness of the *E*-vector finite element analysis to obtain the solution of the nonlinear guided wave problems. The algorithm adopted here includes the normalization, the acceleration techniques for reaching the eigenvalue

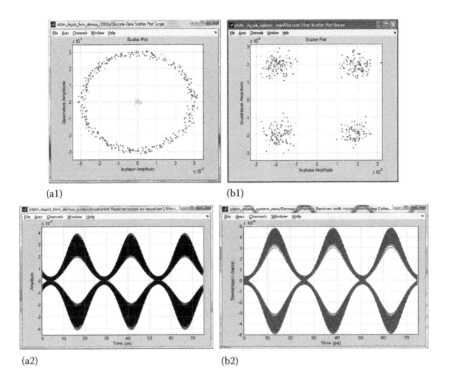

(a1) (b1)

(a2) (b2)

FIGURE 8.51
Constellations of the demultiplexed signal and the eyediagrams of demodulated signals of (a1) incoherent receiver and (b1) coherent receiver; and recovered eye diagram (a2) incoherent (b2) coherent.

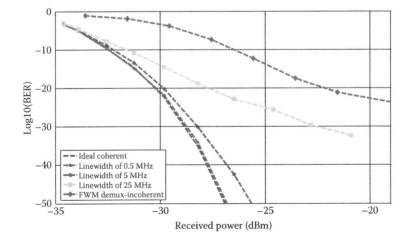

FIGURE 8.52
(See Color Insert) Back to back performance of incoherent receiver and coherent receiver with different phase noise levels of pulsed laser.

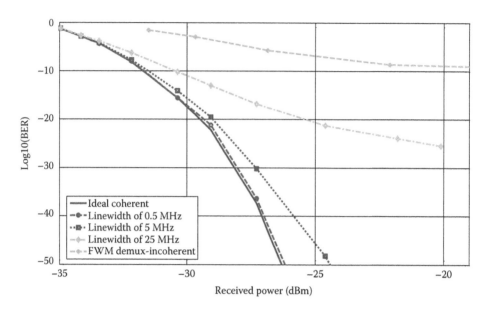

FIGURE 8.53
(See Color Insert) Transmission performance of incoherent receiver and coherent receiver with different phase noise levels of pulsed laser.

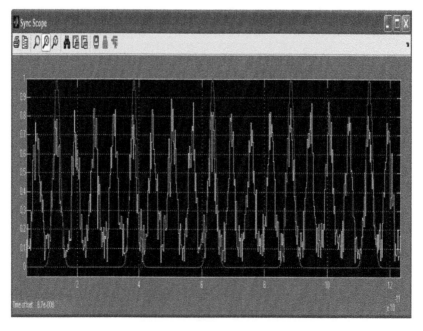

FIGURE 8.54
(See Color Insert) Synchronization between the control and the OTDM pulses.

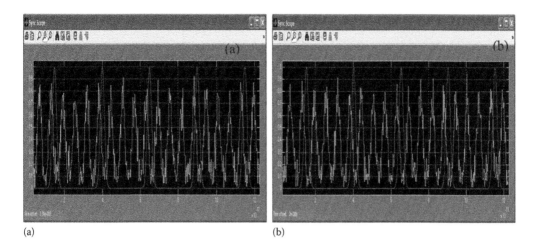

(a)   (b)

FIGURE 8.55
The time traces of the control and the OTDM signals at different synchronization delays: (a) ~19%, and (b) ~31% of the pulse period.

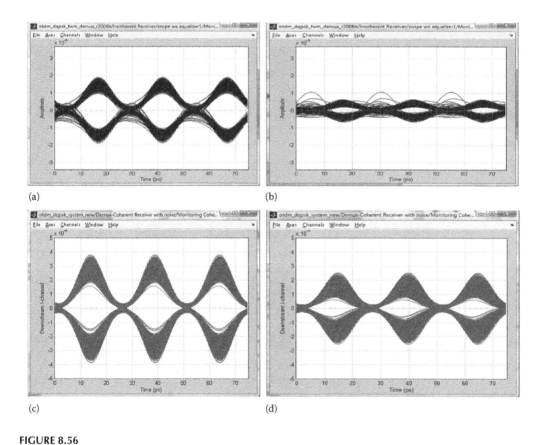

(a)   (b)

(c)   (d)

FIGURE 8.56
Eyediagram of the demodulated signal at two synchonization delays ~19% (a and c) and ~31% (b and d) pulse-period of the incoherent (above) and coherent (below) receivers.

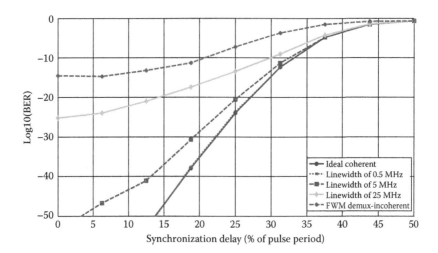

FIGURE 8.57
(See Color Insert) BER curves versus the synchronization delay.

solutions, and a nonlinear iteration scheme. Switching of the lightwave beam can be observed and produced.

A nonlinear optical waveguide can also be used as an optical signal processor such as time division demultiplexing and triple correlation via the FWM nonlinear effects. The details of the use of nonlinearity in guided wave photonic devices will be described in Chapter 10.

8.7 Problems

1. a. What do we mean by "nonlinear optical waveguide"? What region of the wave-guide should be nonlinear? What is the type of nonlinearity of the materials?

   b. Describe the switching behavior in such nonlinear optical waveguides.

2. Refer to Figure 8.4. What do we mean by the "effective refractive index" in such nonlinear waveguide structures? At what power level does the nonlinear switching occur?

References

1. A. Konrad, "Higher order triangular finite elements for electromagnetic waves in anisotropic media," *IEEE Microw. Th. Tech.*, MTT-25, 253–360, 1977.
2. B.M.A. Rahman and J.D. Davies, "Finite element analysis of optical and microwave waveguide problems," *IEEE Trans. Microw. Th. Tech.*, MTT-32, 20–28, 1984.
3. M. Koshiba, K. Hayata, and M. Suzuki, "Finite element solution of anisotropic waveguides with tensor permittivity," *IEEE J. Lightw. Tech.*, LT-4, 121–126, 1986.

4. J.P. Webb, "Finite element analysis of dispersion in waveguides with sharp metal edges," *IEEE Trans. Microw. Th. Tech.*, MTT-36, 1819–1824, 1988.

5. W.J. English and F.J. Young, "An E vector variational formulation of the Maxwell equations for cylindrical waveguide problems," *IEEE Trans. Microw. Th. Tech.*, MTT-19, 40–46, 1971.

6. M. Koshiba, K. Hayata, and M. Suzhuki, "Finite element formulation in terms of the dielectric field vector for electromagnetic waveguide problems," *IEEE Trans. Microw. Th. Tech.*, MTT-33, 900–905, 1985.

7. G.P. Agrawal, *"Nonlinear Fiber Optics,"* Academic Press, New York, 2002.

8. L.A. Hageman and D.M. Young, *"Applied Iterative Method,"* Academic Press, New York, 1981.

9. Jennings, *"Matrix Computation for Engineers and Scientists,"* John Wiley and Sons, NY, 1977.

10. H.R. Schwarz, H. Rutishauser, and E. Stiefel, *"Numerical Analysis of Symmetric Matrices,"* Prentice Hall, NJ, 1973.

11. C.B. Moler, *"Finite Difference Methods for the Eigenvalues of Laplace's Operator,"* Computer Science Department, Standford University, Stanford, CA, Tech report, 1965.

12. J.H. Wilkinson, *"The Algebraic Eigenvalue Problem,"* Oxford University Press, London, 1965.

13. G.F. Carey and J.T. Oden, *"Finite Elements: Computational Aspectsm,"* Vol. III, Prentice Hall, NJ, 1984.

14. A.C. Aitkin, "On the iterative solution of a system of linear equations," *Proc. Royal Soc., Edinburg*, 63, 52–60, 1950.

15. K. Hayata and M. Koshiba, "Full vectorial analysis of nonlinear optical waveguides," *J. Opt. Soc. Am. B*, 2491–2501, 1988.

16. K. Hayata and M. Koshiba, "Three-dimensional simulation of guided-wave second-harmonic generation in the form of coherent Čerenkov radiation", *Optics Letters*, 16(20), 1563–1565, 1991.

17. P.P. Silvester and R.L. Ferrari, *"Finite Elements for Electrical Engineers,"* Cambridge University Press, 1983.

18. G.P. Agrawal, *"Nonlinear Fiber Optics,"* Academics, 2004.

19. J. Hansryd, P.A. Andrekson, M. Westlund, L. Jie, and P. Hedekvist, "Fiber-based optical parametric amplifiers and their applications," *IEEE J. Sel. Topics Quant. Elect.*, 8(3), 506–520, 2002.

20. H.G. Weber, S. Ferber, M. Kroh, C. Schmidt-Langhorst, R. Ludwig, V. Marembert, C. Boerner, F. Futami, S. Watanabe, and C. Schubert, "Single channel 1.28 Tbit/s and 2.56 Tbit/s DQPSK Transmission," *Electron. Lett.*, 42(3), 178–179, 2006.

21. C. Zhang, Y. Mori, K. Igarashi, K. Katoh, and K. Kikuchi, "Ultrafast operation of digital coherent receivers using their time-division demultiplexing function," *J. Lightw. Technol.*, 27(3), 224–232, 2009.

22. D. Ly-Gagnon, S. Tsukamoto, K. Katoh, and K. Kikuchi, "Coherent detection of optical quadrature phase-shift keying signals with carrier phase estimation," *J. Lightw. Technol.*, 24(1), 12–21, Jan 2006.

23. L.N. Binh, *"Optical Fiber Communications Systems: Theory and Practice with Matlab and Simulink Models,"* CRC Press, 2010.

24. H.C.H. Mulvad, L.K. Oxenlwe, M. Galili, A.T. Clausen, L. Gruner-Nielsen, and P. Jeppesen, "1.28 Tbit/s single-polarization serial OOK optical data generation and demultiplexing," *Electron. Lett.*, 45(5), 280–281, Feb 2009.

25. T. Morioka, H. Takara, S. Kawanishi, K. Uchiyama, and M. Saruwatari, "Polarization independent all-optical demultiplexing up to 200 Gbit/s using four-wave mixing in a semiconductor laser amplifier," *Electron. Lett.*, 32(9), 840–842, 1996.

26. S. Kodama, T. Yoshimatsu, and H. Ito, "320Gbit/s error-free demultiplexing using ultrafast optical gate monolithically integrating a photodiode and electroabsorption modulator," *Electron. Lett.*, 39(17), 1269–1270, 2003.

27. A.P. Agrawal, *"Nonlinear fiber optics,"* Academic Press, 3rd Ed., 2001.

28. M.D. Pelusi, V.G. Ta'eed, Libin Fu, E. Magi, M.R.E. Lamont, S. Madden, Duk-Yong Choi, D.A.P. Bulla, B. Luther-Davies, and B.J. Eggleton, "Applications of highly-nonlinear chalcogenide glass devices tailored for high-speed all-optical signal processing," *IEEE J. Sel. Topics Quantum Electron.*, 14, 529–539, 2008.

29. T.D. Vo, H. Hu, M. Galili, E. Palushani, J. Xu, L.K. Oxenløwe, S.J. Madden, D.Y. Choi, D.A. Bulla, M.D. Pelusi, J. Schröder, B. Luther-Davies, and B.J. Eggleton, *"Photonic Chip Based 1.28 Tbaud Transmitter Optimization and Receiver OTDM Demultiplexing,"* Proc. OFC 2010, PDP?, OFC2010, San Diego.
30. P. Winzer, "Advanced optical modulation formats," *Proc. IEEE*, 94(5), 952–985, May 2006.
31. L.N. Binh, *"Digital Optical Communications"* CRC Press, Boca Raton, 2008.

9

Integrated Guided-Wave Photonic Transmitters

Modern optical fiber communications can operate at ultra-high speed of several gigabits per second and over very long distances. This is possible due to the following contributions: (i) single mode guiding in optical fibers; (ii) preservation of single longitudinal mode lightwave source by conducting the modulation of the lightwaves via an external modulator which is normally implemented in an integrated planar structure; (iii) in line optical amplification in the photonic domain; and (iv) an advanced modulation format to generate spectral efficient optical signals to combat various impairments of the transmission medium.

This chapter deals with modulation of the lightwaves using an external modulator of integrated structure, the guided wave photonic modulators, especially the Ti:diffused LiNbO$_3$ type. The fabrication of these kind of modulators can be found in Appendix 8.

9.1 Introduction

A photonic transmitter consists of a single or multiple lightwave source which can be either modulated directly by manipulating the driving current of the laser diode or externally via an integrated optical modulator. This chapter presents the techniques for modulation of lightwaves externally, not directly manipulating the stimulated emission from inside the laser cavity, via the use of electro-optic (EO) effects. Advanced modulation formats [50] have recently attracted much attention for enhancement of the transmission efficiency since the mid 1980s for coherent optical communications. Hence the preservation of the narrow linewidth of the laser source is critical for operation bit rates in the range of several tens of gigabits per second. Thus the external modulation is essential.

A photonic transmitter consists of a structure of optical waveguides in which some sections are incorporated electrodes for biasing and also for launching the microwave electrical signal in order to modulate the phase of the optical carriers. This phase modulation (PM) would result in the depletion or constructive interference of the lightwave at the output of an interference optical guide wave structure. Thus the design of the guided wave structure and that of the electrode, normally a traveling wave type, is very important in order to maximize the PM effect.

Three typical types of optical modulators presented in this chapter use the three-dimensional (3D) optical waveguides, especially the diffused optical waveguides described in Chapter 3. The lithium niobate (LiNbO$_3$) EO modulators are extensively employed. Other typical types such as the electro-absorption (EA) structures are also briefly described. Their operating principles, devices physical structures, device parameters and their applications and driving condition for generation of different modulation formats as well as their impacts on system performance are described. Basic optical effects, such as the interference of two lightwave beams, are extensively used in order to establish a constructive or destructive effect on the intensity of the guided optical wave, hence intensity modulation by the use of the PM of the lightwaves via one path of the interference structure.

Thus this chapter is structured as follows: Section 9.2 gives a general introduction to optical modulators, thence the uses of optical modulators to generate advanced modulation formats for long-haul optically amplified communications systems. Section 9.3 gives a modeling method and fabrication of ultra-broadband electrodes whereby the PM of the lightwaves can be implemented for ultra-band signal generation. Section 9.4 gives a brief overview of the modulators and demands on the fabrication of the electrodes and optical waveguides, including the effects of a non-ideal brick-wall like structure of the traveling wave electrodes. Section 9.6 gives an overview of modulation techniques to manipulate the amplitude, phase or frequency of the lightwave carrier using EO effects. The details of the fabrication of the optical modulators are given in Appendix 9.

9.2 Optical Modulators

The modulation of lightwaves via an external optical modulator can be classified into three types depending on the special effects that alter the lightwaves' property [1], especially the intensity or the phase of the lightwave carrier. In an external modulator the intensity is normally manipulated by manipulating the phase of the carrier lightwaves guided in one path of an interferometer. A Mach-Zehnder interferometric structure using integrated optical waveguide is the most common type [1–25].

An EA modulator employs the Franz and Keldysh effect which is observed as lengthening the wavelength of the absorption edge of a semiconductor medium under the influence of an electric field [22–24]. In a quantum structure such as the multi-quantum well structure this effect is called the Stark effect, or the EA effect. The EA modulator can be integrated with a laser structure on the same integrated circuit chip. For the LiNbO$_3$ modulator, the device is externally connected to a laser source via an optical fiber.

The total insertion loss of a semiconductor intensity modulator is about 8–10 dB including fiber-waveguide coupling loss which is rather high. However, this loss can be compensated by a semiconductor optical amplifier (SOA) that can be integrated on the same circuit. Comparing with LiNbO$_3$, its total insertion loss is about 3–4 dB which can be affordable as Er-doped fiber amplifier (EDFA) is now readily available.

The driving voltage for the EA modulator is usually lower than that required for LiNbO$_3$. However the extension ratio is not as high as that of the LiNbO$_3$ type which is about 25 dB as compared to 10 dB for the EA modulator. This feature contrasts the operating characteristics of the LiNbO$_3$ and EA modulators. Although the driving voltage for the EA modulator is about 2–3 V and 5–7 V for LiNbO$_3$, the former type would be preferred for intensity or PM formats due to this extinction ratio that offers much lower "zero" noise level and hence high quality factor.

9.2.1 Phase Modulators

The phase modulator is a device which manipulates the "phase" of optical carrier signals under the influence of an electric field created by an applied voltage. When voltage is not applied to the radio frequency (RF) electrode, the number of periods of the lightwaves, n, exists in a certain path length. When voltage is applied to the RF-electrode, one or a fraction of one period of the wave is added, which now means $(n + 1)$ waves exist in the same length. In this case, the phase has been changed by 2π and the half voltage of this is called the driving voltage. In the case of long distance optical transmission, waveform is susceptible to degradation due to nonlinear effects, such as self-phase modulation (SPM), etc. A phase modulator can be used to alter the phase of the carrier to compensate for this degradation. The magnitude of the change of the phase depends on the change of the refractive index created via the EO effect that in turn depends on the orientation of the crystal axis with respect to the direction of the established electric field by the applied signal voltage.

An integrated optic phase modulator operates in a similar manner except that the lightwave carrier is guided via an optical waveguide which a diffused or ion-exchanged guiding regions for LiNbO$_3$, and rib-waveguide structures for semiconductor type. Two electrodes are deposited so that an electric field can be established across the waveguiding cross section (see Chapter 3) so that a change of the refractive index via the EO or EA effect occurs, as shown in Figure 9.1. For ultra-fast operation one of the electrodes is a traveling wave type, or hot electrode, and the other is a ground electrode. The traveling wave electrode must be terminated with matching impedance at the end so as to avoid wave

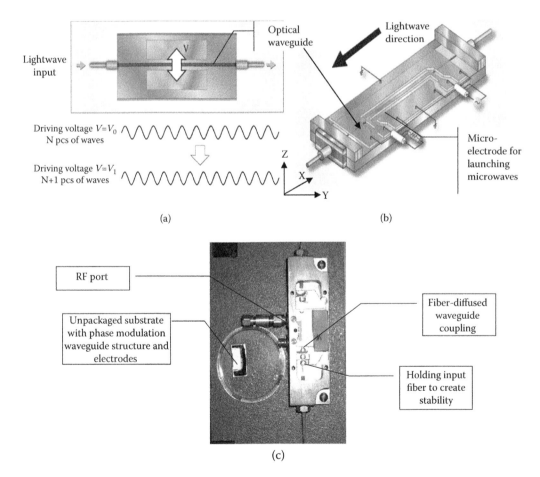

(a)

(b)

(c)

FIGURE 9.1
(See Color Insert) EO-PM in an integrated modulator using LiNbO$_3$. Electrode impedance matching is not shown. (a) Schematic diagram; (b) integrated optic structure; and (c) photograph of a packaged modulator.

reflection. Usually a quarter wavelength impedance is used to match the impedance of the traveling wave electrode to that of the 50 Ω transmission line.

A phasor representation of a phase-modulated lightwave can be by the circular rotation at a radial speed of ω_c. Thus the vector with an angle ϕ represents the magnitude and phase of the lightwave.

9.2.2 Intensity Modulators

A basic structured LiNbO$_3$ modulator comprises of: (i) two waveguides; (ii) two Y-junctions; and (iii) RF/DC traveling wave electrodes. Optical signals coming from the lightwave source are launched into the LiNbO$_3$ modulator through the polarization maintaining fiber, it is then equally split into two branches at the first Y-junction on the substrate. When no voltage is applied to the RF electrodes, the two signals are recombined constructively at the second Y-junction and coupled into a single output. In this case output signals from the LiNbO$_3$ modulator are recognized as "one." When voltage is applied to the RF electrode, due to the EO effects of the LiNbO$_3$ crystal substrate, the waveguide refractive index is changed, and hence the carrier phase in one arm is advanced though retarded in the other arm. Thence the two signals are recombined destructively at the second Y-junction, they are transformed into a higher order mode and radiated at the junction (Figure 9.2). If the phase retarding is in a multiple odd factor of p, the two signals are completely out of phase, the combined signals are radiated into the substrate and the output signal from the LiNbO$_3$ modulator is recognized as "zero." The voltage difference which induces this "zero" and "one" is called the driving voltage of the modulator, and is one of the important parameters in deciding the modulator's performance. The design of traveling wave electrodes for broadband operation is given in Section 9.6 of this chapter.

9.2.2.1 Phasor Representation and Transfer Characteristics

Consider that an interferometric intensity modulator consists of an input waveguide then split into two branches and recombined to a single output waveguide. If the two electrodes are initially biased with voltages V_{b1} and V_{b2} then the initial phases exerted on the lightwaves would be $\phi_1 = \pi V_{b1}/V_\pi = -\phi_2$ which are indicated by the bias vectors shown in Figure 9.3b.

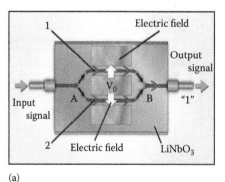
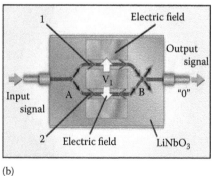

(a)   (b)

FIGURE 9.2

(See Color Insert) Intensity modulation using interferometric principles in guide wave structures in LiNbO$_3$: (a) ON—constructive interference mode; (b) destructive interference mode—OFF. Optical guided wave paths 1 and 2. Electric field is established across the optical waveguide.

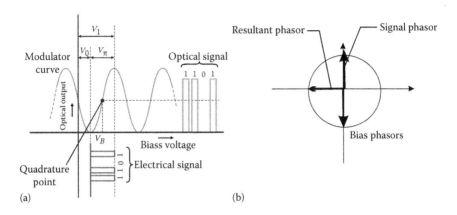

FIGURE 9.3
Electrical to optical transfer curve of an interferometric intensity modulator.

From these positions the phasors are swinging according to the magnitude and sign of the pulse voltages applied to the electrodes. They can be switched to the two positions that can be constructive or destructive. The output field of the lightwave carrier can be represented by

$$E_0 = \frac{1}{2} E_{iRMS} e^{j\omega_c t} \left(e^{j\phi_1(t)} + e^{j\phi_2(t)} \right) \tag{9.1}$$

where ω_c is the carrier radial frequency, E_{iRMS} is the root mean square value of the magnitude of the carrier and $\phi_1(t)$ and $\phi_2(t)$ are the temporal phase generated by the two time-dependent pulse sequences applied to the two electrodes. With the voltage levels varying according to the magnitude of the pulse sequence one can obtain the transfer curve as shown in Figure 9.3a. This phasor representation can be used to determine exactly the biasing conditions and magnitude of the RF or digital signals required for driving the optical modulators to achieve 50%, 33% or 67% bit period pulse shapes.

The power transfer function of the Mach-Zehnder modulator is expressed as[*]

$$P_0(t) = \alpha P_i \cos^2 \frac{\pi V(t)}{V_\pi} \tag{9.2}$$

where $P_0(t)$ is the output transmitted power, α is the modulator total insertion loss, P_i is the input power (usually from the laser diode), $V(t)$ is the time-dependent signal applied voltage, V_π is the driving voltage so that a π phase shift is exerted on the lightwave carrier. It is necessary to set the static bias on the transmission curve through the bias electrode. It is common practice to set the bias point at 50% transmission point or a $\pi/2$ phase difference between the two optical waveguide branches, the quadrature bias point. As shown in Figure 9.3, electrical digital signals are transformed into optical digital signals by switching voltage to both ends of the quadrature points on the positive and negative sides of the transfer characteristic.

---

[*] Note this equation is representing for single drive MZIM—it is the same for dual drive MZIM provided that the bias voltages applied to the two electrodes are equal and opposite in signs. The transfer curve of the field representation would have half the periodic frequency of the transmission curve shown in Figure 9.3.

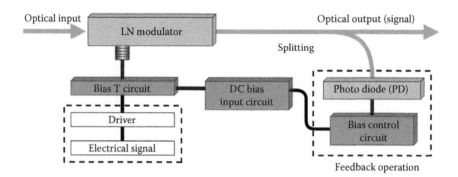

FIGURE 9.4
Arrangement of bias control of integrated optical modulators to achieve stable operating point.

9.2.2.2 Bias Control

One factor that affects the modulator performance is the drift of the bias voltage which happens after a few hours of operation due to the accumulation of charges on the surface of the crystal, thus the electro-effects of a high field. For the Mach-Zehnder interferometric modulator (MZIM) type it is very critical when it is required to bias at the quadrature point or at minimum or maximum locations on the transfer curve. Direct current (DC) drift is the phenomenon that occurs in $LiNbO_3$ due to the build-up of charges on the surface of the crystal substrate. Under this drift the transmission curve gradually shifts in the long term [19,25]. In the case of the $LiNbO_3$ modulator, the bias point control is vital as the bias point will shift long term. To compensate for the drift, it is necessary to monitor the output signals and feed it back into the bias control circuits to adjust the DC voltage so that operating points stay at the same point as shown in Figure 9.4 for example, the quadrature point. It is the manufacturer's responsibility to reduce DC drift so that DC voltage is not beyond the limit throughout the lifetime of the device.

9.2.2.3 Chirp Free Optical Modulators

Due to the symmetry of the crystal refractive index of the uniaxial anisotropy of the class m of $LiNbO_3$, the crystal cut and the propagation direction of the electric field affect both modulator efficiency, denoted as the driving voltage, and the modulator chirp. The uniaxial property of $LiNbO_3$ is shown in Figure 9.5. In the case of the Z-cut structure, as a hot electrode is placed on top of the waveguide, RF field flux is more concentrated, and this results in the improvement of overlap between RF and the optical field as shown in Figure 9.6. However overlap between RF in the ground electrode and waveguide is reduced in the Z-cut structure so that overall improvement of driving voltage for Z-cut structures compared to X-cut is approximately 20%. The different overlapping area for the Z-cut structure results in a chirp parameter of 0.7 whereas X-cut and Z-propagation has almost zero-chirp due to its symmetric structure. A number of commonly arranged electrode and waveguide structures are shown in Figure 9.7 to maximize the interaction between the traveling electric field and the optical guided waves. Furthermore a buffer layer, normally SiO_2 is used to match the velocities between these waves so as to optimize to optical modulation bandwidth.

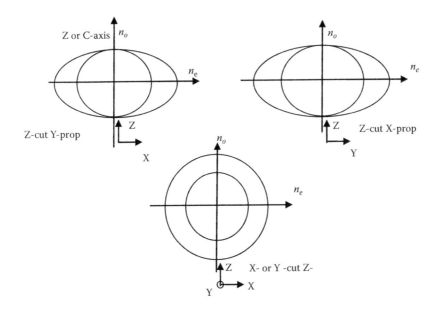

FIGURE 9.5
Refractive index contours of LiNbO$_3$ uniaxial crystal with Z- or C- denoting as the principal axis. (a) Lightwaves propagation in the Z-axis polarized along Y-cut LiNbO$_3$; (b) prop-direction Z-axis and X-cut crystal; (c) Y-prop and Y-cut crystal.

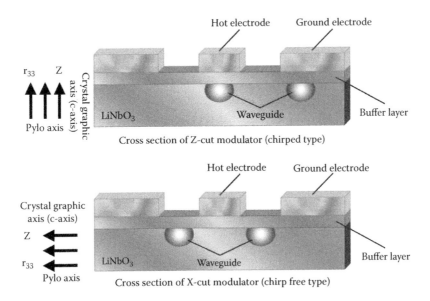

FIGURE 9.6
Different crystal cuts of 3D LiNbO$_3$ diffused waveguide integrated structures. (a) Integration of electrodes and optical waveguides in Z-cut; (b) X-cut crystal structure and orientation of electrodes and waveguides.

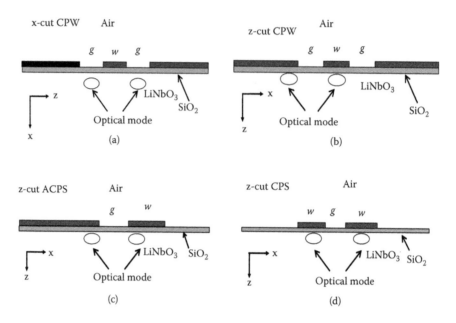

FIGURE 9.7
Cross sectional structure of commonly used electrode structure and crystal orientation for intetferometric modulation to maximize the use of the overlap integral between the optical guided mode and the electric field distribution for largest EO coefficients.

9.2.2.4 Structures of Photonic Modulators

Figure 9.8a and b show the structure of an MZ intensity modulator using single and dual electrode configurations, respectively. The thin line electrode is called the "hot" electrode or traveling wave electrode. RF connectors are required for launching the RF data signals to establish the electric field required for EO effects. Impedance termination is also required. Optical fibers pigtails are also attached to the end faces of the diffused waveguide. The mode spot size of the diffused waveguide is not symmetric and hence some diffusion parameters are controlled so that maximizing the coupling between the fiber and the diffused or rib waveguide can be achieved. Due to this mismatching between the mode spot sizes of the circular and diffused optical waveguides, coupling loss occurs. Furthermore the difference between the refractive indices of fiber and LiNbO$_3$ is quite substantial and thus Fresnel reflection loss would also incur.

Figure 9.8c shows the structure of a polarization modulator which is essential for the multiplexing of two polarized data sequences so as to double the transmission capacity, e.g., 40 Gb/s to 80 Gb/s. Furthermore this type of polarization modulator can be used as a polarization rotator in a polarization dispersion compensating sub-system [20].

9.2.2.5 Typical Operational Parameters

Typical operating parameters of the guided wave optical modulators are listed in Table 9.1 including the modulator operating speed, the total insertion loss, the ON-OFF distinction ratio etc. These parameters are important for system applications in term of speed and attenuation as well as distinction ratio between the on and off states for different modulation formats employed in transmission systems.

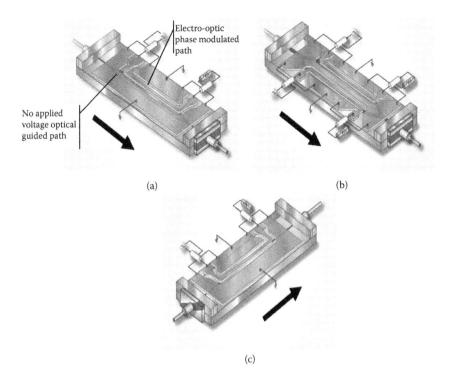

(a)

(b)

(c)

FIGURE 9.8
(See Color Insert) Intensity modulators using LiNbO$_3$: (a) single drive electrode (b) dual electrode structure; and (c) EO polarization scrambler using LiNbO$_3$.

9.3 Traveling Wave Electrodes for Integrated Modulators

An integrated EO modulator is composed of optical waveguides and electrodes. The integrated optical waveguides would follow the structure of 3D waveguides as described in Chapter 4. The electrodes must be in planar structure of either coplanar (CPS) or asymmetric coplanar structure (ACPS) so that ultra-broadband operation can be achieved. This

TABLE 9.1

Typical Operational Parameters of Optical Intensity Modulators

Parameters	Typical Values	Definition/comments
Modulation speed	10 Gb/s	Capability to transmit digital signals
Insertion Loss	Max 5 dB	Defined as the optical power loss within the modulator
Driving Voltage	Max 4 V	The RF voltage required to have a full modulation
Optical Bandwidth	Min 8 GHz	3 dB roll-off in efficiency at the highest frequency in the modulated signal spectrum
ON/OFF Extinction Ratio	Min 20 dB	The ratio of maximum optical power (ON) and minimum optical power (OFF)
Polarization Extinction Ratio	Min 20 dB	The ratio of two polarization states (TM and TE guided modes) at the output

section gives the detailed modeling and design of such electrodes as the modulating electrodes for ultra-high speed broadband optical modulators. The modulation efficiency is also examined with the numerical calculation of the overlap integral between the optical waveguide and the fields generated by the electrodes.

9.3.1 Introduction

Optical transmission at ultra-high bit rate up to 40 Gb/s and multiplexing of several optical channels are becoming the standard deployment of dense division wavelength multiplexing (DWDM) optical networks in certain information transport routes. Various amplitude and phase shift keying (PSK) modulation formats have been investigated over the last decade in order to increase the transmission capacity and mitigate the linear dispersion and nonlinear induced effects. In ultra-high speed transmission systems external modulators play a vital role. Of particular interest are the modulators in LiNbO$_3$ and other compound semiconductor integrated optical devices. The PM can be achieved with single, asymmetric or dual (balanced or symmetric) electrode structures that require precision in the design and fabrication.

Several published works [29] have outlined the important steps and considerations involved to achieve efficient design of traveling wave electrodes [30,31]. The principal part of the design calculations were based upon an adopted empirical model that was derived from the combination of the quasi-transverse electromagnetic modes (TEM) analysis and the finite element method (FEM). The empirical model, despite being impressive in its ability to facilitate the design calculation, does not provide the modulating electric field that is mandatory in calculating the electro-optical overlap integral. More importantly, it does not take into account the more subtle structural factors such as the wall angle of the gold plated electrode. Most analyses such as the conformal mapping (CM) technique, the FEM [32] or the method of images (MoI) [33], assume an infinitely thin electrode structure and a perfect vertical side wall. However when thick electrodes, typically in the range of 10–20 μm, are fabricated to achieve broadband operation of a 3 dB bandwidth of more than 40 GHz, the assumption of a perfectly vertical wall angle is no longer valid. In practical devices, after the gold electroplating stage, the electrode would assume a trapezoidal shape as shown in Figure 9.9.

The wall angle of the electrode in fact has a rather significant influence on the value of the effective microwave index n_m and the electrode characteristic impedance Z. It is therefore

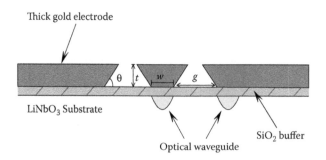

FIGURE 9.9

Cross section structure of an interferometric EO modulator having a tilted wall angle of thick electrodes and a dielectric buffer layer; LiNbO$_3$ substrate and air superstrate.

important to take into consideration such structural effects of the electrodes in the design of broadband traveling wave electrodes. Chung, Chang, and Adler [32] have modeled the electric field of the electrodes using the FEM but their analysis did not take into account the effect of the wall angle. This has led to the limitation of the reported empirical model. Electrode structural factors such as this can only be modeled by a more robust numerical formulation such as the FEM [35,36] or the finite differencing (FD) method [38]. The FD method is selected to solve the anisotropic Laplace equation. The FD method is very computationally effective due to its relatively straightforward analysis. In the PC computing environment that has virtual memory support, this scheme can warrant sufficient accuracy in a relatively short time without tedious mathematical manipulation and programming involved in contrast to the FEM. Non-uniform grid allocation can also be used to economize computer storage capacity.

The next section outlines the numerical formulation of the non-uniform mesh FD scheme to solve the anisotropic Laplace equation under the quasi-TEM assumption of the microwave mode. Based on this numerical model, an application program named FD Traveling Wave Electrodes Analysis (FDTWEA) has been developed to compute Z and n_m, the modulating field, E_x and E_y, and the traveling wave and optical wave overlap integral, Γ. Our simulated results will be compared with published results to verify the validity of our FDTWEA simulator. The simulator is also employed to study the effect of the wall tilted angle on the design of electrode structures.

9.3.2 Numerical Formulation

9.3.2.1 Discrete Fields and Potentials

The traveling wave electrodes are miniature transmissions where a quasi-TEM wave transmission can be assumed, hence, the electric fields can be reduced to a two-dimensional electrostatic field distribution which can be solved by applying the Laplace equation. Since the LiNbO$_3$ substrate and the SiO$_2$ dielectric are involved, we have an anisotropic case. Assuming a Z-cut orientation of the LiNbO$_3$ crystal, then the permittivity tensor is diagonal. Thence the electrostatic potential V is the solution of the anisotropic Laplace equation written as:

$$\frac{\partial}{\partial x}\left(\varepsilon_x \frac{\partial V}{\partial x}\right) + \frac{\partial}{\partial y}\left(\varepsilon_y \frac{\partial V}{\partial y}\right) = 0 \tag{9.3}$$

This equation can be numerically solved by an FD method using a non-uniform mesh allocation scheme. Denser mesh is allocated at the edges of the electrode and also at the buffer layer area so that the effect of the edge field can be modeled more accurately. Points that are further away from the electrode can be modeled with a coarser mesh. Difference equations can then be formulated.

Consider a general electrode structure as shown in Figure 9.10a. The grid points are placed along the dielectric boundary. The electrode structures for our consideration involve only three dielectric medium, namely, the air, the SiO$_2$ thin coated layer and the LiNbO$_3$ substrate. With LiNbO$_3$ being the anisotropic medium, the transition of different dielectric medium occurs only in the y direction. Therefore, it suffices to analyze the two layers of dielectric medium as shown in Figure 9.10b to formulate the difference equations.

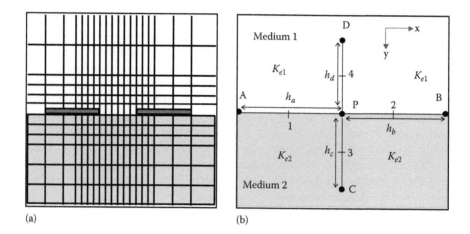

(a)   (b)

FIGURE 9.10
(a) A simple illustration of the non-uniform grid allocation scheme. (b) The grid points for finite differencing.

Considering a mesh point P along the boundary and four other points surrounding it, namely, A, B, C, and D, the point located between AP, BP, CP and DP can be labeled 1, 2, 3 and 4 with their respective grid sizes of h_a, h_b, h_c and h_d. The electrical potentials at each point are V_P, V_A, V_B, V_C and V_D, respectively. K_{e1} and K_{e2} are the dielectric constants of the media.

At point 1 we have:

$$\frac{\partial V}{\partial x} \approx \frac{V_P - V_A}{h_a}; \text{ at } 2 \frac{\partial V}{\partial x} \approx \frac{V_B - V_P}{h_b}; \text{ at } 3 \frac{\partial V}{\partial y} \approx \frac{V_C - V_P}{h_c}; \text{ and at } 4 \frac{\partial V}{\partial y} \approx \frac{V_P - V_D}{h_d} \quad (9.4)$$

With the assumption that half of the flux flowing in medium 1 and the other half in medium 2, we have

$$\left(\varepsilon_x \frac{\partial V}{\partial x} \right)_{PB} = \frac{\frac{1}{2}(Ke_{1x} + Ke_{2x}) \cdot (V_B - V_P)}{h_b} \text{ along segment BP} \quad (9.5)$$

$$\left(\varepsilon_x \frac{\partial V}{\partial x} \right)_{AP} = \frac{\frac{1}{2}(Ke_{1x} + Ke_{2x}) \cdot (V_P - V_A)}{h_a} \text{ along segment AP} \quad (9.6)$$

$$\left(\varepsilon_y \frac{\partial V}{\partial y} \right)_{PC} = \frac{Ke_{2y} \cdot (V_C - V_P)}{h_c} \text{ along segment PC} \quad (9.7)$$

$$\left(\varepsilon_y \frac{\partial V}{\partial y} \right)_{DP} = \frac{Ke_{1y} \cdot (V_P - V_D)}{h_d} \text{ along segment DP} \quad (9.8)$$

Using Equation 9.3, the five point FD relationship between the fields at the grid points can be expressed as

$$K_A V_A + K_B V_B + K_C V_C + K_D V_D - K_P V_P = 0$$

with $\quad K_A = \dfrac{Ke_{1x} + Ke_{2x}}{hb \cdot (ha + hb)}; K_B = \dfrac{Ke_{1x} + Ke_{2x}}{ha \cdot (ha + hb)}$

$$K_C = \dfrac{Ke_{2y}}{\dfrac{1}{2} hc \cdot (hc + hd)}; K_D = \dfrac{Ke_{1y}}{\dfrac{1}{2} hd \cdot (hc + hd)}$$ (9.9)

and $\quad K_P = K_A + K_B + K_C + K_D$

Similarly corresponding coefficients and FD relationships can be obtained for the air and the SiO$_2$ buffer layer and the substrate LiNbO$_3$ regions. Using a sufficiently large problem space, approximately 400×400 μm, we can assume the electric field along the boundary lines to be infinitesimal. For the potential, a Neumann boundary condition [32,39] can be used to obtain the spatial relationships between the fields at the grid points. The remaining points of the windows are just the permutation of the boundary conditions outlined above. Incorporating the boundary conditions into the FD equations, a set of difference equations can be derived from the mesh grid points in form of

$$A \cdot u = b$$ (9.10)

where A is the coefficient matrix, u is the vector that contains the potential V of each grid point, while b is the vector that assumes the right-hand side of the Laplace equation which is mostly zero except for the grid points on the electrodes which take up the value of the potential on the relevant electrodes. A typical matrix equation for a problem space can then be generated whose coefficient matrix is a tri-diagonal matrix with fringes. The matrix elements p, a, b, c and d for each point correspond to the coefficient K_p, K_A, K_B, K_c and K_d, respectively. Take note that for points 7, 8 and 9 which fall on the electrode, $K_p = 1$ while K_A, K_B, K_c and K_d are all zero. The difference equation can be solved by means of the more conventional successive over relaxation method (SOR) [40,41]. This method, however, requires a good initial guess and a good estimate of the relaxation factor in order to achieve a reasonable rate of convergence.

9.3.2.2 Electrode Line Capacitance, Characteristic Impedance and Microwave Effective Index

The above numerical formulation can then be applied to the computation of the operational parameters of the traveling wave electrodes such as the characteristic impedance Z and the microwave effective index n_m which are given as

$$Z = \dfrac{1}{c\sqrt{CC_0}} \text{ and } n_m = \sqrt{\dfrac{C_0}{C}}$$ (9.11)

where C is the capacitance per unit length of the electrode transmission line with the dielectric medium, while C_0 is the capacitance per unit length for the air filled medium, with c being the speed of light in vacuum. The capacitances C and C_0 can be determined by calculating the charges on the conductors. The Gauss theorem can be applied to determine the charges and hence it requires the integration of the normal component of the electric flux over a surface enclosing the "hot" electrode. Forming this surface by lines joining the nodal points parallel to the coordinate directions, at any point P on this surface, we have

$$D_n = \varepsilon E_n = -\varepsilon \frac{\partial V}{\partial n} \tag{9.12}$$

where D_n is the normal component of the electric flux, E_n is the normal component of electric intensity, and the subscript n indicates the orthogonal coordinate. The potential at P may be expressed numerically in terms of the known potentials V_A and V_b on each side of it. For irregular mesh as shown in we have

$$\frac{\partial V}{\partial x} = \frac{V_B - V_A}{h_b + h_a} \text{ and } \frac{\partial V}{\partial y} = \frac{V_C - V_D}{h_c + h_d} \tag{9.13}$$

Thus, if using the enclosing surface of a square box surround the hot conductor of s straight line segments each containing r nodes, the charge per unit length normal to the cross section is then given by

$$Q = \varepsilon_r \varepsilon_o l \sum_s \sum_{P=1}^{4} {}' \left(\frac{\partial V}{\partial n} \right)_P \tag{9.14}$$

The apostrophe sign indicates the first and last terms in the summation are halved, which is seen to be equivalent to integration by the trapezoidal rule, l is the length of the segment of the integration path. For uniform discretization, l is essentially the grid size h. For our non-uniform scheme, l is assumed to be either $(h_a + h_b)/2$ or $(h_c + h_d)/2$ depending on either a horizontal or vertical line segment over which the summation is taken. The value of the relative permittivity, ε_r depends on which dielectric medium point P falls onto. For instance, when we sum the derivative along the first horizontal line segment ($s = 1$), the segment falls completely in the air, so $\varepsilon_r = 1.0$. If the summation is done on line segment number 3, which is in the LiNbO$_3$ crystal, $\varepsilon_r = \varepsilon_z = 43$. The vertical summation ($s = 2, 4$) however needs to be dealt with care because it involves dielectric medium transition. For points located entirely in the air, SiO$_2$ and LiNbO$_3$, the relative permittivities are $\varepsilon_a = 1.0$, $\varepsilon_b = 3.9$ and $\varepsilon_x = 28$, respectively. However, for points fallen on the buffer air interface, we assume half of the flux passing through each medium. So for the air-SiO$_2$ interface, $\varepsilon_r = (1 + \varepsilon_b)/2$, whereas for the SiO$_2$-LiNbO$_3$ interface, $\varepsilon_r = (\varepsilon_b + \varepsilon_x)/2$. The charge capacity is defined as

$$C = \frac{Q}{V_t} \tag{9.15}$$

where V_t is the potential difference between the electrode arms. Similarly for C_o, the Laplace equation can be solved for the transmission line in the air filled medium without the dielectric medium. The charge capacity and hence the capacitance can then be deduced.

9.3.2.3 Electric Fields E_x and E_y and the Overlap Integral

The electric field generated by the traveling wave electrode can be derived as

$$E_x = \frac{\partial V}{\partial x} = \frac{V_B - V_A}{(h_a + h_b)} \text{ and } E_y = \frac{\partial V}{\partial y} = \frac{V_C - V_D}{(h_c + h_d)} \tag{9.16}$$

The overlap integral due to the EO effects, Γ is defined as [42]:

$$\Gamma = \frac{g}{V} \iint |E_o(x, y)|^2 E_m(x, y) dx dy \tag{9.17}$$

where $|E_o(x, y)|^2$ and E_m are the square of the normalized total electric field of the guided lightwaves and the electric field generated the electric traveling wave, respectively. The choice of E_x or E_y depends on the crystal orientation and the polarization of the optical field so as to maximize the EO interaction. The normalized optical field intensity profile assumes a Hermitian-Gaussian profile given as

$$|E_o(x, y)|^2 = \frac{4y^2}{w_x w_y^3 \pi} \exp\left[-\left(\frac{x - p}{w_x}\right)^2\right] \cdot \exp\left[-\left(\frac{y}{w_y}\right)^2\right] \tag{9.18}$$

where the $1/e$ intensity width and depth are $2w_x$ and $1.376w_y$, respectively, and p is the peak position of the optical field in the lateral direction. w_x and w_y are strongly dependent on waveguide fabrication parameters. They can be determined either experimentally or by numerical modeling. For modeling purposes, we use w_x and w_y of 2.5 µm and 2.2 µm, respectively [42]. From the calculated Γ one could determine the voltage V_π defined as the applied voltage to the electrode for a π phase shift change on the lightwave carrier passing through an arm of the MZIM.

9.3.3 Electrode Simulation and Discussions

9.3.3.1 Grid Allocation and Modeling Performance

Figure 9.11 shows contour plots of the potentials for coplanar waveguide (CPW), ACPS and CPS electrode structures obtained from the solution of the Laplace equations (Equations 9.1 through 9.5) in which $w = 10$ µm, $g = 15$ µm, $t = 3$ µm and $t_b = 1.2$ µm are considered. The electric fields generated by these electrodes are obtained by solving Equation 9.6 and are illustrated in Figure 9.12.

The horizontal field E_x is strongest in between the gaps, while the vertical field E_y is strongest along the hot electrodes in all structures. From these plots, we could see that the push-pull operation can be achieved most efficiently by the horizontal field, E_x of a CPW structure for the X-cut Y-propagating device. Another configuration of the CPS electrode structure would allow us to place the waveguides directly underneath for the vertical field, E_y to exert maximum EO effects. This, of course, would correspond to a Z-cut device that in turn may suffer frequency chirping on the light wave carrier. However the CPS structure exhibits a high propagation loss and is therefore seldom employed. A common configuration for a Z-cut device is to place one waveguide under the hot electrodes and the other at

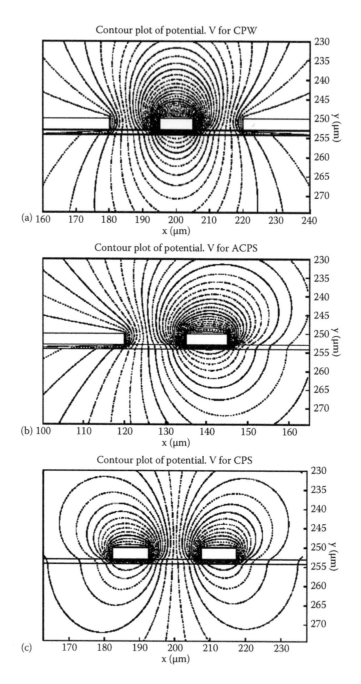

FIGURE 9.11
Contour plot of the Laplace equation solution for electrode structures: (a) CPW; (b) ACPS; and (c) CPS.

the edge of the ground plane in either the CPW or ACPS. The vertical field E_y is employed in such a configuration. This cannot be considered as a full push-pull operation because the waveguide underneath the ground plane sees only one half of the field seen by the waveguide under the hot electrode. We can observe that from the contour plots and the field plots, the strongest field exists around the edges of the electrodes, within the SiO_2

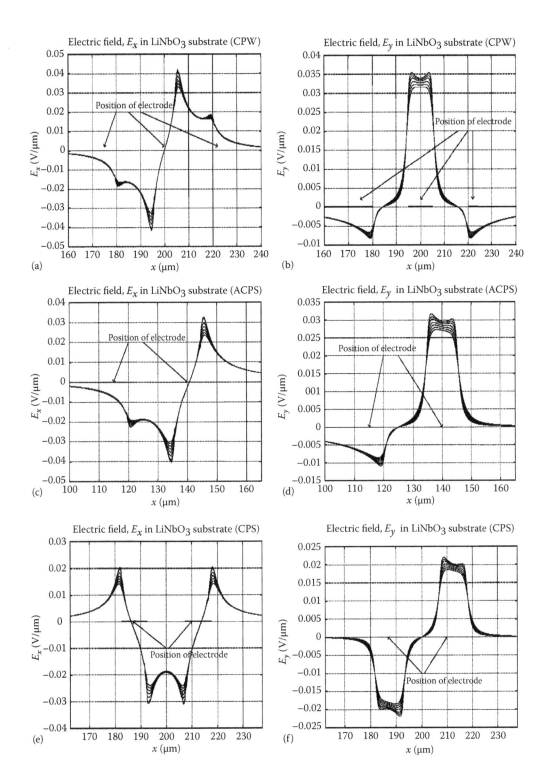

FIGURE 9.12
Electric field distributions in the region of the first 1 µm of LiNbO$_3$ just beneath the SiO$_2$ buffer layer [(a) E_x and (b) E_y] of the CPW electrode; [(c) E_x and (d) E_y] of ACPS electrode; and [(e) E_x and (f) E_y] of CPS electrode.

buffer layer underneath the hot electrodes and also in the gap between the electrodes. Therefore denser grids should be allocated around these areas. From our experience, a grid size as small as 0.05 0.2 μm at around the electrode edges, the gap and buffer region, up to a larger grid size of 8–10 μm for points where the electric field has decayed substantially warrants an accurate calculation result. If the file is not specified, FDTWEA generates one automatically using its default grid allocation scheme suitable for the relevant electrode structure. A problem space of around 400×400 μm is sufficient to assume a metallic box boundary condition.

9.3.3.2 Model Accuracy

9.3.3.2.1 General Comparison

To assess the validity of our modeling results, the values of Z and n_m are compared with published experimental results. The comparisons are tabulated in Table 9.2. The simulated results of FDTWEA for both the CPW and ACPS electrodes corroborate well with published results. In this modeling, we determine the Z and n_m values for the buffer layer varying between 0.6–0.85 μm which are determined from the fabrication tolerances of several fabricated devices [43]. There seems to be a discrepancy in our calculated Z value of the ACPS structure compared to the one reported by Ref. [44] with a Z value of 35 Ω. We have thus assumed that the calculations in Gee, Thunmond, and Yen (1983) [45] have not taken into account the thickness of the electrodes. The recalculated impedance, with the assumption of an infinitely thin electrode, is about 33.15 Ω, which agrees well with their reported value. The third modeling is performed based on experimental results reported by Ref. [33]. The simulated results agree closely with their theoretical calculations which are based on the MoI. However, the measured results in Ref. [33] differ slightly from that of the theoretical predictions. The characteristic impedance measured by time domain reflectometry (TDR) and scattering parameter network analysis (NA) were 49.8 Ω and 47.1 Ω, respectively. The small discrepancy is possibly due to a slightly thicker SiO_2 buffer layer coated in the experiment.

TABLE 9.2

FDTWEA Calculated Values as Compared with Published Results. W, G, T and t Units are in μm

						Published		FDTWEA	
Ref.	Structure	W	G	T	t_b	n_m	$Z(\Omega)$	n_m	$Z(\Omega)$
Chung et al. 1991 [32]	CPW	20	5	3	0.6–0.85	2.7	27	2.867–2.703	27–27.3
—	CPW	48	10	3	0.6–0.85	3.3	24.5	3.358–3.224	24.13–21.4
Korotky et al. 1987 [44]	ACPS	15	5	4	0	—	~35	3.661	29.83
—		15	5	0	0	—	—	4.226	33.15
Chuang et al. 1993 [33]	ACPS	10	10	1.5	0.1	—	45 (cal) 47.1 (TDR) 49.8 (NA)	3.781	42.7

9.3.3.2.2 Finite Differencing Traveling Wave Electrodes Analysis versus Spectral Domain Analysis (SDA)

Our simulations are then compared further by emulating the calculations by the SDA as reported in Ref. [46]. Figure 9.13a and b show the comparison of the two simulations. Initially, we calculated the parameter based on the electrode parameters of $t = 4$ µm, $t_b = 0\text{–}1.5$ µm, $w = 8$ µm and $g = 30$ µm. Our results, however, have seemed to underestimate both the values of Z and n_m. It was later realized that the SDA may not have taken the thickness of the electrode into account. So we recalculate Z and n_m with the assumption of a zero thickness electrode. This has clearly implied that the SDA does not take into account the effect of the electrode thickness. Such limitation does not exist in our present FD analysis.

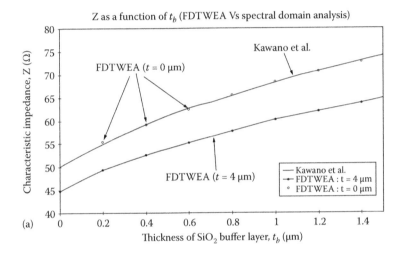

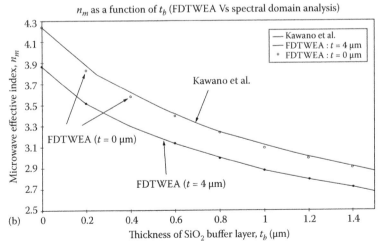

FIGURE 9.13
(a) FDTWEA calculated impedance compared to that of the SDA, and (b) calculation of FDTWEA n_m compared to that of SDA.

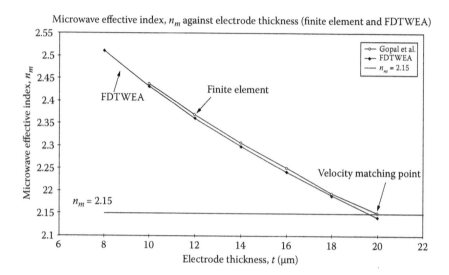

FIGURE 9.14
FDTWEA's calculation of n_m compared with FEM.

9.3.3.2.3 *Finite Differencing Traveling Wave Electrodes Analysis versus Finite Element Method*

We further verify our calculation program by comparing the value of n_m with the simulation result of very thick electrodes of the FEM. The comparison is shown in Figure 9.14. The calculations are intended to show how the thickness of the electrodes can significantly improve the electrical and optical velocity matching. Thick electrodes that are in the range of 10–20 μm are employed. From Figure 9.14, the plot of the microwave refractive indices versus the thickness of the electrode, we observe that our calculations agree to within 0.5% with its FEM counterpart. In our calculations a grid size of 0.125 μm is used for points of meshes along the wall of the thick electrodes. Each calculation takes approximately 3–4 min to complete on a standard PC Windows 2000 operating system. Our FD scheme has achieved a similar level of accuracy achieved by FEM in a relatively straightforward numerical performance.

So far it has been shown that the FD scheme offers potential and reasonably accurate modeling of the traveling wave electrode parameters, Z and n_m. They are consistent with both the simulations and measured data reported in previous works. It has simply verified the validity of our numerical model. FDTWEA, with its verified precision would thus be a useful tool to provide a good quantitative measure in the design and analysis of traveling wave electrodes.

9.3.4 Electro-Optic Overlap Integral, Γ

In this section, we show some of the calculation results and discuss a few issues on the interaction between the electrical traveling wave and the guide optical waves, the electrical-optical overlap integral. To compute the overlap integral we need the normalized optical intensity field profile, which is a Hermitian-Gaussian profile as

$$|E_o(x, y)|^2 = \frac{4y^2}{w_x w_y^3 \pi} \exp\left[-\left(\frac{x-p}{w_x}\right)^2\right] \cdot \exp\left[-\left(\frac{y}{w_y}\right)^2\right] \tag{9.19}$$

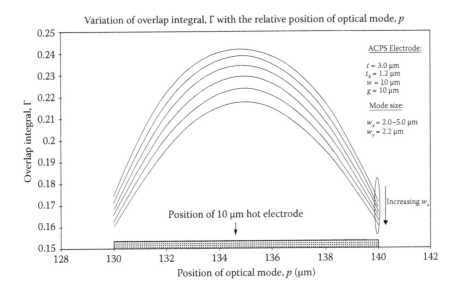

FIGURE 9.15
Variation of Γ as the peak position of the optical mode shift from one end of the hot electrode to another for increasingly wider optical mode.

The optical mode sizes, w_x and w_y, play very important roles in maximizing the overlap integral Γ. The mode size is defined as twice the 1/e modal width and 1.376 times the 1/e modal depth. The parameters w_x and w_y can be modeled using the effective index method or the finite difference or element numerical technique, or even by the experimental measurement of the mode size as described in Chapter 3. Furthermore Γ is also influenced by the relative position of the optical mode with respect to the "hot" electrodes. All the attributes that affect the value of Γ can be easily modeled by FDTWEA. For illustration purposes, we will calculate Γ of the ACPS electrodes and study the effects of various factors that can influence its value.

Figure 9.15 shows the variations of the overlap integral, Γ as the peak position of the optical mode, p shift from one end of the electrodes to the other end. As expected, a tighter confined mode would give a higher value of Γ.

It is thus very critical to fabricate a Ti:LiNbO₃ that guide lightwaves with optimum mode spot sizes to maximize the fiber-diffused waveguide coupling and the EO overlapping. Apart from having a tightly confined optical mode, the relative position of the optical waveguide with respect to the electrodes is also critical to provide an optimum EO effect. When narrow electrodes whose width is comparable to the size of the waveguide are designed, it is preferable to place the waveguide directly underneath the electrode to utilize the strong edge field from both sides of the electrode. This has been shown in Figure 9.15 in which the maximum Γ is obtained when the position of the optical mode is centered underneath the electrode. However, for much wider electrode structures, the preferred position would be just inside the end of the electrode closer to the gap. When a 30 μm wide electrode is employed, a maximum overlap integral can be achieved by exploiting the higher edge field by placing the waveguide at the center, that is $x \sim 135$ μm. Positioning the optical waveguide right at the center of the electrode would give inefficient EO interaction.

Figure 9.16 shows how the thicker buffer layer can impede the EO effect. Although a thicker buffer layer has the advantage of a lower conductor loss and also significantly

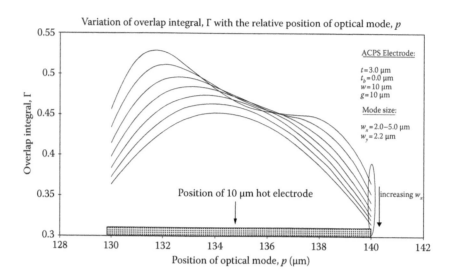

FIGURE 9.16
Variation of Γ as the peak position of the optical mode shift from one end of the hot electrode to another for increasing wider optical mode for electrode with no buffer layer, $t_b = 0$.

improves the velocity match, there is a trade off involved because the overlap integral will be lower, which can therefore lead to a higher V_π. We can see that without the buffer layer, Γ assumes a much higher value. Unfortunately, having no buffer layer would imply a much higher optical-electrical velocity mismatch and hence reduce the effective operational bandwidth. Essentially, the position of the optical waveguide with respect to the electrode should be considered before fabricating the device. It is also important to have a quantitative measure of the effect SiO_2 thickness on the overall performance of the device.

Figure 9.17 confirms that the FDTWEA can provide a measure that would greatly facilitate the design of traveling wave electrodes.

9.3.5 Tilted Wall Electrode

In previous sections, we have used FDTWEA to analyze various rudimentary design parameters of traveling wave electrodes. In this section, we demonstrate how it can handle the problem of greater complexity of practically fabricated traveling electrodes.

The fabrication of traveling wave electrodes is not straightforward. One way of extending the bandwidth of the device is to employ very thick electrodes that are in the range of 10–20 μm. Such thick electrodes, however, do not always assume a rectangular shape. They are more likely to take up a trapezoidal shape and the contour of the distributed electric fields as shown in Figure 9.18.

Such geometrical factors cannot be ignored because they certainly have a subtle effect on both the values of Z and n_m. We can see that the wall angle certainly has a substantial effect on the values of Z and n_m as illustrated in Figure 9.19 and 9.20. For the trapezoidal shape electrodes both Z and n_m are reduced. The effect of the wall angle is less severe for thinner electrodes as we can see from the graphs that the difference of the plots for different θ converges as the electrodes become thinner. For thick electrodes, especially those that are greater than 10 μm, the wall angle effect should not be ignored. For

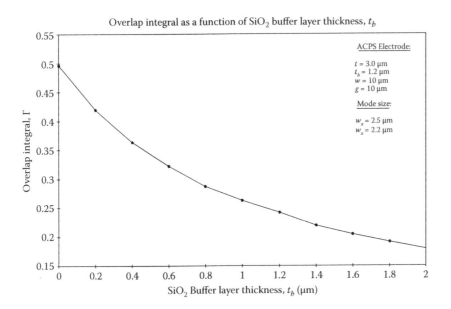

FIGURE 9.17
Γ as a function of the thickness of SiO$_2$ buffer, t_b.

example, if we base our design on the assumption of a $\theta = 90°$ rectangular electrode, then we would use a 20 µm thick electrode for the best velocity match. However, if the fabricated electrodes actually assume a trapezoidal shape with $\theta = 80°$, then we would have overestimated the value of n_m for best velocity matching by about 0.2. This corresponds to almost a 20% loss of the bandwidth that could have been achieved by

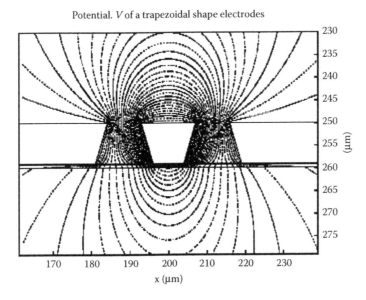

FIGURE 9.18
Potential field distribution of the trapezoidal shape electrode with the contour shown in curves wrapping around the electrodes.

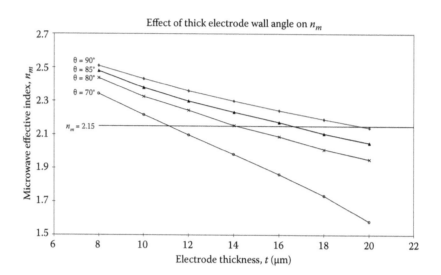

FIGURE 9.19

Electrode microwave impedance n_m as a function of the electrode thickness with the wall angle as a parameter.

a mismatching of 0.01 between the microwave and optical indices. The plots shown in Figure 9.20 suggest that we need only an electrode thickness of around 14 μm to achieve maximum bandwidth, based on the assumption of an 82° electrode wall angle and an electrode thickness of 15 μm. Again this demonstrates another potential application of the FDTWEA numerical modeling scheme. It thus offers much improved analytical

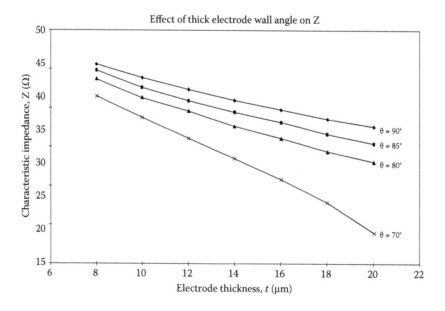

FIGURE 9.20

Characteristic impedance Z (Ω) as a function of electrode thickness with the wall angle as a parameter.

capability beyond most analytical techniques. Undoubtedly, an analytical capability would certainly imply a better design in traveling wave electrodes for ultra-wideband optical modulators, especially for the employment of advanced modulation formats in long-haul optically amplified fiber communications systems.

9.3.6 Frequency Responses of Phase Modulation by Single Electrode

In practice the design and estimation of the frequency response of the microwave electrode and its optical modulation response is very important. Even more important is the measurement technique to determine whether the mismatching of the microwave signals and the lightwave, especially when the optical waves are modulated under the modulation format using amplitude, discrete phase, continuous phase or frequency. Advanced modulation formats have recently emerged as the most effective method in ultra-high speed optical communications to combat the residual dispersion nonlinear effects [47–50]. The role of the PM of the carrier is critical especially when continuous PM is employed [51]. Thus in this section the frequency response of phase modulators is briefly investigated both in theory and experiment.

The velocity mismatch between the microwave signals and the optical waves comes from the difference in the effective refractive indices of the optical guided mode and that of the traveling microwave wave along the electrode length. This mismatching reduces the optical bandwidth of the phase modulator. The essential characteristics of the phase modulator are its frequency response which is the phase variation of the optical carrier under the modulation of the traveling wave. Over the years the measurement of the frequency response of an optical phase modulator has been done in the electrical domain [52], until recently a novel method using a Sagnac interferometer to determine the frequency response of phase modulator was used [53]. This allows us to adapt this measurement technique to characterize the phase responses of a number of optical phase modulators with the set up shown in Figure 9.21, in which the device under test is a phase modulator inserted in a Sagnac interferometer. The length of the ring determines the null of the frequency response. The PM is then converted to

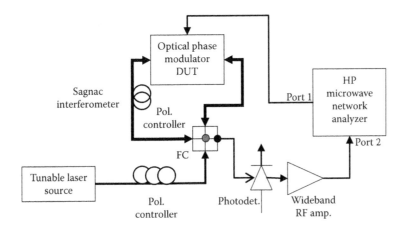

FIGURE 9.21
Schematic of a Sagnac interferometer for the measurement of the frequency response of optical phase modulators. Optical fiber line__ (bold) dashed; electrical line continuous. DUT = device under test. FC = fiber coupler – 4 ports.

amplitude response due to the interference or phase comparison of the optical path in clockwise and counterclockwise directions. Thus any modulation index variation in the modulator would be easily measured. A microwave network analyzer is used to scan the exciting sinusoidal RF wave into the electrode of the optical phase modulator. The optical signals are then detected by the photodiode and a wideband microwave amplifier then fed into port 2 of the network analyzer. The scattering parameter S_{21} is then obtained.

The detected current output of the photodetector is given by

$$i_d = t_f \Re K \sin\left[\sin c\left(2\pi f_{RF}\tau_{PM}\right)v_{RF}(t+\tau) - v_{RF}(t)\right] \tag{9.20}$$

with

$$\sin c\left(2\pi f_{RF}\tau_{PM}\right) = \frac{\sin\left(2\pi f_{RF}\tau_{PM}\right)}{2\pi f_{RF}\tau_{PM}}$$

with f_{RF} is the frequency of the RF electrical signal launched into the phase modulator. τ_{PM} is the delay time difference between the traveling times of the lightwaves in the two waveguide sections before entering to the phase modulator. t is the traveling time of the traveling wave through the electrode. The polarization is assumed to be 1 when the polarization controller aligns the polarized mode launched into the lithium niobate optical waveguide.

Equation 9.42 is obtained by getting the optical field components of the clockwise and counterclockwise with a delay difference depending on the length difference of the fibers. These fields are then modulated with the phase shift via the optical phase modulator. These fields then enter the fiber coupler. The output field is the interfered field at the output and coupled to the photodetector. This field is then followed a square law and then a current is obtained. Thus the responsivity $\Re$ of the detector is included in Equation 9.42.

Furthermore the amplitude conversion from the PM of the two paths can be considered as similar to the case of asymmetric Mach-Zehnder interferometer and the optical amplitude of the interfered optical field does also follow [54]:

$$E_o(t) = \frac{E_{in}}{2}\left[J_o\left(\frac{\pi v_{RF}}{V_{\pi,CW}}\right) - J_o\left(\frac{\pi v_{RF}}{V_{\pi,CCW}}\right)\right]\cos\left(2\pi f_{RF}t\right) \tag{9.21}$$

And the modulation half wave voltage V_π for the CW and CCW direction is given by

$$V_{\pi,CW} = V_{\pi,CCW}\frac{2\pi f_{RF}\tau}{\sin\left(2\pi f_{RF}\tau\right)} \tag{9.22}$$

A number of optical phase modulators are tested in which the buffer exists or not. We observe that the null and maxima of the frequency response is periodic and due to the time

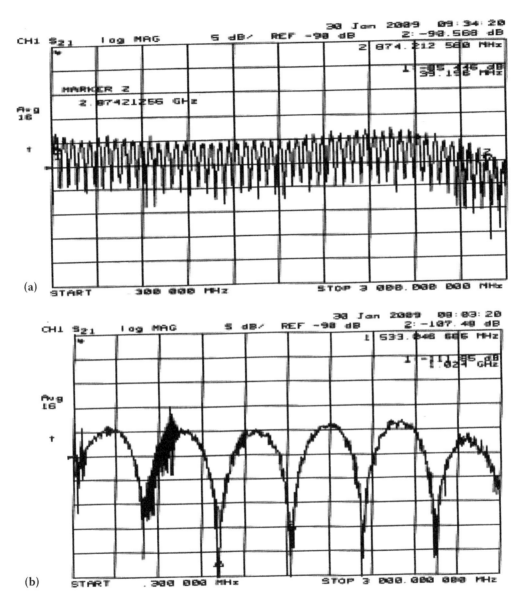

FIGURE 9.22
Frequency response of a non-buffered layer traveling wave electrode: (a) general response, (b) detailed side lobes.

difference of the optical waves before entering the two inputs of the optical phase modulators. No observation of the frequency responses is given in Equation 9.44 but only that of Equation 9.42 (Figure 9.22). That means that the mismatch of the velocities of the traveling wave and optical wave is quite high. While in Figure 9.23 when a SiO_2 buffer layer is used then the traveling wave can offer a wideband operation and the frequency response is a combined response of Equations 9.42 and 9.44.

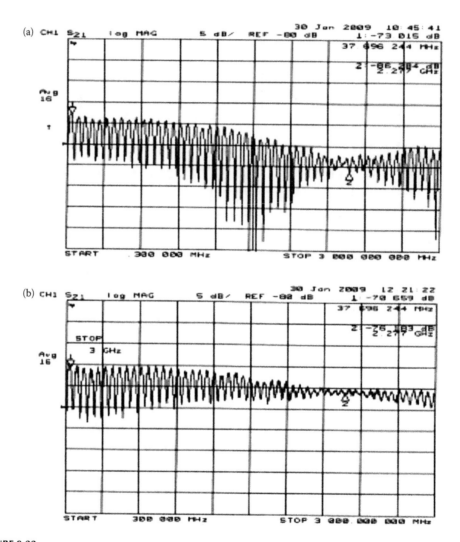

FIGURE 9.23
Frequency response of a SiO$_2$ buffered layer traveling wave electrode: (a) 40 GHz bandwidth, (b) 10 GHz bandwidth.

9.3.7 Remarks

In this section we have formulated a non-uniform FD model that offers an accurate analytical and modeling tool for estimation of the ultra-broadband properties of traveling wave electrodes. An anisotropic Laplace equation solved using the FDTWEA gives accurate microwave properties of symmetric and asymmetric electrode structures. A non-uniform grid allocation scheme is implemented for maximizing the memory computing usage without jeopardizing the model accuracy. It is demonstrated that the calculated characteristic impedance Z and microwave effective index n_m are consistent and an improvement in accuracy as compared to published results. The FDTWEA is also employed to evaluate the EO overlap integral. Principal features are obtained for optimization of the EO interaction overlap integral. The optimization of Γ assures a low half wave voltage V_π, hence lower driving microwave power. Our calculations have demonstrated consistency with their simulated and measured data. The FDTWEA has also been applied to study the impact of the trapezoidal shape tilted-wall thick electrode (>10 μm) on the electrode characteristic impedance and effective index. The effects of the wall angle are not negligible and should not be ignored when a thick electroplated electrode is required to achieve ultra-broadband operation. Quantitatively and qualitatively, our presented model has been shown to be consistent and offers more computationally efficient solutions as compared with other numerical models.

9.4 Lithium Niobate Optical Modulators: Devices and Applications

Optical modulators, using acousto-optic, magneto-optic or EO effects, as the principal components for external modulation of lightwaves, have presently played the important role in modern long-haul ultra-high speed optical communications and photonic signal processing systems. $LiNbO_3$ is an ideal host material for such modulators. In this paper we present: (i) a brief overview of the EO interaction in $LiNbO_3$ optical waveguides and the traveling electric waves for modulation; (ii) fabrication techniques of optical waveguides and electrodes for the excitation of microwave and millimeter waves, including the prospects of the uses of ultra-thick "brick-wall like" type by synchrotron deep x-ray LIGA process; (iii) a comprehensive modeling of the traveling wave electrodes for the design of interferometric optical modulators, including different electrode structures. Tilted and thick and "brick-wall" like electroplated electrodes are modeled and confirmed with implemented modulators operating up to 26 GHz in diffused $LiNbO_3$ optical interferometric waveguide structures. (iv) The design and implementation of $LiNbO_3$ single or dual drive as modulators for modulating lightwaves in modern transmission systems as ultra-high speed pulse carvers, generators of modulation formats (non-return-to-zero [NRZ], return-to-zero [RZ], carrier-suppressed RZ [CS-RZ], differential phase shift keying [DPSK], differential quadrature phase shift keying [DQPSK], etc.) as outlined in Section 9.2 and photonic signal processors.

9.4.1 Mach-Zehnder Interferometric Modulator and Ultra-High Speed Advanced Modulation Formats

Since the invention of EDFA the moderate insertion loss of lightwave coupling, propagation loss in diffused waveguides are no longer the limiting factor for using $LiNbO_3$ MZIM for system applications, especially for ultra-long-haul and ultra-high speed (40 Gb/s and higher) optical fiber communications. Recently several advanced modulation formats have been implemented and demonstrated employing single drive, dual drive and pairs of MZIMs. There are four typical types of modulation techniques: digital amplitude modulation, commonly known as amplitude shift keying (ASK) with NRZ or RZ pulse formats [9], digital differential PM or DPSK or multilevel M-ary PSK, frequency modulation frequency shift keying (FSK) or continuous phase FSM or minimum shift keying (MSK). The operational principles of these modulators will be described in Section 9.6. This section briefly outlines the demands of these transmitters on the integrated modulators and practical fabrication of such modulators.

9.4.1.1 Amplitude Modulation

Amplitude modulation of the lightwave carrier can be easily implemented by applying RF signals into the traveling wave electrodes. The bias voltage is critical whether it is positioned at the phase quadrature, minimum or maximum transmission depending on specific applications. If analog modulation is required, such as in the field of microwave photonics and photonic signal processing, the MZIM is normally biased at the phase quadrature point. While if a suppression of the carrier is required then a different phase shift of π between the two arms must be enforced. For single drive MZIM, bias at V_π while for a dual drive case one could bias both sides with $\pi/2$ and $-\pi/2$ (or alternatively phase quadrature at $\pi/2$ and $3\pi/2$). The biasing at these two operating points can also be used to produce negative gain coefficients in photonic signal processors. For the RZ format, a pulse carver is required to generate RZ pulse trains and then followed by a MZIM data modulator. CS-RZ can thence be generated if the pulse carver is biased at the minimum transmission point.

9.4.1.2 Phase Modulation

PM can offer better performance than amplitude modulation provided that the detection of the phase of the carrier under the modulation format can be recovered error free. A previously coherent technique was used for these types of PM using LiNbO$_3$ MZIM, but it suffers significantly as this technique requires the phase locking of the carriers and hence the linewidth limitation of the source would deteriorate the signals. However this can be overcome by the use of DPSK techniques. DPSK can be implemented using parallel MZIMs and M-ary DPSK can be implemented using only one dual drive LiNbO$_3$ MZIM.

9.4.1.3 Frequency Modulation

Optical FSK can be generated using a pair of parallel LiNbO$_3$ MZIM as shown in Figure 9.24. The RF frequencies can be switched in the time domain according to the coding of the data pulses. Again NRZ or RZ formats can be added if a LiNbO$_3$ MZIM pulse carver is added in tandem to the FSK modulators. Novel continuous phase FSM and MSK as an extension of FSK can be developed using MZIMs.

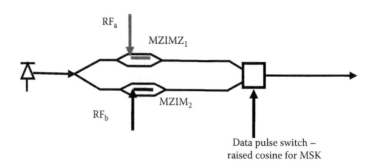

FIGURE 9.24
Schematic of an integrated structure employing LiNbO$_3$ MZIMs for FSK modulation.

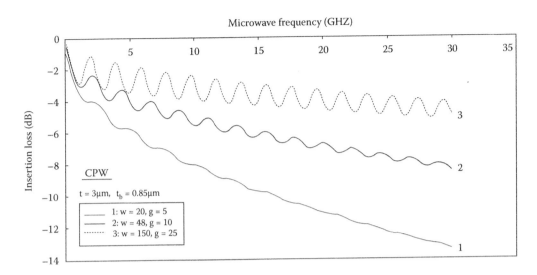

FIGURE 9.25
Insertion loss S_{21} of the RF electrode as a function of the microwave frequency.

9.4.2 LiNbO$_3$ MZIM Fabrication

Most steps of the fabrication of MZIM can be implemented using standard optical lithography. Optical waveguides are fabricated by using Ti in-diffusion of about 50 nm thick at 1050°C for 7 h. The SiO$_2$ layer is then RF sputtered with thickness variable from 0.6–2.0 µm. Electrodes are formed by two main stages. A 20 nm chrome layer is first deposited and the electrode pattern is then created using photographic techniques. A 100 nm thick Au layer is then deposited on top of the Cr temporary electrode pattern. The gold layer is then etched to the pattern of the electrode structures. Electroplating of thick Au pattern is then conducted. The thickness of electroplated Au can reach 7–10 µm with a wall angle of about 12–20° without shorting the electrodes. Note that prior to the fabrication of the electrode, optical channel waveguides are tested using the straight optical channels pattern next to the Mach-Zehnder structures in order to determine the important parameters of the electrodes as designed and simulated. Microwave coupling, fiber pigtails and mounting are then completed.

The insertion loss S_{21} of our typical fabricated electrodes are simulated and measured as shown in Figure 9.25 which shows the measured RF insertion loss with the upper curve showing the calibrated S_{21} of the microwave set up and the lower curve showing the results for the case when the RF electrodes are included in the measured system. Several modulators have been fabricated. The obtained results indicate that the electrical and optical bandwidths reach a consistent range of about 26–30 GHz. Optical measurement of the packaged modulators have also been obtained and the optical transfer bandwidths are compatible with those obtained in the electrical domain. This concludes that the optical loss is minimum and there is good velocity matching between the optical waves and the traveling microwaves. A packaged 26 GHz 3dB bandwidth MZIM is shown in Figure 9.26.

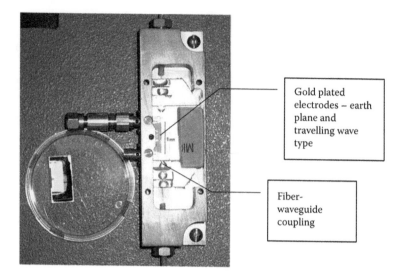

FIGURE 9.26

(See Color Insert) Fabricated and packaged optical modulator: the substrate and fabricated electrode shown on the left side of the photograph and the packaged 26 GHz modulator with fiber pigtail shown on the right side.

9.4.3 Effects of Angled-Wall Structure on RF Electrodes

Since the proposed dielectric optical waveguides in 1966 and thence integrated optics in 1969, the field integrated photonics have progressed tremendously in parallel with the global optical fiber optic communication networks. Optical communication systems have reached an increasing fast pace in both speed and capacity reaching several terabytes per second with 40G OC-192 to 80G OC-768 multiwavelength transport over ultra-long transmission haul. External modulation is the most important technique for modulating lightwaves at these ultra-high speed modulators as external modulation of lightwaves [1]. EO modulators based on $LiNbO_3$ waveguiding structures have been the most popular and feasible devices in the technological evolution of optical fiber communications after the invention of optical amplification in the early 1990s due to its intrinsic high-speed property, its large EO coefficient, its transparency over the 1550 nm communication spectral window, zero or adjustable frequency chirp, efficient butt-coupling and most importantly, relatively simple fabrication technique. Recently several advanced modulation formats such as NRZ, RZ ASK, CS-RZ ASK and DPSK can be generated by employing two single drive interferometric modulators in tandem or a dual electrode drive device. Figure 9.27 shows a typical optical transmission system employing parallel $LiNbO_3$ MZIM for the generation of DQPSK signal. Thus the modulators can be used as a pulse carver to generate periodic pulse shape trains, data modulation and generation and modulation of different phase states of the lightwave carrier. Therefore it demands the modulators have properties of low drive voltage, low insertion loss and wideband. Therefore it requires efficient and accurate modeling of the traveling wave electrodes as well as a method to fabricate thick and vertical "brick-wall" like electroplated electrodes, the optical channel waveguides and the interaction of the launched traveling waves in such combined EO modulation system is developed.

The modeling of ultra-broadband $LiNbO_3$ MZIM incorporating traveling wave electrodes has been briefly described in Section 9.3 as they play a crucial role in the optical

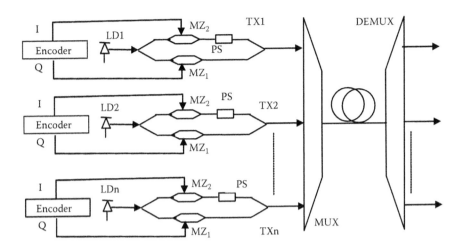

FIGURE 9.27
An advanced modulation format transmitter using MZIMs for multi-carrier multiplexed optical fiber transmission system. PS = phase shifter, LD = laser diode—different wavelength, TX = transmitter, Mux = wavelength multiplexer, DeMUX = wavelength demultiplexer.

bandwidth of the devices. The modulating electric field plays critical roles in the EO overlap integral. More importantly, the more subtle structural factors such as the wall angle of the gold plated electrode can be tailored. Most analyses, e.g., the conformal mapping technique, the Green function method or the MoI, either assume an infinitely thin electrode or assume the wall angle of the electrode to be 90°. However when very thick electrodes, typically in the range of 10–20 μm [6], are employed to achieve ultra-broadband property (40 GHz and higher), the 90° wall angle assumption is no longer valid. Indeed after the electroplating process, the trapezoidal shape of the electrode is illustrated in Figure 9.28b. Figure 9.28a shows the schematic plane view of the EO interferometric optical modulators. The wall angle of the electrode significantly alters the effective microwave index n_m and the characteristic impedance Z. It is thus critical to take into consideration such structural defects in the design of traveling wave electrodes. The comparison of the bandwidth of the tilted electrode to that of the "brick-wall" like structure is important for increasing the operational bandwidth.

Our FDTWEA is used to study the effect of the electrode wall angle that has the most practical effects due to fabrication processes. In extending the bandwidth of the device, very thick electrodes of thickness ranges from 10–20 μm are used. If the conventional Ti-Au electroplating technique is used, such thick electrodes, however, do not assume a perfect "brick wall" shape, but a trapezoidal shape as shown in Figure 9.28b. Such a geometrical factor exerts a subtle effect on both the values of Z and n_m. The wall-angle of tilted electrodes significantly affects the effective values of Z and n_m. The effect of the wall angle is illustrated in Figure 9.28. The bandwidth difference of the plots for different θ converges as the electrodes become thinner. For thick electrodes, especially when greater than 10 μm, the wall angle effect should not be ignored. For example, if we base our design on the assumption of a θ = 90° rectangular electrode, then we would use a 20 μm thick electrode for the best velocity match. However, if the fabricated electrodes actually assume a trapezoidal shape with θ = 80°, we would have overestimated the value of n_m for best velocity matching by about 0.2. This corresponds to an almost 20% reduction of the bandwidth that could have been achieved by a mismatch of 0.01 between the microwave index and

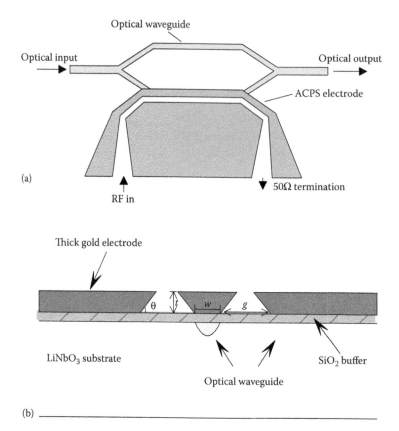

FIGURE 9.28
(a) Schematic diagram of the EO interferometric modulator. (b) The cross sectional view of wall-angle tilted thick electrodes on optical waveguides.

the optical propagation velocity. Modelled results indicate that an electrode thickness of around 14 μm can be used to achieve maximum bandwidth. If the deep x-ray lithographic process is employed with a thick polymer layer (e.g., SU-8), a "brick wall" like polymeric slot structure can be formed. A thick gold layer can then be electroplated and an ideal electrode structure can be fabricated. If such a 20 μm thick gold electrode can be fabricated in this way to give an n_m of 2.15 that matches the optical index, hence a near infinitely-wide-bandwidth modulator can be achieved.

9.4.4 Integrated Modulators and Modulation Formats

Since the proposal of dielectric waveguide and then the advent of optical circular waveguide, the employment of modulation techniques has only recently been extensively exploited since the availability of optical amplifiers. The modulation formats allow the transmission efficiency, and hence the economy of ultra-high capacity information telecommunications. Optical communications have evolved significantly through several phases from single mode systems to coherent detection and modulation which was developed with the main aim to improve on the optical power. The optical amplifiers defeated that main objective of modulation formats and allowed the possibility of incoherent and all possible formats employing the modulation of the amplitude, the phase and frequency of the lightwave carrier.

Currently photonic transmitters play a principal part in the extension of the modulation speed into several gigahertz range and make possible the modulation of the amplitude, the phase and frequency of the optical carriers and their multiplexing. Photonic transmitters using LiNbO$_3$ have been proven in laboratory and installed systems. The principal optical modulator is the MZIM which can be a single or a combined set of these modulators whereby the electrical signals can be applied to the electrodes to form binary or multilevel amplitude or PM and even more effective for discrete or continuous PSK modulation schemes.

Spectral properties of the optical 80 Gb/s dual-level MSK, 40 Gb/s MSK and 40 Gb/s DPSK with various RZ pulse shapes are compared. The spectral properties of the first two formats are similar. Compared to the optical DPSK, the power spectra of optical MSK and dual-level MSK modulation formats have more attractive characteristics. These include the high spectral efficiency for transmission, higher energy concentration in the main spectral lobe and more robustness to inter-channel crosstalk in DWDM due to greater suppression of the side lobes. In addition, the optical MSK offers the orthogonal property, which may offer a great potential in coherent detection, in which the phase information is reserved via I- and Q-components of the transmitted optical signals. In addition the multilevel formats would permit the lowering of the bit rate and hence substantial reduction of the signal effective bandwidth and the possibility of reaching the highest speed limit of the electronic signal processing, the digital signal processing, for equalization and compensation of distortion effects. The demonstration of electrical time division multiplexing (ETDM) receiver at 80 Gb/s and higher speed [27] would make these modulation format schemes attracting significant applications in ultra-high speed transmission.

High speed operation can only be possible if the guided lightwaves can be modulated by high speed broadband electrodes (Figure 9.29). Thus the next section describes the modeling and fundamental characteristics of the microwave traveling wave electrodes for such integration with optical channel waveguide to maximize the bandwidth of optical modulators.

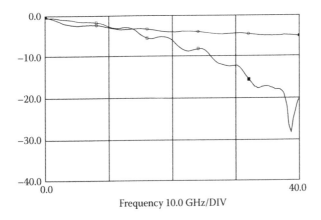

FIGURE 9.29
Electrical (S_{21})/optical response of the implemented interferometric modulator.

9.4.5 Remarks

An effective numerical model has been briefly presented for modeling of LiNbO$_3$ MZIM, especially the effects of the properties of the traveling electrodes on the EO modulation. The diffused optical waveguides can be designed with ease using the non-uniform FE approach. The characteristic impedance Z and the microwave effective index n_m have been developed and implemented in a 26 GHz bandwidth LiNbO$_3$ MZIM with a single drive. The effects of the trapezoidal shape electrode are significant with the wall angle θ, created by the thick electrode (>10 μm) fabricated by electroplating. The fabrication technique for ultra-wideband MZIM has been described. A near 30 n GHz 3 dB BW has been demonstrated. We also propose a thick electrode can be fabricated if the deep x-ray synchrotron lithography is used so that an extremely wideband MZIM can be produced. We have shown that LiNbO$_3$ MZIMs have great potential in the generation of digital ASK, DPSK and FSK as well as negative gain coefficient photonic processors. Other potential and novel modulation formats such as MSK, continuous phase FSK (CPFSK), etc., can be generated using MZIMs and are described in Section 9.6.

9.5 Generation and Modulation of Optical Pulse Sequences

9.5.1 Return-to-Zero Optical Pulses

9.5.1.1 Generation

Figure 9.30 shows the conventional structure of an RZ-ASK transmitter in which two external LiNbO$_3$ MZIMs can be used. The MZIM shown in this transmitter can be either a single or dual drive (push-pull) type. The optical on-off keying (OOK or ASK) transmitter would normally consist of a narrow linewidth laser source to generate lightwaves whose wavelength satisfies the international telecommunication union (ITU) standard for wavelength grid.

The first MZIM, commonly known as the pulse carver, is used to generate the periodic pulse trains with a required RZ format. The suppression of the lightwave carrier can also be carried out at this stage if necessary, commonly known as the CS-RZ. Compared

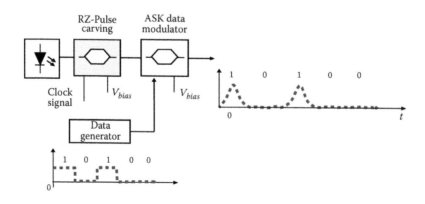

FIGURE 9.30
Conventional structure of an OOK optical transmitter utilizing two MZIMs.

to other RZ types, the CS-RZ pulse shape is found to have attractive attributes for long-haul WDM transmissions including the π phase difference of adjacent modulated bits, suppression of the optical carrier component in optical spectrum and narrower spectral width.

Different types of RZ pulses can be generated depending on the driving amplitude of the RF voltage and the biasing schemes of the MZIM. The equations governing the RZ pulses electric field waveforms are

$$
E(t) = \begin{cases}
\sqrt{\dfrac{E_b}{T}} \sin\left[\dfrac{\pi}{2} \cos\left(\dfrac{\pi t}{T}\right)\right] & \text{67\% duty-ratio RZ pulses or CS-RZ} \\[3ex]
\sqrt{\dfrac{E_b}{T}} \sin\left[\dfrac{\pi}{2}\left(1 + \sin\left(\dfrac{\pi t}{T}\right)\right)\right] & \text{33\% duty-ratio RZ pulses or RZ33}
\end{cases}
\tag{9.25}
$$

where E_b is the pulse energy per a transmitted bit and T is one bit period.

The 33% duty-ratio RZ pulse are denoted as return-to-zero with 33% pulse period width (RZ33) pulse whereas the 67% duty cycle RZ pulse is known as the CS-RZ type. The art in generation of these two RZ pulse types stays at the difference of biasing point on the transfer curve of an MZIM.

The bias voltage conditions and the pulse shape of these two RZ types, the carrier suppression and non-suppression of maximum carrier can be implemented with the biasing points at the minimum and maximum transmission point of the transmittance characteristics of the MZIM, respectively. The peak-to-peak amplitude of the RF driving voltage is $2V_\pi$ where V_π is the required driving voltage to obtain a π phase shift of the lightwave carrier. Another important point is that the RF signal is operating at only half of the transmission bit rate. Hence, pulse carving is actually implementing the frequency doubling. The generation of RZ33 and CS-RZ pulse trains are demonstrated in Figure 9.31a and b.

The pulse carver can also utilize a dual drive MZIM which is driven by two complementary sinusoidal RF signals. This pulse carver, assumed to be a dual drive MZIM, is then biased at $-V_{\pi/2}$ and $+V_{\pi/2}$ with the peak-to-peak amplitude of $V_{\pi/2}$. Thus a π phase shift is created between the state "1"and "0" of the pulse sequence and hence the RZ with alternating phase 0 and π. If carrier suppression is required then the two electrodes are applied with voltages V_π and swing voltage amplitude of V_π.

Although RZ modulation offers improved performance, RZ optical systems usually require more complex transmitters than those in the NRZ ones. Compared to only one stage for modulating data on the NRZ optical signals, two modulation stages are required for generation of RZ optical pulses.

9.5.1.2 Phasor Representation

Recalling the interference of the two guided lightwaves from the two branches of an interferometric structure, the resulting field is the sum of the two lightwave oscillating fields written with the phase variation and removing the carrier angular frequency:

$$
E_o = \frac{E_i}{2}\left[e^{j\varphi_1(t)} + e^{j\varphi_2(t)}\right] = \frac{E_i}{2}\left[e^{j\pi v_1(t)/V_\pi} + e^{j\pi v_2(t)/V_\pi}\right]
\tag{9.24}
$$

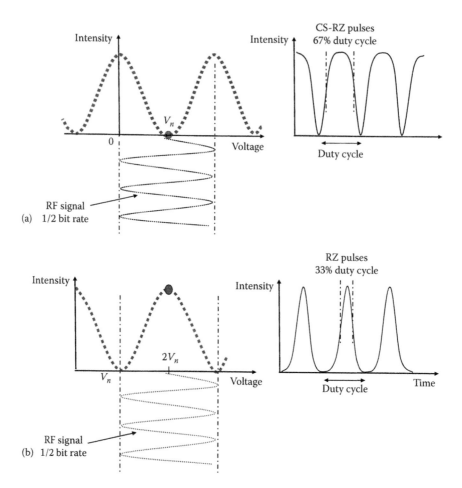

FIGURE 9.31
Bias point and RF driving signals for generation of: (a) CS-RZ, and (b) RZ33 pulses.

It can be seen that the modulating process for the generation of RZ pulses can be represented by a phasor diagram as shown in Figure 9.31. This technique gives a clear understanding of the superposition of the fields at the coupling output of two arms of the MZIM. Here, a dual-drive MZIM is used, that is the data driving signals $[V_1(t)]$ and inverse data ($\overline{data} : V_2(t) = -V_1(t)$) are applied into each arm of the MZIM, respectively, and the RF voltages swing in inverse directions. Applying the phasor representation, vector addition and simple trigonometric calculus, the process of generation RZ33 and CS-RZ is explained in detail and verified (Figure 9.32).

The width of these pulses are commonly measured at the position of full-width half maximum (FWHM). It is noted that the measured pulses are intensity pulses whereas we are considering the addition of the fields in the MZIM. Thus, the normalized E_o field vector has the value of $\pm 1/\sqrt{2}$ at the FWHM intensity pulse positions and the time interval between these points gives the FWHM values.

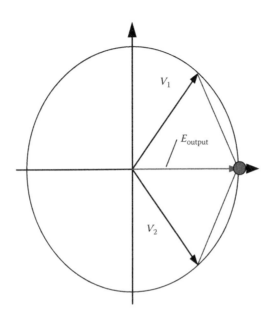

FIGURE 9.32
Phasor representation of the lightwaves propagating through two paths of the interferometers for generation of output field in dual-drive MZIM.

9.5.1.2.1 Phasor Representation of Carrier Suppressed-return-to-zero Pulses

Key parameters including the V_{bias}, the amplitude of the RF driving signal are shown in Figure 9.33a. Accordingly, its initialized phasor representation is demonstrated in Figure 9.33b.

The values of the key parameters are outlined as follows: (i) V_{bias} is $\pm V_\pi/2$; (ii) swing voltage of driving RF signal on each arm has the amplitude of $V_\pi/2$ (i.e., $V_{p-p} = V_\pi$); (iii) RF signal operates at half of bit rate ($B_R/2$); (iv) at the FWHM position of the optical pulse, the $E_{out} = \pm 1/\sqrt{2}$ and the component vectors V_1 and V_2 form with vertical axis a phase of $\pi/4$ as shown in Figure 9.34.

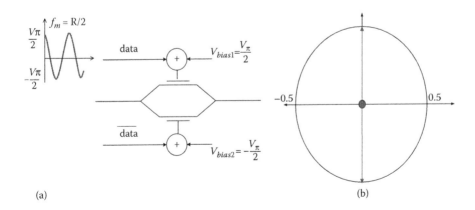

(a)   (b)

FIGURE 9.33
Initialized stage for generation of CS-RZ pulse: (a) RF driving signal and the bias voltages. (b) Initial phasor representation.

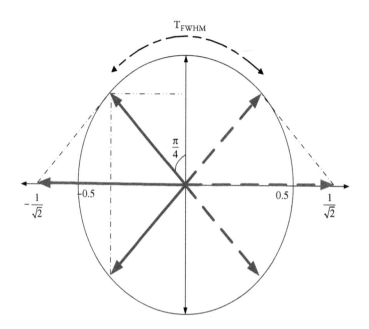

FIGURE 9.34
Phasor representation of CS-RZ pulse generation using dual-drive MZIM.

Considering the scenario for the generation of 40 Gb/s CS-RZ optical signal, the modulating frequency is f_m ($f_m = 20$ GHz $= B_R/2$). At the FWHM positions of the optical pulse, the phase is given by the following expressions:

$$\frac{\pi}{2}\sin(2\pi f_m) = \frac{\pi}{4} \Rightarrow \sin 2\pi f_m = \frac{1}{2} \Rightarrow 2\pi f_m = \left(\frac{\pi}{6}, \frac{5\pi}{6}\right) + 2n\pi \qquad (9.25)$$

Thus, the calculation of T_{FWHM} can be carried out and hence, the duty cycle of the RZ optical pulse can be obtained as given in the following expressions:

$$T_{FWHM} = \left(\frac{5\pi}{6} - \frac{\pi}{6}\right)\frac{1}{R2\pi} = \frac{1}{3}\pi \times \frac{1}{R} \Rightarrow \frac{T_{FWHM}}{T_{BIT}} = \frac{1.66\times10^{-4}}{2.5\times10^{-11}} = 66.67\% \qquad (9.26)$$

The result obtained in Equation 9.29 clearly verifies the generation of CS-RZ optical pulses from the phasor representation.

9.5.1.2.2 *Phasor Representation of Return-to-zero with 33% Pulse Period Width*

Key parameters including the V_{bias}, the amplitude of driving voltage and its correspondent initialized phasor representation are shown in Figure 9.35a and b, respectively.

The values of the key parameters are: (i) V_{bias} is V_π for both arms; (ii) swing voltage of driving RF signal on each arm has the amplitude of $V_\pi/2$ (i.e., $V_{p-p} = V_\pi$); (iii) RF signal operates at half of bit rate ($B_R/2$).

At the FWHM position of the optical pulse, the $E_{output} = \pm1/\sqrt{2}$ and the component vectors V_1 and V_2 form with the horizontal axis a phase of $\pi/4$ as shown in Figure 9.36.

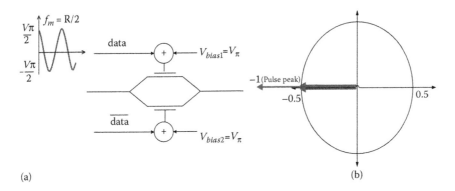

(a)   (b)

FIGURE 9.35
Initialized stage for generation of RZ33 pulse: (a) RF driving signal and the bias voltage; (b) initial phasor representation.

Considering the scenario for the generation of 40 Gb/s CS-RZ optical signal, the modulating frequency is f_m ($f_m = 20$ GHz $= B_R/2$). At the FWHM positions of the optical pulse, the phase is given by the following expressions:

$$\frac{\pi}{2}\cos(2\pi f_m t) = \frac{\pi}{4} \Rightarrow t_1 = \frac{1}{6f_m} \tag{9.28}$$

$$\frac{\pi}{2}\cos(2\pi f_m t) = -\frac{\pi}{4} \Rightarrow t_2 = \frac{1}{3f_m} \tag{9.29}$$

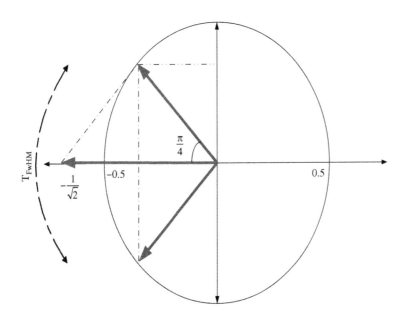

FIGURE 9.36
Phasor representation of RZ33 pulse generation using dual-drive MZIM.

Thus, the calculation of TFWHM can be carried out and hence, the duty cycle of the RZ optical pulse can be obtained as given in the following expressions:

$$T_{\text{FWHM}} = \frac{1}{3f_m} - \frac{1}{6f_m} = \frac{1}{6f_m} \quad \therefore \quad \frac{T_{\text{FWHM}}}{T_b} = \frac{1/6f_m}{1/2f_m} = 33\% \tag{9.30}$$

The result obtained in Equation 9.31 clearly verifies the generation of RZ33 optical pulses from the phasor representation.

9.5.2 Differential Phase Shift Keying

9.5.2.1 Background

Digital encoding of data information by modulating the phase of the lightwave carrier is referred to as optical PSK. In early days, optical PSK was studied extensively for coherent photonic transmission systems. This technique requires the manipulation of the absolute phase of the lightwave carrier. Thus, precise alignment of the transmitter and demodulator center frequencies for the coherent detection is required. These coherent optical PSK systems face severe obstacles such as broad linewidth and chirping problems of the laser source. Meanwhile, the DPSK scheme overcomes those problems, since the DPSK optically modulated signals can be detected incoherently. This technique only requires the coherence of the lightwave carriers over one bit period for the comparison of the differentially coded phases of the consecutive optical pulses.

A binary "1" is encoded if the present input bit and the past encoded bit are of opposite logic whereas a binary 0 is encoded if the logic is similar. This operation is equivalent to an XOR logic operation. Hence, an XOR gate is employed as a differential encoder. NOR can also be used to replace XOR operation in differential encoding as shown in Figure 9.37a. In DPSK, the electrical data "1" indicates a π phase change between the consecutive data bits in the optical carrier, while the binary "0" is encoded if there is no phase change between the consecutive data bits. Hence, this encoding scheme gives rise to two points located exactly at π phase difference with respect to each other in a signal constellation diagram. For continuous PSK, such as MSK, the phase evolves continuously over a quarter of the section, thus a phase change of $\pi/2$ between one phase state to the other. This is indicated by the inner bold circle as shown in Figure 9.37b.

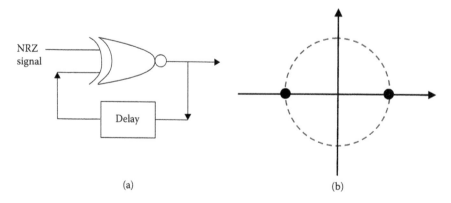

(a)   (b)

FIGURE 9.37
(a) DPSK pre-coder. (b) Signal constellation diagram of DPSK [1,50].

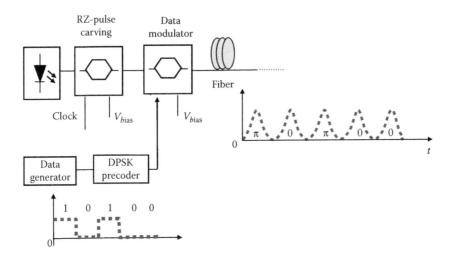

FIGURE 9.38
DPSK optical transmitter with RZ pulse carver.

9.5.2.2 Optical Differential Phase Shift Keying Transmitter

Figure 9.38 shows the structure of a 40 Gb/s DPSK transmitter in which two external LiNbO$_3$ MZIM are used. Operational principles of a MZIM were presented above. The MZIMs shown in Figure 9.38 can be either of single or dual drive type. The optical DPSK transmitter also consists of a narrow linewidth laser to generate a lightwave whose wavelength conforms to the ITU grid.

The RZ optical pulses are then fed into the second MZIM through which the RZ pulses are modulated by the pre-coded binary data to generate RZ-DPSK optical signals. Electrical data pulses are differentially pre-coded in a pre-coder using the XOR coding scheme. Without a pulse carver the structure shown in Figure 9.38 is an optical NRZ-DPSK transmitter. In data modulation for DPSK format, the second MZIM is biased at the minimum transmission point. The pre-coded electrical data has peak-to-peak amplitude equal to $2V_\pi$ and operates at the transmission bit rate. The modulation principles for the generation of optical DPSK signals are demonstrated in Figure 9.39.

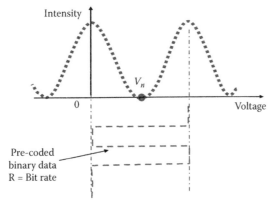

FIGURE 9.39
Bias point and RF driving signals for generation of optical DPSK format.

The EO phase modulator (EO-PM) might also be used for generation of DPSK signals instead of MZIM. Using the optical phase modulator, the transmitted optical signal is chirped whereas using MZIM, especially the X-cut type with Z-propagation, chirp-free signals can be produced. However, in practice, a small amount of chirp might be useful for transmission [23].

9.6 Generation of Modulation Formats

Modulation is the process facilitating the transfer of information over a medium, for example, a wireless or optical environment. Three basic types of modulation techniques are based on the manipulation of a parameter of the optical carrier to represent the information digital data. These are ASK, PSK and FSK. In addition to the manipulation of the carrier, the occupation of the data pulse over a single period would also determine the amount of energy concentrates and the speed of the system required for transmission. The pulse can remain constant over a bit period or RZ level within a portion of the period. These formats would be named NRZ or RZ. They are combined with the modulation of the carrier to form various modulation formats which are presented in this section. Figure 9.40 shows the base band signals of the NRZ and RZ formats and its corresponding block diagram of a photonic transmitter.

9.6.1 Amplitude Shift Keying

9.6.1.1 Amplitude–Modulation Amplitude Shift Keying-Non-Return-to-Zero and Amplitude Shift Keying-Return-to-Zero

The amplitude of the electrical data stream (Figure 9.40a) is fed to the integrated optical intensity modulators to switch the lightwaves on and off as shown in Figure 9.40b. For an operating bit rate above 10 Gb/s the laser linewidth must be preserved by using this external modulation. Thus the bandwidth of the ASK signals contribute to the dispersion effects, that is the interference and broadening of the ASK pulse due to the phase response of the single mode fibers.

However when the data sequence follows a NRZ format, the amplitude of the optical pulse may not be that high so the transmission distance would be limited. One way to enhance the height of the amplitude of the pulse for the same optical power is to use the RZ pulse shape (Figure 9.40c), that is with the bit period the pulse is switched on and then return to the zero level for a "1." This would thus double the pulse amplitude for the same power contained in a pulse period as compared with the NRZ pulse shape provided that the RZ pulse width is half of that of the NRZ. The disadvantage of RZ pulse shaping is that the speed of detection circuitry must be twice as fast as that of NRZ. This wider bandwidth of the receiving circuit would also contribute more noise. A summary of RZ pulse shaping is tabulated in Table 9.3 for a different scenario in which the lightwave carrier can be retained, suppressed by positioning the biasing points. For example, if the suppression of the carrier is required then a dual drive MZIM must be biased at $V_\pi/2$ and $-V_\pi/2$ to give a total π difference in the phase of the lightwaves in the two paths of the interferometer, thus depletion of the carrier at the output of the MZIM.

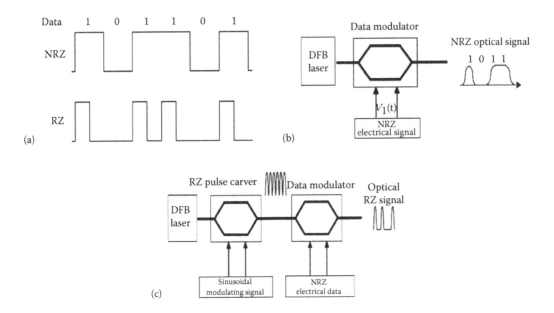

FIGURE 9.40
(a) Baseband NRZ and RZ line coding for 101101 data sequence. (b) Block diagram of NRZ photonics transmitter, and (c) RZ photonics transmitter incorporating a pulse carver.

9.6.1.2 Amplitude–Modulation on-off Keying Return-to-Zero Formats

There are a number of advanced formats used in advanced optical communications, based on the intensity of the pulse this may include NRZ, RZ and duo-binary. These ASK formats can also be integrated with the PM to generate discrete or continuous phase NRZ or RZ formats. Currently the majority of 10 Gb/s installed optical communication systems have been developed with NRZ due to its simple transmitter design and bandwidth efficient characteristic. However, RZ format has higher robustness to fiber nonlinearity and polarization mode dispersion (PMD). In this section, the RZ pulse is generated by MZIM commonly known as *pulse carver* as arranged in Figure 9.41. There are a number of variations in the RZ format based on the biasing point in the transmission curve shown in Table 9.3 in which the phasor representation of the biasing and driving signals are given.

CS-RZ has been found to have more attractive attributes in long-haul WDM transmissions compared to the conventional RZ format due to the possibility of reducing the upper level of the power contained in the carrier that serve no purpose in the transmission but only increase the total energy level so approaching the nonlinear threshold level faster. CS-RZ pulse has optical phase difference of π in adjacent bits, removing the optical carrier component in optical spectrum and reducing the spectral width. This offers an advantage in compact WDM channel spacing.

9.6.1.3 Amplitude–Modulation Carrier-Suppressed Return-to-Zero Formats

The suppression of the carrier can be implemented by biasing the MZ interferometer in such a way so that there is a π phase shift between the two arms of the interferometer. The magnitude of the sinusoidal signals applied to an arm or both arms would determine the

TABLE 9.3

Summary of RZ Format Generation and Characteristics of Single Drive MZIM Based on Biasing Point, Drive Signal Amplitude and Frequency

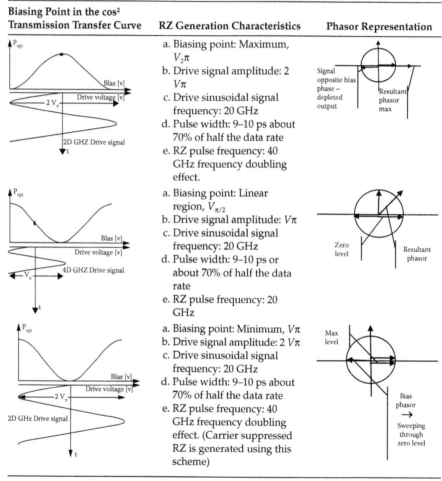

Biasing Point in the cos² Transmission Transfer Curve	RZ Generation Characteristics	Phasor Representation
	a. Biasing point: Maximum, $V_{2\pi}$ b. Drive signal amplitude: 2 $V\pi$ c. Drive sinusoidal signal frequency: 20 GHz d. Pulse width: 9–10 ps about 70% of half the data rate e. RZ pulse frequency: 40 GHz frequency doubling effect.	
	a. Biasing point: Linear region, $V_{\pi/2}$ b. Drive signal amplitude: $V\pi$ c. Drive sinusoidal signal frequency: 20 GHz d. Pulse width: 9–10 ps or about 70% of half the data rate e. RZ pulse frequency: 20 GHz	
	a. Biasing point: Minimum, $V\pi$ b. Drive signal amplitude: 2 $V\pi$ c. Drive sinusoidal signal frequency: 20 GHz d. Pulse width: 9–10 ps about 70% of half the data rate e. RZ pulse frequency: 40 GHz frequency doubling effect. (Carrier suppressed RZ is generated using this scheme)	

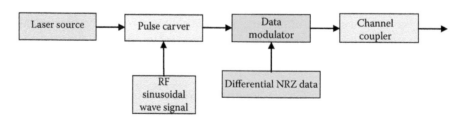

FIGURE 9.41
Block diagrams of RZ-DPSK transmitter.

width of the optical output pulse sequence. The driving conditions and phasor representation are shown in Table 9.3.

9.6.2 Discrete Phase–Modulation Non-Return-to-Zero Formats

The term discrete PM is referred to as DPSK, whether DPSK or DQPSK, to indicate the states of the phases of the lightwave carrier are switched from one distinct location on the phasor diagram to the other state, for example, from 0 to π, or $-\pi/2$ to $\pi2$ for binary PSK (BPSK) or even more evenly spaced PSK levels as in the case of M-ary PSK.

9.6.2.1 Differential Phase Shift Keying

Information encoded in the phase of an optical carrier is commonly referred to as optical PSK. In early days, PSK requires precise alignment of the transmitter and demodulator center frequencies. Hence, the PSK system is not widely deployed. With the DPSK scheme introduced, coherent detection is not critical since DPSK detection only requires source coherence over one bit period by comparison of two consecutive pulse intervals.

A binary "1" is encoded if the present input bit and the past encoded bit are of opposite logic and a binary 0 is encoded if the logic is similar. This operation is equivalent to XOR logic operation. Hence, an XOR gate is usually employed in a differential encoder. NOR can also be used to replace XOR operation in differential encoding as shown in Figure 9.42.

In optical application, electrical data "1" is represented by a π phase change between the consecutive data bits in the optical carrier, while state "0" is encoded with no phase change between the consecutive data bits. Hence, this encoding scheme gives rise to two points located exactly at π phase difference with respect to each other in the signal constellation diagram as indicated in Figure 9.42b.

A RZ-DPSK transmitter consists of an optical source, pulse carver, data modulator, differential data encoder and a channel coupler. A channel coupler model is not developed in simulation by assuming no coupling losses when an optical RZ-DPSK modulated signal is

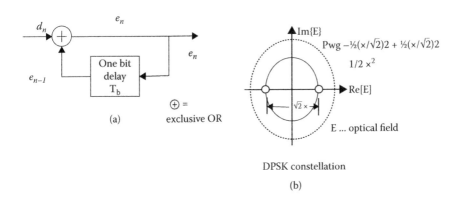

FIGURE 9.42
(a) The encoded differential data are generated by $e_n = d_n \oplus e_{n-1}$. (b) Signal constellation diagram of DPSK.

launched into the optical fiber. This modulation scheme has combined the functionality of a dual drive MZIM modulator of pulse carving and PM.

The pulse carver, usually a MZ interferometric intensity modulator, is driven by a sinusoidal RF signal for single drive MZIM and two complementary electrical RF signals for dual drive MZIM, to carve pulses out from the optical carrier signal forming RZ pulses. These optical RZ pulses are fed into a second MZ intensity modulator where RZ pulses are modulated by differential NRZ electrical data to generate RZ-DPSK. This data PM can be performed using a straight line phase modulator but the MZ waveguide structure has several advantages over a phase modulator due to its chirpless property. Electrical data pulses are differentially pre-coded in a differential pre-coder as shown in Figure 9.42a. Without the pulse carver and sinusoidal RF signal, the output pulse sequence follows NRZ-DPSK format that is the pulse would occupy 100% of the pulse period and there is no transition between the consecutive "1s."

9.6.2.2 Differential Quadrature Phase Shift Keying

This differential coding is similar to DPSK except that each symbol consists of two bits which are represented by the two orthogonal axial discrete phase at $(0, \pi)$ and $(-\pi/2, +\pi/2)$ as shown in Figure 9.42b or two additional orthogonal phase positions are located on the imaginary axis.

9.6.2.2.1 Non-return-to-zero-Differential Phase Shift Keying

Figure 9.43 shows the block diagram of a typical NRZ-DPSK transmitter. Differential pre-coder of electrical data is implemented using the logic explained in the previous section. In phase modulating of the optical carrier, an MZ modulator known as a data phase modulator is biased at minimum point and driven by a data swing of $2V_\pi$. The modulator showed an excellent behavior that the phase of the optical carrier will be altered by π exactly when the signal transiting the minimum point of the transfer characteristic.

9.6.2.2.2 Return-to-Zero-Differential Phase Shift Keying

The arrangement of the RZ-DPSK transmitter is essentially similar to RZ-ASK as shown in Figure 9.44 with the data intensity modulator replaced with the data phase modulator.

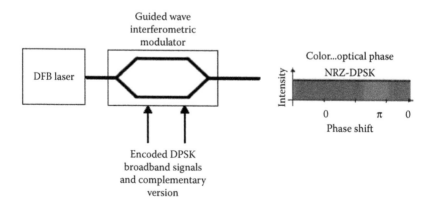

FIGURE 9.43
Block diagram of NRZ-DPSK photonics transmitter. Phase shift shown by the intensity under the intensity line.

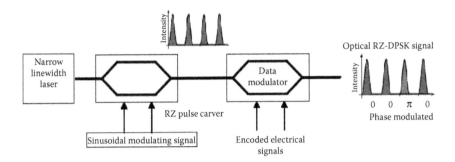

FIGURE 9.44
(See Color Insert) Block diagram of RZ-DPSK photonics transmitter. Phase modulated shown by the intensity of the shaded region under the pulses.

The difference between them is the biasing point and the electrical voltage swing. Different RZ formats can also be generated.

9.6.2.3 M-Ary Amplitude Differential Phase Shift Keying

As an example a 16-ary MADPSK signal can be represented by a constellation shown in Figure 9.45. It is, indeed, a combination of a 4-ary ASK and a DQPSK scheme. At the transmitting end, each group of four bits $[D_3D_2D_1D_0]$ of user data are encoded into a symbol, among them the two least significant bits $[D_1D_0]$ are encoded into four phase states $[0, \pi/2, \pi, 3\pi/2]$ and the other two most significant bits $[D_3D_2]$ are encoded into four amplitude levels. At the receiving end, as MZ delay interferometers (DI) are used for phase comparison and detection, a diagonal arrangement of the signal constellation shown in Figure 9.45a is preferred. This simplifies the design of the transmitter and receiver, minimizes the number of phase detectors, hence leading to high receiver sensitivity.

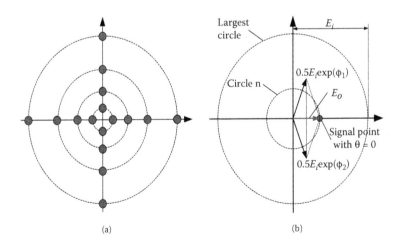

(a) (b)

FIGURE 9.45
Signal constellation of 4-ary ADPSK format and phasor representation of a point on the constellation point for driving voltages applied to dual dive MZIM.

In order to balance the bit error rate (BER) between ASK and DQPSK components, the signal levels corresponding to four circles of the signal space should be adjusted to a reasonable ratio which depends on the noise power at the receiver. As an example, if this ratio is set to $[I_0/I_1/I_2/I_3] = [1/1.4/2/2.5]$, where I_0, I_1, I_2 and I_3 are the intensity of the optical signals corresponding to circle 0, circle 1, circle 2, and circle 3, respectively, then by selecting E_i equal to the amplitude of circle 3 and V_π equal to 1, the driving voltages should have the values given in Table 9.4. Inversely speaking one can set the outer most level such that its peak value is below the nonlinear SPM threshold, the voltage level of the outer most circle would be determined. The inner most circle is limited to the condition that the largest signal to noise ratio (SNR) should be achieved. This is related to the optical SNR (OSNR) required for a certain BER. Thus from the largest amplitude level and smallest amplitude level we can then design the other points of the constellation.

Furthermore, to minimize the effect of inter-symbol interference, 66%-RZ and 50%-RZ pulse formats are also used as alternatives to the traditional NRZ counterpart. These RZ pulse formats can be created by a pulse carver that precedes or follows the dual-drive MZIM modulator. Mathematically, waveforms of NRZ and RZ pulses can be represented by the following equations, where E_{on}, $n=0, 1, 2, 3$ are peak amplitude of the signals in the circle 0, circle 1, circle 2, and circle 3 of the constellation, respectively:

$$p(t) = \begin{cases} E_{on} & \text{for NRZ} \\[2ex] E_{on} \cos\left(\dfrac{\pi}{2}\cos^2\left(\dfrac{1.5\pi t}{T_s}\right)\right) & \text{for 66\%-RZ} \\[2ex] E_{on} \cos\left(\dfrac{\pi}{2}\cos^2\left(\dfrac{2\pi t}{T_s}\right)\right) & \text{for 50\%-RZ} \end{cases} \qquad (9.31)$$

A typical arrangement of the signals of the pre-coder and driving signals for the MZIM is shown in Figure 9.46.

9.6.3 Continuous Phase-Modulation (PM)-Non-Return-to-Zero Formats

In the previous section the optical transmitters for discrete PSK modulation formats have been described. Obviously the phase of the carrier has been used to indicate the digital states of the bits or symbols. These phases are allocated in a non-continuous manner around a circle corresponding to the magnitude of the wave. Alternatively, the phase of the carrier can be continuously modulated and the total phase changes at the transition instants, usually at the end of the bit period, would be the same as those for discrete cases. Since the phase of the carrier is continuously varied during the bit period, this can be considered as a FSK modulation technique, except that the transition of the phase at the end of one bit and the beginning of the next bit would be continuous. The continuity of the carrier phase at these transitions would reduce the signal bandwidth and hence more tolerable to dispersion effects and higher energy concentration for effective transmission over the optical guided medium. One of the examples of the reduction of the phase at the transition is the offset DQPSK which is a minor but important variation on the QPSK or DQPSK. In OQPSK the Q-channel is shifted by half a symbol period so

TABLE 9.4

Driving Voltages for 16-ary MADPSK Signal Constellation

	Circle 0			Circle 1			Circle 2			Circle 3		
Phase	$V_i(t)$ volt	$V_2(t)$ volt	Phase	$V_i(t)$ volt	$V_2(t)$ volt	Phase	$V_i(t)$ volt	$V_2(t)$ volt	Phase	$V_i(t)$ volt	$V_2(t)$ volt
0	0.38	−0.38	0	0.30	−0.30	0	0.21	−0.21	0	0.0	0.0
$\pi/2$	0.88	0.12	$\pi/2$	0.80	0.20	$\pi/2$	0.71	0.29	$\pi/2$	0.5	0.5
π	−0.62	0.62	π	−0.7	0.70	π	−0.79	0.79	π	−1.0	1.0
$3\pi/2$	−0.12	−0.88	$3\pi/2$	−0.20	−0.8	$3\pi/2$	−0.29	−0.71	$3\pi/2$	−0.5	−0.5

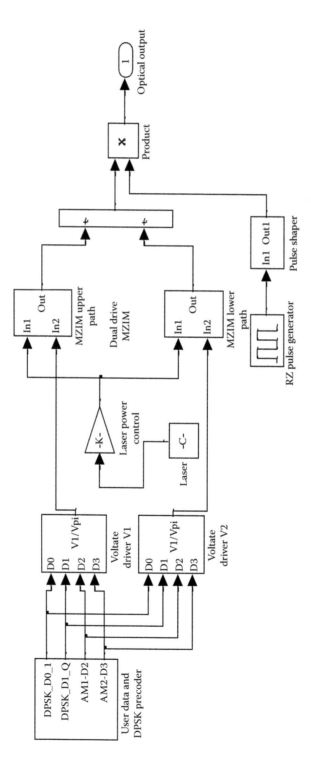

FIGURE 9.46
Matlab Simulink model of a MADPSK photonic transmitter. The MZIM is represented by two phase shifter blocks.

that the transition instants of I- and Q-channel signals do not happen at the same time. The result of this simple change is that the phase shifts at any one time are limited and hence the offset QPSK is a more constant envelope than that of the "conventional" QPSK.

The enhancement of the efficiency of the bandwidth of the signals can be further improved if the phase changes at these transitions are continuous. In this case the change of the phase during the symbol period is continuously changed by using a half-cycle sinusoidal driving signal with the total phase transition over a symbol period is a fraction of π, depending on the levels of this PSK modulation. If the change is $\pi/2$ then we have a MSK scheme. The orthogonality of the I- and Q-channels will also reduce further the bandwidth of the carrier-modulated signals. In this section we describe the basic principles of optical MSK and the photonic transmitters for these modulation formats. Ideally the driving signal to the phase modulator should be a triangular wave so that a linear phase variation of the carrier in the symbol period is linear. However, when a sinusoidal function is used there are some nonlinear variation, we term this type of MSK as a nonlinear MSK format. This nonlinearity contributes to some penalty in the OSNR (see ref. [50]). Furthermore the MSK as a special form of offset differential quadrature phase shift keying (ODQPSK) is also described for optical systems.

9.6.3.1 Linear and Nonlinear Minimum Shift Keying

MSK is a special form of CPFSK signal in which the two frequencies are spaced in such a way that they are orthogonal, hence there is minimum spacing between them, defined by

$$s(t) = \sqrt{\frac{2E_b}{T_b}} \cos\left[2\pi f_1 t + \theta(0)\right] \quad \text{for symbol 1} \tag{9.32}$$

$$s(t) = \sqrt{\frac{2E_b}{T_b}} \cos\left[2\pi f_2 t + \theta(0)\right] \quad \text{for symbol 0} \tag{9.33}$$

As shown by the equations above, the signal frequency change corresponds to higher frequency for data-1 and lower frequency for data-0. Both frequencies, f_1 and f_2, are defined by

$$f_1 = f_c + \frac{1}{4T_b} \tag{9.34}$$

$$f_2 = f_c - \frac{1}{4T_b} \tag{9.35}$$

Depending on the binary data, the phase of signal changes; data-1 increases the phase by $\pi/2$, while data-0 decreases the phase by $\pi/2$. The variation of phase follows paths of sequence of straight lines in phase trellis (Figure 9.47), in which the slopes represent frequency changes. The change in carrier frequency from data-0 to data-1, or vice versa, is equal to half the bit rate of incoming data [5]. This is the minimum frequency spacing that allows the two FSK signals representing symbols 1 and 0, to be coherently orthogonal in the sense that they do not interfere with one another in the process of detection.

A MSK signal consists of both I- and Q-components including the carrier frequency, which can be written as

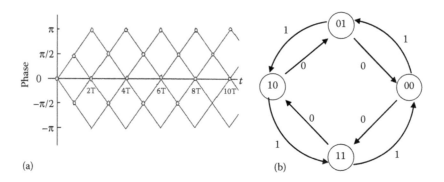

FIGURE 9.47
(a) Phase trellis for MSK. (b) State diagram for MSK.

$$s(t) = \sqrt{\frac{2E_b}{T_b}} \cos[\theta(t)]\cos[2\pi f_c t] - \sqrt{\frac{2E_b}{T_b}} \sin[\theta(t)]\sin[2\pi f_c t] \qquad (9.36)$$

The in-phase component consists of a half-cycle cosine pulse defined by

$$s_I(t) = \pm\sqrt{\frac{2E_b}{T_b}} \cos\left(\frac{\pi t}{2T_b}\right), \quad -T_b \le t \le T_b \qquad (9.37)$$

while the quadrature component would take the form

$$s_Q(t) = \pm\sqrt{\frac{2E_b}{T_b}} \sin\left(\frac{\pi t}{2T_b}\right), \quad 0 \le t \le 2T_b \qquad (9.38)$$

During even bit interval, the I-component consists of positive cosine waveform for phase of 0, while negative cosine waveform for phase of π; during odd bit interval, the Q- component consists of positive sine waveform for phase of $\pi/2$, while negative sine waveform for phase of $-\pi/2$ (as shown in Figure 9.47). Any of four states can arise: 0, $\pi/2$, $-\pi/2$, π. However, only state 0 or π can occur during any even bit interval and only $\pi/2$ or $-\pi/2$ can occur during any odd bit interval. The transmitted signal is the sum of I- and Q-components and its phase is continuous with time.

Two important characteristics of MSK are each data bit is held for two bit period, meaning the symbol period is equal two bit period ($h = 1/2$) and the I- and Q-components are interleaved. I- and Q-components are delayed by one bit period with respect to each other. Therefore, only I- or Q-components can change at a time (when one is at zero-crossing, the other is at maximum peak). The pre-coder can be a combinational logic as shown in Figure 9.48.

A truth table can be constructed based on the logic state diagram and combinational logic diagram shown in Figure 9.47. For positive half cycle cosine wave and positive half cycle sine wave, the output is 1; for negative half cycle cosine wave and negative half cycle sine wave, the output is 0. Then, a K-map can be constructed to derive the logic gates of the pre-coder, based on the truth table. The following three pre-coding logic equations are derived:

$$S_0 = \overline{b_n}\overline{S_0}'S_1' + b_n\overline{S_0}'S_1' + \overline{b_n}S_0'S_1' + b_nS_0'\overline{S_1}' \qquad (9.39)$$

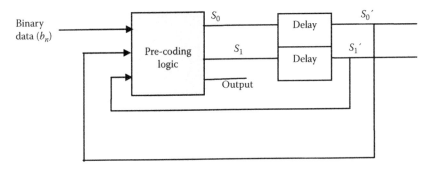

FIGURE 9.48
Combinational logic, the basis of the logic for constructing the pre-coder in electronic domain.

$$S_1 = \overline{S_1}' = \overline{b_n}\,\overline{S_1}' + b_n\,\overline{S_1}' \tag{9.40}$$

$$\text{Output} = \overline{S_0} \tag{9.41}$$

9.6.3.2 Minimum Shift Keying as a Special Case of Continuous Phase Frequency Shift Keying

CPFSK signals are modulated in upper side band (USB) and lower side band (LSB) frequency carriers f_1 and f_2 as expressed in Equations 9.34 and 9.35.

$$s(t) = \sqrt{\frac{2E_b}{T_b}}\,\cos\left[2\pi f_1 t + \theta(0)\right] \quad \text{for symbol "1"} \tag{9.42}$$

$$s(t) = \sqrt{\frac{2E_b}{T_b}}\,\cos\left[2\pi f_2 t + \theta(0)\right] \quad \text{for symbol "0"} \tag{9.43}$$

where $f_1 = f_c + 1/4T_b$ and $f_2 = f_c - 1/4T_b$.with T_b is the bit period.
  The phase slope of lightwave carrier changes linearly or nonlinearly with the modulating binary data. In the case of linear MSK, the carrier phase linearly change by $\pi/2$ at the end of the bit slot with data "1", while it linearly decreases by $\pi/2$.with data "0." The variation of phase follows paths of well-defined phase trellis in which the slopes represent frequency changes. The change in carrier frequency from data-0 to data-1, or vice versa, equals half the bit rate of incoming data [13]. This is the minimum frequency spacing that allows the two FSK signals representing symbols 1 and 0, to be coherently orthogonal in the sense that they do not interfere with one another in the process of detection. MSK carrier phase is always continuous at bit transitions. The MSK signal in Equations 9.37 and 9.38 can be simplified as:

$$s(t) = \sqrt{\frac{2E_b}{T_b}}\,\cos[2\pi f_c t + d_k\,\frac{\pi t}{2T_b} + \Phi_k], \quad kT_b \le t \le (k+1)T_b \tag{9.44}$$

and the base-band equivalent optical MSK signal is represented as:

$$\tilde{s}(t) = \sqrt{\frac{2E_b}{T_b}} \exp\left\{ j\left[d_k \frac{\pi t}{2T} + \Phi(t,k) \right] \right\}, \quad kT \le t \le (k+1)T$$

$$= \sqrt{\frac{2E_b}{T_b}} \exp\left\{ j\left[d_k 2\pi f_d t + \Phi(t,k) \right] \right\}$$

(9.45)

where $d_k = \pm 1$ are the logic levels; f_d is the frequency deviation from the optical carrier frequency and $h = 2f_d T$ is defined as the frequency modulation index. In case of optical MSK, $h = 1/2$ or $f_d = 1/(4T_b)$.

9.6.3.3 Minimum Shift Keying as Offset Differential Quadrature Phase Shift Keying

Equation 9.21 can be rewritten to express MSK signals in form of I-Q components as:

$$s(t) = \pm\sqrt{\frac{2E_b}{T_b}} \cos\left(\frac{\pi t}{2T_b} \right) \cos[2\pi f_c t] \pm \sqrt{\frac{2E_b}{T_b}} \sin\left(\frac{\pi t}{2T_b} \right) \sin[2\pi f_c t]$$

(9.46)

During even bit intervals, the in-phase component consists of positive cosine waveform for phase of 0, while negative cosine waveform for phase of π; during odd bit interval, the Q-component consists of positive sine waveform for phase of $\pi/2$, while negative sine waveform for phase of $-\pi/2$. Any of four states can arise: 0, $\pi/2$, $-\pi/2$, π. However, only state 0 or π can occur during any even bit interval and only $\pi/2$ or $-\pi/2$ can occur during any odd bit interval. The transmitted signal is the sum of I- and Q-components and its phase is continuous with time.

Two important characteristics of MSK are each data bit is held for a two bit period, meaning the symbol period is equal two bit period ($h = 1/2$) and the I-component and Q-component are interleaved. I- and Q-components are delayed by 1 bit period with respect to each other. Therefore, only I- or Q-component can change at a time (when one is at zero-crossing, the other is at maximum peak).

9.6.3.4 Configuration of Photonic Minimum Shift Keying Transmitter Using Two Cascaded Electro-Optic Phase Modulators

EO-PMs and interferometers operating at high frequency using resonant-type electrodes have been studied and proposed in Refs. [14,26,27]. In addition, high-speed electronic driving circuits evolved with the ASIC technology using 0.1 μm GaAs p-HEMT or InP HEMTs [25] enables the feasibility in realization of the optical MSK transmitter structure. The base-band equivalent optical MSK signal is represented in Equation 9.46.

The first EO-PM enables the frequency modulation of data logic into USBs and LSBs of the optical carrier with frequency deviation of f_d. Differential phase pre-coding is not necessary in this configuration due to the nature of the continuity of the differential phase trellis. By alternating the driving sources $V_d(t)$ to sinusoidal waveforms for simple implementation or combination of sinusoidal and periodic ramp signals which was first proposed by Amoroso [16], different schemes of linear and nonlinear phase shaping MSK transmitted sequences can be generated [50].

The second EO-PM enforces the phase continuity of the lightwave carrier at every bit transition. The delay control between the EO-PMs is usually implemented by the phase shifter shown in Figure 9.49. The driving voltage of the second EO-PM is pre-coded to fully compensate the transitional phase jump at the output $E_{o1}(t)$ of the first EO-PM. Phase continuity characteristic of the optical MSK signals is determined by the algorithm in Equations 9.42 through 9.44.

$$\Phi(t,k) = \frac{\pi}{2}\left(\sum_{j=0}^{k-1} a_j - a_k I_k \sum_{j=0}^{k-1} I_j\right) \tag{9.47}$$

where $a_k = \pm 1$ are the logic levels; $I_k = \pm 1$ is a clock pulse whose duty cycle is equal to the period of the driving signal $V_d(t)$; f_d is the frequency deviation from the optical carrier frequency and $h = 2f_d T$ is previously defined as the frequency modulation index. In the case of optical MSK, $h = 1/2$ or $f_d = 1/(4T)$. The phase evolution of the continuous phase optical MSK signals are explained in Figure 9.49. In order to mitigate the effects of unstable stages of rising and falling edges of the electronic circuits, the clock pulse $V_c(t)$ is offset with the driving voltages $V_d(t)$.

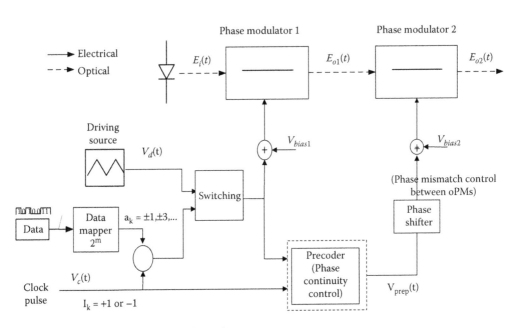

FIGURE 9.49
Block diagram of optical MSK transmitter employing two cascaded/integrated optical phase modulators.

9.6.3.5 Configuration of Optical Minimum Shift Keying Transmitter Using Mach-Zehnder Intensity Modulators: I-Q Approach

The conceptual block diagram of optical MSK transmitter is shown in Figure 9.50. The transmitter consists of two dual-drive EO MZIM modulators generating the chirpless I- and Q-components of MSK modulated signals which is considered as a special case of staggered or offset QPSK. The binary logic data is pre-coded and de-interleaved into even and odd bit streams which are interleaved with each other by one bit duration offset.

Two arms of the dual-drive MZIM are biased at $V_\pi/2$ and $-V_\pi/2$ and driven with *data* and $\overline{data}$. Phase shaping driving sources can be a periodic triangular voltage source in case of linear MSK generation or simply a sinusoidal source for generating a nonlinear MSK-like signal which also obtain linear phase trellis property but with small ripples introduced in the magnitude. The magnitude fluctuation level depends on the magnitude of the phase shaping driving source. High spectral efficiency can be achieved with tight filtering of the driving signals before modulating the EO MZIMs.

9.6.4 Single Side Band (SSB) Optical Modulators

An SSB modulator can be formed using a primary interferometer with two secondary MZIM structures, the optical Ti-diffused waveguide paths that form a nested primary MZ structure as shown in Figure 9.51. Each of the two primary arms contains a MZ structures. Two RF ports are for RF modulation and three DC ports are for biasing the two secondary MZIMs and one primary MZIM. The modulator consists of X-cut Y-propagation LiNbO$_3$ crystal, where you can produce an SSB modulation just by driving each MZ. DC voltage is supplied to produce the π phase shift between upper and lower arms. DC bias voltages are also supplied from DC$_B$ to produce the phase shift between 3rd and 4th arms. A DC bias voltage is supplied from DC$_C$ to achieve a $\pi/2$ phase shift between MZIM$_A$ and MZIM$_B$. The RF voltage applied given as $\pi_1(t) = \phi \cos \omega_m t$ and $\phi_1(t) = \phi \sin \omega_m t$ are inserted from RF$_A$ and RF$_B$, respectively, by using a wideband $\pi/2$ phase shifter. ϕ is the modulation level, ω_m is the RF angular frequency.

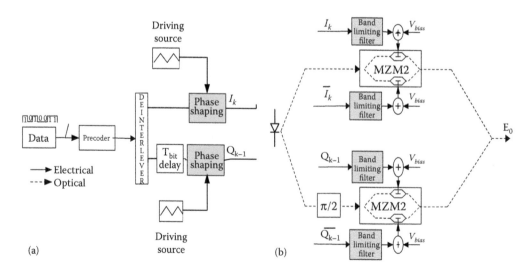

FIGURE 9.50
Block diagram configuration of band limited phase shaped optical MSK.

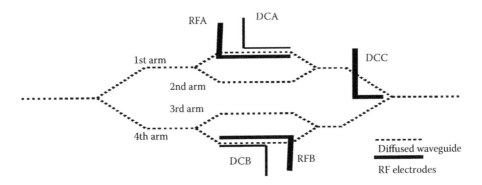

FIGURE 9.51
Schematic diagram (not to scale) of a SSB FSK optical modulator formed by nested MZ modulators.

9.7 Problems

1. A carrier-suppressed RZ optical transmitter is shown below.

    i. Propose a guided wave structure having a Mach-Zehnder interferometer that would permit the modulation of the intensity and phase of the lightwaves guided from the input through the interferometric branches and then combined at the output. You may search for a commercial modulator and extract the operational characteristic parameters such as operating wavelength, total insertion loss, Vpi, DC drift of bias voltage, monitored signals, and so on.

    ii. Now suppose an integrated optical modulator is used as the lightwave modulation device. The electrical data source generates electrical data sequence. Sketch the time-domain pulse sequence over a 10 bit period for a bit rate of 40 Gb/s at the output of the pulse pattern generator.

    iii. Give a brief description of the principles of the suppression of the carrier. Which component of the transmitter would implement the suppression?

    iv. What are the functions of the laser, the modulator, and the push-pull modulator? If the V_π for the two modulators are 5 volts, sketch the transfer characteristics of the modulators—that is, the output power versus the input driving voltage. Make sure that you set appropriate biasing voltages for the modulators. The output power of the laser is 10 dBm and the total insertion loss for each modulator is 4 dB. For the pulse pattern generator, the output power at the output port data is 10 dBm and that at the clock output port is 2 Vp-p. All line impedance are 50 ohms.

    v. Is it necessary to use a booster optical amplifier to increase the total average power launching into an optical fiber for transmission? If so, then what is the gain and noise of the optical amplifier? Note that the nonlinear limit of a SSMF is around 5 dBm.

    vi. Sketch the spectra at the outputs of the laser, the modulator, and the push-pull modulator.

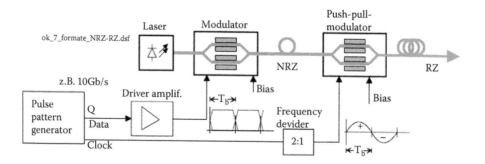

2. Optical modulator and phasor diagram—dual-drive MZIM.

   a. Sketch the transfer characteristics of a MZIM—that is, the output intensity or field as a function of applied voltage V. Now, with a bias voltage of $V_b = V_\pi/2$ and V_π, and a time-varying signal of $v_s(t) = V_{\pi/2} \cos\omega_s t$ with $\omega_s = 2\pi f_s$ and $f_s = 20$ GHz, sketch this applied electrical source on the transfer curve and then obtain the output optical envelop. You may assume that the MZIM of this part is a single-drive type.

   b. Now the modulator is a dual-drive MZM, repeat Problem 2a with a bias voltage of $V_b = V_{\pi/2}$ and V_π and a time-varying signal of $v_s(t) = V_{\pi/2} \cos\omega_s t$ with $\omega_s = 2\pi f_s$ and $f_s = 20$ GHz.

3. An optical fiber communication system consists of an optical transmitter using a 1550 nm DFB laser with a line width of 10 pm (pico-meters), an external optical modulator whose bandwidth is 20 GHz, and a total insertion loss of 5 dB. The modulator is driven with a bit-pattern-signal generator with a 10 dBm electrical power output into a 50 ohm line. A microwave amplifier is used to boost the electrical data pulse to an appropriate level for driving the optical modulator. The data bit rate is 10 Gb/s and its format is non-return-to-zero (NRZ) and an 80 km SSMF is used for the transmission of the modulated signals.

   a. Sketch the block diagram of the transmission system, especially the optical transmitter employing a guided wave optical modulator using MZIM as an external modulation device.

   b. If the V_π of the external modulator is 5 V, what is the gain of the microwave amplifier so that an extension ration of 20 dB can be achieved for the output pulses of "1" and "0" at the output of the modulator? Make sure that you sketch the amplitude and power output of the modulator versus the driving voltage into the modulator. What type of connector would you use to connect the microwave amplifier to the modulator and that to the bit-pattern-generator?

   c. If the DFB laser emits 0 dBm optical power at its pig tail output, then what is the average optical power contained in the signal spectrum? You may assume that the pulse sequence generated at the output of the bit-pattern-generator is a perfect rectangular shape. What is the effective 3 dB bandwidth of the signal power spectrum? Estimate the total pulse broadening of the pulse sequence at the end of the 80 km fiber length. Similarly, estimate the pulse sequence if the bit rate is 40 Gb/s.

   d. Now if a dispersion-compensating fiber of 20 km is used to compensate for the signal distortion in the 80 km fiber, what is the required dispersion factor of this fiber so that there would be no distortion? If the loss of the dispersion-compensating fiber is 1.0 dB/km at 1550 nm, estimate the average optical power of the signal at the output of the dispersion-compensating fiber.

e. Based on the dispersion limit given below, plot the dispersion length as a function of the bit rate for NRZ format.

NOTE: The dispersion limit, under linear regime operation, can be estimated in the following equation (ref. Forgheti et al. 1997)

$$L_D = \frac{c}{\lambda} \frac{\rho}{B_R^2 D}$$

where is the bit rate, D is the dispersion factor (s/m²), ρ is the duty cycle ration— that is, the ration between the "ON" and "OFF" in a bit period and L_D is in meters.

4. Repeat Problem 3 for return-to-zero (RZ) format and ASK modulation. Sketch the structure of the RZ optical transmitter. Note that an extra optical modulator must be used and coupled with the data modulator of Problem 1, the optical pulse carver. Give details of the pulse carver, including: driving voltage, driving signal, and synchronization with the data generator.

5. Spectral efficiency

a. A DWDM optical transmission system can transmit optical channels whose channel spacing is 100 GHz. What is the spectral efficiency if the bit rate of each channel is 40 Gb/s and the modulation is NRZ-ASK?

b. Repeat (a) for RZ-ASK modulation format.

c. Repeat (a) and (b) for the channel spacing of 50 GHz.

6. Give the structure of an optical transmitter for the generation of RZ-ASK modulation format. Make sure that you assign the optical power of lightwaves generated from the light source and at the output of the optical modulators so that a maximum of 10 dBm of optical power is launched into the SSMF so that it is below the nonlinear SPM effect limit.

Describe the operation of the optical modulator, the pulse carver, so that it can generate periodic pulse sequences before feeding into the data generator. Make sure that you provide the amplitude and intensity versus the driving signal voltage levels which are used to drive the optical modulators.

7. Nonlinear SPM Effect

The nonlinear refractive index coefficient of silica-based SSMF is $n_2 = 2.5 \times 10^{-20}$ m²/W.

What is the effective area of the SSMF? (You can refer to the technical specification of the Corning SMF-28 and its mode field diameter to estimate this area.)

Estimate the change of the refractive index as a function of the average optical power. Estimate the total phase change due to this nonlinear effect after propagating though a length L (in km) of this fiber.

Estimate the maximum length L of the SSMF that the lightwaves can travel so that a no higher than 0.1 rad of the phase change on this lightwave carrier would be suffered.

Show how you can generate a format that would have a RZ format and a suppression of the lightwave carrier. Show that the width of the RZ pulse in this case is 67% of the bit period. Hint: you may represent the lightwaves in the path of the optical modulator, an optical interferometer by using phasors. First, sketch the phasor of the input lightwave, then those of the two paths, and then the phase applied onto these paths. Then sum up at the output to give the resultant output. For the pulse width, you can estimate the width over which the amplitude fall to 1/sqrt(2) of its maximum.

8. Revision Problems About Structures of Optical Modulators

   a. Sketch a structure of an integrated optical modulator including: (i) the substrate, (ii) the diffused waveguide if LiNbO3 material is used, (iii) the electrode structure. Give a brief description of the modulation of the phase of the guided lightwave carrier. Give an estimate of the total insertion loss and which part of the modulator would suffer the greatest loss of power.

   b. Use the phase modulation structure of (a) to form an interferometric structure of the symmetric path of the interferometer.

      i. Sketch the interferometer with only one electrode applied to one path of the guided wave structure.

      ii. Similar to (i) but using two electrodes applied to each path of the interferometer, give advantages and disadvantages of this dual-electrode driving optical modulator in terms of driving voltage level (applied to the electrodes) and the difficulty in the control of the phase paths of the electrical signals.

      iii. Search the Internet for some commercial devices on optical phase modulation, amplitude modulators, modulators for generating modulation formats BDPSK, QPSK, DQPSK, M-ary QAM (e.g., 16-QAM), and I-Q modulators. Extract the important parameters of the modulators.

References

1. (a) R.C. Alferness, "Optical guided-wave devices," *Science*, 234(4778), 825–829, 14 November 1986. (b) M. Rizzi and B. Castagnolo, "Electro-Optic Intensity Modulator for Broadband Optical Communications," *Fiber and Integrated Optics*, 21, 243–251, 2002. (c) H. Takara, "High-speed optical time-division-multiplexed signal generation," *Optical and Quantum Electronics*, 33(7–10), 795–810, July 2001. (d) E.L. Wooten, K.M. Kissa, A. Yi-Yan, E.J. Murphy, D.A. Lafaw, P.F. Hallemeier, D. Maack, D.V. Attanasio, D.J. Fritz, G.J. McBrien, and D.E. Bossi, "A review of lithium niobate modulators for fiber-optic communications systems," *IEEE J. Sel. Topics Quant. Elect.*, 6(1), 69–80, January/February 2000. (e) K. Noguchi, O. Mitomi, H. Miyazawa, and S. Seki, "A broadband Ti:LiNbO$_3$ optical modulator with a ridge structure," *IEEE J. Lightwave Tech.*, 13(6), 1164, June 1995.

2. J. Noda, "Electro-optic modulation method and device using the low-energy oblique transition of a highly coupled super-grid," *IEEE J. Lightwave Tech.*, LT-4, 1986.

3. M. Suzuki, Y. Noda, H. Tanaka, S. Akiba, Y. Kuahiro, and H. Isshiki, "Monolithic integration of InGaAsP/InP distributed feedback laser and electroabsorption modulator by vapor phase epitaxy," *IEEE J. Lightwave Tech.*, 5(9), 1277–1285, September 1987.

4. P.K. Tien, "Integrated optics and new wave phenomena in optical waveguides," *Rev. Mod. Phys.*, 49, 361–420, 1977.
5. A. Yariv, C.A. Mead, and J.V. Parker, *IEEE J. Quantum Electron.*, QE-2, 243, 1966.
6. M. Suzuki, Y. Noda, H. Tanaka, S. Akiba, Y. Kuahiro, and H. Isshiki, "Monolithic integration of InGaAsP/InP distributed feedback laser and electroabsorption modulator by vapor phase epitaxy", *IEEE J. Light Tech.*, 5(9), 1277–1285, September 1987.
7. H. Nagata, Y. Li, W.R. Bosenberg, and G.L. Reiff, "DC Drift of X-Cut LiNbO₃ Modulators," *IEE Photonics Tech.Lett.*, 16(10), 2233–2335, October 2004.
8. H. Nagata, "DC Drift Failure Rate Estimation on 10 Gb/s X-Cut Lithium Niobate Modulators," *IEEE Photonics Tech Lett.*, 12(11), 1477–1479, November 2000.
9. R.E. Epworth, K.S. Farley, and D. Watley, "Polarization mode dispersion compensation," US patent 398/152, 398/202, 398/65.
10. J.G. Proahkis, "*Digital Communications,*" 3rd Ed., McGraw-Hill, NY, USA, 1994.
11. K.K. Pang, "*Digital Transmission,*" Melbourne: Mi-Tec Publishing, 58, 2002.
12. R. Redner and H. Walker, *SIAM Review,* 26(2), 1984.
13. T. Kawanishi, S. Shinada, T. Sakamoto, S. Oikawa, K. Yoshiara, and M. Izutsu, "Reciprocating optical modulator with resonant modulating electrode," *Electr. Lett.*, 41(5), 271–272, 2005.
14. R. Krahenbuhl, J.H. Cole, R.P. Moeller, and M.M. Howerton, "High-speed optical modulator in LiNbO₃ with cascaded resonant-type electrodes," *J. Light. Technol.*, 24(5), 2184–2189, 2006.
15. I.P. Kaminow and T. Li, *Optical Fiber Communications, Volume IVA* (Chapter 16), Elsevier Science, 2002.
16. F. Amoroso, "Pulse and Spectrum Manipulation in the Minimum Frequency Shift Keying (MSK) format," *IEEE Trans. Commun.*, 24, 381–384, March 1976.
17. E. Lach and K. Schuh, "Recent advances in ultrahigh bit rate ETDM transmission systems", *IEEE J. Lightw. Tech.*, 24(12), 4455–4467, 2006.
18. W.S. Lee, "80+Gb/s ETDM systems implementation: An overview of current technology," in *Proc. OFC* 2006, Paper no. OTuB3.
19. B. Ivan, Djordjevic, and Bane Vasic,100-Gb/s transmission using orthogonal frequency-division multiplexing," *IEEE Photonics Tech. Lett.*,18(15), August 2006.
20. A.J. Lowery, L. Du, and J. Armstrong, "Orthogonal frequency division multiplexing for adaptive dispersion compensation in long Haul WDM systems," OFC 2006 Postdeadline Sessions, Anaheim, CA, USA, 9th March 2006, Paper PDP39.
21. A.H. Gnauck, X. Liu, X. Wei, D.M. Gill, and E.C. Burrows, "Comparison of modulation formats for 42.7-gb/s single-channel transmission through 1980 km of SSMF," *IEEE Photonics Tech. Lett.*,16(3), 909–911, 2004.
22. G.P. Agrawal, "*Fiber-Optic Communication Systems,*" 3rd ed., New York: Wiley, 2002.
23. A. Hirano, Y. Miyamoto, and S. Kuwahara, "Performances of CSRZ-DPSK and RZ-DPSK in 43-Gbit/s/ch DWDM G.652 single-mode-fiber transmission," *in Proceedings of OFC'03*, 2, 454–456, 2003.
24. A.H. Gnauck, G. Raybon, P.G. Bernasconi, J. Leuthold, C.R. Doerr, and L.W. Stulz, "1-Tb/s (6/spl times/170.6 Gb/s) transmission over 2000-km NZDF using OTDM and RZ-DPSK format," *IEEE Photonics Tech. Lett.*, 15(11), 1618–1620, 2003.
25. Y. Yamada, H. Taga, and K. Goto, "Comparison Between VSB, CS-RZ and NRZ format in a Conventional DSF based long Haul DWDM system," *in Proceedings of ECOC'02*, 4, 1–2, 2002.
26. O. Painter, P.C. Sercel, K.J. Vahala, D.W. Vernooy, and G.H. Hunziker, "Resonant Optical Modulators," US Patent No. WO/2002/050575, 27 June 2002.
27. C. Schubert, R.H. Derksen, M. Möller, R. Ludwig, C.-J. Weiske, J. Lutz, S. Ferber, A. Kirstädter, G. Lehmann, and C. Schmidt-Langhorst, "Integrated 100-Gb/s ETDM Receiver," *IEEE J. Lightw. Tech.*, 25(1), January 2007.
28. O.G. Ramer, "Integrated optic electro-optic modulator electrode analysis," *IEEE J. Quant. Elect.*, QE-18(3), 386–392, 1982.
29. H.J.M. Bélanger and Z. Jakubezyk, "General analysis of electrodes in integrated optics electro-optic Devices," *IEEE J. Quant. Elect.*, 27(2), 243–251, 1991.

30. D. Marcuse, "Optimal electrode design for integrated optics modulators," *IEEE J. Quant. Elect.*, QE-18(3), 393–398, March 1982.
31. C. Sabatier and E. Caquot, "Influence of a dielectric buffer layer on the field distribution in an electro-optic guided wave device," *IEEE J. Quant. Elect.*, QE-22(1), 32–37, 1986.
32. H. Chung, W.S.C. Chang, and E.L. Adler, "Modeling and optimization of travelling-wave LiNbO$_3$ interferometric modulators," *IEEE J. Quant. Elect.*, 27(3), 608–617, 1991.
33. W.-C. Chuang, W.Y. Le, J.H. Lieu, and W.-S. Wang, "A comparison of the performance of LiNbO$_3$ travelling wave phase modulators with various dielectric buffer layers," *J. Opt. Commun.*, 14(4), 142–148, 1993.
34. P.J. Mares and S.L. Chuang, "Modeling of self-electro-optic-effect devices," *J. Appl. Phys.*, 74(2), 1388–1389, July 1993.
35. Z. Pantic and R. Mittra, "Quasi-TEM analysis of microwave transmission lines by finite element method," *IEEE Trans. Microw. Th. Tech.*, MTT-34(11), 1986.
36. G.K. Gopalakrishnan, W.K. Burns, and C.H. Bulmer, "Electrical loss mechanisms in traveling wave LiNbO$_3$ optical modulators," *Elect. Lett.*, 28(2), 207–208, 1992.
37. G.K. Gopalakrishnan and W.K. Burns, "Performance and modeling of broadband LiNbO$_3$ traveling wave optical intensity modulators," *J. Lightw. Tech.*, 12(10), 1807–1818, 1994.
38. H.E. Green, "The numerical solution of some important transmission-line problems," *IEEE Trans. Microw. Th. Tech.*, MTT-13(5), 676–692, 1965.
39. S. Ramo and W. Whinery, *Fields and Waves in Communication Electronics*, 2nd Ed., J. Wiley & Sons, Inc, 1973.
40. T.C. Oppe, "NSPCG user's guide version 1.0 - a package for solving large sparse linear systems by various iterative methods," Center for Numerical Analysis, The University of Texas, Austin, USA.
41. C.M. Kim and R. Ramaswamy, "Overlap integral factors in integrated optic modulators and switches," *J. Lightw. Tech.*, 7(7), 1063–1070, 1989.
42. E.L. Wooten and W.S.C. Chiang, "Test structures for characterization of electro-optic waveguide modulators in lithium niobate," *IEEE J. Quant. Elect.*, 29(1), 161–170, 1993.
43. L.N. Binh "LiNbO$_3$ optical modulators: devices and applications," *J. Crystal Growth*, 180–187, January 2006.
44. S.K. Korotky, G. Eisenstein, R.S. Tucker, J.J. Veselka, and G. Raybon, "Optical intensity modulation to 40GHz using a waveguide electro-optic switch," *Appl. Phys. Lett.*, 50(23), 1631–1633, 1987.
45. C.M. Gee, G.D. Thurmond, and H.W. Yen, "17 GHz bandwidth electro-optic modulator," *Appl. Phys. Lett.*, 43(11), December 1983.
46. T. Kitoh and K. Kawano, "Modeling and design of Ti:LiNbO$_3$ optical modulator electrodes with a buffer layer," *Elect. Comm. in Japan*, Part 2, 76(1), 25–34, 1993.
47. P.J. Winzer and R.-J. Essiambre, "Advanced modulation formats for high-capacity optical transport networks," *J. Lightw. Technol.*, 24(12), 4711–4728.
48. I. Morita, "Advanced modulation format for 100-Gbit/s transmission," Paper TuE3.3, 252–253, OFC 2008.
49. K. Petermann, "Modulation formats for optical fiber transmission," in *Optical Communication Theory and Techniques*, Springer, Berlin, 2006.
50. L.N. Binh, *"Digital Optical Communications,"* CRC Press, Taylor & Francis Group, Boca Raton, FL, 2009.
51. L.N. Binh, T.L. Huynh, and K.K. Pang, "Direct detection frequency discrimination optical receiver for minimum shift keying format transmission," *IEEE J. of Lightw. Technol.*, 26(18), 3234–3247, September 2008.
52. W.R. Leeb, A.L. Schotz, and E. Bonek, "Measurement of velocity mismatch in travelling wave electro-optic modulators," *IEEE J. Quant. Elect.*, QE-18(1) 14–16, 1982.
53. E.H.W. Chan and R.A. Minassian, "A new optical phase modulator dynamic response measurement technique," *IEEE J. Lightwave Tech.*, 26(16), 2008, 2882–2888, 2008.
54. M.Y. Frankel and R.D. Esman, "Optical single sideband suppression carrier for wideband signal processing," *IEEE J. Lightw Tech.*, 16(5), May 1989.

10

Nonlinearity in Guided Wave Devices

This chapter describes the applications of nonlinear (NL) effects in guided wave devices, especially the channel and rib-waveguide structure, so that the converted lightwave beam can be coupled with optical transmission systems to generate phase conjugation for compensation of distorted data pulses, or as a bispectrum photonic pre-processing, or in a fiber ring laser to generate mode-locked pulse sequences.

10.1 Nonlinear Effects in Integrated Optical Waveguides for Photonic Signal Processing

10.1.1 Introductory Remarks

With increasing demand for high capacity, communication networks are facing several challenges, especially in signal processing at the physical layer at ultra-high speed. When the processing speed is over that of the electronic limit or requires massive parallel and high speed operations, the processing in the optical domain offers significant advantages. Thus all-optical signal processing is a promising technology for future optical communication networks. An advanced optical network requires a variety of signal processing functions including optical regeneration, wavelength conversion, optical switching and signal monitoring. An attractive way to realize these processing functions in transparent and high speed mode is to exploit the third-order nonlinearity in optical waveguides, particularly parametric processes.

Nonlinearity is a fundamental property of optical waveguides including channel, rib integrated structures or circular fibers. The origin of nonlinearity comes from the third-order NL polarization in optical transmission media [1]. It is responsible for various phenomena such as self-phase modulation (SPM), cross-phase modulation (XPM) and four-wave mixing (FWM) effects. In these effects, the parametric FWM process is of special interest because it offers several possibilities for signal processing applications [2–16]. To implement all-optical signal processing functions, high nonlinearity of optical waveguides is required. Therefore the highly NL fibers (HNLF) are commonly employed for this purpose because the NL coefficient of HNLF is about ten times higher than that of standard transmission fibers. Indeed, the third-order nonlinearity of conventional fibers is often very small to prevent the degradation of the transmission signal from NL distortion. Recently, NL chalcogenide and tellurite glass waveguides have emerged as a promising device for ultra-high speed photonic processing [2,17]. Because of their geometries these waveguides are called planar waveguides. A planar waveguide can confine a high intensity of light, within an area comparable to the wavelength of light, over a short distance of a few centimeters. Hence they are compatible with optical integrated circuits which can produce very compact photonic devices for signal processing.

In this paper we demonstrate a number of important applications in optical signal processing using third-order nonlinearity, especially parametric FWM processes by simulation. This section is organized as follows: Section 10.1.2 gives us a mathematical review of third-order nonlinearity in optical waveguides. We particularly focus on the parametric process FWM. The basic propagation equations, which describe the propagation of optical signals as well as interactions between optical waves in optical waveguides, are also given in this section. Then a MATLAB model of NL waveguides is developed and integrated into the Simulink platform as described in Section 10.1.3 enabling our investigation of the parametric processes in NL waveguides. This model is used as a functional block in all-optical signal processing applications. System applications and performance evaluations based on parametric processes are demonstrated in Section 10.1.4. Besides important applications such as optical amplification, ultra-high speed switching and distortion compensation, we have also demonstrated the potential of a triple correlation as a high-order spectrum estimation for ultra-sensitive optical receivers based on FWM.

10.1.2 Third-Order Nonlinearity and Parametric Four-Wave Mixing Process

10.1.2.1 Nonlinear Wave Equation

In optical waveguides including optical fibers, the third-order nonlinearity is of special importance because it is responsible for all NL effects. Lightwaves confinement and propagation in optical waveguides are generally governed by the NL wave equation (NLE) which can be derived from Maxwell's equations under the coupling of the NL polarization. The NL wave propagation in NL waveguide in the time-spatial domain in vector form can be expressed as [1] (see also the Appendices of this book):

$$\nabla^2 \vec{E} - \frac{1}{c^2} \frac{\partial^2 \vec{E}}{\partial t^2} = \mu_0 \left(\frac{\partial^2 \overrightarrow{P_L}}{\partial t^2} + \frac{\partial^2 \overrightarrow{P_{NL}}}{\partial t^2} \right) \tag{10.1}$$

where $\vec{E}$ is the electric field vector of the optical wave, μ_0 is the vacuum permeability assuming a non-magnetic waveguiding medium, c is the speed of light in vacuum, $\overrightarrow{P_L}, \overrightarrow{P_{NL}}$, are, respectively, the linear and NL polarization vectors which are formed as:

$$\overrightarrow{P_L}(\vec{r}, t) = \varepsilon_0 \chi^{(1)} \cdot \vec{E}(\vec{r}, t) \tag{10.2}$$

$$\overrightarrow{P_{NL}}(\vec{r}, t) = \varepsilon_0 \chi^{(3)} \vdots \vec{E}(\vec{r}, t) \vec{E}(\vec{r}, t) \vec{E}(\vec{r}, t) \tag{10.3}$$

where $\chi^{(3)}$ is the third-order susceptibility. Thus the linear and NL coupling effects in optical waveguides can be described by Equation 10.1. The second term on the right hand side (RHS) is responsible for the NL processes including interaction between the optical waves through the third-order susceptibility.

In most telecommunication applications, only the complex envelope of optical signal is considered in analysis because the bandwidth of the optical signal is much smaller than the optical carrier frequency. To model the evolution of the light propagation in optical waveguides, it requires that Equation 10.1 can be further modified and simplified by some assumptions which are valid in most telecommunication applications [1]. Hence the electrical field $\vec{E}$ can be written as

$$\vec{E}(\vec{r},t) = \tfrac{1}{2}\hat{x}\left\{F(x,y)A(z,t)\exp[i(kz-\omega t)]+c.c.\right\} \tag{10.4}$$

where $A(z,t)$ is the slowly varying complex envelope propagating along z in the waveguide, k is the wave number. After some algebra using a method of separating variables, the following equation for propagation in the optical waveguide is obtained as:

$$\frac{\partial A}{\partial z}+\frac{\alpha}{2}A-i\sum_{n=1}^{\infty}\frac{i^{n}\beta_{n}}{n!}\frac{\partial^{n}A}{\partial t^{n}}=i\gamma\left(1+\frac{i}{\omega_{0}}\frac{\partial}{\partial t}\right)\times A\int_{-\infty}^{\infty}g(t')\left|A(z,t-t')\right|^{2}dt' \tag{10.5}$$

where the effect of propagation constant β around ω_{0} is Taylor-series expanded, $g(t)$ is the NL response function including the electronic and nuclear contributions. For the optical pulses wide enough to contain many optical cycles, Equation 10.5 can be simplified as

$$\frac{\partial A}{\partial z}+\frac{\alpha}{2}A+\frac{i\beta_{2}}{2}\frac{\partial^{2}A}{\partial\tau^{2}}-\frac{\beta_{3}}{6}\frac{\partial^{3}A}{\partial\tau^{3}}=i\gamma\left[\left|A\right|^{2}A+\frac{i}{\omega_{0}}\frac{\partial\left(\left|A\right|^{2}A\right)}{\partial\tau}-T_{R}A\frac{\partial\left(\left|A\right|^{2}\right)}{\partial\tau}\right] \tag{10.6}$$

where a frame of reference moving with the pulse at the group velocity v_{g} is used by making the transformation $\tau = t-z/v_{g} \equiv t-\beta_{1}z$, and A is the total complex envelope of propagation waves, α, β_{k} are the linear loss and dispersion coefficients, respectively, $\gamma = \omega_{0}n_{2}/cA_{\text{eff}}$ is the NL coefficient of the guided wave structure and the first moment of the NL response function is defined as

$$T_{R}\equiv\int_{0}^{\infty}tg(t')dt' \tag{10.7}$$

Equation 10.6 is the basic propagation equation, commonly known as the NL Schroedinger equation (NLSE) which is very useful for investigating the evolution of the amplitude of the optical signal and the phase of the lightwave carrier under the effect of third-order nonlinearity in optical waveguides. The left hand side (LHS) in Equation 10.6 contains all linear terms, while all NL terms are contained on the RHS. In this equation the first term on the RHS is responsible for the intensity dependent refractive index effects, including FWM.

10.1.2.2 Four-Wave Mixing Coupled-Wave Equations

FWM is a parametric process through the third-order susceptibility $\chi^{(3)}$. In the FWM process, the superposition and generation of the propagating of the waves with different complex amplitudes, the frequencies ω_{n} and wave numbers k_{n} through the waveguide can be represented as

$$A=\sum_{n}A_{n}e^{\left[j(k_{n}z-\omega_{n}\tau)\right]}\quad\text{with }n=1,\ldots,4 \tag{10.8}$$

It is noted that the complex amplitudes are used as the envelope of the waves, the complex parts of these amplitudes represent the phases of the lightwaves of different frequencies. By ignoring the linear and scattering effects and with the introduction of Equation 10.8

into Equation 10.6, the NLSE can be separated into coupled differential equations, each of which is responsible for one distinct wave in the waveguide

$$\frac{\partial A_1}{\partial z} + \frac{\alpha}{2} A_1 = i\gamma A_1 \left[|A_1|^2 + 2\sum_{n\neq1} |A_n|^2 \right] + i\gamma 2 A_3 A_4 A_2^* \exp\left(-i\Delta k_1 z\right)$$

$$\frac{\partial A_2}{\partial z} + \frac{\alpha}{2} A_2 = i\gamma A_2 \left[|A_2|^2 + 2\sum_{n\neq2} |A_n|^2 \right] + i\gamma 2 A_3 A_4 A_1^* \exp\left(-i\Delta k_2 z\right)$$

$$\frac{\partial A_3}{\partial z} + \frac{\alpha}{2} A_3 = i\gamma A_3 \left[|A_3|^2 + 2\sum_{n\neq3} |A_n|^2 \right] + i\gamma 2 A_1 A_2 A_4^* \exp\left(-i\Delta k_3 z\right)$$

$$\frac{\partial A_4}{\partial z} + \frac{\alpha}{2} A_4 = i\gamma A_4 \left[|A_4|^2 + 2\sum_{n\neq4} |A_n|^2 \right] + i\gamma 2 A_1 A_2 A_3^* \exp\left(-i\Delta k_4 z\right)$$

(10.9)

where $\Delta k = k_1 + k_2 - k_3 - k_4$ is the wave vector mismatch. The equation system, Equation 10.9, thus describes the interaction between different waves in NL waveguides. The interaction which is represented by the last term in Equation 10.9 can generate new waves. For three waves with different frequencies, a fourth wave can be generated at frequency $\omega_4 = \omega_1 + \omega_2 - \omega_3$. The waves at frequencies ω_1 and ω_2 are called pump waves, whereas the wave at frequency ω_3 is the signal and the generated wave at ω_4 is called the idler wave as shown in Figure 10.1a. If all three waves have the same frequency ($\omega_1 = \omega_2 = \omega_3$) the interaction is called a degenerate FWM with the new wave at the same frequency. If only two of the three waves are at the same frequency ($\omega_1 = \omega_2 \neq \omega_3$) the process is called partly degenerate FWM which is important for some applications like the wavelength converter and parametric amplifier.

10.1.2.3 Phase Matching

In parametric NL processes such as FWM, the energy conservation and momentum conservation must be satisfied to obtain a high efficiency of the energy transfer as shown in Figure 10.1a. The phase matching condition for the new converted wave requires

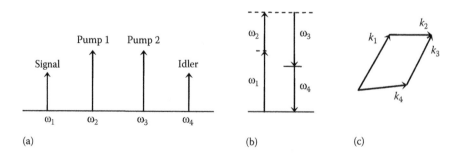

(a)   (b)   (c)

FIGURE 10.1
(a) Position and notation of the distinct waves, (b) diagram of energy conservation, and (c) diagram of momentum conservation in FWM.

$$\Delta k = k_1 + k_2 - k_3 - k_4 = \frac{1}{c}(n_1\omega_1 + n_2\omega_2 - n_3\omega_3 - n_4\omega_4)$$

$$= 2\pi\left(\frac{n_1}{\lambda_1} + \frac{n_2}{\lambda_2} - \frac{n_3}{\lambda_3} - \frac{n_4}{\lambda_4}\right) \tag{10.10}$$

During propagation in optical waveguides, the relative phase difference $\theta(z)$ between four involved waves is determined by [3,5]:

$$\theta(z) = \Delta kz + \varphi_1(z) + \varphi_2(z) - \varphi_3(z) - \varphi_4(z) \tag{10.11}$$

where $\varphi_k(z)$ relates to the initial phase and the NL phase shift during propagation. An approximation of a phase matching condition can be given as follows [4]:

$$\frac{\partial\theta}{\partial z} \approx \Delta k + \gamma(P_1 + P_2 - P_3 - P_4) = \kappa \tag{10.12}$$

where P_k is the power of the waves and κ is the phase mismatch parameter. Thus the FWM process has maximum efficiency for $\kappa = 0$. The mismatch comes from the frequency dependence of the refractive index and the dispersion of optical waveguides. Depending on the dispersion profile of the NL waveguides, it is very important in the selection of pump wavelengths to ensure that the phase mismatch parameter is minimized.

10.1.3 Transmission Models and Nonlinear Guided Wave Devices

To model the parametric FWM process between multi-waves, the basic propagation equations described in Section 10.1.2 are used. There are two approaches to simulate the interaction between waves. The first approach, named as the separating channel technique, is to use the coupled equations system, Equation 10.9, in which the interactions between different waves are obviously modeled by certain coupling terms in each coupled equation. Thus each optical wave considered as one separated channel is represented by a phasor. The coupled equations system is then solved to obtain the solutions of the FWM process. The outputs of the NL waveguide are also represented by separated phasors, hence the desired signal can be extracted without using a filter.

The second or alternating approach is to use the propagation equation, Equation 10.6 that allows us to simulate all evolutionary effects of the optical waves in the NL waveguides. In this technique a total field is used instead of individual waves. The superimposed complex envelope A is represented by only one phasor which is the summation of individual complex amplitudes of different waves given as

$$A = \sum_k A_k e^{\left[j(\omega_k - \omega_0)\tau\right]} \tag{10.13}$$

where ω_0 is the defined angular central frequency, A_n, ω_n are the complex envelope and the carrier frequency of individual waves, respectively. Hence, various waves at different frequencies are combined into a total signal vector which facilitates integration of the NL waveguide model into the Simulink® platform (Figure 10.2). Equation 10.6 can

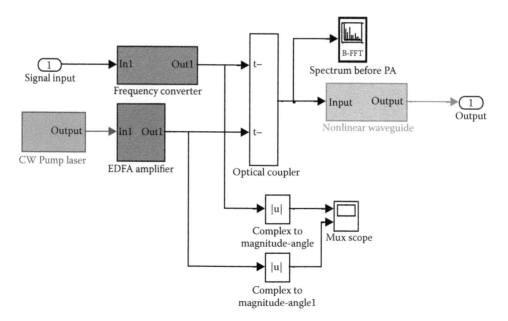

FIGURE 10.2
Typical Simulink setup of the parametric amplifier using the model of NL waveguide.

be numerically solved by the split-step method. The Simulink block representing the NL waveguide is implemented with an embedded MATLAB program. Because only complex envelopes of the guided waves are considered in the simulation, each of the different optical waves are shifted by a frequency difference between the central frequency and the frequency of the wave to allocate the wave in the frequency band of the total field. Then the summation of individual waves, which is equivalent to the combination process at the optical coupler, is performed prior to entering the block of NL waveguide as depicted in Figure 10.3. The output of NL waveguide will be selected by an optical bandpass filter (BPF). In this way, the model of NL waveguide can be easily connected to other Simulink blocks which are available in the platform for simulation of optical fiber communication systems [18].

10.1.4 System Applications of Third-Order Parametric Nonlinearity in Optical Signal Processing

In this section, a range of signal processing applications are demonstrated through simulations which use the model of NL waveguide to model the wave mixing process.

10.1.4.1 Parametric Amplifiers

One of the important applications of the $\chi^{(3)}$ nonlinearity is parametric amplification. The optical parametric amplifiers (OPA) offer a wide gain bandwidth, high differential gain and optional wavelength conversion and operation at any wavelength [4–7]. These important features of OPA are obtained because the parametric gain process does not rely on energy transitions between energy states, but it is based on highly efficient FWM in which two photons at one or two pump wavelengths interact with a signal photon. The 4th photon, the idler, is formed with a phase such that the phase difference between the pump

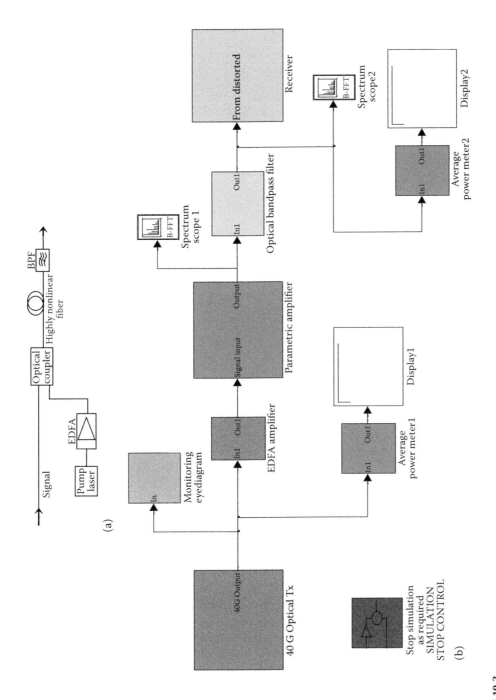

FIGURE 10.3

(a) A typical setup of an OPA. (b) Simulink model of OPA.

photons and the signal and idler photons satisfies the phase matching condition (Equation 10.10). The schematic of the fiber-based parametric amplifier is depicted in Figure 10.3a.

For the parametric amplifier using one pump source, from the coupled equations, Equation 10.9, with $A_1 = A_2 = A_p$, $A_3 = A_s$ and $A_4 = A_i$, it is possible to derive three coupled equations for complex field amplitude of the three waves $A_{p,s,i}$

$$\frac{\partial A_p}{\partial z} = -\frac{\alpha}{2} A_p + i\gamma A_p \left[|A_p|^2 + 2\left(|A_s|^2 + |A_i|^2 \right) \right] + i2\gamma A_s A_i A_p^* \exp(-i\Delta kz)$$

$$\frac{\partial A_s}{\partial z} = -\frac{\alpha}{2} A_s + i\gamma A_s \left[|A_s|^2 + 2\left(|A_p|^2 + |A_i|^2 \right) \right] + i\gamma A_p^2 A_i^* \exp(-i\Delta kz) \qquad (10.14)$$

$$\frac{\partial A_i}{\partial z} = -\frac{\alpha}{2} A_i + i\gamma A_i \left[|A_i|^2 + 2\left(|A_s|^2 + |A_p|^2 \right) \right] + i\gamma A_p^2 A_s^* \exp(-i\Delta kz)$$

The analytical solution of these coupled equations determines the gain of the amplifier [1,3]:

$$G_s(L) = \frac{|A_s(L)|^2}{|A_s(0)|^2} = 1 + \left[\frac{\gamma P_p}{g} \sinh(gL) \right]^2 \qquad (10.15)$$

with L is the length of the HNLF/waveguide, P_p is the pump power, and g is the parametric gain coefficient:

$$g^2 = -\Delta k \left(\frac{\Delta k}{4} + \gamma P_p \right) \qquad (10.16)$$

where the phase mismatch Δk can be approximated by extending the propagation constant in a Taylor series to around ω_0:

$$\Delta k = -\frac{2\pi c}{\lambda_0^2} \frac{dD}{d\lambda} (\lambda_p - \lambda_0)(\lambda_p - \lambda_s)^2 \qquad (10.17)$$

Here, $dD/d\lambda$ is the slope of the dispersion factor $D(\lambda)$ evaluated at the zero-dispersion of the guided wave component, i.e., at the optical wavelength, $\lambda_n = 2\pi c/\omega_n$.

Figure 10.4b shows the Simulink® setup of the 40 Gb/s return-to-zero (RZ) transmission system using a parametric amplifier. The setup contains a 40 Gb/s optical RZ transmitter, an optical receiver for monitoring, a parametric amplifier block and a BPF that filters the desired signal from the total field output of the amplifier and the parameters of the system are given in Table 10.1. Details of the parametric amplifier block can be seen in Figure 10.3. The block setup of a parametric amplifier consists of a continuous wave (CW) pump laser source, an optical coupler to combine the signal and the pump, a HNLF block which contains the embedded MATLAB model for NL propagation. The important simulation parameters of the system are listed in Table 10.2.

Figure 10.5 shows the signals before and after the amplifier in the time domain. The time trace indicates the amplitude fluctuation of the amplified signal as a noisy source from the wave mixing process. Their corresponding spectra are shown in Figure 10.6. The noise

(a)

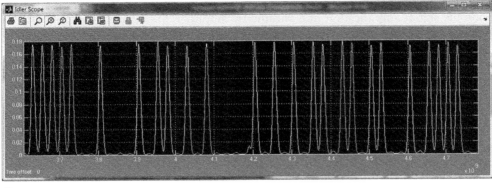

(b)

FIGURE 10.4
Time traces of the 40 Gb/s signal before (a) and, (b) after the parametric amplifier.

floor of the output spectrum of the amplifier shows the gain profile of OPA. Simulated dependence of OPA gain on the wavelength difference between the signal and the pump is shown in Figure 10.7 together with theoretical gain using Equation 10.15. The plot shows an agreement between the theoretical and the simulated results. The peak gain is achieved at phase matched conditions where the linear phase mismatch is compensated by the NL phase shift.

TABLE 10.1

Critical Parameters of the Parametric Amplifier in 40 Gb/s System

RZ 40 Gb/s Transmitter

$\lambda_s = 1520$ nm–1600 nm, $\lambda_0 = 1559$ nm
Modulation: RZ-OOK, $P_s = 0.01$ mW (peak), $B_r = 40$ Gb/s

Parametric Amplifier

Pump source: $P_p = 1$ W (after EDFA), $\lambda_p = 1560.07$ nm
HNLF: $L_f = 500$ m, $D = 0.02$ ps/km/nm, $S = 0.09$ ps/nm²/km, $\alpha = 0.5$ dB/km, $A_{eff} = 12$ μm², $\gamma = 13$
  1/W/km
BPF: $\Delta\lambda_{BPF} = 0.64$ nm

Receiver

Bandwidth $B_e = 28$ GHz, $i_{eq} = 20$ pA/Hz$^{1/2}$, $i_d = 10$ nA

TABLE 10.2

Critical Parameters of the Parametric Amplifier in 40 Gb/s Transmission System

RZ 40 Gb/s Signal

$\lambda_0 = 1559$ nm, $\lambda_s = \{1531.12, 1537.4, 1543.73, 1550.12\}$ nm,
$P_s = 1$ mW (peak), $B_r = 40$ Gb/s

Parametric Amplifier

L pump source: $P_p = 100$ mW (after EDFA), $\lambda_p = 1560.07$ nm
HNLF: $L_f = 200$ m, $D = 0.02$ ps/km/nm, $S = 0.03$ ps/nm²/km, $\alpha = 0.5$ dB/km, $A_{eff} = 12$ μm²,
$\gamma = 13$ 1/W/km
BPF: $\Delta\lambda_{BPF} = 0.64$ nm, $\lambda_i = \{1587.91, 1581.21, 1574.58, 1567.98\}$ nm

10.1.4.2 Wavelength Conversion and Nonlinear Phase Conjugation

Beside the signal amplification in a parametric amplifier, the idler is generated after the wave mixing process. Therefore, this process can also be applied to wavelength conversion. Due to the very fast response of the third-order nonlinearity in optical waveguides, the wavelength conversion based on this effect is transparent to the modulation format and the bit rate of signals. For a flat wideband converter which is a key device in wavelength division multiplexing (WDM) networks, a short length HNLF with a low dispersion slope is required in design. By a suitable selection of the pump wavelength, the wavelength converter can be optimized to obtain a bandwidth of 200 nm [7]. Therefore, the wavelength conversion between bands such as C- and L- bands can be performed in WDM networks. Figure 10.7 shows an example of the wavelength conversion for four WDM channels at C-band. The important parameters of the wavelength converter are shown in Table 10.2. The WDM signals are converted into L-band with the conversion efficiency of –12 dB.

Another important application with the same setup is the nonlinear phase conjugation (NPC). A phase conjugated replica of the signal wave can be generated by the FWM process. From Equation 10.8, the idler wave is approximately given in the case of degenerate FWM for simplification: $E_i \sim A_p^2 A_s^* e^{-j\Delta kz}$ or $E_i \sim r A_s^* e^{[j(-kz-\omega t)]}$ with the signal wave $E_s \sim A_s e^{[j(kz-\omega t)]}$. Thus the idler field is the complex conjugate of the signal field. In appropriate conditions,

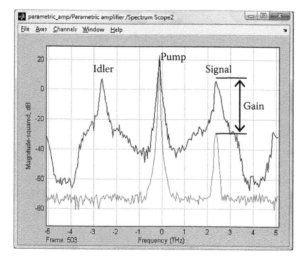

FIGURE 10.5
Optical spectra at the input (red—lower) and the output (black—higher) of the OPA.

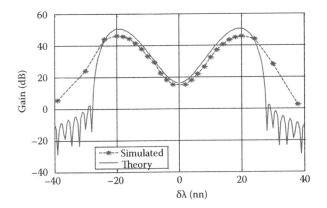

FIGURE 10.6
Calculated and simulated gain of the OPA at $P_p = 30$ dBm.

optical distortions can be compensated by using NPC and optical pulses propagating in the fiber link can be recovered. The basic principle of distortion compensation with NPC refers to spectral inversion. When an optical pulse propagates in an optical fiber, its shape will be spread in time and distorted by the group velocity dispersion. The phase conjugated replica of the pulse is generated in the middle point of the transmission link by the NL effect. On the other hand, the pulse is spectrally inverted where spectral components in the lower frequency range are shifted to the higher frequency range and vice versa. If the pulse propagates in the second part of the link with the same manner in the first part, it is inversely distorted again and that can cancel the distortion in the first part to recover the pulse shape at the end of the transmission link. When using NPC for distortion compensation, a 40%–50% increase in transmission distance compared to a conventional transmission link can be obtained [8–11]. Figure 10.8 shows the setup of a long-haul 40 Gb/s transmission system demonstrating the distortion compensation using NPC. The fiber transmission link of the system is divided into two sections by an NPC based on a

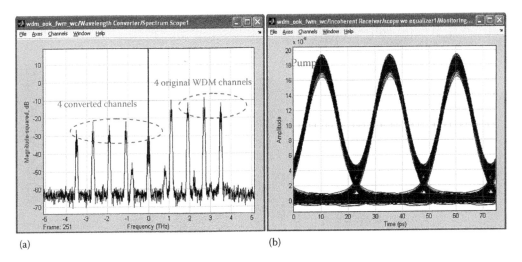

(a)   (b)

FIGURE 10.7
(a) The wavelength conversion of 4 WDM channels. (b) Eyediagram of the converted 40 Gb/s signal after BPF.

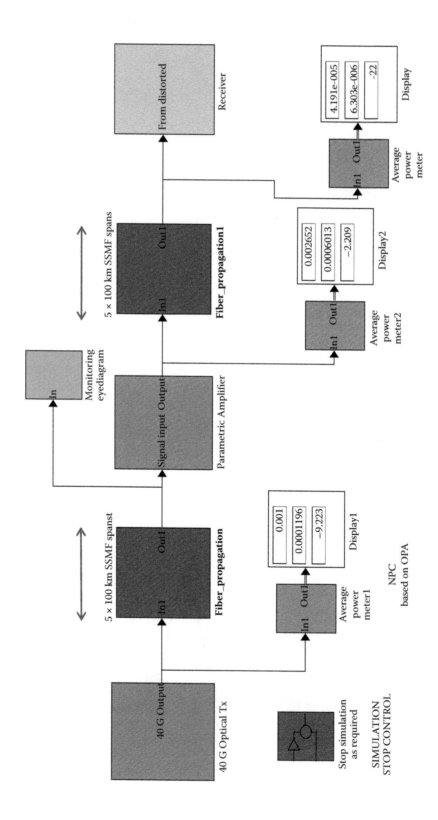

FIGURE 10.8
Simulink setup of a long-haul 40 Gb/s transmission system using NPC for distortion compensation.

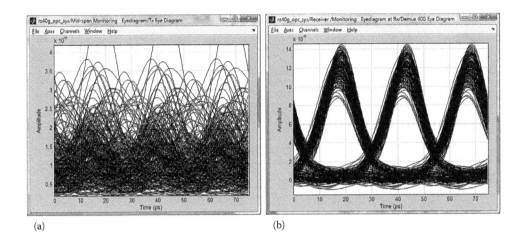

(a)   (b)

FIGURE 10.9
Eyediagrams of the 40 Gb/s signal at the end (a) of the first section and (b) of the transmission link.

parametric amplifier. Each section consists of five spans with 100 km standard single mode fiber (SSMF) in each span. Figure 10.9a shows the eyediagram of the signal after propagating through the first fiber section. After the parametric amplifier at the midpoint of the link, the idler signal, a phase conjugated replica of the original signal, is filtered for transmission in the next section. The signal in the second section suffers the same dispersion as in the first section. At the output of the transmission system the optical signal is regenerated as shown in Figure 10.9b. Due to a change in wavelength of the signal in NPC, a tunable dispersion compensator can be required to compensate the residual dispersion after transmission in real systems (Table 10.3).

10.1.4.3 High-Speed Optical Switching

When the pump is an intensity modulated signal instead of the CW signal, the gain of the OPA is also modulated due to its exponential dependence on the pump power in a phase

TABLE 10.3

Critical Parameters of the Long-haul Transmission System using NPC for Distortion Compensation

RZ 40 Gb/s Transmitter
λ_s = 1547 nm, λ_0 = 1559 nm
Modulation: RZ-OOK, P_s = 1 mW (peak), B_r = 40 Gb/s
Fiber Transmission Link
SMF: L_{SMF} = 100 km, D_{SMF} = 17 ps/nm/km, α = 0.2 dB/km
EDFA: Gain = 20 dB, NF = 5dB;
Number of spans: 10 (5 in each section), L_{link} = 1000 km
NPC Based on OPA
Pump source: P_p = 1 W (after EDFA), λ_p = 1560.07 nm
HNLF: L_f = 500 m, D = 0.02 ps/km/nm, S = 0.09 ps/nm²/km, α = 0.5 dB/km, A_{eff} = 12 μm², γ = 13 1/W/km
BPF: $\Delta\lambda_{BPF}$ = 0.64 nm
Receiver
Bandwidth B_e = 28 GHz, i_{eq} = 20 pA/Hz$^{1/2}$, i_d = 10 nA

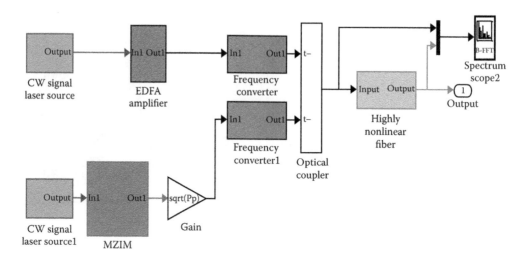

FIGURE 10.10
Simulink setup of the 40 GHz short-pulse generator.

matched condition. The width of gain profile in time domain is inversely proportional to the product of the gain slope (S_p) or the NL coefficient and the length of the NL waveguide (L) [3]. Therefore an OPA with high gain or large S_pL operates as an optical switch with an ultra-high bandwidth which is very important in some signal processing applications such as pulse compression or short-pulse generation [12,13]. A Simulink® setup for a 40 GHz short-pulse generator is built with the configuration as shown in Figure 10.10. In this setup, the input signal is a CW source with low power and the pump is amplitude modulated by a Mach-Zehnder interferometric intensity modulator (MZIM) which is driven by a RF sinusoidal wave at 40 GHz. The waveform of the modulated pump is shown in Figure 10.11a. Important parameters of the FWM-based short-pulse generator are shown in Table 10.4. Figure 10.11b shows the generated short-pulse sequence with the pulse width of 2.6 ps at the signal wavelength after the optical BPF.

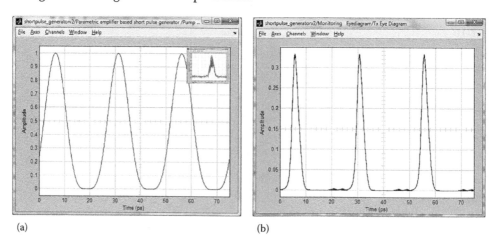

(a)　　　　　　　　　　　　　　　　　　　　(b)

FIGURE 10.11
Time traces of: (a) the sinusoidal amplitude modulated pump; and (b) the generated short-pulse sequence. (Inset: The pulse spectrum.)

TABLE 10.4

Parameters of the 40 GHz Short-pulse Generator

Short-pulse Generator

Signal: $P_s = 0.7$ mW, $\lambda_s = 1535$ nm, $\lambda_0 = 1559$ nm
Pump source: $P_p = 1$W (peak), $\lambda_s = 1560.07$ nm, $f_m = 40$ GHz
HNLF: $L_f = 500$ m, D = 0.02 ps/km/nm, S = 0.03 ps/nm^2/km, $\alpha = 0.5$ dB/km, $A_{eff} = 12$ μm^2,
$\gamma = 13$ 1/W/km
BPF: $\Delta\lambda_{BPF} = 3.2$ nm

Another important application of the optical switch based on the FWM process is the demultiplexer, a key component in ultra-high speed optical time division multiplexing (OTDM) systems. OTDM is a key technology for terabit per second Ethernet transmission which can meet the increasing demand of traffic in future optical networks. A typical scheme of OTDM demultiplexer in which the pump is a mode-locked laser (MLL) to generate short pulses for control is shown in Figure 10.12a. The working principle of the FWM-based demultiplexing is described as follows: The control pulses generated from a MLL at tributary rate are pumped and co-propagated with the OTDM signal through the NL waveguide. The mixing process between the control pulses and the OTDM signal during propagation through the NL waveguide converts the desired tributary channel to a new idler wavelength. Then the demultiplexed signal at the idler wavelength is extracted by a BPF before going to a receiver as shown in Figure 10.12a.

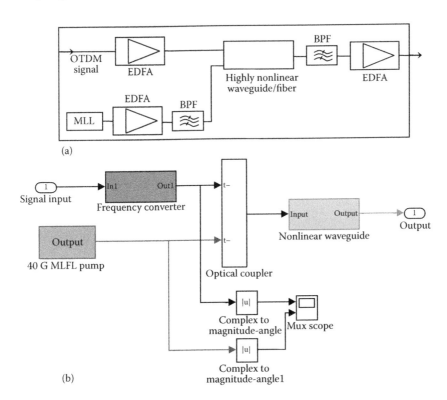

FIGURE 10.12
(a) A typical setup of the FWM-based OTDM demultiplexer. (b) Simulink model of the OTDM demultiplexer.

TABLE 10.5

Important Parameters of the FWM-based OTDM Demultiplexer using a Nonlinear Waveguide

---

OTDM Transmitter

MLL: $P_0 = 1$ mW, $T_p = 2.5$ ps, $f_m = 40$ GHz
Modulation formats: OOK and DQPSK; OTDM multiplexer: 4 × 40 Gsymbols/s

FWM-based Demultiplexer

Pumped control: $P_p = 500$ mW, $T_p = 2.5$ ps, $f_m = 40$ GHz, $\lambda_p = 1556.55$ nm
Input signal: $P_s = 10$ mW (after EDFA), $\lambda_s = 1548.51$ nm
Waveguide: $L_w = 7$ cm, $D_w = 28$ ps/km/nm, $S_w = 0.003$ ps/nm²/km
$\alpha = 0.5$ dB/cm, $\gamma = 10^4$ 1/W/km
BPF: $\Delta\lambda_{BPF} = 0.64$ nm

Using HNLF is relatively popular in structures of the OTDM demultiplexer [6,14]. However its stability, especially the walk-off problem is still a serious obstacle. Recently, planar NL waveguides have emerged as promising devices for ultra-high speed photonic processing [15,16]. These NL waveguides offer a lot of advantages such as no free-carrier absorption, stability at room temperature, no requirement of quasi-phase matching and the possibility of dispersion engineering. With the same operational principle, planar waveguide-based OTDM demultiplexers are very compact and suitable for photonic integrated solutions. Figure 10.12b shows the Simulink setup of the FWM-based demultiplexer of the on-off keying (OOK) 40 Gb/s signal from the 160 Gb/s OTDM signal using a highly NL waveguide instead of HNLF. Important parameters of the OTDM system in Table 10.5 are used in the simulation. Figure 10.13a shows the spectrum at the output of the NL waveguide. Then the demultiplexed signal is extracted by the BPF as shown in Figure 10.13b. Figure 10.14 shows the time traces of the 160 Gb/s OTDM signal, the control signal and the 40 Gb/s demultiplexed signal, respectively. The red dots in Figure 10.14a indicate the timeslots of the desired tributary signal in the OTDM signal. The developed model of OTDM demultiplexer can be applied, not only to the conventional OOK format but also to advanced modulation formats such as differential quadrature phase shift keying (DQPSK) which increases the data load of the OTDM system without an increase in bandwidth of the signal. By using available blocks developed for the DQPSK system [18], a Simulink model of DQPSK-OTDM system is also setup for demonstration. The bit rate of the OTDM system is doubled to 320 Gb/s with the same pulse repetition rate. Figure 10.15 shows the

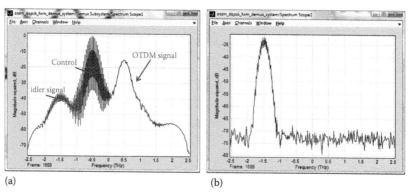

(a)   (b)

FIGURE 10.13

Spectra at the outputs: (a) of NL waveguide, and (b) of BPF.

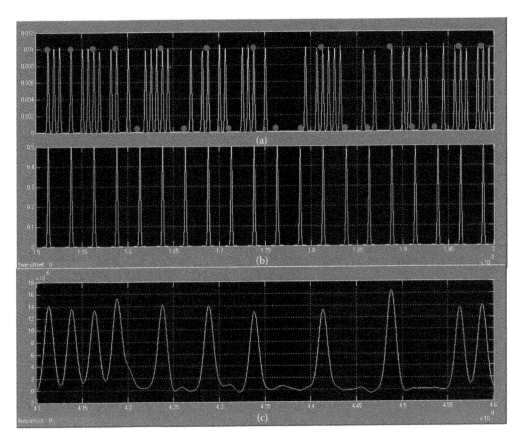

FIGURE 10.14
(See Color Insert) Time traces of: (a) the 160 Gb/s OTDM signal, (b) the control signal, and (c) the 40 Gb/s demultiplexed signal.

simulated performance of the demultiplexer in both 160 Gb/s OOK- and 320 Gb/s DQPSK-OTDM systems. The bit error rate (BER) curve in the case of the DQPSK-OTDM signal shows a low error floor which may be resulted by the influence of NL effects on phase-modulated signals in the waveguide.

10.1.4.4 Triple Correlation

One of promising applications exploiting the $\chi^{(3)}$ nonlinearity is the implementation of triple correlation in the optical domain. Triple correlation is a higher order correlation technique and its Fourier transform, called bispectrum, is very important in signal processing, especially in signal recovery [19,20]. The triple correlation of a signal $s(t)$ can be defined as

$$C^3(\tau_1, \tau_2) = \int s(t)s(t-\tau_1)s(t-\tau_2)dt \qquad (10.18)$$

where τ_1, τ_2 are time-delay variables. To implement the triple correlation in the optical domain, the product of three signals including different delayed versions of the original signal need to be generated, and then detected by an optical photodiode to perform the integral operation. From the representation of the NL polarization vector (see Equation

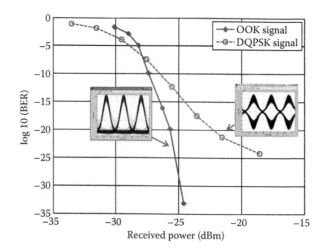

FIGURE 10.15
Simulated performance of the demultiplexed signals for 160 Gb/s OOK- and 320 Gb/s DQPSK-OTDM systems. (Insets: Eyediagrams at the receiver.)

10.3), this triple product can be generated by the $\chi^{(3)}$ nonlinearity. One way to generate the triple correlation is based on third harmonic generation (THG) where the generated new wave containing the triple product is at a frequency three times that of the original carrier frequency. Thus if the signal wavelength is in the 1550 nm band, the new wave needs to be detected at around 517 nm. The triple-optical autocorrelation based on single stage THG has been demonstrated in direct optical pulse shape measurement [21]. However, this way is hard to obtain high efficiency in the wave mixing process due to the difficulty of phase matching between three signals. Moreover, the triple product wave is in the 517 nm band where wideband photo-detectors are not available for high-speed communication applications. Therefore a possible alternative to generate the triple product is based on other NL interactions such as FWM. From Equation 10.9, the fourth wave is proportional to the product of three waves $A_4 \sim A_1 A_2 A_3^* e^{-j\Delta kz}$. If A_1 and A_2 are the delayed versions of the signal A_3, the mixing of three waves results in the fourth wave A_4 which is obviously the triple product of three signals. As mentioned in Section 10.1.2, all three waves can take the same frequency, however, these waves should propagate into different directions to possibly distinguish the new generated wave in a diverse propagation direction which requires a strict arrangement of the signals in the spatial domain. An alternative method we propose is to convert the three signals into different frequencies (ω_1, ω_2 and ω_3). Then the triple product wave can be extracted at the frequency $\omega_4 = \omega_1 + \omega_2 - \omega_3$ which is still in the 1550 nm band.

Figure 10.15a shows the Simulink® model for the triple correlation based on FWM in NL waveguide. The structural block consists of two variable delay lines to generate delayed versions of the original signal as shown in Figure 10.16 and frequency converters to convert the signal into three different waves before combining at the optical coupler to launch into the NL waveguide. Then the fourth wave signal generated by FWM is extracted by the bandpass filter. To verify the triple product based on FWM, another model shown in Figure 10.16b to estimate the triple product by using Equation 10.18 is also implemented for comparison. The integration of the generated triple product signal is then performed at the photo-detector in the optical receiver to estimate the triple correlation of the signal. A repetitive signal, which is a dual-pulse sequence with unequal amplitude, is generated for investigation. Important parameters of the setup are shown in Table 10.6. Table 10.6

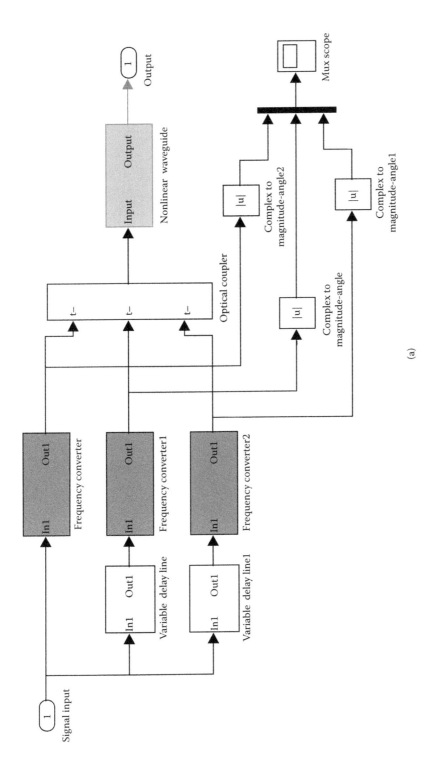

FIGURE 10.16

(a) Simulink setup of the FWM-based triple product generation. (b) Simulink setup of the theory-based triple product generation.

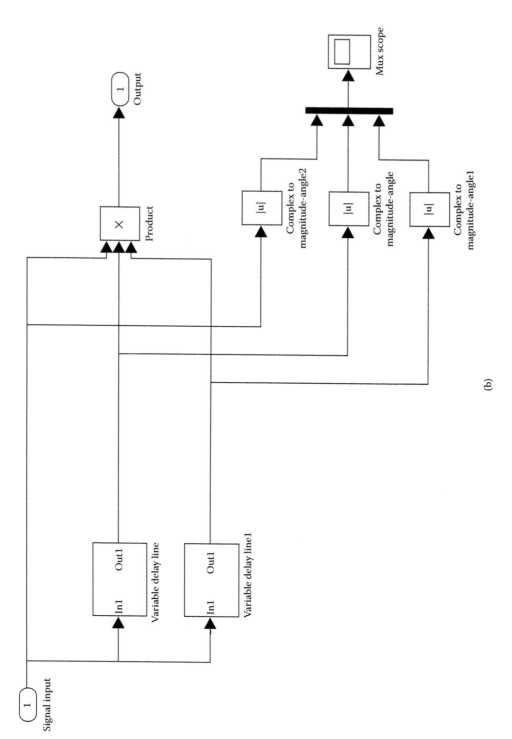

(b)

FIGURE 10.16 (Continued)

TABLE 10.6

Important Parameters of the FWM-based OTDM Demultiplexer using a NL Waveguide

Signal Generator

Single-pulse: $P_0 = 100$ mW, $T_p = 2.5$ ps, $f_m = 10$ GHz
Dual-pulse: $P_1 = 100$ mW, $P_2 = 2/3P_1$, $T_p = 2.5$ ps, $f_m = 10$ GHz

FWM-based Triple-product Generator

Original signal: $\lambda_{s1} = 1550$ nm, $\lambda_{s1} = 1552.52$ nm
Delayed τ_1 signal: $\lambda_{s2} = 1552.52$ nm, delayed τ_2 signal: $\lambda_{s3} = 1554.13$ nm
Waveguide: $L_w = 7$ cm, $D_w = 28$ ps/km/nm, $S_w = 0.003$ ps/nm²/km
$\alpha = 0.5$ dB/cm, $\gamma = 10^4$ 1/W/km
BPF: $\Delta\lambda_{BPF} = 0.64$ nm

show the waveform of the dual-pulse signal and the spectrum at the output of the NL waveguide. The wavelength spacing between three waves is unequal to reduce the noise from other mixing processes. The triple product waveforms estimated by theory and the FWM process are shown in Table 10.6. In case of the estimation based on FWM, the triple product signal is contaminated by the noise generated from other mixing processes. Figure 10.17 shows the triple correlations of the signal after processing at the receiver in both cases based on theory and FWM. The triple correlation is represented by a three dimensional (3D) plot which is displayed by the image. The x and y axes of the image represent the time-delay variables (τ_1 and τ_2) in terms of samples with a step-size of $T_m/32$ where T_m is the pulse period. The intensity of the triple correlation is represented by colors with scale specified by the color bar (Figure 10.18). Although the FWM-based triple correlation result is noisy, the triple correlation pattern is still distinguishable as compared to the theory (Figure 10.19). Another signal pattern of the single pulse which is simpler has also been investigated as shown in Figures 10.20 and 10.21.

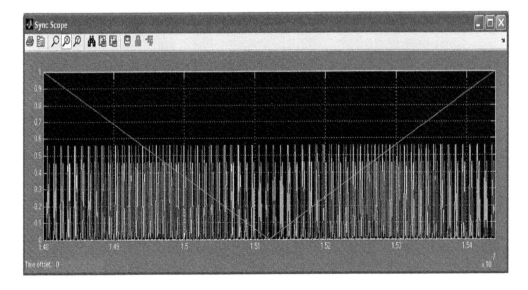

FIGURE 10.17
(See Color Insert) The variation in time domain of the time delay (cyan), the original signal (violet) and the delayed signal (yellow).

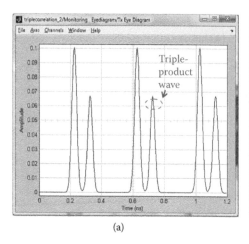
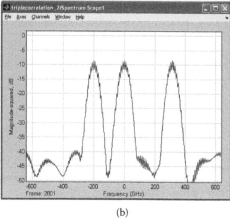

(a)　　　　　　　　　　　　　　　　(b)

FIGURE 10.18
(a) Time trace of the dual-pulse sequence for investigation. (b) Spectrum at the output of the NL waveguide.

10.1.5 Application of Nonlinear Photonics in Advanced Telecommunications

This section looks at the uses of NL effects and applications in modern optical communication networks in which 100 Gb/s optical Ethernet is expected to be deployed.

A typical performance of a photonic signal pre-processor employing no linear FWM is given and that of an advanced processing of such received signals in the electronic domain

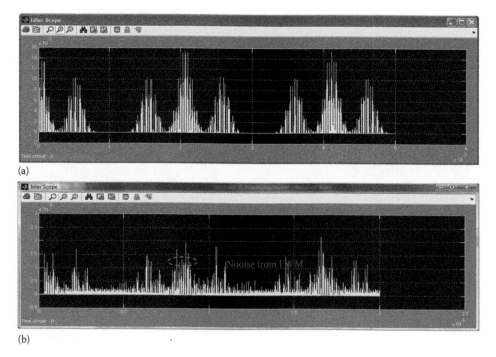

(a)

(b)

FIGURE 10.19
Generated triple product waves in time domain of the dual-pulse signal based on: (a) theory, and (b) FWM in NL waveguide.

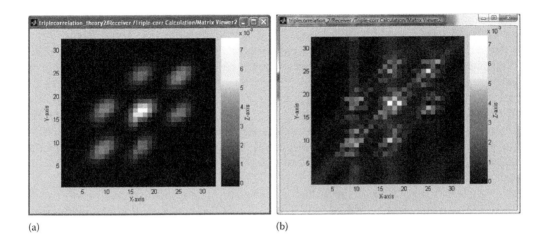

(a)   (b)

FIGURE 10.20
(See Color Insert) Triple correlation of the dual-pulse signal based on: (a) theoretical estimation, and (b) FWM in NL waveguide.

processed by a digital triple correlation system. At least 10 dB improvement is achieved on the receiver sensitivity.

Regarding the NL effects, the nonlinearity of the optical fibers hinder and limit the maximum level of the total average power of all the multiplexed channels for maximizing the transmission distance. These are due to the change of the refractive index of the guided medium as a function of the intensity of the guided waves. This in turn creates the phase changes and hence different group delays, thence distortion. Furthermore, other associate NL effects such as the FWM, Raman scattering, Brillouin scattering, and inter-modulation also create jittering and distortion of the received pulse sequences after a long transmission distance.

However, recently we have been able to use these NL optical effects to our advantage as a pre-processing element before the optical receiver to improve its sensitivity. A higher

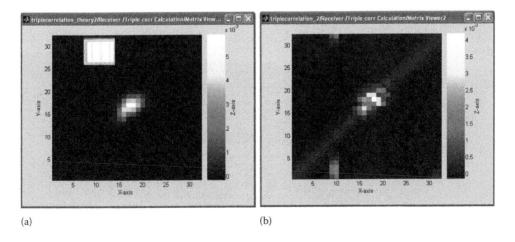

(a)   (b)

FIGURE 10.21
(See Color Insert) Triple correlation of the single-pulse signal based on: (a) theoretical estimation, and (b) FWM in NL waveguide. (Inset: the single-pulse pattern.)

order spectrum technique is employed with the triple correlation implemented in the optical domain via the use of the degenerate FWM effects in a high NL optical waveguide. However, this may add additional optical elements and filtering in the processor and hence complicate the receiver structure. We can overcome this by bringing this NL higher order spectrum processing to after the opto-electronic conversion and in the digital processing domain after a coherent receiving and electronic amplification sub-system.

In this section, we illustrate some applications of NL effects in demultiplexing of ultra-fast pulse sequence, phase conjugation and parametric amplification and NL processing algorithms to improve the sensitivity of optical receivers. NL optical processing can be implemented at the front end of the photo-detector and NL processing algorithm in the electronic domain.

The spectral distribution of the FWM and the simulated spectral conversion can be achieved. There is a degeneracy of the frequencies of the waves so that efficient conversion can be achieved by satisfying the conservation of momentum (Figure 10.22). This is detailed in another paper in the special session of this workshop [2]. The detected phase states and bispectral properties are depicted in Figure 10.23 in which the phases can be distinguished based on the diagonal spectral lines. Under noisy conditions these spectral distributions can be observed in Figure 10.24.

Alternating to the optical processing described above, a NL processing technique using a high order spectrum (HOS) technique can be implemented in the electronic domain. This is implemented after the analog-to-digital converter (ADC) which samples the incoming electronic signals produced by the coherent optical receiver as shown in Figure 10.25 [26]. The operation of a third-order spectrum analysis is based on the combined interference of three signals (in this case the complex signals produced at the output of the ADC) two of which are the delayed version of the original. Thence the amplitude and phase distribution of the complex signals are obtained in 3D graphs which allow us to determine the signal and noise power and the phase distribution. These distributions allow us to perform several functions necessary for the evaluation of the performance of optical transmission systems. Simultaneously the processed signals allow us to monitor the health of the transmission systems such as the effects due to NL effects, the distortion due to chromatic dispersion of the fiber transmission lines, the noises contributed by in-line amplifiers, etc. A typical curve that compares the performance of this innovative processing with convention detection techniques is shown in Figure 10.25. If we project the error rate of the receivers employing conventional techniques and our high order spectral receiver employing digital signal processing techniques at a BER of 10^{-9} then at least 1000 times lower. This is equivalent to at least one unit improvement on the quality factor of the eye opening. This is in turn equivalent to about 10 dB in the signal-to-noise ratio (SNR).

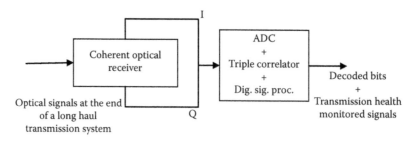

FIGURE 10.22
Schematic diagram of a high order spectral optical receiver and electronic processing.

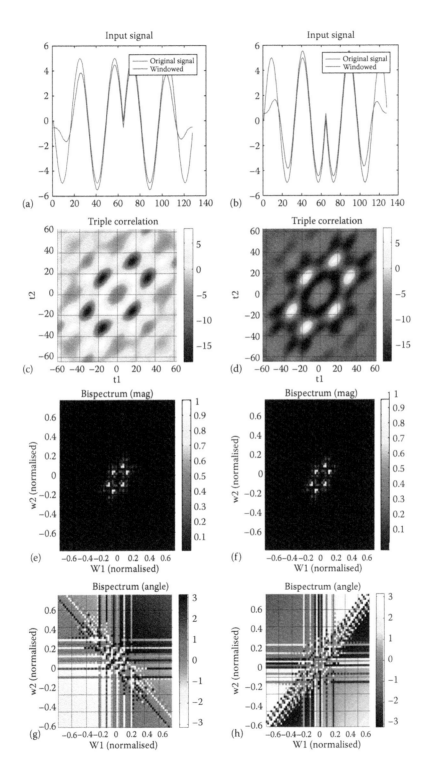

FIGURE 10.23
(See Color Insert) Input waveform with phase changes at the transitions (a and b), triple correlation and bispectrum (c–h) of both phase and amplitude.

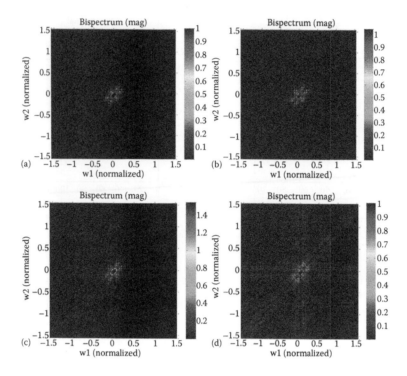

FIGURE 10.24
(See Color Insert) Effect of Gaussian noise on the bispectrum (a and c) amplitude distribution in two dim (b and d) phase spectral distribution.

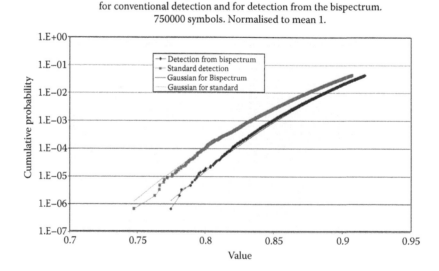

FIGURE 10.25
(See Color Insert) Error estimation version detection level of the HOS processor.

These results are very exciting for network and system operators as significant improvement of the receiver sensitivity can be achieved, this allows significant flexibility in the operation and management of the transmission systems and networks. Simultaneously the monitored signals produced by the high order spectral techniques can be used to determine the distortion and noises of the transmission line and thus the management of the tuning of the operating parameters of the transmitter, the number of wavelength channels and the receiver or in-line optical amplifiers.

The effect of additive white Gaussian noise on the bispectrum magnitude is shown in Figure 10.24, the sequence of figures are as indicated in Figure 10.23. The uncorrupted bispectrum magnitude is shown in Figure 10.23a while b, c and d were generated using SNRs of 10 dB, 3 dB and 0 dB, respectively. This provides a method of monitoring the integrity of a channel and illustrates another attractive attribute of the bispectrum. It is noted that the bispectrum phase is more sensitive to Gaussian white noise then is the magnitude and quickly becomes indistinguishable below 6 dB.

Although the algorithm employing the NL processing, albeit optical pre-processing or electronic processing of the triple correlation and thence bispectrum, will involve hardware and soft implementation. From the point of view of industry, it needs to deliver to the market at the right time for systems and networks operating in the terabits per second speed. One can thus be facing the following dilemma

- Optical pre-processing requires an efficient NL optical waveguide which must be in the integrated structure whereby efficient coupling and interaction can be achieved. If not then the gain of about 3 dB in SNR would be defeated by this loss. Furthermore, the integration of the linear optical waveguiding section and a NL optical waveguide is not matched due to the difference in the waveguide structures of both regions. For the linear waveguide structure to be efficient for coupling with circular optical fibers, silica on silicon would be best suited due to the small refractive index difference and the technology of burying such waveguides to form an embedded structure whose optical spot size would match that of a single guided mode fiber. This silica on silicon would not match with an efficient NL waveguide made by As_2S_3 on silicon.

- On the other hand if electronic processing is employed then it requires an ultra-fast ADC and then fast electronic signal processors. Currently a 56 GSamples/s ADC is available from Fujitsu as shown in Figure 10.26. It is noted that the data output samples of the ADC are structured in parallel forms with the referenced clock rate of 1.75 GHz. Thus all processing of the digital samples must be in parallel form and thus parallel processing algorithms must be structured in parallel. This is the most challenging problem which we must overcome in the near future.

- An application specific integrated circuit (ASIC) must also be designed for this processor.

- A hard decision that must be made also is the fitting of such ASIC and associate optical and opto-electronic components into international standard compatible size. Thus all design and components must meet this requirement.

- Finally the laboratory and field testing must be demonstrated for market delivery.

These challenges will be met and we are currently progressing toward the final target for the delivery of such sensitive receivers for 100 Gb/s optical internet.

56 GSamples/s ADC Two-channel version
using CHAIS architecture

(a)

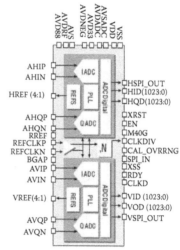

56 GSamples/s ADC Four-channel version

(b)

FIGURE 10.26
(See Color Insert) Plane view of the Fujitsu ADC operating at 56 GSamples/s: (a) integrated view, and (b) operation schematic.

10.1.6 Remarks

In this section we have demonstrated a range of signal processing applications exploiting the parametric process in NL waveguides. A brief mathematical description of the parametric process through third-order nonlinearity has been reviewed. A Simulink model of NL waveguide has been developed to simulate interaction of multi-waves in optical waveguides including optical fibers. Based on the developed Simulink modeling platform, a range of signal processing applications exploiting parametric FWM process has been investigated through simulation. With a CW pump source, the applications such as parametric amplifier, wavelength converter and optical phase conjugator have been implemented for demonstration. The ultra-high speed optical switching can be implemented by using an intensity modulated pump to apply in the short-pulse generator and the OTDM demultiplexer. Moreover, the FWM process has been proposed to estimate the triple correlation which is very important in signal processing. The simulation results showed the possibility of the FWM-based triple correlation using the NL waveguide with different pulse patterns. Although the triple correlation is contaminated by noise from other FWM processes, it is possibly distinguishable. The wavelength positions as well as the power of three delayed signals need to be optimized to obtain the best results.

Furthermore, we have also addressed the important issues of nonlinearity and its uses in optical transmission systems, the management of networks if the signals which indicate the health of the transmission system are available. There is no doubt that the NL phenomena play several important roles in the distortion effects of signals transmitted, but also allow us to improve the transmission quality of the signals. This has been briefly described in this paper on the optical processing using FWM effects and NL signal processing using high order spectral analysis and processing in the electronic domain. This ultra-high speed optical pre-processing and/or electronic triple correlation and bispectrum receivers are the first system using NL processing for 100 Gb/s optical Internet.

10.2 Nonlinear Effects in Actively Mode-locked Fiber Lasers

10.2.1 Introductory Remarks

Fiber lasers have several interesting characters that distinguish them from other solid state or gas lasers. The low loss of the fiber (0.2 dB/km attenuation) makes it possible to form a low loss cavity and thus low threshold laser. Since the intensity in the fiber is very high due to its small core area, a NL cavity can be easily obtained. With the introduction of nonlinearity into the cavity, lasers with advance quality and performance can be obtained by optimizing the laser parameters.

In fact, the effects of dispersion and nonlinearity on an optical pulse when it propagates through a fiber have been studied extensively using NLSE [22–28]. By solving the NLSE using the inverse scattering method [29], the authors obtained the solution in the form of soliton, whose shape is kept unchanged if the dispersion and NL effects are balanced. Although soliton in optical fiber was first predicted in 1973 [26], it was not experimentally observed until 1980 [30] due to the lack of a short optical pulse source [24]. Numerically solving the NLSE using the split-step Fourier [31] or finite-difference methods shows that the pulse can be compressed by NL effect or be broadened by dispersion effect.

Since pulses are formed and travel in actively mode-locked fiber lasers, dispersion and nonlinearity of the fiber cavity would play an important role in this mode-locking process. By introducing the master equation for mode-locking and numerical simulation, Haus and Silberberg showed that SPM can shorten the mode-locked pulse by a factor of 2 [32]. Later, Kartner, Kopf, and Keller added that a further shortening factor could be obtained when abnormal dispersion is introduced into the cavity [33]. The pulse shortening is the result of a soliton pulse forming in the cavity and has been reported in many papers [33–42]. However, the results were obtained with the assumption that mode-locking conditions are satisfied and pulse propagation in the cavity can be approximately described by the master equation. The performance of the laser and characteristics of the pulse when the laser is detuned were not discussed.

In this section, we present a detailed study on the NL effects on the performance of actively mode-locked fiber laser. Characteristics of the mode-locked pulses are discussed not only when the laser is exactly tuned but also when the laser is detuned. We show that there is a trade-off between pulse shortening and stability of the laser. Nonlinearity should be optimized to obtain the shortest pulse while the laser is still stable within a certain amount of detuning.

10.2.2 Laser Model

Figure 10.27 shows the laser model which consists of an optical amplifier, an optical BPF, a fiber and an intensity modulator. The optical wave is amplified in the optical amplifier and then filtered by the BPF. After filtering, the lightwave propagates through the fiber, where it experiences the dispersion and NL effects, and finally is modulated by the optical modulator. The output of the modulator is fed back into the amplifier to create a loop.

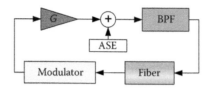

FIGURE 10.27
General schematic model of the actively mode-locked fiber laser, nonlinearity is employed from the fiber section with high peak power pulses amplified by the optical amplifiers of gain G. ASE = amplified stimulated emission.

10.2.2.1 Modeling of the Fiber

The fiber can be modeled by the propagation equation [24]

$$\frac{\partial A}{\partial z} + \frac{j}{2}\beta_2 \frac{\partial^2 A}{\partial T^2} + \frac{\alpha}{2}A = j\gamma|A^2|A \tag{10.19}$$

where A is the signal envelope, z is the axial distance, T is the delayed time ($T = t - z/v_g$), v_g is the group velocity, α is the linear attenuation factor of the fiber and accounts for the loss, γ is the NL coefficient which accounts for the SPM effect, β_2 is the second-order derivative of the propagation constant β and can be calculated from the fiber dispersion parameter as

$$\beta_2 = -\frac{\lambda^2 D}{2\pi c} \tag{10.20}$$

in which λ is the operating wavelength in the vacuum, D is the dispersion factor of the fiber, c is the speed of light.

10.2.2.2 Modeling of the Er:Doped Fiber Amplifiers

Since the bandwidth of the optical amplifier is much larger than the filter bandwidth, it can be modeled by a wavelength independent gain with saturated equation:

$$g_0 = \frac{g_{ss}}{1 + P_{av} / P_{sat}} \tag{10.21}$$

where g_{ss} is small signal gain factor, P_{av} is the signal average power, P_{sat} is the saturation power level.

10.2.2.3 Modeling of the Optical Modulator

The modulator can be modeled from the transmission function

$$T = \alpha_m \cos^2\left(\frac{\pi(v_m - V_{sh})}{2V_\pi}\right) \tag{10.22}$$

where V_π is the voltage applied into the modulator causes a π phase shift in one arm of the integrated optical interferometer, V_{sh} accounts for the DC drift of the modulator, α_m is the insertion loss, v_m is the modulating voltage signal and can be given by

$$v_m = V_m \cos \omega_m t + V_b \tag{10.23}$$

in which V_m is the amplitude of the modulating signal, ω_m is the modulating frequency, V_b is the bias voltage.

Substituting Equation 10.23 into 10.22 and noting that V_{sh} can be assumed to be zero without any affect to the final result, the transmission function of the modulator can be written as

$$T = \alpha_m \cos^2\left(\frac{\pi}{4}\Delta_m \cos(\omega_m t) + \frac{\pi}{2}\frac{V_b}{V_\pi}\right) \tag{10.24}$$

where $\Delta m = 2V_m/V_\pi$ is the modulation depth. When the modulation is biased at the quadrature point $V_b = V_\pi/2$, Equation 10.24 becomes

$$T = \alpha_m \cos^2\left[\frac{\pi}{2}\left(\Delta_m \cos\left(\omega_m t\right) + 1\right)\right] \tag{10.25}$$

10.2.2.4 Modeling of the Optical Filter

The optical filter BPF can be described by the following transfer function following a Gaussian profile as

$$H(f) = \alpha_F \exp\left(-\frac{1}{2}\left(\frac{f}{B_0}\right)^2\right) \tag{10.26}$$

where α_F is the insertion loss, B_0 is half of the e^{-1} bandwidth of the filter.

10.2.3 Nonlinear Effects in Actively Mode-Locked Fiber Lasers

10.2.3.1 Zero Detuning

The cavity nonlinearity is determined as

$$\gamma = \frac{2\pi L n_2}{\lambda_0 A_{eff}} \tag{10.27}$$

where L is the fiber length, n_2 is the fiber NL-index coefficient ($\sim 2.3 \times 10^{-23}$ m^2/W for GeO$_2$: SiO$_2$), λ_0 is the operating wavelength, and A_{eff} is the effective area of the fiber which is the area of the Gaussian mode guided in single mode optical fibers given as πr_0^2; $r_0 \triangleq$ mode spot size of the single mode fiber (SMF). For convenience, we normalize the cavity nonlinearity using

$$\gamma_n = \frac{\gamma P_{sat}}{1.5\tau_G f_m k_{sat}} \tag{10.28}$$

in which P_{sat} is saturation power of the Er:doped fiber amplifier (EDFA), f_m is the modulation frequency, k_{sat} is the saturation coefficient, and τ_G is the pulse width.

Figure 10.28 shows the pulse buildup process in an actively MLL with cavity nonlinearity of 0.2. Steady-state pulse is obtained after about 200 round trips and its pulse width is shorter than that without nonlinearity effect. Comparison between steady-state pulses of different nonlinearity values and the dependence of pulse width on the cavity nonlinearity are given in Figures 10.29 and 10.30, respectively. As the nonlinearity increases the pulse becomes shorter and shorter. Since the nonlinearity introduces NL phase variation to the optical pulse, which in turn generates new frequencies, the spectrum of the optical pulse is broadened. When the spectrum becomes wider, more modes can be locked and hence the pulse becomes shorter. This is verified in Figure 10.31 which shows that the spectrum of the pulse with high nonlinearity is wider than that without nonlinearity effect.

It can be seen from Figures 10.29 and 10.31 that when the nonlinearity increases from zero to 0.2, the pulse width is shortened by a half while the pulse bandwidth is increased four times. This indicates that the pulse is chirped when the nonlinearity increases.

The dependence of the pulse chirp on the cavity nonlinearity is shown in Figure 10.32. The pulse is up-chirped with low frequency at the leading edge and high frequency at the trailing edge. The new frequency generated due to the nonlinearity effect of the fiber can be determined from

$$\delta\omega(T) = -\frac{\partial\phi_{NL}}{\partial T} = -\left(\frac{L}{L_{NL}}\right)\frac{\partial|U(0,T)|^2}{\partial T} \tag{10.29}$$

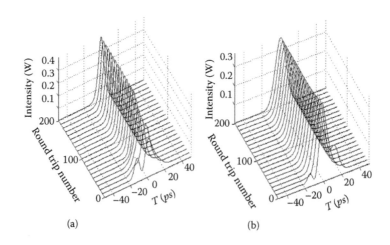

(a)   (b)

FIGURE 10.28
Pulse buildup in an actively mode-locked fiber laser with: (a) normalized cavity nonlinearity of 0.2, and (b) zero nonlinearity.

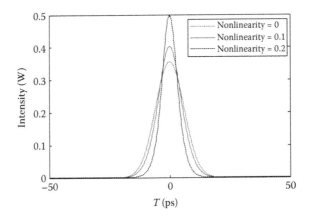

FIGURE 10.29
(See Color Insert) Pulses with different cavity nonlinearity values.

For the leading edge $dU^2/dT > 0 \rightarrow d\omega < 0$, low frequency is generated. While for the trailing edge, $d\omega > 0$, high frequency is generated. Therefore, the carrier under the pulse is up-chirped.

10.2.3.2 Detuning in Actively Mode-locked Fiber Laser with Nonlinearity Effect

Figure 10.33a shows the steady-state pulses of an actively mode-locked fiber laser having a cavity nonlinearity of 0.1 with different frequency detuning values. Similar to the case of linear cavity, the pulse position is shifted away from the transmission peak (the center) of the modulation function. For a positive detune ($f_m > 0$) the pulse is delayed and the leading edge experiences less loss than the trailing edge when passing the modulator. This pushes the pulse toward the center and hence compensates for the delay caused by the detuning and hence, steady-state pulse is formed. The position shifting also increases linearly with the detuning as shown in Figure 10.34a.

The spectra of the pulses are plotted in Figure 10.33b. It is seen that the lasing wavelength is shifted toward the lower frequency for positive detuning as in the case of normal

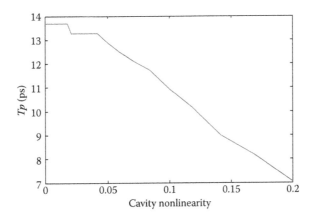

FIGURE 10.30
Pulse shortening when nonlinearity increases.

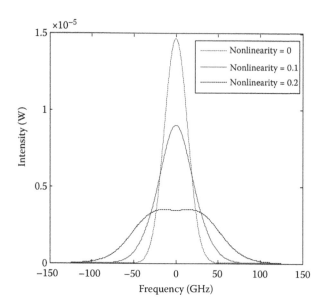

FIGURE 10.31
(See Color Insert) Pulse spectra with different cavity nonlinearity values.

dispersion. Although the pulse is also up-chirped as the case of normal dispersion, the principle of wavelength shifting is different. The wavelength shifting in the NL case is the result of the filtering effect of the modulator. Since the pulse is up-chirped, the low frequency is at the leading edge and hence it experiences lower loss than the high frequency when passing the modulator. Therefore, the spectrum is shifted toward low frequency as shown in Figure 10.34b. On the other hand, when detuning is negative, the pulse lags behind. Its trailing edge, where the high frequency is distributed, passes the modulator at the transmission peak. The modulator now acts as a filter which is in favor of the high frequency and hence shifts the spectrum toward the high frequency region.

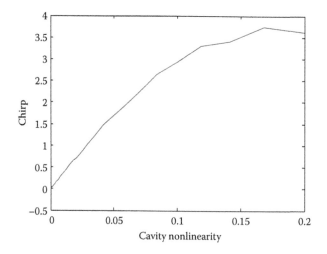

FIGURE 10.32
The nonlinearity causes the pulse up-chirp and the chirp increases as nonlinearity increases.

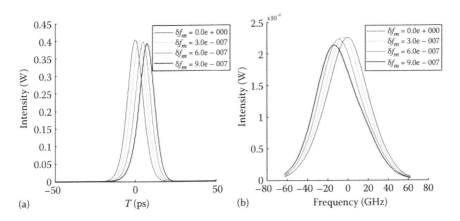

(a)   T (ps)   (b)   Frequency (GHz)

FIGURE 10.33
(See Color Insert) Steady-state pulses and spectra of an actively mode-locked fiber laser having cavity nonlinearity of 0.1 with different detuning values.

While the wavelength shifting in the dispersion cavity helps to compensate and balance for detuning, the wavelength shifting in the NL cavity does not give any help in compensating for the detuning. The wavelength shifting is just the result of the modulator filtering effect on the up-chirp distribution of the frequency components of the pulse. The only effect that reacts and compensates for the detuning is the pushing or pulling effect when the pulse position is shifted. However, these pushing/pulling effects are weaker for the narrower pulse under the NL effect. Therefore, the laser is more sensitive to the detuning. Figure 10.35 shows the locking range of the laser when the nonlinearity is changed. The locking range starts dropping dramatically as the nonlinearity reaches 0.04 where the nonlinearity starts taking its shortening effect on the pulse as shown in Figure 10.30.

10.2.3.3 Pulse Amplitude Equalization in Harmonic Mode-locked Fiber Laser

In harmonic MLLs, the modulation frequency is not equal to the fundamental frequency of the cavity but is a harmonic of it. Not only one pulse but N pulses circulate in sequence in the cavity. This results in an N times increase of the repetition rate. However, the pulse

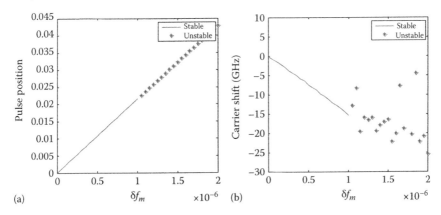

(a) δf_m $\times 10^{-6}$ (b) δf_m $\times 10^{-6}$

FIGURE 10.34
Shifting of pulse position and lasing wavelength in detuned actively mode-locked fiber laser with nonlinearity effect.

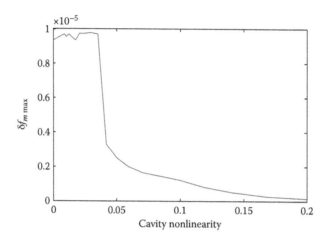

FIGURE 10.35
Detuning range versus cavity nonlinearity.

amplitude may be different from one to another due to the super-mode noise [43–45] which causes the laser modes to jump from one group to the other groups.

The amplitude difference between two pulses in a harmonic MLL without nonlinearity can be clearly observed in Figure 10.36a, but their amplitudes become equalized when nonlinearity is added in the cavity as shown in Figure 10.36b. Nonlinearity helps to equalize the pulse amplitude in harmonic MLLs through the SPM and filtering effect. The higher energy pulse experiences higher SPM effect and hence its spectrum becomes wider. Therefore it will experience more loss when passing through the filter. This results in equalization of the pulse energy.

10.2.4 Experiments

10.2.4.1 Experimental Setup

Figure 10.37 shows the experimental setup of an actively mode-locked fiber laser using Er:doped fiber (EDF) as a gain medium. The EDF is 5.6 m long and pumped by a 980 nm

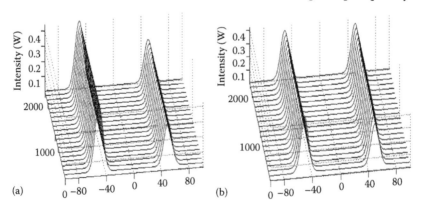

FIGURE 10.36
Harmonic mode-locked pulses (a) with and (b) without nonlinearity effect.

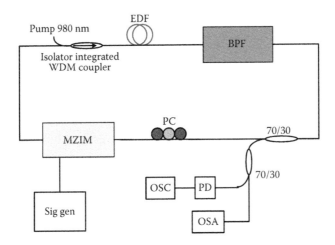

FIGURE 10.37
Actively mode-locked fiber laser setup; EDF: Er:doped fiber; BPF: bandpass filter; PC: polarization controller; MZIM: Mach-Zehnder intensity modulator; PD: photodiode; OSC: oscilloscope; OSA: optical spectrum analyzer; Sig Gen: signal generator.

laser diode through a WDM coupler. An isolator integrated in the WDM ensures unidirectional lasing. A LiNbO₃ MZIM is inserted in the ring for mode-locking. Mode-locking is obtained by introducing a periodical loss to the signal when driving the modulator by a sinusoidal signal extracted from an RF signal generator. The polarization controller (PC) is used to adjust the polarization state of the lightwave signal traveling in the ring. The lasing wavelength can be tuned by adjusting the central wavelength of the thin-film tunable optical BPF with a 3 dB bandwidth of 1.2 nm. The optical signal in the ring is coupled to the output port through a 70:30 fiber coupler.

10.2.4.2 Mode-Locked Pulse Train with 10 GHz Repetition Rate

The total length of the cavity is about 20.7 m which corresponds to a fundamental frequency of 9.859 MHz. When the modulation frequency is set at 10.006885 GHz, which corresponds to the 1015th harmonic of the fundamental frequency, a stable mode-locked pulse train is obtained as shown in Figure 10.38. The full-width at half-maximum (FWHM) of the pulse measured from the oscilloscope is about 21 ps. It is noted that this value includes the rise time of the photodiode and the oscilloscope. Using an auto-correlator for measuring the pulse width to minimize the rise time effects of those electrical components, we obtain the pulse width of 14.3 ps.

The optical spectrum of the pulse train is shown in Figure 10.39. It can be seen from Figure 10.39 that the pulse train spectrum has side lobes with a separation of 0.08 nm, which corresponds to a longitudinal mode separation of 10 GHz. This spectrum profile is typical for the MLL and usually referred as the mode-locked structure spectrum [46–48].

The above spectrum profile can be explained by taking the Fourier transform of the periodic Gaussian pulse train:

$$p(t) = e^{-t^2/2T_0^2} * \text{III}\left(\frac{t}{T}\right) \tag{10.30}$$

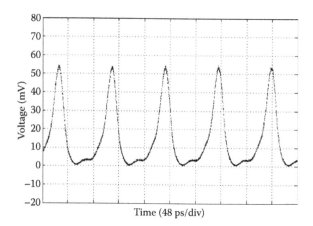

FIGURE 10.38
10 GHz repetition rate pulse train generated from the actively mode-locked fiber laser when the modulator is driven by a sinusoidal RF signal.

where T_0 is the width of the Gaussian pulse, T is the pulse repetition period and $|||(t/T)$ is the comb function given by

$$||| \left(\frac{t}{T} \right) = \sum_{n=-\infty}^{\infty} \delta(t - nT)$$

(10.31)

$$P(\omega) = F(p(t)) = T_0 \sqrt{2\pi} e^{-T_0^2 \omega^2 / 2} \sum_{n=-\infty}^{\infty} \delta(\omega - n\Omega)$$

(10.32)

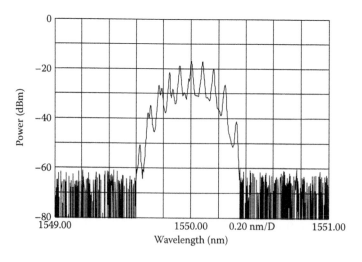

FIGURE 10.39
Spectrum of the mode-locked pulse train; the 10 GHz (0.08 nm) spacing modes are clearly observed.

where $\Omega = 2\pi/T$. $P(\omega)$ has the structure of a train of Dirac function pulses separated by the repetition frequency with a Gaussian envelope. The laser wavelength can be tunable over the whole C-band by tuning the central wavelength of the thin film BPF placed in the loop.

10.2.4.3 Pulse Shortening and Spectrum Broadening under Nonlinearity Effect

The nonlinearity effect is explored by increasing the optical power pumped into the EDF. As the power is increased the NL effect is enhanced and thus contributes to the forming of the mode-locked pulse. Figure 10.40 shows the oscilloscope trace of the pulse train when the pumped power is increased to 190 mW. The pulse width has been shortened to 12.3 ps, a 24% decrease compared to that of the dispersion cavity without NL effect. The pulse will be compressed more if higher optical power is pumped into the EDF.

The optical spectrum of the pulse is shown in Figure 10.41. As discussed in the previous section, the NL effect generates new frequencies when it introduces NL phase change to the pulse and thus broadens its spectrum. The 3 dB bandwidth of the pulse has been increased from 0.26 nm to 0.5 nm when the pump power increases from 40 mW to 190 mW.

10.2.5 Remarks

We have presented the nonlinearity effects on the actively mode-locked fiber lasers. A laser model which includes optical amplifier, BPF, modulator and optical fiber has been developed for studying the dispersion and NL effects. It is found that nonlinearity compresses the pulse and reduces the locking range. The mode-locked pulse is positive chirped due to the generation of new frequencies through the SPM effect and thus the pulse bandwidth is increased. Therefore, the pulse is compressed and a shorter pulse is generated. However, there is a trade-off between short pulse width and large locking range. If the laser is designed to have a shorter pulse width, the locking range of the laser is narrower and vice versa. Therefore, our laser model can be used to optimize the laser parameters to obtain the required pulse width while maintaining the laser stability over a certain locking range.

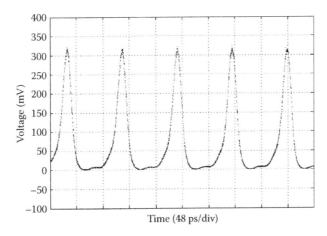

FIGURE 10.40
Oscilloscope trace of the mode-locked pulse train when the pump power increases to 190 mW; pulse shortened when the pumped power increases.

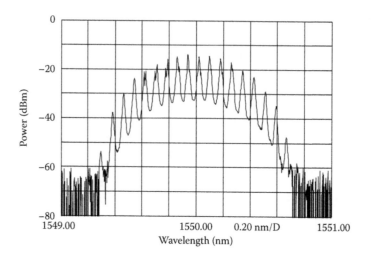

FIGURE 10.41
Nonlinearity causes the pulse spectrum to broaden when the pump power increases to 190 mW.

10.3 Nonlinear Photonic Pre-Processing for Bispectrum Optical Receivers

In this section, we present the processing of optical signals before the opto-electronic detection in the optical domain in a NL optical waveguide as a NL signal processing technique for digital optical receiving system for long-haul optically amplified fiber transmission systems. The algorithm implemented is a HOS technique in which the original signals and two delayed versions are correlated via the FWM or third harmonic conversion process. The optical receivers employing higher order spectral photonic pre-processor and a very large scale integrated circuit (VLSI) electronic system for the electronic decoding and evaluation of the BER of the transmission system are presented. A photonic signal pre-processing system, developed to generate the triple correlation via the third harmonic conversion in a NL optical waveguide, is employed as the photonic pre-processor to generate the essential part of a triple correlator. The performance of an optical receiver incorporating the HOS processor is given for long-haul phase modulated fiber transmission.

10.3.1 Introductory Remarks

Recently, tremendous efforts have been made to reach higher transmission bit rates and longer haul for optical fiber communication systems [49–52]. The bit rate can reach several hundreds of gigabits per second and toward terabit per second. In this extremely high speed operational region, the limits of electronic speed processors have been surpassed and optical processing is assumed to play an important part of the optical receiving circuitry. Furthermore, novel processing techniques are required in order to minimize the bottlenecks of electronic processing, noises and distortion due to the impairment of the transmission medium, the linear and NL distortion effects.

This paper deals with the photonic-processing of optically modulated signals prior to the electronic receiver for long-haul optically amplified transmission systems. NL optical waveguides in planar or channel structures are studied and employed as a third harmonic converter so as to generate a triple product of the original optical waves and its two delayed copies. The triple product is then detected by an opto-electronic receiver. Thence the detected current would be electronically sampled and digitally processed to obtain the bispectrum of the data sequence and a recovery algorithm is used to recover the data sequence. The generic structures of such a high order spectral optical receiver are shown in Figure 10.42. For the NL photonic processor (Figure 10.42a) the optical signals at the input are delayed and then coupled to a NL photonic device in which the NL conversion process is implemented via the use of a third harmonic conversion or degenerate FWM. While in the NL digital processor (see Figure 10.42b) the optical signals are detected coherently and then sampled and processed using the NL triple correlation and decoding algorithm.

We propose and simulate this NL optical pre-processor optical receiver under the MATLAB Simulink platform for differentially coded phase shift keying, the DQPSK modulation scheme.

This section is organized as follows: Section 10.3.2 gives a brief introduction of the triple correlation and bispectrum processing techniques. Section 10.3.3 introduces the simulation platform for the long-haul optically amplified optical fiber communication systems. Section 10.3.4 then gives the implementation of the NL optical processing and its association with the optical receiver with digital signal processing in the electronic domain so as to recover the data sequence. The performance of the transmission system is given with evaluation of the BER under linear and NL transmission regimes.

10.3.2 Bispectrum Optical Receiver

Figure 10.42 shows the structure of a bispectrum optical receiver in which there are three main sections: an all-optical pre-processor, an opto-electronic detection, and amplification including an ADC to generate sampled values of the triple correlated product. This section is organized as follows: the next subsection gives an introduction to bispectrum and associate noise elimination as well as the benefits of the bispectrum techniques, then the details of the bispectrum processor are given and then some implementation aspects of the bispectrum processor using VLSI are stated.

10.3.3 Triple Correlation and Bispectra

10.3.3.1 Definition

The power spectrum is the Fourier transform of the autocorrelation of a signal. The bispectrum is the Fourier transform of the triple correlation of a signal. Thus both the phase and amplitude information of the signals are embedded in the triple correlated product.

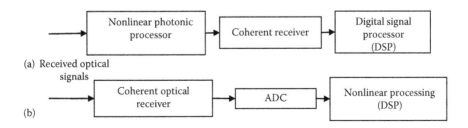

FIGURE 10.42
Generic structure of HOS optical receiver: (a) photonic pre-processor, and (b) NL digital signal processing in electronic domain.

While autocorrelation and its frequency domain power spectrum does not contain the phase information of a signal, the triple correlation contains both, due to the definition of the triple correlation,

$$c(\tau_1, \tau_2) = \int S(t)S(t+\tau_1)S(t+\tau_2)dt \qquad (10.33)$$

where $S(t)$ is the continuous time domain signals to be recovered. τ_1; τ_2 are the delay time intervals. For the special case where $\tau_1 = 0$ or $\tau_2 = 0$, the triple correlation is proportional with the autocorrelation. It means that the amplitude information is also contained in the triple correlation. The benefit of the holding phase and amplitude information is that it gives the potential to recover the signal back from its triple correlation. In practice the delays τ_1 and τ_2 indicate the path difference between the three optical waveguides. These delay times correspond to the frequency regions in the spectral domain. Thus different time intervals would determine the frequency lines in the bispectrum.

10.3.3.2 Gaussian Noise Rejection

Given a deterministic sampled signal $S(n)$, the sampled version of the continuous signals $S(t)$, corrupted by Gaussian noise $w(n)$, with n as the sampled time index, the observed signal takes the form $Y(n) = S(n) + w(n)$. The poly-spectra of any Gaussian process are zero for any order greater than two [53]. The bispectrum is the third-order poly-spectrum and offers a significant advantage for signal processing over the second-order poly-spectrum, commonly known as the power spectrum, which is corrupted by Gaussian noise. Theoretically speaking, the bispectral analysis allows us to extract a non-Gaussian signal from the corrupting affects of Gaussian noise.

Thus for a signal arrived at the optical receiver, the steps to recover the amplitude and phase of the lightwave modulated signals are [20]: (i) estimate the bispectrum of $S(n)$ based on observations of $Y(n)$; (ii) from the amplitude and phase bispectra form an estimate of the amplitude and phase distribution in one dimensional frequency of the Fourier transform of $S(n)$, this forms the constituents of the signal $S(n)$ in the frequency domain; and (iii) thus by taking the inverse Fourier transform to recover the original signal $S(n)$. This type of receiver is termed as the bispectral optical receiver.

10.3.3.3 Encoding of Phase Information

The bispectra contains almost complete information about the original signal (magnitude and phase). Thus if the original signal $x(n)$ is real and finite it can be reconstructed except for a shift a. Equivalently the Fourier transform can be determined except for a linear shift factor of $e^{-j2\pi\omega a}$. By determining two adjacent pulses any differential phase information will be readily available [54]. In other words, the bispectra, hence the triple correlation, contain the phase information of the original signal allowing it to "pass through" the square law photodiode which would otherwise destroy this information. The encoded phase information can then be recovered up to a linear phase term thus necessitating a differential coding scheme.

10.3.3.4 Eliminating Gaussian Noise

For any processes that have zero mean and the symmetrical probability density function (pdf), their third-order cummulant are equaled to zero. Therefore, in a triple correlation, those symmetrical processes are eliminated. Gaussian noise is assumed to affect signal quality. Mathematically, the third cummulant is defined as

$$c_3(\tau_1, \tau_2) = m_3(\tau_1, \tau_2) - m_1 \begin{bmatrix} m_2(\tau_1) + m_2(\tau_2) \\ + m_2(\tau_1 - \tau_2)] + 2(m_1)^3 \end{bmatrix} \tag{10.34}$$

where m_k is the kth order moment of the signal. Thus for the zero mean and symmetrical pdf, its third-order cummulant becomes zero [19]. Theoretically, considering the signal as $u(t) = s(t) + n(t)$, where $n(t)$ is an additive Gaussian noise, the triple correlation of $u(t)$ will reject Gaussian noise affecting the $s(t)$ [55,56].

10.3.4 Bispectral Optical Structures

Figure 10.43 shows the generic and detailed structure of the bispectral optical receiver, respectively, which consists of: (i) an all-optical pre-processor front-end; followed by (ii) a photo-detector and electronic amplifier to transfer the detected electronic current to a voltage level appropriate for sampling by an ADC, thus the signals at this stage is in sampled form; (iii) the sampled triple correlation product is then transformed to the Fourier domain using the fast Fourier transform (FFT). The product at this stage is the row of the matrix of the bispectral amplitude and phase plane (see Figure 10.44). A number of parallel structures may be required if passive delay paths are used. (iv) A recovery algorithm is used to derive the one-dimensional distribution of the amplitude and phase as a function of the frequency which are the essential parameters required for taking the inverse Fourier transform to recover the time domain signals.

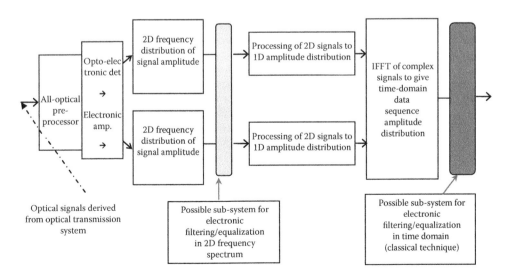

FIGURE 10.43
Generic structure of an optical pre-processing receiver employing bispectrum processing technique.

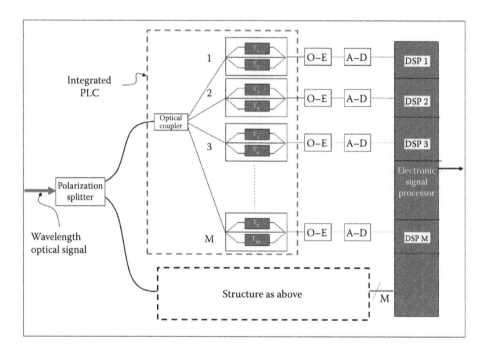

FIGURE 10.44
Parallel structures of photonic pre-processing to generate triple correlation product in the optical domain.

10.3.4.1 Principles

The physical process of mixing the three waves to generate the fourth wave whose amplitude and phase are proportional to the product of the three input waves is well known in literatures of NL optics.

This process requires: (i) highly NL medium so as to efficiently convert the energy of the three waves to the that of the fourth wave; and (ii) phase matching conditions of the three input waves of the same frequency (wavelength) to satisfy the conservation of momentum.

10.3.4.2 Technological Implementation

10.3.4.2.1 Nonlinear Optical Waveguides

In order to satisfy the condition of FWM we must use: (i) a rib-waveguide for guiding whose NL refractive index coefficient is about 100,000 times greater than that of silica. The material used in this waveguide is chalcogenide glass type (e.g., AS_2S_3) or TeO_2. The three waves are guided in this waveguide structure. Their optical fields are overlapped. The cross-section of the waveguide is in the order of 4 µm × 0.4 µm. (ii) The waveguide cross-section is designed such that the dispersion is "flat" over the spectral range of the input waves, ideally from 1520 nm to 1565 nm. This can be done by adjusting the thickness of the rib structure.

10.3.4.2.2 Mixing and Integrating

The fourth wave generated from the FWM waveguide is then detected by the photo-detector which acts as an integrating device. Thus the output of this detector is the triple correlation product in the electronic domain that we are looking for.

10.3.4.2.3 Equalization and Filtering

If equalization or filtering is required, then these functional blocks can be implemented in the bispectral domain as shown in Figure 10.1. Figure 10.2 shows the parallel structures of the bispectral receiver so as to obtain all rows of the bispectral matrix. The components of the structure are almost similar except for the delay time of the optical pre-processor.

10.3.5 Four-Wave Mixing in Highly Nonlinear Media

In the NL channel waveguide fabricated using TeO_2 (tellurium oxide) on silica the interaction of the three waves, one original and two delayed beams, happens via the electronic processes with highly NL coefficient $[\chi^3]$ will convert to the 4th wave.

When the three waves are co-propagating, the conservation of the momentum of the three waves and the fourth wave is satisfied to produce efficient FWM. Indeed phase matching can also be satisfied by one forward wave and two backwards propagating waves (delayed version of the first wave), leading to almost 100% conversion efficiency to generate the fourth wave.

The interaction of the three waves via the electronic process and the $\chi^{(3)}$ gives rise to the polarization vector P which in turn couples to the electric field density of the light waves and then to the NLSE. By solving and modeling this wave equation with the FWM term on the RHS of the equation one can obtain the wave output (the fourth wave) at the output at the end of the NL waveguide section.

10.3.6 Third Harmonic Conversion

Third harmonic conversion may happen but at extremely low efficiency, at least one thousand times less than that of FWM due to the non matching of the effective refractive indices of the guided modes at 1550 nm (fundamental wave) and 517 nm (third harmonic wave).

Note: The common term for this process is the matching of the dispersion characteristics, i.e., k/omega with omega is the radial frequency of the waves at 1550 nm and 517 nm; versus the thickness of the waveguide.

10.3.7 Conservation of Momentum

The conservation of momentum and thus phase matching condition for the FWM is satisfied without much difficulty as the wavelengths of the three input waves are the same. The optical NL channel waveguide is to be designed such that there is mismatching of the third harmonic conversion and it is most efficient for FWM. It is considered that single polarized mode, either transverse electronic (TE) or transverse magnetic (TM), will be use to achieve efficient FWM. Thus the dimension of the channel waveguide would be estimated at about 0.4 μm (height) × 4 μm (width).

10.3.8 Estimate of Optical Power Required for Four-Wave Mixing

In order to achieve the most efficient FWM process the NL coefficient n_2 which is proportional to $\chi^{(3)}$ by a constant ($8n/3$, with n as the refractive index of the medium or approximately the effective refractive index of the guided mode). This NL coefficient is then multiplied by the intensity of the guided waves to give an estimate of the phase change and thence estimation of the efficiency of the FWM. With the cross-section estimated in Equation 10.3 and the well confinement of the guided mode, the effective area of the guided waves is very close to the cross-section area. Thus an average power of the guided waves would be about 3–5 mW or about 6 dBm.

With the practical data of the loss of the linear section (section of multimode interference and delay split—similar to array waveguide grating technology) is estimated at 3 dB. Thus the input power of the three waves required for efficient FWM is about 10 dBm (maximum).

10.3.9 Mathematical Principles of Four-Wave Mixing and the Wave Equations

10.3.9.1 Phenomena of Four-Wave Mixing

The origin of FWM comes from the parametric processes that lie in the NL responses of bound electrons of a material to applied optical fields. More specifically the polarization induced in the medium is not linear in the applied field but contains NL terms whose magnitude is governed by the NL susceptibilities [24,57,58]. The 1st, 2nd and 3rd order parametric processes can occur due to these NL susceptibilities [$\chi^1 \chi^2 \chi^3$]. The coefficient χ^3 is responsible for the FWM which is exploited in this work. Simultaneously with this FWM there is also a possibility of generating third harmonic waves by mixing the three waves and parametric amplification. The THG is normally very small due the phase mismatching of the guided wave number (the momentum vector) between the fundamental waves and the third harmonic waves. FWM in a guided wave medium such as single mode optical fibers has been extensively studied due to its efficient mixing to give the fourth wave [59,60]. The exploitation of the FWM processes has not been extensively exploited yet in channel optical waveguides. In this work we demonstrate theoretically and experimentally the application of the spectrum analysis technique in optical signal processing.

The three lightwaves are mixed to generate the polarization vector $\vec{P}$ due to the NL third-order susceptibility given as

$$\vec{P}_{NL} = \varepsilon_0 \chi^{(3)} \vec{E} \cdot \vec{E} \cdot \vec{E} \tag{10.35}$$

where ε_0 is the permittivity in vacuum, $\vec{E}_1, \vec{E}_2, \vec{E}_3$ are the electric field components of the lightwaves, $\vec{E} = \vec{E}_1 + \vec{E}_2 + \vec{E}_3$ is the total field entering the NL waveguide, and $\chi^{(3)}$ is the third-order susceptibility of the NL medium. P_{NL} is the product of the three total optical fields of the three optical waves here that gives the triple product of the waves required for the bispectrum receiver in which the NL waveguide acts as a multiplier of the three waves which are considered as the pump waves in this section. The mathematical analysis of the coupling equations via the wave equation is complicated but straightforward. ω_1, ω_2, ω_3 and ω_4 are the angular frequencies of the four waves of the FWM process and are linearly polarized along the horizontal y-direction of the channel waveguides and propagating along the z-direction. The total electric field vector of the four waves is given by:

$$\vec{E} = \frac{1}{2} \vec{a}_y \sum_{i=1}^{4} \vec{E}_i e^{j(k_i z - \omega_i t)} + c.c. \tag{10.36}$$

with $\vec{a}_y$= unit vector along the y axis; and $c.c.$ = complex conjugate. The propagation constant can be obtained by $k_i = (n_{eff,i}\omega_i)/c$ with $n_{eff,i}$ the effective index of the ith guided waves E_i ($i = 1,...4$) which can be either TE or TM polarized guided mode propagating along the channel NL optical waveguide and all four waves are assumed to be propagating along the same direction. Substituting Equation 10.36 into Equaiton 10.35, we have

$$\vec{P}_{NL} = \frac{1}{2}\vec{a}_y \sum_{i=1}^{4} P_i e^{j(k_i z - \omega_i t)} + c.c.$$ (10.37)

where P_i ($i = 1, 2...4$) consists of a large number of terms involving the product of three electric fields of the optical guided waves, e.g., the term P_4 can be expressed as:

$$P_4 = \frac{3\varepsilon_0}{4}\chi_{xxxx}^{(3)}\left\{\begin{array}{l}|E_4|^2 E_4 + 2(|E_1|^2 + |E_2|^2 + |E_3|^2)E_4 \\ +2E_1 E_2 E_3 e^{j\varphi^+} + 2E_1 E_2 E_3^* e^{j\varphi^-} + c.c.\end{array}\right\}$$

with

$$\varphi^+ = (k_1 + k_2 + k_3 + k_4)z - (\omega_1 + \omega_2 + \omega_3 + \omega_4)t$$
$$\varphi^- = (k_1 + k_2 - k_3 - k_4)z - (\omega_1 + \omega_2 - \omega_3 - \omega_4)t$$ (10.38)

The first four terms of Equation 10.38 represents the SPM and XPM effects which are dependent on the intensity of the waves. The remaining terms results in FWM. Thus the question is which terms are the most effective components resulting from the parametric mixing process? The effectiveness of the parametric coupling depends on the phase matching terms governed by φ^+ and φ^- or a similar quantity.

It is obvious that significant FWM would occur if the phase matching is satisfied. This requires the matching of both the frequency as well as the wave vectors as given in Equation 10.38. From Equation 10.38 we can see that the term φ^+ corresponds to the case in which three waves are mixed to give the fourth wave whose frequency is three times that of the original wave, this is the THG. However, the matching of the wave vector would not normally be satisfied due to the dispersion effect or the wave vectors of a channel optical waveguide of the fundamental and third-order harmonic are largely different and only minute THG might occur.

The conservation of momentum derived from the wave vectors of the four waves requires that:

$$\Delta k = k_1 + k_2 - k_3 - k_4 = \frac{n_{eff,1}\omega_1 + n_{eff,2}\omega_2 - n_{eff,3}\omega_3 - n_{eff,4}\omega_4}{c} = 0$$ (10.39)

as the effective refractive indices of the guided modes of the three waves E_1, E_2 and E_3 are the same so are their frequencies. This condition is automatically satisfied provided that the NL waveguide is designed such that it supports only a single polarized mode TE or TM and with minimum dispersion difference within the band of the signals.

10.3.9.2 Coupled Equations and Conversion Efficiency

To derive the wave equations to represent the propagation of the three waves to generate the fourth wave, we can resort to the Maxwell equations. It is lengthy to write down all the steps involved in this derivation so we summarize the standard steps usually employed to derive the wave equations as follows: First, add the NL polarization vector given in Equation 10.35 into the electric field density vector **D**. Then, take the curl of Maxwell's first equation and use the second equation of Maxwell's four equations and, substituting the electric field density vector and using the fourth equation, one would then come up with the vectorial wave equation.

For the FWM process occurring during the interaction of the three waves along the propagation direction of the NL optical channel waveguide, the evolution of the amplitudes, A_1 to A_4, of the four waves, E_1 to E_4, is given by (only A_1 term is given)

$$\frac{dA_1}{dz} = \frac{jn_2\omega_1}{c}\left[\left(\Gamma_{11}|A_1|^2 + 2\sum_{k\neq 1}\Gamma_{1k}|A_k|^2\right)A_1 + 2\Gamma_{1234}A_2^*A_3A_4e^{j\Delta kz}\right] \tag{10.40}$$

where the wave vector mismatch Δk is given in Equation 10.39; the * denotes the complex conjugation. Note that the coefficient n_2 in Equation 10.40 is the NL coefficient related to the NL susceptibility coefficient, and defined as:

$$n_2 = \frac{3}{8n\,\mathrm{Re}(\chi_{xxxx}^3)} \tag{10.41}$$

10.3.9.3 Evolution of Four-Wave Mixing along the Nonlinear Waveguide Section

Once the fourth wave is generated, the interaction of the four waves along the section of the waveguide continues happening, thus the NLSE must be used to investigate the evolution of the waves. The NLSE is well known and presented in Agrawal, 2001 [24] and is given for the temporal amplitude of the waves as

$$\frac{dA_j}{dz} \rightarrow \frac{dA_j}{dz} + \beta_1\frac{dA_j}{dz} + \frac{j}{2}\beta_2\frac{d^2A_j}{dz^2} + \frac{1}{2}\alpha_j A_j \tag{10.42}$$

This makes the four equations complicated and only numerical simulations can offer the evolution of the complex amplitude and the power of the fourth wave at the output of the NL waveguide. This takes into account the dispersion of the waveguide and material of the waveguide under chromatic dispersion.

10.3.10 Transmission and Detection

10.3.10.1 Optical Transmission Route and Simulation Platform

Figure 10.3 shows the schematic of the transmission link over a total length of 700 km, with sections from Melbourne City (Victoria, Australia) to Gippsland, the inland section in Victoria, Australia, thence an undersea section of more than 300 km crossing the Bass Strait to George Town, Tasmania and then inland transmission to Hobart, Tasmania. The Gippsland/George Town link is shown in Figure 10.45. Other inland sections in Victoria and Tasmania are structured with optical fibers and lumped optical amplifiers (EDFA). Raman distributed optical amplification (ROA) is used by pump sources located at both ends of the Melbourne, Victoria to Hobart, Tasmania link including the 300 km undersea

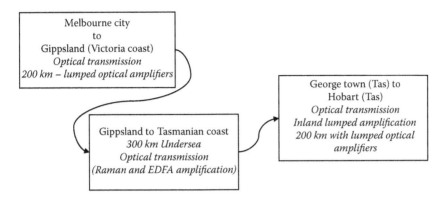

FIGURE 10.45
Schematic of the transmission link including inland and undersea sections between Melbourne (Victoria) and Hobart (Tasmania) of Australia.

section which consists of only the transmission and dispersion compensating fibers, no active sub-systems are included. Only Raman amplification is used with pump sources located at both sides of the section and installed inland. This 300 km distance is a fairly long distance and only Raman distributed gain is used. Simulink models of the transmission system include the optical transmitter, the transmission line and the bispectrum optical receiver.

10.3.10.2 Four-Wave Mixing and Bispectrum Receiving

We have also integrated the photonic signal processing using the nonlinear interactions into the MATLAB Simulink of the transmission system to enable us to investigate the NL parametric conversion system very close to practice. The spectra of the optical signals before and after this amplification are shown in Figures 10.46 through 10.48, indicating the conversion efficiency. This indicates the performance of the bispectrum optical receiver.

10.3.10.3 Performance

We implement the models for both techniques for the binary phase shift keying (BPSK) modulation format for serving as a guideline for phase modulation optical transmission systems using NL pre-processing. We note the following

1. The arbitrary white Gaussian noise (AWGN) block in the Simulink platform can be set in different operating modes. This block then accepts the signal input and assumes the sampling rate of the input signal, then estimates the noise variance based on Gaussian distribution and the specified SNR. This is then superimposed on the amplitude of the sampled value. Thus we believe at that stage, the noise is contributed evenly across the entire band of the sampled time (converted to spectral band).

2. The ideal curve SNR versus BER plotted in the graph provided is calculated using the commonly used formula in several textbooks on communication theory. This is evaluated based on the geometrical distribution of the phase states and then the noise distribution over those states. That means that all the modulation and demodulation are assumed to be perfect. However, in the digital system simulation

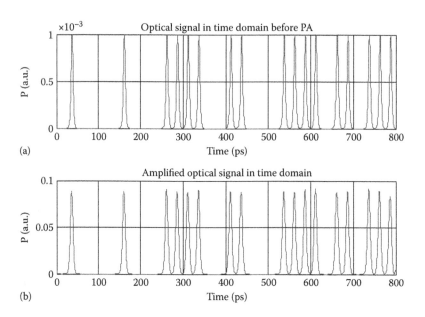

(a)

(b)

FIGURE 10.46

Time traces of the optical signal (a) before, and (b) after the parametric amplifier.

the signals must be sampled. This is even more complicated when a carrier is embedded in the signal, especially when the phase shift keying modulation format is used.

3. One can thus re-setup the models of: (i) AWGN in a complete BPSK modulation format with both the ideal coherent modulator and demodulator and any necessary filtering required; and (ii) AWGN blocks with the coherent modulator and demodulator incorporating the triple correlator and the necessary signal processing block. This is done in order to make a fair comparison between the two pressing systems.

4. In the former model, the AWGN block was being used incorrectly in that it was being used in "SNR" mode which applies the noise power over the entire bandwidth of the channel which, of course, is larger than the data bandwidth, meaning that the amount of noise in the data band was a fraction of the total noise applied. We accept that this was an unfair comparison to the theoretical curve which is given against E_b/N_0 as defined in Proakis, 2002 [61].

5. In the current model we provide, a fair comparison noise was added to the modulated signal using the AWGN block in E_b/N_0 mode (E_0 is the energy per bit and N_0 are the noises contained within the bit period) with the "symbol period" set to the carrier period, in effect this set the "carrier to noise" ratio (CNR). Also, the triple correlation receiver was modified a little from the original, namely, the addition on the BPF and some tweaking of the triple correlation delays, this resulted in the BER curve shown in Figure 10.49. Also an ideal homodyne receiver model was constructed with noise added and measured in the exact same method as the triple correlation model. This provides a benchmark to compare the triple correlation receiver against the conventional detection system. Furthermore, we can compare the simulated BER values with the theoretical limit set by

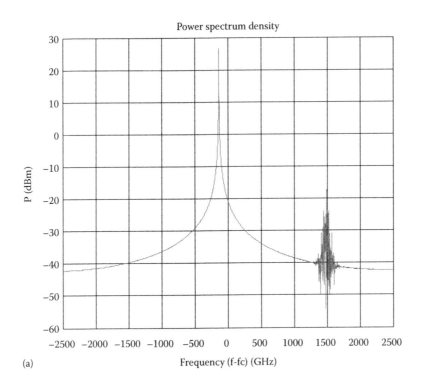

(a)

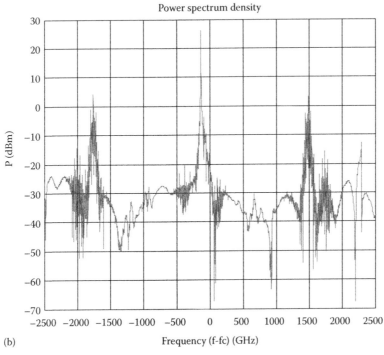

(b)

FIGURE 10.47
Corresponding spectra of the optical signal (a) before, and (b) after the parametric amplifier.

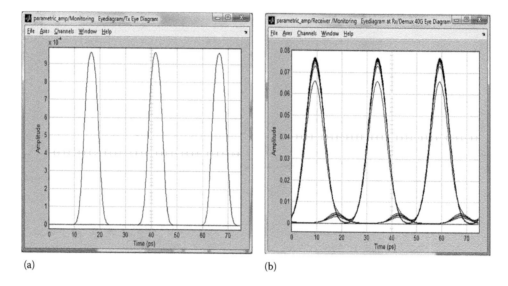

(a)   (b)

FIGURE 10.48

(a) Input data sequence. (b) Detected sequence processed using triple correlation NL photonic processing and recovery scheme bispectrum receiver.

$$P_b = \frac{1}{2} erfc \sqrt{\frac{E_b}{N_0}} \tag{10.43}$$

by relating the CNR to E_b/N_0 like so

$$\frac{E_b}{N_0} = CNR \frac{B_W}{f_s} \tag{10.44}$$

where the channel bandwidth BW is 1600 Hz set by the sampling rate and f_s is the symbol rate, in our case a symbol is one carrier period (100 Hz) as we are adding noise to the carrier. These frequencies are set at the normalized level so as to scale to wherever the spectral regions would be of interest. As can be seen in Figure 10.49 the triple correlation receiver matches the performance of the ideal homodyne case and closely approaches the theoretical limit of BPSK (approximately 3 dB at BER of 10^{-10}). As discussed the principle benefit from the triple correlation over the ideal homodyne case will be the characterization of the noise of the channel which is achieved by analysis of the regions of symmetry in the 2D bispectrum. Finally, we still expect possible performance improvement when symbol identification is performed directly from the triple correlation matrix as opposed to the traditional method which involves recovering the pulse shape first. It is not possible at this stage to model the effect of the direct method.

10.3.11 Remarks

This section demonstrates the employment of a NL optical waveguide and associated NL effects, such as parametric amplification, FWM and THG, for the implementation of the

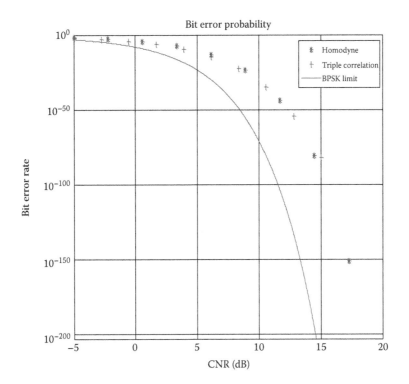

FIGURE 10.49
BER versus CNR for NL triple correlation, ideal BPSK·under coherent detection and ideal BPSK limit.

triple correlation and thence the bispectrum creation and signal recovery techniques to reconstruct the data sequence transmitted over a long-haul optically amplified fiber transmission link.

10.4 Raman Effects in Microstructure Optical Fibers or Photonic Crystal Fibers

10.4.1 Introductory Remarks

The development of optical fiber technology has been a critical milestone in global telecommunications and information technology since the 1970s [62]. As we have seen in Chapters 4 and 5, the guiding of the lightwaves as a signal carrier inside the optical fiber is a leading character in the information revolution due to its extremely high frequency range (~10^{14} Hz). Conventional single mode transmission fiber consists of glass fiber, 125 μm in cladding diameter covered by a protective polymer to an outer diameter of 250 μm, and the central core of about 8 μm diameter region is doped with germanium to raise the refractive index by about 0.3%. The higher refractive index of the core traps the light and weakly guides it as a single transverse mode by "total internal reflection." Attenuation, dispersion, nonlinearity and birefringence are major factors of optical fibers related to the applications. As for the developing interests and requirements in optical waveguide, integration, and signal processing, we need that fibers can be designed to carry much more power, have higher nonlinearities, be used as sensors or acquire widely engineered dispersion and birefringence to satisfy versatile fields besides optical telecommunications, in which only the standard step index fiber did well.

Recently, great excitement has been generated by the demonstration of air-silica photonic crystal fibers (PCFs), or holey fibers (HFs), microstructure fibers (MOFs), which have a large variation of refractive index within the light guiding region [63] as described in Chapter 5. Unlike conventional optical fibers, PCFs can be made from a single material. Waveguiding in these structures can be due to an effective index difference between a defected region, which forms the core, and the arrangement of air holes surrounding the core to act as the cladding, as shown in Figure 10.50a. The added feature of air holes has opened up exciting new possibilities for controlling and guiding light, including making fiber as a photonic bandgap (PBG) to guide light predominantly in air [64]. PCFs are particularly attractive for photonic components because the optical properties can be engineered during fabrication. For example, the large index difference between glass and air can be explored to push the zero-dispersion wavelength (ZDW) below 800 nm [65] and to maintain rigorously single mode [2,63]. The anomalous or near-zero dispersion in a wideband wavelength region is essential to many NL device applications, such as soliton formation [66] and supercontinuum generation [67]. Furthermore, through the fiber design, the mode size can be tailored by an appropriate air-hole-core geometrical arrangement, which in turn changes the light intensity inside the fiber to alter the effective nonlinearity. An effective mode area as small as 1.3 μm^2 has successfully been fabricated with an effective nonlinearity coefficient as high as 70 W^{-1} km^{-1} 1550 nm [68], i.e., around 70 times more NL than SSMF. The combination of highly NL material composition and small core/high numerical aperture (NA) allows a dramatic increase of the fiber nonlinearity. Recently, a highly NL PCF with the propagation loss as low as 2.6 dB/m at 1550 nm and with a NL coefficient as high as 640 W^{-1} km^{-1} has been reported [69] that provides a promising future for the NL fiber devices based on PCF.

In this section we briefly present the Raman effects in PCF or MOF and other Raman-assisted NL parametric processes such as self-frequency shift and continuum generation. The stimulated Raman scattering (SRS) is caused by molecular vibrations in the light waveguiding material, typically in the multi-terahertz range, which interact with the light

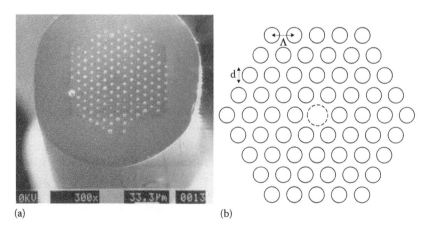

(a)   (b)

FIGURE 10.50
(a) Photomicrograph of a photonic crystal fiber. (b) Transverse section of a PCF with triangular lattice.

photon simultaneously (femtosecond range) [70]. The SRS shift the original light frequency both up (anti-Stokes) and down (Stokes). When the Stokes light exceeds a certain threshold value, the major fraction of pump power will be converted to the Stokes frequency. Raman amplification based on SRS effects offers the necessary low-noise amplification for economical field-deployed systems [71] and thus becomes one of the most competitive candidates for the next generation fiber amplifiers. With the maturity of high power semiconductor pump lasers, the fiber Raman amplifiers (FRAs) overwhelm the EDFAs in performances such as: (i) amplification can be achieved at any frequency band by simply changing the pump wavelength; (ii) simple upgrade scheme; and (iii) reducing the equivalent noise figure while keeping a low NL penalty. Since the Raman effect is also a manifestation of light-material interaction mediated by $\chi^{(3)}$ (third-order susceptibility tensor), it is natural to explore the possibility of using a high NL PCF to construct a compact FRA with enhanced amplification performance, which is expected to overcome the typical drawbacks of using conventional fiber: long fiber length (tens of kilometers) and related Rayleigh scattering [72]. Firstly, the Raman effect of high NL PCF is experimentally investigated, and then the study of the effective Raman coefficient in PCF, by changing the geometrical parameters, the hole diameter d and the spacing Λ between air holes (Figure 10.50b). Thence it is possible to alter the effective index of the cladding and thus the field distribution. The effective Raman gain coefficient can be modified by the variation of the modal area of the PCF. Since the Kerr nonlinearity (parametric process) is also dependent on the effective mode area, we discuss the optical signal transmission quality considering the interaction between Raman amplification and the NL phase shift, where the dispersion property of PCF must be included.

10.4.2 Raman Gain in Photonic Crystal Fibers

10.4.2.1 Measurement of Raman Gain

The schematic experimental setup is shown in Figure 10.51. A tunable laser source (TLS) is used through the circulator to scan the Raman gain spectrum. A length of 1.5 m of high NL PCF is used as the Raman gain medium, which has an effective core area of 3.7 μm^2 and a NL coefficient $\gamma = 52$ W^{-1} km^{-1}. 1455 nm Raman fiber laser is injected through

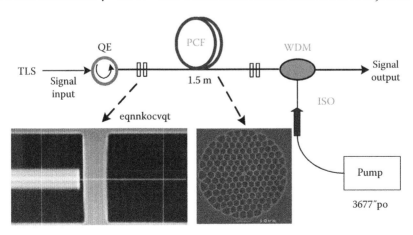

FIGURE 10.51
Experimental setup for Raman amplification in PCF. TLS: tunable laser source, OC: optical circulator, ISO: isolator, WDM: wavelength division multiplexer.

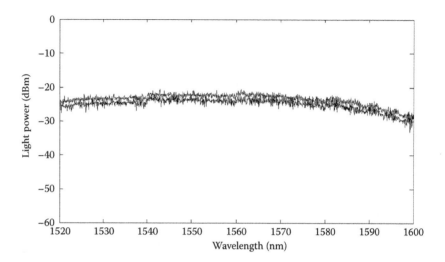

FIGURE 10.52
Signal power spectrum with- and without pump power. The maximum pump power is 1 W.

the WDM to act as Raman pump. The light is coupled in or out of the PCF by using the fiber collimator (shown in Figure 10.51). Since the ZDW of PCF is at about 810 nm and the dispersion slope is 0.84 ps/nm^{-2}/km^{-1}, the Raman amplification process occurs in a highly dispersive environment.

The signal power spectrum is measured with the pump ON/OFF when the TLS scans from 1520 nm to 1600 nm. The results are shown in Figure 10.52. Although the amplification is very weak due to the short gain medium length (1.5 m) compared to Yusoff, 2002 [73], one can observe the signal level increase with pump ON. The average gain around the 1550 nm region is about 1.04 dB. Neglecting pump depletion in low signal input, the ON-OFF Raman gain can be expressed as $G = \exp(g_R P_p L_{eff}/A_{eff})$, where P_p is the input pump power, g_R is the Raman gain coefficient, A_{eff} is the effective area, and the effective length $L_{eff} = [1 - \exp(-\alpha L)]/\alpha$, respectively. Here the effective Raman gain coefficient is defined as $g = g_R/A_{eff}$ (neglect the polarization scrambling), thus the effective Raman gain coefficient can be estimated as 0.015 W^{-1} m^{-1}, which is about 30 times larger than that of the SSMF. Thus the PCF promises to be an optimal choice to practical FRA.

10.4.2.2 Effective Area and Raman Gain Coefficient

Since the fabrication material of PCF comes from conventional silica or Ge-doped silica, the enhanced nonlinearity in PCF is mainly due to the fact that the fibers feature a small core area that can be changed from the air-hole arrangement. From the definition of effective gain coefficient g, the much stronger Raman process inside the PCF can also be caused by the effective core area.

In standard single mode optical fibers, the growth of stimulated Raman scattered signal intensity is proportional to the product of the pump and signal intensities and the gain coefficient g_R, such that

$$\frac{dI}{dz} = g_R I_P I_S \qquad (10.45)$$

where I_s and I_p are the intensities of the Stokes-shifted and pump waves, respectively. In order to generate stimulated emission, it is strictly necessary that the Stokes and pump waves overlap spatially and temporally. In circular SMFs, the power P is found by integrating the intensity over the cross-section of the fiber

$$P = \int_0^{2\pi} \int_0^{\infty} I(r,\theta) \cdot r \, dr \, d\theta \tag{10.46}$$

By integrating Equation 10.45 over the cross-section of the fiber, the power evolution of Raman Stokes light is given by

$$\frac{dP_S(z)}{dz} = P_P(z)P_S(z) \frac{\int_0^{\infty} g_R I_S I_P \cdot r \, dr}{2\pi \int_0^{\infty} I_S \cdot r \, dr \int_0^{\infty} I_P \cdot r \, dr} = P_P(z)P_S(z)g_0 \tag{10.47}$$

Here we assume that the g_R is uniform with the radius, thus a new parameter A_{eff}^R that describes the spatial overlap between the signal and pump modes can be given as:

$$g_0 = \frac{\int_0^{\infty} g_R I_S I_P \cdot r \, dr}{2\pi \int_0^{\infty} I_S \cdot r \, dr \int_0^{\infty} I_P \cdot r \, dr} = \frac{g_R}{A_{eff}^R} \tag{10.48}$$

$$A_{eff}^R = \frac{2\pi \int_0^{\infty} I_S \cdot r \, dr \int_0^{\infty} I_P \cdot r \, dr}{\int_0^{\infty} I_S I_P \cdot r \, dr} \tag{10.49}$$

where the Raman effective core area A_{eff}^R is similar to the effective area of SPM [66]. In order to modify A_{eff}^R in the case of PCF where triangular lattice is often used, we can rewrite the effective area in the xy plane as

$$A_{eff}^R = \frac{\iint_{Area} I_S(x,y) \cdot dx \, dy \iint_{Area} I_P(x,y) \cdot dx \, dy}{\iint_{Area} I_S(x,y) I_P(x,y) \cdot dx \, dy} = \frac{1}{\iint_{Area} i_S(x,y) i_P(x,y) \cdot dx \, dy} \tag{10.50}$$

where i_S, i_P are the normalized signal and pump intensities. The value of the Raman effective area usually presents values between those of pump mode and signal mode. This parameter provides an insight into the Raman interaction. The Raman effects depends not only on the small fiber core area but also on the spectral separation between the pump and signal wavelength, so that a fiber with small A_{eff}^R for a given pump source performs a high gain. For this reason, we will analyze the mode area of the PCF according to the air-hole arrangement to calculate the Raman effective area A_{eff}^R in order to get optimal Raman gain. The normalized intensities appearing in Equation 10.50 can be evaluated by means of a numerical simulator for effective index approach [74]. The procedure is repeated twice, first for the signal field and then for pump mode.

The PCF structure considered here is illustrated in Figure 10.50b. The transverse section of a triangular PCF consists of a regular hexagonal array of circular air holes in silica glass with a defect, the absence of a hole, located at the center. The geometrical parameters describing the PCF are the air-hole diameter d and the space between the holes, Λ. The number of rings is variable and all the considered PCFs in this paper have at least 10 rings to give accurate results.

First triangular PCFs with bulk silica, relative hole diameter $d/\Lambda = 0.6, 0.7, 0.8, 0.9$ and pitch that varies from 0.7 to 2 μm are simulated. The obtained Raman effective area as a function of the pitch Λ is depicted in Figure 10.53, while the Raman gain coefficient for different PCFs as a function of the pitch Λ is presented in Figure 10.54. The effective coefficient g_0 is calculated using Equation 10.48 and the peak Raman gain coefficient g_R is given as $3.34 \cdot 10^{-14}$ m/W [75]. Noting that fixing the ratio d/Λ is equivalent to considering different PCFs with the same air-filling fraction, the ratio of air to silica in the cross-section, while varying the pitch Λ is equivalent to changing the core size since the core radius of the defect core region is given as $\Lambda - d/2$. We can observe that there is an optimal value of Λ that minimize A_{eff}^R and maximize g_0. For the same Λ, A_{eff}^R decreases with the increasing relative hole diameter d/Λ, which is due to the increasing refractive-index difference between the core and air-filling cladding. Actually, the mean refractive index of the microstructure

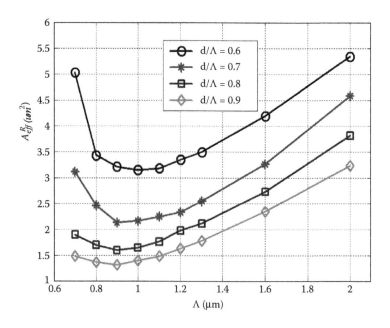

FIGURE 10.53
Raman effective area A_{eff}^R for different PCFs as a function of the pitch Λ.

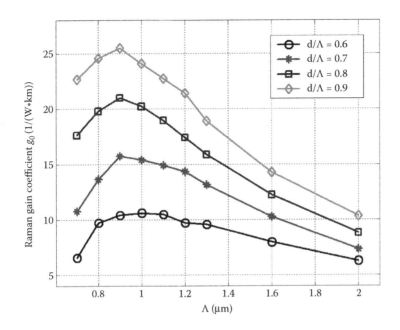

FIGURE 10.54
Raman gain coefficient for different PCFs as a function of the pitch Λ.

cladding depends only on the air-filling ratio. Whenever changing the pitch Λ to 0 or ∞ (fixing the ratio d/Λ), the corresponding PCF has a core radius Λ − d/2 → 0 or ∞. Therefore, the field cannot be guided and A_{eff}^R → ∞. Since the fiber is un-doped, the Raman gain coefficient is inversely proportional to the effective area.

Next, we study the Raman amplification in PCF with silica bulk and Germania-doped core. In normal fiber consists of silica hosts (SiO_2), adding Germania (GeO_2) (i.e., germane-silica), would increase the refractive index. The definition of Raman gain coefficient can be given as [76]:

$$g_0 = g_{R-Si} \iint_{Area} [1 - 2m(x, y)] i_S(x, y) i_P(x, y) \cdot dx dy$$

$$+ g_{R-Ge} \iint_{Area} 2m(x, y) i_S(x, y) i_P(x, y) \cdot dx dy \tag{10.51}$$

In the expression, $m(x,y)$ is the Germania concentration, g_{R-Si} and g_{R-Ge} is the Raman gain spectrum relative to the bounds Si-O-Si and Ge-O-Si, respectively. The peak value of g_{R-Si} is given before and that of g_{R-Ge} is given as $11.8 \cdot 10^{-14}$ m/W. Thus the Raman gain efficiency is dependent on the effective area and the doped region and concentration.

We first examine the PCFs with fixed geometric characteristics d/Λ = 0.4, 0.6, 0.8, and Λ = 1.6 μm, and a Ge-doped region of radius R_d equals to Λ/2 were considered. Since the refractive index of core increases with the Germania concentration, we plot the effective area of doped PCF as a function of the refractive index (equivalent to the doping concentration) in Figure 10.55. Only the small relative hole diameter case has a significant core area variation with different doping concentration. At the large relative hole diameter (large air-filling fraction), the effective area is almost constant. This phenomenon suggests that

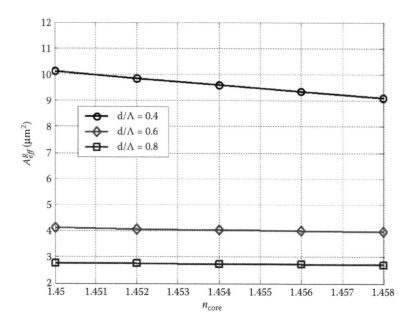

FIGURE 10.55
Raman effective area of Ge-doped PCFs, with $d/\Lambda = 0.4, 0.6, 0.8$, and $\Lambda = 1.6$ μm, Rd equals to $\Lambda/2$, as a function of Germania concentration.

the refractive index different in doped PCF is mainly dependent on the air-filling fraction (mean refractive index of microstructure cladding). Since the PCFs with high d/Λ ratio guarantee highly NL application as well as Raman amplification, a combination of large d/Λ ratio and high Ge-doping is expected to own better Raman gain performances.

The optimal dimension of the doped region of PCF is considered. PCFs with $d/\Lambda = 0.6$, 0.7, 0.8, 0.9 are investigated. The GeO_2 concentration remains at 20% mol meaning constant refractive index of 1.47 at 1550 nm wavelength. For different fixed pitch Λ, the doped core radius R_d increases from the original value close to $\Lambda/2$. Such doped structure is chosen in order to analyze the conditions as much as possible near to physically feasible fibers. Figure 10.56 illustrates the effective Raman area as a function of the mean doping radius for PCF with different hole geometries. The corresponding gain coefficient is shown in Figure 10.57, which is calculated using Equation 10.51.

The optimal positions can be found to get the minimum effective area A_{eff}^R and the maximum gain coefficient g_0, while the best values happen at the different position of R_d. We note that the maximum g_0 occurs at the position of $R_d = \Lambda - d/2$, where the Ge-doping fill in the entire core region. This can be explained that in the doping cases, the Raman gain coefficient depends on the effective area and the guided field lies in the Ge-doped region. The more the effective area lies in the doped region, the greater g_0. From Figure 10.56, we observe that the mode area is not confined very well for doped PCFs. When the doped core radius $R_d < \Lambda - d/2$, the core can be assumed equal to the Ge-doped region. When $R_d = \Lambda - d/2$, the core is equal to $\Lambda - d/2$ like in the silica-bulk PCF. When the doped region extends over to the cladding rings, the field confinement is mainly due to the refractive index difference between the doped core and the air-hole cladding. The light field can leak out partly through the doped cladding ring, thus there is a weak increase of the effective area when R_d increases (shown in Figure 10.56). The

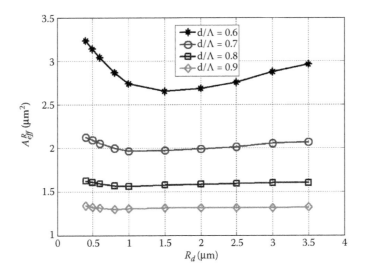

FIGURE 10.56
Raman effective area for Ge-doped PCFs.

Raman gain coefficient always occurs for $R_d = \Lambda - d/2$. Further increasing the doping region does not improve the gain performance because of the effective area increase and the additional part of the doped region in the air rings. Thus, we can expect that the best condition is to fill the Ge-doping inside the solid core region $R_d = \Lambda - d/2$, which is also a meaningful result for the fabrication process. With careful design of the core area and doped region, a maximum Raman gain coefficient >50 W^{-1} km^{-1} can be obtained, which is 100 times higher than the standard single mode transmission fiber. The gain performance can be further enhanced if the Germania concentration is increased or by using other NL doping materials.

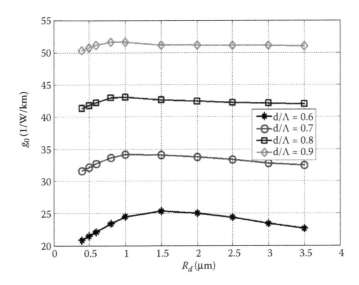

FIGURE 10.57
Raman gain coefficient for Ge-doped PCFs.

10.4.3 Remarks

In this section, the Raman amplification properties of the MOFs are analyzed in detail. A 1.5 m highly NL PCF is used to test the Raman effects in HF. After that, we define the parameter Raman effective area to describe the enhancement of Raman gain efficiency in PCF. By using the numerical simulation, the effective Raman area has been varied by changing the geometrical parameters, the relative air-hole diameter d/Λ, and the pitch Λ. The minimal A_{eff}^R can be obtained for different PCFs thus the maximal Raman gain coefficient is achieved. Furthermore, we study the PCFs with a Ge-doping region in order to provide an optimal solution for Raman amplification. Similarly, the effective area can be minimized by varying the structure parameters but the best gain performance can only be obtained by combing the effective area and the Ge-doped area. A maximal Raman gain efficiency can be achieved when the doping core area is internally tangent to the first ring of air holes. All these results suggest that a Ge-doped PCF with a carefully designed core-cladding structure is promising for the efficient Raman amplification of optical signal.

10.5 Raman Gain of Segmented Core Profile Fibers

In Chapter 4 the design of optical fibers with segmented profile is given, this kind of fiber offers a number of advantages in effective mode spot size, dispersion property for compensating dispersion, etc. In this section we employ this design of the segmented core fiber to tailor its property/profile so as to achieve the best Raman amplification gain. The Raman gain of this kind of fibers can be contrasted with those of MOFs described in the previous section which support single guided modes over extremely wideband.

The fiber design can be initiated by considering the selection of either. Since n_1 is known, setting Δ_1 will change n_2. If n_2 is also set, the tuning of Δ_1 or δ will lead to the variation of n. Similarly, a change in the ratio b/a can lead to the variation of b for a known radius. However, it is best (as described in Chapter 5) to design on the $Vd^2(Vb)/dV^2$ curve firstly based on a careful selection of δ and b/a. This combination also needs to be considered for the cutoff V values of LP_{01} and LP_{11} modes. The V_{C01} and V_{C11} are shown on the curve of the waveguide factor with the asterisk and the square marker, respectively. When the LP_{11} or LP_{01} cutoffs are properly determined to fall in the window of the wavelength spectrum being considered, L-band, this then leads to the curve of the waveguide factor versus the range of operating wavelengths. The fiber properties including dispersion slope, attenuation, cutoff wavelength, spot-size, the FOM, the RDS and the critical curvature for bending loss evaluation can all be obtained. The drawbacks of W-fiber are the high attenuation, high sensitivity due to micro-bending loss and the shortage of freedom design control parameters, which gives W-fiber a low design flexibility and tolerance. Segmented core fibers such as triple-clad fiber can cope with the problems. They provide higher design freedom degree by providing a larger number of design parameters. The analysis and design steps of optical fibers with triple-cladding regions is presented in this section so that a complete design platform for dispersion compensation based on in-line optical fibers is available for DWDM system designers. A new and simple method has been developed for the approximation of waveguide dispersion parameter curves (see also designs described in Chapter 5).

10.5.1 Segmented-Core Fiber Design for Raman Amplification

SRS causes power to be transmitted from lower-wavelength channels to higher-wavelength channels, known as Stoke signals. The gain coefficient is a function of wavelength spacing. SRS occurs when the transmitted power of the channel exceeds a certain level, which is known as the threshold power. For a single-channel lightwave system, it has been shown that the threshold power P_{th} is given by [18]

$$P_{th} = \frac{16 A_{eff}}{k_P L_{eff} g_{R\,max}} \tag{10.52}$$

k_P is the polarization factor having a value between 1, when the polarizations between Stokes signal and pump are preserved, and 1/2 when the polarizations are completely scrambled. $L_{eff} = [1 - \exp(-\alpha_P L)]/\alpha_P$ is the effective length of the fiber [77]. In the case of CW and quasi-CW conditions, the NL interaction between the pump and Stokes waves of SRS is governed by the following set of two coupled equations:

$$\frac{dI_S}{dz} = g_R I_P I_S - \alpha_S I_S$$

$$\frac{dI_P}{dz} = -\frac{\omega_P}{\omega_S} g_R I_P I_S - \alpha_P I_P \tag{10.53}$$

where I_P, I_S, α_P, α_S, ω_P, ω_S are the intensity, the attenuation and frequency of pump and signal, respectively.

The gain provided by Raman amplification can be obtained from the above equations. With the common assumption that the intensity of signal remains much smaller than the pump intensity, the Raman amplification gain can be expressed:

$$G_A = \exp(g_R P_0 L_{eff} / A_{eff}) \tag{10.54}$$

where $P_0 I_0 A_{eff}$ is the pump power. One of the most important parameters for Raman amplification in any application is the Raman effective gain coefficient, which is defined as $g_{eff}^R = g_R / A_{eff}$ where g_R is the Raman gain coefficient. The effective Raman gain coefficient depends not only on the Raman gain coefficient itself but also on the effective area of the fiber. Hence it leads to the significance of designing a fiber with a small effective area in order to achieve high gain. In addition, flattening the Raman gain coefficient as much as possible is also a desirable need. DCF fiber offers excellent gain medium for discrete Raman amplifiers [78] which have already been started to be widely deployed in the long-haul transmission systems.

10.5.2 Advantages of Dispersion Compensating Fiber as a Lumped/Discrete Raman Amplifier (DRA)

DCF offers advantageous operations in transmission systems: (i) due to high anomalous negative dispersion as designed given in Chapter 5, the deployed DCF can be just a few kilometers long, which also meets the requirement of DRA; (ii) small effective area and high Ge concentration gives higher Raman gain efficiency, which results in less threshold power to excite Raman amplification; (iii) offering additional flexibility in the system; and (iv) saving space and potentially low cost.

The threshold power for scattering is linearly proportional to the effective area as

$$P_{th} \propto \frac{\lambda A_{eff}}{2\pi n_{NL} L_{eff}} \tag{10.55}$$

where n_0 is the refractive index of the fiber core at low optical power levels; n_{NL} is the NL refractive index coefficient which varies from 2.35×10^{-20} to approximately 3.2×10^{-20} m²/W and P_{th} is the optical threshold power in Watts. Thus, a smaller value of A_{eff} gives smaller threshold power, i.e., less power required to excite the SRS for amplification, hence less cost since the high power laser sources are expensive. The effective area has been shown to be one of the most important properties of the DRAs. It is therefore of concern to point out the assumption and the equation implemented to find the effective area. The transverse modal profile was assumed to be approximately equal to that of the laser pump source. Besides, in the case of Gaussian assumption for the radial intensity distribution of the mode, the effective area for Raman amplification can be approximated as [79,80]:

$$A_{eff}(\Delta v, \lambda_P) = \frac{A_{eff}(\lambda_S) + A_{eff}(\lambda_P)}{2} \tag{10.56}$$

where λ_S and λ_p are the signal and pump wavelength, respectively.

$$A_{eff} = \pi \omega_{eff}^2 \tag{10.57}$$

10.5.3 Spectrum of Raman Amplification

The Stoke signals generated by SRS are shifted toward the longer wavelengths. The peak gain occurring at the first Stoke signal is about 100 nm shift from the pump wavelength or about 440 cm⁻¹ in wavelength number and approximately 13 THz shift in frequency domain. The shape and the peak gain value, varying with different amounts of Germanium doping concentrations, are illustrated in Figure 10.58. It is shown that the Raman gain spectrum extends over 20–30 THz.

The Raman gain spectrum for pure bulk silica with pump signal at 1.55 μm was achieved by segmenting the experimental graph obtained in Andre and Correia, 2003 [81]. However, the actual silica Raman gain spectrum used in the simulation is at the pump wavelength of 1.46 μm, which is utilized to amplify the signals operating in C-band. Thus, scaling of the Raman gain is required. It is important to note that the Raman gain spectrum is inversely proportional to the pump wavelength.

10.5.4 Key Equations for Deducing the Raman Gain of Ge-Doped Silica

Obtaining a precise Raman gain is very critical in order to obtain accurate computational results of the discrete/lumped Raman amplifier using DCF fiber. From Refs. [82–84] the method of predicting Raman gain spectrum with a certain amount of Ge-doping concentrations are obtained. The linear regression curve has been introduced. The linear relation

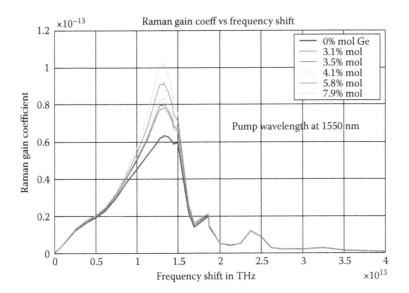

FIGURE 10.58
(See Color Insert) Spectrum of Raman gain coefficient with various Germanium doping concentrations.

between the RI and the Germanium doping concentrations has also been compared with [85]. In Figure 10.59 the RI, corresponding to different Ge-doped concentrations, follows a linear equation and verifies with the actual RIs obtained from Sellmeier's coefficients. The spectral profile of Raman gain spectrum $g(\Delta f)$ can be expressed as [83]

$$g(\Delta f) = \frac{\sigma_0 (\Delta f) \lambda^3}{c^2 h n (\Delta f)^2} \tag{10.58}$$

where Δf is the frequency-shift between the pump and the first Stoke signal; n is the refractive index and $\sigma_0(\Delta f)$ is zero-Kelvin Raman cross-section [84,86], which is defined as

$$\sigma_0 (x_{GeO_2}, \Delta f) = \left[1 + C(\Delta f) \times x_{GeO_2}\right] \times \sigma_0 (x_{SiO_2}, \Delta f) \tag{10.59}$$

where $C(\Delta f)$ is the linear regression coefficient.

From the above equations, the Raman gain coefficients of a particular Ge-doping concentration can be obtained from the Raman spectrum of pure silica

$$g(x_{GeO_2}, \Delta f) = \left[1 + C(\Delta f) x_{GeO_2}\right] g(x_{SiO_2}, \Delta f) \left(\frac{n_{Si}}{n_{Si-Ge}}\right)^2 \tag{10.60}$$

for a relatively small amount of Ge-doping concentration in pure silica, the difference between two refractive indices is negligible. Therefore, the Raman gain spectrum utilized in the simulation can be implemented as

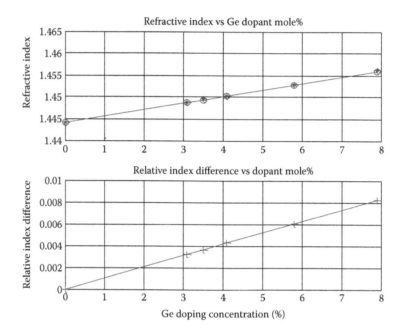

FIGURE 10.59
Linearity of refractive index with Germanium doping concentration.

$$g\left(x_{GeO_2}, \Delta f\right) = \left[1 + C\left(\Delta f\right) x_{GeO_2}\right] g\left(x_{SiO_2}, \Delta f\right) \tag{10.61}$$

where

$$\Delta = \frac{n_{core} - n_{clad}}{n_{core}}$$

This approach for the Raman gain is implemented in the simulation.

10.5.5 Design Methodology for Dispersion Compensating Fiber – Discrete Raman Amplifiers

The technique focuses on a single pump to avoid the interactive interferences between the pumps with the asymptotic approximation for Raman gain within C-band (30 nm bandwidth). The insight of this technique is discussed in detail. Defining the normalized far-field RMS spot size as $\varpi = \omega_0/a$ with

$$\varpi = \frac{2\int\limits_0^\infty \psi^2\left(R\right) R dR}{\int\limits_0^\infty \left[\psi'\left(R\right)\right]^2 R dR}$$

and $R = r/a$ is the normalized distance with respect to the core radius, ω places a lower bound on the microbending loss [87] and is useful for calculations of the splice loss due to small offsets. More importantly, it was proven to be directly related to the measurable root-mean square (RMS) width of the far-field spot size [78].

The relation existing between this normalized far-field RMS spot size and the waveguide dispersion has been presented in Chung, 1987 [88]:

$$\frac{d(bV)}{dV} - b = \frac{4}{V^2\omega^2} \tag{10.62}$$

where $b(V)$ is the normalized propagation constant. The equation is then rearranged as:

$$\omega^2 = \frac{4}{V^2\left[\dfrac{d(bV)}{dV} - b\right]} = \frac{4}{V^2\left[\left(V\dfrac{db(V)}{dV} + b\dfrac{dV}{dV}\right) - b\right]} \tag{10.63}$$

Hence, the equation can be simplified to be:

$$\omega^2 = \frac{4}{V^3\,\dfrac{db(V)}{dV}} \tag{10.64}$$

Moreover,

$$b(V) = \left(1.1428 - \frac{0.996}{V}\right)^2 \tag{10.65}$$

The relative error is less than 2% for $1 < V < 3$.

$$\frac{db(V)}{dV} = 2\left(1.1428 - \frac{0.996}{V}\right)\frac{0.996}{V^2} \tag{10.66}$$

From Equations 10.64 through 10.66, it can be derived that

$$\omega^2 = \frac{4}{2.27646V - 1.984} \tag{10.67}$$

Therefore, the spot size can be calculated as:

$$\omega_0^2 = \frac{4a_{eff}^2}{2.27646V - 1.984} \tag{10.68}$$

Thus the effective area is

$$A_{eff} = \pi\omega_0^2 = \frac{4\pi a_{eff}^2}{2.27646V - 1.984} \tag{10.69}$$

In the case of Gaussian assumption for the radial intensity distribution of the mode, the effective area for Raman amplification can be approximated as [89]:

$$A_{eff}\left(\Delta f, \lambda_P\right) = \frac{A_{eff}\left(\lambda_S\right) + A_{eff}\left(\lambda_P\right)}{2} \tag{10.70}$$

The effective Raman gain

$$g_{eff} = \frac{g_{Raman}}{A_{eff}\left(\Delta f, \lambda_P\right)} \tag{10.71}$$

and V is given by

$$V = \frac{2\pi}{\lambda} an_2 \sqrt{\frac{n_1 - n_2}{n_2}} = \frac{2\pi}{\lambda} an_2 \sqrt{2\Delta} \tag{10.72}$$

Defining the variable:

$$X = an_2\sqrt{2\Delta}$$

Thus,

$$V = \frac{2\pi}{\lambda} X \tag{10.73}$$

The X-variable is the key parameter which gives the options for the design of DCF-DRAs. The solutions of the X-variable can be easily obtained.

In order to control the flatness of effective Raman gain over 30 nm bandwidth of C-band, we may simply control the values of the two ends which are located at 1530 nm and 1560 nm. In this method, the Raman gain within the C-band operating wavelength is asymptotically approximated (Figure 10.60).

On the other hand, since the known curves of the spot size and hence the effective area are hyperbolic, the effective Raman gain cannot be perfectly flattened and a level of gain tilt is expected to exist and considered to be the ripple created by the single pump. The value of the tilt or the curvature of the effective gain is optimally determined depending on the feasibility of the design solutions for DCF.

Defining the gain tilt to be K_{tilt} and manipulating the two ends of the effective Raman gain, given as:

$$g_{eff\,1560} = K_{tilt}\,g_{eff\,1530} \tag{10.74}$$

$$\frac{g_{R1560}}{A_{eff}\left(\Delta f\right)_{1560}} = K_{tilt}\frac{g_{R1530}}{A_{eff}\left(\Delta f\right)_{1530}} \tag{10.75}$$

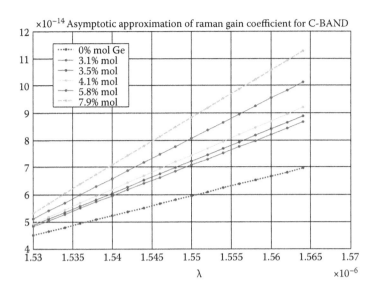

FIGURE 10.60
Asymptotic approximation of Raman gain within C-band.

The equation showing the control technique for the effective Raman gain of DRAs over the C-band is formulated as follows:

$$\frac{g_{R1560}\left(\dfrac{2.27646\times 2\pi}{\lambda_{1560}}X - 1.984\right)}{\left(\dfrac{2.27646\times 2\pi}{\lambda_{1560}} + \dfrac{2.27646\times 2\pi}{\lambda_{pump}}\right)X - 2\times 1.984}$$

$$= K_{tilt}\frac{g_{R1530}\left(\dfrac{2.27646\times 2\pi}{\lambda_{1530}}X - 1.984\right)}{\left(\dfrac{2.27646\times 2\pi}{\lambda_{1530}} + \dfrac{2.27646\times 2\pi}{\lambda_{pump}}\right)X - 2\times 1.984} \tag{1.0.76}$$

Rearranging the above equations, a quadratic equation is obtained and hence the solutions can be derived. These solutions are correspondent to various amounts of doping concentrations of Germanium in the core. The solution determines the gain tilt of the effective single-pump Raman gain and is then used for the design of an appropriate DCF profile of this DCF-DRA.

10.5.6 Design Steps

Since $x = an_2(2\Delta)^{1/2}$, it can be seen that a solution of x may give various options for the design according to different combinations of the values of the core radius (a) and the inner cladding RI (n_2). In our design n_1 is known as it is the core material with a specific Ge-doping concentration. The core radius and the RI of the core and the inner-cladding can be found. These parameters of DCF are utilized for controlling and estimating the effective Raman gain of the DRAs. However, these design solutions are only selected when they

TABLE 10.7

Profile for Designing DCF-DRAs

Core Radius a (μm)	Inner-cladding Radius b (μm)	Core RI (n_1)	Inner-cladding RI (n_2)	Outer-cladding RI (n)
1.31	2.751	1.4487 (3.1% Ge-doped)	1.4208 (5.05% F-doped)	1.42554

also show a feasible design to achieve a good DCF profile. This can be obtained based on the consideration of the normalized frequency V-value. The unknown design parameters are then optimized to achieve a good compensation scheme for simultaneously compensating the dispersion and the dispersion slope.

The method can also be inversely used to estimate the gain tilt when the profile of the DCF fiber has already been aware. By knowing the values of normalized frequencies V-values at 1.53 μm and 1.56 μm, the derived equation above can now be utilized for the calculation of the gain tilt.

10.5.7 Sampled Profile Design

The index profile given in Table 10.7 gives a good design for a DRA using DCF fiber. The gain tilt obtained is within the acceptable range and the DCF with high RDS and FOM is capable of compensation of high RDS transmission fibers. The simulated results are tabulated in Table 10.8.

The graphical results are demonstrated in Figures 10.61 through 10.68. The achieved results such as the dispersion value, RDS, FOM, attenuation, etc., are compared to those reported in Refs. [89,90].

The gain tilt of 1.382 dB is obtained as the result of the design of DCF whereas the predicted value K_{tilt} for solving the equation is 1.615 dB. The achieved results are compared to those reported in Refs. [89] in which the reported gain ripple of the DRAs are 2.3 dB and 1.2 dB, respectively. The slight difference in the values between the predicted and the simulated gain tilt has validated this fast, simple and relatively accurate way of designing DRAs using DCF fiber with a single pump scheme. The DRAs utilizing a multiple pump scheme can be developed from this method by considering the gain tilt caused by each individual pump and superimposing all the gains. It is believed that the gain ripple in the multi-pump DRAs when properly designed is considerably reduced.

Comparing the properties of the designed DCF to those reported in Refs. [89,90] the obtained dispersion and dispersion slope, hence RDS value, are comparable and even relatively higher.

TABLE 10.8

Summary of the Designed Fiber

Gain Tilt (dB)	Dispersion (ps/nm/km)	Dispersion Slope (ps/nm²/km)	RDS (nm⁻¹)	Attenuation (dB/km)	FOM	λ_{c11} (μm)	Effective Area (μm²)	Critical Curvature (mm)
1.382	−300.6	−3.237	0.0107	1.167	257.5	0.803	6.9617	11.77

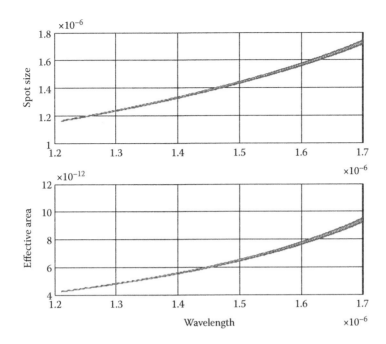

FIGURE 10.61
Effective area of profile of Table 10.7.

10.5.8 Remarks

In summary, a new and fast method for predicting and controlling the effective Raman gain of the segmented core fiber is described. The value of the effective Raman gain and especially its gain tilt caused by a single pump can be fast predicted. This design significantly

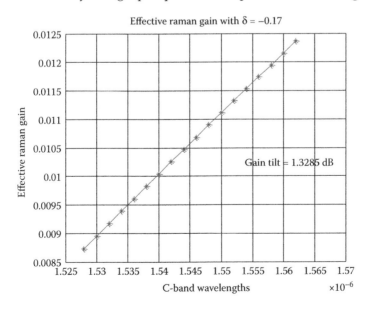

FIGURE 10.62
Effective Raman gain of profile of Table 10.7.

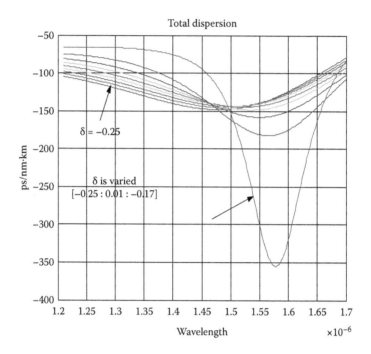

FIGURE 10.63
Total chromatic dispersion of profile of Table 10.7.

reduces the complexity of numerical simulation and enables a simple but accurate second-order approximation for the effective Raman gain. However, further research is needed to improve the currently developed technique so as to involve the effects of multi-pump sources on the effective Raman gain and predict its gain ripple which is created by the superimposition of those sources. The Raman amplification phenomena and its coefficients in dispersion compensating fibers have been investigated leading to an optimum design of multi-cladding optical fibers. The mode spot size of the fibers can be optimized for Raman amplification and still gives a high dispersion factor and slopes for dispersion compensating applications. Various types of doped materials can also be used to exploit its combination with the garmented index profile for the design.

10.6 Summary

In this chapter, a number of NL wave propagation equations that describe the evolution as well as the parametric interaction of multi-optical waves in NL waveguides are developed for photonic signal processing. A range of signal processing applications based on the parametric conversion process of third-order nonlinearity are demonstrated by simulation, such as HOS analysis for ultra-sensitive receivers, time division demultiplexing, amplification and wavelength conversion and phase conjugation. Models of NL waveguides are integrated into the MATLAB Simulink platform for processing performance simulations. FWM is the central NL physical process.

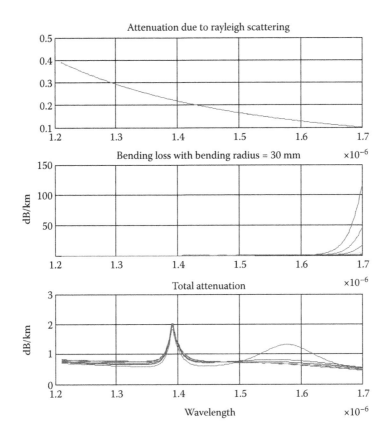

FIGURE 10.64
Total attenuation of fibers of profile of Table 10.7.

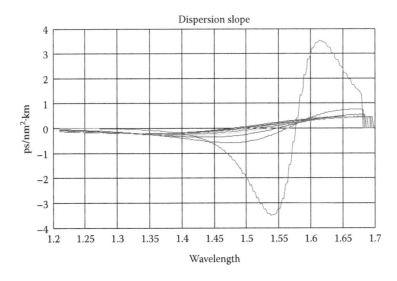

FIGURE 10.65
(See Color Insert) Dispersion slope of profile of Table 10.7.

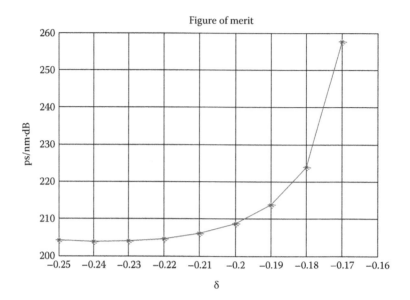

FIGURE 10.66
Figure of merit of profile of materials of the core given in Table 10.7.

Furthermore, NL effects in actively mode-locked fiber lasers are described. A laser model including a gain media, a BPF, optical fiber and an intensity modulator is introduced to investigate the performance of the laser under NL effects. Lasers in both working conditions, non-detuning and detuning, are studied. We find that the NL effects compress the pulse through the SPM effect. However, there is a trade-off between pulse shortening

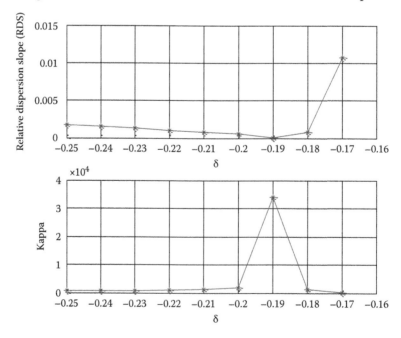

FIGURE 10.67
Relative dispersion slope and Kappa of fibers with profile no. 5 (see profiles given in Chapter 5).

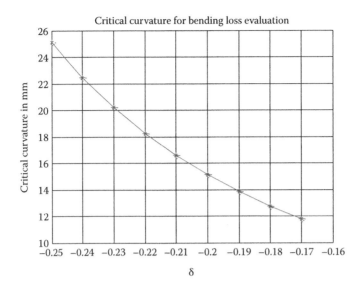

FIGURE 10.68
Radius of critical curvature of fiber with profile 5 (see profiles given in Chapter 5).

and a large detuning range. A laser with a shorter pulse has a narrower detuning range and vice versa.

Lastly, we present the Raman amplification properties of MOF or PCFs in order to explore the application of PCF as a Raman gain medium. We examine, by experiment, the Raman effects in 1.5 m highly NL PCF. Raman amplification occurs inside the quite short length fiber with 1-W pump power. The Raman gain enhancement in PCF is studied according to the effective Raman core area by changing the geometrical parameters of PCF. By numerical simulation, a minimal effective area can be acquired with proper air-hole diameter and arrangement. Further, PCFs with a Ge-doped region are analyzed to increase the Raman gain performance with the combination of Ge-doping and a small effective area. The results show that a promising Raman gain medium can be made with a doped solid core region and modified minimum effective area to perform a significant improvement compared to the conventional fibers.

Furthermore, we illustrate the design of a segmented core fiber to obtain the optimum Raman gain. This is described in the last section of this chapter.

References

1. G.P. Agrawal, *"Nonlinear Fiber Optics"*, 4th ed., Academic Press, Boston, 2007.
2. F. Luan, M.D. Pelusi, M.R.E. Lamont, D.-Y. Choi, S. Madden, B. Luther-Davies, and B.J. Eggleton, "Dispersion engineered As2S3 planar waveguides for broadband four-wave mixing based wavelength conversion of 40 Gb/s signals," *Opt. Express*, 17(5), 3514–3520, 2009.
3. M.E. Marhic, N. Kagi, T.-K. Chiang, and L.G. Kazovsky, "Broadband fiber optical parametric amplifiers," *Opt. Lett.*, 21, 573–575, 1996.
4. R. Stolen, and J. Bjorkholm, "Parametric amplification and frequency conversion in optical fibers," *IEEE J. Sel. Topics Quantum Electron.*, 18, 1062–1072, 1982.

5. J. Hansryd, P.A. Andrekson, M. Westlund, L. Jie, P. Hedekvist, "Fiber-based optical parametric amplifiers and their applications," *IEEE J. Sel. Topics Quantum Electron.*, 8(3), 506–520, 2002.

6. P.O. Hedekvist, M. Karlsson, and P.A. Andrekson, "Fiber four-wave mixing demultiplexing with inherent parametric amplification," *J. Lightw. Technol.*, 19, 977–981, 2001.

7. M.C. Ho, K. Uesaka, M. Marhic, Y. Akasaka, and L.G.Kazovsky, "200-nm-bandwidth fiber optical amplifier combining parametric and Raman gain," *IEEE J. Lightw. Technol.*, 19, 977–981, Jul 2001.

8. R.M. Jopson, A.H. Gnauck, and R.M. Derosier, "Compensation of fiber chromatic dispersion by spectral inversion," *Electron. Lett.*, 29(7), 576–578, Apr 1993.

9. S.L. Jansen, D. van den Borne, P.M. Krummrich, S. Spalter, G.-D. Khoe, and H. de Waardt, "Long-haul DWDM transmission systems employing optical phase conjugation," *IEEE J. Sel. Topics Quantum Electron.*, 12, 505–520, 2006.

10. S. Watanabe, S. Takeda, G. Ishikawa, H. Ooi, J.G. Nielsen, and C. Sonne, "Simultaneous wavelength conversion and optical phase conjugation of 200 Gb/s (5x40Gb/s) WDM signal using a highly nonlinear fiber four-wave mixer," in Proc. ECOC'97, 1997, 1–4.

11. R.M. Jopson, S. Radic, C.J. McKinstrie, A.H. Gnauck, and S. Chandrasekhar, "*Wavelength division multiplexed transmission over standard single mode fiber using polarization insensitive signal conjugation in highly nonlinear optical fiber,*" in Proc. OFC'03, Atlanta, 2003, Paper PD12.

12. T. Yamamoto and M. Nakazawa, "Active optical pulse compression with a gain of 29 dB by using four-wave mixing in an optical fiber," *IEEE Photon. Technol. Lett.*, 9, 1595–1597, Dec 1997.

13. J. Hansryd and P.A. Andrekson, "Wavelength tunable 40 GHz pulse source based on fiber optical parametric amplifier," *Electron. Lett.*, 37, 584–585, Apr 2001.

14. H.C.H. Mulvad, L.K. Oxenlwe, M. Galili, A.T. Clausen, L. Gruner-Nielsen, and P. Jeppesen, "1.28 Tbit/s single-polarization serial OOK optical data generation and demultiplexing," *Electron. Lett.*, 45(5), 280–281, Feb 2009.

15. M.D. Pelusi, V.G. Ta'eed, E. Libin Fu Magi, M.R.E. Lamont, S. Madden, D.A.P. Duk-Yong Choi Bulla, B. Luther-Davies, B.J. Eggleton, "Applications of highly-nonlinear chalcogenide glass devices tailored for high-speed all-optical signal processing," *IEEE J. Sel. Topics Quantum Electron.*, 14, 529–539, 2008.

16. T.T.D. Vo, H. Hu, M. Galili, E. Palushani, J. Xu, L.K. Oxenløwe, S.J. Madden, D.-Y. Choi, D.A.P. Bulla, M.D. Pelusi, J. Schröder, B. Luther-Davies, and B.J. Eggleton, "*Photonic chip based 1.28 Tbaud transmitter optimization and receiver OTDM demultiplexing,*" Proceedings of OFC 2010, San Diego, 2010, Paper PDPC5.

17. S. J. Madden, D-Y. Choi, D.A. Bulla, A.V. Rode, B. Luther-Davies, V.G. Ta'eed, M.D. Pelusi, and B.J. Eggleton, "Long, low loss etched As2S3 chalcogenide waveguides for all-optical signal regeneration", *Optics Express*, 15(22), 14414–14421, 2007.

18. L.N. Binh, "*Optical Fiber Communications Systems: Theory and Practice with Matlab and Simulink models,*" CRC Press, Boca Raton, 2010.

19. C.L. Nikias and J.M. Mendel, "Signal processing with higher-order spectra," *IEEE Sign. Proc. Mag.*, 10(3), 10–37, Jul 1993.

20. G. Sundaramoorthy, M.R. Raghuveer, and S.A. Dianat, "Bispectral reconstruction of signals in noise: Amplitude reconstruction issues," *IEEE Trans. Acoust. Speech Sign. Proc.*, 38(7), 1297–1306, Jul 1990.

21. T.M. Liu, Y.C. Huang, G.W. Chern, K.H. Lin, C.J. Lee, Y.C. Hung, and C.K. Sun, "Triple-optical autocorrelation for direct optical pulse-shape measurement," *App. Phys. Lett.*, 81(8), 1402–1404, 2002.

22. J.K. Yang and D.J. Kaup, "Stability and evolution of solitary waves in perturbed generalized nonlinear Schrodinger equations," *Siam J. Appl. Math.*, 60, 967–989, 2000.

23. J.C. Bronski, "Nonlinear scattering and analyticity properties of solitons," *J. Nonlinear Sci.*, 8, 161–182, 1998.

24. G.P. Agrawal, "*Applications of Nonlinear Fiber Optics,*" Academic Press, New York, 2001.

25. G.P. Agrawal, "*Fiber-optic communication systems,*" 3rd ed., Wiley-Interscience, New York, 2002.

26. A. Hasegawa and F. Tappert, "Transmission of stationary nonlinear optical pulses in dispersive dielectric fibers. I. Anomalous dispersion," *Appl. Phys. Lett.*, 23, 142–144, 1973.

27. L.F. Mollenauer, R. H. Stolen, J.P. Gordon, and W.J. Tomlinson, "Extreme picosecond pulse narrowing by means of Soliton effect in single-mode optical fibers," *Opt. Lett.*, 8, 289–291, 1983.

28. M.S. Ozyazici and M. Sayin, "Effect of loss and pulse width variation on soliton propagation," *J. Optoelectr. Adv. Mat.*, 5, 447–477, 2003.

29. C.S. Gardner, J.M. Greene, M.D. Kruskal, and R.M. Miura, "Method for solving the Korteweg-deVries equation," *Phys. Rev. Lett.*, 19, 1095–1097, 1967.

30. L.F. Mollenauer, R.H. Stolen, and J.P. Gordon, "Experimental observation of Picosecond pulse narrowing and Solitons in optical fibers," *Phys. Rev. Lett.*, 45, LP-1095–LP-1098, 1980.

31. R.A. Fisher and W. Bischel, "The role of linear dispersion in plane-wave self-phase modulation," *Appl. Phys. Lett.*, 23, 661–63, 1973.

32. H.A. Haus and Y. Silberberg, "Laser mode-locking with addition of nonlinear index," *IEEE J. Quantum Electron.*, 22, 325–31, 1986.

33. F.X. Kartner, D. Kopf, and U. Keller, "Solitary-pulse stabilization and shortening in actively mode-locked lasers," *J. Opt. Soc. Am. B: Opt. Phys.*, 12, 486–496, 1995.

34. F.X. Kartner, U. Morgner, S.H. Cho, J. Fini, J.G. Fujimoto, E.P. Ippen, V. Scheuer, M. Tilsch, and T. Tschudi, *"Advances in short pulse generation,"* Proc. Lasers and Electro-Optics Society Annual Meeting, 1998, LEOS '98, 1998.

35. D. Foursa, P. Emplit, R. Leners, and L. Meuleman, "18 GHz from a σ-cavity Er-fiber laser with dispersion management and rational harmonic active mode-locking," *Electron. Lett.*, 33, 486–488, 1997.

36. B. Bakhshi, P.A. Andrekson, and X. Zhang, "10GHz modelocked, dispersion-managed and polarization-maintaining erbium fibre ring laser with variable output coupling," *Electron. Lett.*, 34, 884–885, 1998.

37. B. Bakhski and P.A. Andrekson, "40 GHz actively modelocked polarization-maintaining erbium fiber ring laser", *Electron. Lett.*, 36, 411–413, 2000.

38. M. Horowitz, C.R. Menyuk, T.F. Carruthers, and I.N. Duling, III, "Theoretical and experimental study of harmonically modelocked fiber lasers for optical communication systems," *IEEE J. Lightwave Technol.*, 18, 1565–1574, 2000.

39. M. Nakazawa, E. Yoshida, and Y. Kimura, "Ultrastable harmonically and regeneratively mode-locked polarisation-maintaining erbium fibre ring laser," *Electron. Lett.*, 30, 1603–1605, 1994.

40. M. Nakazawa, M. Yoshida, and T. Hirooka, "Ultra-stable regeneratively mode-locked laser as an opto-electronic microwave oscillator and its application to optical metrology," *IEICE Trans. Electron.*, E90C, 443–449, 2007.

41. M. Nakazawa, *"Ultrafast optical pulses and solitons for advanced communications,"* Lasers and Electro-Optics, 2003, CLEO/Pacific Rim 2003, Proc. 5th Pacific Rim Conference on, Taipei, Taiwan, 2003.

42. K. Tamura and M. Nakazawa, "Dispersion-tuned harmonically mode-locked fiber ring laser for self-synchronization to an external clock," *Optics Lett.*, 21, 1984–1986, 1996.

43. O. Pottiez, O. Deparis, R. Kiyan, M. Haelterman, P. Emplit, P. Megret, and M. Blondel, "Supermode noise of harmonically mode-locked erbium fiber lasers with composite cavity," *IEEE J. Quantum Electron.*, 38, 252–259, 2002.

44. M. Nakazawa, K. Tamura, and E. Yoshida, "Supermode noise suppression in a harmonically mode-locked fibre laser by selfphase modulation and spectral filtering," *Electron. Lett.*, 32, 461–463, 1996.

45. Y.H. Li, C.Y. Lou, J. Wu, B.Y. Wu, and Y.Z. Gao, "Novel method to simultaneously compress pulses and suppress supermode noise in actively mode-locked fiber ring laser," *IEEE Photon. Technol. Lett.*, 10, 1250–1252, 1998.

46. M. Nakazawa and E. Yoshida, "A 40-GHz 850-fs regeneratively FM mode-locked polarization-maintaining erbium fiber ring laser," *IEEE Photon. Technol. Lett.*, 12, 1613–1615, 2000.

47. T. Pfeiffer and G. Veith, "40 GHz pulse generation using a widely tunable all-polarization preserving erbium fibre ring laser," *Electron. Lett.*, 29, 1849–1850, 1993.

48. H. Takara, S. Kawanishi, M. Saruwatari, and K. Noguchi, "Generation of highly stable 20 GHz transform-limited optical pulses from actively mode-locked Er^{+3}:doped fiber lasers with an all-polarization maintaining ring cavity," *Electron. Lett.*, 28, 2095–2096, 1992.

49. C.T. Seaton, Xu Mai, G.I. Stegeman, and H.G. Winful, "Nonlinear guided wave applications," *Opt. Eng.*, 24, 593–599, 1985.
50. G.I. Stegeman, E.M. Wright, N. Finlayson, R. Zanoni, and C.T. Seaton, "Third order NL integrated optics," *IEEE J. Lightwave Technol.*, 6, 953–970, 1988.
51. G.I. Stegeman and R. H. Stolen, "Waveguides and fibers for nonlinear optics," *J. Opt. Soc. Am. B*, 6, 652–662, 1989.
52. K. Hayata and M. Koshiba, "Full vectorial analysis of nonlinear optical waveguides," *J. Opt. Soc. Am. B*, 5, 2494–2501, 1988.
53. D.R. Brillinger, "Introduction to Polyspectra," *Ann. Math. Stat.*, 36, 1351–1374, 1965.
54. H. Bartelt, A.W. Lohmann, and B. Wirnitzer, "Phase and amplitude recovery from bispectra," *Appl. Opt.*, 23, 3121–3129, 1984.
55. B.M. Sadler, G.B. Giannakis, and D.J. Smith, "Acousto-optic estimation of correlations and spectra using triple correlations and bispectra," *Opt. Eng.*, 31(10), 2139–2147, 1992.
56. J.M. Mendel, "Tutorial on high-order statistic (spectra) in signal processing and system theory: Theoretical results and some applications," *Proc. IEEE*, 79(3), 278–305, 1991.
57. Y.R. Shen, "*Principles of Nonlinear Optics*," John Wiley, New York, 1984.
58. M. Schubert and B. Wilhemi, "Nonlinear Optics and Quantum Electronics," Wiley, New York, 1986.
59. R.H. Stolen, J.E. Bjorkholm, and A. Ashkin, "Phase-matched three-wave mixing in silica fiber optical waveguides" *Appl. Phys. Lett.*, 24, 308–310, 1974.
60. F. Matera, M. Settembre, M. Tamburini, M. Zitelli and S.K. Turitsyn, "Reduction of the four wave mixing in optically amplified links by reducing pulse overlapping," *Opt. Commun.*, 181, 407–411, 2000.
61. J.G. Proakis, "*Digital Communications*," John Wiley, 2002. Also http://en.wikipedia.org/wiki/Phase-shift_keying#Binary_phase_shift_keying_.28BPSK.29. Access date: June 2010.
62. M. Alastair Glass, D.J. DiGiovanni, T.A. Strasser, A.J. Stentz, R.E. Slusher, A.E. White, A. Refik Kortan, B.J. Eggleton, "Advances in fiber optics", *Bell. Labs. Tech. J.*, 168–187, Jan–Mar 2000.
63. T.A. Birks, J.C. Knight, and P.St.J. Russell, "Endlessly single-mode photonic crystal fiber," *Opt. Lett.*, 961–963, Jul 1997.
64. R.F. Cregan, B.J. Mangan, J.C. Knight, T.A. Birks, P.St.J. Russell, P.J. Roberts, and D.C. Allan, "Single-mode photonic bandgap guidance of light in air," *Science*, 285(5433), 1537–1539, Sept 1999.
65. J.C. Knight, J. Arriaga, T.A. Birks, A.O. Blanch, W.J. Wadsworth, and P.St.J. Russell, "Anomalous dispersion in photonic crystal fiber," *IEEE Photon. Technol. Lett.*, 12, 807–809, 2000.
66. G.P. Agrawal, "*Nonlinear Fiber Optics*," 3rd ed., Academic, San Diego, CA, 2001.
67. T. Uamamoto, H. Kubota, and S. Kawanishi, "Supercontinuum generation at 1.55 μm in a dispersion-flattened polarization-maintaining photonic crystal fiber," *Opt. Expr.*, 11, 1537–1540, 30 June 2003.
68. J.H. Lee, W. Belargi, K. Furusawa, P. Petropoulos, Z. Yusoff, T. M. Monro, and D.J. Richardson, "Four-wave mixing based 10-Gb/s tunable wavelength conversion using a holey fiber with a high SBS threshold," *IEEE Photon. Technol. Lett.*, 15, 440–442, 2003.
69. P. Petropoulos, H.E. Heidepriem, V. Finazzi, R.C. Moore, K. Frampton, D.J. Richardson, and T.M. Monro, "Highly nonlinear and anomalously dispersive lead silicate glass holey fibers," *Opt. Expr.*, 11, 3568–3573, 29 Dec 2003.
70. S. Namiki and Y. Emori, "Ultrabroad-band Raman amplifiers pumped and gain equalized by wavelength-division-multiplexed high-power laser diodes," *IEEE Sel. Topics Quant. Electron.*, 7(1), 3–16, 2001.
71. M.N. Islam, "Raman amplifiers for telecommunications," *IEEE J. Sel. Topics Quant. Electron.*, 8(3), 2002.
72. A.K. Srivastava, Y. Sun, "*Advances in Erbium-Doped Fiber Amplifiers*," Chapter 4, Optical Fiber Telecommunications, IVA, (eds.) I.P. Kaminow and Tingye Li, Academic Press, 2002.
73. Z. Yusoff, J.H. Lee, W. Belardi, T.M. Monro, P.C. The, and D.J. Richardson, "Raman effects in a highly nonlinear holey fiber: Amplification and modulation," *Opt. Lett.*, 27, 424–426, 2002.

74. J. Broeng, D. Mogilevstev, S.E. Barkou, and A. Bjarklev, "Photonic crystal fibers: A new class of optical waveguides," *Opt. Fiber Technol.*, 5, 305–330, 1999.

75. M. FuoChi, F. Poli, S. Selleri, A, Cucinotta, and L. Vincetti, "Study of Raman amplification properties in triangular photonic crystal fibers," *J. Lightwave. Technol.*, 21, 2247–2254, Oct 2003.

76. J. Bromage, K. Rottwitt, and M.E. Lines, "A method to predict the Raman gain spectra of germanosilicate fibers with arbitrary index profile," *IEEE Photon. Technol. Lett.*, 14, 24–26, Jan 2002.

77. H.D. Rudolph and E.G. Neumann, "Approximations for the eigenvalue of the fundamental mode of a step index glass fiber waveguide," *Nachrichten Techn. Z.*, 29, 328–329, 1976.

78. C. Pask, "Physical interpretation of Petermann's strange spot size for single-mode fibers," *Elect. Lett.*, 20(3), 144–145, 1984.

79. W.P.L. Urquhart and P.J. Laybourn, "Effective core area for stimulated Raman scattering in single-mode optical fibers," *IEE Proc.*, 132(4), 201–204, Aug 1985.

80. K. Rottwitt and J. Bromage, "Scaling for the Raman gain coefficient: Applications to germano-silicate fibers," *J. Lightwave Tech*, 21, 1652–1662, Jul 2003.

81. P.S. Andre, R. Correia, L.M. Borghesi Jr., A.L.J. Teixeira, R.N. Nogueira, M.J.N. Lima, H.J. Kalinowski, F. Da Rocha, J.L. Pinto, "Raman gain characterization in standard single mode optical fibers for optical simulation purposes," *Optica Applicata*, 33(4), 560–573, 2003.

82. K. Yuhong, *"Calculations and measurements of Raman Gain Coefficients of Different Fiber Types,"* in Department of Electrical Engineering: Virginia Polytechnic Institute and State University, Dec 2002.

83. S.T. Davey, D.L. Williams, B.J. Ainslie, W.J.M. Rothwell, and B. Wakefield, "Optical gain spectrum of GeO_2-SiO_2 Raman fiber amplifiers," *IEE Proc.*, 136, 301–304, Dec 1989.

84. F.L. Galeener, J.C. Mikkelsen, R.H. Geils, and W.J. Mosby, "The relative Raman cross sections of vitreous SiO_2, GeO_2,B_2O_3 and P_2O_5," *Appl. Phys. Lett.*, 32, 34, 1978.

85. B.J. Ainslie and C.R. Day, "A review of single-mode fibers with modified dispersion characteristics," *J. Lightwave Tech.*, LT-4, 967–979, Aug 1986.

86. N. Shibata, M. Horigudhi, and T. Edhario, "Raman spectra of binary high-silica glasses and fibers containing GeO_2, P_2O_5 and B_2O_3," *J. Non-Crystalline Solids*, 45, 115–125, 1981.

87. K. Petermann and R. Kuhne, "Upper and lower limits for the microbending loss in arbitrary single-mode fibers," *J. Lightwave Tech.*, LT-4, 2–7, 1986.

88. S.V. Chung, *"Simplified Analysis and Design of Single Mode Optical Fibers,"* Monash University, Australia, 1987.

89. (a) L. Gruner-Nielsen, Y. Qian, B. Palsdottir, P.B. Gaarde, S. Dyrbol, and T. Veng, "Module for simultaneous C+L-band dispersion compensation and Raman amplification," *Opt. Soc. Am.*, OCIS codes: (060.2360); (060.2320), 2002. (b) M. Wandel, P. Kristensen, T. Veng, Y. Qian, Q. Le, and L. Gruner-Nielsen, "Dispersion compensating fibers for non-zero dispersion fibers," *Opt. Soc. Am.*, 327–329, 2001.

90. K. Aikawa, R. Suzuki, S. Shimizu, K. Suzuki, M. Kenmotsu, M. Nakayama, K. Kaneda, and K. Himeno, "High performance dispersion and dispersion slope compensating fiber modules for non-zero dispersion shifted fibers," *Fujikura Tech. Rev.*, 32, 5–10, 2003.

Appendix 1   Coordinate System Transformations

A1.1 Right Handed Coordinate System

A vector should be represented in a right handed coordinate system as shown in Figure A1.1 in which the labeling of the axes follows a right hand rule, or if we rotate the x axis to the y axis in a clockwise direction of the right hand, then the screwing direction must be the z direction.

A1.1.1 Sum and Difference of Two Vectors

Given that the vector $\underset{\sim}{A}$ is specified by

$$\underset{\sim}{A} = \begin{pmatrix} A_x & A_y & A_z \end{pmatrix} = A_x \underset{\sim}{a}_x + A_y \underset{\sim}{a}_y + A_z \underset{\sim}{a}_z = \underset{\sim}{i} A_x + \underset{\sim}{j} A_y + \underset{\sim}{k} A_z$$

(A1.1)

$\underset{\sim}{a}_x, \underset{\sim}{a}_y, \underset{\sim}{a}_z$ or $\underset{\sim}{i}, \underset{\sim}{j}, \underset{\sim}{k} \triangleq$ unit vector in x, y, z direction

Then for vector $\underset{\sim}{B}$ the summation of these two vectors $\underset{\sim}{A}, \underset{\sim}{B}$ is given as

$$\underset{\sim}{A} + \underset{\sim}{B} = \begin{pmatrix} A_x & A_y & A_z \end{pmatrix} + \begin{pmatrix} B_x & B_y & B_z \end{pmatrix} = \begin{pmatrix} A_x + B_x & A_y + B_y & A_z + B_z \end{pmatrix}$$

(A1.2)

A1.1.2 Multiplication of Scalar and a Vector and Dot Products

A vector can be scaled by the multiplication of a constant.

A scalar product of two vectors $\underset{\sim}{A}, \underset{\sim}{B}$ is expressed by the dot product of the two vectors to give a scalar quantity given as

$$\underset{\sim}{A} \cdot \underset{\sim}{B} = \begin{pmatrix} A_x & A_y & A_z \end{pmatrix} \begin{pmatrix} B_x \\ B_y \\ B_z \end{pmatrix} = A_x B_x + A_y B_y + A_z B_z = |\underset{\sim}{A}||\underset{\sim}{B}| \cos \theta = \underset{\sim}{B} \times \underset{\sim}{A}$$

(A1.3)

where θ is the angle between the two vectors.

The cross product of the two vectors is a vector presented by

$$\underset{\sim}{A} \times \underset{\sim}{B} = \begin{vmatrix} i & j & k \\ A_x & A_y & A_z \\ B_x & B_y & B_z \end{vmatrix} = \begin{pmatrix} A_y B_z - A_z B_y \\ A_z B_x - A_x B_z \\ A_x B_y - A_y B_x \end{pmatrix} \rightarrow \underset{\sim}{A} \times \underset{\sim}{B} = -\underset{\sim}{B} \times \underset{\sim}{A}$$

(A1.4)

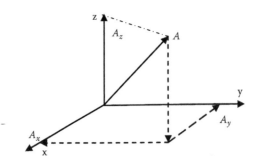

FIGURE A1.1
Right-handed coordinate system—that is if rotating x to y in a clockwise direction then the screwing direction must be z using the right hand.

A1.2 Curl, Divergence, and Gradient Operations

If V is a scalar function then the gradient $\underset{\sim}{V}$ is given by

$$\nabla \underset{\sim}{V} = \begin{pmatrix} \dfrac{\partial V}{\partial x} \\[1mm] \dfrac{\partial V}{\partial y} \\[1mm] \dfrac{\partial V}{\partial z} \end{pmatrix} = \underset{\sim}{i}\,\frac{\partial V}{\partial x} + \underset{\sim}{j}\,\frac{\partial V}{\partial y} + \underset{\sim}{k}\,\frac{\partial V}{\partial z} \tag{A1.5}$$

If $\underset{\sim}{V}(x, y, z)$ is a vector function then the divergence $\nabla \cdot \underset{\sim}{A}$ is given as

$$\nabla \cdot \underset{\sim}{V} = \frac{\partial V_x}{\partial x} + \frac{\partial V_y}{\partial y} + \frac{\partial V_z}{\partial z} \equiv \operatorname{div} \underset{\sim}{V} \tag{A1.6}$$

If $\underset{\sim}{V}(x, y, z)$ is a vector function then the curl of $\underset{\sim}{V}$, $\nabla \times \underset{\sim}{V}$ is given as

$$\nabla \times \underset{\sim}{V} = \begin{vmatrix} \dfrac{\partial}{\partial x} & \dfrac{\partial}{\partial y} & \dfrac{\partial}{\partial z} \\[2mm] V_x & V_y & V_z \\[2mm] \underset{\sim}{i} & \underset{\sim}{j} & \underset{\sim}{k} \end{vmatrix} = \begin{pmatrix} \dfrac{\partial V_z}{\partial y} - \dfrac{\partial V_y}{\partial z} \\[2mm] -\dfrac{\partial V_z}{\partial x} + \dfrac{\partial V_x}{\partial z} \\[2mm] \dfrac{\partial V_y}{\partial x} - \dfrac{\partial V_x}{\partial y} \end{pmatrix} \tag{A1.7}$$

This is the curl operation. Note on the differential variation in one of the planes of the system leads to differential variation in the direction orthogonal to the plane.

A1.2.1 Identity

Given that $\underset{\sim}{A}, V$ are the vector and scalar functions, respectively, we have the following identities:

$$\operatorname{div}\left(\operatorname{curl} \underset{\sim}{A}\right) = \nabla\left(\nabla \times \underset{\sim}{A}\right) = 0$$

$$\operatorname{curl}\left(\operatorname{grad} \underset{\sim}{A}\right) = \left(\nabla \times \nabla \underset{\sim}{V}\right) = 0 \tag{A1.8}$$

$$\operatorname{div}\left(\operatorname{grad} \underset{\sim}{A}\right) = \left(\nabla \cdot \nabla \underset{\sim}{V}\right) = \nabla^2 \underset{\sim}{V} = \frac{\partial^2 V_x}{\partial x^2} + \frac{\partial^2 V_y}{\partial y^2} + \frac{\partial^2 V_z}{\partial z^2} \triangleq \text{Laplacian}$$

A1.2.2 Physical Interpretation of Gradient, Divergence and Curl

Gradient: The gradient of any scalar function is the maximum spatial rate of change of that function. If the scalar function represents the temperature the grad(V) is the temperature gradient or the rate of change of the temperature with distance. It is evident that although the temperature is a scalar function the grad(V) is a vector quantity with its direction being that in which the temperature changes most quickly.

Divergence: As a mathematical tool, vector analysis finds great usefulness in simplifying the expressions of the relations that exists in a three dimensional field. A consideration of fluid flow motion gives a direct interpretation of divergence and curl. First consider an incompressible fluid, e.g., water, then the rectangular parallel pipe shown in Figure A1.2.
   If the fluid is flowing through this small volume then, due to the non-compression of the fluid, it is expected that the same amount of the volume of the fluid at the output of the small volume. Thus there is no divergence of the fluid after flowing through the volume. Hence the divergence of the fluid is zero. However, if the fluid can be compressed then the rate of flow of the fluid through this infinitesimal volume would change and thus the rate of flow would be different at the output of the volume as compared with that at the input. Thence the divergence of the flow is finite.

Curl: The concept of curl or rotation of a vector quantity can be illustrated in a stream of flow problems, for example, a leaf flowing on the surface of a water as shown in Figure A1.3. If the leaf rotates about an axis parallel to the z-axis then the velocity of flow of the leaf would be different, and the curl of the velocity $\underset{\sim}{v}$ denoted as $\nabla \times \underset{\sim}{v}$ is given as

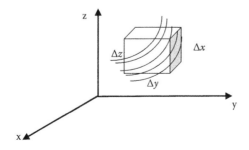

FIGURE A1.2
An infinitesimal volume rectangular parallel piped within a fluid medium.

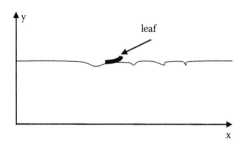

FIGURE A1.3
Flow of a leaf on the surface of water.

$$\nabla \times \underset{\sim}{v} = \begin{vmatrix} \dfrac{\partial}{\partial x} & \dfrac{\partial}{\partial y} & \dfrac{\partial}{\partial z} \\ v_x & v_y & v_z \\ \underset{\sim}{i} & \underset{\sim}{j} & \underset{\sim}{k} \end{vmatrix} = \begin{pmatrix} \dfrac{\partial v_z}{\partial y} - \dfrac{\partial v_y}{\partial z} \\ -\dfrac{\partial v_z}{\partial x} + \dfrac{\partial v_x}{\partial z} \\ \dfrac{\partial v_y}{\partial x} - \dfrac{\partial v_x}{\partial y} \end{pmatrix} \tag{A1.9}$$

Thus the rotation velocity of the leaf about the z axis is given as the change of the velocity of the leaf flow in the x minus that of the y direction.

A1.3 Vector Relation in Other Coordinate Systems

A1.3.1 Cylindrical Coordinates

Referring to the labels of the cylindrical coordinates assigned in Figure A1.4, we have the following expressions for the gradient, divergence and curl operators:

$$\nabla \underset{\sim}{V} = \begin{pmatrix} \dfrac{\partial V}{\partial \rho} \\ \dfrac{1}{\rho}\dfrac{\partial V}{\partial \phi} \\ \dfrac{\partial V}{\partial z} \end{pmatrix} \tag{A1.10}$$

If $\underset{\sim}{A}(x,y,z)$ is a vector function then the divergence $\nabla \cdot \underset{\sim}{A}$ is given as

$$\nabla \cdot \underset{\sim}{A} = \frac{1}{\rho}\frac{\partial(\rho A_\rho)}{\partial \rho} + \frac{1}{\rho}\frac{\partial A_\phi}{\partial y} + \frac{\partial A_z}{\partial z} \equiv \mathrm{div}\,\underset{\sim}{A} \tag{A1.11}$$

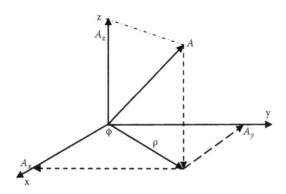

FIGURE A1.4
A cylindrical coordinate system.

If $V(x, y, z)$ is a vector function then the curl(V), with a general curl(A) is given as

$$\nabla \times A = \begin{pmatrix} \left(\dfrac{1}{\rho} \dfrac{\partial (A_z)}{\partial \phi} - \dfrac{\partial A_\phi}{\partial z} \right) a_\rho \\[3mm] \left(-\dfrac{\partial A_z}{\partial \rho} + \dfrac{\partial A_\rho}{\partial z} \right) a_\phi \\[3mm] \dfrac{1}{\rho} \left(\dfrac{\partial (\rho A_\phi)}{\partial \rho} - \dfrac{\partial A_\rho}{\partial \phi} \right) a_z \end{pmatrix} \tag{A1.12}$$

Then the Laplacian in the cylindrical coordinates is given by

$$\nabla^2 V = \frac{1}{\rho} \frac{\partial}{\partial \rho} \left(\rho \frac{\partial V}{\partial \rho} \right) + \frac{1}{\rho^2} \frac{\partial^2 V}{\partial \phi^2} + \frac{\partial^2 V}{\partial z^2} \triangleq \text{Laplacian} \tag{A1.13}$$

A1.3.2 Spherical Coordinates

$$\nabla V = a_r \frac{\partial V}{\partial z} + a_\theta \frac{1}{r} \frac{\partial V}{\partial \theta} + a_\phi \frac{1}{r \sin \theta} \frac{\partial V}{\partial \phi} \tag{A1.14}$$

If $A(x, y, z)$ is a vector function then the divergence $\nabla \cdot A$ is given as

$$\nabla \cdot A = \frac{1}{r^2} \frac{\partial (r^2 A_r)}{\partial r} + \frac{1}{r \sin \theta} \frac{\partial (A_\theta \sin \theta)}{\partial \theta} + \frac{1}{r \sin \theta} \frac{\partial A_\phi}{\partial \phi} \tag{A1.15}$$

If $V(x, y, z)$ is a vector function then the curl(V), with a general curl(A) is given as

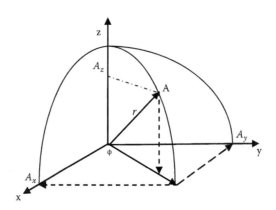

FIGURE A1.5
A spherical coordinate system.

$$\nabla \times \underset{\sim}{A} = \begin{pmatrix} \text{curl}_r \; \underset{\sim}{A} \\ \text{curl}_\theta \; \underset{\sim}{A} \\ \text{curl}_\phi \; \underset{\sim}{A} \end{pmatrix} = \begin{pmatrix} \dfrac{1}{r\sin\theta}\left[\dfrac{\partial(A_\phi \sin\theta)}{\partial\theta} - \dfrac{\partial A_\theta}{\partial\phi}\right] \\[3mm] \left[\dfrac{1}{r\sin\theta}\dfrac{\partial(A_r)}{\partial\phi} - \dfrac{1}{r}\dfrac{\partial(rA_\phi)}{\partial r}\right] \\[3mm] \left[\dfrac{1}{r}\dfrac{\partial(A_\theta)}{\partial r} - \dfrac{\partial A_r}{\partial\theta}\right] \end{pmatrix} \tag{A1.16}$$

The Laplacian of a scalar function V in the spherical coordinate system is given as (see Figure A1.5)

$$\nabla^2 \underset{\sim}{V} = \frac{1}{r^2}\frac{\partial}{\partial r}\left(r^2\frac{\partial V}{\partial r}\right) + \frac{1}{r^2\sin\theta}\frac{\partial}{\partial\theta}\left(\sin\theta\frac{\partial V}{\partial\theta}\right) + \frac{1}{r^2\sin\theta}\frac{\partial^2 V}{\partial\phi^2} \triangleq \text{Laplacian} \tag{A1.17}$$

Appendix 2   Models for Couplers in FORTRAN

The following are four computer programs (FORTRAN 77) developed, along with their variations, in this work. For clarity and brevity, they are presented in the form of four main modules (including the data blocks), followed by four sets of subroutines. The special mathematical functions can be calculated using the method of Press et al. [1].

Main Program (1)

```
    PROGRAM a₀_bₙ.for

C    ************************************************************    C
C    Calculation of a₀(z) and bₙ(z), the expansion coefficients      C
C    for the total field of straight or bended (s=s(z),              C
C    a=a(z),b=b(z)) fibre-slab couplers. R-the fibre bend            C
C    radius, VXR-the upper limit of the slab-mode No.                C
C    ************************************************************    C

     INTEGER M,VX,VXR,D,ZN,VM,N,N1
     PARAMETER(N=202)
     REAL R,ZI,ZF,NC
     REAL WL,DELTA,NA,NO,NS,A,DIA_CL,T,S,K,NF,VF,KF,BFO,UF,N_F
     REAL VSC,AS,KS,VO,VC,DEFF,N_S,P(0:200),BSV(0:200),QF001
     REAL KFOV(0:200),QSUV(0:200,0:200),KSU0(0:200),QF00
     COMMON /BLK1/VX,VXR,D,R,ZI,ZF,ZN,/BLK2/A,/BLK3/NO,NS,T,S
     COMMON /BLK4/K,NC,/BLK5/NF,VF,KF,BFO,UF,N_F
     COMMON /BLK6/AS,VSC,KS,VO,VC,N_S,P,BSV
     COMMON /BLK7/QF00,KFOV,QSUV,KSU0,/BLK8/QF001
     COMPLEX*16 AB(N),DYDX(N),DYT(N),DYM(N)
     COMPLEX*16 YTEMP(N),YSAV(N),DYSAV(N)
     CALL INPUT(M,WL,DELTA,NA,DIA_CL,FN_DATA)
     CALL PAR_F(WL,DELTA,NA)
     CALL PAR_S(M,N1,WL,DEFF,VM)
     CALL A0B(N,N1,AB,DYDX,DYT,DYM,YTEMP,YSAV,DYSAV)
     STOP
     END
     BLOCK DATA
     INTEGER VX,VXR,D,ZN
     REAL R,ZI,ZF,K,NC,R_CM
     COMMON /BLK1/VX,VXR,D,R,ZI,ZF,ZN,/BLK4/K,NC
     DATA VX,D,ZN,R,ZI,ZF,NC/200,500,500,2000E30,0,6000,1.46E0/
     END
```

Main Program (2)

```
    PROGRAM Field.for

C    ************************************************************    C
C    Calculation of the compound-mode equations (e.g. 3.61)         C
C    which allows output of both the eigenvalues (i.e. the          C
C    compound-mode propagation constants) and the eigen-            C
C    vectors (i.e. 2-D or 3-D compound-mode field profiles          C
C    such as E(x), E(y) and E(x,y).                                 C
C    ************************************************************    C
```

```
      INTEGER M,VX,VXR,NP,D,IMAX,VM
      PARAMETER(VX=200,NP=202)
      REAL WL,DELTA,NA,NO,NS,A,DIA_CL,T,S,K,NF,NC,VF,KF,BF0,UF,N_F
      REAL VS,AS,DPH,BS,KS,VO,VC,DEFF,N_S,P(0:VX),BSV(0:VX),QF00
      REAL XI1(0:VX),KF0V(0:VX),QSUV(0:VX,0:VX),KSU0(0:VX),KSVO
      CHARACTER*20 FN_DATA
      REAL Q(NP,NP),DG(NP),E(NP),QI(NP)
      DATA NC/1.46/
      CALL INPUT(M,WL,DELTA,NA,NO,NS,A,DIA_CL,T,S,D,FN_DATA)
      CALL PAR_F(WL,DELTA,NA,A,K,NF,NC,VF,KF,BF0,UF,N_F)
      CALL PAR_S(M,VX,VXR,D,T,NS,NC,WL,VS,NO,AS,K,DPH,BS,KS,VO,
     *,VC,DEFF,N_S,P,BSV,VM,NP,BF0)
      CALL QF00_C(NP,A,T,S,VF,KF,BF0,UF,VS,AS,QF00,Q)
      CALL KF0V_C(VX,VXR,NP,A,T,S,VF,KF,BF0,UF,KS,VC,P,N_S,VS,
     *,XI1,KF0V,Q)
      CALL QSUV_C(VX,VXR,NP,A,T,S,VF,VS,KS,VC,N_S,P,BSV,QSUV,Q)
      CALL KSU0_C(VX,VXR,NP,A,T,S,K,NO,NS,NC,VF,KF,UF,KS,VO,VC,
     *,BSV,N_S,P,AS,KSU0,Q)
      CALL TRED2(Q,VXR+2,NP,DG,E)
      CALL TQLI(Q,IMAX,VXR+2,NP,DG,E)
      CALL FD(Q,IMAX,D,VX,VXR,NP,A,T,S,KF,BF0,UF,N_F,KS,VO,VC,
     *,N_S,P,BSV,QF00,KF0V,DG,QI)
      STOP
      END
```

Main Program (3)

PROGRAM loss.for

```
c    ********************************************************    c
c    Calculation of the loss (attenuation) coefficient of the    c
c    fibre mode due to the radiation loss to the 'unbound'    c
c    slab modes. It can be modified to include a full    c
c    expression of the loss coefficient for two additional    c
c    cases; The four cases now become:    c
c    (1) Alfa1: using the simple formula and coupling coefficients;    c
c    (2) Alfa2: using the simple formula but full coupling coeff.;    c
c    (3) Alfa3: using the full formula but simple coupling coeff.;    c
c    (4) Alfa4: using the full formula and coupling coeeficents.'    c
c    ********************************************************    c
      INTEGER M,VX,VXR,D,VM
      REAL NC
      PARAMETER(VX=160,NC=1.46)
      REAL WL,DELTA,NA,NO,NS,A,DIA_CL,T,S,K,NF,VF,KF,BF0,UF,N_F
      REAL VS,AS,DPH,BS,KS,VO,VC
      REAL DEFF,N_S,P(0:VX),BSV(0:VX),QF00
      REAL XI1(0:VX),KF0V(0:VX),QSUV(0:VX,0:VX)
      REAL XE(0:VX),KSV0(0:VX),KSVO
      REAL XI2(0:VX,0:VX),XI3(0:VX,0:VX)
      REAL BF0V(0:VX),BSV0(0:VX)
      CHARACTER*20 FN_DATA
      CALL INPUT(M,WL,DELTA,NA,NO,NS,A,DIA_CL,T,S,D,FN_DATA)
      CALL PAR_F(WL,DELTA,NA,A,K,NF,NC,VF,KF,BF0,UF,N_F)
      CALL PAR_S(M,VX,VXR,D,T,NS,NC,WL,VS,NO,AS,K,DPH,BS,KS,VO,
     *,VC,DEFF,N_S,P,BSV,VM,BF0)
```

```
      CALL QF00_C(A,T,S,VF,KF,BF0,UF,VS,AS,QF00)
      CALL KF0V_C(VX,VXR,A,T,S,VF,KF,BF0,UF,KS,VC,P,N_S,VS,XI1,
     *,KF0V)
      CALL QSUV_C(VX,VXR,A,T,S,VF,VS,KS,VC,N_S,P,BSV,XI2,XI3,QSUV)
      CALL KSV0_C(VX,VXR,A,T,S,K,NO,NS,NC,VF,KF,UF,KS,VO,VC,BSV,
     *,N_S,P,XE,KSV0)
      CALL COEF(VX,VXR,VM,D,BF0,BSV,QF00,KF0V,QSUV,KSV0,BF0V,BSV0)
      STOP
      END
```

Main Program (4)

PROGRAM Coe.for

```
C     ************************************************************     C
C     Calculation of the coupling coefficients to see the             C
C     dependence of the coupling coefficients on the order            C
C     of the slab modes, etc.                                         C
C     ************************************************************     C
      INTEGER M,VX,VXR,D,ZN,VM,N,N1
      PARAMETER(N=202)
      REAL WL,DELTA,NA,NO,NS,A,DIA_CL,T,S,K,NF,VF,KF,BF0,UF,N_F,NC
      REAL VSC,AS,KS,VO,VC,DEFF,N_S,P(0:200),BSV(0:200),QF001
      REAL KF0V(0:200),QSUV(0:200,0:200),KSU0(0:200),QF00
      COMMON /BLK1/VX,VXR,D,/BLK2/A,/BLK3/NO,NS,T,S
      COMMON /BLK4/K,NC,/BLK5/NF,VF,KF,BF0,UF,N_F
      COMMON /BLK6/AS,VSC,KS,VO,VC,N_S,P,BSV
      COMMON / BLK7/QF00,KF0V,QSUV,KSU0,/BLK8/QF001
      CALL INPUT(M,WL,DELTA,NA,DIA_CL,FN_DATA)
      CALL PAR_F(WL,DELTA,NA)
      CALL PAR_S(M,N1,WL,DEFF,VM)
      STOP
      END
   BLOCK DATA
      INTEGER VX,VXR,D
      REAL K,NC
      COMMON /BLK1/VX,VXR,D,/BLK4/K,NC
      DATA VX,D,NC/200,500,1.46E0/
      END


   SUBROUTINE SET (1)
      SUBROUTINE INPUT(M,WL,DELTA,NA,DIA_CL,FN_DATA)
C  Input of waveguide and coupler parameters etc via data             C
C  files or manual entries.                                           C
      INTEGER VX,VXR,D,M,ZN,NUM
      REAL WL,DELTA,NA,NO,NS,A,DIA_CL,T,S,R,ZI,ZF
      COMMON /BLK1/VX,VXR,D,R,ZI,ZF,ZN,/BLK2/A,/BLK3/NO,NS,T,S
      CHARACTER*20 EDF,FN_DATA
      WRITE(*,*)'Enter data by a file ? (y/n):'
      READ(5,10)EDF
10    FORMAT(A10)
      IF(EDF.EQ.'y'.OR.EDF.EQ.'Y')THEN
        WRITE(*,*)'Enter name of the file(fn):'
        READ(5,20)FN_DATA
20      FORMAT(A20)
```

```
      OPEN(1,FILE=FN_DATA,STATUS='OLD')
        READ(1,*)WL,DELTA,NA,NO,NS,A,DIA_CL,T,S
        NO=1.45
      CLOSE(1)
    ELSE
      WRITE(*,*)'Enter wavelength in micrometers(wl):'
      READ(5,*)WL
      WRITE(*,*)'Enter equivalent step index difference(delta):'
      READ(5,*)DELTA
      WRITE(*,*)'Enter numeric aperture(NA):'
      READ(5,*)NA
      WRITE(*,*)'Enter superstrate refractive index(no):'
      READ(5,*)NO
      WRITE(*,*)'Enter slab refractive index(ns):'
      READ(5,*)NS
      WRITE(*,*)'Enter core radius in micrometers(a):'
      READ(5,*)A
      WRITE(*,*)'Enter cladding diameter in micrometers(dia_cl):'
      READ(5,*)DIA_CL
      WRITE(*,*)'Enter slab thickness in micrometers(t):'
      READ(5,*)T
      WRITE(*,*)'Enter distance between fibre and slab in microme-
     *ters(s):'
      READ(5,*)S
      WRITE(*,*)'Enter a filename to write in these data:'
      READ(5,30)FN_DATA
30      FORMAT(A10)
      OPEN(1,FILE=FN_DATA,STATUS='NEW')
        WRITE(1,*)WL,DELTA,NA,NO,NS,A,DIA_CL,T,S
      CLOSE(1)
    ENDIF
    R_CM=R/10000
35  WRITE(6,40)WL,DELTA,NA,A,R_CM,NO,NS,T,S,D,ZI,ZF,ZN,M
40  FORMAT(' Do you need to change any of the values entered ?'/,
     +'                                        '/,
     +' 1)  Light Wavelength in Free Space   (WL)  : ',F8.4,' micron   '/,
     +' 2)  Profile Height Parameter of Fibre (DELTA): ',F8.4,  /,
     +' 3)  Numerical Aperture of Fibre       (NA)  : ',F8.4, /,
     +' 4)  Radius of Fibre Core      (A)    : ',F8.4,' micron'/,
     +' 5)  Radius of Curvature of Fibre Bend (R) : ',F8.0,' cm'/,
     +'   ****** means R = infinite, i.e. a straight fibre    '/,
     +' 6)  Refractive Index of Superstrate (NO) : ',F8.4,  /,
     +' 7)  Refractive Index of Slab Guide  (NS)  : ',F8.4, /,
     +' 8)  Thickness of Slab Guide     (T) : ',F8.4,' micron'/,
     +' 9)  Fibre-Slab Minimum Distance (S) : ',F8.4,' micron'/,
     +' 10) Half-Width of Slab Where E = 0 (D) : ',I8,' micron'/,
     +' 11) Where Light Propagation Starts (ZI) : ',F8.0,' micron  '/,
     +' 12) Where Light Propagation Ends (ZF) : ',F8.0,' micron'/,
     +' 13) Number of Steps of Propagation (ZN) : ',I8,    /,
     +' 14) Mode-order of the slab waveguide  (M)   : ',I8)
    WRITE(*,*)'Please enter 1-14 to make a change or 0 to proceed:'
    READ(5,*)NUM
    IF(NUM.EQ.0)THEN
      GOTO 190
```

```
        ELSE
          GOTO (50,60,70,80,90,100,110,120,130,140,150,160,170,180)NUM
        ENDIF
50    WRITE(*,*)' Enter the wavelength (in micron): '
      READ(5,*)WL
      GOTO 35
60    WRITE(*,*)' Enter the delta parameter for the fibre:'
      READ(5,*)DELTA
      GOTO 35
70    WRITE(*,*)' Enter the numerical aperture for the fibre:'
      READ(5,*)NA
      GOTO 35
80    WRITE(*,*)' Enter the radius of the fibre core (in micron):'
      READ(5,*)A
      GOTO 35
90    WRITE(*,*)' Enter the radius of curvature of the fibre bend
      * (in cm):'
      READ(5,*)R_CM
      R=10000*R_CM
      GOTO 35
100   WRITE(*,*)' Enter the refractive index of slab superstrate:'
      READ(5,*)NO
      GOTO 35
110   WRITE(*,*)' Enter the refractive index of slab guide:'
      READ(5,*)NS
      GOTO 35
120   WRITE(*,*)' Enter the thickness of Slab Guide (in micron):'
      READ(5,*)T
      GOTO 35
130   WRITE(*,*)' Enter the Minimum Distance between the fibre &
      * Slab (in micron):'
      READ(5,*)S
      GOTO 35
140   WRITE(*,*)' Enter the artificial half-width (in micron) of
      * slab (lyl=D) where E becomes zero:'
      READ(5,*)D
      GOTO 35
150   WRITE(*,*)' Enter the initial point (ZI, in micron) of
      * light propagation:'
      READ(5,*)ZI
      GOTO 35
160   WRITE(*,*)' Enter the final point (ZF, in micron) of light
      *propagation:'
      READ(5,*)ZF
      GOTO 35
170   WRITE(*,*)' Enter the number of steps for light propagation:'
      READ(5,*)ZN
      GOTO 35
180   WRITE(*,*)'Enter a slab-mode order:'
      READ(5,*)M
      GOTO 35
190   RETURN
      END
```

```
      SUBROUTINE PAR_F(WL,DELTA,NA)
c     ********************************************************        c
c     Calculation of fibre parameters                               c
c     N_F:normalization constant of single fibre mode,i.e., LP01     c
c     ********************************************************        c
      REAL WL,K,DELTA,NA,NF,NC,A,VF,KF,XJ,BF0,UF,N_F,BESSJ0,BESSJ1
      DOUBLE PRECISION PI,TWOPI
      COMMON /BLK2/A,/BLK4/K,NC,/BLK5/NF,VF,KF,BF0,UF,N_F
      PI=3.14159265359
      TWOPI=2*PI
      K=TWOPI/WL
      NF=NA/SQRT(2*DELTA)
      VF=K*A*NA
      KF=VF*(1+SQRT(2.))/(A*(1+SQRT(SQRT(4+(VF)**4))))
      UF=SQRT((NA*K)**2-KF**2)
      XJ=KF*A
      BF0=K*SQRT(NC**2+NA**2-(KF/K)**2)
      N_F=UF*BESSJ0(XJ)/(SQRT(PI)*VF*BESSJ1(XJ))
      RETURN
      END

      SUBROUTINE  PAR_S(M,N1,WL,DEFF,VM)
c     ********************************************************        c
c     Calculationof slab parameters.                                c
c     ********************************************************        c
      INTEGER I,J,M,V,VX,VXR,D,VM,N1
      REAL T,NS,NC,WL,VSC,NO,AS,K,BSN,BSX,BTRY,KSTRY,VCTRY
      REAL VOTRY,AC,BF0,AO,PH,DPH,BS,KS,VO,VC,BSTP,DEFF
      REAL N_S,P(0:200),BSV(0:200),EPS,BD,S,NF,VF,KF,UF,N_F,R,A,BDP
      DOUBLE PRECISION PI
      COMMON /BLK1/VX,VXR,D,R,ZI,ZF,ZN,/BLK2/A,/BLK3/NO,NS,T,S
      COMMON /BLK4/K,NC,/BLK5/NF,VF,KF,BF0,UF,N_F
      COMMON /BLK6/AS,VSC,KS,VO,VC,N_S,P,BSV
      CHARACTER*10 EDF
      EXTERNAL QK_C
c     M: order of slab TE modes without artificial reflecting walls.
c     MX, max. order of the above modes.
      PI=3.14159265359
      VSC=PI*T*SQRT(NS**2-NC**2)/WL
      AS=(NO**2-NC**2)/(NS**2-NC**2)
      IF(NO.EQ.NC)AS=0
c     'AS'---An asymmetric measure of the slab waveguide
      IF(NO.GE.NC) THEN
         BSN=K*NO
      ELSE
         BSN=K*NC
      ENDIF
      BSX=K*NS
      I=0
      J=0
      EPS=0.0001
      BSTP=-1.E-05*K
      WRITE(*,*)'Do you like to key in the propagation
```

```
    * constant of the slab mode now ?(y/n):'
     READ(5,40)EDF
 5   FORMAT(A10)
     IF(EDF.EQ.'y'.OR.EDF.EQ.'Y')THEN
       WRITE(*,*)'Enter the propagation constant (Bs):'
       READ(5,*)BS
       KS=SQRT(ABS(BSX**2-BS**2))
       VO=SQRT(ABS(BS**2-(NO*K)**2))
       VC=SQRT(ABS(BS**2-(NC*K)**2))
       GOTO 25
     ENDIF
15   DO 20 BTRY=BSX,BSN,BSTP
       IF(BTRY.EQ.BSX)THEN
         KSTRY=0
       ELSE
         KSTRY=SQRT(ABS(BSX**2-BTRY**2))
       ENDIF
       IF(BTRY.EQ.NC*K)THEN
         VCTRY=0
       ELSE
         VCTRY=SQRT(ABS(BTRY**2-(NC*K)**2))
       ENDIF
       IF(BTRY.EQ.NO*K)THEN
         VOTRY=0
       ELSE
         VOTRY=SQRT(ABS(BTRY**2-(NO*K)**2))
       ENDIF
       IF(KSTRY.EQ.0.OR.VCTRY.GE.KSTRY*3.0E36)THEN
         AC=PI/2
       ELSE
         AC=ATAN(VCTRY/KSTRY)
       ENDIF
       IF(KSTRY.EQ.0.OR.VOTRY.GE.KSTRY*3.0E36)THEN
         AO=PI/2
       ELSE
         AO=ATAN(VOTRY/KSTRY)
       ENDIF
       PH=KSTRY*T-AO-AC
       DPH=ABS(PH-M*PI)
       IF(DPH.LE.EPS) THEN
         I=I+1
         BS=BTRY
         KS=KSTRY
         VO=VOTRY
         VC=VCTRY
       ENDIF
20   CONTINUE
     IF(I.EQ.0)THEN
     J=J+1
     IF(J.GE.5)THEN
        PRINT*,'Bs(',')=0;j','=',J
        WRITE(*,*)'Do you like to key in the propagation
   *constant of the slab mode ?(y/n):'
        READ(5,40)EDF
```

```
40        FORMAT(A10)
         IF(EDF.EQ.'y'.OR.EDF.EQ.'Y')THEN
            WRITE(*,*)'Enter the propagation constant (Bs):'
            READ(5,*)BS
         ELSE
           GOTO 10
         ENDIF
       ENDIF
     BSTP=BSTP/10
     GOTO 15
   ELSE
     IF(I.GE.2)THEN
        EPS=EPS/10
        I=0
        IF(EPS.LE.1.0E-25)THEN
           PRINT*,'I',I
           PRINT*,'EPS',EPS
           EPS=0
        ENDIF
        GOTO 15
     ELSE
25     PRINT*,'I,BS',I,BS
     DEFF=T+1/VO+1/VC
     N_S=SQRT(2/(D*DEFF))
     VXR=IDINT(BS*D*SQRT(1-(K*NC/BS)**2)/PI-5.0E-01)
     N1=VXR+2
     DO 30 V=0,VXR
       P(V)=PI*(V+1/2)/D
       BSV(V)=SQRT(BS**2-P(V)**2)
       BD=ABS(BSV(V)-BF0)
       IF(V.EQ.0)THEN
          BDP=BD
          VM=V
       ELSE
         IF(BD.LT.BDP)THEN
            BDP=BD
            VM=V
         ENDIF
       ENDIF
30     CONTINUE
     ENDIF
   ENDIF
 ENDIF
 WRITE(*,*)'The nunmber of slab transverse modes VXR may be
 *reduced to speed up the calculation with almost no visible
 *effects on the final results. This can be done by comparison
 *with the results of using the original VXR'
 WRITE(*,*)'Do you like to reduce the nunmber of slab
 *transverse modes VXR used in case you need to speed up the
 *calculation?'
 READ(5,50)EDF
50 FORMAT(A10)
 IF(EDF.EQ.'y'.OR.EDF.EQ.'Y')THEN
   WRITE(*,*)'Enter the maximum number of slab transverse modes:'
   READ(5,*)VXR
```

```
      ENDIF
      CALL QK_C
10    RETURN
      END


      SUBROUTINE A0B(N,N1,AB,DYDX,DYT,DYM,YTEMP,YSAV,DYSAV)
      INTEGER N,N1,VX,VXR,D,I,ZN
      REAL ZI,ZF
      REAL Z,ZNEXT,EPS,H1,HMIN,NOK,NBAD,R
      REAL*8 AA
      COMPLEX*16 AB(N),DYDX(N),DYT(N),DYM(N),A0STA,BSTA
      COMPLEX*16 YTEMP(N),YSAV(N),DYSAV(N)
      CHARACTER*20 A0ZDAT
      COMMON /BLK1/VX,VXR,D,R,ZI,ZF,ZN
      EXTERNAL DERIVS,RKQC
      WRITE(*,*)'Enter a file-name for a0zdat:'
      READ(5,1)A0ZDAT
 1    FORMAT(A20)
      OPEN(3,FILE=A0ZDAT)
      I=1
      A0STA=(1.D0,0)
      BSTA=(0.D0,0)
      AB(I)=A0STA
      AA=DCONJG(AB(I))*AB(I)
      WRITE(3,*)ZI,AA
      DO 5 I=2,N1
        AB(I)=BSTA
 5    CONTINUE
      WRITE(*,*)'Enter the required accuracy for the calculation(Eps):'
      READ(5,*)EPS
      H1=(ZF-ZI)/ZN
      HMIN=1.E-10
      IF(ZI.LE.-R)ZI=-R
      IF(ZF.GE.R)ZF=R
      DO 10 Z=ZI,ZF,H1
       ZNEXT=Z+H1
      CALL ODEINT(N,N1,AB,DYDX,DYT,DYM,Z,ZNEXT,EPS,H1,HMIN,NOK,NBAD,
     *DERIVS,RKQC,YTEMP,YSAV,DYSAV)
        I=1
        AA=DCONJG(AB(I))*AB(I)
        WRITE(3,*)ZNEXT,AA
10     CONTINUE
      CLOSE(3)
      RETURN
      END


      SUBROUTINE ODEINT(NVAR,N1,Y,DYDX,DYT,DYM,X1,X2,EPS,H1,HMIN,
     *NOK,NBAD,DERIVS,RKQC,YTEMP,YSAV,DYSAV)
      INTEGER NVAR,N1,NOK,NBAD,MAXSTP,NSTP
      REAL X1,X,X2,EPS,H1,HMIN,H,TWO,ZERO,TINY,HDID,HNEXT
      COMPLEX*16 Y(NVAR),DYDX(NVAR),DYT(NVAR)
      COMPLEX*16 DYM(NVAR),YTEMP(NVAR),YSAV(NVAR),DYSAV(NVAR)
      PARAMETER (MAXSTP=500,ONE=1.0,TWO=2.0,ZERO=0.0,TINY=1.E-30)
      EXTERNAL DERIVS,RKQC,RK4
```

```
        X=X1
        H=SIGN(H1,X2-X1)
        NOK=0
        NBAD=0
        DO 16 NSTP=1,MAXSTP
          CALL DERIVS(NVAR,N1,X,Y,DYDX)
          IF((X+H-X2)*(X+H-X1).GT.ZERO) H=X2-X
          IF(H.NE.0)CALL RKQC(Y,DYDX,DYT,DYM,NVAR,N1,X,H,EPS,HDID,HNE
       *XT,RK4,DERIVS,YTEMP,YSAV,DYSAV)
            IF(HDID.EQ.H)THEN
              NOK=NOK+1
            ELSE
              NBAD=NBAD+1
            ENDIF
          IF((X-X2)*(X2-X1).GE.ZERO)THEN
          RETURN
          ENDIF
        IF(ABS(HNEXT).LT.HMIN)print*,'Stepsize smaller than minimum.'
        IF(ABS(HNEXT).LT.HMIN)H=HMIN
        H=HNEXT
16      CONTINUE
        print*,'Too many steps.'
        RETURN
        END
        SUBROUTINE RKQC(Y,DYDX,DYT,DYM,N,N1,X,HTRY,EPS,HDID,HNEXT,
          *RK4,DERIVS,YTEMP,YSAV,DYSAV)
        INTEGER N,N1,I
        REAL X,HTRY,EPS,HDID,HNEXT,FCOR,ONE,SAFETY,ERRCON
        REAL PGROW,PSHRNK,XSAV,H,HH,ERRMAX,DELTA1,YSCAL
        PARAMETER (FCOR=.0666666667,ONE=1.,SAFETY=0.9,ERRCON=6.E-4)
        COMPLEX*16 Y(N),DYDX(N),DYT(N),DYM(N),YTEMP(N),YSAV(N)
        COMPLEX*16 DYSAV(N)
        EXTERNAL RK4,DERIVS
        PGROW=-0.20
        PSHRNK=-0.25
        XSAV=X
        DO 11 I=1,N1
          YSAV(I)=Y(I)
          DYSAV(I)=DYDX(I)
11      CONTINUE
          H=HTRY
1       HH=0.5*H
        CALL RK4(YSAV,DYSAV,DYT,DYM,N,N1,XSAV,HH,YTEMP,DERIVS)
        X=XSAV+HH
        CALL DERIVS(N,N1,X,YTEMP,DYDX)
        CALL RK4(YTEMP,DYDX,DYT,DYM,N,N1,X,HH,Y,DERIVS)
        X=XSAV+H
        IF(X.EQ.XSAV)print*,'Stepsize not significant in RKQC.'
        CALL RK4(YSAV,DYSAV,DYT,DYM,N,N1,XSAV,H,YTEMP,DERIVS)
        ERRMAX=0
        DO 12 I=1,N1
          DELTA1=SNGL(REAL(YTEMP(I)-Y(I)))
          YSCAL=SNGL(CDABS(YSAV(I))+CDABS(H*DYSAV(I)))
          IF(YSCAL.NE.0)ERRMAX=AMAX1(ERRMAX,ABS(DELTA1/YSCAL))
```

```
12   CONTINUE
   ERRMAX=ERRMAX/EPS
   IF(ERRMAX.GT.ONE) THEN
     H=SAFETY*H*(ERRMAX**PSHRNK)
     GOTO 1
   ELSE
     HDID=H
     IF(ERRMAX.GT.ERRCON)THEN
       HNEXT=SAFETY*H*(ERRMAX**PGROW)
     ELSE
       HNEXT=4.*H
     ENDIF
   ENDIF
   RETURN
   END

   SUBROUTINE RK4(Y,DYDX,DYT,DYM,N,N1,X,H,YOUT,DERIVS)
   INTEGER N,N1,I
   REAL X,H,HH,H6,XH
   COMPLEX*16 Y(N),DYDX(N),YOUT(N),DYT(N),DYM(N)
   EXTERNAL DERIVS
   HH=H*0.5
   H6=H/6.
   XH=X+HH
   DO 11 I=1,N1
     YOUT(I)=Y(I)+HH*DYDX(I)
11   CONTINUE
   CALL DERIVS(N,N1,XH,YOUT,DYT)
   DO 12 I=1,N1
     YOUT(I)=Y(I)+HH*DYT(I)
12   CONTINUE
   CALL DERIVS(N,N1,XH,YOUT,DYM)
   DO 13 I=1,N1
     YOUT(I)=Y(I)+H*DYM(I)
     DYM(I)=DYT(I)+DYM(I)
13   CONTINUE
   CALL DERIVS(N,N1,X+H,YOUT,DYT)
   DO 14 I=1,N1
     YOUT(I)=Y(I)+H6*(DYDX(I)+DYT(I)+2.*DYM(I))
14   CONTINUE
   RETURN
   END

   SUBROUTINE DERIVS(N,N1,Z,AB,DABDZ)
   INTEGER I,U,V,VX,VXR,D,ZN,N,N1
   REAL A,S,T,UF,BF0,VF,KF,AS,NO,NS,K,NC,NF,N_F,VSC,KS,VO,VC,N_S
   REAL P(0:200),BSV(0:200),QF001,KF0V(0:200),R,SZ,ZI,ZF,QF002,
   QF003
   REAL QSUV(0:200,0:200),KSU0(0:200),Z,XE,ERF,QF00
   COMPLEX*16 AB(N),DABDZ(N)
   COMMON /BLK1/VX,VXR,D,R,ZI,ZF,ZN,/BLK2/A,/BLK3/NO,NS,T,S
   COMMON /BLK4/K,NC,/BLK5/NF,VF,KF,BF0,UF,N_F
   COMMON /BLK6/AS,VSC,KS,VO,VC,N_S,P,BSV
   COMMON /BLK7/QF00,KF0V,QSUV,KSU0,/BLK8/QF001
```

```fortran
   IF(R.LE.1.E30)THEN
     SZ=S+Z**2/(2*R)
     QF002=ERF(SQRT(2*UF*(A+SZ+T)))-ERF(SQRT(2*UF*(A+SZ)))
     QF003=AS*(1-ERF(SQRT(2*UF*(A+SZ+T))))
     QF00=(QF002+QF003)*QF001
     I=1
     V=0
     DABDZ(I)=(0,-1.D0)*QF00*AB(V+1)+(0,-1.D0)*KF0V(V)*EXP
   *(-VC*Z**2/(2*R))*AB(V+2)*CDEXP((0,1.D0)*(BF0-BSV(V))*Z)
     DO 10 V=1,VXR
     DABDZ(I)=DABDZ(I)+(0,-1.D0)*KF0V(V)*EXP(-VC*Z**2/
   *(2*R))*AB(V+2)*CDEXP((0,1.D0)*(BF0-BSV(V))*Z)
10   CONTINUE
     DO 20 I=2,N1
       U=I-2
       V=0
       XE=SQRT(UF**2+P(U)**2)
       DABDZ(I)=(0,-1.D0)*KSU0(U)*EXP(-XE*Z**2/(2*R))*AB(V+1)
   *CDEXP((0,1.D0)*(BSV(U)-BF0)*Z)+(0,-1.D0)*QSUV(U,V)*EXP
   *(-VC*Z**2/R)*AB(V+2)*CDEXP((0,1.D0)*(BSV(U)-BSV(V))*Z)
         DO 30 V=1,VXR
       DABDZ(I)=DABDZ(I)+(0,-1.D0)*QSUV(U,V)*EXP(-VC*Z**2/R)*AB(
   *V+2)*CDEXP((0,1.D0)*(BSV(U)-BSV(V))*Z)
30     CONTINUE
20   CONTINUE
   ELSE
   QF002=ERF(SQRT(2*UF*(A+S+T)))-ERF(SQRT(2*UF*(A+S)))
   QF003=AS*(1-ERF(SQRT(2*UF*(A+S+T))))
   QF00=QF001*(QF002+QF003)
   I=1
   V=0
     DABDZ(I)=(0,-1.D0)*QF00*AB(V+1)+(0,-1.D0)*KF0V(V)
   *AB(V+2)*CDEXP((0,1.D0)*(BF0-BSV(V))*Z)
     DO 40 V=1,VXR
     DABDZ(I)=DABDZ(I)+(0,-1.D0)*KF0V(V)*AB(V+2)*CDEXP((0,1.D0)
   *(BF0-BSV(V))*Z)
40   CONTINUE
   DO 50 I=2,N1
     U=I-2
     V=0
     DABDZ(I)=(0,-1.D0)*KSU0(U)*AB(V+1)*CDEXP((0,1.D0)
   *(BSV(U)-BF*0)*Z)+(0,-1.D0)*QSUV(U,V)*AB(V+2)
   *CDEXP((0,1.D0)*(BSV(U)-BSV(V))*Z)
     DO 60 V=1,VXR
     DABDZ(I)=DABDZ(I)+(0,-1.D0)*QSUV(U,V)*AB(V+2)*CDEXP((0,1.
   *D0)*(BSV(U)-BSV(V))*Z)
60     CONTINUE
50   CONTINUE
   ENDIF
   RETURN
   END


   SUBROUTINE QK_C
   INTEGER U,V,VX,VXR
   REAL A,S,T,UF,BF0,XK,VF,KF,NF,N_F,XJ,VSC,AS,QF001
```

```
      REAL BSV(0:200),QF00,BESSJ0,BESSJ1,BESSK0,BESSI0,BESSI1,XI1
      REAL KF0V(0:200),QSUV(0:200,0:200),KSU0(0:200),R,KS1,KS2,KS3
      REAL N_S,KS,VC,P(0:200),XI2,XI3,K,NO,NS,NC,XE,VO,VSO,BS
      DOUBLE PRECISION PI
      COMMON /BLK1/VX,VXR,D,R,ZI,ZF,ZN,/BLK2/A,/BLK3/NO,NS,T,S
      COMMON /BLK4/K,NC,/BLK5/NF,VF,KF,BF0,UF,N_F
      COMMON /BLK6/AS,VSC,KS,VO,VC,N_S,P,BSV
      COMMON /BLK7/QF00,KF0V,QSUV,KSU0,/BLK8/QF001
      PI=3.14159265359
      VSO=K*T*SQRT(NS**2-NO**2)/2
      BS=VSC/VSO
      XJ=KF*A
      XK=UF*A
      QF001=(PI*N_F*VSC/(UF*T*BESSK0(XK)))**2/(SQRT(2.)*BF0)
      QF001=(PI*VSC**2*BESSJ0(XJ)**2)/(SQRT(2.)*BF0*T**2*VF**2
     *BESSK0(XK)**2*BESSJ1(XJ)**2)
C     *********************************************************       C
C     This unit is to calculate the coupling coefficient KF0(V).      C
C     XJ(=KF*A) and XI1(=A*SQRT(VC**2-P(V)**2)) are dummy             C
C     variables for BESSJ0 BESSJ1, BESSI0 AND BESSI1 functions.       C
C     P(V)=pi*(2*v+1)/(2*D)
      DO 10 V=0,VXR
        XI1=A*SQRT(VC**2-P(V)**2)
        KF0V(V)=PI*N_F*N_S*VF**2*KS*T*EXP(-VC*(A+S))*(XI1*BESSJ0(XJ)
     **BESSI1(XI1)+XJ*BESSJ1(XJ)*BESSI0(XI1))/(2*BF0*A**2*VSC*
     **BESSJ0(XJ)(KF**2+VC**2-P(V)**2))
10    CONTINUE
C     *********************************************************       C
C     This unit is to calculate the coupling coefficient Qsuv.        C
C     IX1(U,V)(=A*SQRT(4*VC**2-(B(U)+B(V))**2)),dummy variables for I1; C
C     IX2(U,V)(=A*SQRT(4*VC**2-(B(U)-B(V))**2)),dummy variables for I1. C
C     *********************************************************       C
20    DO 30 U=0,VXR
        DO 40 V=0,VXR
          XI2=A*SQRT(4*VC**2-(P(U)+P(V))**2)
          XI3=A*SQRT(4*VC**2-(P(U)-P(V))**2)
          QSUV(U,V)=PI*(N_S*KS*VF*T)**2*EXP(-2*VC*(A+S))*(BESSI1
     *(XI2)/XI2+BESSI1(XI3)/XI3)/(8*BSV(U)*VSC**2)
40      CONTINUE
30    CONTINUE
      DO 60 U=0,VXR
        XE=SQRT(UF**2+P(U)**2)
        KS1=PI*N_F*N_S*VSC*KS*EXP(-XE*(A+S))/(BSV(U)*BESSK0(XK)*
     *T*XE*)
        KS2=(XE+VC-BS*(XE-VO)*EXP(-XE*T))/(KS**2+XE**2)
        KS3=AS*BS*EXP(-XE*T)/(VO+XE)
        KSU0(U)=KS1*(KS2+KS3)
60    CONTINUE
      RETURN
      END


SUBROUTINE SET (2)
      SUBROUTINE INPUT(M,WL,DELTA,NA,NO,NS,A,DIA_CL,T,S,D,FN_DATA)
C     as in set (1) with some modifications, e.g. parameter and data
      declarations etc.
```

```
      SUBROUTINE PAR_F(WL,DELTA,NA,A,K,NF,NC,VF,KF,BF0,UF,N_F)
C     as in set (1) with some modifications, e.g. parameter and data
      declarations etc.


      SUBROUTINE  PAR_S(M,VX,VXR,D,T,NS,NC,WL,VS,NO,AS,K,DPH,BS,KS,
     *VO,VC,DEFF,N_S,P,BSV,VM,NP,BF0)
C     as in set (1) with some modifications, e.g. parameter and data
      declarations etc.


      SUBROUTINE QF00_C(NP,A,T,S,VF,KF,BF0,UF,VS,AS,QF00,Q)
C     This unit is to calculate the coupling coefficient Qf00.              C
      INTEGER NP,V
      REAL A,S,T,UF,BF0,XK,VF,KF,XJ,VS,AS,ERF,ERFCC
      REAL QF001,QF002,QF003,QF00,BESSJ0,BESSJ1,BESSK0,Q(NP,NP)
      DOUBLE PRECISION PI
      PI=3.14159265359
      XJ=KF*A
      XK=UF*A
      QF001=(PI*VS**2*BESSJ0(XJ)**2)/(SQRT(2.)*BF0*T**2*VF**2*BES
     *SK0(XK)**2*BESSJ1(XJ)**2)
      QF002=ERF(SQRT(2*UF*(A+S+T)))-ERF(SQRT(2*UF*(A+S)))
      QF003=AS*(1-ERF(SQRT(2*UF*(A+S+T))))
      QF00=QF001*(QF002+QF003)
      V=1
      Q(V,V)=QF00+BF0
      RETURN
      END


      SUBROUTINE KF0V_C(VX,VXR,NP,A,T,S,VF,KF,BF0,UF,KS,VC,P,N_S,
     *VS,XI1,KF0V,Q)
C     ****************************************************************    C
C     This unit is to calculate the coupling coefficient Kf0(V).          C
C     XJ(=KF*A) and XI1(=A*SQRT(VC**2-P(V)**2)) are dummy                  C
C     variables for BESSJ0 BESSJ1, BESSI0 AND BESSI1 functions.           C
C     P(V)=pi*(2*v+1)/(2*D)                                               C
C     ****************************************************************    C
      INTEGER V,VX,VXR,NP,V1
      REAL A,T,S,VF,KF,XJ,BF0,UF,KS,VC,P(0:VX),XI1(0:VX),N_S,VS
      REAL BESSJ0,BESSI1,BESSJ1,BESSI0
      REAL KF0V(0:VX),Q(NP,NP)
      CHARACTER*10 V_KF0V
      DOUBLE PRECISION PI,SP
      PI=3.14159265359
      SP=DSQRT(PI)
      XJ=KF*A
      WRITE(*,*)'Enter a file-name for data of (n,Kf0v(n)):'
      READ(5,1)V_KF0V
1     FORMAT(A20)
      OPEN(1,FILE=V_KF0V)
      DO 20 V=0,VXR
        XI1(V)=A*SQRT(VC**2-P(V)**2)
        KF0V(V)=SP*UF*VF*KS*T*N_S*EXP(-VC*(A+S))*(XI1(V)*BESSJ0
     **(XJ)BESSI1(XI1(V))+XJ*BESSJ1(XJ)*BESSI0(XI1(V)))/(2*
```

```
     **BF0*A**2*VS*(KF*2+VC**2-P(V)**2)*BESSJ1(XJ))
     WRITE(1,*)V,KF0V(V)
     V1=1
     Q(V1,V+2)=KF0V(V)
20   CONTINUE
     CLOSE(1)
     RETURN
     END


     SUBROUTINE QSUV_C(VX,VXR,NP,A,T,S,VF,VS,KS,VC,N_S,P,BSV,QSUV,Q)
C    *********************************************************      C
C    This unit is to calculate the coupling coefficient Qsuv.      C
C    IX1(U,V)(=A*SQRT(4*VC**2-(B(U)+B(V))**2)),dummy variables      C
C    for I1;                                                       C
C    IX2(U,V)(=A*SQRT(4*VC**2-(B(U)-B(V))**2)),dummy               C
C    variables for I1.                                             C
C    *********************************************************      C
     INTEGER U,V,VX,VXR,NP
     REAL N_S,VF,A,T,S,VS,KS,VC,P(0:VX),BSV(0:VX)
     REAL QSUV(0:VX,0:VX),XI2,BESSI1
     REAL XI3,Q(NP,NP)
     CHARACTER*10 UV_QSUV
     DOUBLE PRECISION PI
     PI=3.14159265359
     WRITE(*,*)'Enter a file-name for data of (m,n,Qsuv(m,n)):'
     READ(5,2)UV_QSUV
2    FORMAT(A20)
     OPEN(2,FILE=UV_QSUV)
     DO 20 U=0,VXR
       DO 30 V=0,VXR
         XI2=A*SQRT(4*VC**2-(P(U)+P(V))**2)
         XI3=A*SQRT(4*VC**2-(P(U)-P(V))**2)
         QSUV(U,V)=PI*(VF*KS*T*N_S)**2*EXP(-2*VC*(A+S))*(BESSI1
    *(XI2)/XI2+BESSI1(XI3)/XI3)/(8*BSV(U)*VS**2)
         if(U.eq.0)WRITE(2,*)V,QSUV(U,V)
         IF(U.EQ.V)THEN
            Q(U+2,V+2)=QSUV(U,V)+BSV(U)
         ELSE
            Q(U+2,V+2)=QSUV(U,V)
         ENDIF
30   CONTINUE
20   CONTINUE
     CLOSE(2)
     RETURN
     END


     SUBROUTINE KSU0_C(VX,VXR,NP,A,T,S,K,NO,NS,NC,VF,KF,UF,KS,
    *VO,VC,BSV,N_S,P,AS,KSU0,Q)
     INTEGER U,VX,VXR,NP,C1
     REAL A,T,S,K,NO,NS,NC,VF,KF,UF,KS,BSV(0:VX),N_S,P(0:VX),AS
     REAL XJ,XK,XE,VO,VC,KSU0(0:VX),BESSJ0,BESSJ1,BESSK0,Q(NP,NP)
     CHARACTER*10 U_KSU0
     DOUBLE PRECISION PI,SP
```

```
      PI=3.14159265359
      SP=DSQRT(PI)
      C1=1
      WRITE(*,*)'Enter a file-name for data of (n,Ksu0(n)):'
      READ(5,3)U_KSU0
3     FORMAT(A20)
      OPEN(3,FILE=U_KSU0)
      DO 20 U=0,VXR
        XJ=KF*A
        XK=UF*A
        XE=SQRT(UF**2+P(U)**2)
      KSU0(U)=SP*((KS*UF*N_S*EXP(-XE*(A+S))*BESSJ0(XJ))/(2*BSV
     *(U)*XE*VF*BESSJ1(XJ)*BESSK0(XK)))*((K*(NS**2-NC**2)/(KS**
     **2+XE**2))*((XE+VC)/SQRT(NS**2-NC**2)-(XE-VO)*EXP(-XE*T)/
     *SQRT(NS**2-NO**2))+EXP(-XE*T)*AS*(NS**2-NC**2)/(SQRT
     *(NS**2-NO**2)*(VO+XE)))
      WRITE(3,*)U,KSU0(U)
        Q(U+2,C1)=KSU0(U)
20    CONTINUE
      CLOSE(3)
      RETURN
      END


      SUBROUTINE TRED2(Q,N,NP,DG,E)
C     **********************************************************    C
C     Householder reduction of a real N by N matrix Q to            C
C     tridiagonal form with DG returning the diagonal elements      C
C     of the new matrix. See Press et al. (1992).                   C
C     **********************************************************    C
      INTEGER N,NP,I,J,L,K,M
      REAL Q(NP,NP),DG(NP),E(NP),F,G,H,HH,SCALE,QT
      DO 5 L=1,N
       DO 10 M=1,N
         QT=Q(L,M)
10     CONTINUE
5      CONTINUE
      IF(N.GT.1)THEN
       DO 18 I=N,2,-1
        L=I-1
        H=0.
        SCALE=0.
        IF(L.GT.1)THEN
         DO 11 K=1,L
           SCALE=SCALE+ABS(Q(I,K))
11         CONTINUE
         IF(SCALE.EQ.0.)THEN
           E(I)=Q(I,L)
         ELSE
           DO 12 K=1,L
              Q(I,K)=Q(I,K)/SCALE
              H=H+Q(I,K)**2
12    CONTINUE
      F=Q(I,L)
      G=-SIGN(SQRT(H),F)
```

```
      E(I)=SCALE*G
      H=H-F*G
       Q(I,L)=F-G
       F=0.
       DO 15 J=1,L
        Q(J,I)=Q(I,J)/H
        G=0.
        DO 13 K=1,J
         G=G+Q(J,K)*Q(I,K)
13     CONTINUE
       IF(L.GT.J)THEN
        DO 14 K=J+1,L
         G=G+Q(K,J)*Q(I,K)
14     CONTINUE
       ENDIF
       E(J)=G/H
       F=F+E(J)*Q(I,J)
15     CONTINUE
       HH=F/(H+H)
       DO 17 J=1,L
        F=Q(I,J)
        G=E(J)-HH*F
        E(J)=G
        DO 16 K=1,J
         Q(J,K)=Q(J,K)-F*E(K)-G*Q(I,K)
16     CONTINUE
17     CONTINUE
       ENDIF
      ELSE
        E(I)=Q(I,L)
       ENDIF
       DG(I)=H
18     CONTINUE
       ENDIF
       DG(1)=0.
       E(1)=0.
       DO 23 I=1,N
        L=I-1
        IF(DG(I).NE.0.)THEN
         DO 21 J=1,L
          G=0.
          DO 19 K=1,L
           G=G+Q(I,K)*Q(K,J)
19     CONTINUE
        DO 20 K=1,L
         Q(K,J)=Q(K,J)-G*Q(K,I)
20      CONTINUE
21      CONTINUE
       ENDIF
       DG(I)=Q(I,I)
       Q(I,I)=1.
       IF(L.GE.1)THEN
        DO 22 J=1,L
         Q(I,J)=0.
```

```fortran
      Q(J,I)=0.
22    CONTINUE
      ENDIF
23    CONTINUE
      RETURN
      END


      SUBROUTINE TQLI(Q,IMAX,N,NP,DG,E)
C     ***********************************************************     C
C     Tridiagonal QL implicit algorithm to determine the            C
C     eigenvalues and eigenvectors of tridiagonal matrix reduced    C
C     above by TRED2 subroutine. See Press et al. (1992).           C
C     ***********************************************************     C
      INTEGER N,NP,I,J,K,L,M,ITER,IMAX,ROW
      REAL DG(NP),E(NP),Q(NP,NP),B,C,DD,F,G,P,R,S,QT
      IF (N.GT.1) THEN
       DO 11 I=2,N
        E(I-1)=E(I)
11    CONTINUE
      E(N)=0.
      J=0
      DO 15 L=1,N
         ITER=0
1     DO 12 M=L,N-1
      DD=ABS(DG(M))+ABS(DG(M+1))
      IF (ABS(E(M))+DD.EQ.DD) GO TO 2
12    CONTINUE
      M=N
2     IF(M.NE.L)THEN
       IF(ITER.EQ.100) pause'too many iterations'
       ITER=ITER+1
       G=(DG(L+1)-DG(L))/(2.*E(L))
       R=SQRT(G**2+1.)
       G=DG(M)-DG(L)+E(L)/(G+SIGN(R,G))
       S=1.
       C=1.
       P=0.
       DO 14 I=M-1,L,-1
        F=S*E(I)
        B=C*E(I)
        IF(ABS(F).GE.ABS(G))THEN
         C=G/F
         R=SQRT(C**2+1.)
         E(I+1)=F*R
         S=1./R
         C=C*S
        ELSE
         S=F/G
         R=SQRT(S**2+1.)
         E(I+1)=G*R
         C=1./R
         S=S*C
        ENDIF
        G=DG(I+1)-P
```

```
      R=(DG(I)-G)*S+2.*C*B
      P=S*R
      DG(I+1)=G+P
      G=C*R-B
      DO 13 K=1,N
       F=Q(K,I+1)
       Q(K,I+1)=S*Q(K,I)+C*F
       Q(K,I)=C*Q(K,I)-S*F
13    CONTINUE
14    CONTINUE
     DG(L)=DG(L)-P
     IF(DG(L).EQ.0)J=J+1
     IF(J.EQ.1)IMAX=L-1
     E(L)=G
     E(M)=0.
     GO TO 1
    ENDIF
15    CONTINUE
     IF(J.EQ.0)IMAX=L-1
     ENDIF
     DO 20 L=1,N
       ROW=0
       DO 30 M=1,N
        QT=Q(M,L)
        IF(QT.LE.0.)ROW=ROW+1
30     CONTINUE
       IF((NP-ROW).LE.10)THEN
         DO 40 M=1,N
          Q(M,L)=-Q(M,L)
40     CONTINUE
       ELSE
       ENDIF
20    CONTINUE
     RETURN
     END


     SUBROUTINE FD(Q,IMAX,D,VX,VXR,NP,A,T,S,KF,BF0,UF,N_F,KS,
    *VO,VC,N_S,P,BSV,QF00,KF0V,DG,QI)
c    To prepare for field plot.
     INTEGER I,IMAX,V,D,VX,VXR,NP,L,M,IBM
c    IBM: the mode No. which has Max. propagation constant.
     REAL A,T,S,KF,JA,JR,BF0,UF,KA,KR,U,N_F,KS,VO,VC,N_S,P(0:VX)
     REAL BSV(0:VX),X,XMAX,Y,R,RMAX,RSTP,F0,SVX,BESSJ0,BESSK0
     REAL QF00,KF0V(0:VX),Q(NP,NP),QM,FXY,DG(NP),FP,RP,DF,DR,DPDR
     REAL BMI,BM,QI(NP),XY,QN
     DATA XMAX/1.5E01/
     CHARACTER*20 FXM,FYM,MINUS,PLOT2,F01,S0,BCM
     write(*,*)'Enter a file-name for B(m):'
     read(5,14)BCM
14    format(A20)
     open(14,FILE=BCM)
     print*,'The compund-mode propagation constants(eigvalues)are:'
     DO 2 M=1,IMAX-1
     IBM=M
```

```fortran
      DO 3 L=IBM,IMAX
        IF(L.EQ.IBM)THEN
          BM=DG(L)
          IBM=L
      ELSE
       IF(BM.LT.DG(L))THEN
         BM=DG(L)
         IBM=L
       ENDIF
      ENDIF
3     CONTINUE
      IF(IBM.NE.M)THEN
       BMI=DG(M)
       DG(M)=BM
       DG(IBM)=BMI
       DO 5 N=1,VXR+2
        QI(N)=Q(N,M)
        Q(N,M)=Q(N,IBM)
        Q(N,IBM)=QI(N)
5     CONTINUE
      ENDIF
2     CONTINUE
      DO 4 L=1,IMAX
        PRINT*,L,DG(L)
        write(14,*)L,DG(L)
4     CONTINUE
      DO 11 L=1,VXR+2
        DO 12 M=1,VXR+2
          IF(M.EQ.1)THEN
            QN=Q(M,L)**2
          ELSE
            QN=QN+Q(M,L)**2
          ENDIF
12    CONTINUE
      QN=SQRT(QN)
      DO 13 V=1,VXR+2
        QI(L)=Q(V,L)
        IF(QN.NE.0.AND.QI(L)/QN.GE.1.E-30)Q(V,L)=Q(V,L)/QN
13    CONTINUE
11    CONTINUE
1     WRITE(*,*)'Enter a mode-number for the compound-modes(I:1,2,...):'
      READ(5,*)I
      WRITE(*,*)'Enter a file-name for the compound-mode x-plot(fx):'
      READ(5,6)FXM
6     FORMAT(A20)
      WRITE(*,*)'Enter a file-name for the compound-mode y-plot(fy):'
      READ(5,7)FYM
7     FORMAT(A20)
      WRITE(*,*)'Do you need to change the sign of the field? (y/n)'
      READ(5,8)MINUS
8     FORMAT(A20)
      WRITE(*,*)'Enter a file-name for the LP01 fibre mode:'
      READ(5,18)F01
```

```
18    FORMAT(A20)
      WRITE(*,*)'Enter a file-name for the 1st Slab Mode (n=0):'
      READ(5,19)S0
19    FORMAT(A20)
      OPEN(8,FILE=FXM)
      OPEN(9,FILE=FYM)
      OPEN(18,FILE=F01)
      OPEN(19,FILE=S0)
      JA=KF*A
      KA=UF*A
      M=0
      Y=0
      R=XMAX
      RMAX=XMAX
      FXY=0
      RSTP=0.2
10    DO 20 WHILE(R.GE.-RMAX)
         RP=R
         FP=FXY
         IF(M.EQ.0)X=R
         IF(M.NE.0)Y=R
         XY=SQRT(X**2+Y**2)
         IF(XY.LE.A)THEN
          JR=KF*XY
          F0=N_F*BESSJ0(JR)/BESSJ0(JA)
          write(18,*)X,F0
          SVX=N_S*KS*EXP(VC*(X-A-S))/SQRT(KS**2+VC**2)
          write(19,*)X,SVX
         ELSE
          KR=UF*XY
          F0=N_F*BESSK0(KR)/BESSK0(KA)
          write(18,*)X,F0
          IF(X.LT.(A+S))THEN
           SVX=N_S*KS*EXP(VC*(X-A-S))/SQRT(KS**2+VC**2)
          write(19,*)X,SVX
          ELSE
          IF(X.LE.(A+S+T))THEN
            SVX=N_S*(KS*COS(KS*(X-A-S-T))-VO*SIN(KS*(X-A-S-T)))
*/SQRT(KS**2+VO**2)
            write(19,*)X,SVX
          ELSE
            SVX=N_S*KS*EXP(-VO*(X-A-S-T))/SQRT(KS**2+VO**2)
            write(19,*)X,SVX
         ENDIF
        ENDIF
       ENDIF
       V=0
       L=I
       FXY=Q(V+1,L)*F0+Q(V+2,L)*SVX*COS(P(V)*Y)
IF(VXR.EQ.0)THEN
       IF(M.EQ.0.)THEN
        IF(MINUS.EQ.'y'.or.MINUS.EQ.'Y')THEN
         WRITE(8,*)X,-FXY
```

```
      ELSE
        WRITE(8,*)X,FXY
      ENDIF
     ELSE
      IF(MINUS.EQ.'y'.or.MINUS.EQ.'Y')THEN
        WRITE(9,*)Y,-FXY
      ELSE
        WRITE(9,*)Y,FXY
      ENDIF
     ENDIF
    ELSE
     DO 40 V=1,VXR
        FXY=FXY+Q(V+2,L)*SVX*COS(P(V)*Y)
40    CONTINUE
     IF(M.EQ.0.)THEN
      IF(MINUS.EQ.'y'.or.MINUS.EQ.'Y')THEN
       WRITE(8,*)X,-FXY
      ELSE
       WRITE(8,*)X,FXY
      ENDIF
     ELSE
      IF(MINUS.EQ.'y'.or.MINUS.EQ.'Y')THEN
        WRITE(9,*)Y,-FXY
      ELSE
        WRITE(9,*)Y,FXY
      ENDIF
     ENDIF
    ENDIF
    DPDR=100*ABS(FXY-FP)/ABS(RSTP)
    DO 50 L=100,10,-10
     IF(DPDR.GE.L*1.E-02)THEN
       RSTP=RSTP/L
        GOTO 55
     ELSE
        IF(M.EQ.0.)RSTP=0.2
        IF(M.NE.0.)RSTP=1
     ENDIF
50   CONTINUE
55   IF(RSTP.LT.1.E-02)RSTP=1.E-02
    R=R-RSTP
20  ENDDO
    IF(M.EQ.0)THEN
     M=1
     X=4.5
     RMAX=D
     R=D
     RSTP=1
     GOTO 10
    ENDIF
    CLOSE(19)
    CLOSE(18)
    CLOSE(9)
    CLOSE(8)
```

```
      WRITE(*,*)'Do you like to have another compound-mode field plot?
     *(y/n or Y/N)'
      READ(5,100)PLOT2
100   FORMAT(A10)
      IF(PLOT2.EQ.'y'.OR.PLOT2.EQ.'Y')GOTO 1
      CLOSE(14)
      RETURN
      END

                         SUBROUTINE SET (3)
      SUBROUTINE INPUT, PAR_F, PAR_S, QF00_C, KF0V_C, QSUV_C, KSV0_C
      are the same as those in set (2).
      SUBROUTINE COEF(VX,VXR,VM,D,BF0,BSV,QF00,KF0V,QSUV,KSV0,
     *BF0V,BSV0)
C     ***********************************************************     C
C     To prepare for the data files of the loss coefficients.        C
C     ***********************************************************     C
      INTEGER U,V,VX,VXR,VM,D
      REAL BF0,BSV(0:VX),QF00,KF0V(0:VX),KSV0(0:VX),QSUV(0:VX,
      0:VX)
      REAL BF0V(0:VX),BSV0(0:VX),IB,BSVV,ALFA_O,ALFA,PM,Z,
      A_O,A_N
      REAL ALFA3,ALFA4
      COMPLEX*16 ALFA3C,ALFA4C
      CHARACTER*20 A1,A2,A3,A4
      DOUBLE PRECISION PI
      PI=3.14159265359
      WRITE(*,*)'You will be asked to enter four data-file names
     *(1) Alfa1: using the simple formula and coupling coefficients;
     *(2) Alfa2: using the simple formula but full coupling coeff.;
     *(3) Alfa3: using the full formula but simple coupling coeff.;
     *(4) Alfa4: using the full formula and coupling coeeficents.'
      WRITE(*,*)'Enter a filename for the case (1):'
      READ(5,6)A1
6     FORMAT(A10)
      WRITE(*,*)'Enter a filename for the case (2):'
      READ(5,7)A2
7     FORMAT(A10)
      WRITE(*,*)'Enter a filename for the case (3):'
      READ(5,8)A3
8     FORMAT(A10)
      WRITE(*,*)'Enter a filename for the case (4):'
      READ(5,9)A4
9     FORMAT(A10)
      OPEN(1,FILE=A1)
      OPEN(2,FILE=A2)
      OPEN(3,FILE=A3)
      OPEN(4,FILE=A4)
      V=VM
      PM=PI*(2*V+1)/(2*D)
      ALFA_O=2*ABS(KF0V(V)*KSV0(V)*BF0*D/PM)
      DO 10 V=0,VXR
        BSVV=BSV(V)
```

```
    IF(BF0.NE.BSVV)THEN
       BF0V(V)=2*BSV(V)*(BF0*KF0V(V)-BSV(V)*KSV0(V))/(BF0*(BF0**
   *2-BSV(V)**2))
       BSV0(V)=2*BF0*(BF0*KF0V(V)-BSV(V)*KSV0(V))/(BSV(V)*(BF0**
   *2-BSV(V)**2))
     ELSE
      BF0V(V)=2*KF0V(V)/BF0
      BSV0(V)=2*KSV0(V)/BSV(V)
     ENDIF
     IF(V.EQ.0)THEN
      IB=1-BF0V(V)*BSV0(V)
     ELSE
      IB=IB-BF0V(V)*BSV0(V)
     ENDIF
10   CONTINUE
    DO 20 V=0,VXR
      QF00=QF00-BF0V(V)*KSV0(V)
      DO 25 U=0,VXR
        KF0V(V)=KF0V(V)-BF0V(U)*QSUV(U,V)
25   CONTINUE
    IF(V.EQ.VM)KF0V(V)=KF0V(V)/IB
20   CONTINUE
    QF00=QF00/IB
    U=VM
    KSV0(U)=KSV0(U)-BSV0(U)*QF00
    ALFA=2*ABS(KF0V(U)*KSV0(U)*BF0*D/PM)
    DO 50 Z=0,6000,10
      A_O=EXP(-ALFA_O*Z)
      WRITE(1,*)Z,A_O
      A_N=EXP(-ALFA*Z)
      WRITE(2,*)Z,A_N
50   CONTINUE
    CLOSE(4)
    CLOSE(3)
    CLOSE(2)
    CLOSE(1)
    RETURN
    END


                        SUBROUTINE SET (4)
    SUBROUTINE INPUT, PAR_F,PAR_S are the same as those in set (1).
    SUBROUTINE QK_C
c   To prepare for data files of the coupling coefficients.
    INTEGER U,V,VX,VXR
    REAL A,S,T,UF,BF0,XK,VF,KF,NF,N_F,XJ,VSC,AS,QF001
    REAL BSV(0:200),QF00,BESSJ0,BESSJ1,BESSK0,BESSI0,BESSI1,XI1
    REAL KF0V(0:200),QSUV(0:200,0:200),KSU0(0:200),KS1,KS2,KS3
    REAL N_S,KS,VC,P(0:200),XI2,XI3,K,NO,NS,NC,XE,VO,VSO,BS
    DOUBLE PRECISION PI
    CHARACTER*10 D_KF0N,D_KSM0,D_QSMN,Y_N
    COMMON /BLK1/VX,VXR,D,/BLK2/A,/BLK3/NO,NS,T,S
    COMMON /BLK4/K,NC,/BLK5/NF,VF,KF,BF0,UF,N_F
    COMMON /BLK6/AS,VSC,KS,VO,VC,N_S,P,BSV
    COMMON /BLK7/QF00,KF0V,QSUV,KSU0,/BLK8/QF001
```

```
      WRITE(*,*)'Do you want to save the data of coeffs in a file?(y/n)'
      READ(5,2)Y_N
2     FORMAT(A10)
      IF(Y_N.EQ.'y'.OR.Y_N.EQ.'Y')THEN
       WRITE(*,*)'Enter a name for the data-file Kf0n:'
       READ(5,3)D_KF0N
3      FORMAT(A10)
       WRITE(*,*)'Enter a name for the data-file Ksm0:'
       READ(5,4)D_KSM0
4      FORMAT(A10)
       WRITE(*,*)'Enter a name for the data-file Qsmn(m,n):'
       READ(5,6)D_QSMN
6      FORMAT(A10)
      ENDIF
      OPEN(11,FILE=D_KF0N)
      OPEN(12,FILE=D_KSM0)
      OPEN(13,FILE=D_QSMN)
      PI=3.14159265359
      VSO=K*T*SQRT(NS**2-NO**2)/2
      BS=VSC/VSO
      XJ=KF*A
      XK=UF*A
      QF001=(PI*N_F*VSC/(UF*T*BESSK0(XK)))**2/(SQRT(2.)*BF0)
      QF001=(PI*VSC**2*BESSJ0(XJ)**2)/(SQRT(2.)*BF0*T**2*VF**2*
     *BESSK0(XK)**2*BESSJ1(XJ)**2)
      DO 10 V=0,VXR
        XI1=A*SQRT(VC**2-P(V)**2)
        KF0V(V)=PI*N_F*N_S*VF**2*KS*T*EXP(-VC*(A+S))*(XI1*BESSJ0
     *(XJ)
     **BESSI1(XI1)+XJ*BESSJ1(XJ)*BESSI0(XI1))/(2*BF0*A**2*VSC*
     *BESSJ0(XJ)
     **(KF**2+VC**2-P(V)**2))
        WRITE(11,*)V,KF0V(V)
10    CONTINUE
      I=0
      J=0
20    DO 30 U=0,VXR,4
        DO 40 V=0,VXR,4
          XI2=A*SQRT(4*VC**2-(P(U)+P(V))**2)
          XI3=A*SQRT(4*VC**2-(P(U)-P(V))**2)
          QSUV(U,V)=PI*(N_S*KS*VF*T)**2*EXP(-2*VC*(A+S))*(BESSI1
     *(XI2)/XI2+BESSI1(XI3)/XI3)/(8*BSV(U)*VSC**2)
          WRITE(13,*)U,V,QSUV(U,V)
40      CONTINUE
30    CONTINUE
      DO 60 U=0,VXR
        XE=SQRT(UF**2+P(U)**2)
        KS1=PI*N_F*N_S*VSC*KS*EXP(-XE*(A+S))/(BSV(U)*BESSK0(XK)*T*XE
     *)
        KS2=(XE+VC-BS*(XE-VO)*EXP(-XE*T))/(KS**2+XE**2)
        KS3=AS*BS*EXP(-XE*T)/(VO+XE)
        KSU0(U)=KS1*(KS2+KS3)
        WRITE(12,*)U,KSU0(U)
60    CONTINUE
      CLOSE(13)
```

```
CLOSE(12)
CLOSE(11)
RETURN
END
```

Reference

1. W.H. Press, S.A. Teukolvsky, W.T. Vetterling, and B.P. Flannery, "Numerical recipes" in "Fortran: The art of scientific computing," 2nd ed., Chapter 16. Cambridge University Press, 1992.

Appendix 3   Overlap Integral

A3.1 Solution of the Field-overlap Integral (Butt-coupling) Coefficients

By definition of Equation 6.30 of Chapter 6, we have

$$P_{f0n} = \int_{A_\infty} F_0 S_n dA \tag{A3.1}$$

where $dA=dxdy$. This integral can be implemented by integrating over every part of the infinite cross-section area where the index of refraction is a constant. Let $A_f = A_\infty|_{n(x,y)=n_f}$, $A_s = A_\infty|_{n(x,y)=n_s}$, $A_o = A_\infty|_{n(x,y)=n_o}$, and $A_c = A_\infty|_{n(x,y)=n_c}$, then we have $A_\infty = A_f + A_s + A_o = A_c$. Thus

$$P_{f0n} = I_{fn} + I_{sn} + I_{oc} + I_{cn} \tag{A3.2}$$

where

$$I_{fn} = \int_{A_f} F_0 S_n dA \tag{A3.3a}$$

$$I_{sn} = \int_{A_s} F_0 S_n dA \tag{A3.3b}$$

$$I_{on} = \int_{A_o} F_0 S_n dA \tag{A3.3c}$$

$$I_{cn} = \int_{A_c} F_0 S_n dA \tag{A3.3d}$$

First, from Equations 6.29a, 6.63a and 6.65a of Chapter 6, we have immediately $I_{fn}=a^2 C_{f0n}/V_f^2$

$$I_{fn} = a^2 C_{f0n}/V_f^2 = \frac{\pi a N_f N_s k_s t e^{-\gamma_c(a+s)}}{V_{sc} J_0(k_f a)(k_f^2 + \gamma_c^2 - \sigma_n^2)}$$

$$\times [\sqrt{\gamma_c^2 - \sigma_n^2} J_0(k_f a) I_1(\sqrt{\gamma_c^2 - \sigma_n^2}\,a) + k_f J_1(k_f a) I_0(\sqrt{\gamma_c^2 - \sigma_n^2}\,a)] \tag{A3.4}$$

Next, from Equations 6.29b, 6.63b and 6.65b, let $C_{sn0}=k^2[(n_s^2-n_c^2)C_s + (n_o^2-n_c^2)C_o]$, we obtain

$$I_{sn} + I_{on} = C_s + C_o$$

$$I_{sn}+I_{on} = C_s + C_o = \frac{\pi N_f N_s k_s t e^{-\sqrt{\gamma_f^2+\sigma_n^2}\,(a+s)}}{2V_{sc}K_0(\gamma_f a)\sqrt{\gamma_f^2+\sigma_n^2}} \tag{A3.5}$$

$$\times\left\{\frac{\sqrt{\gamma_f^2+\sigma_n^2}+\gamma_c - B_s(\sqrt{\gamma_f^2+\sigma_n^2}-\gamma_o)e^{\sqrt{\gamma_f^2+\sigma_n^2}\,t}}{k_s^2+\gamma_f^2+\sigma_n^2} + \frac{B_s e^{-\sqrt{\gamma_f^2+\sigma_n^2}\,t}}{\gamma_o+\sqrt{\gamma_f^2+\sigma_n^2}}\right\}$$

Finally, from Equation 3.54, let $N_{f0}=N_f/K_0(\gamma_f a)$, we have

$$I_{cn} = N_{f0}\int_{A_c} K_0 S_n dA = I_{cf}-I_f \tag{A3.6}$$

where

$$I_{cf} = N_{f0}\int_{A_c+A_f} K_0 S_n dA \tag{A3.7a}$$

$$I_f = N_{f0}\int_{A_f} K_0 S_n dA \tag{A3.7b}$$

$$I_{cn} = \frac{\pi N_{f0} N_s k_s t}{V_{sc}}\left\{\frac{e^{-\gamma_c(a+s)}}{\gamma_f^2+\sigma_n^2-\gamma_c^2}[\sqrt{\gamma_c^2-\sigma_n^2}\,aK_0(\gamma_f a)I_0(\sqrt{\gamma_c^2-\sigma_n^2}\,a)\right.$$

$$\left.+\gamma_f aK_1(\gamma_f a)I_0(\sqrt{\gamma_c^2-\sigma_n^2}\,a)]+\frac{e^{-\sqrt{\gamma_f^2+\sigma_n^2}\,(a+s)}}{2\sqrt{\gamma_f^2+\sigma_n^2}(\gamma_c-\sqrt{\gamma_f^2+\sigma_n^2})}\right\} \tag{A3.8}$$

A3.2 Vector Formulation of Power Conservation in Coupled-mode Theory

The conservation of the total guided-power along the z-axis P requires:

$$\frac{\partial P}{\partial z} = \frac{1}{4}\frac{\partial}{\partial z}\iint_{A_\infty} (E\times H^*+E^*\times H)\cdot\hat{z}dA = 0 \tag{A3.9}$$

and, utilizing the following field expressions

$$E = \sum_{n=0}^{N_f} a_n e_{fn} + \sum_{n=0}^{N_s} b_n e_{sn} \tag{A3.10a}$$

$$H = \sum_{n=0}^{N_f} a_n h_{fn} + \sum_{n=0}^{N_s} b_n h_{sn} \tag{A3.10b}$$

we have

$$P = \sum_{m,n=0}^{N_f} P_{fmn} a_m^* a_n + \sum_{m,n=0}^{N_s} P_{smn} b_m^* b_n + \sum_{m,n}^{N_f,N_s} P_{fsmn} a_m^* b_n + \sum_{m,n}^{N_f,N_s} P_{sfnm} a_m b_n^* \qquad (A3.11)$$

where

$$P_{fmn} = \frac{1}{4} \int_{A_\infty} (e_{fm}^* \times h_{fn} + e_{fn} \times h_{fm}^*) \cdot \hat{z} dA \qquad (A3.12a)$$

$$P_{smn} = \frac{1}{4} \int_{A_\infty} (e_{sm}^* \times h_{sn} + e_{sn} \times h_{sm}^*) \cdot \hat{z} dA \qquad (A3.12b)$$

$$P_{fsmn} = \frac{1}{4} \int_{A_\infty} (e_{fm}^* \times h_{sn} + e_{sn} \times h_{fm}^*) \cdot \hat{z} dA \qquad (A3.12c)$$

$$P_{sfmn} = \frac{1}{4} \int_{A_\infty} (e_{sm}^* \times h_{fn} + e_{fn} \times h_{sm}^*) \cdot \hat{z} dA \qquad (A3.12d)$$

$$P_{sfnm}^* = P_{fsmn} \qquad (A3.13)$$

or, in matrix form

$$\frac{\partial (\mathbf{A}^+ \mathbf{P} \mathbf{A})}{\partial z} = 0 \qquad (A3.14)$$

where

$$\mathbf{P} = \begin{pmatrix} P_{ff00} & P_{fs00} & P_{fs01} & \cdots & P_{fs0N} \\ P_{sf00} & P_{ss00} & P_{fs11} & \cdots & P_{fs1N} \\ P_{sf10} & P_{sf11} & P_{ss11} & \vdots & P_{fs2N} \\ \vdots & \vdots & \cdots & \ddots & \vdots \\ P_{sfN0} & P_{sfN1} & P_{sfN2} & \cdots & P_{ssNN} \end{pmatrix} \qquad (A3.15)$$

$$A = \begin{pmatrix} a_0 \\ b_0 \\ b_1 \\ \vdots \\ b_N \end{pmatrix} \qquad (A3.16)$$

Appendix 4 Coupling Coefficients

Listed below are the new additional coupling coefficients computed using Equations 7.7 and 7.8 of Chapter 7 exactly, except for Equation A4.3a where a large-argument, asymptotic approximation of the modified Bessel function K_0 is made.

$$V_{f00} = \frac{\pi N_f{}^2 K_f a J_1(k_f a)}{\beta_{f0} J_0(k_f a)} \ln\left(\frac{n_c}{n_f}\right) + \frac{N_f{}^2 \gamma_f}{\beta_{f0} K_0(\gamma_f a)}$$

$$\left[(a+s)\ln\left(\frac{n_s}{n_c}\right) f(a+s) + (a+s+t)\ln\left(\frac{n_o}{n_s}\right) f(a+s+t) \right] \tag{A4.1a}$$

where

$$f(x) = \frac{\pi}{\gamma_f x}\left[E_2(2\gamma_f x) + \frac{1}{4\gamma_f x} E_3(2\gamma_f x) + \left(\frac{1}{2} - \frac{3}{64\gamma_f{}^2 x^2}\right) E_4(2\gamma_f x) \right.$$

$$\left. + \frac{1}{8\gamma_f x} E_5(2\gamma_f x) - \frac{3}{128\gamma_f{}^2 x^2} E_6(2\gamma_f x) \right] \tag{A4.1b}$$

$$V_{smn} = \frac{\pi(N_s k_s t \gamma_c)^2 a}{2\beta_{sm} V_{sc}{}^2} \ln\left(\frac{n_f}{n_c}\right)$$

$$e^{-2\gamma_c(a+s)} \left\{ \frac{I_1\left[\sqrt{4\gamma_c{}^2 - (\sigma_m + \sigma_n)^2}\, a\right]}{\sqrt{4\gamma_c{}^2 - (\sigma_m + \sigma_n)^2}} + \frac{I_1\left[\sqrt{4\gamma_c{}^2 - (\sigma_m - \sigma_n)^2}\, a\right]}{\sqrt{4\gamma_c{}^2 - (\sigma_m - \sigma_n)^2}} \right\} \tag{A4.2}$$

$$V_{fs0n} = \frac{\pi N_f N_s \gamma_f a k_s t K_1(\gamma_f a)}{\beta_{f0} V_{sc} K_0(\gamma_f a)} \ln\left(\frac{n_f}{n_c}\right) e^{-\gamma_c(a+s)} \left\{ \frac{\gamma_c{}^2 + \sigma_n{}^2{}_f}{(\gamma_c{}^2 - \sigma_n{}^2{}_f)^{\frac{3}{2}} a} I_1\left(\sqrt{\gamma_c{}^2 - \sigma_n{}^2}\right) \right.$$

$$- \frac{\gamma_c{}^2}{\gamma_c{}^2 - \sigma_n{}^2} I_0\left(\sqrt{\gamma_c{}^2 - \sigma_n{}^2}\right) \right\} + \frac{\sqrt{\pi} N_f N_s k_s t}{\beta_{f0} V_{sc} K_0(\gamma_f a)} \left(\gamma_c{}^2 + \sigma_n{}^2\right)^{\frac{1}{4}}$$

$$\left\{ \frac{\sqrt{a+s}}{V_{sc}} \ln\left(\frac{n_s}{n_c}\right) K_{\frac{1}{2}}\left[\sqrt{\gamma_c{}^2 + \sigma_n{}^2}(a+s)\right] \right.$$

$$- \frac{\sqrt{a+s+t}}{V_{so}} \ln\left(\frac{n_s}{n_o}\right) K_{\frac{1}{2}}\left[\sqrt{\gamma_c{}^2 + \sigma_n{}^2}(a+s+t)\right] \sqrt{\gamma_c{}^2 - \sigma_n{}^2}$$

$$- \frac{\gamma_c{}^2}{\gamma_c{}^2 - \sigma_n{}^2} I_0\left(\sqrt{\gamma_c{}^2 - \sigma_n{}^2}\right) \right\} \tag{A4.3a}$$

637

$$V_{sfm0} = \frac{\pi N_f N_s k_s t \gamma_c^2 a}{\beta_{sm} V_{sc}^2 \sqrt{\gamma_c^2 - \sigma_m^2}} \ln\left(\frac{n_f}{n_c}\right) e^{-\gamma_c(a+s)} I_1\left[\sqrt{\gamma_c^2 - \sigma_m^2}\, a\right]$$

$$- \frac{\pi N_f N_s k_s t \gamma_c^2 a}{2\beta_{sm} K_0(\gamma_f a)\sqrt{\gamma_f^2 + \sigma_m^2}} \left\{\frac{\gamma_c}{V_{sc}} \ln\left(\frac{n_s}{n_c}\right) e^{-\sqrt{\gamma_f^2 + \sigma_m^2}\,(a+s)}\right.$$

$$\left. + \frac{\gamma_o}{V_{so}} \ln\left(\frac{n_s}{n_o}\right) e^{-\sqrt{\gamma_f^2 + \sigma_m^2}\,(a+s+t)} \right\}$$

(A4.3b)

Appendix 5 Additional Coupling Coefficients

Listed below are the analytical solutions of the additional coupling coefficients

$$\hat{K}^+_{f0n} = \frac{\pi A_0 N_s V_{sc}{}^2 e^{-\sqrt{\gamma_f{}^2+\sigma_n{}^2}(a+s-d)}}{\beta_{f0} V_{so} \sqrt{\gamma_f{}^2+\sigma_n{}^2} \, t(k_s{}^2+\gamma_f{}^2+\sigma_n{}^2)}$$

$$\{k_s(\gamma_o - \sqrt{\gamma_f{}^2+\sigma_n{}^2})[\cos(2k_st) - e^{-\sqrt{\gamma_f{}^2+\sigma_n{}^2}h}] - (k_s{}^2 + \gamma_o\sqrt{\gamma_f{}^2+\sigma_n{}^2})\sin(2k_st)\} \tag{A5.1}$$

$$\hat{K}^-_{f0n} = \frac{\pi A_0 N_s V_{sc}{}^2 e^{-\sqrt{\gamma_f{}^2+\sigma_n{}^2}(a+s-d)}}{\beta_{f0} V_{so} \sqrt{\gamma_f{}^2+\sigma_n{}^2} \, t(k_s{}^2+\sigma_n{}^2-\gamma_o{}^2)}$$

$$\{k_s(\gamma_o - \sqrt{\gamma_f{}^2+\sigma_n{}^2})e^{-\gamma_o(a+s)}[e^{-(\gamma_o+\frac{\sqrt{\gamma_f{}^2+\sigma_n{}^2}}{2})d} - e^{\gamma_o d}]\} \tag{A5.2}$$

$$\hat{Q}^+_{smn} = \frac{N_s{}^2 V_{sc}{}^2 \delta_{mn}}{4\beta_{sm} V_{so}{}^2}$$

$$\{2\gamma_o \sin(k_sd/2)\sin(2k_st) + \frac{1}{k_s}(\gamma_o{}^2 - k_s{}^2)\sin(k_sd/2)\cos(2k_st) - (k_s{}^2 + \gamma_o{}^2)\} \tag{A5.3}$$

$$\hat{Q}^-_{smn} = \frac{\varepsilon_0 N_s{}^2 k_s{}^2 V_{sc}{}^2 \delta_{mn} e^{2\gamma_o t} sh(\gamma_o h)}{2\beta_{sm} \gamma_o V_{so}{}^2} \tag{A5.4}$$

$$\hat{K}^+_{sm0} = \frac{\beta_{f0}}{\beta_{sm}} \hat{K}^+_{f0m} \tag{A5.5}$$

$$\hat{K}^-_{sm0} = \frac{\beta_{f0}}{\beta_{sm}} \hat{K}^-_{f0m} \tag{A5.6}$$

Appendix 6   Elliptic Integral

The solution of the elliptic integral in Equation 7.30 can be expressed in the odd and even Jacobian elliptic functions, i.e., sn and cn (Byrd and Friedman 1954; Meng and Okamoto 1991), respectively, we have

$$P_r = \left\{\alpha_2 \operatorname{sn}^2\left[\kappa z \mid m\right]\right\} / \left\{1 - (\alpha_2 / \alpha_1)\operatorname{cn}^2\left[\kappa z \mid m\right]\right\} \tag{A6.1}$$

where κ and m are a constant and the modulus of the elliptic functions sn and cn, respectively, defined as follows

$$\kappa \equiv 2Q_{ba}(1 + 4\zeta\eta)^{1/4} / (1 - 2P_{ab}\zeta) \tag{A6.2a}$$

$$m = \alpha_2(\beta_2 - \alpha_1) / \left[\beta_2(\alpha_2 - \alpha_1)\right] \tag{A6.2b}$$

In general cases, $\zeta\eta \ll 1$ and, when the coupling is not too strong, $P_{ab}^2 \ll 1$, Equation 7.33a becomes

$$m \approx \zeta^2\left[1 + 2P_{ab}(\zeta - \eta)\right] \tag{A6.3}$$

Appendix 7 Integrated Photonics: Fabrication Processes for LiNbO₃ Ultra-Broadband Optical Modulators

Mach-Zehnder interferometric modulators (MZIM) currently attract much attention in the modulation of lightwaves of different modulation formats such as RZ, NRZ, CS-RZ phase, amplitude or frequency schemes. This chapter thus presents the design and fabrication of the traveling wave electrodes and optical waveguide structures for ultra-broadband operations. A program for the design of the traveling wave electrodes of coplanar waveguide (CPW), coplanar structures (CPS) and asymmetric coplanar structures (ACPS) are described with the procedures for the execution of the design package.

The fabrication processes for optical waveguides using Ti diffusion and gold plated electrodes are described in detail. The test results of the MZIMs and traveling wave electrodes are illustrated to demonstrate that we have implemented our design package and the fabrication of 30 GHz 3 dB bandwidth MZIM.

We further propose the extension of the operational bandwidth of optical modulators into the 100 GHz with an ultra-thick metal layer of several hundreds of micrometers using the deep x-ray lithography fabrication process.

A7.1 Introduction

Integrated photonics has evolved over the last thirty years from the proposal of optical waveguides in 1966 [1] and then to guided wave optics and integrated optics in 1969 [2] to the present generation of photonics, the integrated photonics. Concurrently optical fiber communications have progressed tremendously from a mere 10 Mb/s in 1974 to 40 Gb/s, 80 Gb/s and then 160 Gb/s per wavelength channels. The multiplexing of several wavelength channels has allowed the increase of the total transmission capacity over a single mode fiber to a few tens of terahertz over the S-, C- and L-bands of the 1550 nm spectral windows. We also note that several modulation formats have also been generated* by these external modulators.

The principal technique that allows the development of such enhancement of the transmission speed is the external modulation of lightwaves, thus the minimization of the broadening of linewidth of the laser sources. The external modulation is usually performed by photonic transmitters. Presently the most popular modulator is the lithium niobate (LN) interferometric types which are formed by an interferometric waveguide structure along which the lightwaves interact with the electrical traveling waves via wideband traveling wave electrodes.

---

* Readers are referred to several papers presented in the IEEE Workshop on Advanced Modulation Formats, San Francisco, 2004.

This chapter is the second part of a series of technical articles to present the fabrication techniques for MZIM in LN material systems [3]. The processes for the design of the electrodes, optical waveguides using optical lithography are described including all chemical and packaging stages. We further extend this work with the proposal of x-ray lithography for fabrication of photonic transmitters that could operate into the millimeter-wave region (>100 GHz bandwidth).

This chapter is organized as follows: In the next section we give the details of the design and execution of the design package of the traveling wave electrodes and the processes for fabrication of such electrode structures. Section A7.3 details the fabrication processes for optical waveguides, optical modulator structures and the integration of the electrodes and waveguide structures. We then propose the use of x-ray lithography for thick electroplating of the gold electrodes so as to produce optical modulators operating into the 100 GHz region.

A7.2 Design and Fabrication Processes

The design and fabrication of the interferometric modulators for broadband modulation and generation of advanced modulation formats require: (i) accurate and detailed design of the traveling wave electrodes, the guided optical waveguides, the interferometric guided wave structures, the electro-optic interaction and its effective overlap integral; (ii) fabrication of the optical waveguides and the traveling wave electrodes; (iii) packaging of fiber-optical waveguide coupling; and (iv) testing of optical and microwave properties of the MZIM. The flow charts in Figure A7.1 show the design and fabrication of these broadband optical modulators.

Figure A7.1 shows the flowchart of the design steps for the design of the traveling wave electrodes—all processes (step by step) for fabrication of the MZIM. Four fundamental stages are

- Stage 1: Numerical modeling and design of Mach-Zehnder (MZ) waveguides and electrode structures using the finite difference and conformal methods—specification of modulator parameters including the design of the traveling wave electrodes and the design of optical waveguides and interferometric structures.
- Stage 2: Mask generation for electrode structures and interferometric optical waveguiding structures using AutoCAD.
- Stage 3: Fabrication processes of a wideband optical modulator.
- Stage 4: Modulator packaging including fiber-diffused optical waveguide coupling, pigtailing and testing.

A7.2.1 Design Software Packages—FDTWE

All programs used for the design of ultra-broadband integrated optical modulators have been developed for gigahertz optical modulation devices written by Chua [4] in which the finite difference method (FDM) and conformal mapping have been extensively exploited to minimize the computing time and maximize the accuracy.

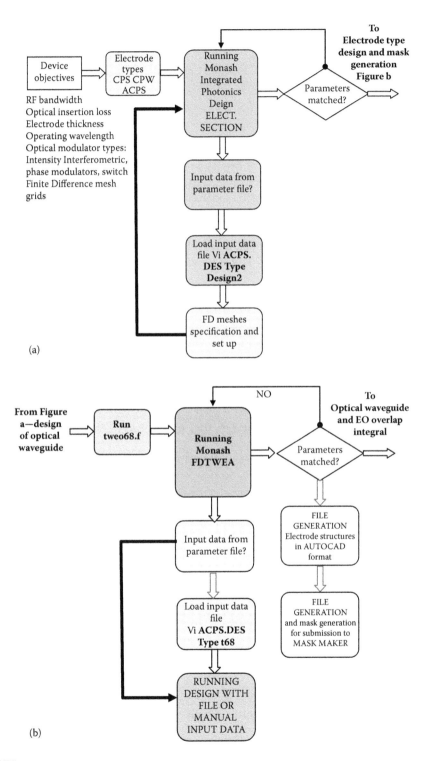

FIGURE A7.1
(a) Traveling wave electrode design for ultra-wideband interferometric modulation. (b) Generation of AutoCAD data file for mask generation.

A7.2.1.1 Design of Traveling Wave Electrode Structures

Figure A7.1b shows the flowchart of the steps of the design of the ultra-broadband traveling wave electrode types such as the CPW, the CPS and ACPS. The details of the steps and execution procedures for the program are shown in the next section.

A7.2.1.2 Hardware Requirement

The programs used for designing electrode devices must be run in a Unix or PC system which can be **mdw, mda,** and **redyn** or **fawlty** machine.[*] Due to the size of the matrix that is used to store large results (500 × 500), a memory of 1–10 G RAM may be required to solve the sparse matrix of 10,000 × 10,000 elements. Naturally the lower memory capacity demands longer execution time. It is expected that the running time would be around 4–5 min of CPU for the design of a typical traveling wave electrode and MZIM structure.

The two main files of the design of a traveling wave electrode are: **design2.f** and **twe068.f** as indicated in Figure A7.1. The external files linked by the two main programs are

- **Chargeo23.f, cpwo4.f, cpso3.f, conformal, acpso9.f**
- **Chargeo23.o, cpso23.o, twe058.o, acpso9.o, cpwo4.o** then linking to programs for the generation of the executable files

The names of all input data files are

- **acps.des, cpw.des, and cps.des** (input data files for design2.f program)
- **acps.in, cpw.in, and cps.in** (input data files for twe068.f program)

The command for compiling all these FORTAN 77 packages is

f77 -o filename filename.f (the filename is the name execution file)

The following commands are for compiling a program applicable for the mda machine (alpha machine)

mk f1 & (f1 file with extension f (filename.f))

then **mk1 f1 f2 f3. fe** (f1-f3 are files with extension **o** (filename.o) and **fe** is the name of the link file, used for execution). Then call **t68** for the execution file (fe = **t68**).

All the **NSPCG** files are the FORTRAN 77 math files, so there no need to compile them, but they must be included in a directory called **nspcg.o**.

A7.2.2 Execution of Simulation Programs

Two main programs are to be used for the design of the traveling wave electrodes

- **design2.f**
- **twe068.f**

---

[*] A number of these workstations may not be supported by Monash IT Computer Centre.

The **design2**.f program generates *approximate* results for use as the initial guest data to ensure the convergence of the matrix solver. This gives the range of the electrode gap and the width of electrodes which would give the best approximation for bandwidth, V_π, etc. Using these data outputs one can run the **tweo68.f program** to double check and obtain more accurate parameters which can then be checked with the specifications. The **tweo68.f** program gives us very accurate results with about 5% differences compared to experimental results. Furthermore, another two steps must be followed for designing different types of electrodes: the CPS and ACPS. For ACPS the **acps.des** and **acps.in** are used as the input data files for the **desgn2.f** and **tweo68.f** programs.

A7.2.2.1 Step 1: Running the design.f Program

This program prompts questions requesting all information required for calculations: (i) calculation mode (individual design or design curve); (ii) structure type of electrodes: ACPS, CPW or CPS; (iii) name of files to which the data output can be saved; (iv) the length of the electrode (cm); (v) the range of the operating frequencies (f1,f2) (GHz); (vi) the thickness of the electrode (ts) (μm); (vii) The range of the electrode width (w1,w2) and gap (g1,g2) (μm) that may satisfy the requirement; and (viii) the thickness of the buffer layer (SiO$_2$ thin film) (μm).

These parameters can be loaded directly from an input data file or entered manually. To run the program and input data from the file then type

   design2 < acps.des

The **design2.f** program imports all input data from the input (**acps.des**) file instead of reading data from the keyboard. Alternatively one can instruct the program to read data from the keyboard. In this case type

   design2

The program will prompt all of the above questions. The output results are stored in two files. One contains all the data about characteristics of the device and the other all the values of the scattering parameters **S$_{21}$**. The value of **S$_{21}$** allows us to evaluate the electrical insertion loss of the device.

All output data can be plotted out by using **Gnuplot** or **Matlab** software.

Using **Gnuplot** software to plot all data you want to see. This software can plot many curves in one graph. To use **Gnuplot** software, type

   gnuplot <enter> (run software)

User will now be in **Gnuplot**, and then plots of all results are stored in the output files. Gnuplot can be used to plot the results

   gnuplot> plot "filename" using xcol:ycol, 'filename' using xcol: ycol

(where **xcol, ycol** are columns of plotted results, and filename is the name of the result file to be plotted).

From the graph or the output data file, we could inspect all possible ranges of data that may satisfy requirements such as gap, width, thickness of gold layer, thickness of SiO$_2$ layer. Then use the **tweo68.f** program to determine the set that are closed or matched with the specified parameters.

A7.2.2.2 Step 2: Running the tweo68.f Program

The running procedures for this program are the same as for the **design2.f** program. The input data can be imported from the input file or from the keyboard. The input data files corresponding to three types of electrodes are **acps.in, cps.in,** and **cpw.in**. At the prompt the input data file requests the following data: (i) type of electrode structures: (ACPS, CPW, or CPS); (ii) type of LiNbO$_3$ wafer (crystal orientation); (iii) thickness of air region (μm) (choose 250 μm); (iv) thickness of gold layer (μm); (v) width of electrode (μm); (vi) gap of electrode (μm); (vii) thickness of buffer layer (μm); (viii) thickness of LiNbO$_3$ substrate (mm); (ix) number of grid points in x-dimension (xdim) and in y-dimension (ydim); (x) Bandwidth (GHz); (xi) operating wavelength (μm); (xii) type of results you want to get (1: compute everything, 2: Z and nm, 3: V_π and γ only); and (xiii) name of output data file.

All the above data are stored in the files **acps.in, cpw.in** or **cps.in,** which are the input data files for running the **tweo68.f** program.

To run **tweo68.f**, type

> **t68 < acps.in** (if you want the program to get data from the file and the electrode device is ACPS)

The program automatically imports all data values from the input file, calculates and outputs results.

> or **t68** (if users intend for the program to import data from the keyboard/manual entries)

The program will prompt all questions asking for the device information. All results are stored in a file. To see the result, type

> **vi filename**

If all the results in this file satisfy the device requirements, then print out this file or record them for drawing and fabrication. If the results do not satisfy, use other values or adjust the input data then run the program again until the results match with the device's parameters.

A7.2.3 Method of Drawing Devices

A7.2.3.1 Flowchart for MASK Drawing

There are many steps in drawing a mask using AutoCAD software. Figure A7.5 summarizes the steps for the use of AutoCAD for drawing a mask. Details of all steps are described in Section A7.2.3.3 (Using AutoCAD).

A7.2.3.2 Design Rules for ASM 2500

The following rules only apply with the ASM 2500 software of the mask generator as specified by the Defense Science and Technology Organization (DSTO). Note that this may be different to other mask manufacturers. AutoCAD software is a general purpose drafting program we can use for electronic mask layouts. However there are limitations to using AutoCAD. Some commands are strictly used in drawing.

Some rules for drawing the mask are

1. The ASM 2500 works only with DXF files so a drawing mask file drawn in AutoCAD must be converted to a DXF file and converted back to DWG file then reconverted back to DXF file for checking.
2. The drawing grid size unit can be **mm**, **μm**, or **inch**. But often we use μm because the dimensions of our devices are in **μm**).
3. The area of the drawing mask is not higher than the maximum, $6' \times 9'$, but our mask area is $3' \times 3'$.
4. Set Grid Snap mode and set it as small as possible depending on your requirements. Normally a snap size equal to 0.1 is used.
5. Set **Orth ON** whenever it is possible. A pattern generator is most efficient when imaging orthogonal rectangles.
6. The **LINES command**: the endpoint of a line must touch precisely and one line cannot share by more than one closed boundary.
7. The **PLINE command**: the width of the poly-lines can be zero or finite. The poly-line has zero width must close (starting and end points must touch). The poly-line with finite width must not touch or cross upon itself.
8. **All blocks** must be completed before insertion into the master file.
9. **Islands**: this is used when you draw alignment marks or there is an island in a device. The AutoCAD's PLINE command with zero width is used for the best result.
10. Any gap in a device **must not be less** than 2.5 μm.

Before starting the drawing, the following values must be known or determined

- The area of the each pattern (width and height)
- The area of the mask (master file) you are going to use (width and height)

These data will be used to set the drawing area.

A7.2.3.3 Using AutoCAD

AutoCAD can accept commands in lower case or capital letters. To run AutoCAD, log onto the drive or directory containing the AutoCAD program file and enter:

ACAD

Then the main menu appears on the display screen. It has eight options and the main menu would be:

Main Menu

0. Exit AutoCAD
1. Beginning a NEW drawing
2. Edit an EXISTING drawing

   3. *Plot a drawing*

   4. *Print plot a drawing*

   5. *Configure AutoCAD*

   6. *File utilities*

   7. *Compile shape/font description file*

   8. *Convert old drawing file*

Enter selection

If a new file is to be open for drawing then enter 1 (new drawing), and if you want to open an existing drawing then enter 2 (an existing file). Then AutoCAD prompts:

   Enter NAME of drawing: **<name of the file for saving> < enter >**

The user is now in the AutoCAD window and can start to draw. Because our device's area is greater than the area of the drawing windows, we must set up the drawing environment such as unit, drawing limits in the first instance. The following commands instruct users how to set up a drawing window. At the command area, do the following steps:

   command: LIMITS   <enter>
   ON/OFF/<Lower left corner> <0.0000,0.0000>: type 0,0 <enter>
   Upper right corner <12.0000,9.0000>: x,y <enter>

(where x and y are the values of a point at the top right corner of the drawing window)

   command: ZOOM   <enter>
   All/Center/Dynamic/Extents/Left/Previous/Windows/<Scale>: All <enter>

(Choose All for full window)

   Now the dimension of the drawing window is equal (or larger) than the device's area.

   To get the coordinate of a point, press **ctrl-d** (at the same time), then you will see two numbers on the top right corner of the window that indicates the coordinates (row, column) of that point. If you move the mouse then the numbers also change. Turn ORTH ON by pressing **ctrl-o** (at the same time)

   command: GRID   <enter>
   Grid spacing(x) or ON/OFF/Snap/Aspect <10.000>: < size of grid lines> <enter>
   command: SNAP <enter>
   Snap spacing or ON/OFF/Snap/Aspect <1.0000>: < size of the snap lines> <enter>

The snap size is as small as possible, that makes the drawing more accurate (I used 0.1 for the size of snap lines).

   After you set-up the drawing windows you can start to draw.

Device drawing

Our devices consist of two types

- MZ waveguides
- Electrodes

There is a special function for drawing the MZ waveguides, and the electrode structures can be drawn to match with the waveguides structures.

Drawing Mach-Zehnder Waveguides

Move the mouse to the top of the window, the main menu will appear, then click special function. Users will see many types of waveguide pattern, then choose MZ. After the selection you will see the following questions at the command lines (one by one) and users must enter the values for each question

Start point? (the point where you want to start)

Do you want to move the object yourself <Y/N>?:

Orientation?

| **Length 1?** | **Length 2?** | **Length 3?** | **Length 4?** | **Radius?** |
| **Angle 1?** | **Angle 2?** | **Width 1?** | **Width 2?** | |

Note: The commands used in this special function are AutoCAD's PLINE with finite width value.

The meaning of all the symbols in the special function can be found in the AutoCAD User Guides. The gap between one waveguide to the other should be greater than 20 μm to avoid interference between the two waveguides.

Drawing Electrode Structures

- The two commands that can be used are: LINE or PLINE.
- The PLINE command is very complicated. You must know the beginning width and end width of a line.
- Using the LINE command is easier than the PLINE command if all alignment marks are outside the area of the pattern. For ease try to place all alignment marks outside the area of the pattern. All the lines must touch exactly (the end point of one line must be the beginning point of the next line) and the lines make a closed figure.
- Drawing the alignment marks, the PLINE with zero width and right wind rule must be used.
- The coordinator of a point can be entered manually or by mouse indicator.
- After drawing all individual device components, then open a file (master file) to arrange all sub-structures including their superposition on each other to generate a drawing mask. This file is to be sent to the mask manufacturer for mask generation.

To further understand the design rules of ASM 2500 and AutoCAD commands, see *Design Rules for ASM 2500* and the AutoCAD manual.

A7.2.4 Fabrication Processes

A7.2.4.1 Flowchart of Fabrication Procedures of a 20 GHz Optical Modulator

The processes for the fabrication of an optical modulator are summarized in Figure A7.1b, the flowchart of the fabrication of an optical modulator. Details of each stage are described in the following sections.

A7.2.4.2 Fabrication Procedures

Throughout all processes, $LiNbO_3$ substrates must be handled with plastic (preferably Teflon) tweezers to avoid damage to the substrates. All procedures are done in a clean room at room temperature and using yellow light, unless otherwise stated. All containers used in all processes are of plastic or Teflon types to avoid substrate damage. Three types of photo-resist (PR) used are: AZ5214, AZ4210 and AZ4620.

   Equipment: Spinner; a mask aligner (KARLSUSS MJB 3); a vacuum oven; an ultrasonic cleaner; an electromagnetic hotplate stirrer; magnetic rods; a 3-L beaker (glass); a microscope; an ultrasonic cleaner.

Process 1: Procedures for cleaning $LiNbO_3$ substrates

1. Use a cotton tip to clean a substrate with acetone (spray) for 1 min
2. Rinse it with deionized (DI) water
3. Clean it with Decon90 (using a cotton tip)
4. Rinse it with DI water
5. Dry a substrate by blowing nitrogen air
6. Check the substrate under a microscope. If it cleans then bake it in a 90°C vacuum oven then store it in a Fluoroware container. Otherwise, repeat steps 3–6 if there are dirt spots (brown color) or repeat all steps if there are oil spots (multicolor)

Process 2: Ti thin-film evaporation and deposition

All clean substrates are placed on the metal plate. The evaporation electron beam current and evaporation time are set according to the calibration curves of the Balzer e-beam evaporation machine. For the best result of evaporation, all the substrates must be cleaned and baked (in the vacuum oven), and evaporated on the same day.

Process 3: PR layer coating

1. Turn on the Spinner machine
2. Place the substrate on the substrate holder
3. Vacuum on
4. Blow the substrate with nitrogen air
5. Drop PR on
6. Adjust the time and speed meter as required

- Time: 40 s
- Speed: 5000 rpm if using AZ4210 and adjust to about 1.9 μm thick

    5000 rpm if using AZ5214 and get about 1 μm thick

    6000 rpm if using AZ4620 and get 6 μm thick

    The thickness of PR has been measured and plotted as a calibrated curve.

7. Press the operation button
8. Vacuum off
9. Take out the substrate
10. Turn off the Spinner machine
11. Place the substrate in the 90°C vacuum oven
    - About 15 min if using AZ5214 or AZ4210
    - About 30 min if using AZ4620
12. Remove the substrate out of the oven and let it cool before proceeding to the next process. The cooling time depends on the thickness of the PR layer, roughly about 10 min/1 μm.

Process 4: **Photolithography**

The KARLSUSS MJB 3 mask aligner is used for this process. The machine **must be turned on an hour** before the lithographic processing.

Exposure Time

- AZ4620: 75 s
- AZ4210: 45 s
- AZ5214: 5 s

Developer

- AZ5214E and AZ4210: using one part of AZ400K mixed with four parts of DI water
- AZ4620: using one part of AZ400K mixed with three parts of DI water

The following steps tell you how to use the mask aligner to get the PR pattern used for gold plating

1. Place the mask on the mask holder of the mask aligner (device to be taken must stay in the center of the hole) then press the vacuum button
2. Place substrate onto the metal plate then slide it in.

    (see operation manual for detail)
3. PR development
    - Place the substrate on an island of the cleaner then vacuum on to hold the substrate
    - Spray the developer into it (around) until PR removed
    - Water it in about 1 or 2 min

- Dry the substrate with nitrogen air (blowing)
- Bake it in the 90°C oven for 10 min before going on to the next process

Note: (i) time to develop is about 50 s; (ii) the pattern must be checked carefully before doing further processes. If it is not as good as we expected then remove the PR layer and repeat process 3 and 4.

Process 5: Titanium etching

Repeat processes 3 and 4 to get the PR pattern of waveguides. The waste Ti surrounding the waveguides is removed by etching in a fast Ti etching solution consisting of two parts of HF and 98 parts of DI water (HF 2%).

1. Pour the Ti etching solution into the plastic beaker (small)
2. Place the substrate on the substrate holder
3. Dip it into the solution and wait until the Ti is starting to dissolve

   (normally it takes about 10 s), then immediately take the substrate out of the etchant and rinse a substrate with DI water for about 2 min.
4. Dry substrate with nitrogen air (blow)
5. Check Ti waveguides, if all waveguides are good and if
   - the substrate is clean (especially the waveguide's area) then go to step 6
   - the substrate is dirty then repeat process 1 (do it lightly) until the substrate is clean then go to step 6, otherwise, remove the Ti layer
6. Bake it in an oven at 90°C for 15–30 min then the substrate must be stored in a clean Fluoroware container for diffusion.

Process 6: Titanium diffusion

Ti film diffuses into $LiNbO_3$ substrate in a diffusion furnace under high temperature. Only three substrates can be diffused at a time due to the small size of the substrate holder boat. The conditions of diffusion are

- Time: 7 h
- Diffusion temp: 1050°C
- Flow rate of O_2: ~0.5 L/min
- Temp. of water: 25°C
- Wet atmosphere with dry oxygen air bubbling through 1/2 bottle of warm DI water (set temperature = 25°C)

(See the manual of the furnace for more details of furnace operations).
Note: Place the substrates on a pure silica boat (three at a time), and only move the boat into the diffusion tube when the temperature of the furnace reaches 1050°C. Move in and take it very slowly out of the furnace (about 20–30 min) so as to avoid a sudden change of temperature on the substrate. Otherwise cracks will occur. After the diffusion, all the waveguides must be examined under a microscope before polishing. If all waveguides are not good (i.e., not smooth), that is all Ti materials are not fully diffused, then do not polish but continue the diffusion steps again to ensure that the diffusion is completed.

Process 7: **Polishing**

There are about three main steps in this process and the first step of preparation is very important. If this step is well done the two edges of the substrate are very sharp, therefore low coupling loss could be achieved. Only one substrate can be polished at a time and several preparation steps must be conducted before polishing the substrate:

- Cut the LiNbO$_3$ substrate (21 × 11 × 1.0 µm) into small pieces (about 4 × 11 × 1.0 µm). Then clean them and store them in a container, these will be used for the polishing process.
- Clean the substrate holder and cut the **Post-it** paper (glue part) into small pieces (about 4 × 11 mm).

Step 1:   Preparation of the substrate

1. Place a small piece of Post-it paper to the substrate holder.
2. Place the substrate (back side) in contact with the paper.
3. Place the small LiNbO$_3$ in contact with the substrate.
4. Stick the small piece of Post-it paper on the small piece of metal then place it back to the hold. Using a Philip key to draw gently two bolts in until you see the COLOR RING between the substrate and the small piece of LiNbO$_3$.
5. If you think they are not well in contact, then unpack and repack them again until you think it is good contact (see the color ring) for polishing.

Step 2: Polishing

1. Use the sand paper: the P.1200 or P.800 will be used in this step. Cut the sand paper into a small circle with a diameter equal to about half the diameter of the polishing disk. Then place it on the disk (however, the size of sand paper is up to you).
2. Turn on the machine. Hold the substrate holder by hand to polish until the unwanted edge of LiNbO$_3$ is removed.
3. Turn off the machine and clean it with acetone (using cotton tips). Observe it under the microscope, if the gap between the substrate and the small piece of LiNbO$_3$ is very small then clean it with Decon90 by using ultrasonic cleaner for about 15 min, and then go to step 4 if it is very clean. Otherwise, repeat steps 1–3.
4. Diamond paste for polishing
   a. **Using a 9 µm polishing disk**
      - Place the 9 µm disk into the polishing machine.
      - Place the substrate holder onto the 9 µm disk then lower the arm of the machine for holding it, then turn the machine on.
      - Drop three drops of Karosene onto the surface of the polishing disk before running the machine. Running time is about 20 min.
      - Turn off the machine and wash the substrate holder as before, then observe it under the microscope. If the gap between the substrate and the small piece of LiNbO$_3$ is smaller, then clean them for 15 min (using ultrasonic cleaner and Decon90). If it is not then re-polish it again until the gap is

smaller. If the edges are very clean after using the ultrasonic cleaner then go to next step (step b), otherwise re-clean it

b.   **Using a 3 µm polishing disk**

  - Place the 3 µm into the polishing machine then repeat all things in step (a).

If the two edges are good then repeat step (a) with a 1 µm disk and again with a 1/4 µm disk. However, the time to run the machine is about 8 min (about 12 min less than when you used a 9 µm and a 3 µm disk).

Step 3: Inspection

If the edge is good (i.e., no chips, not broken) then you will see the bright spots along the edge of the edge which is the end points of waveguides. If you do not see any bright spots then there are several things that one must do

  - Re-clean it, then observe it again, if you see them then go to step 4

Continue polishing for about 5 min then recheck it, if you see the bright spots then go to step 4. If you still do not see any bright spots, then unpack all and repeat step 1 to step 3 until you see a number of bright spots along the edge of the substrate.

Step 4: Finish polishing of the other end

Unpack everything out of the substrate holder. Clean the substrate then repeat the above steps 1 to 3 to polish the other end of the substrate.

After polishing the two ends of the waveguides, check the waveguides by applying the white light into one end so you can see the light is out at other end with the aid of a microscope. The two end points of the waveguides must not be chipped and it must be flat for low and minimum coupling loss. If the above features are not satisfied then unpack them (take the substrate out of the substrate holder for cleaning), then polish the other end (repeating steps 1 to 3).

This is a very time consuming process and it would take at least 1.5 days to finish polishing one substrate.

Process 8: Testing and determination of guided optical modes

After polishing all substrates must be subject to a test of optical insertion loss before going on to further steps. All waveguides in the substrate must be numbered before starting the test.

Firstly use the 633 nm laser for alignment between the fiber and the waveguide, and then use the 1300 nm or 1550 nm laser for measurement of the optical coupling loss (insertion loss). The results that must be taken are: (i) input power from laser (Pin); (ii) output power from one end of the testing waveguide (Pout); (iii) calculate optical insertion loss (Ploss); and (iv) the Ploss = 10 log (Pout/Pin).

Write down the results for each waveguide then choose the one that has the least optical insertion loss. This waveguide will be used to place the electrode on.

Process 9: SiO_2, Chrome (Cr) and Gold (Au) evaporation

  - The SiO_2 can be deposited on the substrate by the Balzer e-beam evaporation machine (not very good, bubble) or PCVD system (Plasma System). The thickness of the SiO$_2$ layer varies from 0.6–2 µm, the thicker SiO$_2$ layer the higher V_π produced.

- Thickness of Cr: 200 Å.
- Thickness of Au: 1000 Å.

Cr and Au layers are evaporated by the Balzer e-beam evaporation machine.

A7.2.4.3 Preparation of Electrode Pattern Used for Gold Plating

Repeat processes 3 and 4 to get the PR electrode pattern. PR AZ4620 and the reverse mask are used this time. The electrode pattern must be checked carefully to insure that there is no short circuit between the hot-line and the ground plant of an electrode. Also check under-etch of the PR pattern, if the difference of the width of the PR pattern and the width of the original pattern is about 2 µm then proceed with gold plating. If the pattern does not satisfy the above conditions, then use acetone to remove the PR layer and repeat processes 3 and 4. Record the ratio between the gap and the width of the electrode after developing. Then bake it in the 90°C vacuum oven for about 20 min, take it out and let it cool for about 5 min before electroplating.

Process 10: Electroplating

Equipment: (i) 1 3-L beaker (glass); (ii) 2 L of gold solution; (iii) a magnetic hotplate stirrer; (iv) a magnetic rod; (v) a digital multi-meter (it can measure $R > 1\ \Omega$, $I > 4$ mA, and $V > 1V$); (vi) 15V DC power supply.

Procedures

1. Prepare the plating bath: Pour 2 L of gold plating solution into the beaker then place onto a heating plate and heat it to 60°C. Turn the speed to 5; set up the DC current source (without load) to about 5 mA.

2. Using the tweezers to hold the substrate hang it to the hook. Ensure the tweezers is in contact with the gold layer of the device by using the multi-meter to measure the resistance. Only dip the substrate into the gold plating solution when the temperature of the solution is 60°C.

3. Turn the power on; turn the voltage knob to get the limited current.The plating current is about 5–5.5 mA. It depends on the plating area.

4. Turn off the power when the time is up and take the substrate out of the bath, cool it for about 1 min before wetting it with water then drying it with nitrogen air. Plating time depends on the thickness of the gold layer required and the plating area. Normally
   - 1:0 → 4 µm
   - 1:20 → 5 µm

5. After this process, the device must be carefully checked under the microscope. If everything is good (i.e., no short circuit, uniform plating, etc.), then bake it in the 90°C vacuum oven for about 10 min.

6. Measure the thickness of the gold and PR layers using Talystep.

7. Lift off the PR layer by acetone (ultrasonic cleaner **must** be used).

8. Bake it in the 90°C oven for about 10 min before doing Au and Cr etch

Process 11: Chrome and Gold etch

1. Repeat process 3 using PR AZ4620
2. Repeat process 4 using electrode pattern and reverse mask
3. Au etch (using the substrate holder to hold the substrate)
   a. *Making Au etchant*

   The waste gold around the electrode is removed by etching in a very fast Au etching solution. It consists of one part HNO_3 and three parts of HCl.

   The gold etching solution **must be stored** in an open container due to the possibility of a gas explosion. The solution should be left for about 30 min or until the color of the solution turns red-yellowish before use.
   b. *Gold etching* (i) dip the substrate holder with the substrate into the etching solution, and immediately take it out when the gold starts to dissolve. Rinse it with DI water. Etching time varies, normally it takes about 6–8 s per 1000 Å; (ii) dry the substrate with nitrogen air; (iii) check the electrode pattern, if it is good then go to Cr etch, otherwise it is depending on the electrode pattern, then *step b* can be repeated, or remove the Au and Cr layers.
4. Cr etching
   a. *Preparing the Cr etchant: The unwanted Cr around the electrode can be removed by using the following solution*
      - Ammonium cerium nitrate $((NH_4)_2Ce(NO_3)_6)$: 109.64 g
      - Hydrochloric acid 70% $(HClO_4)$: 85 ml
      - DI water: 1 L

      Mix them in a 1.5 L beaker stirring with a magnetic rod. Use the magnetic hotplate stirrer to mix them for about half an hour, and then store it in a glass bottle.
   b. *Cr etching*
      - Dip the substrate holder with the substrate into the etching solution and take it out when the Cr is dissolved (it takes about 8–10 s per 200 Å) then rinse it with DI water.
      - Dry it with nitrogen air
      - Check the electrode (i.e., under-etch, over-etch, short circuit, etc.)

      If everything is good, then go to step 5, otherwise remove all.
5. Bake it in a vacuum oven for about 5 min.
6. Use Talystep to measure the thickness of the electrode layers.

Process 12: Etching Pads

1. Repeat process 3
2. Repeat process 4 by using cpw reverse mask, about 500 μm distance from each corner of the electrode pattern
3. Repeat process 10

After this process, the finished optical modulator devices are ready for packaging. All the devices must be checked under the microscope for the following things before going to fiber pigtail

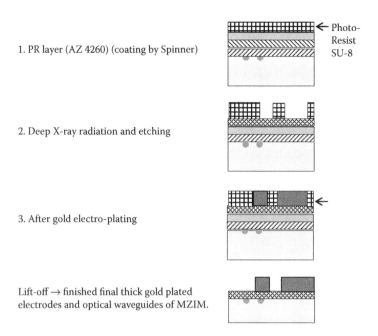

1. PR layer (AZ 4260) (coating by Spinner)

← Photo-
   Resist
   SU-8

2. Deep X-ray radiation and etching

3. After gold electro-plating

Lift-off → finished final thick gold plated
electrodes and optical waveguides of MZIM.

FIGURE A7.2
Electroplating of electrodes using the thick "guiding slot" patterned by deep x-ray radiation. *Note:* Observe the "brick-wall" structures.

- Is there any Au or Cr on the top of the end points the waveguides?
- Is the surface of the two end points perfectly clean?
- Is there any chip at the end points? If it is then continue to the next processing step.

If the device is clean then store it in a container for pig tailing. If it is not then carefully clean the device by using a cotton tip with acetone and DI water until everything is clear and clean. A cross-section view of the device and $LiNbO_3$ substrate for each fabrication process is shown in Figures A7.2 and A7.3 (from the beginning to the end).

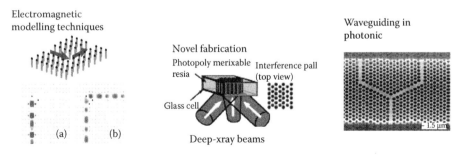

Electromagnetic
modelling techniques

Novel fabrication

Photopoly merixable   Interference pall
resia   (top view)

Glass cell

(a)   (b)

Deep-xray beams

Waveguiding in
photonic

1.5 μm

Synchrotron Radiation Assisted Lithography (LIGA)

FIGURE A7.3
Synchrotron radiation process for fabrication of microwave traveling electrode using SU-8 polymeric structures and electroplating.

A7.2.5 Fiber Pig Tailing and Optical Testing

A7.2.5.1 Fiber-channel Waveguide Pig Tailing

This process is very critical. If the waveguide had low coupling loss obtained from wave-guide testing, then the value of the optical insertion loss is now dependent on the method of coupling the fiber to the MZ waveguide. A schematic diagram of this process is shown in Figure A7.4. These steps developed for pigtailing are

1. Cut the paper (thick) into small pieces (about 2×20 mm). These will be used to apply glue to the edges of the end points of the MZ waveguide.
2. Wax the substrate into the small hole in the substrate holder.
3. Cut the fiber and curve the end face of the fiber by using the slicing machine.
4. Arrange the fiber and the substrate to be pigtailed into position.
5. Move the fiber in close to the edge of the waveguide.
6. Align the fiber until you get the best result of optical insertion loss then move the fiber out (about 10 mm from the waveguide's end) very slowly. Note, do not touch anything on the stage.
7. Using the small piece of paper and a small amount of glue, gently apply it to the edge of the substrate and close to the waveguide position. Try to use the smallest amount of glue possible.
8. Move the fiber in and align it with the waveguide to get the same or higher output power as before. Then shine the UV light into that position for about 1–2 min.
9. Again do not touch anything while using the UV light (i.e., do not vibrate), otherwise the optical coupling loss is very high and it is impossible to observe the guided mode at the other end of the waveguide.
10. Repeat all above steps for the other end of the waveguide.
11. Remove the substrate from the stage and place it in a box.

A7.2.5.2 Electrical Microwave and Optical Testing

The measured parameters are: (i) microwave index, (ii) bandwidth, (iii) V_π, and (iv) optical coupling loss. Several devices have been fabricated and tested. An illustration of the

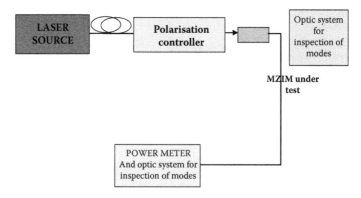

FIGURE A7.4
Schematic diagram of optical system for testing coupling loss of waveguides after polishing.

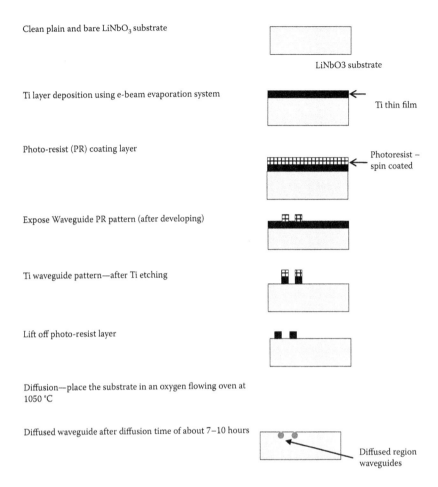

Clean plain and bare LiNbO$_3$ substrate

LiNbO3 substrate

Ti layer deposition using e-beam evaporation system

Ti thin film

Photo-resist (PR) coating layer

Photoresist – spin coated

Expose Waveguide PR pattern (after developing)

Ti waveguide pattern—after Ti etching

Lift off photo-resist layer

Diffusion—place the substrate in an oxygen flowing oven at 1050 °C

Diffused waveguide after diffusion time of about 7–10 hours

Diffused region waveguides

FIGURE A7.5
Cross-section view of device of all steps for fabricating Ti:LiNbO$_3$ MZ waveguide.

lithography and fabrication of electrodes, waveguide structures and MZIM is shown in Figures A7.5 and A7.6.

A7.3 Deep x-ray Fabrication Process for Thick Electrode Electroplating

The problems of electroplating thick electrodes can be observed from the tilted wall angle showCross-section view of device of all steps for fabricating Ti:LiNbO$_3$ MZ waveguide.n in Figure A7.6. This is inevitable as electroplating is conducted with a thin Au film and without "guiding slots" which can be formed with polymeric film structures and etched with x-ray radiation. This process is usually called the LIGA process. In this process a thick polymer film is coated and patterned into a guiding slot for electroplating.

X-ray radiation usually emitted from synchrotron light source can be used as the deep x-ray source. A typical deep x-ray lithography is shown in Figures A7.7 and A7.8. The

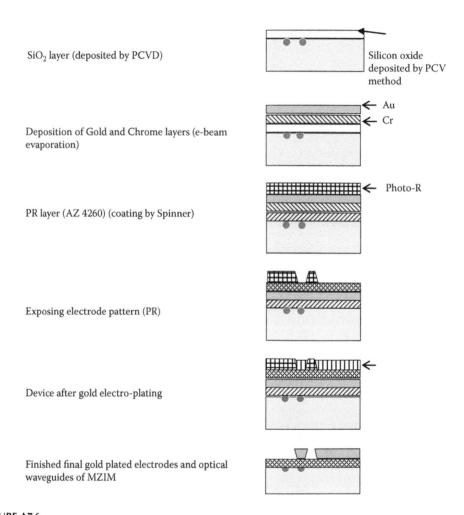

SiO₂ layer (deposited by PCVD)

Silicon oxide
deposited by PCV
method

Deposition of Gold and Chrome layers (e-beam
evaporation)

← Au
← Cr

PR layer (AZ 4260) (coating by Spinner)

← Photo-R

Exposing electrode pattern (PR)

Device after gold electro-plating

←

Finished final gold plated electrodes and optical
waveguides of MZIM

FIGURE A7.6
Electroplating of electrodes on optical waveguide structures. *Note:* Observe the tilted wall structures.

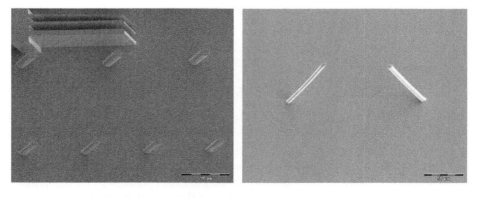

FIGURE A7.7
X-ray lithography: inclination x-ray radiation on SU-8 polymer—single and angled inclination exposure—processed at SSLS as a test for fabrication of integrated photonic devices.

LiMiNT—Lithography for micro- and nano-technology

Bemline LIGA x-ray lithography   Clean room processing

Deep x-ray LIGA lithography beam line arrangement

FIGURE A7.8
Facilities at LiMiNT for lithography for micro- and nano-technology. The proposed techniques for fabrication of integrated photonic devices including the ultra-broadband modulators MZIM.

facility is the lithography beam line of the Singapore Synchrotron Light Source. We have fabricated some test patterns as shown in Figure A7.7. This demonstrates that thick guiding slot polymer can be formed for electroplating of thick electrodes for ultra-wideband optical modulation.

A7.4 Concluding Remarks

We have described the fabrication process using optical lithography and deep x-ray lithography for the ultra-wideband optical modulation devices with interferometric optical waveguides and the asymmetric or symmetric electrodes to generate appropriate fields for electro-optic interaction. The processes are simple and should be implemented in a standard laboratory for microelectronic and x-ray lithography. The procedures for the modeling and design of the optical waveguides and traveling wave electrode structures are described further.

The x-ray deep lithography using synchrotron radiation would be a potential technique for producing "brick wall" like thick electrodes for mm-wave region optical modulation.

References

1. C.K. Kao and G.A. Hockham, "Dielectric-fibre surface waveguides for optical frequencies," *Optoelectronics, IEE Proceedings J.*, 133(3), 191–198, 1966.
2. S.E. Miller, "Integrated optics," *Bell Syst. Tech. J.*, 48, 2059–2069, 1969.
3. L.N. Binh *"Modelling of symmetric and asymmetric travelling wave electrodes for ultra-broadband optical modulation,"* ECSE Monash University, Technical Report MECSE-8-2005, 1–38.
4. T.W. Chua, "Numerical modelling for the design and analysis of high speed integrated optical modulators," M. Eng. Sc. dissertation, 1996, Monash University.

Appendix 8 Planar Waveguides by Finite Difference Method— FORTRAN PROGRAMS

A8.1 Straight Waveguides

A8.1.1 FORTRAN Language Program of the FD1 for Straight Waveguide

```
*$NOEXTENSIONS
  PROGRAM FD

C*****************************************************************
C*                                      *
C* This program uses the Finite Different (FD) method to compute   *
C* the electric field as a light wave propagates through a slab   *
C* waveguide. The waveguide structure implemented in this program *
C* is a straight waveguide with and without abrupt change of the   *
C* refractive index.              *
C* WG -- WAVEGUIDE              *
C* CACL-- CALCULATE              *
C* IDX -- REFRACTIVE INDEX              *
C* INIT-- INITIALIZED              *
C* MTRX-- MATRIX              *
C* ABRP-- ABRUPT CHANGE              *
C* CMF -- COMFIRM              *
C*              *
C*****************************************************************

    PARAMETER (MAX=512)
    COMMON/BLK1/PRF
    COMMON/BLK2/INDEX
    COMMON/BLK4/R
    REAL DX,DZ,X,Y,Z,LAMBDA,WAVE,WIDTH,LENGTH,S,DIST
    REAL INDEX(MAX)
    REAL DN,HF,H,K0,KL,NREF,BETA
    DOUBLE PRECISION PRF(MAX)
    INTEGER XGRID,NPRO,STEP
    COMPLEX*16 R(MAX)

    CALL INTRO
    CALL WG(XGRID,WAVE,NREF,DN,WIDTH,LENGTH,HF,DIST)
    CALL CALC(WAVE,KL,K0,NREF,WIDTH,DX,DZ,NPRO,XGRID,LENGTH)
    CALL CMF(WAVE,NREF,DN,WIDTH,LENGTH,HF,DX,DZ,XGRID,NPRO,K0,S)
```

```
      CALL IDX(XGRID,WIDTH,HF,DN,NREF,KL,DX)
      CALL INIT(Z,XGRID,WIDTH,HF,DN,NREF,K0,DX,KL,NPRO,DZ,S,STEP)
      CALL MTRX2(K0,KL,DZ,DX,NREF,XGRID,NPRO,STEP,DIST)
      STOP
      END


C-------------------------------------------------------------

      SUBROUTINE INTRO

C*************************************************************
C*               *
C* This section is to give the user some knowledge on the program.  *
C*               *
C*************************************************************
      WRITE(6,5)
    5 FORMAT(
     +'                    '/,
     +'*************************************************'/,
     +'*             *'/,
     +'* This program uses finite different method to compute  *'/,
     +'* the electric field as light propagates through a    *'/,
     +'* optical waveguide. Suggestions for various parameters  *'/,
     +'* are shown with the input questions. Mistakes can be  *'/,
     +'* corrected after all the values are fed at the input.  *'/,
     +'* After the computation is completed, a 3-d view of the  *'/,
     +'* field will be plotted.      *'/,
     +'*             *'/,
     +'*************************************************',/)

      RETURN
      END


C-------------------------------------------------------------

      SUBROUTINE WG(XGRID,WAVE,NREF,DN,WIDTH,LENGTH,HF,DIST)

C*************************************************************
C*                  *
C* This part of the program is to prompt for the propagation
C* condition.*
C* width -- Waveguide width in x-dir.            *
C* length-- Waveguide length in z-dir.            *
C* xgrid -- no of section along x-dir.          *
C* wave -- wavelength in vacuum.        *
C* nref -- referenced reflective index.        *
C* dn -- change of reflective index.      *
C* dec -- decision           *
C*                  *
C*************************************************************

      REAL WAVE,HF,NREF,DN,LENGTH,DIST
      INTEGER XGRID
```

```
 WRITE (6,5)
5 FORMAT ('1. Please enter wavelength of lightwaves in vacuum:'/,
 +' in micron. (eg 0.633)',/)
 READ (5,*) WAVE


 WRITE (6,10)
10 FORMAT ('2. Enter total width of slab waveguide system W in micron:'/,
 +' (eg: 20)',/)
 READ (5,*) WIDTH


 WRITE (6,20)
20 FORMAT ('3. Please enter the waveguide thickness h in micron:'/,
 +' (eg:1)',/)
 READ (5,*) HF


 WRITE (6,30)
30 FORMAT ('4. Please enter the no of slices in X-dir:'/,
 +'   note the value must be in integer and must < 256.',/)
 READ (5,*) XGRID


 WRITE (6,40)
40 FORMAT ('5. Enter the waveguide length for beam propagation in micron:'/,
 +'   (eg:500)',/)
 READ (5,*) LENGTH


 WRITE (6,50)
50 FORMAT ('6. Please enter the refractive index of the substrate:'/,
 +'   (eg:2.2)',/)
 READ (5,*) NREF


 WRITE (6,60)
60 FORMAT ('7. Please enter the refractive index change delta (n) :'/,
 +'   (eg:0.01)',/)
 READ (5,*) DN


 WRITE (6,70)
70 FORMAT ('8. Is there any abrupt change in refractive index?'/,
 +' type (1) for [YES] and (0) for [NO]',/)
 READ (5,*) DEC
 IF (DEC.EQ.1) THEN
  WRITE (6,80)
80 FORMAT ('9. What is the length from the edge of the waveguide'/,
 + ' does the abrupt change occurs ?',/)
 READ (5,*) DIST
 ELSE
  DIST=0.0
 END IF
 RETURN
 END
```

```
C------------------------------------------------------------------

    SUBROUTINE CALC(WAVE,KL,K0,NREF,WIDTH,DX,DZ,NPRO,XGRID,LENGTH)

C****************************************************************
C* This section is to calculate the wave constant and to prompt    *
C* the input of delta-z.                          *
C*                                 *
C* KL -- wave number                        *
C* K0 -- KL/NREF                           *
C* DX -- step size in x-dir.                    *
C* DZ -- step size in z-dir.                    *
C* NPRO- no of propagation step in z-dir.              *
C*                              *
C****************************************************************


    REAL WAVE,WAVE1,KL,NREF,K0,WIDTH,DX,DZ,LENGTH
    INTEGER XGRID,NPRO,DEC
    DOUBLE PRECISION PI,TWOPI

    PI=3.14159265359
    TWOPI=2.0*PI
    WAVE1=WAVE/NREF
    KL=TWOPI/WAVE1
    K0=KL/NREF
    WRITE(6,10)
 10 FORMAT('Please input the propagation step size (DZ) in micron.'/)
    READ(5,*)DZ
    RETURN
    END



C------------------------------------------------------------------

    SUBROUTINE CMF(WAVE,NREF,DN,WIDTH,LENGTH,HF,DX,DZ,XGRID,NPRO,K0,S)

C****************************************************************
C*                            *
C* This section is to confirm to the user that the values that have  *
C* been input. If any value that wish to be change, it can be done.  *
C* DEC  --- decision                       *
C*                             *
C****************************************************************


    REAL WAVE,NREF,DN,WIDTH,LENGTH,HF,DX,DZ,K0
    INTEGER XGRID,DEC,NPRO


  5 NPRO=NINT(LENGTH/DZ)
    DX=WIDTH/XGRID
    V=K0*HF*SQRT(2.0*NREF*DN)
    S=0.5*(SQRT(V*V+1)-1)
```

```
      WRITE(6,10)WAVE,WIDTH,HF,LENGTH,DZ,XGRID,NREF,DN,DX,NPRO,S
   10 FORMAT(
     +'                              '/,
     +'                              '/,
     +'   The values entered are:                  '/,
     +'  (1) Wavelength       =',F5.3,' micron          '/,
     +'  (2) Waveguide thickness =',F6.3,' micron          '/,
     +'  (3) Film thickness    =',F6.3,' micron          '/,
     +'  (4) Waveguide length   =',F6.1,' micron          '/,
     +'  (5) Delta-z        =',F6.3,' micron        '/,
     +'  (6) Grid points in X-dir =',I6,'             '/,
     +'  (7) Substrate ref. index =',F6.3,'            '/,
     +'  (8) Delta-n        =',F6.3,'          '/,
     +'                              '/,
     +'   The value of delta-x is :',F6.3,'             '/,
     +'   The number of propagation step is :',I7,'         '/,
     +'   The mode number is :',F6.3,'            '/,
     +'                              '/)


   15 WRITE(6,20)
   20 FORMAT(
     +'Do you wish to change any values ?             '/,
     +'Press any number, (1) to (8), of the parameters shown above '/,
     +'that you wish to change. Otherwise press (0) to continue    '/)
      READ(5,*)DEC
      IF (DEC.EQ.0) THEN
         GOTO 65
       ELSE
         GOTO (25,30,35,40,45,50,55,60) DEC
       END IF


   25 WRITE(6,26)
   26 FORMAT('New wavelength is :'/)
      READ(5,*)WAVE
      GOTO 5


   30 WRITE(6,31)
   31 FORMAT('New thickness is :'/)
      READ(5,*)WIDTH
      GOTO 5


   35 WRITE(6,36)
   36 FORMAT('New film thickness is :'/)
      READ(5,*)HF
      GOTO 5


   40 WRITE(6,41)
   41 FORMAT('New waveguide length is :'/)
      READ(5,*)LENGTH
      GOTO 5
```

```
 45 WRITE(6,46)
 46 FORMAT('New delta-z is :'/)
    READ(5,*)DZ
    GOTO 5


 50 WRITE(6,51)
 51 FORMAT('New x-grid is :'/)
    READ(5,*)XGRID
    GOTO 5


 55 WRITE(6,56)
 56 FORMAT('New substrate refractive index is :'/)
    READ(5,*)NREF
    GOTO 5


 60 WRITE(6,61)
 61 FORMAT('New refractive index difference is :'/)
    READ(5,*)DN
    GOTO 5


 65 CONTINUE
    RETURN
    END
C--------------------------------------------------------------

    SUBROUTINE PROF(XGRID,WIDTH,HF,DN,NREF,DX,S)

C***************************************************************
C*                                  *
C* This subroutine compute the profile distribution.      *
C* PRF  -- PROFILE                        *
C* XC,XC1,J  -- LOCAL VAR.                    *
C*                                  *
C***************************************************************

    PARAMETER (MAX=512)
    COMMON/BLK1/PRF
    REAL WIDTH,DX,HF,NREF,S
    DOUBLE PRECISION XC,XC1,PRF(MAX)
    INTEGER J,XGRID

    XC=WIDTH/2.0-DX/2.0
    DO 10 J=1,XGRID
     PRF(J)=0.0
     XC1=COSH(2.0*XC/HF)
     PRF(J)=1.0+0.01*EXP(-S*LOG(XC1))
     PRF(J)=(PRF(J)-1.0)/0.01
     XC=XC-DX
 10 CONTINUE
    RETURN
    END
```

```
C-------------------------------------------------------------

   SUBROUTINE IDX(XGRID,WIDTH,HF,DN,NREF,KL,DX)

C*************************************************************
C*                              *
C* THIS SECTION IS TO SET UP THE REFRACTIVE INDEX FOR THE MATRIX    *
C* COMPUTATION.                          *
C*                              *
C* PERT -- PERTURDBATION                      *
C*                              *
C*************************************************************

   PARAMETER (MAX=512)
   COMMON/BLK2/INDEX
   REAL WIDTH,DX,XC1,XC,PERT,KL,HF,NREF,S,DN,INDEX(MAX)
   INTEGER J,XGRID


   S=2.0
   XC=(WIDTH/2.0)-DX/2.0
   DO 10 J=1,XGRID
    INDEX(J)=0.0
    XC1=COSH(2.0*XC/HF)
    PERT=DN*EXP(-S*LOG(XC1))
    INDEX(J)=NREF+PERT
    XC=XC-DX
  10 CONTINUE
   RETURN
   END



C-------------------------------------------------------------

   SUBROUTINE INIT(Z,XGRID,WIDTH,HF,DN,NREF,K0,DX,KL,NPRO,DZ,S,STEP)

C*************************************************************
C*                              *
C* THIS SECTION IS TO STORE THE INITIAL FIELD INTENSITY AND TO CREAT *
C* AN OUTPUT FILE FOR THE RESULT.                 *
C*                              *
C* INTEN  -- INTENSITY                 *
C* PROF   -- PROFILE                 *
C*                              *
C*************************************************************

   PARAMETER (MAX=512)
   REAL Z,WIDTH,HF,DN,NREF,K0,DX,KL,LENGHT,S,DZ
   INTEGER XGRID,NPRO,STEP


   CHARACTER*20 FNAME
   WRITE(6,10)
```

```
10 FORMAT('PLEASE ENTER THE NAME OF THE THE OUTPUT FILE:',/)
   READ(5,20)FNAME
20 FORMAT(A)
   OPEN(3,FILE=FNAME,STATUS='OLD')


   WRITE(6,30)
30 FORMAT('PLEASE ENTER THE NO OF PLANE TO BE PLOTTED.')
   READ(5,*)STEP


   OPEN(UNIT=4,FILE='a:PAM.DAT')
   WRITE(4,35)XGRID,STEP
35 FORMAT(1X,I5,1X,I5)
   CLOSE(4)

   Z=0.0

   CALL PROF(XGRID,WIDTH,HF,DN,NREF,DX,S)

   CALL INTEN(KL,NPRO,XGRID,DZ)

   RETURN
   END



C--------------------------------------------------------------

   SUBROUTINE INTEN(KL,NPRO,XGRID,DZ)

C*************************************************************
C*                                                         *
C* THIS SECTION OUTPUT THE FIELD INTENSITY FOR THE FIRST PLANE.   *
C*                                                         *
C*************************************************************

   PARAMETER (MAX=512)
   COMMON/BLK1/PRF
   COMMON/BLK4/R
   REAL T,KL,DZ
   INTEGER J,K,XGRID
   DOUBLE PRECISION ZERO,PRF(MAX),R2(MAX)
   COMPLEX*16 R(MAX)


   ZERO=0.0
   DO 20 J=1,XGRID
    R(J)=DCMPLX(PRF(J),ZERO)
    R2(J)=ABS(R(J))
    WRITE(3,10)R2(J)
10   FORMAT(F10.8)
20 CONTINUE
   RETURN
   END
```

```
C---------------------------------------------------------------

   SUBROUTINE MTRX2 (K0,KL,DZ,DX,NREF,XGRID,NPRO,STEP,DIST)

C***************************************************************
C*                              *
C* THIS SECTION IS TO COMPUTE LIGTH INTENSITY AS IT PROPAGATES   *
C* THROUGH. COMPUTATION IS USING THE FINITE DIFFERENT METHOD     *
C* WHICH USE TRIDIOGANAL MATRIX. ELEMENTS IN THE MATRIX IS COMPUTED  *
C* IN THE SUBROUTINE 'COEFF.'                     *
C*                              *
C* CK    --- CHECK                    *
C* COEFF   --- COEFFICIENT                   *
C* SKIP   --- NO OF TIME TO SKIP BEFORE THE INTENSITY IS OUTPUTED   *
C* OPUT   --- OUTPUT.                  *
C*                              *
C***************************************************************

   PARAMETER (MAX=512)
   REAL K0,KL,DZ,DX,NREF,Z,DIST
   INTEGER I,J,K,L,SKIP,XGRID,NPRO,STEP
   DOUBLE PRECISION EOUT(MAX),A(MAX),AN(MAX),ZERO
   COMMON/BLK4/R
   COMPLEX*16 B(MAX),C(MAX),R1(MAX),BN(MAX)
   COMPLEX*16 R(MAX),E(MAX),TEM(MAX)


   ZERO=0.0
   Z=0.0
   SKIP=INT(NPRO/STEP)
   L=SKIP


   CALL COEFF(K0,DZ,DX,NREF,XGRID,A,B,C)
   DO 50 K=2,NPRO


    CALL CK(Z,DZ,DIST,XGRID,NREF,K0,DX,A,B,C,K)
    CALL MTRX1(A,C,XGRID,R1)


    IF (B(1).EQ.0)PAUSE
    BN(1)=B(1)
    AN(1)=-A(1)
    DO 10 J=2,XGRID
      AN(J)=-A(J)
      TEM(J)=AN(J)/BN(J-1)
      BN(J)=B(J)-TEM(J)*AN(J-1)
      R1(J)=R1(J)-TEM(J)*R1(J-1)
  10 CONTINUE


    E(XGRID)=R1(XGRID)/BN(XGRID)
    R(XGRID)=E(XGRID)
```

```
      DO 30 J=XGRID-1,1,-1
        E(J)=(R1(J)-AN(J)*E(J+1))/BN(J)
        R(J)=E(J)
  30    CONTINUE

        CALL OPUT(E,XGRID,K,L,STEP,NPRO,SKIP)

        WRITE(6,40)K
  40    FORMAT('DOING STEP #',I4)
  50  CONTINUE
      CLOSE(3)
      RETURN
      END


C------------------------------------------------------------

      SUBROUTINE OPUT(E,XGRID,K,L,STEP,NPRO,SKIP)

C*************************************************************
C*                              *
C* This section in charge of the output of the field computed to the  *
C* result file. Note that not all planes are outputted to the result  *
C* file.                        *
C*                              *
C*************************************************************

      PARAMETER(MAX=512)
      DOUBLE PRECISION EOUT(MAX)
      INTEGER XGRID,J,K,L,N,STEP,SKIP,NPRO
      COMPLEX*16 E(MAX)

      N=K-L
      IF (N.EQ.0) THEN
        DO 20 J=1,XGRID
         EOUT(J)=ABS(E(J))
         WRITE(3,10)EOUT(J)
  10     FORMAT(F10.8)
  20    CONTINUE
        L=K+SKIP
      ELSE
       ENDIF
      RETURN
      END


C------------------------------------------------------------

      SUBROUTINE CK(Z,DZ,DIST,XGRID,NREF,K0,DX,A,B,C,K)

C*************************************************************
C*                              *
C* THIS SECTION IS TO CHECK THE DISTANCE OF PROPAGATION IN Z-DIR .   *
C* IF ANY ABRUPT IS SPECIFIED BEFORE, IT WILL BRANCH TO A ROUTINE   *
C* WHICH COMPUTE THE NEW REFRACTIVE INDEX.              *
C*                              *
C*************************************************************
```

```
   PARAMETER(MAX=512)
   REAL Z,Z1,ZE,DZ,NREF,K0,DX,DIST
   DOUBLE PRECISION A(MAX)
   INTEGER XGRID,K,TPRO
   COMPLEX*16 B(MAX),C(MAX)

   Z=Z+DZ
   TPRO=INT(DIST/DZ)
   ZE=K-TPRO
   IF (ZE.EQ.0) THEN
     CALL ABRP(XGRID,NREF)
     CALL COEFF(K0,DZ,DX,NREF,XGRID,A,B,C)
     END IF
     RETURN
     END

C-------------------------------------------------------------

   SUBROUTINE COEFF(K0,DZ,DX,NREF,XGRID,A,B,C)

C*****************************************************************
C*                               *
C* THIS SECTION COMPUTES THE ELEMENTS OF THE TRIDIAGONAL MATRIX.   *
C*                               *
C*****************************************************************

   PARAMETER (MAX=512)
   REAL DX,DZ,NREF,K0,INDEX(MAX)
   INTEGER J,K,XGRID
   DOUBLE PRECISION A(MAX),B1,B2,C1,C2
   COMMON/BLK2/INDEX
   COMPLEX*16 B(MAX),C(MAX)

   DO 10 J=1,XGRID
   A(J)=DZ/(2.0*(DX*DX))
   B1=DZ/(DX*DX)-((DZ*(K0*K0))/2.0)*(INDEX(J)*INDEX(J)-NREF*NREF)
   B2=2.0*K0*NREF
   C1=-(DZ/(DX*DX))+((DZ*(K0*K0))/2.0)*(INDEX(J)*INDEX(J)-NREF*NREF)
   C2=2.0*K0*NREF
   B(J)=DCMPLX(B1,B2)
   C(J)=DCMPLX(C1,C2)
  10 CONTINUE
   RETURN
   END

C-------------------------------------------------------------

   SUBROUTINE MTRX1(A,C,XGRID,R1)

C*****************************************************************
C*                               *
C* THIS ROUTINE IS TO COMPUTE THE RHS OF EQUATION (6).      *
C*                               *
C*****************************************************************
```

```fortran
      PARAMETER(MAX=512)
      INTEGER J,XGRID
      DOUBLE PRECISION A(MAX)
      COMMON/BLK4/R
      COMPLEX*16 C(MAX),R1(MAX),R(MAX)


      R1(1)=C(1)*R(1)+A(1)*R(2)
      DO 10 J=2,XGRID-1
      R1(J)=A(J)*R(J-1)+C(J)*R(J)+A(J)*R(J+1)
   10 CONTINUE
      R1(XGRID)=A(XGRID)*R(XGRID-1)+C(XGRID)*R(XGRID)
      RETURN
      END



C-------------------------------------------------------------


      SUBROUTINE ABRP(XGRID,NREF)


C*************************************************************
C*                                *
C* THIS SECTION IS TO SET THE NEW REF. INDEX FOR THE ABRUPT CHANGE   *
C* IN REF. INDEX.                              *
C*                                *
C* X1,X2 ---- LOCAL VAR                      *
C*                                *
C*************************************************************


      PARAMETER(MAX=512)
      REAL INDEX(MAX),NREF,X1
      INTEGER X2,XGRID,J
      COMMON/BLK2/INDEX


      X1=XGRID/2
      X2=NINT(X1)
      DO 10 J=1,X2
        INDEX(J)=NREF
   10 CONTINUE
      RETURN
      END
```

A8.1.2 MATLAB Program to Plot

```
% This is the matlab file for the Fortran program fd1.for. It is to
% load the result file from "RES.DAT" and plot the 3-D filed intensity
% pattern of the whole waveguide. The contour of the intensity will
% also be plotted.

%NOTE: results obtained by the FORTRAN Program must be stored in a
%file named Res.dat. The dimensions of such data set must be stored
```

```
%in PAM.dat. These can be changed accordingly in the FORTRAN program
%set up.

clear all

load ('RES.DAT')
load ('PAM.DAT')
jmax=PAM(:,1);
imax=PAM(:,2);


k=0;
for i=1:imax;
 for j=1:jmax;
  k=k+1;
  z(i,j)=RES(k);
 end
end


m=[-37.5,80];


figure(1)
mesh(z,m)
title('3-D VIEW OF THE FIELD INTENSITY.')
xlabel('SLICED INTERVALS IN X-DIR')
ylabel('PROP. STEPS')
zlabel('Field intensity')
pause


figure(2)
contour(z)
title('INTENSITY CONTOUR PLOT ALONG Z-PROP DIRECTION')
xlabel('SLICED INTERVALS IN X-DIR')
ylabel('PROP. STEPS')
pause
```

A8.2 Bend Waveguides

```
*$NOEXTENSIONS
   PROGRAM FD

C***************************************************************
C*                                *
C* This program uses the Finite Different (FD) method to compute  *
C* the electric field as a light wave propagates through a slab    *
C* waveguide. The waveguide structure implemented in this program  *
C* is a straight waveguide with and without abrupt change of the   *
```

```
C* refractive index.                        *
C* WG -- WAVEGUIDE                         *
C* CACL-- CALCULATE                        *
C* IDX -- REFRACTIVE INDEX                   *
C* INIT-- INITIALIZED                        *
C* MTRX-- MATRIX                          *
C* ABRP-- ABRUPT CHANGE                     *
C* CMF -- COMFIRM                         *
C*                              *
C****************************************************************

   PARAMETER (MAX=512)
   COMMON/BLK1/PRF
   COMMON/BLK2/INDEX
   COMMON/BLK4/R
   REAL DX,DZ,X,Y,Z,LAMBDA,WAVE,WIDTH,LENGTH,S,DIST
   REAL INDEX(MAX)
   REAL DN,HF,H,K0,KL,NREF,BETA
   DOUBLE PRECISION PRF(MAX)
   INTEGER XGRID,NPRO,STEP
   COMPLEX*16 R(MAX)


   CALL INTRO
   CALL WG(XGRID,WAVE,NREF,DN,WIDTH,LENGTH,HF,DIST)
   CALL CALC(WAVE,KL,K0,NREF,WIDTH,DX,DZ,NPRO,XGRID,LENGTH)
   CALL CMF(WAVE,NREF,DN,WIDTH,LENGTH,HF,DX,DZ,XGRID,NPRO,K0,S)
   CALL IDX(XGRID,WIDTH,HF,DN,NREF,KL,DX)
   CALL INIT(Z,XGRID,WIDTH,HF,DN,NREF,K0,DX,KL,NPRO,DZ,S,STEP)
   CALL MTRX2(K0,KL,DZ,DX,NREF,XGRID,NPRO,STEP,DIST)
   STOP
   END


C ------------------------------------------------------------

   SUBROUTINE INTRO

C****************************************************************
C*                              *
C* This section is to give the user some knowledge on the program.  *
C*                              *
C****************************************************************

   WRITE(6,5)
 5 FORMAT(
  +'                        '/,
  +'****************************************************'/,
  +'*                        *'/,
  +'* This program uses finite different method to compute   *'/,
  +'* the electric field as light propagates through a    *'/,
  +'* optical waveguide. Suggestions for various parameters  *'/,
  +'* are shown with the input questions. Mistakes can be   *'/,
  +'* corrected after all the values are inputted. After   *'/,
  +'* the computation is completed, a 3-d view of the field  *'/,
```

```
 +'* will be plotted.                         *'/,
 +'*                              *'/,
 +'*****************************************************',/)


   RETURN
   END


C-------------------------------------------------------------

   SUBROUTINE WG(XGRID,WAVE,NREF,DN,WIDTH,LENGTH,HF,DIST)

C*****************************************************************
C*                              *
C* This part of the program is to prompt for the propagation  *
C* condition.*
C* width -- Waveguide width in x-dir.                *
C* length-- Waveguide length in z-dir.                *
C* xgrid -- no of section along x-dir.               *
C* wave -- wavelength in vacuum.                *
C* nref -- reference reflective index.               *
C* dn   -- change of reflective index.              *
C* dec  -- decision                    *
C*                              *
C*****************************************************************

   REAL WAVE,HF,NREF,DN,LENGTH,DIST
   INTEGER XGRID


   WRITE(6,5)
 5 FORMAT('1. Please enter wavelength of lightwaves in vacuum :'/,
  +'  in micron. (e.g. 0.633)',/)
   READ(5,*)WAVE


   WRITE(6,10)
 10 FORMAT('2. Enter total width of slab waveguide system W in micron:'/,
  +'  (eg: 20)',/)
   READ(5,*)WIDTH


   WRITE(6,20)
 20 FORMAT('3. Please enter the waveguide thickness h in micron:'/,
  +'  (eg:1)',/)
   READ(5,*)HF


   WRITE(6,30)
 30 FORMAT('4. Please enter the no of slices in X-dir:'/,
  +'  note the value must be in integer and must < 256.',/)
   READ(5,*)XGRID


   WRITE(6,40)
 40 FORMAT('5. Enter the waveguide length for beam propagation in micron:'/,
  +'  (eg:500)',/)
   READ(5,*)LENGTH
```

```
    WRITE(6,50)
 50 FORMAT('6. Please enter the refractive index of the substrate:'/,
   +'   (eg:2.2)',/)
    READ(5,*)NREF


    WRITE(6,60)
 60 FORMAT('7. Please enter the refractive index change delta(n) :'/,
   +'  (eg:0.01)',/)
    READ(5,*)DN


    WRITE(6,70)
 70 FORMAT('8. Is there any abrupt change in refractive index ?'/,
   +' type(1) for [YES] and (0) for [NO]',/)
    READ(5,*)DEC
    IF (DEC.EQ.1) THEN
     WRITE(6,80)
 80 FORMAT('9. What is the length from the edge of the waveguide'/,
   + ' does the abrupt change occurs ?',/)
    READ(5,*)DIST
    ELSE
     DIST=0.0
    END IF
    RETURN
    END




C-----------------------------------------------------------------

    SUBROUTINE CALC(WAVE,KL,K0,NREF,WIDTH,DX,DZ,NPRO,XGRID,LENGTH)

C********************************************************************
C* This section is to calculate the wave constant and to prompt    *
C* the input of delta-z.                          *
C*                                *
C* KL -- wave number                        *
C* K0 -- KL/NREF                         *
C* DX -- step size in x-dir.                     *
C* DZ -- step size in z-dir.                     *
C* NPRO- no of propagation step in z-dir.             *
C*                                *
C********************************************************************


    REAL WAVE,WAVE1,KL,NREF,K0,WIDTH,DX,DZ,LENGTH
    INTEGER XGRID,NPRO,DEC
    DOUBLE PRECISION PI,TWOPI


    PI=3.14159265359
    TWOPI=2.0*PI
    WAVE1=WAVE/NREF
    KL=TWOPI/WAVE1
```

```
   K0=KL/NREF
   WRITE(6,10)
 10 FORMAT('Please input the propagation step size (DZ) in micron.'/)
   READ(5,*)DZ
   RETURN
   END


C------------------------------------------------------------

   SUBROUTINE CMF(WAVE,NREF,DN,WIDTH,LENGTH,HF,DX,DZ,XGRID,NPRO,K0,S)

C**************************************************************
C*                              *
C* This section is to confirm to the user that the values that have  *
C* been input. If any value that wish to be change, it can be done.   *
C* DEC  --- decision                        *
C*                              *
C**************************************************************

   REAL WAVE,NREF,DN,WIDTH,LENGTH,HF,DX,DZ,K0
   INTEGER XGRID,DEC,NPRO


 5 NPRO=NINT(LENGTH/DZ)
   DX=WIDTH/XGRID
   V=K0*HF*SQRT(2.0*NREF*DN)
   S=0.5*(SQRT(V*V+1)-1)


   WRITE(6,10)WAVE,WIDTH,HF,LENGTH,DZ,XGRID,NREF,DN,DX,NPRO,S
 10 FORMAT(
  +'                        '/,
  +'                        '/,
  +'   The values entered are:                '/,
  +' (1) Wavelength       =',F5.3,' micron         '/,
  +' (2) Waveguide thickness  =',F6.3,' micron          '/,
  +' (3) Film thickness     =',F6.3,' micron         '/,
  +' (4) Waveguide length    =',F6.1,' micron          '/,
  +' (5) Delta-z          =',F6.3,' micron          '/,
  +' (6) Grid points in X-dir =',I6,'             '/,
  +' (7) Substrate ref. index =',F6.3,'             '/,
  +' (8) Delta-n          =',F6.3,'          '/,
  +'                        '/,
  +'   The value of delta-x is :',F6.3,'          '/,
  +'   The number of propagation step is :',I7,'       '/,
  +'   The mode number is :',F6.3,'           '/,
  +'                        '/)


 15 WRITE(6,20)
 20 FORMAT(
  +'Do you wish to change any values ?          '/,
```

```
+'Press any number, (1) to (8), of the parameters shown above '/,
+'that you wish to change. Otherwise press (0) to continue    '/)
 READ(5,*)DEC
 IF (DEC.EQ.0) THEN
     GOTO 65
  ELSE
     GOTO (25,30,35,40,45,50,55,60) DEC
  END IF


25  WRITE(6,26)
26 FORMAT('New wavelength is :'/)
 READ(5,*)WAVE
 GOTO 5


30  WRITE(6,31)
31 FORMAT('New thickness is :'/)
 READ(5,*)WIDTH
 GOTO 5


35  WRITE(6,36)
36 FORMAT('New film thickness is :'/)
 READ(5,*)HF
 GOTO 5


40  WRITE(6,41)
41 FORMAT('New waveguide length is :'/)
 READ(5,*)LENGTH
 GOTO 5


45  WRITE(6,46)
46 FORMAT('New delta-z is :'/)
 READ(5,*)DZ
 GOTO 5


50  WRITE(6,51)
51 FORMAT('New x-grid is :'/)
 READ(5,*)XGRID
 GOTO 5


55  WRITE(6,56)
56 FORMAT('New substrate refractive index is :'/)
 READ(5,*)NREF
 GOTO 5


60  WRITE(6,61)
61 FORMAT('New refractive index difference is :'/)
 READ(5,*)DN
 GOTO 5
```

```
 65 CONTINUE
    RETURN
    END


C-------------------------------------------------------------

    SUBROUTINE PROF(XGRID,WIDTH,HF,DN,NREF,DX,S)

C****************************************************************
C*                                  *                          *
C* This subroutine compute the profile distribution.        *
C* PRF  -- PROFILE                        *
C* XC,XC1,J  -- LOCAL VAR.                   *
C*                                  *
C****************************************************************


    PARAMETER (MAX=512)
    COMMON/BLK1/PRF
    REAL WIDTH,DX,HF,NREF,S
    DOUBLE PRECISION XC,XC1,PRF(MAX)
    INTEGER J,XGRID


    XC=WIDTH/2.0-DX/2.0
    DO 10 J=1,XGRID
     PRF(J)=0.0
     XC1=COSH(2.0*XC/HF)
     PRF(J)=1.0+0.01*EXP(-S*LOG(XC1))
     PRF(J)=(PRF(J)-1.0)/0.01
     XC=XC-DX
 10 CONTINUE
    RETURN
    END



C-------------------------------------------------------------

    SUBROUTINE IDX(XGRID,WIDTH,HF,DN,NREF,KL,DX)

C****************************************************************
C*                                  *                          *
C* THIS SECTION IS TO SET UP THE REFRACTIVE INDEX FOR THE MATRIX    *
C* COMPUTATION.                          *
C*                                  *
C* PERT -- PERTURDBATION                       *
C*                                  *
C****************************************************************


    PARAMETER (MAX=512)
    COMMON/BLK2/INDEX
    REAL WIDTH,DX,XC1,XC,PERT,KL,HF,NREF,S,DN,INDEX(MAX)
    INTEGER J,XGRID
```

```
      S=2.0
      XC=(WIDTH/2.0)-DX/2.0
      DO 10 J=1,XGRID
        INDEX(J)=0.0
        XC1=COSH(2.0*XC/HF)
        PERT=DN*EXP(-S*LOG(XC1))
        INDEX(J)=NREF+PERT
        XC=XC-DX
     10 CONTINUE
      RETURN
      END



C----------------------------------------------------------------

      SUBROUTINE  INIT(Z,XGRID,WIDTH,HF,DN,NREF,K0,DX,KL,NPRO,DZ,S,STEP)

C*******************************************************************
C*                                 *
C* THIS SECTION IS TO STORE THE INITIAL FIELD INTENSITY AND TO CREAT *
C* AN OUTPUT FILE FOR THE RESULT.                    *
C*                                 *
C* INTEN  -- INTENSITY                   *
C* PROF   -- PROFILE                  *
C*                            *
C*******************************************************************


      PARAMETER  (MAX=512)
      REAL Z,WIDTH,HF,DN,NREF,K0,DX,KL,LENGHT,S,DZ
      INTEGER XGRID,NPRO,STEP


      CHARACTER*20  FNAME
      WRITE(6,10)
     10 FORMAT('PLEASE ENTER THE NAME OF THE THE OUTPUT FILE:',/)
      READ(5,20)FNAME
     20 FORMAT(A)
      OPEN  (3,FILE=FNAME,STATUS='OLD')


      WRITE(6,30)
     30 FORMAT('PLEASE ENTER THE NO OF PLANE TO BE PLOTTED.')
      READ(5,*)STEP


      OPEN(UNIT=4,FILE='a:PAM.DAT')
      WRITE(4,35)XGRID,STEP
     35 FORMAT(1X,I5,1X,I5)
      CLOSE(4)


      Z=0.0
```

```
   CALL PROF(XGRID,WIDTH,HF,DN,NREF,DX,S)

   CALL INTEN(KL,NPRO,XGRID,DZ)

   RETURN
   END




C-------------------------------------------------------------

   SUBROUTINE INTEN(KL,NPRO,XGRID,DZ)

C*************************************************************
C*                           *
C* THIS SECTION OUTPUT THE FIELD INTENSITY FOR THE FIRST PLANE.   *
C*                           *
C*************************************************************

   PARAMETER (MAX=512)
   COMMON/BLK1/PRF
   COMMON/BLK4/R
   REAL T,KL,DZ
   INTEGER J,K,XGRID
   DOUBLE PRECISION ZERO,PRF(MAX),R2(MAX)
   COMPLEX*16 R(MAX)


   ZERO=0.0
   DO 20 J=1,XGRID
    R(J)=DCMPLX(PRF(J),ZERO)
    R2(J)=ABS(R(J))
    WRITE(3,10)R2(J)
 10  FORMAT(F10.8)
 20 CONTINUE
   RETURN
   END




C-------------------------------------------------------------

   SUBROUTINE MTRX2(K0,KL,DZ,DX,NREF,XGRID,NPRO,STEP,DIST)

C*************************************************************
C*                           *
C* THIS SECTION IS TO COMPUTE LIGTH INTENSITY AS IT PROPAGATES    *
C* THROUGH. COMPUTATION IS USING THE FINITE DIFFERENT METHOD      *
C* WHICH USE TRIDIOGANAL MATRIX. ELEMENTS IN THE MATRIX IS COMPUTED *
C* IN THE SUBROUTINE 'COEFF.'                    *
C*                           *
C* CK   --- CHECK                      *
C* COEFF  --- COEFFICIENT                  *
C* SKIP   --- NO OF TIME TO SKIP BEFORE THE INTENSITY IS OUTPUTED  *
C* OPUT   --- OUTPUT.                    *
C*                           *
C*************************************************************
```

```
      PARAMETER (MAX=512)
      REAL K0,KL,DZ,DX,NREF,Z,DIST
      INTEGER I,J,K,L,SKIP,XGRID,NPRO,STEP
      DOUBLE PRECISION EOUT(MAX),A(MAX),AN(MAX),ZERO
      COMMON/BLK4/R
      COMPLEX*16 B(MAX),C(MAX),R1(MAX),BN(MAX)
      COMPLEX*16 R(MAX),E(MAX),TEM(MAX)


      ZERO=0.0
      Z=0.0
      SKIP=INT(NPRO/STEP)
      L=SKIP


      CALL COEFF(K0,DZ,DX,NREF,XGRID,A,B,C)
      DO 50 K=2,NPRO


       CALL CK(Z,DZ,DIST,XGRID,NREF,K0,DX,A,B,C,K)
       CALL MTRX1(A,C,XGRID,R1)


       IF (B(1).EQ.0)PAUSE
       BN(1)=B(1)
       AN(1)=-A(1)
       DO 10 J=2,XGRID
         AN(J)=-A(J)
         TEM(J)=AN(J)/BN(J-1)
         BN(J)=B(J)-TEM(J)*AN(J-1)
         R1(J)=R1(J)-TEM(J)*R1(J-1)
   10 CONTINUE


       E(XGRID)=R1(XGRID)/BN(XGRID)
       R(XGRID)=E(XGRID)


      DO 30 J=XGRID-1,1,-1
       E(J)=(R1(J)-AN(J)*E(J+1))/BN(J)
       R(J)=E(J)
   30  CONTINUE


       CALL OPUT(E,XGRID,K,L,STEP,NPRO,SKIP)

       WRITE(6,40)K
   40  FORMAT('DOING STEP #',I4)
   50 CONTINUE
      CLOSE(3)
      RETURN
      END


C-------------------------------------------------------------

      SUBROUTINE OPUT(E,XGRID,K,L,STEP,NPRO,SKIP)
```

```
C******************************************************************
C*                                  *                             *
C* This section in charge of the output of the field computed to the  *
C* result file. Note that not all planes are outputted to the result  *
C* file.                                           *
C*                                  *                             *
C******************************************************************


   PARAMETER(MAX=512)
   DOUBLE PRECISION EOUT(MAX)
   INTEGER XGRID,J,K,L,N,STEP,SKIP,NPRO
   COMPLEX*16  E(MAX)


   N=K-L
   IF (N.EQ.0) THEN
     DO 20 J=1,XGRID
       EOUT(J)=ABS(E(J))
       WRITE(3,10)EOUT(J)
10       FORMAT(F10.8)
20   CONTINUE
     L=K+SKIP
   ELSE
    ENDIF
   RETURN
   END




C------------------------------------------------------------

   SUBROUTINE  CK(Z,DZ,DIST,XGRID,NREF,K0,DX,A,B,C,K)

C******************************************************************
C*                                  *                             *
C* THIS SECTION IS TO CHECK THE DISTANCE OF PROPAGATION IN Z-DIR .   *
C* IF ANY ABRUPT IS SPECIFIED BEFORE, IT WILL BRANCH TO A ROUTINE     *
C* WHICH COMPUTE THE NEW REFRACTIVE INDEX.                        *
C*                                  *                             *
C******************************************************************


   PARAMETER(MAX=512)
   REAL Z,Z1,ZE,DZ,NREF,K0,DX,DIST
   DOUBLE PRECISION A(MAX)
   INTEGER XGRID,K,TPRO
   COMPLEX*16  B(MAX),C(MAX)


   Z=Z+DZ
   TPRO=INT(DIST/DZ)
   ZE=K-TPRO
   IF (ZE.EQ.0) THEN
     CALL ABRP(XGRID,NREF)
     CALL COEFF(K0,DZ,DX,NREF,XGRID,A,B,C)
```

```
      END IF
      RETURN
      END



C------------------------------------------------------------

   SUBROUTINE  COEFF(K0,DZ,DX,NREF,XGRID,A,B,C)

C***********************************************************
C*                             *
C* THIS SECTION COMPUTES THE ELEMENTS OF THE TRIDIAGONAL MATRIX.   *
C*                             *
C***********************************************************


   PARAMETER  (MAX=512)
   REAL DX,DZ,NREF,K0,INDEX(MAX)
   INTEGER J,K,XGRID
   DOUBLE PRECISION A(MAX),B1,B2,C1,C2
   COMMON/BLK2/INDEX
   COMPLEX*16  B(MAX),C(MAX)


   DO 10  J=1,XGRID
   A(J)=DZ/(2.0*(DX*DX))
   B1=DZ/(DX*DX)-((DZ*(K0*K0))/2.0)*(INDEX(J)*INDEX(J)-NREF*NREF)
   B2=2.0*K0*NREF
   C1=-(DZ/(DX*DX))+((DZ*(K0*K0))/2.0)*(INDEX(J)*INDEX(J)-NREF*NREF)
   C2=2.0*K0*NREF
   B(J)=DCMPLX(B1,B2)
   C(J)=DCMPLX(C1,C2)
  10 CONTINUE
   RETURN
   END



C------------------------------------------------------------

   SUBROUTINE  MTRX1(A,C,XGRID,R1)

C***********************************************************
C*                             *
C* THIS ROUTINE IS TO COMPUTE THE RHS OF EQUATION (6).       *
C*                             *
C***********************************************************


   PARAMETER(MAX=512)
   INTEGER J,XGRID
   DOUBLE PRECISION A(MAX)
   COMMON/BLK4/R
   COMPLEX*16  C(MAX),R1(MAX),R(MAX)
```

```
   R1(1)=C(1)*R(1)+A(1)*R(2)
   DO 10 J=2,XGRID-1
   R1(J)=A(J)*R(J-1)+C(J)*R(J)+A(J)*R(J+1)
 10 CONTINUE
   R1(XGRID)=A(XGRID)*R(XGRID-1)+C(XGRID)*R(XGRID)
   RETURN
   END



C------------------------------------------------------------

   SUBROUTINE ABRP(XGRID,NREF)

C*************************************************************
C*                             *
C* THIS SECTION IS TO SET THE NEW REF. INDEX FOR THE ABRUPT CHANGE   *
C* IN REF. INDEX.                        *
C*                             *
C* X1,X2 ---- LOCAL VAR                    *
C*                             *
C*************************************************************


   PARAMETER(MAX=512)
   REAL INDEX(MAX),NREF,X1
   INTEGER X2,XGRID,J
   COMMON/BLK2/INDEX


   X1=XGRID/2
   X2=NINT(X1)
   DO 10 J=1,X2
     INDEX(J)=NREF
 10 CONTINUE
   RETURN
   END
```

A8.3 Taper Waveguides: FORTRAN PROGRAM

```
*$NOEXTENSIONS
   PROGRAM FD


C*************************************************************
C*                             *
C* This program uses the Finite Different(FD) method to compute   *
C* the electric field as a light wave propagates through a slab   *
C* waveguide. The waveguide structure that we implement in this   *
C* program is a taper structure.                 *
C* WG -- WAVEGUIDE                      *
C* CACL-- CALCULATE                      *
```

```
C* IDX -- REFRACTIVE INDEX                    *
C* INIT-- INITIALIZED                        *
C* MTRX-- MATRIX                        *
C* ABRP-- ABRUPT CHANGE                      *
C* CMF -- COMFIRM                        *
C*                              *
C*****************************************************************


      PARAMETER (MAX=512)
      COMMON/BLK1/PRF
      COMMON/BLK2/INDEX
      COMMON/BLK4/R
      REAL DX,DZ,X,Y,Z,LAMBDA,WAVE,WIDTH,LENGTH,S,DIST
      REAL INDEX(MAX),ANG
      REAL DN,HF,H,K0,KL,NREF,BETA
      DOUBLE PRECISION PRF(MAX)
      INTEGER XGRID,NPRO,STEP
      COMPLEX*16 R(MAX)


      CALL INTRO
      CALL WG(XGRID,WAVE,NREF,DN,WIDTH,LENGTH,HF,DIST)
      CALL CALC(WAVE,KL,K0,NREF,WIDTH,DX,DZ,NPRO,XGRID,LENGTH)
      CALL CMF(WAVE,NREF,DN,WIDTH,LENGTH,HF,DX,DZ,XGRID
     +,NPRO,K0,S,DIST)
      CALL IDX(XGRID,WIDTH,HF,DN,NREF,KL,DX)
      CALL INIT(XGRID,WIDTH,HF,DN,NREF,K0,DX,KL,NPRO,DZ,S,STEP)
      CALL MTRX2(K0,KL,DZ,DX,NREF,XGRID,NPRO,STEP,DIST,LENGTH,WIDTH)
      STOP
      END




C  ----------------------------------------------------------


      SUBROUTINE INTRO

C*****************************************************************
C*                              *
C* This section is to give the user some knowledge on the program.  *
C*                              *
C*****************************************************************


      WRITE(6,5)
    5 FORMAT(
     +'                        '/,
     +'****************************************************'/,
     +'*                      *'/,
     +'* This program uses finite different method to compute  *'/,
     +'* the electric field as light propagates through a   *'/,
     +'* optical waveguide. Suggestions for various parameters  *'/,
     +'* are shown with the input questions. Mistakes can be  *'/,
```

```
+'* corrected after all the values are inputted. After  */,
+'* the computation is completed, a 3-d view of the field  */,
+'* will be plotted.                     */,
+'*                      */,
+'***************************************************',/)


   RETURN
   END


C------------------------------------------------------------------

   SUBROUTINE  WG(XGRID,WAVE,NREF,DN,WIDTH,LENGTH,HF,DIST)

C**************************************************************
C*                          *
C* This part of the program is to prompt for the propagation condition.*
C* width -- Waveguide width in x-dir.                *
C* length-- Waveguide length in z-dir.               *
C* xgrid -- no of section along x-dir.               *
C* wave  -- wavelength in vacuum.               *
C* nref  -- reference reflective index.               *
C* dn    -- change of reflective index.               *
C* dec   -- decision                  *
C*                          *
C**************************************************************


   REAL  WAVE,HF,NREF,DN,LENGTH
   REAL ANG,DIST
   INTEGER XGRID


   WRITE(6,5)
 5 FORMAT('1. Please enter wavelength of light in vacuum :'/,
   +'  in micron. (e.g. 0.633)',/)
   READ(5,*)WAVE


   WRITE(6,10)
10 FORMAT('2. Please enter thickness of slab waveguide in micron:'/,
   +'  (eg: 20)',/)
   READ(5,*)WIDTH


   WRITE(6,20)
20 FORMAT('3. Please enter the film thickness in micron:'/,
   +' (eg:1)',/)
   READ(5,*)HF


   WRITE(6,30)
30 FORMAT('4. Please enter the no of slice in X-dir:'/,
   +'  note the value must be in integer and must < 256.',/)
   READ(5,*)XGRID
```

```
   WRITE(6,40)
40 FORMAT('5. Please enter the waveguide length in micron:'/,
  +'  (eg:500)',/)
   READ(5,*)LENGTH


   WRITE(6,50)
50 FORMAT('6. Please enter the refractive index of the substract:'/,
  +'  (eg:2.2)',/)
   READ(5,*)NREF


   WRITE(6,60)
60 FORMAT('7. Please enter the refractive index change:'/,
  +'  (eg:0.01)',/)
   READ(5,*)DN


   WRITE(6,70)
70 FORMAT('8. Please enter the length from the edge of the wg'/,
  +'  where the taper starts.',/)
   READ(5,*)DIST


   RETURN
   END
C-----------------------------------------------------------------

   SUBROUTINE CALC(WAVE,KL,KO,NREF,WIDTH,DX,DZ,NPRO,XGRID,LENGTH)

C*****************************************************************
C* This section is to calculate the wave constant and to prompt    *
C* the input of delta-z.                        *
C*                              *
C* KL -- wave number                       *
C* KO -- KL/NREF                        *
C* DX -- step size in x-dir.                   *
C* DZ -- step size in z-dir.                   *
C* NPRO- no of propagation step in z-dir.              *
C*                              *
C*****************************************************************


   REAL WAVE,WAVE1,KL,NREF,KO,WIDTH,DX,DZ,LENGTH
   INTEGER XGRID,NPRO,DEC
   DOUBLE PRECISION PI,TWOPI


   PI=3.14159265359
   TWOPI=2.0*PI
   WAVE1=WAVE/NREF
   KL=TWOPI/WAVE1
   KO=KL/NREF
   WRITE(6,10)
```

```
   10 FORMAT('Please input the propagation step size (DZ) in micron.',/)
      READ(5,*)DZ
      RETURN
      END


C-------------------------------------------------------------

      SUBROUTINE CMF(WAVE,NREF,DN,WIDTH,LENGTH,HF,DX,DZ,XGRID
     +,NPRO,K0,S,DIST)

C**************************************************************
C*                                 *
C* This section is to confirm to the user that the values that have  *
C* been input. If any value that wish to be change, it can be done.  *
C* DEC   --- decision                          *
C*                                 *
C**************************************************************


      REAL WAVE,NREF,DN,WIDTH,LENGTH,HF,DX,DZ,K0,DIST
      INTEGER XGRID,DEC,NPRO


    5 NPRO=NINT(LENGTH/DZ)
      DX=WIDTH/XGRID
      V=K0*HF*SQRT(2.0*NREF*DN)
      S=0.5*(SQRT(V*V+1)-1)


      WRITE(6,10)WAVE,WIDTH,HF,LENGTH,DZ,XGRID,NREF,DN,DIST
     +,DX,NPRO,S
   10 FORMAT(
     +'                          '/,
     +'                          '/,
     +'   The values entered are:              '/,
     +' (1) Wavelength      =',F5.3,' micron         '/,
     +' (2) Waveguide thickness =',F6.3,' micron         '/,
     +' (3) Film thickness    =',F6.3,' micron       '/,
     +' (4) Waveguide length   =',F6.1,' micron       '/,
     +' (5) Delta-z      =',F6.3,' micron       '/,
     +' (6) Grid points in X-dir =',I6,'          '/,
     +' (7) Substrate ref. index =',F6.3,'          '/,
     +' (8) Delta-n      =',F6.3,'        '/,
     +' (9) Distance (taper)   =',F6.1,'         '/,
     +'                          '/,
     +'   The value of delta-x is :',F6.3,'          '/,
     +'   The number of propagation step is :',I7,'       '/,
     +'   The mode number is :',F6.3,'         '/,
     +'                      '/)


   15 WRITE(6,20)
   20 FORMAT(
```

```
      +'Do you wish to change any values?                  '/,
      +'Press any number, (1) to (9), of the parameters shown above '/,
      +'that you wish to change. Otherwise press (0) to continue   '/)
      READ(5,*)DEC
      IF (DEC.EQ.0) THEN
          GOTO 75
        ELSE
          GOTO (25,30,35,40,45,50,55,60,65) DEC
        END IF


   25 WRITE(6,26)
   26 FORMAT('New wavelength is :'/)
      READ(5,*)WAVE
      GOTO 5


   30 WRITE(6,31)
   31 FORMAT('New thickness is :'/)
      READ(5,*)WIDTH
      GOTO 5


   35 WRITE(6,36)
   36 FORMAT('New film thickness is :'/)
      READ(5,*)HF
      GOTO 5


   40 WRITE(6,41)
   41 FORMAT('New waveguide length is :'/)
      READ(5,*)LENGTH
      GOTO 5


   45 WRITE(6,46)
   46 FORMAT('New delta-z is :'/)
      READ(5,*)DZ
      GOTO 5


   50 WRITE(6,51)
   51 FORMAT('New x-grid is :'/)
      READ(5,*)XGRID
      GOTO 5


   55 WRITE(6,56)
   56 FORMAT('New substrate refractive index is :'/)
      READ(5,*)NREF
      GOTO 5
```

```
60  WRITE(6,61)
61 FORMAT('New refractive index different is :'/)
   READ(5,*)DN
   GOTO 5



65  WRITE(6,66)
66 FORMAT('New distance where taper starts is:'/)
   READ(5,*)DIST
   GOTO 5



75  CONTINUE
   RETURN
   END



C---------------------------------------------------------------

   SUBROUTINE PROF(XGRID,WIDTH,HF,DN,NREF,DX,S)

C*************************************************************
C*                                       *
C* This subroutine compute the profile distribution.      *
C* PRF  -- PROFILE                          *
C* XC,XC1,J  -- LOCAL VAR.                    *
C*                                     *
C*************************************************************



   PARAMETER (MAX=512)
   COMMON/BLK1/PRF
   REAL WIDTH,DX,HF,NREF,S
   DOUBLE PRECISION XC,XC1,PRF(MAX)
   INTEGER J,XGRID



   XC=WIDTH/2.0-DX/2.0
   DO 10 J=1,XGRID
    PRF(J)=0.0
    XC1=COSH(2.0*XC/HF)
    PRF(J)=1.0+0.01*EXP(-S*LOG(XC1))
    PRF(J)=(PRF(J)-1.0)/0.01
    XC=XC-DX
10 CONTINUE
   RETURN
   END
```

```
C--------------------------------------------------------------

    SUBROUTINE IDX(XGRID,WIDTH,HF,DN,NREF,KL,DX)

C*************************************************************
C*                                   *
C* THIS SECTION IS TO SET UP THE REFRACTIVE INDEX FOR THE MATRIX    *
C* COMPUTATION.                                *
C*                                   *
C* PERT -- PERTURDBATION                      *
C*                                   *
C*************************************************************


    PARAMETER (MAX=512)
    COMMON/BLK2/INDEX
    REAL WIDTH,DX,XC1,XC,PERT,KL,HF,NREF,S,DN,INDEX(MAX)
    INTEGER J,XGRID


    S=2.0
    XC=(WIDTH/2.0)-DX/2.0
    DO 10 J=1,XGRID
     INDEX(J)=0.0
     XC1=COSH(2.0*XC/HF)
     PERT=DN*EXP(-S*LOG(XC1))
     INDEX(J)=NREF+PERT
     XC=XC-DX
  10 CONTINUE
    RETURN
    END

C--------------------------------------------------------------

    SUBROUTINE INIT(XGRID,WIDTH,HF,DN,NREF,K0,DX,KL,NPRO,DZ,S,STEP)

C*************************************************************
C*                                   *
C* THIS SECTION IS TO STORE THE INITIAL FIELD INTENSITY AND TO CREAT *
C* AN OUTPUT FILE FOR THE RESULT.                  *
C*                                   *
C* INTEN  -- INTENSITY                      *
C* PROF   -- PROFILE                        *
C*                                   *
C*************************************************************


    PARAMETER (MAX=512)
    REAL Z,WIDTH,HF,DN,NREF,K0,DX,KL,LENGHT,S,DZ
    INTEGER XGRID,NPRO,STEP


    CHARACTER*20 FNAME
    WRITE(6,10)
  10 FORMAT('PLEASE ENTER THE NAME OF THE THE OUTPUT FILE:',/)
```

```
   READ(5,20)FNAME
 20 FORMAT(A)
   OPEN  (3,FILE=FNAME,STATUS='OLD')


   WRITE(6,30)
 30 FORMAT('PLEASE ENTER THE NO OF PLANE TO BE PLOTTED.',/)
   READ(5,*)STEP


   OPEN(UNIT=4,FILE='PAM.DAT')
   WRITE(4,35)XGRID,STEP
 35 FORMAT(1X,I5,1X,I5)
   CLOSE(4)

   CALL  PROF(XGRID,WIDTH,HF,DN,NREF,DX,S)


   CALL  INTEN(KL,NPRO,XGRID,DZ)


   RETURN
   END



C-------------------------------------------------------------

   SUBROUTINE  INTEN(KL,NPRO,XGRID,DZ)

C*************************************************************
C*                              *
C* THIS SECTION OUTPUT THE FIELD INTENSITY FOR THE FIRST PLANE.   *
C*                              *
C*************************************************************


   PARAMETER (MAX=512)
   COMMON/BLK1/PRF
   COMMON/BLK4/R
   REAL T,KL,DZ
   INTEGER J,K,XGRID
   DOUBLE PRECISION ZERO,PRF(MAX),R2(MAX)
   COMPLEX*16 R(MAX)


   ZERO=0.0
   DO 20 J=1,XGRID
    R(J)=DCMPLX(PRF(J),ZERO)
    R2(J)=ABS(R(J))
    WRITE(3,10)R2(J)
 10   FORMAT(F10.8)
 20 CONTINUE
   RETURN
   END
```

```
C-----------------------------------------------------------------

      SUBROUTINE MTRX2(K0,KL,DZ,DX,NREF,XGRID,NPRO,STEP,DIST
     +,LENGTH,WIDTH)

C*****************************************************************
C*                                *
C* THIS SECTION IS TO COMPUTE LIGTH INTENSITY AS IT PROPAGATES   *
C* THROUGH. COMPUTATION IS USING THE FINITE DIFFERENT METHOD     *
C* WHICH USE TRIDIOGANAL MATRIX. ELEMENTS IN THE MATRIX IS COMPUTED *
C* IN THE SUBROUTINE 'COEFF.'                      *
C*                                *
C* CK    --- CHECK                         *
C* COEFF --- COEFFICIENT                     *
C* SKIP  --- NO OF TIME TO SKIP BEFORE THE INTENSITY IS OUTPUTED  *
C* OPUT  --- OUTPUT.                        *
C*                                *
C*****************************************************************


      PARAMETER (MAX=512)
      REAL K0,KL,DZ,DX,NREF,Z,Z1,DIST
      REAL M,LENGTH,ANG,WIDTH,W1,L1
      INTEGER I,J,K,L,SKIP,XGRID,NPRO,STEP
      DOUBLE PRECISION EOUT(MAX),A(MAX),AN(MAX),ZERO
      COMMON/BLK4/R
      COMPLEX*16 B(MAX),C(MAX),R1(MAX),BN(MAX)
      COMPLEX*16 R(MAX),E(MAX),TEM(MAX)


      ZERO=0.0
      Z=0.0
      Z1=0.0
      SKIP=INT(NPRO/STEP)
      L=SKIP
      W1=WIDTH/2.0
      L1=LENGTH-DIST
      M=W1/L1
C     ANG=TAN(W1/L1)


      CALL COEFF(K0,DZ,DX,NREF,XGRID,A,B,C)
      DO 50 K=2,NPRO
       Z=Z+DZ
       CALL CK(Z1,DZ,DIST,XGRID,NREF,LENGTH,K0,W1,DX,A,B,C,K,WIDTH,M)
       CALL MTRX1(A,C,XGRID,R1)


       IF (B(1).EQ.0)PAUSE
       BN(1)=B(1)
       AN(1)=-A(1)
       DO 10 J=2,XGRID
         AN(J)=-A(J)
         TEM(J)=AN(J)/BN(J-1)
```

```
   BN(J)=B(J)-TEM(J)*AN(J-1)
   R1(J)=R1(J)-TEM(J)*R1(J-1)
10 CONTINUE


  E(XGRID)=R1(XGRID)/BN(XGRID)
  R(XGRID)=E(XGRID)


 DO 30 J=XGRID-1,1,-1
   E(J)=(R1(J)-AN(J)*E(J+1))/BN(J)
   R(J)=E(J)
30  CONTINUE


  CALL OPUT(E,XGRID,K,L,STEP,NPRO,SKIP)
  WRITE(6,40)K
40  FORMAT('DOING STEP #',I4)
50 CONTINUE
 CLOSE(3)
 RETURN
 END



C--------------------------------------------------------------

  SUBROUTINE OPUT(E,XGRID,K,L,STEP,NPRO,SKIP)

C*************************************************************
C*                              *
C* This section in charge of the output of the field computed to the *
C* result file. Note that not all planes are outputted to the result *
C* file.                         *
C*                              *
C*************************************************************


  PARAMETER(MAX=512)
  DOUBLE PRECISION EOUT(MAX)
  INTEGER XGRID,J,K,L,N,STEP,SKIP,NPRO
  COMPLEX*16 E(MAX)


 N=K-L
 IF(N.EQ.0) THEN
   DO 20 J=1,XGRID
    EOUT(J)=ABS(E(J))
    WRITE(3,10)EOUT(J)
10   FORMAT(F10.8)
20  CONTINUE
   L=K+SKIP
 ELSE
  ENDIF
 RETURN
 END
```

```
C-------------------------------------------------------------

   SUBROUTINE CK(Z1,DZ,DIST,XGRID,NREF,LENGTH,K0,W1
  +,DX,A,B,C,K,WIDTH,M)


C*************************************************************
C*                              *
C* THIS SECTION IS TO CHECK THE DISTANCE OF PROPAGATION IN Z-DIR .  *
C* IF ANY ABRUPT IS SPECIFIED BEFORE, IT WILL BRANCH TO A ROUTINE   *
C* WHICH COMPUTE THE NEW REFRACTIVE INDEX.              *
C*                              *
C*************************************************************


   PARAMETER(MAX=512)
   REAL Z,Z1,ZE,DZ,NREF,K0,DX,DIST,ANG,WIDTH,LENGTH,M
   DOUBLE PRECISION A(MAX)
   INTEGER XGRID,K,TPRO
   COMPLEX*16 B(MAX),C(MAX)


   TPRO=INT(DIST/DZ)
   ZE=K-TPRO
   IF (ZE.GT.0) THEN
     Z1=Z1+DZ
     CALL TAP(Z1,DZ,LENGTH,WIDTH,XGRID,NREF,DIST,DX,M,W1)
     CALL COEFF(K0,DZ,DX,NREF,XGRID,A,B,C)
   ENDIF
   RETURN
   END


C-------------------------------------------------------------

   SUBROUTINE COEFF(K0,DZ,DX,NREF,XGRID,A,B,C)


C*************************************************************
C*                              *
C* THIS SECTION COMPUTES THE ELEMENTS OF THE TRIDIAGONAL MATRIX.   *
C*                              *
C*************************************************************


   PARAMETER (MAX=512)
   REAL DX,DZ,NREF,K0,INDEX(MAX)
   INTEGER J,K,XGRID
   DOUBLE PRECISION A(MAX),B1,B2,C1,C2
   COMMON/BLK2/INDEX
   COMPLEX*16 B(MAX),C(MAX)


   DO 10 J=1,XGRID
   A(J)=DZ/(2.0*(DX*DX))
   B1=DZ/(DX*DX)-((DZ*(K0*K0))/2.0)*(INDEX(J)*INDEX(J)-NREF*NREF)
   B2=2.0*K0*NREF
```

```
   C1=-(DZ/(DX*DX))+((DZ*(K0*K0))/2.0)*(INDEX(J)*INDEX(J)-NREF*NREF)
   C2=2.0*K0*NREF
   B(J)=DCMPLX(B1,B2)
   C(J)=DCMPLX(C1,C2)
 10 CONTINUE
   RETURN
   END




C--------------------------------------------------------------

   SUBROUTINE MTRX1(A,C,XGRID,R1)

C**************************************************************
C*                              *
C* THIS ROUTINE IS TO COMPUTE THE RHS OF EQUATION (6).       *
C*                              *
C**************************************************************


   PARAMETER(MAX=512)
   INTEGER J,XGRID
   DOUBLE PRECISION A(MAX)
   COMMON/BLK4/R
   COMPLEX*16 C(MAX),R1(MAX),R(MAX)


   R1(1)=C(1)*R(1)+A(1)*R(2)
   DO 10 J=2,XGRID-1
   R1(J)=A(J)*R(J-1)+C(J)*R(J)+A(J)*R(J+1)
 10 CONTINUE
   R1(XGRID)=A(XGRID)*R(XGRID-1)+C(XGRID)*R(XGRID)
   RETURN
   END




C--------------------------------------------------------------

    SUBROUTINE TAP(Z1,DZ,LENGTH,WIDTH,XGRID,NREF,DIST,DX,M,W1)

C**************************************************************
C*                              *
C* THIS SECTION IS TO SET THE NEW REF. INDEX FOR THE ABRUPT CHANGE   *
C* IN REF. INDEX.                        *
C*                              *
C* X1,X2 ---- LOCAL VAR                   *
C*                              *
C**************************************************************


   PARAMETER(MAX=512)
   REAL INDEX(MAX),Z,Z1,DZ,NREF,DIST,W1,M
   REAL WIDTH,X,DX,X1,NAIR,LENGTH,ANG
C    DOUBLE PRECISION M
   INTEGER XGRID,J,N
   COMMON/BLK2/INDEX
```

```
      NAIR=1.0
      X=-M*Z1+W1
      X1=W1-X
      N=NINT(X1/DX)
C     PRINT*,X,X1,N
      IF (N.GT.0) THEN
        DO 10 J=1,N
         INDEX(J)=NAIR
         INDEX(XGRID-J)=NAIR
   10 CONTINUE
      ENDIF


      RETURN
      END
```

A8.4 Y-Junction: FORTRAN PROGRAM

```
*$NOEXTENSIONS
      PROGRAM FD


C*************************************************************
C*                                    *
C* This program uses the Finite Different(FD) method to compute   *
C* the electric field as a light wave propagates through a slab   *
C* waveguide. This program is for calculating the field of a Y-   *
C* junction.                              *
C* WG -- WAVEGUIDE                        *
C* CACL-- CALCULATE                         *
C* IDX -- REFRACTIVE INDEX                     *
C* INIT-- INITIALIZED                        *
C* MTRX-- MATRIX                        *
C* ABRP-- ABRUPT CHANGE                       *
C* CMF -- COMFIRM                         *
C*                                 *
C*************************************************************


      PARAMETER (MAX=512)
      COMMON/BLK1/PRF
      COMMON/BLK2/INDEX
      COMMON/BLK4/R
      REAL DX,DZ,X,Y,Z,WAVE,WIDTH,S,TL
      REAL INDEX(MAX),AL,BE,LT
      REAL DN,HF,K0,KL,NREF,D(7)
      DOUBLE PRECISION PRF(MAX)
      INTEGER XGRID,NPRO,STEP
      COMPLEX*16 R(MAX)


      CALL INTRO
      CALL WG(XGRID,WAVE,NREF,DN,WIDTH,HF,AL,BE,LT)
```

```
      CALL CALC(WAVE,KL,K0,NREF,WIDTH,DX,DZ,NPRO,XGRID)
      CALL CMF(WAVE,NREF,DN,WIDTH,HF,DX,DZ,XGRID
     +,NPRO,K0,S,AL,BE,LT)
      CALL PROP(XGRID,NPRO,DZ,DX,KL,K0,WIDTH,HF,DN,NREF
     +,STEP,TL,S,LT,BE,AL)
      STOP
      END


C  ------------------------------------------------------------


      SUBROUTINE INTRO

C*****************************************************************
C*                               *
C* This section is to give the user some knowledge on the program.   *
C*                               *
C*****************************************************************


      WRITE(6,5)
    5 FORMAT(
     +'                            '/,
     +'*****************************************************'/,
     +'*                         *'/,
     +'* This program uses finite different method to compute   *'/,
     +'* the electric field as light propagates through a    *'/,
     +'* optical waveguide. Suggestions for various parameters  *'/,
     +'* are shown with the input questions. Mistakes can be   *'/,
     +'* corrected after all the values are inputted. After   *'/,
     +'* the computation is completed, a 3-d view of the field  *'/,
     +'* will be plotted.                 *'/,
     +'*                         *'/,
     +'*****************************************************',/)


      RETURN
      END



C---------------------------------------------------------------


      SUBROUTINE WG(XGRID,WAVE,NREF,DN,WIDTH,HF,AL,BE,LT)

C*****************************************************************
C*                               *
C* This part of the program is to prompt for the propagation condition.*
C* width -- Waveguide width in x-dir.              *
C* le  -- Waveguide lenrth in z-dir.             *
C* xgrid -- no of section along x-dir.            *
C* wave -- wavelength in vacuum.               *
C* nref -- reference reflective index.             *
C* dn  -- change of reflective index.            *
C* dec  -- decision                   *
C*                               *
C*****************************************************************
```

```
      REAL WAVE,HF,NREF,DN
      REAL AL,BE,LT
      INTEGER XGRID


      WRITE(6,5)
    5 FORMAT('1. Please enter wavelength of light in vacuum :'/,
     +'  in micron. (e.g. 0.633)',/)
      READ(5,*)WAVE


      WRITE(6,10)
   10 FORMAT('2. Please enter thickness of slab waveguide in micron:'/,
     +'  (eg: 20)',/)
      READ(5,*)WIDTH


      WRITE(6,20)
   20 FORMAT('3. Please enter the film thickness in micron:'/,
     +'  (eg:1)',/)
      READ(5,*)HF


      WRITE(6,30)
   30 FORMAT('4. Please enter the no of slice in X-dir:'/,
     +'  note the value must be in integer and must < 256.',/)
      READ(5,*)XGRID


      WRITE(6,40)
   40 FORMAT('6. Please enter the refractive index of the substrate:'/,
     +'  (eg:2.2)',/)
      READ(5,*)NREF


      WRITE(6,50)
   50 FORMAT('7. Please enter the refractive index change:'/,
     +'  (eg:0.01)',/)
      READ(5,*)DN


      WRITE(6,60)
   60 FORMAT('8. Please enter angle of the bend in degree:'/,
     +'  (eg. 0.6)',/)
      READ(5,*)AL


      WRITE(6,70)
   70 FORMAT('9. Please enter the length of the beginning '/,
     +'  horizontal section in micron (eg.100) :',/)
      READ(5,*)BE


      WRITE(6,80)
   80 FORMAT('10. Please enter the total length of the waveguide'/,
     +'  in micron (eg.100)',/)
      READ(5,*)LT
```

```
      RETURN
      END



C------------------------------------------------------------

      SUBROUTINE CALC(WAVE,KL,K0,NREF,WIDTH,DX,DZ,NPRO,XGRID)

C*************************************************************
C* This section is to calculate the wave constant and to prompt    *
C* the input of delta-z.                          *
C*                              *
C* KL -- wave number                      *
C* K0 -- KL/NREF                        *
C* DX -- step size in x-dir.                  *
C* DZ -- step size in z-dir.                  *
C* NPRO- no of propagation step in z-dir.              *
C*                          *
C*************************************************************


      REAL WAVE,WAVE1,KL,NREF,K0,WIDTH,DX,DZ,LENGTH
      INTEGER XGRID,NPRO,DEC
      DOUBLE PRECISION PI,TWOPI


      PI=3.14159265359
      TWOPI=2.0*PI
      WAVE1=WAVE/NREF
      KL=TWOPI/WAVE1
      K0=KL/NREF
      WRITE(6,10)
   10 FORMAT('Please input the propagation step size (DZ) in micron.'/)
      READ(5,*)DZ
      RETURN
      END


C------------------------------------------------------------

      SUBROUTINE CMF(WAVE,NREF,DN,WIDTH,HF,DX,DZ,XGRID
     +,NPRO,K0,S,AL,BE,LT)

C*************************************************************
C*                              *
C* This section is to confirm to the user that the values that have  *
C* been input. If any value that wish to be change, it can be done.   *
C* DEC  --- decision                    *
C*                              *
C*************************************************************


      REAL WAVE,NREF,DN,WIDTH,S
      REAL HF,DX,DZ,K0,AL,BE,LT
      INTEGER XGRID,DEC,NPRO
```

```
5 NPRO=NINT(LT/DZ)
  DX=WIDTH/XGRID
  V=K0*HF*SQRT(2.0*NREF*DN)
  S=0.5*(SQRT(V*V+1)-1)


  WRITE(6,10)WAVE,WIDTH,HF,DZ,XGRID,NREF,DN,AL,BE,LT
 +,DX,NPRO,S
10 FORMAT(
 +'   The values entered are:                  '/,
 +'  (1)  Wavelength        =',F5.3,' micron        '/,
 +'  (2)  Waveguide thickness  =',F6.3,' micron         '/,
 +'  (3)  Film thickness     =',F6.3,' micron       '/,
 +'  (4)  Delta-z          =',F6.3,' micron      '/,
 +'  (5)  Grid points in X-dir =',I6,'          '/,
 +'  (6)  Substrate ref. index =',F6.3,'          '/,
 +'  (7)  Delta-n          =',F6.3,'       '/,
 +'  (8)  Bend-junction angle  =',F6.3,'          '/,
 +'  (9)  Beginning length     =',F6.1,'        '/,
 +'  (10) Total length       =',F6.1,'        '/,
 +'   The value of delta-x is :',F6.3,'          '/,
 +'   The number of propagation step is :',I7,'         '/,
 +'   The mode number is :',F6.3,'          '/)


15 WRITE(6,20)
20 FORMAT(
 +'Do you wish to change any values ?              '/,
 +'Press any number, (1) to (10), of the parameters shown above '/,
 +'that you wish to change.Otherwise press (0) to continue    '/)
  READ(5,*)DEC
  IF (DEC.EQ.0) THEN
     GOTO 75
   ELSE
     GOTO (25,30,35,40,45,50,55,60,65,70) DEC
   END IF



25 WRITE(6,26)
26 FORMAT('New wavelength is :'/)
  READ(5,*)WAVE
  GOTO 5



30 WRITE(6,31)
31 FORMAT('New thickness is :'/)
  READ(5,*)WIDTH
  GOTO 5



35 WRITE(6,36)
36 FORMAT('New film thickness is :'/)
  READ(5,*)HF
  GOTO 5
```

```
 40 WRITE(6,41)
 41 FORMAT('New delta-z is :'/)
  READ(5,*)DZ
  GOTO 5


 45 WRITE(6,46)
 46 FORMAT('New x-grid is :'/)
  READ(5,*)XGRID
  GOTO 5


 50 WRITE(6,51)
 51 FORMAT('New substract refractice index is :'/)
  READ(5,*)NREF
  GOTO 5


 55 WRITE(6,56)
 56 FORMAT('New refractive index different is :'/)
  READ(5,*)DN
  GOTO 5


 60 WRITE(6,61)
 61 FORMAT('New angle is:'/)
  READ(5,*)AL
  GOTO 5


 65 WRITE(6,66)
 66 FORMAT('New beginning length is:'/)
  READ(5,*)BE
  GOTO 5


 70 WRITE(6,71)
 71 FORMAT('New total length is:'/)
  READ(5,*)LT
  GOTO 5


 75 CONTINUE
  RETURN
  END


C-------------------------------------------------------------

  SUBROUTINE PROP(XGRID,NPRO,DZ,DX,KL,K0,WIDTH,HF,DN,NREF
 +,STEP,TL,S,LT,BE,AL)

C***************************************************************
C*                                *
C* This routine is to check the propagation distance of the light.  *
C* When
C***************************************************************
```

```
PARAMETER(MAX=512)
REAL Z,DX,DZ,KL,K0,WIDTH,HF,NREF,TL,DS
REAL INDEX(MAX),S,BE,AL,LT
DOUBLE PRECISION A(MAX),PRF(MAX),PI,M,DH
INTEGER XGRID,NPRO,K,L,STEP,SKIP,T,DH1,XX
LOGICAL FLAG
COMMON/BLK1/PRF
COMMON/BLK2/INDEX
COMMON/BLK4/R
COMPLEX*16  R(MAX),B(MAX),C(MAX)


PI=3.141592654
Z=0.0
DH=0.0
M=TAN((AL*PI)/180)
XX=0
T=0
CALL INIT(XGRID,WIDTH,HF,DN,NREF,K0,DX,KL,NPRO,DZ,S,STEP)
SKIP=INT(NPRO/STEP)
L=SKIP
DO 20 K=2,NPRO
Z=Z+DZ
IF (Z.GT.BE) THEN
 DH=M*(Z-BE)
ENDIF


CALL  INDX(XGRID,WIDTH,HF,DN,NREF,KL,DX,DS,DH)
CALL COEFF(K0,DZ,DX,NREF,XGRID,A,B,C)
CALL MTRX2(K0,KL,DZ,DX,NREF,XGRID,NPRO,STEP,A,B,C,K,L,SKIP,T)


 20 CONTINUE
 CLOSE(3)
 RETURN
 END



C-------------------------------------------------------------

  SUBROUTINE  PROF(XGRID,WIDTH,HF,DN,NREF,DX,S)

C**************************************************************
C*                               *
C* This subroutine compute the profile distribution.     *
C* PRF  -- PROFILE                     *
C* XC,XC1,J  -- LOCAL VAR.                *
C*                          *
C**************************************************************


  PARAMETER (MAX=512)
  COMMON/BLK1/PRF
```

```
   REAL WIDTH,DX,HF,NREF,S
   DOUBLE PRECISION XC,XC1,PRF(MAX)
   INTEGER J,XGRID


  XC=WIDTH/2.0-DX/2.0
  DO 10 J=1,XGRID
   PRF(J)=0.0
   XC1=COSH(2.0*XC/HF)
   PRF(J)=1.0+0.01*EXP(-S*LOG(XC1))
   PRF(J)=(PRF(J)-1.0)/0.01
   XC=XC-DX
  10 CONTINUE
   RETURN
   END



C-----------------------------------------------------------------

   SUBROUTINE INDX(XGRID,WIDTH,HF,DN,NREF,KL,DX,DS,DH)

C****************************************************************
C*                              *
C* THIS SECTION IS TO SET UP THE REFRACTIVE INDEX FOR THE MATRIX   *
C* COMPUTATION.                         *
C*                              *
C* PERT -- PERTURBATION                     *
C*                              *
C****************************************************************


   PARAMETER (MAX=512)
   COMMON/BLK2/INDEX
   REAL WIDTH,DX,XC1,XC,PERT,KL,HF,NREF,U
   REAL S,DN,INDEX(MAX),DS,M,AL,LT,BE,Z,XC2,XC3,XC4
   DOUBLE PRECISION DH
   INTEGER J,XGRID,X,HXGRID,Y


  S=2.0
  U=MOD(XGRID,2)
  XC2=((WIDTH/2.0)+DH)-DX/2.0
  XC3=(-(WIDTH/2.0)-DH)+DX/2.0
  HXGRID=INT(XGRID/2)+U
  DO 10 J=1,XGRID
   INDEX(J)=0.0
   XC1=COSH(2.0*XC2/HF)
   PERT=DN*EXP(-S*LOG(XC1))
   IF((J.EQ.HXGRID+1).OR.(J.GT.HXGRID+1)) THEN
   INDEX(J)=NREF+PERT
C    PRINT*,J,INDEX(J)
   ENDIF
   XC2=XC2-DX
  10 CONTINUE
```

```
      DO 20 J=XGRID,1,-1
        XC4=COSH(2.0*XC3/HF)
        PERT=DN*EXP(-S*LOG(XC4))
        IF((J.LT.HXGRID).OR.(J.EQ.HXGRID))THEN
        INDEX(J)=NREF+PERT
C       PRINT*,J,INDEX(J)
        ENDIF
        XC3=XC3+DX
   20 CONTINUE


      RETURN
      END



C----------------------------------------------------------------

      SUBROUTINE INIT(XGRID,WIDTH,HF,DN,NREF,K0,DX,KL,NPRO,DZ,S,
     STEP)

C***************************************************************
C*                                  *
C* THIS SECTION IS TO STORE THE INITIAL FIELD INTENSITY AND TO CREAT C*  AN OUTPUT
FILE FOR THE RESULT.                            *
C*                                  *
C* INTEN  -- INTENSITY                      *
C* PROF   -- PROFILE                     *
C*                                  *
C***************************************************************


      PARAMETER (MAX=512)
      REAL Z,WIDTH,HF,DN,NREF,K0,DX,KL,LENGHT,S,DZ
      INTEGER XGRID,NPRO,STEP


      CHARACTER*20 FNAME
      WRITE(6,10)
   10 FORMAT('PLEASE ENTER THE NAME OF THE THE OUTPUT FILE:',/)
      READ(5,20)FNAME
   20 FORMAT(A)
      OPEN (3,FILE=FNAME,STATUS='OLD')


      WRITE(6,30)
   30 FORMAT('PLEASE ENTER THE NO OF PLANE TO BE PLOTTED.')
      READ(5,*)STEP


      OPEN(UNIT=4,FILE='PAM.DAT')
      WRITE(4,35)XGRID,STEP
   35 FORMAT(1X,I5,1X,I5)
      CLOSE(4)
```

```
   CALL PROF(XGRID,WIDTH,HF,DN,NREF,DX,S)

   CALL INTEN(KL,NPRO,XGRID,DZ)

   RETURN
   END




C-------------------------------------------------------------

   SUBROUTINE INTEN(KL,NPRO,XGRID,DZ)

C**************************************************************
C*                             *
C* THIS SECTION OUTPUT THE FIELD INTENSITY FOR THE FIRST PLANE.   *
C*                             *
C**************************************************************


   PARAMETER (MAX=512)
   COMMON/BLK1/PRF
   COMMON/BLK4/R
   REAL T,KL,DZ
   INTEGER J,K,XGRID
   DOUBLE PRECISION ZERO,PRF(MAX),R2(MAX)
   COMPLEX*16 R(MAX)


   ZERO=0.0
   DO 20 J=1,XGRID
    R(J)=DCMPLX(PRF(J),ZERO)
    R2(J)=ABS(R(J))
    WRITE(3,10)R2(J)
 10   FORMAT(F10.8)
 20 CONTINUE
   RETURN
   END




C-------------------------------------------------------------

   SUBROUTINE MTRX2(K0,KL,DZ,DX,NREF,XGRID,NPRO
  +,STEP,A,B,C,K,L,SKIP,T)



C**************************************************************
C*                             *
C* THIS SECTION IS TO COMPUTE LIGTH INTENSITY AS IT PROPAGATES   *
C* THROUGH. COMPUTATION IS USING THE FINITE DIFFERENT METHOD    *
C* WHICH USE TRIDIOGANAL MATRIX. ELEMENTS IN THE MATRIX IS COMPUTED  *
C* IN THE SUBROUTINE 'COEFF.'                     *
```

```
C*                              *
C* CK    --- CHECK                    *
C* COEFF  --- COEFFICIENT                  *
C* SKIP   --- NO OF TIME TO SKIP BEFORE THE INTENSITY IS OUTPUTED   *
C* OPUT  --- OUTPUT.                    *
C*                        *
C***************************************************************


   PARAMETER (MAX=512)
   REAL K0,KL,DZ,DX,NREF,Z,DIST
   INTEGER I,J,K,L,SKIP,XGRID,NPRO,STEP,T
   DOUBLE PRECISION EOUT(MAX),A(MAX),AN(MAX),ZERO
   COMMON/BLK4/R
   COMPLEX*16 B(MAX),C(MAX),R1(MAX),BN(MAX)
   COMPLEX*16 R(MAX),E(MAX),TEM(MAX)


   ZERO=0.0

   CALL MTRX1(A,C,XGRID,R1)
   IF (B(1).EQ.0)PAUSE
   BN(1)=B(1)
   AN(1)=-A(1)
   DO 10 J=2,XGRID
    AN(J)=-A(J)
    TEM(J)=AN(J)/BN(J-1)
    BN(J)=B(J)-TEM(J)*AN(J-1)
    R1(J)=R1(J)-TEM(J)*R1(J-1)
  10 CONTINUE


   E(XGRID)=R1(XGRID)/BN(XGRID)
   R(XGRID)=E(XGRID)


  DO 30 J=XGRID-1,1,-1
   E(J)=(R1(J)-AN(J)*E(J+1))/BN(J)
   R(J)=E(J)
  30 CONTINUE


  CALL OPUT(E,XGRID,K,L,STEP,NPRO,SKIP,T)

  WRITE(6,40)K
 40 FORMAT('DOING STEP #',I4)
 50 CONTINUE
  RETURN
  END



C-------------------------------------------------------------

   SUBROUTINE OPUT(E,XGRID,K,L,STEP,NPRO,SKIP,T)
```

```
C*****************************************************************
C*                                      *
C* This section in charge of the output of the field computed to the *
C* result file. Note that not all planes are outputed to the result   *
C* file.                                *
C*                                      *
C*****************************************************************


   PARAMETER(MAX=512)
   DOUBLE PRECISION EOUT(MAX)
   INTEGER XGRID,J,K,L,N,STEP,SKIP,NPRO,T
   COMPLEX*16 E(MAX)


   N=K-L
   IF (N.EQ.0) THEN
   T=T+1
    IF ((T.LT.STEP).OR.(T.EQ.STEP-1)) THEN
     DO 20 J=1,XGRID
      EOUT(J)=ABS(E(J))
      WRITE(3,10)EOUT(J)
10       FORMAT(F10.8)
20   CONTINUE
     L=K+SKIP
    ENDIF
   ELSE
    ENDIF
   RETURN
   END


C----------------------------------------------------------------

   SUBROUTINE COEFF(K0,DZ,DX,NREF,XGRID,A,B,C)

C*****************************************************************
C*                                      *
C* THIS SECTION COMPUTES THE ELEMENTS OF THE TRIDIAGONAL MATRIX.    *
C*                                      *
C*****************************************************************


   PARAMETER (MAX=512)
   REAL DX,DZ,NREF,K0,INDEX(MAX)
   INTEGER J,K,XGRID
   DOUBLE PRECISION A(MAX),B1,B2,C1,C2
   COMMON/BLK2/INDEX
   COMPLEX*16 B(MAX),C(MAX)


   DO 10 J=1,XGRID
   A(J)=DZ/(2.0*(DX*DX))
   B1=DZ/(DX*DX)-((DZ*(K0*K0))/2.0)*(INDEX(J)*INDEX(J)-NREF*NREF)
```

```
   B2=2.0*K0*NREF
   C1=-(DZ/(DX*DX))+((DZ*(K0*K0))/2.0)*(INDEX(J)*INDEX(J)-NREF*NREF)
   C2=2.0*K0*NREF
   B(J)=DCMPLX(B1,B2)
   C(J)=DCMPLX(C1,C2)
 10 CONTINUE
   RETURN
   END


C------------------------------------------------------------------


   SUBROUTINE MTRX1(A,C,XGRID,R1)


C******************************************************************
C*                                    *
C* THIS ROUTINE IS TO COMPUTE THE RHS OF EQUATION (6).     *
C*                                *
C******************************************************************


   PARAMETER(MAX=512)
   INTEGER J,XGRID
   DOUBLE PRECISION A(MAX)
   COMMON/BLK4/R
   COMPLEX*16 C(MAX),R1(MAX),R(MAX)


   R1(1)=C(1)*R(1)+A(1)*R(2)
   DO 10 J=2,XGRID-1
   R1(J)=A(J)*R(J-1)+C(J)*R(J)+A(J)*R(J+1)
 10 CONTINUE
   R1(XGRID)=A(XGRID)*R(XGRID-1)+C(XGRID)*R(XGRID)
   RETURN
   END
```

A8.5 Mach-Zehnder Waveguide Structures: FORTRAN Program

```
*$NOEXTENSIONS
   PROGRAM FD


C******************************************************************
C*                                    *
C* This program uses the Finite Different (FD) method to compute   *
C* the electric field as a light wave propagates through a slab   *
C* waveguide. A Mach-Zhender interferometer structure is used    *
C* in this program.                     *
C* WG -- WAVEGUIDE                        *
```

```
C* CACL-- CALCULATE                         *
C* IDX -- REFRACTIVE INDEX                     *
C* INIT-- INITIALIZED                        *
C* MTRX-- MATRIX                          *
C* ABRP-- ABRUPT CHANGE                      *
C* CMF -- COMFIRM                         *
C*                                  *
C***************************************************************

      PARAMETER (MAX=512)
      COMMON/BLK1/PRF
      COMMON/BLK2/INDEX
      COMMON/BLK4/R
      REAL DX,DZ,X,Y,Z,WAVE,WIDTH,S,TL
      REAL INDEX(MAX),AL,LE,BE,EN,SEP
      REAL DN,HF,K0,KL,NREF,D(7)
      DOUBLE PRECISION PRF(MAX)
      INTEGER XGRID,NPRO,STEP
      COMPLEX*16 R(MAX)


      CALL INTRO
      CALL WG(XGRID,WAVE,NREF,DN,WIDTH,LE,HF,AL,BE,EN,SEP)
      CALL CALC(WAVE,KL,K0,NREF,WIDTH,DX,DZ,NPRO,XGRID)
      CALL DIM(AL,LE,BE,EN,SEP,HF,D,TL)
      CALL CMF(WAVE,NREF,DN,WIDTH,HF,DX,DZ,XGRID
     +,NPRO,K0,S,AL,LE,BE,EN,SEP,D)
      CALL PROP(D,Z,XGRID,NPRO,DZ,DX,KL,K0,WIDTH,HF,DN,NREF,STEP,TL,S)
      STOP
      END



C -------------------------------------------------------------

      SUBROUTINE INTRO

C***************************************************************
C*                                  *
C* This section is to give the user some knowledge on the program.   *
C*                                  *
C***************************************************************


      WRITE(6,5)
    5 FORMAT(
     +'                          '/,
     +'***********************************************************'/,
     +'*                        *'/,
     +'* This program uses finite different method to compute   *'/,
     +'* the electric field as light propagates through a    *'/,
     +'* optical waveguide. Suggestions for various parameters  *'/,
     +'* are shown with the input questions. Mistakes can be   *'/,
     +'* corrected after all the values are inputted. After   *'/,
```

```
+'* the computation is completed, a 3-d view of the field  */,
+'* will be plotted.                        */,
+'*                              */,
+'*************************************************',/)


    RETURN
    END




C------------------------------------------------------------------

    SUBROUTINE WG(XGRID,WAVE,NREF,DN,WIDTH,LE,HF,AL,BE,EN,SEP)

C***********************************************************
C*                              *
C* This part of the program is to prompt for the propagation  *
C* condition.*
C* width -- Waveguide width in x-dir.                *
C* le  -- Waveguide length in z-dir.                *
C* xgrid -- no of section along x-dir.                *
C* wave -- wavelength in vacuum.                *
C* nref -- reference reflective index.                *
C* dn  -- change of reflective index.                *
C* dec  -- decision                      *
C*                              *
C***********************************************************


    REAL WAVE,HF,NREF,DN
    REAL AL,BE,LE,EN,SEP
    INTEGER XGRID


    WRITE(6,5)
  5 FORMAT('1. Please enter wavelength of light in vacuum :'/,
   +'  in micron. (eg 0.633)',/)
    READ(5,*)WAVE


    WRITE(6,10)
 10 FORMAT('2. Please enter thickness of slab waveguide in micron:'/,
   +'  (eg: 20)',/)
    READ(5,*)WIDTH


    WRITE(6,20)
 20 FORMAT('3. Please enter the film thickness in micron:'/,
   +'  (eg:1)',/)
    READ(5,*)HF


    WRITE(6,30)
 30 FORMAT('4. Please enter the no of slice in X-dir:'/,
```

```
+'   note the value must be in integer and must < 256.',/)
  READ(5,*)XGRID


  WRITE(6,40)
40 FORMAT('6. Please enter the refractive index of the substrate:'/,
  +'   (eg:2.2)',/)
  READ(5,*)NREF


  WRITE(6,50)
50 FORMAT('7. Please enter the refractive index change:'/,
  +'  (eg:0.01)',/)
  READ(5,*)DN


  WRITE(6,60)
60 FORMAT('8. Please enter Y-junction half angle in degree:'/,
  +'   (eg. 0.6)',/)
  READ(5,*)AL


  WRITE(6,70)
70 FORMAT('9. Please enter the length of the beginning '/,
  +'   horizontal section in micron (eg.100):',/)
  READ(5,*)BE


  WRITE(6,80)
80 FORMAT('10. Please enter the length of the centre section'/,
  +'   in micron (eg. 500):',/)
  READ(5,*)LE


  WRITE(6,90)
90 FORMAT('Please enter the length of the ending horizontal'/,
  +'   section in micron (eg.100)',/)
  READ(5,*)EN


  WRITE(6,100)
100 FORMAT('Please enter the waveguide separation in micron'/,
  +'   (eg. 10)',/)
  READ(5,*)SEP


  RETURN
  END


C-------------------------------------------------------------


  SUBROUTINE DIM(AL,LE,BE,EN,SEP,HF,D,TL)
```

```
C***************************************************************
C*                                  *                          *
C* This section is the calculate the length of each section of the  *
C* waveguide.                                *                 *
C*                                  *                          *
C***************************************************************


      REAL X,SL,TL,AL,BE,EN,SEP,LE,DE,HF,D(7)
      DOUBLE PRECISION PI,TWOPI


      PI=3.14159265359
      TWOPI=2*PI
      TL=TAN(PI*AL/180.0)
      X=HF/(2.0*TL)
      SL=SEP/(2.0*TL)
      D(1)=BE
      D(2)=D(1)+X
      D(3)=D(1)+SL
      D(4)=D(3)+LE
      D(5)=D(4)+SL-X
      D(6)=D(5)+X
      D(7)=D(6)+EN
      RETURN
      END




C-------------------------------------------------------------


      SUBROUTINE CALC(WAVE,KL,K0,NREF,WIDTH,DX,DZ,NPRO,XGRID)

C***************************************************************
C* This section is to calculate the wave constant and to prompt    *
C* the input of delta-z.                         *            *
C*                                  *                          *
C* KL -- wave number                         *                *
C* K0 -- KL/NREF                         *                    *
C* DX -- step size in x-dir.                     *            *
C* DZ -- step size in z-dir.                     *            *
C* NPRO- no of propagation step in z-dir.              *       *
C*                                  *                          *
C***************************************************************


      REAL WAVE,WAVE1,KL,NREF,K0,WIDTH,DX,DZ,LENGTH
      INTEGER XGRID,NPRO,DEC
      DOUBLE PRECISION PI,TWOPI


      PI=3.14159265359
      TWOPI=2.0*PI
      WAVE1=WAVE/NREF
      KL=TWOPI/WAVE1
```

```
   K0=KL/NREF
   WRITE(6,10)
 10 FORMAT('Please input the propagation step size (DZ) in micron.'/)
   READ(5,*)DZ
   RETURN
   END


C--------------------------------------------------------------

   SUBROUTINE CMF(WAVE,NREF,DN,WIDTH,HF,DX,DZ,XGRID
  +,NPRO,K0,S,AL,LE,BE,EN,SEP,D)

C*****************************************************************
C*                              *                               *
C* This section is to confirm to the user that the values that have  *
C* been input. If any value that wish to be change, it can be done.   *
C* DEC   --- decision                          *
C*                                   *
C*****************************************************************


   REAL WAVE,NREF,DN,WIDTH,D(7),S
   REAL HF,DX,DZ,K0,AL,LE,BE,EN,SEP
   INTEGER XGRID,DEC,NPRO


  5 NPRO=NINT(D(7)/DZ)
   DX=WIDTH/XGRID
   V=K0*HF*SQRT(2.0*NREF*DN)
   S=0.5*(SQRT(V*V+1)-1)


   WRITE(6,10)WAVE,WIDTH,HF,DZ,XGRID,NREF,DN,AL,BE,LE,EN,SEP
  +,DX,NPRO,S
 10 FORMAT(
  +'  The values entered are:              '/,
  +' (1) Wavelength       =',F5.3,' micron       '/,
  +' (2) Waveguide thickness =',F6.3,' micron        '/,
  +' (3) Film thickness    =',F6.3,' micron      '/,
  +' (4) Delta-z        =',F6.3,' micron      '/,
  +' (5) Grid points in X-dir =',I6,'            '/,
  +' (6) Substrate ref. index =',F6.3,'          '/,
  +' (7) Delta-n       =',F6.3,'        '/,
  +' (8) Y-junction angle   =',F6.3,'          '/,
  +' (9) Beginning length   =',F6.1,'          '/,
  +' (10) Centre length    =',F6.1,'        '/,
  +' (11) Ending length    =',F6.1,'        '/,
  +' (12) Centre separation  =',F6.1,'         '/,
  +'  The value of delta-x is :',F6.3,'          '/,
  +'  The number of propagation step is :',I7,'          '/,
  +'  The mode number  is :',F6.3,'          '/)


 15 WRITE(6,20)
 20 FORMAT(
```

```
+'Do you wish to change any values ?                     '/,
+'Press any number, (1) to (12), of the parameters shown above '/,
+'that you wish to change. Otherwise press (0) to continue   '/)
 READ(5,*)DEC
 IF (DEC.EQ.0) THEN
    GOTO 85
  ELSE
    GOTO (25,30,35,40,45,50,55,60,65,70,75,80) DEC
  END IF


25 WRITE(6,26)
26 FORMAT('New wavelength is :'/)
 READ(5,*)WAVE
 GOTO 5


30 WRITE(6,31)
31 FORMAT('New thickness is :'/)
 READ(5,*)WIDTH
 GOTO 5


35 WRITE(6,36)
36 FORMAT('New film thickness is :'/)
 READ(5,*)HF
 GOTO 5


40 WRITE(6,41)
41 FORMAT('New delta-z is :'/)
 READ(5,*)DZ
 GOTO 5


45 WRITE(6,46)
46 FORMAT('New x-grid is :'/)
 READ(5,*)XGRID
 GOTO 5


50 WRITE(6,51)
51 FORMAT('New substrate refractive index is :'/)
 READ(5,*)NREF
 GOTO 5


55 WRITE(6,56)
56 FORMAT('New refractive index different is :'/)
 READ(5,*)DN
 GOTO 5


60 WRITE(6,61)
61 FORMAT('New angle is:'/)
```

```
   READ(5,*)AL
   CALL DIM(AL,LE,BE,EN,SEP,HF,D,TL)


65 WRITE(6,66)
66 FORMAT('New beginning length is:'/)
   READ(5,*)BE
   CALL DIM(AL,LE,BE,EN,SEP,HF,D,TL)


70 WRITE(6,71)
71 FORMAT('New centre length is:'/)
   READ(5,*)LE
   CALL DIM(AL,LE,BE,EN,SEP,HF,D,TL)


75 WRITE(6,76)
76 FORMAT('New ending length is:'/)
   READ(5,*)EN
   CALL DIM(AL,LE,BE,EN,SEP,HF,D,TL)



80 WRITE(6,81)
81 FORMAT('New separation is :'/)
   READ(5,*)SEP
   CALL DIM(AL,LE,BE,EN,SEP,HF,D,TL)



85 CONTINUE
   RETURN
   END


C-------------------------------------------------------------

   SUBROUTINE PROP(D,Z,XGRID,NPRO,DZ,DX,KL,K0,WIDTH
  +,HF,DN,NREF,STEP,TL,S)

C*************************************************************
C*                                            *
C* This routine is to check the propagation distance of the light.   *
C* When
C*************************************************************


   PARAMETER(MAX=512)
   REAL D(7),Z,DX,DZ,KL,K0,WIDTH,HF,NREF,DW,TL
   REAL INDEX(MAX),DH,S
   DOUBLE PRECISION A(MAX),PRF(MAX)
   INTEGER XGRID,NPRO,K,L,STEP,SKIP,T
   COMMON/BLK1/PRF
   COMMON/BLK2/INDEX
   COMMON/BLK4/R
   COMPLEX*16 R(MAX),B(MAX),C(MAX)
```

```
      Z=0.0
      XX=0.0
      T=0
      CALL INIT(XGRID,WIDTH,HF,DN,NREF,K0,DX,KL,NPRO,DZ,S,STEP)
      SKIP=INT(NPRO/STEP)
      L=SKIP
      DO 20 K=2,NPRO
      Z=Z+DZ
      IF((Z.LT.D(1)).OR.(Z.GT.D(6)))THEN
        DH=0.0
      ELSE
        ENDIF
      IF((Z.GT.D(1)).AND.(Z.LT.D(3)))THEN
        DH=TL*(Z-D(1))
      ELSE
        ENDIF
      IF((Z.GT.D(3)).AND.(Z.LT.D(4)))THEN
        DH=DH
      ELSE
        ENDIF
      IF((Z.GT.D(4)).AND.(Z.LT.D(6)))THEN
        DH=TL*(D(6)-Z)
      ELSE
        ENDIF


      CALL INDX(XGRID,WIDTH,HF,DN,NREF,KL,DX,DH)
      CALL COEFF(K0,DZ,DX,NREF,XGRID,A,B,C)
      CALL MTRX2(K0,KL,DZ,DX,NREF,XGRID,NPRO,STEP,A,B,C,K,L,SKIP,T)

   20 CONTINUE
      CLOSE(3)
      RETURN
      END



C-------------------------------------------------------------

      SUBROUTINE PROF(XGRID,WIDTH,HF,DN,NREF,DX,S)

C*****************************************************************
C*                               *
C* This subroutine compute the profile distribution.       *
C* PRF  -- PROFILE                       *
C* XC,XC1,J  -- LOCAL VAR.                 *
C*                               *
C*****************************************************************


      PARAMETER (MAX=512)
      COMMON/BLK1/PRF
      REAL WIDTH,DX,HF,NREF,S
      DOUBLE PRECISION XC,XC1,PRF(MAX)
      INTEGER J,XGRID
```

```
  XC=WIDTH/2.0-DX/2.0
  DO 10 J=1,XGRID
   PRF(J)=0.0
   XC1=COSH(2.0*XC/HF)
   PRF(J)=1.0+0.01*EXP(-S*LOG(XC1))
   PRF(J)=(PRF(J)-1.0)/0.01
   XC=XC-DX
 10 CONTINUE
  RETURN
  END




C-------------------------------------------------------------

  SUBROUTINE INDX(XGRID,WIDTH,HF,DN,NREF,KL,DX,DH)

C*****************************************************************
C*                              *
C* THIS SECTION IS TO SET UP THE REFRACTIVE INDEX FOR THE MATRIX    *
C* COMPUTATION.                        *
C*                           *
C* PERT -- PERTURBATION                      *
C*                        *
C*****************************************************************


  PARAMETER (MAX=512)
  COMMON/BLK2/INDEX
  REAL WIDTH,DX,XC1,XC2,XC3,XC4,PERT,KL,HF,NREF
  REAL DW,S,DN,INDEX(MAX),DH,HF1
  INTEGER J,XGRID,U,HXGRID


  S=2.0
  U=MOD(XGRID,2)
  XC2=((WIDTH/2.0)+DH)-DX/2.0
  XC3=(-(WIDTH/2.0)-DH)+DX/2.0
  HXGRID=INT(XGRID/2)+U
  DO 10 J=1,XGRID
   INDEX(J)=0.0
   XC1=COSH(2.0*XC2/HF)
   PERT=DN*EXP(-S*LOG(XC1))
   IF((J.EQ.HXGRID+1).OR.(J.GT.HXGRID+1)) THEN
   INDEX(J)=NREF+PERT
   ENDIF
   XC2=XC2-DX
 10 CONTINUE
  DO 20 J=XGRID,1,-1
  XC4=COSH(2.0*XC3/HF)
  PERT=DN*EXP(-S*LOG(XC4))
  IF((J.LT.HXGRID).OR.(J.EQ.HXGRID)) THEN
  INDEX(J)=NREF+PERT
  ENDIF
```

```
      XC3=XC3+DX
   20 CONTINUE
      RETURN
      END



C----------------------------------------------------------------

      SUBROUTINE INIT(XGRID,WIDTH,HF,DN,NREF,K0,DX,KL,NPRO,DZ,S,STEP)


C***************************************************************
C*                               *
C* THIS SECTION IS TO STORE THE INITIAL FIELD INTENSITY AND TO CREAT *
C* AN OUTPUT FILE FOR THE RESULT.                 *
C*                               *
C* INTEN  -- INTENSITY                    *
C* PROF   -- PROFILE                     *
C*                               *
C***************************************************************


      PARAMETER (MAX=512)
      REAL Z,WIDTH,HF,DN,NREF,K0,DX,KL,LENGHT,S,DZ
      INTEGER XGRID,NPRO,STEP


      CHARACTER*20 FNAME
      WRITE(6,10)
   10 FORMAT('PLEASE ENTER THE NAME OF THE THE OUTPUT FILE:',/)
      READ(5,20)FNAME
   20 FORMAT(A)
      OPEN (3,FILE=FNAME,STATUS='OLD')


      WRITE(6,30)
   30 FORMAT('PLEASE ENTER THE NO OF PLANE TO BE PLOTTED.')
      READ(5,*)STEP


      OPEN(UNIT=4,FILE='PAM.DAT')
      WRITE(4,35)XGRID,STEP
   35 FORMAT(1X,I5,1X,I5)
      CLOSE(4)


      Z=0.0

      CALL PROF(XGRID,WIDTH,HF,DN,NREF,DX,S)

      CALL INTEN(KL,NPRO,XGRID,DZ)

      RETURN
      END
```

```
C-------------------------------------------------------------

   SUBROUTINE  INTEN(KL,NPRO,XGRID,DZ)

C***************************************************************
C*                                *
C* THIS SECTION OUTPUT THE FIELD INTENSITY FOR THE FIRST PLANE.    *
C*                                *
C***************************************************************


   PARAMETER  (MAX=512)
   COMMON/BLK1/PRF
   COMMON/BLK4/R
   REAL T,KL,DZ
   INTEGER J,K,XGRID
   DOUBLE PRECISION ZERO,PRF(MAX),R2(MAX)
   COMPLEX*16  R(MAX)


   ZERO=0.0
   DO 20 J=1,XGRID
    R(J)=DCMPLX(PRF(J),ZERO)
    R2(J)=ABS(R(J))
    WRITE(3,10)R2(J)
 10   FORMAT(F10.8)
 20 CONTINUE
   RETURN
   END




C-------------------------------------------------------------

   SUBROUTINE  MTRX2(K0,KL,DZ,DX,NREF,XGRID,NPRO
  +,STEP,A,B,C,K,L,SKIP,T)

C***************************************************************
C*                                *
C* THIS SECTION IS TO COMPUTE LIGTH INTENSITY AS IT PROPAGATES    *
C* THROUGH. COMPUTATION IS USING THE FINITE DIFFERENT METHOD    *
C* WHICH USE TRIDIOGANAL MATRIX. ELEMENTS IN THE MATRIX IS COMPUTED   *
C* IN THE SUBROUTINE 'COEFF.'                   *
C*                                *
C* CK    --- CHECK                    *
C* COEFF  --- COEFFICIENT                   *
C* SKIP  --- NO OF TIME TO SKIP BEFORE THE INTENSITY IS OUTPUTED    *
C* OPUT   --- OUTPUT.                    *
C*                                *
C***************************************************************


   PARAMETER  (MAX=512)
   REAL K0,KL,DZ,DX,NREF,Z,DIST
   INTEGER I,J,K,L,SKIP,XGRID,NPRO,STEP,T
```

```
DOUBLE PRECISION EOUT(MAX),A(MAX),AN(MAX),ZERO
COMMON/BLK4/R
COMPLEX*16 B(MAX),C(MAX),R1(MAX),BN(MAX)
COMPLEX*16 R(MAX),E(MAX),TEM(MAX)


ZERO=0.0
Z=0.0


CALL MTRX1(A,C,XGRID,R1)
IF (B(1).EQ.0)PAUSE
BN(1)=B(1)
AN(1)=-A(1)
DO 10 J=2,XGRID
 AN(J)=-A(J)
 TEM(J)=AN(J)/BN(J-1)
 BN(J)=B(J)-TEM(J)*AN(J-1)
 R1(J)=R1(J)-TEM(J)*R1(J-1)
10 CONTINUE


 E(XGRID)=R1(XGRID)/BN(XGRID)
 R(XGRID)=E(XGRID)


DO 30 J=XGRID-1,1,-1
 E(J)=(R1(J)-AN(J)*E(J+1))/BN(J)
 R(J)=E(J)
30 CONTINUE

 CALL OPUT(E,XGRID,K,L,STEP,NPRO,SKIP,T)

 WRITE(6,40)K
40 FORMAT('DOING STEP #',I4)
50 CONTINUE
 RETURN
 END



C-------------------------------------------------------------

 SUBROUTINE OPUT(E,XGRID,K,L,STEP,NPRO,SKIP,T)

C*************************************************************
C*                              *
C* This section in charge of the output of the field computed to the  *
C* result file. Note that not all planes are outputted to the result  *
C* file.                         *
C*                              *
C*************************************************************
```

```
   PARAMETER(MAX=512)
   DOUBLE PRECISION EOUT(MAX)
   INTEGER XGRID,J,K,L,N,STEP,SKIP,NPRO,T
   COMPLEX*16 E(MAX)


   N=K-L
   IF (N.EQ.0) THEN
   T=T+1
    IF ((T.LT.STEP).OR.(T.EQ.STEP-1)) THEN
     DO 20 J=1,XGRID
      EOUT(J)=ABS(E(J))
      WRITE(3,10)EOUT(J)
10      FORMAT(F10.8)
20   CONTINUE
     L=K+SKIP
    ENDIF
   ELSE
    ENDIF
   RETURN
   END


C---------------------------------------------------------------

   SUBROUTINE COEFF(K0,DZ,DX,NREF,XGRID,A,B,C)

C****************************************************************
C*                              *
C* THIS SECTION COMPUTES THE ELEMENTS OF THE TRIDIAGONAL MATRIX.  *
C*                              *
C****************************************************************


   PARAMETER (MAX=512)
   REAL DX,DZ,NREF,K0,INDEX(MAX)
   INTEGER J,K,XGRID
   DOUBLE PRECISION A(MAX),B1,B2,C1,C2
   COMMON/BLK2/INDEX
   COMPLEX*16 B(MAX),C(MAX)


   DO 10 J=1,XGRID
   A(J)=DZ/(2.0*(DX*DX))
   B1=DZ/(DX*DX)-((DZ*(K0*K0))/2.0)*(INDEX(J)*INDEX(J)-NREF*NREF)
   B2=2.0*K0*NREF
   C1=-(DZ/(DX*DX))+((DZ*(K0*K0))/2.0)*(INDEX(J)*INDEX(J)-NREF*NREF)
   C2=2.0*K0*NREF
   B(J)=DCMPLX(B1,B2)
   C(J)=DCMPLX(C1,C2)
10 CONTINUE
   RETURN
   END
```

```
C-------------------------------------------------------------------

      SUBROUTINE MTRX1(A,C,XGRID,R1)

C*****************************************************************
C*                                                              *
C*  THIS ROUTINE IS TO COMPUTE THE RHS OF EQUATION (6).         *
C*                                                              *
C*****************************************************************


      PARAMETER(MAX=512)
      INTEGER J,XGRID
      DOUBLE PRECISION A(MAX)
      COMMON/BLK4/R
      COMPLEX*16  C(MAX),R1(MAX),R(MAX)


      R1(1)=C(1)*R(1)+A(1)*R(2)
      DO 10 J=2,XGRID-1
      R1(J)=A(J)*R(J-1)+C(J)*R(J)+A(J)*R(J+1)
   10 CONTINUE
      R1(XGRID)=A(XGRID)*R(XGRID-1)+C(XGRID)*R(XGRID)
      RETURN
      END
```

Appendix 9 Interdependence between Electric and Magnetic Fields and Electromagnetic Waves

In this appendix the time varying electric and magnetic fields are described. That means the interdependence and the dynamics of the electromagnetic (EM) fields, and the formation and propagation of plane waves are assumed which are important for the radiation, guiding and propagation through either wireless or guided media. Essential fundamental understanding of the vector analyses can be found in many textbooks on electromagnetism.

A9.1 Wave Representation

A9.1.1 General Property

We consider in this book the continuous and time-harmonic waves represented by sine waves under steady state conditions. An electric field propagating in the z-direction can be represented by

$$\underset{\sim}{E}(z,t) = E_0 e^{-\alpha z} \cos(\omega t - \beta z + \phi)\underset{\sim}{a_z} \tag{A9.1}$$

where ω is the angular frequency of the wave ($\omega = 2\pi f$), f is the frequency of the wave, β is the propagation of the waves along the z-direction, ϕ is the initial phase, and α is the attenuation constant of the propagation medium. It is noted here that the propagation/phase constant β can be estimated as very straightforward if the waves are propagating in a wireless medium assumed to be air or vacuum. However this is more complex if the waves propagate in a guided medium in which the propagation/phase constant depends on the modes of the waves confined in the transverse directions and how tight the waves are confined in this plane. We will deal with these issues in chapters 8 and 9. Thus this constant indicates how fast or how slow the waves are propagating through the medium. Sometimes this is called the wave number.

Assuming that the initial phase can be adjusted to zero and at the initial instant, i.e., $t = 0$, then in a lossless ($\alpha = 0$) medium the waves can be rewritten as

$$\underset{\sim}{E}(z,0) = E_0 \cos(-\beta z)\underset{\sim}{a_z} \tag{A9.2}$$

This indicates that the phase/propagation constant β is related to a wavelength λ such that

$$\beta = \frac{2\pi}{\lambda} \tag{A9.3}$$

with λ as the wavelength in vacuum if the EM waves propagating in air or vacuum and shorter value when under guided wave or media of a definite permittivity. Note that when the EM waves propagate in a medium the frequency of the waves remains unchanged and only the wavelength changes as the propagating velocity is slowed down by the permittivity constant of the medium. Thus the wavelength varies accordingly. If in a guided medium, then the wave velocity is slowed down by the effective permittivity of the medium as seen by the waves under guiding condition.

Under an attenuated medium with the coefficient along the z-direction we have

$$E(z,0) = E_0 e^{-\alpha z} \cos(-\beta z) a_z \tag{A9.4}$$

Thus we could see that the amplitude of the wave is exponentially decreased with respect to z. That means an oscillatory wave as a function of time with its amplitude E_0 is multiplied by the exponentially decreased coefficient $e^{-\alpha z}$.

Exercise

Plot the amplitude of an EM wave as a function of time represented by

$$E(z,t) = E_0 e^{-\alpha z} \cos(\omega t - \beta z + \phi) a_z$$

under the following conditions

a. $E_0 = 1.0$; $\alpha = 0$ dB/km; $\omega = 2\pi \cdot 10^9$ Hz, $\lambda = 1000$ nm; initial phase $= 0$
b. $E_0 = 10$; $\alpha = 0$ dB/km; $\omega = 2\pi \cdot 10^9$ Hz, $\lambda = 1000$ nm; initial phase $= \pi/2$
c. $E_0 = 10$; $\alpha = 1$ dB/km; $\omega = 2\pi \cdot 10^9$ Hz, $\lambda = 1000$ nm; initial phase $= 0$ with the total traveled distance of 10 km.

A9.1.2 Waves by Phasor Representation

The wave represented in Equation A9.1 can be rewritten in the case that the attenuation constant is negligible, as

$$E(z,t) = \Re\left(E_0 e^{-\alpha z} e^{j\phi} e^{j(\omega t - \beta z)}\right) a_z$$

$$E(z,t) = \Re\left(\vec{E}_s e^{j\omega t}\right) a_z \tag{A9.5}$$

$$E_s = \Re\left(E_0 e^{j\phi}\right) a_z$$

This uses the property of the polar form representation of the sinusoidal wave as follows

$$re^{j\phi} = r\cos\phi + jr\sin\phi = r \prec \phi \tag{A9.6}$$

In general form we have

$$E_s = E_0(x, y, z) e^{j\phi} a_r$$

with a_r is the unit vector along the radial direction of the phasor. That means that the sinusoidal representation is now transformed to complex form. This complex representation

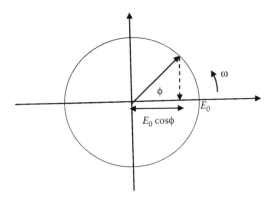

FIGURE A9.1
Phasor representation of the wave of amplitude E_0 and frequency ω with an amplitude E_0. The vector rotates around the circle with an angular frequency ω.

assists the manipulation of the waves using a vectorial operation and hence simplification of the algebraic steps. The waves expressed in Equation A9.5 can be represented in vector form as shown in Figure A9.1 in which the initial phase and angular frequency are represented by the a vector of an initial angle ϕ. This vector is rotating around the circle of amplitude E_0 with an angular frequency of ω. The projection of this vector to the real horizontal axis of the complex plane is the sinusoidal wave. The attenuation factor can be incorporated according to the length of the propagation along the z-direction. This vector representation allows us to add substrate to several other wave vectors provided that they are oscillating at the same frequency.

A9.1.3 Phase Velocity

Now considering a traveling wave in a loss less medium expressed by

$$\underline{E}(z,t) = E_0 \cos(\omega t - \beta z + \phi)\underline{a}_z \tag{A9.7}$$

This wave is propagating in the +z-direction with a constant phase of

$$\omega t - \beta z = C \tag{A9.8}$$

in which we assume a zero initial phase.
   Then the phase velocity or the speed of the wave propagating along the z-direction is given by differentiating Equation A9.8

$$v_p = \frac{dz}{dt} = \frac{\omega}{\beta} = \lambda f \tag{A9.9}$$

Now if the wave angular frequency is not a single frequency but composed of several other spectral components, such as in the case that the wave is modulated by a pulse envelope, then these spectral components are propagating along the z-direction with a group velocity of

$$v_g = \frac{dv_p}{d\omega} \tag{A9.10}$$

Indeed this group velocity allows us to evaluate the dispersion of carrier-modulated signals after propagating through a distance L in a wireless or guided medium.

A9.2 Maxwell's Equations

Maxwell's equations are formed with the unity of the four equations, Ampere's circuital law, Faraday's law, and Gauss Law for the electric and magnetic fields as listed in Table A9.1.

A9.2.1 Faraday's Law

Faraday's law is a fundamental relationship between a generated voltage or electric field in a changing magnetic field. The induced electro-motive voltage V_{emf} is related to the electric field $\underset{\sim}{E}$ and related by the magnetic field strength $\underset{\sim}{B}$ by:

$$\oint \underset{\sim}{E} \cdot d\underset{\sim}{L} = V_{emf} = -\frac{\partial}{\partial t} \oiint_S \underset{\sim}{B} \cdot d\underset{\sim}{S} \tag{A9.11}$$

TABLE A9.1

Fundamental EM Equation

Equations	Differential Form	Integral Form
Gauss's law	$\nabla \cdot \underset{\sim}{D} = \rho$	$\oiint_S \underset{\sim}{D} \cdot d\underset{\sim}{S} = Q_{enc}$
Gauss's magnetic law	$\nabla \cdot \underset{\sim}{B} = 0$	$\oiint_S \underset{\sim}{B} \cdot d\underset{\sim}{S} = 0$
Faraday's law	$\nabla \times \underset{\sim}{E} = -\frac{\partial \underset{\sim}{B}}{\partial t}$	$\oint \underset{\sim}{E} \cdot d\underset{\sim}{L} = -\frac{\partial}{\partial t} \oiint_S \underset{\sim}{B} \cdot d\underset{\sim}{S}$
Ampere's circuital law	$\nabla \times \underset{\sim}{H} = \underset{\sim}{J}_c + \frac{\partial \underset{\sim}{D}}{\partial t}$	$\oint \underset{\sim}{H} \cdot d\underset{\sim}{L} = \oiint_S \underset{\sim}{J}_c \cdot d\underset{\sim}{S} + \frac{\partial}{\partial t} \oiint_S \underset{\sim}{D} \cdot d\underset{\sim}{S}$

Other EM Equations

Lorentz force equation

$$\underset{\sim}{F} = q\left(\underset{\sim}{E} + \underset{\sim}{v} \times \underset{\sim}{B}\right)$$

E = electric field; B is the magnetic field intensity and q is the charge moving with a velocity v through a magnetic field.

Constitutive equations

$$\begin{cases} \underset{\sim}{D} = \varepsilon \underset{\sim}{E} \\ \underset{\sim}{B} = \mu \underset{\sim}{H} \\ \underset{\sim}{J} = \sigma \underset{\sim}{E} \end{cases}$$

ε, μ, σ = permittivity, permeability; conductivity of the medium

Current continuity equation

$$\nabla \cdot \underset{\sim}{J} = -\frac{\partial \rho_v}{\partial t}$$

with $d\underset{\sim}{L}$ is the vectorial differential length within a close loop and is the $d\underset{\sim}{S}$ differential area perpendicular to the surface. This equation involves the interaction of charges and magnetic field. Note that the vectorial directions of the fields and the surface and differential length must follow the right-hand rule.

A9.2.2 Ampere's Law

The continuity of the current flow is the original form of Ampere's circuital law. This relates the relationship between the electric current source and the magnetic field. The law can be written in integral form and its equivalent differential form via the Stoke's theorem as follows:

$$\oint_C \underset{\sim}{H} \cdot d\underset{\sim}{L} = \oiint_S \underset{\sim}{J}_c \cdot d\underset{\sim}{S} + \frac{\partial}{\partial t} \oiint_S \underset{\sim}{D} \cdot d\underset{\sim}{S} \Leftrightarrow \nabla \times \underset{\sim}{H} = \underset{\sim}{J}_c + \frac{\partial \underset{\sim}{D}}{\partial t} \tag{A9.12}$$

where

- $\oint_C$ is the closed line integral around the closed curve C
- $\underset{\sim}{B}$ is the magnetic field in Webers[*]
- $\cdot$ is the vector dot product
- $d\underset{\sim}{L}$ is an infinitesimal element (a differential) of the curve C (i.e., a vector with magnitude equal to the length of the infinitesimal line element, and the direction given by the tangent to the curve C)
- $\oiint_S$ denotes an integral over the surface S enclosed by the curve C (the double integral sign is meant simply to denote that the integral is two-dimensional in nature)
- $\underset{\sim}{J}_C$ is the free current density through the surface S enclosed by the curve C
- $d\underset{\sim}{S}$ is the vectorial differential area of an infinitesimal element of the surface S (that is, a vector with magnitude equal to the area of the infinitesimal surface element, and direction normal to surface S). The direction of the normal must correspond with the orientation of C by the right-hand rule. I_{enc} is the net free current that penetrates through the surface S
- The term

$$\oiint_S \underset{\sim}{J}_c \cdot d\underset{\sim}{S}$$

represents the conduction current and the term

$$\frac{\partial}{\partial t} \oiint_S \underset{\sim}{D} \cdot d\underset{\sim}{S}$$

represents the displacement current.

---

[*] Note: the unit of the magnetic field intensity $\underset{\sim}{B}$ 1 Weber = kg $\cdot$ m^2/C $\cdot$ s.

A9.2.2.1 Gauss's Law for Electric Field and Charges

The electric flux through any closed surface is proportional to the enclosed electric charge Q_{enc}

$$\oiint_S \underset{\sim}{D} \cdot d\underset{\sim}{S} = Q_{enc} \tag{A9.13}$$

A9.2.2.2 Gauss's Law for Magnetic Field

The divergence of the magnetic field $\underset{\sim}{B}$ equates to zero, in other words, that it is a solenoidal vector field. It is equivalent to the statement that magnetic monopoles do not exist.

$$\nabla \cdot \underset{\sim}{B} = 0 \tag{A9.14}$$

A9.3 Current Continuity

Consider a volume of charge Q enclosed in a volume surface S which is varying as a function of time. If the volume is reduced then the only possibility is that the charges are flowing through the surface. This flow of charges under the conservation of charges or energy is the current which must be equal to the rate of change of the contained charge. The flowing current I can be expressed in terms of the differential charge and the current density as:

$$I = \oint_S \underset{\sim}{J} \cdot d\underset{\sim}{S} = -\frac{\partial Q}{\partial t} \tag{A9.15}$$

where $\underset{\sim}{J}$ is the current density in vector form and t is the time variable $d\underset{\sim}{S}$ whose normal direction is the direction of the surface vector which is the infinitesimal surface element and S is the surface area of the volume.

Using the divergence theorem [1] we can write

$$\oint_S \underset{\sim}{J} \cdot d\underset{\sim}{S} = -\oiint_V \left(\nabla \cdot \underset{\sim}{J} \right) dV = \int_V \left(\nabla \cdot \underset{\sim}{J} \right) dV \tag{A9.16}$$

and

$$-\frac{\partial Q}{\partial t} = -\frac{\partial}{\partial t} \oiint_V \rho_V dV \tag{A9.17}$$

In which ρ_V is the charge density enclosed within the surface under consideration. Thus comparing Equations A9.15 through A9.17 we obtain:

$$\nabla \cdot \underset{\sim}{J} = -\frac{\partial \rho_V}{\partial t} \tag{A9.18}$$

This is the point form of the current continuity equation. This equation indicates that the divergence of the current density equals of the differential changes of the charges flowing out or in the volume under consideration. In the steady state condition when there is no change in the charge density then the conservation of charges leads to the fact that the total currents flowing into a node equates that of the total current flowing out of a node. This is Kirchoff's current law which is valid for under-transient or phasor steady state conditions.

A9.4 Lossless Transverse Electric and Magnetic Waves

Consider an x-polarized wave propagating along the +z-direction in a medium characterized by the material constants μ,ε and lossless, that is the attenuation constant is zero. The field can be represented by

$$\underset{\sim}{E}(z,t) = E_0 \cos(\omega t - \beta z)\underset{\sim}{a}_x \tag{A9.19}$$

Then applying Faraday's law given in Table A9.1, we have

$$\nabla \times \underset{\sim}{E} = -\frac{\partial \underset{\sim}{B}}{\partial t} = -\mu\frac{\partial \underset{\sim}{H}}{\partial t}$$

$$\begin{vmatrix} \underset{\sim}{a}_x & \underset{\sim}{a}_y & \underset{\sim}{a}_z \\ \dfrac{\partial}{\partial x} & \dfrac{\partial}{\partial y} & \dfrac{\partial}{\partial z} \\ E_0 \cos(\omega t - \beta z) & 0 & 0 \end{vmatrix} = \beta E_0 \sin(\omega t - \beta z)\underset{\sim}{a}_y = -\mu\frac{\partial \underset{\sim}{H}}{\partial t} \tag{A9.20}$$

Then by taking the integral of the differential component of the magnetic field flux we arrive at

$$\int d\underset{\sim}{H} = \underset{\sim}{H} = \frac{\beta E_0}{\omega\mu}\cos(\omega t - \beta z)\underset{\sim}{a}_y + C_1 \tag{A9.21}$$

$$\therefore \underset{\sim}{H} = \frac{\beta E_0}{\omega\mu}\cos(\omega t - \beta z)\underset{\sim}{a}_y \qquad \because C_1 = 0 \quad \text{initial condition}$$

with C_1 is the integral constant which turns out to be nil by using the initial condition. It is from Equation A9.21 and A9.19 that the $\underset{\sim}{H}$ and $\underset{\sim}{E}$ fields are orthogonal to each other and their magnitudes are differed by a scaling factor due to the characteristic of the medium and the propagation constant and radial frequency of the EM waves.

Now we can use the other Maxwell's equations to find the relationship between the propagation constant and the velocity of light

$$c = \frac{1}{\sqrt{\mu_0\varepsilon_0}}$$

and the medium characteristic μ, ε.

$$\nabla \times \underset{\sim}{H} = \underset{\sim}{J}c + \frac{\partial \underset{\sim}{D}}{\partial t} = \varepsilon \frac{\partial \underset{\sim}{E}}{\partial t} \rightarrow$$

$$\begin{vmatrix} \underset{\sim}{a}_x & \underset{\sim}{a}_y & \underset{\sim}{a}_z \\ \dfrac{\partial}{\partial x} & \dfrac{\partial}{\partial y} & \dfrac{\partial}{\partial z} \\ 0 & \dfrac{\beta}{\omega\mu} E_0 \cos(\omega t - \beta z) & 0 \end{vmatrix} = -\frac{\beta E_0}{\omega\mu} \frac{\partial}{\partial z} \cos(\omega t - \beta z) \underset{\sim}{a}_x \qquad (A9.22)$$

$$= -\frac{\beta^2 E_0}{\omega\mu} \sin(\omega t - \beta z) \underset{\sim}{a}_x$$

Therefore leading

$$\rightarrow \qquad\qquad \frac{\partial \underset{\sim}{E}}{\partial t} = -\frac{\beta^2 E_0}{\omega\mu} \sin(\omega t - \beta z) \underset{\sim}{a}_x$$

$$\qquad\qquad\qquad\qquad\qquad\qquad\qquad\qquad\qquad\qquad (A9.23)$$

$$\rightarrow \qquad\qquad \underset{\sim}{E} = -\frac{\beta^2 E_0}{\omega^2\mu} \sin(\omega t - \beta z) \underset{\sim}{a}_x \qquad \because \frac{\partial}{\partial t} = \omega$$

The additional parameters of the amplitude of the electric field must be equal to unity in order to satisfy the field expression assumed from the beginning. Thus we have

$$\beta = \omega\sqrt{\mu\varepsilon} \qquad\qquad (A9.24)$$

and it follows that the phase velocity of the EM wave is given by

$$v_p = \frac{\omega}{\beta} = \frac{1}{\sqrt{\mu\varepsilon}} = \frac{1}{\sqrt{\mu_0\mu_r\varepsilon_0\varepsilon_r}} = \frac{c}{\sqrt{\mu_r\varepsilon_r}} \qquad\qquad (A9.25)$$

or the phase velocity of an EM wave is that of the velocity of light reduced by a factor related to the relative permeability and permittivity of the medium. It is noted in Equation A9.25 that the propagation constant β is not restricted by the guiding condition but assumed as a plane wave in a medium. Under guiding condition this propagation constant would be restricted by the confinement of the electric and magnetic components by the boundary conditions of the guided medium. We will describe these guiding conditions and the propagation constant or relative refractive indices as seen by the EM waves in such guided media.

Example

A y-polarized plane wave is propagating along the x-direction and the medium is air. The frequency of the wave is 10 GHz with a magnitude of 1.0 V/m. Write down the expression for the electric field of this wave.

Solution

$$\beta = \omega/c = 2\pi \times 10^{10}/3.10^8 = \frac{2\pi}{3} \text{ rad/s.}$$

$$\underset{\sim}{E} = 1.0\cos\left(2\pi \times 10^{10} t - \frac{2\pi}{3}x\right)\underset{\sim}{a_y} \text{ V/m}$$

Example

A wave is propagating in a non-magnetic medium whose magnitude and time-oscillating characteristics are given by

$$\underset{\sim}{E}(x,t) = 10\cos\left(10^8 \pi t + \frac{\pi}{10}x + \frac{\pi}{3}\right)\underset{\sim}{a_y} \text{ V/m}$$

Identify the frequency of the wave, its phase velocity and the permittivity and permeability of the medium.
The flux $\underset{\sim}{H}$.
Sketch the fields $\underset{\sim}{B}$ and $\underset{\sim}{H}$.

Solution

Inspecting the expression of the wave we have

$$f = \frac{10^8 \pi}{2\pi} = 5\times 10^7 \text{ Hz} = 50 \text{ MHz}$$

The propagation constant and the phase velocity of the wave are related by

$$\beta = \frac{2\pi}{3} = \frac{\omega}{v_p} \to v_p = \frac{3}{2\pi}2\pi f \to 3f = 3\times 50\cdot 10^6 = 1.5\times 10^8 \text{m/s}$$

Thus for the non-magnetic medium ($\mu_r = 1$) we have

$$v_p = \frac{c}{\sqrt{\varepsilon_r}} \to \varepsilon_r = \frac{c^2}{v_p} = \frac{3\times 10^8}{1.5\times 10^8} = 2$$

The magnetic field density is then given as

$$\underset{\sim}{H}(x,t) = \frac{\beta E_0}{\omega \mu_0 \mu_r}\cos\left(10^8 \pi t + \frac{\pi}{10}x + \frac{\pi}{3}\right)\underset{\sim}{a_z}$$

$$\underset{\sim}{H}(x,t) = \frac{\frac{2\pi}{3}10}{2\pi 5x10^7 \mu_0}\cos\left(10^8 \pi t + \frac{\pi}{10}x + \frac{\pi}{3}\right)\underset{\sim}{a_z}$$

$$= \frac{2}{3\mu_0}\cos\left(10^8 \pi t + \frac{\pi}{10}x + \frac{\pi}{3}\right)\underset{\sim}{a_z} \text{ A/m}$$

note that the fields are spatially perpendicular.

A9.5 Maxwell's Equations in Time-harmonic and Phasor Forms

Given that the field time varying is sinusoidal, substituting $\partial/\partial t = j\omega$ we have the Maxwell's equations in differential form in which the subscript s stands for the phasor representation

Equations	Differential Form
Gauss's law	$\nabla \cdot \underset{\sim}{D}_s = \rho$
Gauss's magnetic law	$\nabla \cdot \underset{\sim}{B}_s = 0$
Faraday's law	$\nabla \times \underset{\sim}{E}_s = -j\omega B_s$
Ampere's circuital law	$\nabla \times \underset{\sim}{H}_s = \underset{\sim}{J}_c + j\omega \underset{\sim}{D}_s$

A9.6 Plane Waves

EM waves can be radiated from a source. Far away from the source the waves resemble a uniform phase front, this is called plane waves. In the uniform wave field the electric and magnetic components of the waves are orthogonal to the propagation direction, hence the name transverse electric and magnetic (TEM) waves.

A9.6.1 General Wave Equations

In this section it is assumed that the simple medium is linear, isotropic, and time invariant. Then the Maxwell equations can be written for this medium as

$$\nabla \cdot \underset{\sim}{E} = 0 \tag{A9.26}$$

$$\nabla \cdot \underset{\sim}{H} = 0 \tag{A9.27}$$

$$\nabla \times \underset{\sim}{E} = -\mu \frac{\partial \underset{\sim}{H}}{\partial t} \tag{A9.28}$$

$$\nabla \times \underset{\sim}{H} = \sigma \underset{\sim}{E} + \varepsilon \frac{\partial \underset{\sim}{E}}{\partial t} \tag{A9.29}$$

Taking the curl of both sides of Equation A9.28 and the property of an isotropic medium we have

$$\nabla \times \left(\nabla \times \underset{\sim}{E} \right) = -\mu \frac{\partial}{\partial t} \left(\nabla \times \underset{\sim}{H} \right) \tag{A9.30}$$

Then using the identity

$$\nabla \times \nabla \underset{\sim}{A} = \nabla \cdot \underset{\sim}{A} - \nabla^2 \underset{\sim}{A} \tag{A9.31}$$

Then combining Equations A9.29 and A9.30 we arrive at

$$\nabla \cdot \underset{\sim}{E} - \nabla^2 \underset{\sim}{E} = -\mu\sigma \frac{\partial \underset{\sim}{E}}{\partial t} - \mu\varepsilon \frac{\partial^2 \underset{\sim}{E}}{\partial t^2} \tag{A9.32}$$

Under the conditions of charge free, the divergence of $\underset{\sim}{E}$ is zero and Equation A9.32 becomes

$$\nabla^2 \underset{\sim}{E} = +\mu\sigma \frac{\partial \underset{\sim}{E}}{\partial t} + \mu\varepsilon \frac{\partial^2 \underset{\sim}{E}}{\partial t^2} \tag{A9.33}$$

This is the Helmholtz wave equation for the electric field component $\underset{\sim}{E}$, the medium is operating in the linear region, that is the permittivity does not vary with the strength of the intensity of the field. We will see this in chapters 8 and 9 dealing with the nonlinear medium, especially the case of guided waves in optical fibers with high power of the guided modes.

A similar equation can be derived for the magnetic component $\underset{\sim}{H}$ without much difficulty.

Example

Write the wave equation for an EM wave whose field is expressed by

$$\underset{\sim}{E}(z,t) = \begin{pmatrix} E_x(z,t) \\ 0 \\ 0 \end{pmatrix}$$

propagating in a charge free medium.

Solution

$$\frac{\partial^2}{\partial t^2} E_x(z,t) = +\mu\sigma \frac{\partial E_x(z,t)}{\partial t} + \mu\varepsilon \frac{\partial^2 E_x(z,t)}{\partial t^2}$$

A9.6.2 Time-harmonic Wave Equation

When the wave is time-harmonic then the wave equation can be written in terms of the phasor of the electric field component as

$$\nabla^2 \underset{\sim}{E}_s = \left(j\omega\mu\sigma - \omega^2\mu\varepsilon \right) \underset{\sim}{E}_s \tag{A9.34}$$

Or alternatively

$$\nabla^2 \underset{\sim}{E}_s - \gamma^2 \underset{\sim}{E}_s = 0$$

with   (A9.35)

$$\gamma^2 = j\omega\mu\sigma - \omega^2\mu\varepsilon \rightarrow \gamma = \alpha + j\beta$$

For the magnetic field component, we can similarly obtain

$$\nabla^2 \underline{H}_s - \gamma^2 \underline{H}_s = 0 \tag{A9.36}$$

Equations A9.35 and A9.36 are the well-known Helmholtz equations for time-harmonic fields propagating either in free space or in confined and guided structure in the linear regions. When the waves propagate under the nonlinear regime of the medium then these equations would be modified to include a number of terms in the right-hand side. These equations would be termed as the nonlinear Schrodinger equations (NLSE) which will be on optical waveguiding in integrated and circular structures. These equations would then be subject to the boundary conditions so that the eigenvalue equations can be obtained and thence the eigenvalues can be obtained. These values correspond to the propagation constant or wave vector that determines the speed of propagating in such guided structures.

Let us assume an x-polarized plane wave traveling in the z-direction whose electric field is given as:

$$\underline{E}_s(z) = E_{xs}(z)\underline{a}_x \tag{A9.37}$$

The wave amplitude is solely dependent on the z-direction only. Then substituting this into the Helmholtz equation (Equation A9.35) we have:

$$\frac{\partial^2 E_{xs}}{\partial z^2} - \gamma^2 E_{xs} = 0 \tag{A9.38}$$

This equation is a linear second order and homogeneous differential equation. A possible solution for this equation is

$$E_{xs} = Ae^{\lambda z} \tag{A9.39}$$

where A, λ are arbitrary constants. It is straightforward to prove that the differential equation can now be written as

$$\frac{\partial^2 E_{xs}}{\partial z^2} = \lambda^2 A E_{xs} e^{\lambda x} \tag{A9.40}$$

Thence

$$\lambda^2 - \gamma^2 = 0 \tag{A9.41}$$

The only physical value of the value of γ must be negative so that the wave does not grow to infinitive when the propagation distance becomes very large. Thus with the time dependent factor and the frequency component $e^{j\omega t}$, both waves propagating in the forward and backward directions are possible. For the forward propagating waves with $A = E_0^-$ we have

$$E_{xsF} = Ae^{-\gamma z}$$

$$\rightarrow E_{xsF} = E_0^- \cos(\omega t - \beta z) \tag{A9.42}$$

Thus we have

$$E_{xsF} = E_0^- e^{-\alpha z} \cos(\omega t + \beta z) \tag{A9.43}$$

The waves in the backward direction take the form

$$E_{xsB} = E_0^+ e^{\alpha z} \cos(\omega t + \beta z) \tag{A9.44}$$

Thence the superposition of the two waves from the forward and backward directions arrive at

$$E_{xs} = E_0^- e^{-\alpha z} \cos(\omega t - \beta z) + E_0^+ e^{\alpha z} \cos(\omega t + \beta z) \tag{A9.45}$$

By applying Faraday's law to the electric field the magnetic field can be found as

$$\nabla \times \underset{\sim}{E}_s = -j\omega\mu \underset{\sim}{H}_s = \left(-\gamma E_0^- e^{-\gamma z} + \gamma E_0^+ e^{\gamma z}\right) \underset{\sim}{a}_y \tag{A9.46}$$

Thus we can obtain the magnetic field vector by expanding the curl of Equation A9.46 to give

$$\underset{\sim}{H}_s = \left(\frac{\gamma E_0^+}{j\omega\mu} e^{-\gamma z} - \frac{\gamma E_0^-}{j\omega\mu} e^{+\gamma z}\right) \underset{\sim}{a}_y$$

$$= \left(H_0^+ e^{-\gamma z} - H_0^- e^{+\gamma z}\right) \underset{\sim}{a}_y \tag{A9.47}$$

Hence we can define the intrinsic impedance as

$$\eta = \frac{E_0^+}{H_0^+} = \frac{j\omega\mu}{\gamma} = -\frac{E_0^-}{H_0^-} \tag{A9.48}$$

with

$$\eta = \sqrt{\frac{j\omega\mu}{\sigma + j\omega\varepsilon}}$$

when the expression of γ given by Equation 11.35 is used.

Reference

1. E. Kreyszig, "*Advanced Engineering Mathematics (Chapter 9 and 10),*" 8th ed., John Wiley and Son, New York, 2008.

Index

Printed and bound by CPI Group (UK) Ltd, Croydon, CR0 4YY

23/10/2024

01778251-0012